Nonparametric Statistics on Manifolds and Their Applications to Object Data Analysis

Nonparametric Statistics on Manifolds and Their Applications to Object Data Analysis

Victor Patrangenaru

Florida State University

Tallahassee, USA

Leif Ellingson

Texas Tech University

Lubbock, USA

CRC Press is an imprint of the
Taylor & Francis Group, an **informa** business
A CHAPMAN & HALL BOOK

CRC Press
Taylor & Francis Group
6000 Broken Sound Parkway NW, Suite 300
Boca Raton, FL 33487-2742

First issued in paperback 2020

ISBN-13: 978-1-4398-2050-6 (hbk)
ISBN-13: 978-0-367-73782-5 (pbk)

Visit the Taylor & Francis Web site at
http://www.taylorandfrancis.com

and the CRC Press Web site at
http://www.crcpress.com

Contents

Foreword

I am delighted to introduce the first extensive book on nonparametric statistics on manifolds and their applications. When I came to know about this book project undertaken by the most active researcher in the field, I was pleased to know that this book is coming at an early stage in a field that is in need of an early synthesis. In most emerging research fields, a book can play a significant role in bringing some maturity to the field. Research fields advance through research papers. In research papers, however, only a limited perspective can be provided regarding the field, its application potential, the techniques required and already developed in the field. A book presents the chance to provide the foundations, the main themes, the methods, to present the existing results in a systematic manner, to outline the field's horizon, including its open problems. This book succeeds in unifying the field by bringing in disparate topics, already available in several papers, but not easy to understand, under one roof. I was supportive of this book project even before I had seen any material on it. The project is a brilliant and a bold idea by an active researcher, who is now joined in coauthorship by an enthusiastic, hard working and talented younger peer. Now that I have it on my screen, it exceeds all expectations, in particular regarding the extent to which complex differential geometric notions permeate statistics.

Nonparametric Statistics on Manifolds started gaining recognition in 1990s as a field. Image and shape analysis, directional data analysis, bioinformatics, pattern recognition and medical imaging had advanced to a point where statisticians, computer scientists, and engineers started building a geometric methodology aimed at combining information in these various types of data sources, images, video, or text, by representing them as points on higher dimensional curved spaces, or manifolds. Statistics computing and manifold techniques are today recognized for their synergetic effect on any numerical data analysis. However, when applied separately, these methodologies provide insufficient information. By landmark selecting and preprocessing data, such methodologies aspire, much like the human perception system, to create a fused picture of the data rather then using only partial information arrived at from separate methodologies.

Nonparametric statistics on manifolds is a blend of progress in nonparametric multivariate data analysis including large sample theory and nonparametric bootstrap, and differential geometry. When it became possible to store and preprocess huge digital vectors (stored as digital images, shapes, sounds, etc) in data libraries and run nonlinear statistical algorithms to explore all possible matches among the contents provided, the field of nonparametric statistics on manifolds was born as a perfect match for tackling complex questions in modern data analysis. This field has already seen spectacular applications in many diverse domains such as astronomy (providing real evidence, as opposed to voting bodies at international conferences, that Pluto is not a planet!), biology, bioinformatics and proteomics, computer vision, geology, medical imaging, manufacturing, meteorology, surveillance and defense, pattern recognition and neurology.

In fact, new research areas are a direct outgrowth of nonparametric statistics on manifolds, and, since the latter extends multivariate analysis, it is likely to become a much more accurate tool in the statistical analysis of any research area. So far, the main problem with applying nonparametric statistics on manifolds for the practicing statistician has been the daunting task of understanding the formidable apparatus behind it, and the absence of a textbook presenting the right amount of information needed for an adequate understanding. Here it is, that textbook!

Victor Pambuccian
Professor of Mathematics
Arizona State University

Preface

The main objective of this book is to introduce the reader to a **new way of analyzing object data, that primarily takes into account the geometry of the spaces of objects measured on the sample space**. This elementary fact that was too long ignored in statistics becomes a must, given that, in our day, observations extracted from electronic sources, including medical imaging, can not be regarded as linear data points, and even if by some procedure they are embedded in a linear space, the key statistic, the sample mean falls outside the space of objects regarded as a subset of a linear space. Classical Statistics, based on Probability theory, Mathematical Analysis, Linear Algebra and (more recently) Computer Science, is thus fundamentally challenged to find new ways for analyzing object data problems. For this reason, our top priority is to *give an appropriate notion of mean object, as a point that lives on the object space*, like every other observation. When considering the notion of sample mean, we go back the the basic least squares principle, bearing in mind that, on the object space, the distance between two points is no longer a Euclidean distance.

Fortunately, this problem is not new. It was already posed by Cartan (1928) [58] in his *Lectures on the Geometry of Riemannian spaces*, where he considered the notion of the *barycenter of a finite system of points on a Riemannian space* (the notion of manifold was not known at the time). Certainly, there are metrizable object spaces, which do not necessarily have a smooth structure, where the least expected value of the square distance of the random object to a given point was considered by Fréchet (1948) [121]. All the minimizers of this expected value, are being regarded as mean values of the random point. On the other hand, Cartan, to avoid the problem of multiple barycenters of the same finite set, imposed an additional topological restriction on the Riemannian object space by assuming that it *is simply connected and has nonpositive sectional curvature*. Cartan and Fréchet were contemporaries, so we decided to give them both credit for their breakthrough definition of a mean value by naming such values *Fréchet means, and Cartan means* (in case we sample points from a Riemannian manifold).

Computing the value of a Cartan mean turns out to be a time consuming process. When it comes to non-Euclidean data, historically, statisticians considered sample means that are not a computational burden to the data analysis, such as the mean direction for directional data (see Mardia and Jupp (2000) [230]) or the Procrustes mean for similarity shape data (see Dryden and Mardia (1998) [91], and Kent (1992) [181]). Such means appeared as maximum likelihood estimates (m.l.e.'s) in certain directional (respectively shape) parameterized distributions, so the analysis was carried out in the numerical parameter space in those instances. On the other hand, in the general case of an arbitrary manifold there are no known distributional families and a *nonparametric* approach is sought.

An early useful result, making Object Data Analysis (ODA) possible, is the consistency of the Fréchet sample mean set as an estimator of the mean Fréchet set (see Ziezold (1977) [345]). The bad news, though, was that for general metric spaces there was no asymptotic theory, leading to a statistical table-based inference, therefore one could not go further with data analysis. An additional structure was needed, which is where smoothness (manifold structure) of the object space comes in handy. Given the computational issues with the Cartan means, Patrangenaru (1998) [266] considered a nonparametric analysis of *extrinsic means* on manifolds, which are Fréchet means associated with a distance between two observations on the object space and which is equal to the chord distance between their images of the object space under an *embedding* into the Euclidean space.

Inspired by Milnor's work in Differential Topology (see Milnor (1963) [242]), one realized that a random point on an embedded manifold has a unique extrinsic mean when the mean of its push forward distribution in the numerical space is *non focal.* Since not being a focus point of an embedded manifold is a generic property for a distribution of such a random point on a given embedded manifold, the extrinsic mean is the preimage via the embedding of the projection of the mean vector of the push forward distribution; since projections preserve consistency, a simpler view of consistency of the extrinsic mean is obtained. Since consistency is a *local property* on the embedded manifold, one could apply Cramer's delta method to the projection map, and obtain the asymptotic normality of extrinsic sample mean around the extrinsic population mean (see Patrangenaru (1998) [266]).

Slightly earlier, independently, Hendriks and Landsman derived the asymptotic normality of the linearized extrinsic sample means on a submanifold of the Euclidean space. Additionally they derived the first tests for the two extrinsic sample mean problem on manifolds (see Hendriks and Landsman (1996,1998) [152, 154]). Bhattacharya and Patrangenaru (2003, 2005) [42], [43]) introduced the *extrinsic sample covariance matrix* as a descriptive statistic. Patrangenaru (1998) [257] and Hendriks and Landsman (1998) [154] however use different estimation techniques of the extrinsic mean. Hendricks and Landsman used estimators with a sample covariance based on the Weingarten map (see Spivak (1979) [316]), while Patrangenaru (1998) [266] used the technique of *adapted frames,* inspired from Cartan's moving frames (see Spivak(1979) [316]), a technique that at the time was already available in axial data analysis due to Fisher et al. (1996) [116], where it was effective in deriving nonparametric confidence regions for mean axes.

In summary, ODA requires some sort of differentiable structure on the **object space**, that **has to be** consequently either **a manifold, or having some manifold related structure, no matter what the nature of the objects is.**

Therefore we structured our book as follows. In Part I, we introduce the basics for the three "pillars of ODA": (i) examples of object data, (ii) nonparametric multivariate statistics and (iii) geometry and topology of manifolds. This part is a **must read and learn** for anyone new to the area, who genuinely wants to understand what is behind ODA, as opposed to simply using it. Secondly, we develop a general nonparametric analysis on arbitrary manifolds, as well as density estimation on manifolds. In Part II, we "translate" this methodology, in the context of certain sampling manifolds arising in Statistics. In Part III, we apply this methodology to concrete examples of ODA. Finally, in Part IV, we introduce the reader to more recent methodologies for data analysis on manifolds, such as persistence homology and intrinsic PCA on manifolds.

Chapter 1 provides an overview of data on manifolds, presenting a number of examples of such data in practical applications. A manifold is an abstract metric space that looks locally, but not necessarily globally, like a numerical space $\mathbb{R}^m$. Data on manifolds originally arose in astronomy, meteorology, geology, cartography, biology, and physics. A drastic increase in the need of data analysis on manifolds occurred with the computer revolution in digital imagery. Digital images, which arguably account for the largest types of data available, are today the bread and butter of modern sciences, such as bioinformatics, medical imaging, computer vision, pattern recognition, astrophysics, learning, forensics, etc. Although a digital image is a long vector of values representing the brightness of a given color at any specific pixel or voxel location, the marginal distribution of a random image of a given size, at a given pixel, has no relevance for the scene pictured. A standard way of analyzing the contents of a random image, is often by associating a random object on a certain manifold $\mathcal{M}$ (shape space, DTI space, learning space, etc.) and analyzing that random object on $\mathcal{M}$.

Real life data can be best analyzed only if a sufficiently large number of variables have been measured. Therefore, in Chapter 2, the reader is provided with the supporting knowledge, from basic definitions from probability theory through more advanced theory, necessary for the proper study of such data using nonparametric multivariate methods. It begins by introducing random vectors and their mean and covariance parameters, along with their sample estimators and associated sampling distributions. As necessary introductory topics, multivariate normal distributions, inference procedures for mean vectors and covariance matrices, and principal components along with certain related techniques are then introduced. Nonparametric methodologies, for which the assumptions are that nothing is known about the form of the multivariate distribution except, perhaps, that it has certain finite moments, or that it has a continuous and positive joint density over the effective range, are the other primary focus of this chapter. These assumptions suffice to construct reasonably good estimators of population parameters, such as mean vectors, covariance matrices, moments, and spatial quantiles. The Edgeworth expansion (see Edgeworth (1905) [93], Bhattacharya and Denker (1990) [36]) and the nonparametric bootstrap (see Efron (1982, 1979, 2003) [97, 96, 98], Hall (1997) [137]) are the key techniques discussed here for inference from small samples. The chapter continues with a discussion of function and density estimation, and concludes with considering tests of neighborhood hypotheses for the one-and multi-sample problem in a Hilbert space.

Chapter 3 presents needed Differential Geometry and Topology background needed for the study of manifolds. A *manifold* is a metric space that can be locally parameterized by a number of independent parameters, for which the transition between two parameterizations is smooth. The number of parameters needed in such a parametrization is the *dimension* of the manifold. Curves and surfaces of the Euclidean space are basic examples of low dimensional manifolds. Their geometry was thoroughly studied by Gauss. However, often the number of independent parameters is high, so that a manifold arising in data analysis is impossible to visualize. A high dimensional representation of an abstract manifold as a smooth object in a vector space is called an *embedding*. Whenever possible, it is preferred to *equivariantly* embed a manifold so that its representation is left invariant by a group of symmetries of the ambient numerical space, allowing for the comparison of data sets at different locations on the manifold. A more abstract approach to manifolds, due to Riemann, is to endow a manifold with a a geodesic distance depending on a Riemannian structure and, whenever possible, it is preferred to find a Riemannian structure on the manifold with the largest possible group of isometries (transformations that preserve the geodesic distance).

The median and the mean of a probability measure on a Riemannian manifold $(\mathcal{M}, \rho_g)$ were introduced in the case of a random sample by Cartan (1928) [57] and, in general, by Fréchet (1948) [121]. Following Fréchet's original ideas, in Chapter 4 we introduce random objects (r.o.) on metric spaces, their associated Fréchet functions, and related probability Fréchet parameters, such as the median, total variance and mean. Unlike with random vectors, the Fréchet mean (or median) of a r.o. is not necessarily unique; the more appropriate notion of a Fréchet mean set is therefore considered here. The corresponding sample counterparts, Fréchet sample total variance and Fréchet sample mean set are shown to be consistent estimators of the Fréchet total variance and Fréchet mean, provided the metric space $(\mathcal{M}, \rho)$ is complete.

Asymptotic distributions of extrinsic sample means for embeddings of abstract manifolds in an Euclidean space are given in Chapter 5. Here is also given the Central Limit Theorem for the distribution of Fréchet sample means from a probability distribution on a complete manifold. As a corollary, for a large family of probability measures on a complete Riemannian manifold, the Cartan means yield asymptotically a normal distribution in the tangent space of the intrinsic mean of that distribution. In this chapter, one proves this result and, as a consequence, one gives large sample confidence regions for the intrinsic mean as well as nonparametric confidence regions for the intrinsic mean of such distributions using pivotal and nonpivotal bootstrap.

In Chapter 6, we first address the problem of comparing two population total extrinsic variances. Next, we study the two sample extrinsic mean problem for random objects on an embedded manifold. The last part of the chapter is concerned with large sample and nonparametric bootstrap tests for mean change associated with a matched pair of random objects on a Lie group, and with nonparametric tests for the change in the extrinsic mean on a manifold that admits a simply transitive Lie group action.

Chapter 7 focuses in details on density estimation using harmonic analysis methods on Riemannian symmetric spaces (see Cartan (1927) [57], Helgason(1984)[148]), that are due to Kim et al. (2009) [191], or kernel density estimation on manifolds methods (see Pelletier (2005) [278]). Emphasizing the work of Kim et al. (2009) [191], in this chapter the technical details for infinite dimensional linear data analysis on symmetric is given. Note that unlike in Chapters 4–6, the approximations are on a linear space of functions, not on the manifold proper. Here, one is concerned with statistical inverse problems and function estimation on manifolds. After detailing the approximations using harmonic functions on a Riemannian manifold, specific applications are discussed. Kernel density estimation of probability distributions on a Riemannian manifold due to Pelletier (op. cit.) is also detailed here.

Part II of the book focuses on inference procedures based on asymptotics and nonparametric bootstrap for some special manifolds. Chapter 8 deals with such methodology for two-sample tests for means on Riemannian manifolds with a simply transitive group of isometries. In particular, it presents a two-sample procedure for testing the equality of the intrinsic means on a homogeneous Hadamard–Cartan manifold. Although all the Riemannian structures on homogeneous space are locally classified (see Patrangenaru (1994) [264]), Hadamard–Cartan structures are preferred for easing the computations.

In Chapter 9, we present asymptotic results for extrinsic means of submanifolds on the Euclidean space that are due to Hendriks and Landsman (1998) [154]; special attention is paid to Stiefel manifolds, including the orthogonal groups and the special orthogonal groups.

Chapter 10 applies the results from Chapter 5 to (i) real projective spaces $\mathbb{R}^{N-1}$ - the *axial spaces* and (ii) complex projective spaces $\mathbb{C}P^{k-2}$- the *planar shape spaces*. Another application is to the products of real projective spaces $(\mathbb{R}P^m)^{k-m-1}$, or the so-called *projective shape spaces* (see Mardia and

Patrangenaru (2005) [233]). In this chapter, one considers two-sample tests for mean projective shapes based on the Lie group structure of 3D projective shapes in general position and nonparametric bootstrap tests for VW mean 3D projective shape change.

In Chapter 11 the one-sample test for functional data is applied to the problem of identifying the projective shape of a planar curve. This leads to Chapter 12, which presents general theory for nonparametric statistics on Hilbert manifolds, including a one-sample neighborhood hypothesis test for extrinsic means.

Chapter 13 details similarity shape analysis for k-ads that lie in the plane as well as those in higher dimensions. The shape spaces for such configurations are described along with embeddings with which to perform statistical analysis. The chapter concludes with definitions of extrinsic means in this context and the asymptotic distributions of their sample analogues. For medical imaging problems, nonparametric statistics on size-and-reflection manifolds in 2D or 3D are performed for purposes of diagnostics, discrimination, and/or identification.

Chapter 14 focuses on equivariant embeddings of Grassman manifolds. Grassmann manifolds were identified in Chapter 5 as manifold models for affine shape spaces. Their equivariant embeddings are used to derive the extrinsic means and extrinsic covariance matrices for data on Grassmannians.

A variety of applications for the previously presented methodology are presented in Part III. Chapter 15 presents an application in diffusion tensor imaging (DTI), which is a fairly novel modality of MR imaging that allows noninvasive mapping of the brain's white matter. A particular map derived from DTI measurements is a map of water's principal diffusion directions, which are proxies for neural fiber directions. The method presented here naturally leads to an analysis based on Cholesky decompositions of covariance matrices, which helps to decrease computational time and does not increase dimensionality. The resulting nonparametric matrix-valued statistics are used for testing if there is a difference on average between corresponding signals in Diffusion Tensor Images in young children with dyslexia when compared to their clinically normal peers, based on data that was previously analyzed using parametric methods. Both methodologies show a significant difference.

The solar nebula theory hypothesizes that planets are formed from an accretion disk of material that over time condenses into dust, small planetesimals, and planets that should have, on average, coplanar, nearly circular orbits. If the orbit of Pluto has a different origin than the other planets in the solar system, that will have a tremendous effect on modeling the spacecrafts for a mission to Pluto. The nebula theory is tested for Pluto. Chapter 16 uses both parametric and nonparametric methods to test this hypothesis.

Chapter 17 presents applications to medical imaging data analysis for distributions on the complex projective space $\mathbb{C}P^{k-2}$, as a space of planar similarity shapes of k-ads, which are embedded in spaces of matrices via Veronese-Whitney embeddings, as well as a test for increased internal ocular pressure (IOP) detection in animal models, run on the 3-dimensional direct similarity shape space Σ_3^4.

Motivated by the problem of nonparametric inference in high level digital image analysis, Chapter 18 applies the methodology for statistical analysis on Hilbert manifolds presented in Chapter 12 to the analysis of shapes of contours lying in the plane. The space of such shapes, which is a Hilbert manifold, is embedded into a space of Hilbert–Schmidt operators to define extrinsic mean shapes and their sample analogues. Computational restrictions faced when utilizing digital imaging data are also considered.

Chapter 19 focuses on various preprocessing and post-processing steps in a larger project on planning reconstructive surgery in severe skull injuries. These steps are needed in order to perform a reflection size-and-similarity shape based statistical analysis of the human skull based on CT images. The image processing problems associated with extracting the bone structure from the CT images are considered, as well.

In Chapter 20, a study of affine shapes is considered in connection with problems that arise in bioinformatics and pattern recognition. In particular, affine shape analysis is used in retrieving a larger image from aerial images. Linear shape spaces are defined here, and identified with Grasmannians, leading to an analysis of 2D electrophoresis images for protein matching.

In Chapter 21, the problem of identifying the projective shape of a planar curve is considered as a practical application of the neighborhood hypothesis testing in a data-driven example where we determine δ, the smallest radius of the neighborhood hypothesis for which the neighborhood hypothesis is not rejected. The theory is applied to the recognition of the projective shape of a planar curve extracted from digital images of a flat scene.

Following Patrangenaru et al. (2010) [272], in Chapter 22, one develops a nonparametric methodology for analysis of projective shapes of configurations of landmarks on real 3D objects from their regular camera pictures. A fundamental result in computer vision, emulating the principle of human vision in space, claims that, generically, a finite 3D configuration of points can be retrieved from corresponding configurations in a pair of camera images, up to a projective transformation. Consequently, the projective shape of a 3D configuration can be retrieved from two of its planar views, and a projective shape analysis can be pursued from a sample of images. Using large sample and nonparametric bootstrap methodology for extrinsic means on manifolds, one can give confidence regions and tests for the mean projective shape of a 3D configuration from its 2D camera images. In Chapter 22, two examples are given: an example of testing for accuracy of a simple manufactured object using mean projective shape analysis, and a face identification example. Both examples are data driven based on landmark registration in digital images.

In Chapter 23, we extend tests for mean 3D projective shape change in matched pairs to independent samples. We provide a brief introduction of projective shapes of spatial configurations obtained from their digital camera images, building on previous results of Crane and Patrangenaru (2011) [78]. The manifold of projective shapes of k-ads in 3D containing a projective frame at five given landmark indices has a natural Lie group structure, which is inherited from the quaternion multiplication. Here, given the small sample size, one estimates the mean 3D projective shape change in two populations, based on independent random samples of possibly different sizes using Efron's nonparametric bootstrap (see Efron (1979) [96]). This methodology is applied in three relevant applications of analysis of 3D scenes from digital images: visual quality control, face recognition, and scene recognition.

Stereo data of the eye is the most common type of imaging data for eye disease detection and control. Chapter 24 considers an application of projective shape analysis to a set of data from the Louisiana Experimental Glaucoma Study (Burgoyne et al. (2000) [56]). A matched pairs test is presented and applied to detect mean glaucomatous projective shape change for this data.

In Chapter 25, one gives a class of adjusted Pelletier density estimators, on homogeneous spaces, that converge uniformly and almost surely at the same rate as naive kernel density estimators on Euclidean spaces. A concrete example of projective shape density estimation of 6-ads arising from digitized images of the "actor" data set in Chapter 1 is also given here.

Part IV of this book considers some additional topics. Chapter 26, considers the related topic using topological methods for multivariate statistics. This chapter is from Bubenik et al. (2010) [53]. Using *persistent homology* from computational algebraic topology, a random sample is used to construct estimators of persistent homology. The estimation procedure can then be evaluated using the so-called bottleneck distance (see Cohen-Steiner et al. (2005) [75]) between the estimated persistent homology and the true persistent homology. The connection to Statistics comes from the fact that, when viewed as a nonparametric regression problem, the distance is bounded by the sup-norm loss. Consequently, a sharp asymptotic minimax bound is determined under the sup–norm risk over Hölder classes of functions for the nonparametric regression problem on manifolds. This provides good convergence properties for the persistent homology estimator in terms of the expected bottleneck distance.

Chapter 27 considers some additional directions in nonparametric statistics on manifolds and consists of two main sections. The first one discusses some recent developments in this area that are not presented at length in this text, but certainly warrant inclusion. These topics are extensions of principal component analysis and spatial statistics for data on manifolds, shape analysis of surfaces in three dimensions, and the analysis of data on stratified spaces. The second section is concerned with computational issues with calculating Cartan means on manifolds. This section is motivated by the flurry of papers that have appeared in computer vision, statistical learning, pattern recognition, medical imaging, and other computationally intensive applied areas that utilize nonparametric methodology for statistics on manifolds. While pursuing location parameters in these works, scientists have been using intrinsic means almost without exception. However, there are often computational, in addition to theoretical, advantages to utilizing extrinsic means instead. This section presents a number of examples in which this is the case.

This is the first extended monograph on statistics on manifolds and their applications to Object Data Analysis, addressing in detail the blend of geometry, statistics, and their practical case studies from a nonparametric viewpoint. The targeted audience is graduate students in Mathematics, Statistics, Biostatistics and Bioinformatics and Engineering. The book is also addressed to a larger community, including statisticians, mathematicians, computer vision experts, image analysts, bioinformaticians, medical imaging specialists, geologists and geophysicists, and anthropologists, among others.

Acknowledgement

We are very grateful to David Grubbs for inviting us to write this monograph, to Vasken Pambuccian for his kind words in the Foreword, and to the referees for their constructive suggestions that helped improve the quality of the manuscript.

This book would not have come to the light without continuous summer research support during the past ten years from the National Science Foundation and from the National Security Agency, for which we are extremely grateful. Many of the results in this book are from joint research papers with Getulio Amaral, Vladimir Balan, Ananda Bandulasiri, Rabi Bhattacharya, Marius Buibas, Michael Crane, Gordana Derado, Xavier Descombes, Ian L. Dryden, David Groisser, Harrie Hendriks, Thomas Hotz, Stephan Huckemann, John T. Kent, Huiling Le, Xiuwen Liu, Kanti V. Mardia, Steve J. Marron, Ezra Miller, Axel Munk, Megan Owen, Rob L. Paige, Johnny Pang, Vlad P. Patrangenaru, Raymond Pruett, Mingfei Qiu, Frits Ruymgaart, Armin Schwartzman, Sean Skewer, Samanmalee Sughatadasa, Hilary H. Thompson, Paul San Valentin, Andy Wood and Jinfeng Zhang. We thank them for our wonderful collaboration, and for their willingness to bring Statistics on Manifolds to new heights.

We are extremely grateful to Peter Kim for providing LaTeX and image files for Chapter 7 and Chapter 26 from his published work, to Frits Ruymgaart for his notes on asymptotics, and to Rudy Beran for his excellent hints for improving an earlier version of the manuscript.

The second author wishes to thank his Ph.D. students, Dhanamalee Banadara, Chalani Prematilake, and Iresha Premarathna, for thoughtfully reading through sections of the manuscript and providing helpful suggestions for improvement. He also wishes to thank his parents for instilling in him the character, strength, work ethic, and yearning to do what he loves that have brought him to where he is today. Their unconditional support has helped him through both the toughest and greatest of times and the completion of this manuscript.

The first author was very fortunate to meet the late Wijesuriya Dayawansa, whose Applied Mathematics insight helped him realize the role of 3D projective shape in human vision; memories of his bright and unselfish persona and high scientific standard will live forever. He is also thankful to his Differential Geometry PhD adviser Izu Vaisman and to his Mathematical Statistics PhD adviser Rabi N. Bhattacharya.

Above all, the first author would like to thank his wife Ania and his son Vlad, for their unabated support during the preparation of this monograph.

Part I

Nonparametric Statistics on Manifolds

Chapter 1

Data on Manifolds

There are many disciplines in which data arises on object spaces which are manifolds. Among these are anthropology, astronomy, bioinformatics, computer vision, geology, image analysis, medical imaging, meteorology, and statistics. Over the next several chapters, this monograph will present the general theory and methodology underlying a nonparametric statistical analysis of data arising on manifolds. However, each application and type of data will have its own specific problems that need to be taken into consideration in the analysis, which will be addressed in later portions of this book. First, though, we wish to provide a number of examples of data lying on manifolds. In addition to showing applications in which such data arise, these example can help provide context in subsequent chapters focusing on theory.

1.1 Directional and Axial Data

Statistical data analysis on spheres is a relatively old discipline. Watson (1983) [333] points out that one of the first statistical tests ever known is due to D. Bernoulli (1734) [21], who was asking whether the unit normals to orbital planes are uniformly distributed on the celestial sphere. Here, the angle of the orbital plane of a planet is in reference to the ecliptic, which is the apparent path of the Sun around the Earth.

Let i be the inclination of the orbital plane of a planet to the ecliptic and Ω be the angle between a fixed line in the ecliptic (the line joining the Sun and the Earth at the time of the vernal equinox) and the line joining the ascending node of the planet (the point where the orbit of the planet rises to the positive side of the ecliptic). Then each orbit determines one directed unit vector n perpendicular to the orbital plane of the planet with the sense of direction given by the right hand rule,

$$n = (\sin(\Omega)\sin(i), -\cos(\Omega)\sin(i), \cos(i))$$

The University of Uppsala data (Mardia and Jupp (2000), Table 10.2) [230] provides a set of measurements (i, Ω) for the (at the time) nine planets in the solar system. From this data, Patrangenaru (1998) [266] derived the coordinates n_x, n_y, n_z of the unit normals to orbital planes of the planets as of 1998. See Table 1.1 for these coordinates.

Table 1.1: *The normals to the orbital planes of the nine planets in the solar system*

Planet	n_x	n_y	n_z
Mercury	0.001151	0.121864	0.99255
Venus	0.022170	-0.054694	0.99826
Earth	0.000000	0.000000	1.00000
Mars	0.032156	-0.002858	0.99948
Jupiter	0.020454	-0.010471	0.99974
Saturn	0.013473	-0.041487	0.99905
Uranus	0.012596	0.004514	0.99991
Neptune	0.029663	-0.009412	0.99952
Pluto	0.241063	0.170303	0.95545

Additional data on spheres or projective planes can be found in Fisher et al. (1987) [117]. One such example concerning wind directions at a given location on Earth can be found on p. 308 in that reference and is graphically displayed here in Figure 1.1. The data, itself, can be found in Table 1.3 at the conclusion of this chapter.

1.2 Similarity Shape Data and Size and Shape Data

Images arise as data in a number of fields, including anthropology, biology, computer vision, and medicine. In a number of cases, the entire image may not be of interest to researchers. Instead, they may be interested only in describing certain geometric information, commonly called the shape, of key features of the image. Depending on the manner in which the images were obtained, this

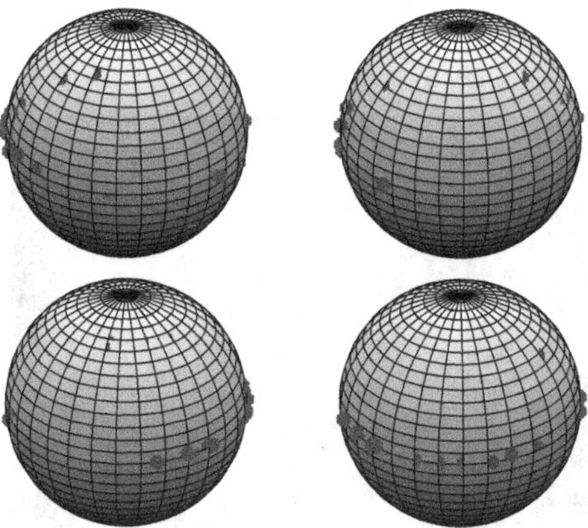

Figure 1.1 *Four views of a sample of wind directions on consecutive days of March 1981, near Singleton, United Kingdom, at a height of 300 meters. (Source: Bhattacharya et al.(2012), Figure 4. Reproduced by permission of John Wiley & Sons LTD).*

data may lend itself to **similarity shape** or **size-and-shape** analysis. We will now consider a few examples of such data.

A useful data library, maintained by James Rohlf at the State University of New York, Stony Brook, can be found on his website [291]. This library includes classical data sets from Bookstein (1991) [51] in electronic format. Such data typically contains observations that consist of lists of coordinates for **landmarks**, which are points of key interest in an image. Two such data sets describe locations in the human skull. For instance, the University School Study data (Bookstein (1991), pp. 400-405) contains landmark coordinates from X-rays of children's midface bones. These observations can be found in Tables 1.4, 1.5, 1.6, and 1.7 at the conclusion of this chapter.

The Apert data set (Bookstein (1991), pp. 405-406), as shown in Table 1.8, consists of a number of landmarks describing children who have Apert syndrome. Apert syndrome is a genetic craniosynostosis, of a markedly deformed tower-shaped head resulting from the premature fusion of all cranial sutures. X-rays of a clinically normal and an Apert syndrome skull are displayed in Figure 1.2. In each of these examples, the data sets describe 2D, or planar, similarity shapes of the finite number of landmarks. For additional examples of 2D similarity shape data, we recommend the data sets in Dryden and Mardia (1998) [91].

While classical planar similarity shape analysis is concerned with the types of data shown in the preceding examples, in recent years, many researchers have begun to focus instead on analyzing the direct similarity shape of outlines or boundaries of objects, which are often referred to as *contours*, as infinite-

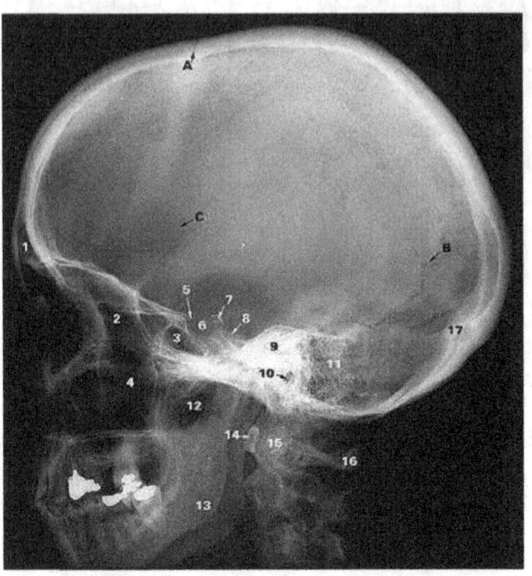

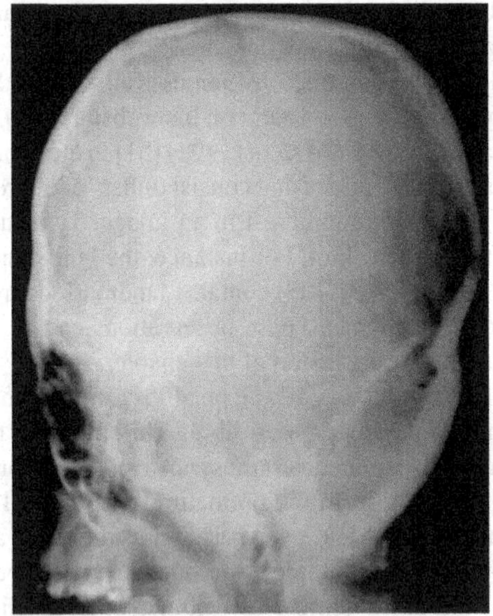

Figure 1.2 *Lateral X-ray of a clinically normal skull (top, with landmarks) and an Apert syndrome skull (bottom). (Source: Bandulasiri et al.(2009), Figure 4. Reproduced by permission of Elsevier).*

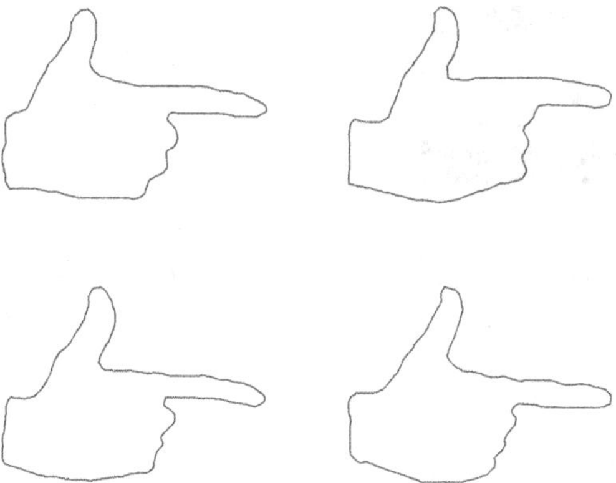

Figure 1.3 *Sample of 4 contours of a hand gesture. (Source: Bhattacharya et al.(2012), Figure 4. Reproduced by permission of John Wiley & Sons LTD).*

dimensional objects rather than finite. For instance, consider the following sample of contours of hand gestures from Sharvit et. al. (1998) [306] shown in Figure 1.3. Here, the contours are represented by evaluating the underlying function at a sufficiently large number of sampling points, as suggested in Ellingson et al. (2013) [104]. A list of the 300 sampling points used for Figure 1.3 is given in Tables 1.9, 1.10, 1.11, 1.12, 1.13, 1.14, 1.15, and 1.16.

Albert Einstein's brain was removed shortly after his death (most likely without prior family consent), weighed, dissected and photographed by a pathologist. Among other pictures, a digital scan of a picture of the General Relativity creator's half brain taken at the autopsy is displayed below; we extracted the contour of the corpus callosum(CC) from this Einstein's brain image, the shape of which would be set as a null hypothesis in our testing problem (see Figure 1.2).

Fletcher (2013) [118] extracted contours of CC midsagittal sections from MRI images to study possible age related changes in this part of the human brain. His study points to certain age related shape changes in the CC. Given that Einstein passed away at 76, we consider a subsample of CC brain contours from Fletcher (2013) [118] in the age group 64-83 to test how far the average CC contour is from Einstein's. The data is displayed in Figure 1.2.

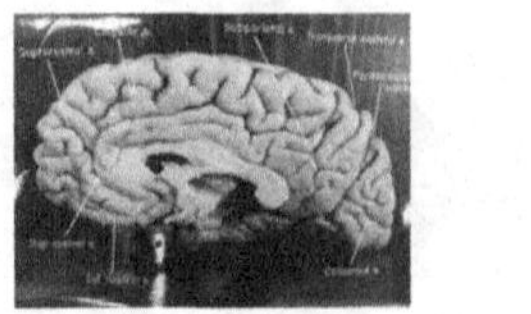

Figure 1.4 *Right hemisphere of Einstein's brain including CC midsagittal section (left) and its contour (right).*

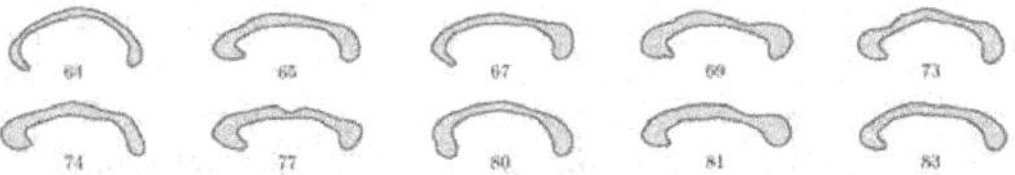

Figure 1.5: *Corpus callosum midsagittal sections shape data, in subjects ages - 65 to 83*

With advancements in imaging technology, **3D-direct similarity shape and 3D size-and-shape data** are also produced in various fields, including the biological sciences and medicine. Our first example of such data arises from medical imaging. In the Louisiana State University Experimental Glaucoma Study (LEGS), the optic nerve head (ONH) region of both eyes of twelve mature Rhesus monkeys were imaged with a Heidelberg Retina Tomograph (HRT) device, also called Scanning Confocal Laser Tomograph (SCLT). The experimental glaucoma was induced in one eye of each animal and the second eye was kept as control. The images are 256×256 arrays of elevation values which represent the "depth" of the ONH and are thus range images. Figure 1.6 shows contour plots for glaucomatous change (see Derado et al. 2004 [82]).

From clinical experience, it is known that the ONH area contains all the relevant information related to glaucoma onset. Figure 1.7 shows the relevant area with four landmarks. Namely, S for the superior aspect of the retina towards the top of the head, N for the nasal or nose side of the retina, T for temporal, the side of the retinal closest to the temple or temporal bone of the skull, and V for the ONH deepest point. The first three are anatomical landmarks and the fourth one is a mathematical landmark. A fifth landmark, called I for inferior, was also recorded.

The data set was obtained from a library of Heidelberg Retina Tomograph (HRT) images of the complicated ONH topography. Those images are so-called *range images*. A range image is, loosely speaking, like a digital camera image, except that each pixel stores a depth rather than a color level. It can also be seen as a set of points in $3D$. The range data acquired by $3D$ digitizers, such as optical scanners, commonly consist of depths sampled on a regular grid. In the mathematical sense, a range image is a $2D$ array of real numbers which represent those depths.

A combination of modules in C++ and SAS took the raw image output and processed it into 256×256 arrays of height values

Another byproduct was a file that we will refer to as the "abxy" file. This file contains the following information: subjects names (denoted by: 1c, 1d, 1e, 1f, 1g, 1i, 1j, 1k, 1l, 1n, 1o, 1p), observation points that distinguish the normal and treated eyes and the 10 or 15 degree fields of view for the imaging. Observation point "03" denotes a 10 degree view of the experimental glaucoma eye, "04" denotes 15 degree view of the experimental glaucoma eye, "11" – 10 degree fellow normal eye, "12"– 15 degree fellow normal eye. Recall that of the two eyes of one animal one was given experimental glaucoma, and the other was left untreated (normal) and imaged over time as a control.

The coordinates of the ellipse that determines the border of the optic nerve head were determined by the software of the HRT as it interacts with the operator of the device.

Two-dimensional coordinates of the center of these ellipses, as well as the sizes of the small and the large axes of the ellipses, are also stored in the "abxy" file. To find out more about the LSU study and the image acquisition, see Burgoyne et al. (2000) [56]. Files names (each file is one observation) were constructed from the information in these so-called "abxy" file. The list of all the observations was then used as an input for the program (created by G. Derado in C++), which determined the three dimensional coordinates of the landmarks for each observation considered in our analysis.

The XY coordinates of the "cardinal" papilla landmarks, were recovered from "abxy" data file in the LEGS library, which further allowed a reading from each of the the Z coordinate from the corresponding 256×256 array files.

The original data was collected in experimental observations on Rhesus monkeys. Given that, after treatment, a healthy eye slowly returns to its original shape, for the purpose of IOP increment detection, only the first set of after-treatment observations of the treated eye were considered. Table 1.17 contains the original sample coordinates, The filenames and their corresponding x_i^r coordinates, in microns, are given for 12 animals.

A large source of 3D size-and-shape data is the RCSB Protein Data Bank (PDB) website at http://www.rcsb.org/pdb/home/home.do, which provides a wealth of information about , including about their physical structures. As of March 1, 2015, there were 106,858 structures posted on the there. One of the many problems of interest in bioinformatics is the relationship of the physical structure and chemical sequence of a protein to its biological function and size-and-shape analysis provides one avenue for exploring this relationship.

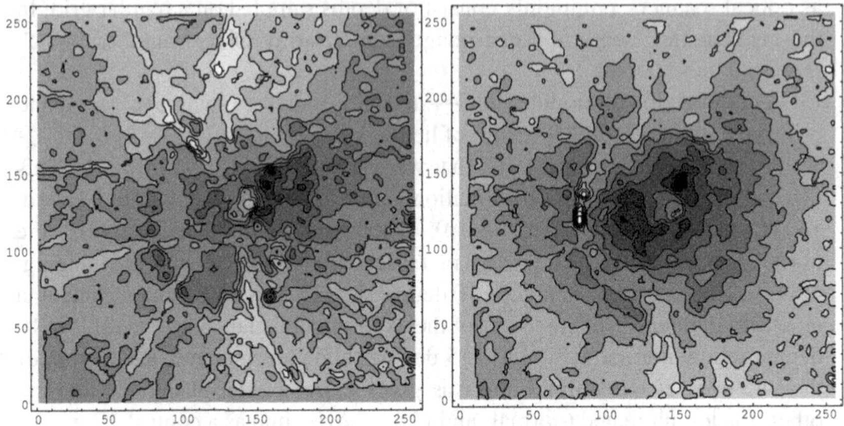

Figure 1.6 *Range images as contour plots for HRT images of a Rhesus monkey ONH area before and after the induced glaucomatous shape change. (Source: Derado et al.(2004), Figure 3. Reproduced by permission of Taylor & Francis).*

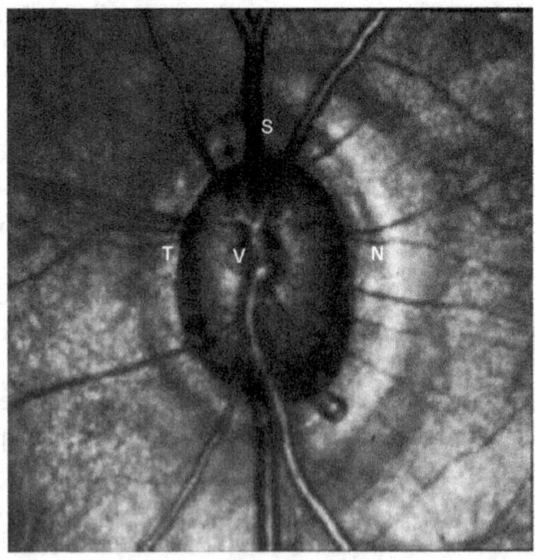

Figure 1.7 *Magnified HRT image of the central region of the retina including ONH, and the four landmarks N, S, T, V. (Source: Derado et al.(2001), Figure 1. Reproduced by permission of Taylor & Francis).*

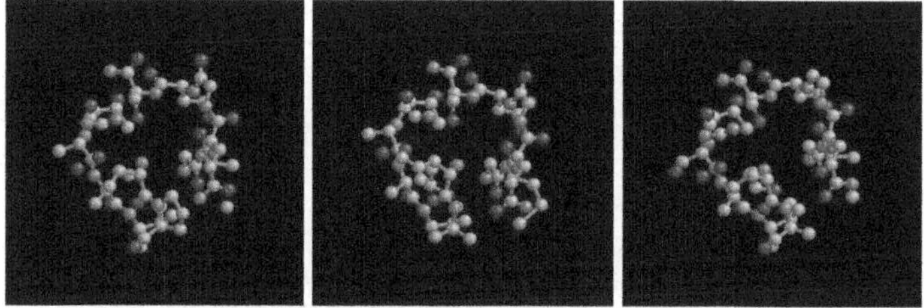

Figure 1.8 *Serine proteinase protein binding site structures.(Source: Bhattacharya et al.(2012), Figure 4. Reproduced by permission of John Wiley & Sons LTD)*

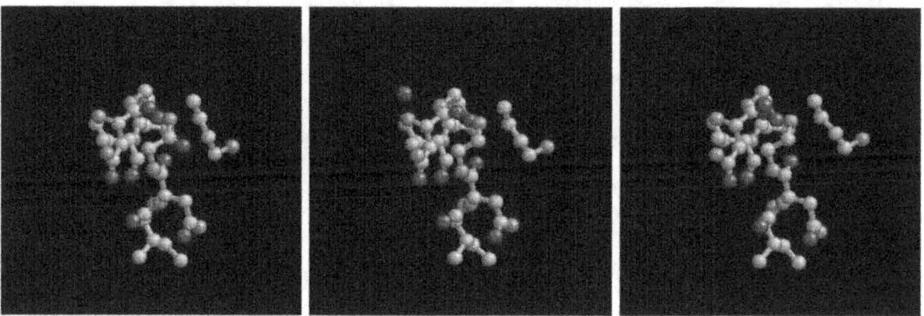

Figure 1.9 *Acid proteinase protein binding site structures. (Source: Bhattacharya et al.(2012), Figure 4. Reproduced by permission of John Wiley & Sons LTD).*

Here, we will consider the 3D atomic structures of two groups of binding sites, which are locations on the surface of a protein where binding activity occurs. In Figure 1.8, we show binding sites from three examples of *hydrolase (serine proteinase)* which were obtained by X-ray diffraction, as displayed using the software Rasmol. Their structure i.d.s on PDB are 1ela, 1eld and 1ele and their primary citation is Mattos et al. (1997) [237]. In Figure 1.8, the atoms are gray level coded where hydrogen is dark gray, oxygen is medium gray, and carbon is light gray. Similarly, Figure 1.9 displays binding sites for three examples of *acid proteinase*. The coordinates and chemical types of atoms from these acid proteinase binding sites are shown in Tables 1.18 and 1.19, where the units are measured in $\mathring{A}$.

1.3 Digital Camera Images

As suggested in the previous section, the manner in which image data is obtained impacts the type of shape information contained in the image. For similarity shape analysis to be suitable, great care must be taken with collecting the data. On the other hand, because image acquisition from regular digital cameras is based on a principle, **projective shape** data is overwhelmingly the least expensive and most available source of digital imaging data. The important

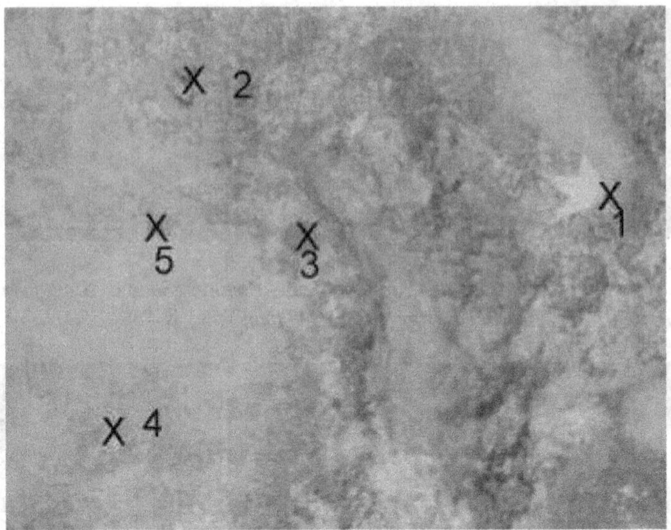

Figure 1.10 *Landmarks - Sope Creek data. (Source Patrangenaru (2001), Figure 1. Reproduced by permission of Taylor & Francis).*

distinction here is that since two digital images of the same flat scene roughly differ by a projective transformation, *in the absence of additional knowledge of the 2D scene pictured*, the only information that can be retrieved from an image is the resulting 2D projective shape.

A set of basic 2D projective shape data is contained in Table 1.20. This set consists of coordinates of five landmarks identified across a random sample of 41 scanned images of a flat river stone. The marked landmarks in one of these images are displayed in Figure 1.10. A more sophisticated example of planar projective shape data is the so-called "Big Foot" data set from Munk et al. (2007) [252]. A subset of this data set is displayed in Figure 1.11. This data involves projective shapes of planar contours, rather than configurations of landmarks.

Another example of projective shape data is in the area of face recognition. Our data consists in fourteen digital images of the same person (an actor posing in different disguises), from the live BBC program "Tomorrow's World". Face appearance in these pictures may be neither frontal or lateral, as seen in Figure 1.12. Table 1.21 contains the coordinates of eight landmarks: the tip of the nose (1), the base of the nose (2), the left end of lips (3), the right end of lips (4), the left end of left eye (5), the right end of left eye (6), the left end of right eye (7) and the right end of right eye (8).

Since scenes are naturally three dimensional, the 2D projective shape of one image of a 3D scene often yields an inaccurate reflection of the true pro-

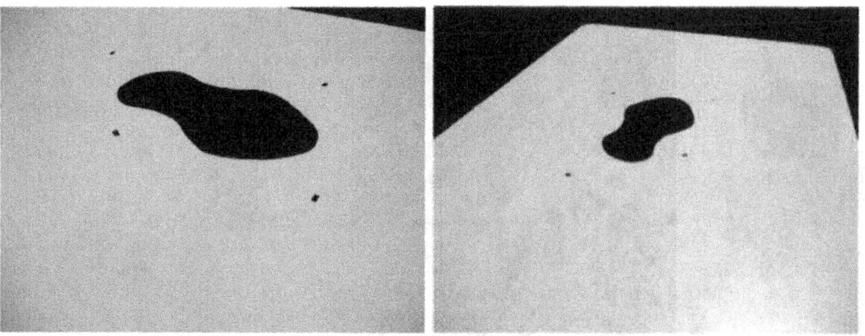

Figure 1.11 *Two views of a planar scene including a natural projective frame (the four points) and a curve (edge of the footprint).*

Figure 1.12 *BBC data:14 views of an actor's face in different disguises and poses. Front views are on the top row.*

jective shape of the scene. Instead, if one has at least two images of the scene, the scene's 3D projective shape can be extracted, as shown in Chapter 22. We consider a case of 3D projective shape in our next example. Here, we consider a 3D polyhedral object manufactured from three cubes to match a blueprint displayed in Figure 1.13. The three cubes sit on the top of each other as shown in the left side of Figure 1.13 and have sides, from top to bottom, of four, six, and ten units. The right side of Figure 1.13 displays a digital image of the object with visible corners, taken as landmarks, numbered from 1 to 19.

Sixteen randomly selected pictures of the object that show all of the selected landmarks were paired into eight pairs of images. The recording of landmark coordinates of camera image pairs was done using the MATLAB commands *imread* and *cpselect*. The sixteen images of the object are displayed in Figure 1.14. The 2D coordinates of the visible corners that were selected as landmarks are listed in Table 1.22 (see from Patrangenaru et al. (2010)[272]).

This sample of images in Figure 1.14 will be considered in Chapter 22, for a one-sample test for a VW-mean 3D projective shape. A second data set of 16 digital images of a related polyhedral scene, that was obtained by a slight modification of the first polyhedral object, is displayed in Figure 1.15. These pictures were taken with another digital camera, and the 2D coordinates of the visible corners selected as landmarks are listed in Table 1.23. The data sets in Tables 1.22 and 1.23 will be used for two-sample tests for VW-means 3D

Figure 1.13 *Blueprint of a 3D object (left) and nineteen landmarks displayed that are used for 3D scene reconstruction, and statistical analysis (right). (Source: Patrangenaru et al.(2010), Figure 1 and Figure 2. Reproduced by permission of Elsevier).*

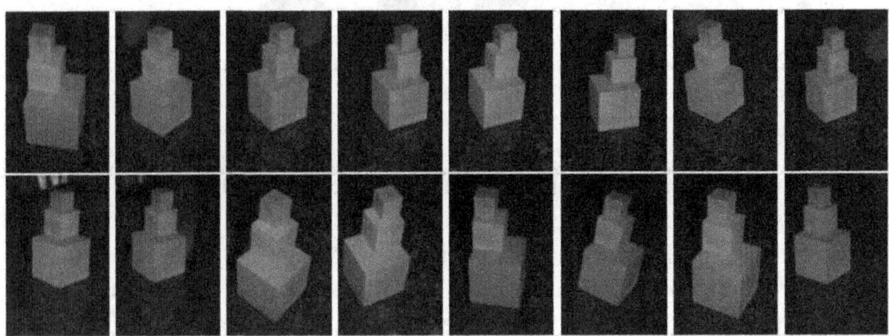

Figure 1.14 *Sixteen digital camera views of a 3D object resembling the blueprint. (Source: Patrangenaru et al.(2010), Figure 3. Reproduced by permission of Elsevier).*

projective shapes in Chapter 23.

In this example the data consists of twenty-four photos taken of the busts of the Greek philosopher Epicurus. These are displayed in Figure 1.16. Sixteen of the images are from a one-headed statue, others, in the third row are from a double-headed statue, including Epicurus and one of his disciples. Nine landmarks, displayed in Figure 1.17 were selected from the right half of the face of the statues. In Figure 1.17 the facial landmarks used in our image analysis are marked from 1 to 9. The registered coordinates are used in Chapter 23 for two sample tests for mean 3D projective shapes extracted from the data.

Figure 1.15 *Top box slightly larger than in Figure 1.14. (Source: Patrangenaru et al.(2014), Figure 3. Reproduced by permission of Springer)*

Figure 1.16 *Epicurus bust images - bottom row shows his disciple the back of the bust. (Source: Patrangenaru et al.(2014), Figure 8. Reproduced by permission of Springer).*

1.4 Stereo Imaging Data of the Eye Fundus

Medical imaging has changed the medical practice in many ways. Nevertheless, some of the medical devices used are prohibitively expensive, or worse, can do harm, when used in excess. For this reason cheaper and less invasive imaging devices are welcome by medical practitioners. In the case of eye imaging, besides the Heidelberg Retina Tomograph (HRT), a cheaper device used is a stereo camera. The data provided here consists of 15 sets of pairs of stereo images of the complicated optic nerve head (ONH) topography in animal models, that are displayed in Figure 1.18. Of the two eyes of one subject, one was given experimental glaucoma, and the other was left untreated (normal) and

Figure 1.17 *Landmarks for Epicurus bust. (Source: Patrangenaru et al.(2014), Figure 10. Reproduced by permission of Springer).*

imaged over time as a control. In Figure 1.18, for each subject, there are four images: one stereo pair for the normal eye, and one stereo pair for the glaucomatous eye, so that each row consists in eye fundus images from two Rhesus monkeys, except for the bottom row that has images from one subject only.

The coordinates of nine landmarks on the approximate elliptic contour that determines the ridge of the ONH are recorded, as well as those of certain blood vessels junctions and the estimated location of the deepest point are tabulated in the tables 1.24, 1.25, 1.26. In addition to the landmarks S(superior), I(inferior), N(nasal), T(templar) and V(vertex), considered for HRT data, in Figure 1.19 we marked four anatomical landmarks SM(mid-superior), IM(mid-inferior), NM(mid-nasal) and TM(mid-templar), that are located at the junction of important blood vessels. Their positions in the ONH cup are schematically displayed in Figure 1.19.

1.5 CT Scan Data

A (CT) scan uses X-rays to make detailed pictures of structures inside of the body. A CT scan can be used to study various parts of the human body, such as the chest, abdomen, pelvis, and an arm or leg. A CT scan can also take pictures of body organs, such as the bladder, liver, lungs, pancreas, intestines, kidneys, and heart. Additionally, CT scans can be used to study the spinal cord, blood vessels, and bones. Figure 1.20 displays a series of CT scans of one person's head. This observation comes from a data set consisting of 34 different patients, as shown in Osborne (2012) [257] and Osborne et al. (2012) [259]. We will return to this data set in Chapter 19.

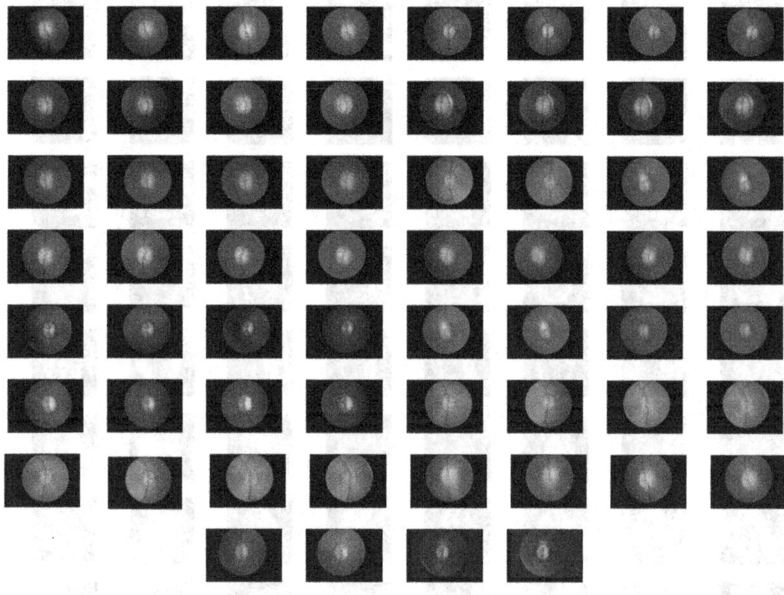

Figure 1.18 *Stereo pairs of the Optic Nerve Head region for a normal and glaucomatous eye in Rhesus monkeys. (Source: Crane and Patrangenaru(2011), Figure 2. Reproduced by permission of Elsevier).*

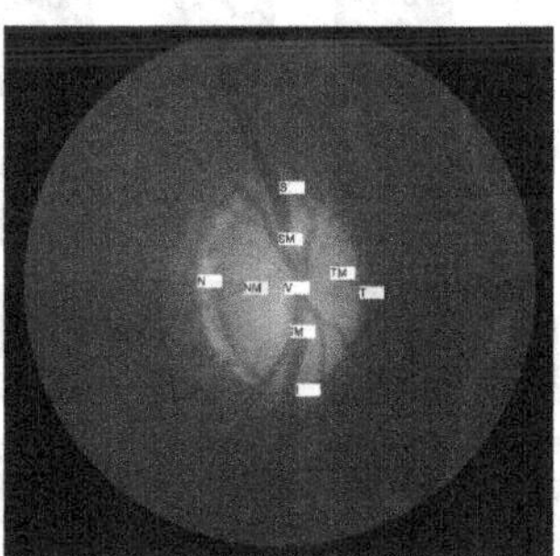

Figure 1.19 *Nine anatomical landmarks of the Optic Nerve Head region. (Source: Crane and Patrangenaru (2011), Figure 3. Reproduced by permission of Elsevier).*

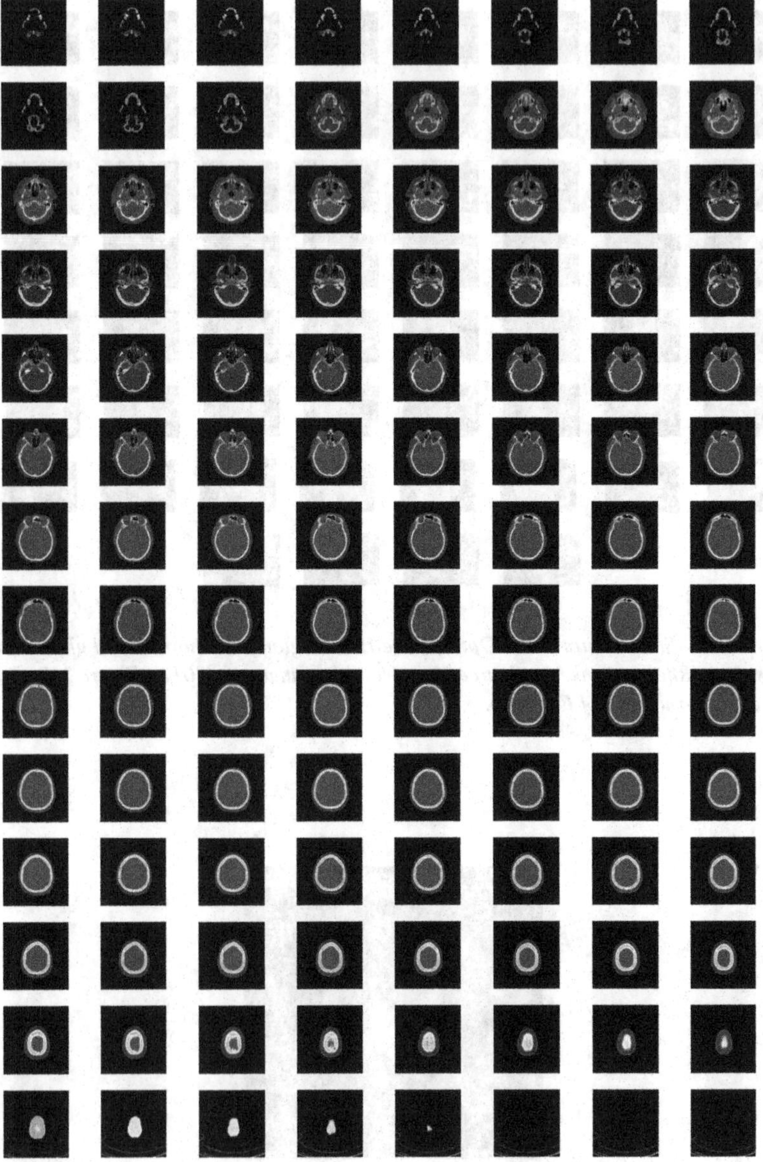

Figure 1.20 *CT scan slices of a human head from the bottom (top left) to the top (bottom right).*

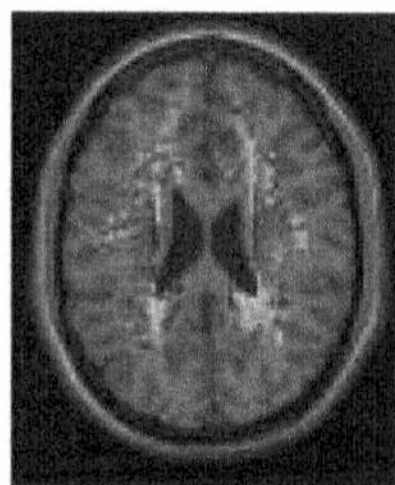
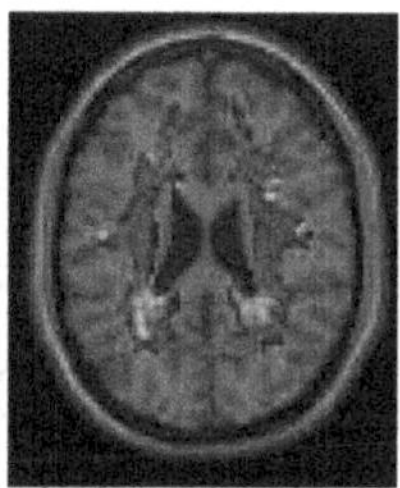

Figure 1.21 *DTI slice images of a control subject (left) and of a dyslexia subject (right). (Source: Osborne at a.(2013), Figure 1, p.171. Reproduced by permission of Elsevier).*

1.6 DTI Data

Diffusion Tensor Imaging (DTI) is a sophisticated magnetic resonance (MR) imaging technique that enables researchers to visualize white matter fibers in the brain. The researchers can then trace subtle changes in the white matter that are associated with either unusual brain wiring, as in conditions like dyslexia and schizophrenia, or brain diseases, such as multiple sclerosis or epilepsy. Diffusion tensor images store information about the diffusion of water molecules in the brain. As an example, Figure 1.21 displays a DTI slice for a control subject and a patient with dyslexia, which is a reading disability characterized by a delay in the age at which a child begins to read.

The diffusion of a water molecule at a certain location in the brain has a probability distribution in 3-dimensional space. Its diffusion D is one-half of the covariance matrix of that distribution, which is a symmetric positive semidefinite matrix. The diffusion matrix

$$D = \begin{pmatrix} d_{11} & d_{12} & d_{13} \\ d_{21} & d_{22} & d_{23} \\ d_{31} & d_{32} & d_{33} \end{pmatrix}$$

at any given voxel may be stored in the form of a column vector $(d_{11}\ d_{22}\ d_{33}\ d_{12}\ d_{13}\ d_{22}\ d_{23})^T$. As a result, the location, orientation, and anisotropy of the white fiber tracts can be measured. The degree of anisotropy, which is the departure from a spherical diffusion, in children's brains may offer insight to the study of dyslexia. Table 15.1 displays D for a voxel chosen from two samples of DTI slices, one consisting of clinically normal subjects and the other patients with dyslexia. The control and dyslexia subjects from Figure 1.21 are, respectively, the first and seventh observations in the table.

We will discuss theory for analyzing such data in Chapter 8 and return to this example, in particular, in Chapter 15. As a final remark, we would like to note that these methodologies may be similarly applied to cosmic background

Table 1.2 *DTI data in a group of control (columns 1 – 6) and dyslexia (columns 7 – 12).*

1	2	3	4	5	6
0.8847	0.6516	0.4768	0.6396	0.5684	0.6519
0.9510	0.9037	1.1563	0.9032	1.0677	0.9804
0.8491	0.7838	0.6799	0.8265	0.7918	0.7922
0.0448	-0.0392	0.0217	0.0229	-0.0427	0.0269
-0.1168	-0.0631	-0.0091	-0.1961	-0.0879	-0.1043
0.0162	-0.0454	-0.1890	-0.1337	-0.1139	-0.0607
7	**8**	**9**	**10**	**11**	**12**
0.5661	0.6383	0.6418	0.6823	0.6159	0.5643
0.7316	0.8381	0.8776	0.8376	0.7296	0.8940
0.8232	1.0378	1.0137	0.9541	0.9683	0.9605
0.0358	-0.0044	-0.0643	0.0309	-0.0929	-0.0635
-0.2289	-0.2229	-0.1675	-0.2217	-0.1713	-0.1307
-0.1106	-0.0449	-0.0192	-0.0925	-0.0965	-0.1791

radiation (CBR) data, which consists of 2×2 symmetric positive definite matrices.

1.7 Data Tables

Table 1.3 *Wind directions at a height of 300 m on consecutive days near Singleton, UK.*

Colatitude (°)	Longitude (°)
90	132.2
90	107.2
90	282.5
87	213.3
87.6	167.7
90	158.3
88.6	182.8
90	135
92.7	300.3
91.3	240.3
88.8	131.5
88.7	108.6
91.8	282.7
90	296.6
92.7	243.4
89.1	115.3
92.2	283.5
91.3	336.3
92.1	155.8
89.4	98.3
89.3	95.8
88.3	112.1
90	225
89.4	120.2
88.6	124.9
89	144.1
88.2	85.4
90	67.8
88.8	83.6

Table 1.4 *Coordinates of landmarks in a normal children growth study measured from X-rays at 8 and 14 years of age – part I.*

Id-months-gender	1x	1y	2x	2y	3x	3y	4x	4y
1872-96-1	6.000	6.000	9.018	6.000	4.941	4.710	5.356	3.229
1872-168-1	6.000	6.000	9.268	6.000	4.876	4.604	5.165	2.772
1890-96-2	6.000	6.000	9.142	6.000	5.082	4.685	5.776	3.414
1890-170-2	6.000	6.000	9.385	6.000	5.036	4.587	5.580	3.128
1953-96-1	6.000	6.000	8.870	6.000	4.817	4.591	5.433	3.407
1953-168-1	6.000	6.000	9.086	6.000	4.607	4.242	5.199	3.037
1956-94-2	6.000	6.000	8.789	6.000	4.800	4.622	5.546	3.175
1956-173-2	6.000	6.000	8.999	6.000	4.654	4.459	5.493	2.713
2000-96-1	6.000	6.000	8.920	6.000	4.872	4.867	5.442	3.560
2000-167-1	6.000	6.000	9.121	6.000	4.696	4.677	5.118	3.240
2002-92-1	6.000	6.000	8.893	6.000	5.028	4.822	5.609	3.359
2002-167-1	6.000	6.000	9.099	6.000	4.933	4.450	5.821	2.812
2007-97-1	6.000	6.000	8.723	6.000	4.984	4.748	5.375	3.554
2007-162-1	6.000	6.000	8.941	6.000	4.877	4.541	5.199	3.234
2026-96-1	6.000	6.000	8.886	6.000	4.975	5.099	5.605	3.829
2026-168-1	6.000	6.000	9.079	6.000	4.805	5.049	5.355	3.314
2101-96-1	6.000	6.000	9.082	6.000	4.969	4.400	5.622	3.042
2101-167-1	6.000	6.000	9.319	6.000	4.808	4.186	5.540	2.434
2102-96-2	6.000	6.000	8.809	6.000	4.828	4.769	5.338	3.495
2102-168-2	6.000	6.000	8.959	6.000	4.511	4.571	5.062	3.193
2108-96-1	6.000	6.000	9.049	6.000	5.000	4.708	5.392	3.438
2108-167-1	6.000	6.000	9.286	6.000	4.809	4.438	5.111	2.802
2109-96-2	6.000	6.000	8.694	6.000	4.992	4.814	5.396	3.624
2109-172-2	6.000	6.000	8.872	6.000	4.762	4.464	5.234	3.130
2113-95-1	6.000	6.000	8.924	6.000	5.029	4.624	5.577	3.275
2113-172-1	6.000	6.000	9.219	6.000	5.005	4.555	5.446	2.668
2123-96-2	6.000	6.000	8.926	6.000	5.122	4.510	5.640	3.345
2123-168-2	6.000	6.000	9.151	6.000	5.007	4.471	5.381	3.209
2124-96-1	6.000	6.000	8.939	6.000	4.809	4.959	5.563	3.302
2124-167-1	6.000	6.000	9.157	6.000	4.714	4.731	5.492	2.902
2135-96-1	6.000	6.000	8.642	6.000	4.881	4.954	5.350	3.535
2135-168-1	6.000	6.000	8.827	6.000	4.840	4.806	5.137	3.097
2190-90-2	6.000	6.000	8.834	6.000	5.012	4.916	5.699	3.557
2190-162-2	6.000	6.000	9.041	6.000	4.864	4.820	5.590	3.046
2191-96-1	6.000	6.000	8.917	6.000	4.954	4.407	5.504	3.294
2191-171-1	6.000	6.000	9.092	6.000	4.739	4.457	5.298	2.967
2192-90-2	6.000	6.000	8.777	6.000	5.021	4.695	5.588	3.565
2192-162-2	6.000	6.000	8.988	6.000	4.922	4.562	5.485	3.136
2196-96-2	6.000	6.000	8.687	6.000	4.917	4.887	5.374	3.578
2196-172-2	6.000	6.000	8.866	6.000	4.872	4.891	5.193	3.182
2197-95-2	6.000	6.000	8.810	6.000	4.929	4.587	5.552	3.588
2197-170-2	6.000	6.000	9.046	6.000	4.774	4.399	5.322	3.200
2198-90-2	6.000	6.000	8.785	6.000	5.040	4.756	5.693	3.538
2198-163-2	6.000	6.000	8.976	6.000	4.928	4.651	5.682	3.154
2245-95-1	6.000	6.000	9.085	6.000	5.043	4.790	5.435	3.274
2245-168-1	6.000	6.000	9.257	6.000	4.901	4.695	5.172	2.757
2257-97-1	6.000	6.000	8.863	6.000	4.816	4.757	5.378	3.546
2257-168-1	6.000	6.000	8.951	6.000	4.699	4.615	5.363	2.995
2259-96-1	6.000	6.000	9.116	6.000	4.748	4.760	5.303	3.407
2259-168-1	6.000	6.000	9.339	6.000	4.739	4.553	5.160	2.939
2271-96-2	6.000	6.000	8.837	6.000	5.091	4.731	5.546	3.320
2271-167-2	6.000	6.000	9.031	6.000	4.919	4.507	5.404	2.783
2276-95-2	6.000	6.000	8.838	6.000	5.016	4.529	5.795	3.115
2276-168-2	6.000	6.000	9.037	6.000	4.844	4.391	5.646	2.664
2279-96-2	6.000	6.000	8.805	6.000	4.773	4.817	5.432	3.586
2279-168-2	6.000	6.000	8.928	6.000	4.719	4.640	5.095	3.358
2286-90-2	6.000	6.000	8.873	6.000	4.770	4.722	5.408	3.502
2286-166-2	6.000	6.000	9.057	6.000	4.658	4.620	5.180	3.220
2367-96-1	6.000	6.000	9.015	6.000	4.779	4.715	5.448	3.086
2367-168-1	6.000	6.000	9.289	6.000	4.675	4.445	5.339	2.495
2373-96-1	6.000	6.000	8.884	6.000	5.113	4.601	5.718	3.362
2373-167-1	6.000	6.000	9.025	6.000	5.187	4.466	5.697	2.873
2377-90-1	6.000	6.000	9.034	6.000	5.099	4.647	5.695	3.348
2377-170-1	6.000	6.000	9.335	6.000	4.954	4.365	5.601	2.713
2378-97-2	6.000	6.000	8.726	6.000	4.953	4.789	5.431	3.627
2378-168-2	6.000	6.000	8.928	6.000	4.859	4.675	5.328	3.173
2392-96-1	6.000	6.000	9.029	6.000	5.065	4.641	5.722	3.166
2392-167-1	6.000	6.000	9.378	6.000	5.002	4.411	5.697	2.567
2398-96-1	6.000	6.000	9.007	6.000	4.918	4.760	5.479	3.324
2398-168-1	6.000	6.000	9.235	6.000	4.627	4.780	5.098	2.791
2399-90-1	6.000	6.000	9.084	6.000	5.000	4.707	5.636	3.206
2399-173-1	6.000	6.000	9.257	6.000	4.791	4.440	5.422	2.425
2400-90-1	6.000	6.000	8.961	6.000	4.822	4.645	5.318	3.332
2400-162-1	6.000	6.000	9.092	6.000	4.680	4.427	5.202	2.817
2406-96-2	6.000	6.000	8.964	6.000	4.838	4.827	5.397	3.217
2406-168-2	6.000	6.000	9.183	6.000	4.662	4.762	5.359	2.749

Table 1.5 *Coordinates of landmarks in a normal children growth study measured from X-rays at 8 and 14 years of age – part II.*

Id-months-gender	5x	5y	6x	6y	7x	7y	8x	8y
1872-96-1	7.528	1.739	7.007	6.069	8.823	4.159	6.740	4.265
1872-168-1	7.836	1.160	7.049	6.121	9.043	3.802	6.826	3.997
1890-96-2	7.441	1.898	7.110	5.961	8.749	3.984	6.883	4.359
1890-170-2	7.549	1.259	7.185	6.017	8.906	3.672	6.816	4.108
1953-96-1	7.494	1.522	6.959	5.945	8.736	4.122	6.754	4.241
1953-168-1	7.676	.905	7.047	6.077	9.050	3.718	6.742	3.954
1956-94-2	7.859	1.917	7.036	5.972	8.599	3.972	6.701	4.293
1956-173-2	8.303	1.269	7.048	5.972	8.776	3.678	6.767	4.086
2000-96-1	7.368	1.779	6.947	6.137	8.678	4.051	6.702	4.228
2000-167-1	7.358	1.220	7.074	6.098	8.908	3.700	6.522	4.014
2002-92-1	7.609	1.714	6.818	5.908	8.781	4.052	6.620	4.349
2002-167-1	8.192	.948	6.873	5.960	9.219	3.730	6.817	3.988
2007-97-1	7.323	1.757	6.897	6.003	8.443	4.230	6.679	4.425
2007-162-1	7.448	1.289	6.928	5.983	8.708	3.922	6.636	4.166
2026-96-1	7.601	2.278	7.059	5.942	8.732	4.250	6.779	4.518
2026-168-1	7.861	1.579	7.086	5.906	9.014	3.880	6.761	4.310
2101-96-1	7.562	1.260	7.020	5.923	9.117	3.959	6.842	3.989
2101-167-1	7.775	.403	7.033	5.960	9.383	3.544	6.884	3.664
2102-96-2	7.298	2.010	6.988	5.933	8.593	4.114	6.333	4.413
2102-168-2	7.311	1.453	6.962	6.000	8.761	3.679	6.398	4.224
2108-96-1	7.066	1.645	7.161	6.066	8.794	3.964	6.689	4.333
2108-167-1	7.193	.750	7.227	6.028	9.058	3.483	6.750	4.048
2109-96-2	7.353	1.970	7.072	6.116	8.415	4.079	6.543	4.417
2109-172-2	7.727	1.319	7.080	6.154	8.622	3.809	6.480	4.197
2113-95-1	7.556	1.823	7.059	5.997	8.820	3.967	6.754	4.270
2113-172-1	8.001	.845	7.056	5.974	9.214	3.686	6.754	3.994
2123-96-2	7.862	2.003	7.109	6.044	8.705	4.109	6.845	4.313
2123-168-2	7.803	1.401	7.205	6.129	8.744	3.864	6.761	4.152
2124-96-1	7.628	1.656	7.157	6.118	8.756	4.138	6.844	4.400
2124-167-1	7.986	1.190	7.261	6.098	9.062	3.880	6.860	4.143
2135-96-1	7.239	1.960	6.947	6.032	8.323	4.172	6.486	4.406
2135-168-1	7.150	1.226	7.009	6.009	8.368	3.757	6.397	4.144
2190-90-2	7.797	1.952	7.122	5.993	8.760	4.150	6.702	4.465
2190-162-2	8.301	1.404	7.186	5.932	9.069	3.842	6.788	4.169
2191-96-1	7.681	1.699	7.110	5.891	8.641	4.194	6.447	4.309
2191-171-1	7.749	1.184	7.154	5.941	8.837	3.925	6.385	4.154
2192-90-2	7.424	2.061	7.005	6.042	8.644	4.196	6.719	4.384
2192-162-2	7.708	1.391	7.053	6.162	8.835	3.892	6.704	4.201
2196-96-2	7.265	2.000	7.067	6.041	8.453	4.241	6.445	4.516
2196-172-2	7.454	1.361	7.098	6.015	8.644	3.938	6.422	4.250
2197-95-2	7.455	2.035	7.161	6.086	8.741	4.236	6.732	4.371
2197-170-2	7.630	1.362	7.203	6.133	8.843	3.877	6.626	4.084
2198-90-2	7.532	2.140	7.125	6.019	8.772	4.117	6.743	4.369
2198-163-2	8.008	1.575	7.099	6.053	9.034	3.790	6.789	4.121
2245-95-1	7.399	1.659	7.119	6.036	8.955	3.911	6.783	4.247
2245-168-1	7.627	.865	7.102	6.065	9.111	3.581	6.604	3.964
2257-97-1	7.379	1.992	7.021	5.956	8.590	4.192	6.446	4.311
2257-168-1	7.901	1.445	7.044	5.985	8.938	4.016	6.552	3.887
2259-96-1	7.196	1.569	7.042	6.019	8.707	4.010	6.590	4.291
2259-168-1	7.306	.781	7.088	5.965	8.960	3.555	6.557	4.055
2271-96-2	7.703	1.834	7.094	6.035	8.773	4.072	6.839	4.331
2271-167-2	7.786	.989	7.095	6.004	9.026	3.699	6.776	4.036
2276-95-2	8.319	1.972	7.078	5.981	8.848	4.218	6.961	4.331
2276-168-2	8.553	1.440	7.146	6.001	9.041	3.954	6.853	4.207
2279-96-2	7.151	1.917	7.070	6.134	8.502	4.029	6.611	4.373
2279-168-2	7.165	1.406	7.072	6.129	8.705	3.820	6.423	4.163
2286-90-2	7.237	1.777	7.019	6.085	8.612	4.086	6.626	4.340
2286-166-2	7.272	1.192	7.034	6.100	8.701	3.780	6.524	4.155
2367-96-1	7.637	1.544	7.046	5.943	8.711	3.954	6.594	4.212
2367-168-1	7.935	.754	7.116	5.963	8.957	3.600	6.570	3.887
2373-96-1	7.799	1.920	7.146	6.102	8.789	4.176	6.812	4.308
2373-167-1	8.320	1.385	7.176	6.105	9.009	4.081	6.863	4.094
2377-90-1	7.631	1.935	7.087	5.995	8.913	4.112	6.855	4.278
2377-170-1	7.991	1.266	7.137	6.020	9.332	3.721	6.847	4.036
2378-97-2	7.618	1.980	6.973	5.989	8.679	4.075	6.759	4.275
2378-168-2	7.990	1.513	7.026	5.979	8.971	3.774	6.755	4.071
2392-96-1	7.788	1.830	7.084	5.923	9.052	4.118	6.842	4.238
2392-167-1	8.319	1.180	7.103	5.972	9.507	3.781	6.954	3.991
2398-96-1	7.451	1.768	7.050	6.104	8.801	4.065	6.781	4.258
2398-168-1	7.530	.827	7.155	6.050	8.935	3.548	6.619	3.970
2399-90-1	7.989	1.798	7.244	6.115	8.973	4.012	6.967	4.319
2399-173-1	8.305	.837	7.225	6.190	9.092	3.710	6.805	4.101
2400-90-1	7.582	1.717	7.052	5.943	8.732	4.024	6.768	4.225
2400-162-1	7.758	1.132	7.083	5.976	8.962	3.819	6.718	4.030
2406-96-2	7.550	1.777	7.211	6.032	8.900	4.048	6.792	4.253
2406-168-2	7.896	1.167	7.122	6.043	9.176	3.773	6.725	4.038

Table 1.6 *Coordinates of landmarks in a normal children growth study measured from X-rays at 8 and 14 years of age – part III.*

Id-months-gender	1x	1y	2x	2y	3x	3y	4x	4y
2407-96-1	6.000	6.000	8.763	6.000	5.024	4.415	5.553	3.099
2407-169-1	6.000	6.000	8.941	6.000	4.744	4.217	5.448	2.482
2410-96-1	6.000	6.000	9.183	6.000	4.894	4.806	5.681	3.314
2410-167-1	6.000	6.000	9.396	6.000	4.766	4.737	5.425	3.001
2411-91-1	6.000	6.000	9.152	6.000	5.178	4.670	5.807	3.211
2411-162-1	6.000	6.000	9.328	6.000	5.068	4.565	5.600	2.910
2429-91-1	6.000	6.000	9.004	6.000	5.028	4.715	5.703	3.381
2429-162-1	6.000	6.000	9.230	6.000	4.966	4.550	5.498	2.878
2437-91-2	6.000	6.000	8.814	6.000	4.843	4.613	5.218	3.421
2437-163-2	6.000	6.000	8.990	6.000	4.632	4.413	5.083	3.026
2449-91-2	6.000	6.000	9.100	6.000	5.069	4.859	5.931	3.259
2449-162-2	6.000	6.000	9.269	6.000	4.888	4.759	5.810	2.996
2539-96-2	6.000	6.000	8.835	6.000	4.844	4.572	5.543	3.081
2539-168-2	6.000	6.000	8.976	6.000	4.631	4.333	5.358	2.650
2541-96-1	6.000	6.000	8.985	6.000	4.882	4.565	5.501	3.151
2541-168-1	6.000	6.000	9.137	6.000	4.816	4.329	5.421	2.702
2545-95-1	6.000	6.000	9.017	6.000	5.053	4.517	5.571	3.131
2545-168-1	6.000	6.000	9.185	6.000	4.977	4.456	5.492	2.790
2548-90-2	6.000	6.000	8.618	6.000	5.291	4.654	5.741	3.004
2548-162-2	6.000	6.000	8.754	6.000	5.178	4.422	5.698	2.409
2549-95-1	6.000	6.000	8.960	6.000	5.015	4.707	5.468	3.137
2549-168-1	6.000	6.000	9.170	6.000	4.781	4.527	5.104	2.711
2561-96-2	6.000	6.000	8.657	6.000	5.045	5.063	5.574	3.581
2561-168-2	6.000	6.000	8.777	6.000	4.810	4.861	5.394	3.227
2572-90-2	6.000	6.000	8.964	6.000	4.894	4.524	5.415	3.406
2572-168-2	6.000	6.000	9.110	6.000	4.906	4.462	5.209	2.991
2578-96-1	6.000	6.000	8.968	6.000	4.801	4.664	5.461	3.423
2578-168-1	6.000	6.000	9.221	6.000	4.590	4.358	5.314	2.680
2580-96-1	6.000	6.000	9.018	6.000	4.991	4.493	5.472	2.890
2580-168-1	6.000	6.000	9.170	6.000	4.844	4.243	5.405	2.396
2594-90-1	6.000	6.000	8.867	6.000	5.009	4.636	5.442	3.272
2594-168-1	6.000	6.000	9.068	6.000	4.839	4.500	5.046	2.726
2595-95-1	6.000	6.000	8.676	6.000	4.780	4.748	5.517	3.233
2595-167-1	6.000	6.000	8.799	6.000	4.475	4.459	5.042	2.585
2596-90-2	6.000	6.000	8.712	6.000	5.162	4.908	5.587	3.512
2596-162-2	6.000	6.000	8.925	6.000	4.995	4.786	5.544	3.045
2603-99-1	6.000	6.000	8.807	6.000	5.120	4.541	5.669	3.150
2603-168-1	6.000	6.000	8.986	6.000	5.009	4.334	5.424	2.617
2680-95-2	6.000	6.000	8.898	6.000	4.585	4.800	5.272	3.517
2680-168-2	6.000	6.000	8.995	6.000	4.563	4.710	5.167	3.090
2726-96-2	6.000	6.000	8.715	6.000	4.925	4.741	5.468	3.446
2726-168-2	6.000	6.000	8.863	6.000	4.708	4.721	5.017	3.051
2779-101-1	6.000	6.000	9.011	6.000	4.822	4.572	5.216	3.504
2779-162-1	6.000	6.000	9.241	6.000	4.589	4.353	5.047	3.008
2781-96-2	6.000	6.000	8.828	6.000	4.866	4.777	5.580	3.792
2781-167-2	6.000	6.000	8.964	6.000	4.763	4.666	5.525	3.448
2802-90-1	6.000	6.000	8.891	6.000	4.952	4.877	5.823	3.300
2802-168-1	6.000	6.000	9.085	6.000	4.796	4.718	5.607	2.898

Table 1.7 *Coordinates of landmarks in a normal children growth study measured from X-rays at 8 and 14 years of age – part IV.*

Id-months-gender	5x	5y	6x	6y	7x	7y	8x	8y
2407-96-1	7.733	1.827	6.966	6.019	8.641	3.995	6.650	4.274
2407-169-1	7.943	1.007	6.914	5.989	8.966	3.610	6.760	3.998
2410-96-1	7.732	2.014	7.328	6.054	8.922	3.996	6.761	4.422
2410-167-1	7.796	1.609	7.376	6.034	9.146	3.639	6.733	4.196
2411-91-1	7.971	1.763	7.126	5.956	8.854	3.999	6.869	4.277
2411-162-1	8.072	1.230	7.186	6.015	9.151	3.651	6.818	4.034
2429-91-1	8.090	1.797	7.022	6.217	8.943	4.242	6.924	4.410
2429-162-1	8.043	1.142	7.085	6.237	9.099	4.054	6.937	4.194
2437-91-2	7.380	1.809	6.876	6.104	8.843	4.106	6.678	4.299
2437-163-2	7.480	1.321	6.884	6.148	8.900	3.718	6.537	4.133
2449-91-2	8.253	1.826	7.205	6.009	9.161	4.172	7.001	4.297
2449-162-2	8.528	1.534	7.193	6.067	9.332	3.986	6.979	4.171
2539-96-2	7.571	1.615	7.047	5.951	8.928	3.971	6.810	4.197
2539-168-2	7.819	.988	7.049	5.974	9.001	3.804	6.701	3.951
2541-96-1	7.910	1.862	7.129	6.007	8.789	4.053	6.847	4.114
2541-168-1	8.235	1.308	7.129	5.995	9.029	3.726	6.773	3.833
2545-95-1	7.738	1.718	7.166	6.019	8.930	3.962	7.017	4.202
2545-168-1	8.058	1.235	7.187	6.040	9.127	3.703	6.954	3.969
2548-90-2	7.764	1.873	7.140	6.248	8.735	4.171	6.761	4.339
2548-162-2	8.160	1.341	7.132	6.242	8.872	3.817	6.801	4.078
2549-95-1	7.721	1.746	7.174	6.045	8.872	4.032	6.765	4.159
2549-168-1	7.796	.980	7.239	6.167	9.024	3.661	6.671	3.930
2561-96-2	7.376	2.094	7.158	6.135	8.586	4.221	6.759	4.537
2561-168-2	7.529	1.510	7.149	6.112	8.735	3.890	6.626	4.263
2572-90-2	7.372	1.500	7.057	5.972	8.946	3.832	6.775	4.159
2572-168-2	7.647	.888	7.015	5.912	9.019	3.586	6.616	3.971
2578-96-1	7.289	1.693	6.969	5.925	8.823	4.080	6.844	4.292
2578-168-1	7.874	.778	7.032	5.924	9.312	3.631	6.957	3.925
2580-96-1	7.747	1.393	7.129	5.989	8.907	3.897	6.866	4.068
2580-168-1	8.113	.736	7.137	6.123	9.103	3.683	6.981	3.816
2594-90-1	7.560	1.705	7.022	6.159	8.885	3.975	6.815	4.183
2594-168-1	7.851	.911	7.077	6.115	8.965	3.480	6.597	3.815
2595-95-1	7.688	1.895	6.920	6.008	8.868	4.048	6.655	4.222
2595-167-1	7.963	1.022	6.930	6.015	8.995	3.618	6.630	3.937
2596-90-2	7.464	2.108	7.083	6.039	8.696	4.213	6.714	4.374
2596-162-2	7.967	1.673	7.102	6.019	9.011	3.928	6.721	4.182
2603-99-1	7.679	1.730	6.939	6.140	8.821	4.081	6.741	4.241
2603-168-1	7.713	1.016	6.961	6.135	8.982	3.700	6.671	3.923
2680-95-2	7.280	1.790	7.077	5.949	8.593	3.970	6.622	4.328
2680-168-2	7.705	1.284	7.099	5.955	8.785	3.742	6.554	4.156
2726-96-2	7.569	1.829	6.981	6.102	8.731	4.104	6.791	4.366
2726-168-2	7.559	1.204	6.999	6.121	8.777	3.801	6.627	4.110
2779-101-1 7.230	1.530	7.104	6.073	8.730	3.802	6.773	4.240	
2779-162-1	7.457	.798	7.192	6.092	8.965	3.426	6.723	3.980
2781-96-2	7.267	2.253	6.975	6.002	8.491	4.143	6.650	4.423
2781-167-2	7.502	1.810	6.967	6.122	8.708	4.005	6.635	4.259
2802-90-1	7.827	1.951	7.078	6.074	8.832	4.150	6.838	4.378
2802-168-1	8.103	1.405	7.034	6.044	9.001	3.808	6.749	4.184

Table 1.8 *Coordinates of landmarks in Apert children, including their age and gender.*

Gender-age(yr)	Ax	Ay	Bx	By	Cx	Cy	Dx	Dy
Fem.-9	12.59	8.90	9.74	9.70	11.07	10.16	12.13	10.67
Male-7	12.34	8.58	10.04	10.04	11.36	9.98	12.69	10.12
Fem.-6	12.19	9.43	9.76	10.31	11.00	10.59	12.12	10.96
Male-6	12.29	8.31	10.53	9.79	11.29	9.75	12.70	9.70
Male1-15	12.11	9.39	9.70	10.24	11.05	10.47	12.21	11.00
Male-13	12.23	9.09	9.57	9.87	10.84	10.18	11.96	10.99
Fem.-11	12.46	8.16	10.12	9.14	11.39	9.31	12.56	9.79
Fem.1-15	10.97	8.49	8.92	9.97	10.13	10.06	11.45	10.01
Fem.-12	11.73	9.17	9.48	10.35	10.66	10.44	12.00	10.62
Fem.2-15	12.66	9.14	10.00	10.33	11.45	10.39	12.77	10.83
Fem.3-15	12.31	8.94	9.75	9.51	10.93	10.09	12.15	10.44
Male-10	12.12	9.36	9.82	10.68	11.16	10.82	12.39	10.77
Male2-15	11.97	8.88	9.86	10.51	10.87	10.48	12.71	10.52
Fem.4-15	11.79	9.71	9.42	11.36	10.80	11.18	12.03	11.49

Gender-age(yr)	**Ex**	**Ey**	**Fx**	**Fy**	**Gx**	**Gy**	**Hx**	**Hy**
Fem.-9	11.85	9.49	11.53	8.72	10.80	8.86	11.08	8.64
Male-7	11.74	9.09	11.30	8.50	10.74	9.01	10.75	8.67
Fem.-6	11.67	9.99	11.28	9.47	10.50	9.52	10.66	9.28
Male-6	12.08	8.95	11.56	8.54	10.89	8.78	11.15	8.68
Male1-15	11.58	9.77	11.26	8.80	10.38	9.03	10.29	8.80
Male-13	11.51	9.46	11.13	8.80	10.34	8.92	10.39	8.69
Fem.-11	11.99	8.43	11.59	7.83	10.73	7.91	10.93	7.90
Fem.1-15	10.65	9.13	10.15	8.38	9.42	8.84	9.57	8.48
Fem.-12	11.38	9.59	10.98	8.99	-0.	-0.	9.99	9.14
Fem.2-15	11.98	9.61	11.59	8.95	10.88	8.90	10.86	9.09
Fem.3-15	11.54	9.16	11.33	8.49	10.43	8.99	10.72	8.71
Male-10	11.62	9.91	11.23	9.27	10.47	9.43	10.49	9.33
Male2-15	11.37	9.52	11.13	8.93	10.46	9.25	10.34	9.17
Fem.4-15	11.32	10.12	10.77	9.65	9.98	9.96	10.07	9.85

Table 1.9: *List A of pseudo-landmarks on the first contour in Figure 1.3.*

Landmarks 1-50		Landmarks 51-100		Landmarks 101-150	
X	Y	X	Y	X	Y
-0.0690195	-0.0607218	-0.0356867	0.00061036	-0.0045762	0.05527602
-0.0690195	-0.0589441	-0.0330201	0.00416585	-0.0036873	0.05438714
-0.0712416	-0.0531664	-0.0316868	0.00594359	-0.0032429	0.05394271
-0.0716861	-0.0518331	-0.030798	0.00683247	-0.0032429	0.05305384
-0.072575	-0.0496109	-0.0303535	0.0072769	-0.002354	0.05127609
-0.072575	-0.0491665	-0.0281313	0.01038796	-0.0014652	0.04949835
-0.072575	-0.0482776	-0.0272425	0.01172127	-0.0014652	0.04905391
-0.072575	-0.0478332	-0.0272425	0.0121657	-0.0005763	0.04594286
-0.0730194	-0.0473888	-0.0254647	0.01438788	-0.0001318	0.04460955
-0.0730194	-0.045611	-0.0254647	0.01483232	-0.0001318	0.04372067
-0.0730194	-0.0451666	-0.0241314	0.0170545	0.00031259	0.04238737
-0.0734638	-0.0438333	-0.0241314	0.01749894	0.00120146	0.04016518
-0.0739083	-0.0416111	-0.023687	0.01794337	0.0016459	0.03972075
-0.0739083	-0.0411666	-0.0219092	0.02327661	0.0016459	0.03927631
-0.0739083	-0.0385	-0.0214648	0.02505435	0.0016459	0.03749857
-0.0739083	-0.0380556	-0.0210204	0.02594323	0.0016459	0.03705413
-0.0739083	-0.0371667	-0.0210204	0.02638766	0.00120146	0.03572082
-0.0739083	-0.0362778	-0.0205759	0.02727653	0.00120146	0.03483195
-0.0739083	-0.0331668	-0.0201315	0.02772097	0.00120146	0.03394308
-0.0739083	-0.0291669	-0.0201315	0.02860984	0.0016459	0.03305421
-0.0743527	-0.0287224	-0.019687	0.02905428	0.0016459	0.03216533
-0.0743527	-0.0260558	-0.019687	0.03038759	0.00209033	0.02994315
-0.0734638	-0.0229448	-0.019687	0.03127646	0.00209033	0.02860984
-0.0734638	-0.0216115	-0.019687	0.0317209	0.00253477	0.02727653
-0.0730194	-0.0202781	-0.019687	0.03216533	0.0029792	0.02594323
-0.0716861	-0.0198337	-0.019687	0.03260977	0.0029792	0.02549879
-0.0699083	-0.0193893	-0.0192426	0.03305421	0.0029792	0.02460992
-0.0676862	-0.0185004	-0.019687	0.0366097	0.00431251	0.0219433
-0.0672417	-0.0185004	-0.019687	0.03705413	0.00520139	0.02105443
-0.0667973	-0.0185004	-0.019687	0.03883188	0.00564582	0.02016555
-0.0663529	-0.0185004	-0.019687	0.04060962	0.00964575	0.01438788
-0.0650195	-0.0185004	-0.019687	0.04105406	0.01009018	0.01394345
-0.0645751	-0.0185004	-0.019687	0.04327624	0.01053462	0.01349901
-0.0641307	-0.018056	-0.019687	0.04638729	0.01097906	0.01305457
-0.0632418	-0.018056	-0.019687	0.04727616	0.01186793	0.01172127
-0.0627974	-0.018056	-0.019687	0.04860947	0.0127568	0.01083239
-0.0610196	-0.0176115	-0.019687	0.04905391	0.01320124	0.01038796
-0.0570197	-0.0167227	-0.019687	0.05038722	0.01453455	0.00949908
-0.0561308	-0.0167227	-0.019687	0.0526094	0.01675673	0.00772134
-0.0539086	-0.0167227	-0.0201315	0.05572045	0.0176456	0.0072769
-0.051242	-0.0167227	-0.0210204	0.05883151	0.01986778	0.00594359
-0.0499087	-0.0162782	-0.0214648	0.05927594	0.02164553	0.00505472
-0.048131	-0.0153893	-0.0214648	0.05972038	0.02564545	0.00461029
-0.0432422	-0.0096117	-0.0214648	0.06060925	0.02653433	0.00461029
-0.0427977	-0.0091672	-0.0210204	0.06060925	0.0274232	0.00505472
-0.0419089	-0.0078339	-0.0201315	0.06060925	0.03186756	0.0072769
-0.0414644	-0.0069451	-0.0152427	0.06105369	0.03453418	0.00816578
-0.04102	-0.0065006	-0.0103539	0.05972038	0.03497861	0.00816578
-0.0379089	-0.0025007	-0.0090206	0.05883151	0.03586749	0.00861021
-0.0356867	0.00016592	-0.006354	0.05705376	0.03631192	0.00861021

Table 1.10: *List B of pseudo-landmarks on the first contour in Figure 1.3.*

Landmarks 151-200		Landmarks 201-250		Landmarks 251-300	
X	Y	X	Y	X	Y
0.03675636	0.0086102	0.07186682	0.02416548	0.01720116	-0.0362778
0.03808967	0.0090546	0.07275569	0.0219433	0.01675673	-0.0367223
0.04164516	0.010388	0.07275569	0.02149886	0.01631229	-0.0367223
0.04208959	0.0117213	0.07320013	0.01972112	0.01453455	-0.0376112
0.04253403	0.013499	0.07320013	0.01927668	0.01320124	-0.0385
0.04253403	0.0139434	0.07320013	0.01838781	0.01186793	-0.0389445
0.04253403	0.0148323	0.074089	0.01572119	0.00831244	-0.0402778
0.04253403	0.0152768	0.07497788	0.01394345	0.00564582	-0.0411666
0.04297847	0.0166101	0.07675562	0.01038796	0.00520139	-0.0411666
0.04431178	0.0183878	0.07675562	0.00905465	0.00431251	-0.0416111
0.04520065	0.02061	0.07586675	0.00505472	0.00386808	-0.0420555
0.04520065	0.0219433	0.07453344	0.00327698	0.00342364	-0.0420555
0.04520065	0.0232766	0.074089	0.00283254	0.0029792	-0.0420555
0.04520065	0.023721	0.06964464	-0.0042784	-0.0036873	-0.045611
0.04520065	0.0250544	0.06964464	-0.0047229	-0.0045762	-0.045611
0.04520065	0.0254988	0.0692002	-0.0051673	-0.0050206	-0.0460554
0.0434229	0.027721	0.06875577	-0.0065006	-0.0072428	-0.0464999
0.04208959	0.0290543	0.06520028	-0.0087228	-0.0099094	-0.0473888
0.04164516	0.0294987	0.06386697	-0.0096117	-0.0103539	-0.0473888
0.04164516	0.030832	0.06075592	-0.0127227	-0.0139094	-0.0491665
0.04164516	0.0312765	0.0580893	-0.0136116	-0.0161316	-0.0500554
0.04120072	0.0330542	0.05675599	-0.0136116	-0.0170204	-0.0500554
0.04120072	0.0352764	0.05631155	-0.014056	-0.0227981	-0.052722
0.04120072	0.0357208	0.05542268	-0.014056	-0.023687	-0.052722
0.04120072	0.037943	0.05497825	-0.0145005	-0.0245758	-0.0531664
0.04120072	0.0383874	0.05364494	-0.0158338	-0.0303535	-0.055833
0.04120072	0.0388319	0.05275606	-0.0167227	-0.0312424	-0.055833
0.04297847	0.0468317	0.05053388	-0.0185004	-0.0347979	-0.0571664
0.0434229	0.048165	0.05008945	-0.0189448	-0.0370201	-0.0580552
0.04431178	0.0490539	0.04608952	-0.0229448	-0.0379089	-0.0584997
0.04520065	0.0499428	0.04431178	-0.0247225	-0.0383534	-0.0584997
0.04608952	0.0503872	0.04253403	-0.0260558	-0.0387978	-0.0589441
0.0483117	0.0512761	0.04120072	-0.0273891	-0.0405755	-0.0593885
0.04920057	0.0517205	0.04075629	-0.0273891	-0.04102	-0.0593885
0.04964501	0.0517205	0.0372008	-0.0291669	-0.0414644	-0.0593885
0.05453381	0.0512761	0.03675636	-0.0296113	-0.0432422	-0.0602774
0.05675599	0.0499428	0.03542305	-0.0300557	-0.0445755	-0.0602774
0.05942261	0.048165	0.03453418	-0.0305002	-0.0476865	-0.0602774
0.06031148	0.0463873	0.03408974	-0.0305002	-0.0485754	-0.0602774
0.06120035	0.0446095	0.032312	-0.0313891	-0.0503531	-0.0607218
0.06608915	0.0397207	0.03142312	-0.0313891	-0.0516865	-0.0611663
0.06831133	0.0361653	0.03097869	-0.0318335	-0.0552419	-0.0624996
0.06875577	0.0343875	0.02920094	-0.0322779	-0.0556864	-0.0624996
0.06875577	0.0330542	0.02831207	-0.0322779	-0.0592419	-0.062944
0.06964464	0.030832	0.02697876	-0.0327224	-0.0619085	-0.0633885
0.06964464	0.0303876	0.02564545	-0.0331668	-0.0627974	-0.0633885
0.07008908	0.0299432	0.0225344	-0.0340557	-0.0650195	-0.062944
0.07053351	0.0286098	0.02164553	-0.0345001	-0.0676862	-0.0616107
0.07097795	0.027721	0.01942335	-0.035389	-0.068575	-0.0611663
0.07142239	0.0259432	0.01809004	-0.0358334	-0.0690195	-0.0607218

Table 1.11: *List A of pseudo-landmarks on the second contour in Figure 1.3.*

Landmarks 1-50		Landmarks 51-100		Landmarks 101-150	
X	Y	X	Y	X	Y
0.09044152	0.01044919	0.03461837	-0.0400804	-0.0457477	-0.0501863
0.08899781	0.00515562	0.03413714	-0.0405616	-0.0462289	-0.0501863
0.08851658	0.00467439	0.03269343	-0.0410428	-0.0486351	-0.0492238
0.08803535	0.00467439	0.0322122	-0.0415241	-0.0491163	-0.0492238
0.08611041	0.00371192	0.03173097	-0.0415241	-0.0495976	-0.0492238
0.08418547	0.00323068	0.02788109	-0.0424865	-0.0500788	-0.0492238
0.07985437	0.00274945	0.02739986	-0.0424865	-0.05056	-0.0492238
0.07792943	0.00323068	0.02643739	-0.0429678	-0.052485	-0.0487426
0.07119215	0.00419315	0.02258752	-0.043449	-0.0534475	-0.0482614
0.07022969	0.00419315	0.02210629	-0.043449	-0.0544099	-0.0482614
0.06926722	0.00419315	0.02162505	-0.043449	-0.0592223	-0.0468177
0.06878598	0.00419315	0.02114382	-0.043449	-0.0601847	-0.0463364
0.06830475	0.00419315	0.02018135	-0.043449	-0.0621097	-0.0463364
0.06734228	0.00419315	0.01777518	-0.0448927	-0.0630721	-0.0463364
0.06686105	0.00419315	0.01536901	-0.0458552	-0.0635534	-0.0458552
0.06637981	0.00419315	0.01344407	-0.0468177	-0.066922	-0.045374
0.06589858	0.00419315	0.0110379	-0.0472989	-0.0678845	-0.045374
0.06445488	0.00419315	0.01007544	-0.0477801	-0.0683657	-0.045374
0.06252994	0.00419315	0.00718803	-0.0482614	-0.0688469	-0.0448927
0.06012377	0.00419315	0.0067068	-0.0482614	-0.0707719	-0.0444115
0.05868007	0.00419315	0.00381939	-0.0497051	-0.0712531	-0.0444115
0.05819884	0.00419315	0.00045076	-0.0506675	-0.0736593	-0.0439302
0.0577176	0.00419315	-0.0005117	-0.0511488	-0.0736593	-0.0424865
0.05242403	0.00419315	-0.0014742	-0.05163	-0.0736593	-0.0405616
0.04953662	0.00467439	-0.0019554	-0.05163	-0.0736593	-0.0395991
0.04857416	0.00419315	-0.0053241	-0.0530737	-0.0736593	-0.037193
0.04520552	0.00371192	-0.0082115	-0.0535549	-0.0736593	-0.0362305
0.04472428	0.00371192	-0.0101364	-0.0540362	-0.073178	-0.035268
0.04424305	0.00323068	-0.0106176	-0.0540362	-0.073178	-0.0328619
0.04231811	-0.000138	-0.0110989	-0.0540362	-0.073178	-0.0294932
0.04183688	-0.0011004	-0.0130238	-0.0545174	-0.073178	-0.0280495
0.04135565	-0.0030254	-0.0173549	-0.0549986	-0.073178	-0.0275683
0.03846824	-0.0068752	-0.0183174	-0.0549986	-0.073178	-0.0261246
0.03558084	-0.0088002	-0.0187986	-0.0549986	-0.073178	-0.0246809
0.03558084	-0.0092814	-0.0197611	-0.0545174	-0.073178	-0.0237184
0.03558084	-0.0097626	-0.0240922	-0.0540362	-0.073178	-0.0203498
0.0350996	-0.0116876	-0.0255359	-0.0535549	-0.0736593	-0.0136125
0.03750577	-0.0160187	-0.0264983	-0.0535549	-0.073178	-0.0112063
0.03798701	-0.0174624	-0.0284233	-0.0535549	-0.0726968	-0.0078377
0.03798701	-0.0179436	-0.0289045	-0.0535549	-0.0722156	-0.0068752
0.03798701	-0.0189061	-0.029867	-0.0535549	-0.0702906	-0.0049503
0.03798701	-0.0203498	-0.0317919	-0.0535549	-0.0688469	-0.0049503
0.03558084	-0.0246809	-0.0337169	-0.0530737	-0.0674032	-0.0054315
0.03558084	-0.0251621	-0.0341981	-0.0530737	-0.0621097	-0.006394
0.03461837	-0.0261246	-0.036123	-0.0525925	-0.0616284	-0.006394
0.03461837	-0.029012	-0.0366043	-0.0525925	-0.060666	-0.006394
0.03606207	-0.0328619	-0.0375667	-0.0521112	-0.0597035	-0.006394
0.03654331	-0.0338243	-0.0399729	-0.05163	-0.0577786	-0.006394
0.03702454	-0.0362305	-0.0409354	-0.0511488	-0.0572973	-0.006394
0.03654331	-0.037193	-0.0447852	-0.0506675	-0.0563349	-0.006394

Table 1.12: *List B of pseudo-landmarks on the second contour in Figure 1.3.*

Landmarks 151-200		Landmarks 201-250		Landmarks 251-300	
X	Y	X	Y	X	Y
-0.0558536	-0.006394	-0.0202423	0.06049753	0.01392531	0.02055511
-0.0544099	-0.006394	-0.0168737	0.06386617	0.01440654	0.02055511
-0.0491163	-0.003507	-0.01543	0.06338494	0.01681271	0.02055511
-0.0486351	-0.002544	-0.0149487	0.06338494	0.01921888	0.02055511
-0.0486351	-0.002063	-0.013505	0.06338494	0.02258752	0.02055511
-0.0471914	0.0022682	-0.0125426	0.0629037	0.02306875	0.02055511
-0.0462289	0.0041932	-0.0120613	0.0629037	0.02354999	0.02055511
-0.0452665	0.0061181	-0.0110989	0.06194123	0.02643739	0.02055511
-0.044304	0.0085243	-0.0101364	0.06146	0.02691863	0.02055511
-0.044304	0.0090055	-0.0096552	0.06097877	0.02739986	0.02055511
-0.0438228	0.0094867	-0.0086927	0.0600163	0.02980603	0.02055511
-0.0438228	0.009968	-0.0082115	0.05905383	0.0322122	0.02055511
-0.0428603	0.0128554	-0.007249	0.05664766	0.03413714	0.02055511
-0.0414166	0.016224	-0.0062865	0.05472272	0.0350996	0.02055511
-0.0414166	0.0171865	-0.0062865	0.05231655	0.03846824	0.02055511
-0.0404541	0.0195926	-0.0058053	0.05039162	0.03894948	0.02055511
-0.0399729	0.0200739	-0.0058053	0.04606051	0.04039318	0.02055511
-0.0394917	0.0215176	-0.0058053	0.04413557	0.04183688	0.02055511
-0.0390104	0.0219988	-0.0058053	0.04365434	0.04231811	0.02055511
-0.0390104	0.02248	-0.0062865	0.04269187	0.04376182	0.02055511
-0.0341981	0.0282549	-0.0062865	0.04221064	0.04424305	0.02055511
-0.0337169	0.0287361	-0.0062865	0.03980447	0.04616799	0.02055511
-0.0337169	0.0292173	-0.0067678	0.03932323	0.04664922	0.02055511
-0.0327544	0.030661	-0.0077302	0.03691706	0.04713045	0.02055511
-0.0313107	0.032586	-0.0082115	0.03499213	0.04761169	0.02055511
-0.0289045	0.0354734	-0.0125426	0.02777362	0.05049909	0.02055511
-0.0284233	0.0364358	-0.0125426	0.02729238	0.05434896	0.02007387
-0.0279421	0.0369171	-0.0120613	0.02055511	0.05819884	0.01959264
-0.0279421	0.0373983	-0.0091739	0.01959264	0.05868007	0.01959264
-0.0274608	0.0378795	-0.0082115	0.01959264	0.05964254	0.01911141
-0.0264983	0.0402857	-0.0077302	0.01959264	0.06060501	0.01911141
-0.0260171	0.0412482	-0.0062865	0.01959264	0.06204871	0.01911141
-0.0250546	0.0431731	-0.0058053	0.01959264	0.06301118	0.01863017
-0.0245734	0.0436543	-0.0053241	0.01959264	0.06397364	0.01863017
-0.0236109	0.045098	-0.0033991	0.01911141	0.06445488	0.01863017
-0.0221672	0.0475042	-0.0029179	0.01911141	0.06589858	0.01863017
-0.021686	0.0484667	-0.0024366	0.01911141	0.06830475	0.01814894
-0.0212048	0.0489479	-0.0019554	0.01911141	0.07119215	0.0176677
-0.0212048	0.0494291	-3.05E-05	0.01911141	0.07311709	0.01718647
-0.0207235	0.0499104	0.00045076	0.01911141	0.07552326	0.01670524
-0.0207235	0.0518353	0.00093199	0.01911141	0.07696696	0.01670524
-0.0207235	0.0523166	0.00189446	0.01959264	0.07792943	0.016224
-0.0207235	0.053279	0.00430063	0.02007387	0.07841067	0.016224
-0.0202423	0.0547227	0.0052631	0.02055511	0.0803356	0.016224
-0.0202423	0.0556852	0.00574433	0.02055511	0.08226054	0.01574277
-0.0202423	0.0571289	0.00622556	0.02055511	0.08514794	0.0147803
-0.0202423	0.0580914	0.0081505	0.02055511	0.08755411	0.0133366
-0.0202423	0.0585726	0.00911297	0.02055511	0.08803535	0.01285536
-0.0202423	0.0590538	0.01007544	0.02055511	0.08851658	0.01237413
-0.0207235	0.0600163	0.01296284	0.02055511	0.08947905	0.01141166

Table 1.13: *List A of pseudo-landmarks on the third contour in Figure 1.3.*

Landmarks 1-50		Landmarks 51-100		Landmarks 101-150	
X	Y	X	Y	X	Y
0.09184302	-0.0002324	0.03448028	-0.0150676	-0.0367286	-0.0516611
0.09134851	-0.0027049	0.03398577	-0.0165511	-0.0382122	-0.0516611
0.09134851	-0.0031995	0.03349127	-0.0175401	-0.0396957	-0.0511666
0.09134851	-0.003694	0.03250226	-0.0185292	-0.0416737	-0.0506721
0.08541444	-0.006661	0.03200775	-0.0190237	-0.0436517	-0.0506721
0.08393092	-0.0061665	0.03151324	-0.0195182	-0.0451352	-0.0506721
0.08145839	-0.005672	0.02706269	-0.0239687	-0.0456298	-0.0506721
0.08046937	-0.0051775	0.02557917	-0.0259467	-0.0466188	-0.0506721
0.07948036	-0.0051775	0.02607367	-0.0279248	-0.0485968	-0.0506721
0.07898585	-0.0051775	0.02656818	-0.0284193	-0.0495858	-0.0501776
0.07700783	-0.004683	0.02706269	-0.0303973	-0.0500803	-0.0501776
0.07354628	-0.0041885	0.02508466	-0.0328698	-0.0530473	-0.0496831
0.07008474	-0.003694	0.02360114	-0.0343534	-0.0535419	-0.0496831
0.06860122	-0.0031995	0.02310663	-0.0353424	-0.0545309	-0.0491885
0.0671177	-0.0027049	0.02211762	-0.0378149	-0.0555199	-0.048694
0.06662319	-0.0027049	0.02162311	-0.0407819	-0.0570034	-0.0481995
0.06563418	-0.0022104	0.02112861	-0.0412764	-0.0574979	-0.0481995
0.06266714	-0.0017159	0.02112861	-0.043749	-0.0579924	-0.0481995
0.06217264	-0.0017159	0.0206341	-0.0452325	-0.0599704	-0.0472105
0.06068912	-0.0017159	0.01964509	-0.0462215	-0.0609594	-0.0472105
0.05871109	-0.0012214	0.01816157	-0.0472105	-0.064421	-0.045727
0.05821659	-0.0012214	0.01766706	-0.047705	-0.0649155	-0.045727
0.05772208	-0.0012214	0.01618354	-0.0481995	-0.0668935	-0.0452325
0.051788	-0.0012214	0.01568904	-0.0481995	-0.067388	-0.044738
0.05079899	-0.0012214	0.01470003	-0.0481995	-0.0703551	-0.0442435
0.04931547	-0.0012214	0.01371101	-0.0481995	-0.0738166	-0.0432545
0.04733745	-0.0007269	0.01222749	-0.0481995	-0.0748056	-0.04276
0.04684294	-0.0007269	0.00678792	-0.048694	-0.0753001	-0.04276
0.04585393	-0.0007269	0.00332638	-0.0496831	-0.0767837	-0.0392984
0.04486491	-0.0007269	0.00184286	-0.0501776	-0.0767837	-0.0388039
0.0438759	-0.0007269	0.00134835	-0.0506721	-0.0767837	-0.0353424
0.04239238	-0.0007269	-0.0031022	-0.0506721	-0.0767837	-0.0343534
0.04140337	-0.0007269	-0.0035967	-0.0506721	-0.0767837	-0.0328698
0.04090886	-0.0007269	-0.0129923	-0.0521556	-0.0762891	-0.0323753
0.03991985	-0.0007269	-0.0139813	-0.0521556	-0.0767837	-0.0274303
0.03893084	-0.0007269	-0.0164539	-0.0526501	-0.0762891	-0.0244632
0.03645831	-0.0007269	-0.0179374	-0.0526501	-0.0757946	-0.0234742
0.03151324	-0.0007269	-0.0189264	-0.0531446	-0.0757946	-0.0224852
0.0285462	-0.0007269	-0.0199154	-0.0531446	-0.0753001	-0.0214962
0.0280517	-0.0031995	-0.0218934	-0.0531446	-0.0753001	-0.0195182
0.0280517	-0.003694	-0.022388	-0.0526501	-0.0753001	-0.0190237
0.02755719	-0.0041885	-0.023377	-0.0526501	-0.0748056	-0.0150676
0.02755719	-0.004683	-0.024366	-0.0526501	-0.0748056	-0.0125951
0.02755719	-0.005672	-0.0278275	-0.0526501	-0.0743111	-0.0116061
0.03002972	-0.00765	-0.0288165	-0.0526501	-0.0743111	-0.009628
0.03052423	-0.00765	-0.0298055	-0.0526501	-0.0743111	-0.0071555
0.03299676	-0.009628	-0.0303001	-0.0526501	-0.0733221	-0.003694
0.03398577	-0.0111116	-0.0322781	-0.0526501	-0.0713441	-0.0007269
0.03448028	-0.0116061	-0.0337616	-0.0521556	-0.0688715	-0.0012214
0.03448028	-0.0121006	-0.0357396	-0.0521556	-0.0668935	-0.0012214

Table 1.14: *List B of pseudo-landmarks on the third contour in Figure 1.3.*

Landmarks 151-200		Landmarks 201-250		Landmarks 251-300	
X	Y	X	Y	X	Y
-0.0629375	-0.0017159	-0.0209044	0.06207538	0.00579891	0.01608629
-0.0619485	-0.0017159	-0.0184319	0.06652594	0.00728243	0.01559179
-0.0599704	-0.0012214	-0.0179374	0.06702045	0.01074397	0.01509728
-0.0589814	-0.0012214	-0.0144758	0.06751495	0.01123848	0.01509728
-0.0579924	-0.0007269	-0.0139813	0.06751495	0.01371101	0.01460277
-0.0560144	-0.0002324	-0.0134868	0.06800946	0.01717256	0.01410827
-0.0540364	0.0012511	-0.0124978	0.06751495	0.01964509	0.01410827
-0.0525528	0.00273462	-0.0120033	0.06702045	0.0201396	0.01460277
-0.0515638	0.00372363	-0.0115088	0.06702045	0.02162311	0.01460277
-0.0505748	0.00471265	-0.0100253	0.06553693	0.02607367	0.01410827
-0.0505748	0.00520715	-0.0090363	0.06454792	0.03200775	0.01262475
-0.0500803	0.00570166	-0.0070583	0.06108637	0.03299676	0.01262475
-0.0495858	0.00619616	-0.0060692	0.05960285	0.03398577	0.01262475
-0.0490913	0.00718518	-0.0060692	0.05910835	0.0359638	0.01262475
-0.0476078	0.00965771	-0.0055747	0.05811933	0.03744732	0.01262475
-0.0461243	0.01213024	-0.0055747	0.05762483	0.03843633	0.01262475
-0.0446407	0.01559179	-0.0050802	0.0556468	0.03893084	0.01262475
-0.0446407	0.0165808	-0.0040912	0.05317427	0.04288689	0.01262475
-0.0436517	0.01954784	-0.0035967	0.05070174	0.04437041	0.01262475
-0.0426627	0.02202037	-0.0035967	0.05020723	0.04733745	0.01262475
-0.0421682	0.02300938	-0.0035967	0.04822921	0.04832646	0.01262475
-0.0411792	0.0244929	-0.0035967	0.0477347	0.04931547	0.01262475
-0.0396957	0.02696543	-0.0035967	0.04724019	0.04980998	0.01262475
-0.0377176	0.02844895	-0.0035967	0.04625118	0.05079899	0.01311925
-0.0357396	0.02993247	-0.0040912	0.04575667	0.0512935	0.01311925
-0.0332671	0.032405	-0.0040912	0.04427315	0.051788	0.01311925
-0.0327726	0.03289951	-0.0055747	0.03932809	0.05327152	0.01311925
-0.0307946	0.03636105	-0.0060692	0.03833908	0.05376603	0.01311925
-0.0298055	0.03833908	-0.0075528	0.03438303	0.05426054	0.01262475
-0.0288165	0.03932809	-0.0100253	0.03092148	0.05722757	0.01213024
-0.0288165	0.0398226	-0.0110143	0.02894346	0.06167813	0.01064672
-0.0268385	0.04229513	-0.0115088	0.02844895	0.06316165	0.01015222
-0.0258495	0.04377865	-0.0115088	0.02795444	0.06563418	0.01015222
-0.0258495	0.04427315	-0.0120033	0.02745994	0.06612869	0.01015222
-0.0248605	0.04575667	-0.0124978	0.02597642	0.06662319	0.01015222
-0.024366	0.04674569	-0.0134868	0.0244929	0.0671177	0.01015222
-0.0238715	0.04724019	-0.0134868	0.02399839	0.06761221	0.01015222
-0.023377	0.04872371	-0.0134868	0.02350389	0.06810671	0.01015222
-0.023377	0.04971272	-0.0139813	0.02300938	0.06959023	0.00965771
-0.0228825	0.05070174	-0.0139813	0.02202037	0.07008474	0.00965771
-0.0228825	0.05119624	-0.0144758	0.02103136	0.07156826	0.0091632
-0.022388	0.05169075	-0.0120033	0.01756981	0.07354628	0.0086687
-0.022388	0.05267976	-0.0100253	0.01608629	0.07799684	0.00817419
-0.0218934	0.05317427	-0.0095308	0.01608629	0.07948036	0.00767968
-0.0218934	0.05366878	-0.0075528	0.01608629	0.08145839	0.00669067
-0.0218934	0.05416328	-0.0070583	0.01608629	0.08442542	0.00570166
-0.0209044	0.05614131	-0.0060692	0.01608629	0.08491993	0.00520715
-0.0209044	0.05713032	-0.0040912	0.01559179	0.08887598	0.00273462
-0.0204099	0.05861384	0.00134835	0.01559179	0.08937049	0.00224011
-0.0209044	0.05910835	0.00233737	0.01559179	0.08986499	0.00174561

Table 1.15: *List A of pseudo-landmarks on the fourth contour in Figure 1.3.*

Landmarks 1-50		Landmarks 51-100		Landmarks 101-150	
X	Y	X	Y	X	Y
0.09563799	0.002356	0.03315564	-0.0161388	-0.035325	-0.0496293
0.09513813	-0.0026426	0.03015648	-0.0196378	-0.0383242	-0.0486296
0.09263884	-0.0036423	0.02865691	-0.0216372	-0.0458221	-0.0461303
0.09163912	-0.0041422	0.02765719	-0.0226369	-0.0473216	-0.0456304
0.09113926	-0.004642	0.02615761	-0.026136	-0.0478215	-0.0456304
0.09013954	-0.004642	0.02615761	-0.0271357	-0.0503208	-0.0446307
0.08664053	-0.0051419	0.02665747	-0.0276355	-0.0518204	-0.0441309
0.08514095	-0.0051419	0.02665747	-0.0291351	-0.0528201	-0.043631
0.08414124	-0.0051419	0.02615761	-0.0301348	-0.0538198	-0.043631
0.08164194	-0.0051419	0.02465804	-0.0326341	-0.0548195	-0.0431312
0.08064223	-0.0051419	0.02115902	-0.0356333	-0.0578187	-0.0421314
0.08014237	-0.0051419	0.01965945	-0.0371328	-0.0593182	-0.0411317
0.0761435	-0.0051419	0.01965945	-0.0381326	-0.060318	-0.0411317
0.07564364	-0.0051419	0.02015931	-0.0386324	-0.0628173	-0.040132
0.07114491	-0.0051419	0.02015931	-0.040132	-0.0648167	-0.040132
0.07014519	-0.0051419	0.01965945	-0.0431312	-0.0653166	-0.040132
0.06964533	-0.0051419	0.01865973	-0.0441309	-0.0668161	-0.0406319
0.06914547	-0.0051419	0.01666029	-0.0461303	-0.0688156	-0.040132
0.06864561	-0.0051419	0.01566058	-0.04713	-0.0733143	-0.0356333
0.06664618	-0.0051419	0.01516072	-0.0476299	-0.0738142	-0.033134
0.0651466	-0.0051419	0.01466086	-0.0476299	-0.0738142	-0.0326341
0.06214745	-0.0051419	0.014161	-0.0481297	-0.0738142	-0.0321343
0.06164759	-0.0051419	0.00666312	-0.050629	-0.0738142	-0.0311345
0.06114773	-0.0051419	0.00466368	-0.0511289	-0.0738142	-0.0306347
0.06064787	-0.0051419	0.00416382	-0.0511289	-0.0738142	-0.0286352
0.05864844	-0.0051419	0.00166453	-0.0516288	-0.0738142	-0.0266358
0.05714886	-0.0051419	-0.0003349	-0.0521286	-0.0738142	-0.026136
0.05614914	-0.0051419	-0.0008348	-0.0526285	-0.0738142	-0.0256361
0.05564928	-0.0051419	-0.0013346	-0.0526285	-0.0738142	-0.0251362
0.05265013	-0.0051419	-0.0028342	-0.0526285	-0.0738142	-0.0236367
0.05215027	-0.0051419	-0.0048336	-0.0526285	-0.0738142	-0.0231368
0.05165041	-0.004642	-0.0063332	-0.0526285	-0.0738142	-0.0211374
0.0506507	-0.004642	-0.0078328	-0.0526285	-0.0738142	-0.0201376
0.04865126	-0.004642	-0.0103321	-0.0526285	-0.0738142	-0.0186381
0.0481514	-0.0041422	-0.0108319	-0.0526285	-0.0738142	-0.0181382
0.04665182	-0.0036423	-0.0118317	-0.0526285	-0.0738142	-0.0176384
0.03815422	-0.0031424	-0.0148308	-0.0526285	-0.0733143	-0.0141393
0.03665465	-0.0031424	-0.0163304	-0.0526285	-0.0738142	-0.01164
0.03565493	-0.0031424	-0.0173301	-0.0526285	-0.0738142	-0.0111402
0.03515507	-0.0031424	-0.01783	-0.0526285	-0.074314	-0.0106403
0.03315564	-0.0041422	-0.0193295	-0.0526285	-0.074314	-0.0096406
0.03265578	-0.004642	-0.0198294	-0.0526285	-0.074314	-0.0091408
0.0311562	-0.0066415	-0.021329	-0.0526285	-0.0748139	-0.008141
0.03065634	-0.008141	-0.0218288	-0.0526285	-0.0748139	-0.0061416
0.0311562	-0.0091408	-0.0263276	-0.0526285	-0.0748139	-0.0021427
0.03165606	-0.0096406	-0.0268274	-0.0526285	-0.0738142	0.00035657
0.03315564	-0.01164	-0.028327	-0.0526285	-0.0733143	0.00085642
0.0336555	-0.0121399	-0.0288269	-0.0521286	-0.0733143	0.00135628
0.0336555	-0.0131396	-0.0308263	-0.0511289	-0.070815	0.00285586
0.03315564	-0.0156389	-0.0328257	-0.050629	-0.0703151	0.00335572

Table 1.16: *List B of pseudo-landmarks on the fourth contour in Figure 1.3.*

Landmarks 151-200		Landmarks 201-250		Landmarks 251-300	
X	Y	X	Y	X	Y
-0.0693154	0.00385558	-0.0148308	0.0618392	0.01516072	0.01435261
-0.0683157	0.00385558	-0.0153307	0.06633793	0.01566058	0.01435261
-0.0658164	0.00385558	-0.0148308	0.06733765	0.01815987	0.01385275
-0.0643168	0.00385558	-0.014331	0.06833737	0.01915959	0.01385275
-0.0623174	0.00385558	-0.0133312	0.0703368	0.02065917	0.01385275
-0.0608178	0.00385558	-0.0128314	0.07083666	0.02365832	0.01385275
-0.0588184	0.00385558	-0.0058334	0.06883723	0.02415818	0.01435261
-0.0583185	0.00385558	-0.0038339	0.06683779	0.02515789	0.01435261
-0.0553194	0.0048553	-0.0033341	0.06633793	0.02615761	0.01435261
-0.0543197	0.00535515	-0.0033341	0.06583807	0.02765719	0.01385275
-0.0528201	0.00585501	-0.0028342	0.0643385	0.02815705	0.0133529
-0.0513205	0.00735459	-0.0023343	0.06283892	0.02865691	0.0133529
-0.0508206	0.00785445	-0.0018345	0.06233906	0.02915677	0.0133529
-0.0503208	0.00835431	-0.0018345	0.06083948	0.02965662	0.0133529
-0.0493211	0.00935402	-0.0013346	0.05784033	0.03015648	0.01285304
-0.0493211	0.00985388	-0.0008348	0.05684061	0.03215592	0.01235318
-0.0478215	0.01135346	-0.0013346	0.05534104	0.03315564	0.01235318
-0.0463219	0.0133529	-0.0013346	0.05484118	0.0336555	0.01185332
-0.0453222	0.01485247	-0.0013346	0.05434132	0.03415535	0.01185332
-0.0448223	0.01535233	-0.0018345	0.05384146	0.03515507	0.01135346
-0.0433228	0.01685191	-0.0033341	0.04784315	0.03715451	0.01135346
-0.0408235	0.0218505	-0.0038339	0.04684344	0.03815422	0.01135346
-0.0403236	0.02285021	-0.0058334	0.04084513	0.03865408	0.01185332
-0.0398238	0.02434979	-0.0063332	0.03934555	0.04065352	0.01185332
-0.0393239	0.02484965	-0.0078328	0.0363464	0.04415253	0.01185332
-0.0393239	0.02534951	-0.0093324	0.03334725	0.04465239	0.01185332
-0.0383242	0.02634923	-0.0098322	0.03134781	0.04665182	0.01185332
-0.0363247	0.02884852	-0.0103321	0.03084795	0.04715168	0.01185332
-0.0358249	0.02934838	-0.0103321	0.02984824	0.05165041	0.01185332
-0.035325	0.02984824	-0.0108319	0.02834866	0.05265013	0.01185332
-0.0343253	0.03084795	-0.0113318	0.0278488	0.05314999	0.01185332
-0.0333256	0.03234753	-0.0113318	0.02734894	0.05564928	0.01185332
-0.0328257	0.03334725	-0.0108319	0.02684908	0.05764872	0.01135346
-0.031826	0.03484683	-0.0103321	0.02534951	0.05864844	0.01135346
-0.0313262	0.03584654	-0.0093324	0.02384993	0.05964815	0.0108536
-0.0308263	0.0363464	-0.0073329	0.0218505	0.0651466	0.00935402
-0.0298266	0.03734612	-0.0048336	0.01985106	0.06564646	0.00935402
-0.0293267	0.03784598	-0.0033341	0.0193512	0.06814575	0.00935402
-0.028327	0.0388457	-0.0013346	0.01785163	0.07064505	0.00935402
-0.0258277	0.04184485	0.00066481	0.01735177	0.0736442	0.00985388
-0.0253278	0.04284457	0.00116467	0.01735177	0.07414406	0.00985388
-0.0238283	0.044844	0.00266425	0.01735177	0.07564364	0.00985388
-0.0233284	0.04584372	0.00316411	0.01735177	0.07664335	0.00985388
-0.0233284	0.04634358	0.00366397	0.01735177	0.08064223	0.00935402
-0.0228286	0.0473433	0.00516354	0.01735177	0.08614067	0.00785445
-0.0223287	0.04784315	0.0056634	0.01685191	0.08714039	0.00785445
-0.0208291	0.05134217	0.00666312	0.01685191	0.09213898	0.00635487
-0.0198294	0.05284174	0.00716298	0.01685191	0.0931387	0.00535515
-0.0168302	0.05834019	0.01116185	0.01585219	0.09463827	0.00385558
-0.0163304	0.05884005	0.01466086	0.01435261	0.09563799	0.002356

Table 1.17 *Coordinates of five landmarks in a normal eye (n) and in a treated eye (0) measured in 12 Rhesus monkeys.*

Filename	Sx	Sy	Sz	Tx	Ty	Tz	Nx	Ny	Nz	Ix	Iy	Iz
1a1pn103.12b	2580	1060	60.31	1360	2660	-78.40	3800	2660	-132.68	2580	4260	126.65
1a1p0103.12b	2580	1060	78.01	1360	2660	39.00	3800	2660	-208.02	2580	4260	110.51
1a1cn103.12b	2540	1180	84.36	1500	2460	-72.31	3580	2460	-210.91	2540	3740	114.50
1a1c0103.12b	2540	1180	-5.79	1500	2460	-144.63	3580	2460	-237.19	2540	3740	75.21
1a1dn103.12b	2620	700	46.32	1440	2500	30.88	3800	2500	-108.09	2620	4300	101.24
1a1d0103.12b	2480	800	-37.89	1280	2500	-86.61	3680	2500	-162.40	2480	4200	108.27
1a1en103.12b	2480	900	5.88	1180	2560	-23.52	3780	2560	-146.97	2480	4220	229.27
1a1e0103.12b	2540	900	53.55	1220	2520	-119.01	3860	2520	-53.55	2540	4140	208.27
1a1fn103.12b	2640	880	-8.97	1260	2640	-71.72	4020	2640	-35.86	2640	4400	53.79
1a1f0103.12b	2640	880	5.67	1260	2640	-39.69	4020	2640	-136.09	2640	4400	90.73
1a1gn103.12b	2520	1060	23.05	1280	2560	-17.29	3760	2560	-149.84	2520	4060	201.71
1a1g0103.12b	2520	1060	-123.44	1280	2560	-101.97	3760	2560	-161.01	2520	4060	53.67
1a1in103.12b	2620	840	140.08	1320	2680	-170.53	3920	2680	-146.17	2620	4520	158.35
1a1i0103.12b	2580	620	26.64	1280	2380	-154.49	3880	2380	-111.87	2580	4140	159.81
1a1jn103.12b	2760	960	296.99	1600	2580	-77.22	3920	2580	-65.34	2760	4200	160.37
1a1j0103.12b	2720	1020	208.38	1520	2620	-191.02	3920	2620	-81.04	2720	4220	-28.94
1a1kn103.12b	2580	840	126.64	1320	2460	-143.16	3840	2460	-99.11	2580	4080	137.66
1a1k0103.12b	2580	840	-32.91	1320	2460	-98.73	3840	2460	-142.61	2580	4080	142.61
1a1ln103.12b	2560	1000	71.99	1420	2580	-95.99	3700	2580	-185.98	2560	4160	233.97
1a1l0103.12b	2560	1000	119.73	1420	2580	-125.17	3700	2580	-163.26	2560	4160	250.34
1a1nn103.12b	2540	860	65.71	1220	2540	-175.24	3860	2540	-147.86	2540	4220	93.10
1a1n0103.12b	2540	860	53.71	1220	2540	-102.05	3860	2540	-139.64	2540	4220	134.27
1a1on103.12b	2620	1120	49.15	1420	2600	-24.57	3820	2600	-147.44	2620	4080	141.30
1a1o0103.12b	2620	1120	95.55	1420	2600	106.16	3820	2600	-159.24	2620	4080	196.40

Filename	Vx	Vy	Vz
1a1pn103.12b	2180	2820	-542.77
1a1p0103.12b	2860	2400	-585.04
1a1cn103.12b	3100	2880	-403.75
1a1c0103.12b	2960	2820	-480.17
1a1dn103.12b	2960	2380	-339.70
1a1d0103.12b	2720	2460	-552.16
1a1en103.12b	2900	2220	-388.00
1a1e0103.12b	2080	2500	-636.70
1a1fn103.12b	2080	3220	-421.36
1a1f0103.12b	2820	2540	-487.65
1a1gn103.12b	2920	2720	-438.00
1a1g0103.12b	2760	2920	-558.16
1a1in103.12b	2820	2660	-347.15
1a1i0103.12b	3020	2420	-447.48
1a1jn103.12b	2480	2700	-497.80
1a1kn103.12b	2340	2240	-390.95
1a1k0103.12b	2300	2260	-422.34
1a1ln103.12b	2680	2900	-419.95
1a1l0103.12b	2800	2600	-353.74
1a1nn103.12b	2960	3220	-498.33
1a1n0103.12b	2860	2700	-386.70
1a1on103.12b	2940	2920	-681.91
1a1o0103.12b	2900	2220	-642.28

Table 1.18: *Acid proteinase atoms coordinates and types – I.*

			Protein A				Protein B
X	Y	Z	Element	X	Y	Z	Element
9.919	28.923	-8.61	N	9.504	27.24	-5.912	O
11.278	29.372	-8.909	C	9.717	29.014	-8.516	N
12.35	28.532	-8.306	C	11.058	29.475	-8.851	C
13.365	28.337	-8.933	O	12.178	28.64	-8.308	C
11.509	30.797	-8.429	C	13.19	28.482	-8.972	O
10.897	31.807	-9.331	C	11.26	30.911	-8.383	C
10.794	32.972	-8.983	O	10.69	31.912	-9.347	C
10.485	31.374	-10.515	N	10.528	33.074	-9.019	O
12.121	28.052	-7.086	N	10.402	31.468	-10.561	N
13.108	27.255	-6.366	C	11.99	28.106	-7.106	N
13.297	27.834	-4.974	C	13.027	27.333	-6.451	C
14.067	29.007	-5.073	O	13.28	27.929	-5.072	C
13.063	29.505	0.84	C	14.093	29.077	-5.186	O
12.749	29.348	-0.652	C	12.917	29.41	0.72	C
12.828	28.238	-1.183	O	12.584	29.33	-0.767	C
14.565	29.741	1.067	C	12.64	28.251	-1.353	O
12.322	30.44	-1.293	N	14.421	29.646	0.928	C
11.988	30.461	-2.721	C	12.163	30.451	-1.351	N
12.577	31.724	-3.313	C	11.821	30.513	-2.77	C
12.23	32.822	-2.89	O	12.428	31.782	-3.338	C
10.477	30.53	-2.975	C	12.11	32.878	-2.884	O
10.218	30.488	-4.467	C	10.304	30.563	-3.015	C
9.747	29.399	-2.268	C	10.042	30.693	-4.506	C
13.761	34.807	-5.994	N	9.612	29.308	-2.465	C
13.011	35.835	-6.723	C	13.707	34.88	-5.997	N
13.091	35.436	-8.181	C	13.047	35.982	-6.698	C
12.32	35.905	-9.014	O	13.091	35.604	-8.17	C
13.716	37.201	-6.736	C	12.26	36.033	-8.972	O
14.463	37.458	-5.443	C	13.864	37.298	-6.597	C
12.72	38.293	-7.009	C	14.267	37.578	-5.151	C
15.544	38.518	-5.604	C	13.072	38.462	-7.186	C
14.13	34.664	-8.482	N	15.416	38.578	-5.017	C
14.434	34.213	-9.827	C	14.111	34.814	-8.497	N
14.194	32.723	-10.067	C	14.393	34.358	-9.851	C
14.456	31.878	-9.197	O	14.139	32.866	-10.095	C
15.896	34.554	-10.12	C	14.503	32.02	-9.265	O
16.213	36.017	-9.858	C	15.863	34.667	-10.149	C
15.843	36.883	-10.645	O	16.248	36.085	-9.761	C
16.867	36.299	-8.735	N	15.887	37.04	-10.447	O
13.676	32.42	-11.255	N	16.958	36.232	-8.642	N
13.407	31.046	-11.677	C	13.529	32.554	-11.24	N
14.744	30.314	-11.634	C	13.264	31.164	-11.64	C
15.752	30.891	-12.012	O	14.626	30.475	-11.669	C
14.798	29.08	-11.152	N	14.752	29.266	-11.14	N
16.469	27.921	-9.734	C	16.463	28.183	-9.749	C
16.892	29.032	-8.753	C	16.781	29.327	-8.768	C
17.468	30.254	-9.47	C	17.599	30.456	-9.415	C
17.911	31.352	-8.512	C	18.204	31.431	-8.39	C
19.119	30.905	-7.76	N	19.413	30.854	-7.681	N

Table 1.19: *Acid proteinase atoms coordinates and types.*

			Protein C
X	Y	Z	Element
10.167	29.245	-8.599	N
11.476	29.668	-9.02	C
12.528	28.755	-8.443	C
13.438	28.32	-9.158	O
11.707	31.11	-8.574	C
10.952	32.098	-9.422	C
11.073	33.296	-9.244	O
10.178	31.597	-10.372	N
12.302	28.339	-7.199	N
13.251	27.499	-6.499	C
13.49	28.082	-5.115	C
14.203	29.294	-5.257	O
13.51	29.73	0.626	C
13.202	29.635	-0.865	C
13.511	28.628	-1.504	O
15.014	29.878	0.879	C
12.51	30.644	-1.391	N
12.136	30.686	-2.798	C
12.68	31.955	-3.418	C
12.25	33.05	-3.097	O
10.615	30.639	-2.987	C
10.296	30.497	-4.451	C
10.014	29.477	-2.215	C
14.053	34.998	-6.056	N
13.422	36.056	-6.848	C
13.723	35.698	-8.305	C
13.185	36.295	-9.24	O
14.099	37.421	-6.659	C
14.701	37.533	-5.27	C
13.101	38.543	-6.927	C
15.748	38.621	-5.185	C
14.653	34.771	-8.492	N
15.045	34.357	-9.824	C
14.798	32.868	-10.073	C
15.201	32.022	-9.276	O
16.506	34.753	-10.039	C
16.738	36.253	-9.801	C
16.53	37.074	-10.7	O
17.12	36.616	-8.577	N
14.018	32.587	-11.118	N
13.678	31.225	-11.536	C
15.018	30.532	-11.645	C
15.169	29.342	-11.075	N
16.806	28.073	-9.747	C
17.239	29.11	-8.723	C
18.193	30.141	-9.324	C
18.779	31.099	-8.272	C
19.95	30.517	-7.495	N

Table 1.20: *Landmark coordinates – Sope Creek data.*

Id	x1	y1	x3	y3	x2	y2	x5	y5	x4	y4
01	46.00	18.3	28.70	20.5	21.80	11.4	20.0	20.0	17.7	31.7
02	17.44	44.3	58.70	31.0	71.50	30.5	64.8	23.0	54.1	17.8
03	48.80	44.5	58.60	31.0	71.70	30.5	65.0	25.0	52.4	17.6
04	55.20	54.0	42.00	35.3	47.00	24.7	36.0	29.0	23.0	35.0
05	17.70	3.0	50.40	16.8	68.65	31.0	65.6	24.3	61.3	16.3
06	18.00	22.0	40.30	20.3	56.20	27.1	54.0	20.0	50.7	12.0
07	13.70	29.3	33.60	21.3	50.85	24.0	43.2	18.3	34.0	11.0
08	17.80	43.0	39.45	32.3	61.70	29.5	49.4	28.0	34.3	26.1
09	53.40	17.3	41.40	32.0	29.40	32.7	34.7	38.7	41.6	46.0
10	29.20	48.3	40.90	37.8	50.60	43.8	47.6	31.9	37.7	20.9
11	46.60	11.2	42.00	21.4	29.00	22.3	37.9	26.2	49.8	30.8
12	24.00	31.5	25.00	18.4	37.70	8.7	27.0	12.2	13.6	16.3
13	24.20	23.1	30.70	11.5	46.10	1.0	34.4	6.1	10.1	15.9
14	61.80	34.1	41.20	33.7	34.80	22.3	34.8	31.9	33.3	44.7
15	29.40	14.0	26.10	21.2	41.90	29.3	48.0	25.5	50.9	20.6
16	54.20	19.3	33.70	17.0	40.10	13.3	14.7	28.2	10.8	16.9
17	53.50	12.2	42.20	10.1	49.80	7.3	24.1	8.0	26.0	8.2
18	49.20	9.7	52.00	26.8	34.70	35.8	51.6	36.8	67.4	36.6
19	62.00	14.4	55.10	30.4	35.10	29.0	40.6	38.7	66.2	49.9
20	46.20	9.3	61.30	22.3	56.60	34.1	68.9	30.2	79.4	25.5
21	30.90	20.5	54.10	24.5	63.20	33.9	66.0	27.7	68.5	18.1
22	19.20	30.7	45.70	22.3	64.00	28.5	58.5	19.1	51.6	8.3
23	65.30	29.9	44.10	32.9	34.90	21.1	34.3	32.1	24.0	46.7
24	34.40	32.3	55.80	28.5	64.70	41.5	65.5	28.9	66.1	12.8
25	42.70	36.0	25.10	29.2	22.60	11.5	18.7	24.4	13.4	39.6
26	23.70	40.3	38.40	19.2	59.30	12.2	46.0	10.4	30.0	8.8
27	51.30	27.2	28.60	35.6	14.10	29.1	14.6	38.2	18.2	58.9
28	36.50	8.8	27.40	32.8	16.50	38.3	20.9	44.1	28.7	52.6
29	56.20	13.7	43.40	18.1	32.00	15.1	35.4	19.4	40.8	24.9
30	40.80	14.9	38.30	30.6	31.30	33.5	35.5	37.5	41.8	42.7
31	60.30	19.7	44.90	16.6	44.70	9.9	38.7	13.9	30.6	19.6
32	53.10	24.9	44.20	18.2	48.50	12.8	41.4	14.6	31.6	17.1
33	37.40	17.1	28.50	24.8	24.80	16.4	24.3	25.9	24.1	39.7
34	52.30	29.9	43.10	21.0	49.10	15.6	40.4	16.6	28.6	17.9
35	40.50	28.3	31.00	20.1	36.50	9.0	28.4	14.9	17.8	23.0
36	46.10	26.1	30.20	23.8	30.10	16.3	23.8	21.3	15.0	28.2
37	53.20	17.2	38.30	28.5	30.70	17.3	30.5	24.0	31.5	33.1
38	49.50	8.9	30.70	10.2	26.40	11.8	24.7	20.7	22.6	31.0
39	63.10	26.0	41.40	28.3	34.70	16.3	32.0	27.2	29.2	43.3
40	65.50	21.1	37.40	32.4	14.00	23.1	25.1	35.2	38.3	48.6
41	30.00	7.1	62.10	18.1	74.20	30.6	79.7	24.2	84.6	13.5

Table 1.21: *Landmark coordinates – BBC "impersonator" data.*

Landmark	1	2	3	4	5	6	7	8
Image								
1	466,403	469,191	350,501	608,501	278,191	397,199	554,200	665,202
2	482,385	483,164	328,552	602,570	254,148	40,161	598,167	730,173
3	511,358	533,170	371,500	600,531	308,136	428,157	616,178	742,188
4	482,334	513,162	365,462	589,480	299,140	412,154	577,169	701,185
5	505,260	511,84	394,422	628,430	302,83	421,89	599,83	719,91
6	530,328	539,142	389,450	605,468	344,133	457,152	619,155	728,173
7	465,264	478,56	338,457	620,454	263,98	385,103	580,113	700,120
8	491,382	503,164	343,494	593,521	287,143	430,172	580,181	727,199
9	208,238	300,59	224,422	421,423	157,94	239,98	391,106	497,109
10	155,295	240,94	139,495	389,515	91,94	193,122	376,128	494,139
11	167,303	244,129	161,449	365,468	127,113	196,128	353,136	475,148
12	196,280	256,91	152,447	401,482	92,101	203,113	373,117	514,130
13	690,484	683,268	412,569	620,662	449,212	568,257	728,310	842,349
14	195,315	240,141	250,477	439,446	152,173	224,155	350,140	454,124

Table 1.22 *Coordinates of landmarks from camera images of 3D object in Figure 1.14.*

Landmark No.		1	2	3	4	5	6	7	8	9	10
Image 1	x	89.50	72.50	142.00	154.00	75.50	144.00	156.50	68.00	171.00	188.50
	y	29.50	66.00	73.00	37.50	125.50	134.50	96.00	146.50	159.50	99.50
Image 2	x	135.00	97.50	137.50	174.50	99.00	138.50	175.00	80.00	138.50	194.00
	y	39.00	54.50	70.50	53.50	105.00	122.00	104.00	116.00	140.50	113.00
Image 3	x	177.50	133.50	165.50	209.50	133.00	164.50	207.50	112.50	157.00	222.50
	y	28.00	39.00	56.50	46.00	89.00	109.50	95.50	96.00	125.50	107.50
Image 4	x	223.00	176.50	201.00	248.50	175.50	199.50	246.50	152.00	187.00	258.50
	y	35.50	44.50	63.50	53.50	94.50	114.50	104.50	100.00	130.00	115.00
Image 5	x	189.00	140.50	162.50	212.50	138.00	159.00	209.00	114.00	144.50	219.00
	y	33.50	41.00	65.00	57.00	89.00	113.00	106.00	94.50	132.00	119.50
Image 6	x	209.50	163.00	178.00	227.00	161.00	175.00	224.50	138.00	157.00	231.50
	y	63.50	67.50	83.00	80.50	115.50	132.50	129.00	119.00	145.00	138.50
Image 7	x	110.00	75.00	112.50	146.50	80.00	117.00	151.00	63.00	118.00	169.00
	y	24.00	40.50	53.50	35.50	88.50	101.50	83.00	99.50	119.00	90.00
Image 8	x	161.50	121.50	147.00	187.50	122.00	147.00	187.00	102.50	139.50	199.50
	y	47.00	56.00	71.50	62.00	101.00	117.00	106.00	107.50	130.00	115.50
Image 9	x	176.50	141.50	176.50	210.50	140.00	175.50	210.50	122.50	174.00	226.50
	y	38.00	44.00	52.50	47.00	94.00	103.00	95.50	99.00	112.50	101.50
Image 10	x	146.50	109.50	136.50	176.00	109.00	136.50	175.00	89.00	131.00	187.50
	y	35.00	41.00	51.50	46.00	88.00	100.50	93.00	93.50	110.50	99.00
Image 11	x	146.50	102.50	159.50	201.00	99.50	153.50	194.00	77.50	159.00	221.00
	y	40.00	71.50	104.00	70.00	128.00	159.00	126.00	145.00	194.50	141.50
Image 12	x	167.00	111.50	143.50	200.50	105.50	136.00	191.00	78.50	123.50	208.00
	y	29.00	42.50	80.50	65.50	96.50	136.50	118.00	104.00	163.00	138.00
Image 13	x	91.00	78.00	142.50	152.00	79.00	142.50	153.50	72.50	168.50	181.50
	y	35.50	63.00	70.00	41.50	122.00	127.50	99.50	138.50	148.00	103.00
Image 14	x	163.50	138.00	190.00	212.00	128.00	178.50	201.50	114.50	190.50	225.50
	y	47.00	62.50	82.00	66.00	115.50	135.00	118.00	126.00	156.00	128.50
Image 15	x	135.00	105.50	167.50	193.00	105.50	166.00	190.50	91.00	180.50	219.00
	y	38.00	64.00	79.50	54.50	124.50	140.50	112.50	140.00	165.00	120.50
Image 16	x	104.00	72.50	112.50	142.00	74.00	114.50	143.50	58.50	119.00	163.50
	y	28.00	36.50	43.50	35.00	87.00	97.00	84.50	94.00	105.50	89.00

Landmark No.		11	12	13	14	15	16	17	18	19
Image 1	x	72.50	170.00	187.50	56.00	225.50	249.50	66.50	222.50	244.50
	y	227.50	238.50	179.50	273.00	295.50	187.00	389.00	412.00	303.50
Image 2	x	81.50	138.50	193.00	43.00	142.00	233.50	50.00	144.00	231.50
	y	187.50	214.00	184.50	208.50	256.00	202.50	319.00	373.00	312.50
Image 3	x	110.50	154.00	219.00	66.00	140.00	251.00	69.50	139.50	246.00
	y	168.00	199.50	179.00	183.00	240.00	200.50	293.00	356.50	312.00
Image 4	x	150.50	184.50	254.50	104.50	159.50	281.00	106.00	158.50	275.50
	y	169.50	202.50	186.00	181.50	240.00	209.50	289.00	354.50	321.00
Image 5	x	110.00	140.00	212.00	62.50	110.50	235.00	61.00	106.00	225.00
	y	162.00	201.50	188.00	173.00	241.50	216.00	276.00	350.50	321.00
Image 6	x	135.00	154.50	228.00	88.00	117.50	245.00	88.00	116.50	238.00
	y	187.50	216.00	209.00	192.00	243.00	230.50	300.50	360.50	344.50
Image 7	x	69.00	122.50	173.50	34.00	125.00	211.50	47.00	136.50	217.50
	y	167.00	187.50	158.50	188.50	227.00	173.00	292.00	336.00	277.00
Image 8	x	103.00	139.00	199.00	63.00	124.00	225.50	68.00	126.50	224.50
	y	171.50	195.50	180.00	184.00	229.00	198.00	283.00	335.00	299.50
Image 9	x	120.00	171.50	224.50	82.50	169.50	259.50	83.00	167.50	256.00
	y	171.50	188.00	174.50	181.50	211.50	185.50	298.00	337.00	303.00
Image 10	x	88.50	128.50	186.50	48.50	118.50	215.50	50.00	117.50	212.00
	y	160.50	180.50	166.50	171.00	206.50	180.50	277.50	321.50	289.50
Image 11	x	75.00	150.00	210.00	30.00	161.50	260.50	32.50	150.00	240.50
	y	218.00	266.00	213.00	256.00	343.50	244.50	356.00	441.00	342.00
Image 12	x	71.00	113.00	193.00	16.00	85.50	225.50	14.00	76.00	204.00
	y	175.50	232.50	206.50	193.00	298.00	248.50	292.00	396.50	345.00
Image 13	x	74.50	166.50	180.50	60.00	220.50	239.00	66.00	216.00	234.50
	y	221.00	229.00	182.00	258.00	276.00	189.50	382.00	400.00	308.50
Image 14	x	99.50	173.50	208.00	71.50	200.50	256.50	52.50	172.50	229.50
	y	200.50	232.00	200.00	223.50	278.50	221.00	340.00	395.00	330.00
Image 15	x	89.00	176.00	212.50	57.50	209.00	270.50	60.00	201.00	260.00
	y	223.50	249.00	203.00	258.00	304.50	219.00	382.50	432.50	338.50
Image 16	x	60.50	120.00	164.00	27.00	131.50	203.00	33.50	133.00	205.00
	y	167.50	181.00	160.50	181.50	207.00	170.50	298.50	329.50	284.50

Table 1.23 *Coordinates of landmarks from camera images of 3D object in Figure 1.15.*

Landmark No.		1	2	3	4	5	6	7	8	9	10	11	12	13	14	15	16	17	18	19
Image 01	x	2147	1737	1864	2279	1717	1839	2249	1636	1788	2279	1616	1747	2244	1302	1519	2345	1286	1494	2279
	y	162	203	425	380	588	815	765	603	876	810	1023	1302	1241	1074	1580	1454	1697	2218	2092
Image 02	x	2289	1879	2310	2687	1854	2264	2649	1763	2254	2730	1737	2213	2659	1428	2228	2973	1403	2142	2851
	y	253	476	775	512	937	1231	983	1003	1347	1038	1023	1302	1241	1074	1580	1454	1697	2218	2092
Image 03	x	2350	1909	2289	2735	1920	2289	2740	1828	2289	2811	1854	2294	2806	1504	2249	3110	1550	2259	3084
	y	198	395	633	436	912	1140	937	962	1251	998	1535	1839	1560	1712	2259	1778	2598	3160	2659
Image 04	x	2304	1925	2360	2740	1920	2345	2715	1839	2355	2786	1839	2325	2755	1535	2375	3084	1540	2320	3003
	y	304	527	775	527	1013	1241	1003	1059	1362	1064	1600	1889	1580	1808	234	178	2613	3145	2573
Image 05	x	2132	1671	1955	2421	1681	1960	2421	1585	1925	2477	1606	1920	2461	1251	1778	2694	1286	1783	2664
	y	152	294	542	395	770	1028	866	810	1119	927	1337	1651	1459	1464	2041	1687	2279	2867	2487
Image 06	x	1940	1560	1925	2304	1570	1925	2294	1499	1925	2355	1514	1914	2340	1231	1894	2608	1266	1899	2548
	y	309	522	775	557	927	1175	942	988	1292	1008	1418	1732	1443	1616	2198	1646	2244	2781	2304
Image 07	x	2477	1970	2269	2781	1965	2264	2765	1864	2223	2826	1879	2208	2801	1489	2046	3049	1519	2051	3003
	y	56	208	486	324	729	1008	836	760	1109	907	1337	1687	1479	1469	2107	1727	2335	2983	2598
Image 08	x	2396	2021	2467	2846	2051	2487	2836	1965	2497	2917	2011	2502	2902	1732	2568	3241	1793	2563	3176
	y	512	820	1069	760	1241	1489	1180	1307	1621	1236	1752	2061	1687	2006	2598	1904	2634	3191	2517
Image 09	x	2436	1914	2112	2649	1869	2082	2603	1768	2011	2644	1732	1950	2588	1332	1687	2755	1276	1631	2659
	y	46	111	375	304	623	891	826	638	972	886	1226	1570	1474	1292	1899	1722	2198	2836	2639
Image 10	x	2340	1980	2269	2624	1970	2249	2608	1889	2228	2659	1889	2213	2634	1611	2147	2857	1611	2122	2796
	y	360	491	699	562	891	1109	957	932	1195	1003	1373	1636	1449	1504	1996	1646	2168	2659	2299
Image 11	x	2396	1864	2006	2553	1808	1950	2492	1697	1864	2517	1646	1808	2446	1231	1464	2568	1160	1388	2456
	y	248	268	547	517	780	1048	1028	785	1124	1089	1347	1707	1661	1393	2011	1920	2274	2927	2826
Image 12	x	2153	1773	1965	2350	1757	1940	2330	1676	1904	2370	1676	1879	2340	1383	1722	2492	1383	1712	2446
	y	263	334	562	466	729	937	856	750	1008	896	1170	1433	1327	1241	1727	1519	1894	2391	2158
Image 13	x	2325	1884	2234	2669	1864	2198	2634	1773	2183	2700	1768	2137	2639	1423	2056	2907	1418	2001	2806
	y	370	557	861	663	1003	1297	1109	1043	1423	1175	1524	1894	1636	1687	2365	1899	2391	3059	2578
Image 14	x	2259	1722	1909	2456	1712	1894	2446	1606	1828	2477	1600	1813	2456	1190	1514	2603	1185	1509	2563
	y	122	198	456	370	714	977	896	734	1059	952	1317	1656	1550	1393	1996	1793	2310	2938	2725
Image 15	x	2624	2193	2502	2943	2158	2451	2897	2071	2436	2953	2036	2380	2892	1697	2269	3130	1651	2198	3029
	y	334	456	704	572	927	1170	1038	962	1266	1099	1484	1798	1621	1606	2168	1849	2421	2998	2659
Image 16	x	2122	1788	2228	2558	1803	2249	2548	1737	2254	2629	1752	2259	2618	1499	2360	2958	1524	2345	2907
	y	506	755	942	689	1200	1398	1135	1276	1499	1190	1763	2006	1676	1980	2446	1854	2745	3196	2598

Table 1.24 *LEGS stereo data: coordinates of the nine ONH anatomical landmarks – I.*

Landmark Id		I	S	N	T	V	IM	SM	NM	TM
1AL	x	3015	2709	2385	3415	2882	3104	2762	2502	3128
	y	2408	1038	1785	1661	1682	2108	1214	1771	1527
1AR	x	3056	2762	2444	3456	2954	3158	2828	2589	3202
	y	2444	1061	1802	1655	1694	2126	1238	1784	1597
1BL	x	3164	2918	2588	3578	3092	3272	2930	2711	3318
	y	2540	1130	1934	1808	1790	2228	1328	1931	1668
1BR	x	3212	2966	2636	3650	3170	3314	3002	2788	3378
	y	2594	1160	1964	1838	1844	2264	1370	1978	1711
2AL	x	2996	2888	2426	3428	3008	2912	2817	2624	3242
	y	2462	1088	1778	1766	1718	2318	1311	1790	1652
2AR	x	3068	2966	2516	3500	3092	2984	2902	2708	3320
	y	2456	1070	1742	1736	1688	2312	1283	1778	1622
2BL	x	3014	2918	2438	3464	3032	2936	2846	2648	3290
	y	2414	1028	1700	1700	1640	2270	1238	1730	1592
2BR	x	3026	2936	2444	3476	3044	2954	2870	2642	3296
	y	2444	1040	1718	1718	1700	2276	1292	1736	1592
3AL	x	2648	2666	2048	3176	2564	2480	2648	2282	2894
	y	2546	1046	1742	1832	1754	2192	1196	1886	1490
3AR	x	2714	2744	2132	3248	2672	2600	2732	2384	2996
	y	2558	1076	1760	1850	1778	2228	1238	1904	1538
3BL	x	2666	2678	2060	3182	2606	2510	2666	2282	2900
	y	2414	1028	1700	1700	1640	2270	1238	1730	1592
3BR	x	2708	2720	2120	3230	2654	2564	2726	2372	2948
	y	2540	1052	1694	1832	1754	2198	1202	1880	1472
4AL	x	3164	3086	2738	3584	3110	3116	3224	2822	3416
	y	2366	1250	1826	1856	1826	2216	1454	1748	1880
4AR	x	3080	2990	2648	3512	3068	3026	3140	2768	3356
	y	2360	1250	1838	1844	1796	2222	1436	1748	1874
4BL	x	3593	3593	3211	4034	3602	3539	3734	3334	3889
	y	2389	1307	1843	1934	1848	2257	1507	1762	1930
4BR	x	3316	3321	2957	3784	3375	3243	3471	3098	3662
	y	2434	1352	1880	1975	1871	2312	1539	1798	1980
5AL	x	2960	2882	2378	3416	2822	2792	2810	2504	3212
	y	2594	1142	1904	1850	1886	2360	1370	1772	1628
5AR	x	3122	3050	2540	3590	3008	2960	2990	2666	3380
	y	2600	1130	1904	1838	1868	2366	1370	1796	1622
5BL	x	3014	3002	2456	3548	2948	2858	2930	2588	3356
	y	2654	1166	1916	1916	1910	2420	1394	1826	1694
5BR	x	3080	3062	2522	3620	3020	2918	2990	2666	3428
	y	2612	1124	1880	1874	1898	2384	1358	1790	1682
6AL	x	3296	3200	2858	3590	3242	3170	3302	2984	3458
	y	2366	1334	1970	1844	1844	2168	1592	1916	1850
6AR	x	3260	3164	2846	3590	3230	3140	3290	2966	3470
	y	2462	1424	2060	1928	1946	2264	1676	2006	1946
6BL	x	3024	2863	2528	3284	2852	2840	2900	2660	3116
	y	2693	1533	2240	2030	2102	2414	1808	2174	2048
6BR	x	2989	2818	2492	3242	2816	2810	2864	2630	3068
	y	2683	1508	2222	1994	2072	2396	1760	2126	2024

Table 1.25 *LEGS stereo data: coordinates of the nine ONH anatomical landmarks – II.*

Landmark Id		I	S	N	T	V	IM	SM	NM	TM
7AL	x	2838	2843	2313	3284	2818	2894	2909	2418	3139
	y	2658	1308	1953	1998	1948	2413	1493	1908	1998
7AR	x	2833	2853	2348	3314	2838	2914	2919	2408	3189
	y	2543	1203	1853	1908	1848	2318	1408	1798	1893
7BL	x	2763	2743	2228	3269	2733	2753	2813	2328	3069
	y	1203	2568	1848	1948	1863	2378	1373	1818	1928
7BR	x	2828	2813	2293	3334	2813	2838	2894	2378	3139
	y	1198	2568	1853	1933	1838	2373	1378	1808	1908
8AL	x	3050	3080	2468	3518	3002	3014	3128	2636	3320
	y	2546	1052	1724	1802	1778	2342	1310	1640	1766
8AR	x	2384	2372	1772	2858	2336	2348	2432	1988	2660
	y	2522	1028	1766	1802	1784	2324	1304	1670	1772
8BL	x	3176	3374	2720	3734	3236	3176	3410	2900	3524
	y	2498	1082	1700	1826	1766	2288	1328	1592	1754
8BR	x	2858	3050	2408	3422	2948	2864	3086	2600	3236
	y	2438	1022	1646	1778	1688	2222	1244	1520	1676
9AL	x	3116	3170	2636	3656	3086	3134	3128	2786	3476
	y	2450	1232	1868	1850	1838	1424	2150	1838	1808
9AR	x	3128	3146	2636	3644	3116	3152	3158	2774	3464
	y	2432	1220	1874	1844	1838	1424	2150	1820	1766
9BL	x	2954	3002	2456	3524	2930	2966	2978	2648	3320
	y	2420	1154	1820	1808	1784	2192	1316	1838	1718
9BR	x	2960	3002	2450	3500	2978	3008	3002	2672	3350
	y	2384	1148	1772	1796	1754	2180	1316	1820	1712
10AL	x	3134	3200	2726	3572	3158	3080	3176	2870	3380
	y	2444	1346	1946	1946	1850	2174	1478	1964	1928
10AR	x	3050	3110	2654	3488	3104	3002	3098	2792	3314
	y	2378	1280	1874	1892	1778	2090	1412	1904	1862
10BL	x	3170	3248	2780	3602	3146	3110	3212	2912	3416
	y	2288	1220	1766	1790	1682	1988	1346	1790	1766
C10BR	x	3080	3146	2690	3506	3074	3026	3104	2816	3320
	y	2276	1196	1778	1778	1688	1970	1328	1796	1754
11AL	x	3170	3050	2678	3560	3098	3170	3254	2834	3404
	y	2258	1160	1730	1616	1724	1970	1484	1748	1694
11AR	x	3152	3044	2702	3566	3098	3146	3248	2864	3416
	y	2216	1106	1700	1550	170	1922	1430	1706	1634
11BL	x	3218	3110	2672	3524	3080	3206	3206	2810	3416
	y	2270	1196	1820	1598	1802	2036	1520	1820	1682
11BR	x	3128	3026	2594	3446	3020	3110	3122	2726	3332
	y	2270	1208	1838	1604	1796	2012	1496	1838	1682
12AL	x	3170	3050	2678	3560	3098	3170	3254	2834	3404
	y	2258	1160	1730	1616	1724	1970	1484	1748	1694
12AR	x	3152	3044	2702	3566	3098	3146	3248	2864	3416
	y	2216	1106	1700	1550	170	1922	1430	1706	1634
12BL	x	3218	3110	2672	3524	3080	3206	3206	2810	3416
	y	2270	1196	1820	1598	1802	2036	1520	1820	1682
12BR	x	3128	3026	2594	3446	3020	3110	3122	2726	3332
	y	2270	1208	1838	1604	1796	2012	1496	1838	1682

Table 1.26 *LEGS stereo data: coordinates of the nine ONH anatomical landmarks – III.*

Landmark Id		I	S	N	T	V	IM	SM	NM	TM
13AL	x	3171	3189	2693	3566	3134	3180	3257	2802	3398
	y	2248	1152	1753	1780	1734	1989	1493	1757	1689
13AR	x	3148	3175	2698	3566	3143	3161	3248	2816	3402
	y	2207	1098	1693	1725	1693	1930	1434	1703	1621
13BL	x	3224	3122	2672	3530	3146	3194	3194	2788	3390
	y	2282	1220	1820	1766	1772	1988	1478	1823	1685
13BR	x	3134	3038	2588	3452	3062	3092	3122	2716	3275
	y	2270	1208	1838	1604	1796	2012	1496	1838	1682
14AL	x	3002	2948	2354	3500	2876	2858	2834	2582	3218
	y	2306	794	1412	1406	1466	1808	944	1472	1430
14AR	x	3122	3044	2474	3614	3014	2984	2930	2702	3320
	y	2288	770	1382	1376	1460	1808	920	1454	1400
14BL	x	2894	2708	2210	3350	2708	2792	2600	2408	3062
	y	2546	1076	1760	1580	1736	2048	1226	1766	1676
14BR	x	3038	2870	2366	3506	2870	2942	2750	2528	3188
	y	2564	1058	1790	1604	1742	2042	1238	1784	1694
15AL	x	2774	2714	2282	3218	2726	2714	2702	2420	3014
	y	2288	1130	1862	1742	1706	2168	1364	1814	1730
15AR	x	2702	2624	2204	3134	2666	2606	2606	2348	2966
	y	2336	1172	1916	1802	1784	2204	1442	1886	1784
15BL	x	2672	2732	2252	3158	2714	2618	2714	2372	2972
	y	2306	1166	1844	1814	1796	2162	1400	1814	1784
15BR	x	2714	2768	2294	3188	2762	2660	2744	2402	3038
	y	2294	1130	1808	1796	1772	2138	1394	1802	1778

Chapter 2

Basic Nonparametric Multivariate Inference

2.1 Basic Probability Theory

Mathematics, including Probability Theory, hinges upon bivalent Logics and its set theoretical derivatives. In *Probability Theory* one assumes that when performing an experiment with more than one possible outcome, these outcomes form a set, the *sample space*, and the *probability* (chance) of a particular **event** (subset of outcomes) to occur, accumulates from probabilities of its disjoint parts (subevents). A more formal definition is:

DEFINITION 2.1.1. *(i) A $\mathcal{A}$ is a set of subsets, called* events, *of a nonempty set Ω, called sample space, defined by the requirements that*

$$\begin{aligned} \Omega &\in \mathcal{A}, \\ \forall E \in \mathcal{A} &\Rightarrow E^c \in \mathcal{A}, \\ \forall E_1, E_2, \dots \in \mathcal{A} &\Rightarrow \cup_{k=1}^{\infty} E_k \in \mathcal{A}. \end{aligned} \tag{2.1}$$

(ii) A **measure** $\mathbb{P}$ *is a function* $\mathbb{P} : \mathcal{A} \to [0,1]$ *with the following properties*

$$\mathbb{P}(\emptyset) = 0. \tag{2.2}$$

$$\mathbb{P}(\cup_{k=1}^{\infty} E_k) = \sum_{k=1}^{\infty} \mathbb{P}(E_k), \forall E_1, E_2, \dots \in \mathcal{A}. \tag{2.3}$$

assuming the events $E_1, E_2, \dots$ are pairwise disjoint (mutually exclusive), and the triple $(\Omega, \mathcal{A}, \mathbb{P})$ is called a measure space. *If in addition*

$$\mathbb{P}(\Omega) = 1, \tag{2.4}$$

$\mathbb{P}$ *is said to be a* probability measure.
(iii) Let $\mathcal{C}$ be a collection of subsets of Ω. The σ-field generated by $\mathcal{C}$, *denoted by $\sigma\{\mathcal{C}\}$, is the smallest σ-field containing $\mathcal{C}$.*

In turn, the σ-field $\sigma\{\mathcal{C}\}$ is the intersection of all σ-fields containing $\mathcal{C}$. Hence a $\sigma-$field is closed under complementation, and under operations of finite or countable union or intersection.

Here are some further definitions and properties related to probability measures. Let $E_1, E_2, \dots \in \mathcal{A}$ and $E_1 \subset E_2 \subset \dots$. Define $E = \cup_{k=1}^{\infty} E_k = \lim_{k\to\infty} E_k$; also write $E_k \uparrow E$. We have that $\mathbb{P}$ is **continuous from below**, meaning that

$$\text{if } E_k \uparrow E, \text{ then } \mathbb{P}(E_k) \uparrow \mathbb{P}(E).$$

Similarly if $E_1 \supset E_2 \supset \dots$ define $E = \cap_{k=1}^{\infty} E_k = \lim_{k\to\infty} E_k$ and write $E_k \downarrow E$. We have

$$\text{if } E_k \downarrow E, \text{ then } \mathbb{P}(E_k) \downarrow \mathbb{P}(E),$$

which expresses the **continuity from above**. The triple $(\Omega, \mathcal{A}, \mathbb{P})$ is a *probability space*. A probability measure is σ- subadditive, meaning that

$$\mathbb{P}(\cup_{k=1}^{\infty} E_k) \leq \Sigma_{k=1}^{\infty} \mathbb{P}(E_k),$$

DEFINITION 2.1.2. *The events, elements of an indexed set* $\{E_j \in \mathcal{A}\}_{j\in J}$, *are said to be independent if for any finite set of indices* $j_1, \ldots, j_a$,

$$\mathbb{P}(\bigcap_{r=1}^{a} E_{j_r}) = \prod_{r=1}^{a} \mathbb{P}(E_{j_r}). \tag{2.5}$$

The notation for the events $\{E_j \in \mathcal{A}\}_{j\in J}$, *being independent is* $\sqcup_{j\in J} E_j$

The notion of independence of events should not be confused with the events being *disjoint*. For example given the events E and F if $E \cap F = \varnothing$ then $E \sqcup F$ only one of them has probability zero.

This idea leads at once to the notion of the *conditional probability* of an event $E \in \mathcal{A}$, given an event $F \in \mathcal{A}$ with positive probability

$$\mathbb{P}(E|F) = \frac{\mathbb{P}(E \cap F)}{\mathbb{P}(F)}, \text{ provided that } \mathbb{P}(F) > 0. \tag{2.6}$$

Fixing F, we can compute $\mathbb{P}(E|F)$ for each $E \in \mathcal{A}$. This means that we have a set-function $E \mapsto \mathbb{P}(E|F), E \in \mathcal{A}$, or $\mathbb{P}(\bullet|F) : \mathcal{A} \to [0,1]$. It is an easy exercise to verify that this set-function has all the properties of a probability measure. Therefore $\mathbb{P}(\bullet|F)$ is called the *conditional probability measure* or **distribution, given** F.

To define the fundamental notion of random vector we need to equip the Euclidean space $\mathbb{R}^m$ with a suitable σ-field of , so that certain events defined in terms of inequalities, have a probability to occur. Consider the classes

$$\mathcal{O}_m = \{\text{all open subsets of } \mathbb{R}^m\},$$

$$\mathcal{R}_m = \{\phi, \text{ and all half-open ``rectangular boxes'' } (a_1, b_1] \times \cdots \times (a_m, b_m],$$

$$-\infty < a_j < b_j < \infty, j = 1, \ldots, p\},$$

$$\mathcal{Q}_m = \{\text{all closed ``quadrants'' } Q(x_1, \ldots, x_m) \tag{2.7}$$

with a quadrant $Q(x), x = (x^1, \ldots, x^m) \in \mathbb{R}^m$ given by

$$Q(x) = (-\infty, x_1] \times \cdots \times (-\infty, x_m]. \tag{2.8}$$

One can show that these collections generate the same σ-field $\mathcal{B}^m$ of *Borel sets* in $\mathbb{R}^m$, that is

$$\mathcal{B}^m = \sigma(\mathcal{O}_m) = \sigma(\mathcal{R}_m) = \sigma(\mathcal{Q}_m). \tag{2.9}$$

This σ-field is very big, although subsets of $\mathbb{R}^m$ can be constructed that are not in $\mathcal{B}^m$. Here we have constructed $\mathcal{B}^m$ directly. It is also possible to obtain $\mathcal{B}^m$ from $\mathcal{B}^1$. In fact we have

$$\mathcal{B}^m = \sigma\{B_1 \times \cdots \times B_m : B_j \in \mathcal{B}^1, j = 1,\ldots,p\}.$$

An important example of a measure is the *Lebesgue measure* λ^m, which is the unique measure on $(\mathbb{R}^m, \mathcal{B}^m)$ that assigns the Euclidean volume $\lambda^m((a_1,b_1] \times \cdots \times (a_m,b_m])) = (b_1 - a_1) \times \cdots \times (b_m - a_m)$, to rectangular boxes in $\mathcal{R}_m$.

A second type of measure is the *counting measure*. Let $D = \{x_1, x_2, \ldots\}' \subset \mathbb{R}^m$ be a countable set, and define the set-function

$$\nu_D(B) = \Sigma_{k=1}^{\infty} 1_B(x_k), B \in \mathcal{B}^m.$$

It is easy to see that ν_D is a measure, so that $(\mathbb{R}^m, \mathcal{B}^m, \nu_D)$ is a measure space. The measure ν_D is called counting measure with respect to D. It should be noted that ν_D can actually be defined on the σ-field of **all** subsets of $\mathbb{R}^m$ and $\nu_D(B)$ equals the number of points from D contained in B.

2.2 Integration on Euclidean Spaces

This section is derived in part from a LaTeX file of class notes in Mathematical Statistics, kindly provided in 2005 by Frits Ruymgaart [297].

A function $f : \mathbb{R}^m \to \overline{\mathbb{R}} = [-\infty, \infty]$ is called $(\mathcal{B}^m, \mathcal{B})$-measurable (or, briefly, *Borel measurable*) if

$$f^{-1}(\{\pm\infty\}) \in \mathcal{B}^m, f^{-1}(B) \in \mathcal{B}^m \ \forall B \in \mathcal{B}.$$

A mapping $f : \mathbb{R}^m \to \overline{\mathbb{R}}^k$ is a $(\mathcal{B}^m, \mathcal{B}^k)$-measurable function if each of its k components is measurable as defined above.

PROPOSITION 2.2.1. *The class of all measurable functions is closed under the algebraic operations of addition, multiplication, and division, under composition, under countable infima and suprema, and under limits. A continuous function is measurable.*

A useful and simple class of measurable functions is the class $\mathcal{S}$ of all step-functions $s : \mathbb{R}^m \to \mathbb{R}$,

$$s(x) = \Sigma_{k=1}^{n} c_k 1_{B_k}(x), x \in \mathbb{R}^m, \tag{2.10}$$

where $n \in \mathbb{N}, c_k \in \mathbb{R}$, and $B_k \in \mathcal{B}^m$. Obviously the representation of a step-function is not unique. The results below, however, are independent of the particular representation used.

For measurable f define

$$f^+ = f \vee 0, f^- = -(f \wedge 0).$$

Clearly f^+, f^- are nonnegative and measurable (see 1. of this subsection). Note that

$$f = f^+ - f^-, |f| = f^+ + f^-.$$

This decomposition allows us to develop most of integration theory for non-negative functions.

PROPOSITION 2.2.2. *Let f be a nonnegative measurable function, and, for $k = 1, \dots, n2^n$, let us set*

$$B_{n,k} = \{x \in \mathbb{R}^m : (k-1)2^{-n} < f(x) \leq k2^{-n}\}, \tag{2.11}$$

$$B_n = \{x \in \mathbb{R}^m : f(x) > n\}, \tag{2.12}$$

$$f_n(x) = \Sigma_{k=1}^{n2^n} \frac{k-1}{2^n} 1_{B_{n,k}}(x) + n\, 1_{B_n}(x), x \in \mathbb{R}^m. \tag{2.13}$$

Then

$$f_n(x) \uparrow f(x),\ as\, n \to \infty, x \in \mathbb{R}^m,$$

thus f can be approximated by step-functions.

Note that these step-functions in the approximation above are obtained by partitioning the **range** (which is the real line and therefore easy to explicitly partition) rather than the **domain** ($\mathbb{R}^m$). This idea is particularly useful for real-valued functions defined on abstract sets, and is a distinction from Riemann integral.

Let $s \in \mathcal{S}$ be a nonnegative step-function, i.e. assume $s(x) = \Sigma_{k=1}^n c_k 1_{B_k}(x), x \in \mathbb{R}^m$, with $0 \leq c_k < \infty$ and $B_k \in \mathcal{B}^m$ for all k. For such a function define the integral with respect to the measure M by

$$0 \leq \int_{\mathbb{R}^m} s(x) dM(x) = \int_{\mathbb{R}^m} s\, dM = \int s\, dM = \Sigma_{k=1}^n c_k M(B_k) \leq \infty. \tag{2.14}$$

For arbitrary **nonnegative** measurable f we define the *integral with respect to the measure M* to be

$$0 \leq \int_{\mathbb{R}^m} f\, dM = \sup\left\{ \int_{\mathbb{R}^m} s\, dM : s \in \mathcal{S}, 0 \leq s \leq f \right\} \leq \infty.$$

The properties

$$0 \leq f \leq g \Rightarrow 0 \leq \int f\, dM \leq \int g\, dM,$$

$$0 \leq f_n \uparrow f \Rightarrow \int f_n\, dM \uparrow \int f\, dM, n \to \infty,$$

are almost immediate (f, g, and $f_1, f_2, \dots$ measurable). The first expresses the

monotonicity of the integral, and the second is called the *monotone convergence theorem* (see Billingsley (1995, Thm16.2) [48]). A property is true M-almost everywhere (a. e.) if it is true on the complement of a set G, with $M(G) = 0$. Assuming that $f \geq 0$ we also have

$$\int f\,dM < \infty \Rightarrow f < \infty, M - \text{a.e.},$$

$$\int f\,dM = 0 \Rightarrow f = 0, M - \text{a.e.}.$$

Let $f \geq 0$ and $g \geq 0$ be measurable. We have $\int (cf+g)dM = c\int f\,dM + \int g\,dM$, They express the *linearity* of the integral (on the cone of nonnegative measurable functions).

Let $f \geq 0$ be measurable and assume M is a measure on $(\mathbb{R}^m, \mathcal{B}^m)$. Then the set function

$$B \mapsto \nu(B) = \int_B f\,dM, B \in \mathcal{B}^m,$$

is a measure on $\mathcal{B}^m$.

Let us now turn to integration of arbitrary measurable functions. If $\int f^{\pm} dM < \infty$, then f is called *M-integrable function*, and its *integral* is defined as

$$\int f dM = \int f^{+} dM - \int f^{-} dM. \tag{2.15}$$

Note that the integral is well-defined. Let $\mathcal{L}_M$ denote the **class of all** M**-integrable functions**. We have

$$f, g \in \mathcal{L}_M \Rightarrow \begin{cases} af + bg \in \mathcal{L}_M, \\ \int (af+bg)dM = a\int f\,dM + b\int g\,dM, \end{cases}$$

for all $a, b \in \mathbb{R}$. Hence

PROPOSITION 2.2.3. *$\mathcal{L}_M$ is a linear space and the integral acts as a linear functional on that space.*

A function on $\mathbb{R}^m$ is *Lebesgue integrable* if it is λ^m-integrable. Let $Q = [a_1, b_1] \times \cdots \times [a_m, b_m]$ be a compact hyper-rectangle and $f : \mathbb{R}^m \to \mathbb{R}$ such that $f = 0$ on Q^c and continuous on Q. Then the *Lebesgue integral* over Q equals the Riemann integral:

$$\int_Q f\,d\lambda^m = \int \ldots \int_Q f(x_1, \ldots, x_m) dx_1 \ldots dx_m.$$

A measurable f is in $\mathcal{L}_{\nu_D}$ if and only if $\Sigma_{k=1}^{\infty} |f(x_k)| < \infty$. For such an f the integral with respect to the counting measure ν_D equals

$$\int f d\,\nu_D = \Sigma_{k=1}^{\infty} f(x_k). \tag{2.16}$$

Any Riemann integrable function is Lebesgue integrable; the converse is false. To conclude this section, recall that the *product measure* $Q_1 \otimes Q_2$ of two measures Q_1, Q_2 is defined on pairs of Borel sets in corresponding Euclidean spaces by

$$Q_1 \otimes Q_2(B_1 \times B_2) = Q_1(B_1)Q_2(B_2) \tag{2.17}$$

Suppose that $f : \mathbb{R}^m \to [0,\infty]$ is Borel measurable. Then we also have ($x \in \mathbb{R}^m, y \in \mathbb{R}^q, (x,y) \in \mathbb{R}^m$)

$$f(\bullet, y) \text{ is Borel measurable on } \mathbb{R}^m \ \forall y \in \mathbb{R}^q$$

$$f(x, \bullet) \text{ is Borel measurable on } \mathbb{R}^q \ \forall x \in \mathbb{R}^m,$$

$$\int_{\mathbb{R}^q} f(\bullet, y) d\,Q(y) \text{ is Borel measurable on } \mathbb{R}^m,$$

$$\int_{\mathbb{R}^m} f(x, \bullet) d\,M(x) \text{ is Borel measurable on } \mathbb{R}^q.$$

We can now formulate the main result, viz.

$$\int_{\mathbb{R}^m} f(x,y) d(M \otimes Q)(x,y) =$$

$$= \int_{\mathbb{R}^m} \left\{ \int_{\mathbb{R}^q} f(x,y) d\,Q(y) \right\} d\,M(x) =$$

$$= \int_{\mathbb{R}^q} \left\{ \int_{\mathbb{R}^m} f(x,y) dM(x) \right\} dQ(y),$$

which is called *Fubini's theorem.*

By way of an example let us apply Fubini's theorem in the situation where $p = q = 1, M = \nu_D$ with $D = \{x_1, x_2, \dots\}$, and $Q = \lambda$. Then we obtain

$$\int_{\mathbb{R}^2} f(x,y) d(\nu_D \otimes \lambda)(x,y) =$$

$$= \Sigma_{k=1}^{\infty} \int_{\mathbb{R}} f(x_k, y) d\lambda(y) = \int_{\mathbb{R}} \{\Sigma_{k=1}^{\infty} f(x_k, y)\} d\lambda(y),$$

apparently an instance where summation and integration can be interchanged.

2.3 Random Vectors

DEFINITION 2.3.1. *Assume* $(\Omega, \mathcal{A}, \mathbb{P})$ *is a probability space. A random vector (r. vec.) is a mapping* $X : \Omega \to \mathbb{R}^m$ *which is* $(\mathcal{A}, \mathcal{B}^m)$ *- measurable, which means that*

$$X^{-1}(B) = \{\omega : X(\omega) \in B\} \in \mathcal{A}, \forall B \in \mathcal{B}^m.$$

If $p = 1, X$ *is said to be a random variable (r.v.).*

The simplest natural example of a family of r. vec.'s is the family of multinomial trials (Johnson and Wichern (2007)[170]), the natural multivariate extension of the Bernoulli trials. A *multinomial trial* with g possible outcomes, $g \geq 2$ is a r. vec. X in $\mathbb{R}^g$, with $\mathbb{P}(X = e_i) = \pi_i > 0, \forall i = 1, \ldots, g$, where $e_1, \ldots, e_g$ is the standard basis of $\mathbb{R}^g$ and $\pi_1 + \cdots + \pi_g = 1$; we will write down $X \sim M_g(\pi)$, where $\pi = (\pi_1 \ldots \pi_g)^T$. A multinomial trial is an example of *discrete* r.vec. X on $\mathbb{R}^g$, that is a r.vec. X for which there is a discrete subset $D \subset \mathbb{R}^g$, such that $\mathbb{P}(X \in D) = 1$. The *probability mass function*(p.m.f.) of a discrete r. vec. X is the function

$$f_X(x) = \mathbb{P}(X = x). \tag{2.18}$$

In the case of a multinomial trial $X \sim M_g(\pi)$ considered above, $D = \{e_1, \ldots, e_g\}$ and if $x \in D, f_X(x) = f_X((x_1 \ldots x_g)^T) = \pi_1^{x_1} \ldots \pi_g^{x_g}$.
The **joint cumulative distribution function (c.d.f.)** of X in definition 2.3.1 is the multivariable function $F_X : \mathbb{R}^m \to [0,1]$ given by

$$F_X(x) = P(X^{-1}(Q(x_1, \ldots, x_m))), \tag{2.19}$$

where $Q(x_1, \ldots, x_m) \in \mathcal{Q}_m$ is the closed lower quadrant $Q(x_1, \ldots, x_m) = (-\infty, x_1] \times \cdots \times (-\infty, x_m]$.

The r. vec. X induces a probability measure Q on $(\mathbb{R}^m, \mathcal{B}^m)$ according to

$$Q(B) = \mathbb{P}_X(B) = \mathbb{P}(X^{-1}(B)), B \in \mathcal{B}^m. \tag{2.20}$$

Naturally $Q = \mathbb{P}_X$ is called the induced or image probability measure (by, respectively under X, on the numerical space $\mathbb{R}^m$).

The are defined by

$$\mathbb{P}_{X^j}(B) = P_j(B) = \mathbb{P}\{X^j \in B\}, B \in \mathcal{B},$$

with the given by

$$F_j(x^j) = \mathbb{P}\{X^j \leq x^j\} = P_j((-\infty, x^j]).$$

Now let $\varphi : \mathbb{R}^m \to \mathbb{R}$ be Borel measurable function. In this section we will assume that for any such function, $\varphi, \int |\varphi| dP < \infty$.

In order to define the *mean vector* and *covariance matrix* of $Q = P_X$ for an arbitrary r.v. X, we should consider an integrable with respect to (w.r.t.) Q. That is $\varphi : \mathbb{R}^m \to \mathbb{R}$ is a measurable function such that $\int_{\mathbb{R}} |\varphi| dQ < \infty$. $\mathcal{L}_Q$ is the space of all integrable functions w.r.t. Q.
Then we define the *expected value* of $\varphi(X^1, \ldots, X^m)$ as

$$E(\varphi(X^1, \ldots, X^m)) = \int_{\mathbb{R}^m} \varphi(x^1, \ldots, x^m) dQ(x^1, \ldots, x^m) = \int \varphi dQ. \tag{2.21}$$

In particular we may again obtain the means and variances of the marginals

$$E(X^j) = \mu^j, \text{Var}(X^j) = \sigma_j^2,$$

and their covariances

$$\mathrm{Cov}(X^i, X^j) = E(X^i - \mu^i)(X^j - \mu^j) = \sigma_{ij},$$

by choosing $\varphi(x^1, \ldots, x^m) = (x^i - \mu^i)(x^j - \mu^j)$.
More generally, the **absolute moment of order** $k \in \mathbb{N}$ is defined as $E(\|X\|^k)$. It follows from standard inequalities that

$$E(\|X\|^m) < \infty \Rightarrow E(\|X\|^k) < \infty, \text{ for all } k \leq m,$$

$k, m \in \mathbb{N}$.
Moments are important parameters of probability distributions. The famous **moment problem** refers to the question whether a probability distribution $Q = P_X$ is uniquely determined by the set of all its moments $E((X^1)^{k_1} \ldots (X^m)^{k_m}), k_1 + \cdots + k_m = k, k \in \mathbb{N}$, assuming these moments exist. Under some additional assumptions this question can be answered in the affirmative.

The *characteristic function of a r. vec. X on* $\mathbb{R}^m$ is the complex function defined on a neighborhood of $0 \in \mathbb{R}^m$ given by

$$\Phi_X(t) = E(\exp(it^T X)). \tag{2.22}$$

The Euclidean *Fréchet function* associated with a r. vec. X is

$$\mathcal{F}_X(y) = E(\|y - X\|^2), \tag{2.23}$$

and, if the Euclidean Fréchet function is finite, its minimizer is the mean vector $E(X) = \mu$, whose components are the means of the marginals $X^j, j = 1, \ldots, p$. Similarly, one may define the *median set* as the set of all vectors, minimizers of the function $\mathcal{F}_1$ given by

$$\mathcal{F}_1(y) = E(\|y - X\|). \tag{2.24}$$

Unlike the mean vector, the median set may have more than one vector.
The mean vector is linear:

$$E(a\varphi(X^1, \ldots, X^m) + b\psi(X^1, \ldots, X^m)) =$$

$$= aE\varphi(X^1, \ldots, X^m) + bE\psi(X^1, \ldots, X^m),$$

for all $a, b \in \mathbb{R}$. In particular we have

$$\mathrm{Cov}(X^i, X^j) = E(X^i X^j) - E(X^i) \cdot E(X^j).$$

The matrix consisting of the covariances $\sigma_{ij}, i, j = 1, \ldots, p$ is called the *covariance matrix* Σ of the random vector $X = (X^1, \ldots, X^m)^T$. Here the "T" operation stands for transpose. Adhering to the convention that the expectation of a matrix of r. v.'s is simply the matrix of expectations, and the **mean vector** is $\mu = (\mu^1 \dots \mu^m)^T$, we can conveniently represent the covariance matrix as

$$\Sigma = Cov(X) = E(X-\mu)(X-\mu)^T. \tag{2.25}$$

If C is a matrix, and X is a r.vec., then $E(CX) = CE(X)$ and $Cov(CX) = CCov(X)C^T$, thus covariance matrix is *symmetric* and *semi-definite positive* corresponding to a symmetric operator on $\mathbb{R}^m$, called the covariance operator. The *total variance* of X is given by

$$\text{tvar}(X) = E(X-\mu)^T(X-\mu) = \text{tr}\Sigma = \sigma_1^2 + \dots + \sigma_m^2 \tag{2.26}$$

and the of X is the determinant $|\Sigma|$ of the covariance matrix of X.

X has a *joint probability density function* (p.d.f.) f_X with respect to the measure M if the c.d.f. F_X can be expressed as the integral

$$F_X(x) = \int_{Q(x)} f_X dM, \tag{2.27}$$

where $Q(x)$ is the quadrant given in equation (2.8). In particular, X has a discrete probability measure with support D if and only if X has a joint probability distribution function (p.d.f.) w.r.t. the counting measure ν_D, and in this case the p.m.f. of X is the p.d.f. w.r.t. ν_D.

The random vector X in $\mathbb{R}^m$ has an **absolutely continuous distribution** if X has a joint probability density function (p.d.f.) w.r.t. the Lebesgue measure λ^m, and in this case the mean vector and covariance matrix of X can be expressed in terms of Lebesgue integrals on $\mathbb{R}^m$.

2.4 Parameters and Sampling Distributions of their Estimators

In *Nonparametric Statistics*, a *parameter* is a vector valued functional $\theta : \mathcal{C} \to \mathbb{R}^m$ on a subset $\mathcal{C}$ of the space $\mathcal{P}(p)$ of all probability measures on $\mathbb{R}^m$. We will often identify the parameter θ with its value $\theta(Q)$ associated with the **unknown** probability measure Q.

DEFINITION 2.4.1. *Assume $X_1, \dots, X_n$ are random vectors, possibly of different dimensions, and let $X = (X_1^T \dots X_n^T)^T$. $X_1, \dots, X_n$ are said to be independent if the probability measure P_X is the product of the probability measures $P_{X_j}, j = 1, \dots, n$.*

We will say that $X_1, \dots, X_n$ are independent, identically distributed r.vec's(**i.i.d.r.vec.'s**) from a population (probability measure) Q on $\mathbb{R}^m$, if $X_1, \dots, X_n$ are independent and $P_{X_i} = Q, \forall i = 1, \dots, n$. A *random sample* of size n from a probability measure Q is a set of values $x_1 = X_1(\omega_1), \dots, x_n = X_n(\omega_n)$, where $X_1, \dots, X_n$ are i.i.d.r.vec's, and $\omega_i \in \Omega$. In Multivariate Analysis, one often has to estimate a parameter $\theta(Q)$ based on a random sample. To this goal, one considers an *estimator* $\hat{\Theta}_n = \hat{\Theta}_n(X_1, \dots, X_n)$, where

$X_1,\dots,X_n$ are i.i.d.r.vec.'s from Q, and $\hat{\Theta}_n : (\mathbb{R}^m)^n \to \mathbb{R}^m$ is a function, such that $\hat{\Theta}_n(x_{\tau(1)},\dots,x_{\tau(n)}) = \hat{\Theta}_n(x_1,\dots,x_n)$, for all τ in the permutation group Σ_n of the indices $\{1,\dots,n\}$. The symmetry is needed, as a sample $x_1,\dots,x_n$ from a probability measure Q can be regarded as a list values of n r.vec.s $X_1,\dots,X_n$ from Q, which are usually assumed to be independent and to provide and equally weighted information. A standard example of estimator of $\theta(Q)$, is $\hat{\Theta}_n = \theta(\hat{Q}_n)$, value of the parameter on the **empirical distribution** $\hat{Q}_n$ which is given by

$$\hat{Q}_n(x_i) = \frac{1}{n}, i = 1,\dots n. \tag{2.28}$$

The goodness of an estimator may be based on its **asymptotic properties**, since the larger the sample, it is assumed the closer $\hat{\Theta}_n(x_1,\dots,x_n)$ to the parameter $\theta(Q)$. The distribution of the estimator $\hat{\Theta}_n(X_1,\dots,X_n)$, is the sampling distribution of $\hat{\Theta}$, which is key to statistical inference for the parameter θ. As a basic example, given a random sample $x_1,\dots,x_n$ of multivariate observations, and its associated data matrix:

$$\mathbf{x} = (x_1 \dots x_n)^T, \tag{2.29}$$

one may **estimate** the mean vector μ a distribution Q, by the mean of the empirical distribution:

$$\hat{\mu} = \sum_{i=1}^{n} \hat{Q}_n(x_i)x_i = \frac{1}{n}\sum_{i=1}^{n} x_i = \bar{x}, \tag{2.30}$$

called the *sample mean vector*. For example, given a multinomial trial $X \sim M_g(\pi)$, its mean vector is $\mu = E(X) = \pi = (\pi_1 \dots \pi_g)^T$. To estimate μ, we consider the estimator $\bar{X}$ in equation (2.30) based on i.i.d.r.vec.'s $X_1,\dots,X_n$ from this multinomial trial. Note that $X_1 + \dots + X_n$ has a *multinomial distribution*, that is the distribution of a random vector on $\mathbb{R}^g$ counting the number of outcomes in each of the g cells in n independent multinomial trials. That is $S(X) = X_1 + \dots + X_g)^T$ has a multinomial $\mathcal{M}_g(n,\pi), \pi = (\pi_1 \dots \pi_g)^T, \pi_i > 0, \pi_1 + \dots + \pi_g = 1$. Then its joint p.d.f. with respect to the counting measure is

$$f_{S(X)}(x) = \binom{n}{x_1 \dots x_g} \pi_1^{x_1} \dots \pi_g^{x_g}. \tag{2.31}$$

Since $f_{\bar{X}}(y) = \mathbb{P}(S(X) = ny)$, we obtain for the sampling distribution of $\bar{X}$, the following p.d.f. with respect to the counting measure:

$$f_{\bar{X}}(y) = \binom{n}{ny_1 \dots ny_g} \pi_1^{ny_1} \dots \pi_g^{ny_g}. \tag{2.32}$$

Define the vector $\mathbf{1_n}$ as the $n \times 1$ column matrix with all entries equal to 1, then the sample mean vector can be expressed in terms of the data matrix as

$$\bar{x} = \frac{1}{n}\mathbf{x^T 1_n}. \tag{2.33}$$

The matrix $\mathbf{x}^T$ will be called the *sampling matrix*.
The *sample covariance matrix estimate* estimate s_n is

$$s_n = \frac{1}{n}\mathbf{x}^{\mathbf{T}}\mathbf{H_n}\mathbf{x}, \tag{2.34}$$

where $\mathbf{H_n}$ is the *centering matrix*

$$\mathbf{H_n} = \mathbf{I_n} - \frac{\mathbf{1}}{\mathbf{n}}\mathbf{1_n}\mathbf{1_n^T}. \tag{2.35}$$

Note that since $\mathbf{H_n}\mathbf{1_n} = \mathbf{0}$, it follows that $rk\mathbf{H_n} = \mathbf{n}-\mathbf{1}$. Also note that the sample covariance matrix estimate is the covariance matrix of the empirical distribution $\hat{Q}_n$.
Since given two matrices A, B for which the product AB is defined, $rkAB \leq min(rkA, rkB)$ and, since $rk\mathbf{x} \leq p, rk\mathbf{H_n} = \mathbf{n}-\mathbf{1}$, from (2.34) we have

$$rk\ s_n \leq min(p, n-1). \tag{2.36}$$

2.5 Consistency and Asymptotic Distributions of Estimators

For corresponding univariate results in this section, see class notes by R. N. Bhattacharya (1997) [31] or Bhattacharya et al. (2015) [40]. A minimal requirement of any reasonable estimator $\hat{\Theta}_n = \hat{\Theta}_n(X_1, \ldots, X_n)$ of a population parameter $\theta : \mathcal{C} \to \mathbb{R}^m$ is that of consistency.

DEFINITION 2.5.1. *A consistent estimator* $\hat{\Theta}_n$ $(n \geq 1)$ *of a parameter* $\theta : \mathcal{C} \to \mathbb{R}^m$, *is an estimator* $\hat{\Theta}_n$ *that converges in probability to* θ: $\hat{\Theta}_n \xrightarrow{P} \theta$, *i.e.,*

$$\mathbb{P}(||\hat{\Theta}_n - \theta|| > \varepsilon) \longrightarrow 0 \quad \textit{as } n \to \infty, \textit{for every } \varepsilon > 0. \tag{2.37}$$

More generally, a sequence of random vectors Y_n *converges in probability to a random vector* Y, $Y_n \xrightarrow{P} Y$, if

$$\mathbb{P}(||Y_n - Y|| > \varepsilon) \longrightarrow 0 \quad \text{as } n \to \infty, \text{ for every } \varepsilon > 0. \tag{2.38}$$

A common method for proving consistency is the following.

PROPOSITION 2.5.1. *(a) If, for some* $q > 0, E||\hat{\Theta}_n - \theta||^q \to 0$, *then* $\hat{\Theta}_n$ *is a consistent estimator of* θ. *(b) If* $\hat{\Theta}_n$ *is an unbiased estimator of* θ *and* tvar$(\hat{\Theta}_n) \to 0$, *then* $\hat{\Theta}_n$ *is a consistent estimator of* θ.

Proof. (a) By Chebyshev's inequality, for every $\varepsilon > 0$,

$$\mathbb{P}(\|\hat{\Theta}_n - \theta\| \geq \varepsilon) \leq \frac{E\|\hat{\Theta}_n - \theta\|^q}{\varepsilon^q} \longrightarrow 0 \quad \text{as } n \to \infty. \tag{2.39}$$

(b) is a special case of (a) with $q = 2$, and

$$\text{tvar}(\hat{\Theta}_n) = E\|\hat{\Theta}_n - \theta\|^2.$$

Note: Let X be a random vector (e.g., $X = \hat{\Theta}_n - \theta$) such that $E\|X\|^q < \infty$ for some $q > 0$, then writing $E(Y : A)$ for the *expectation of* Y *on the set* A, i.e., $E(Y : A) = \int_A Y dP$,

$$\begin{aligned} E\|X\|^q &= E\left[\|X\|^q : \|X\| < \varepsilon\right] + E\left[\|X\|^q : \|X\| \geq \varepsilon\right] \\ &\geq E\left[\|X\|^q : \|X\| \geq \varepsilon\right] \geq \varepsilon^q \mathbb{P}(\|X\| \geq \varepsilon), \end{aligned} \tag{2.40}$$

which gives

$$\mathbb{P}(\|X\| \geq \varepsilon) \leq \frac{E\|X\|^q}{\varepsilon^q}. \tag{2.41}$$

PROPOSITION 2.5.2. *Suppose* U_n *and* V_n *are two sequences of random vectors such that* $U_n \xrightarrow{P} a, V_n \xrightarrow{P} b$. *If* $g(u,v)$ *is a function (of two vector valued variables) which is continuous at* (a,b), *then* $g(U_n, V_n) \xrightarrow{P} g(a,b)$.

Proof. Fix $\varepsilon > 0$. There exists $\delta = \delta(\varepsilon)$ such that if $\|u - a\| \leq \delta$ and $\|v - b\| \leq \delta$ then $\|g(u,v) - g(a,b)\| \leq \varepsilon$. Now

$$\begin{aligned} &\mathbb{P}(\|g(U_n,V_n) - g(a,b)\| > \varepsilon) = \\ = \; &\mathbb{P}(\{\|U_n - a\| > \delta \text{ or } \|V_n - b\| > \delta\} \cap \{\|g(U_n,V_n) - g(a,b)\| > \varepsilon\}) \\ &+ \mathbb{P}(\{\|U_n - a\| \leq \delta \text{ and } \|V_n - b\| \leq \delta\} \cap \{\|g(U_n,V_n) - g(a,b)\| > \varepsilon\}) \\ \leq \; &\mathbb{P}(\|U_n - a\| > \delta) + \mathbb{P}(\|V_n - b\| > \delta) \to 0. \end{aligned} \tag{2.42}$$

Note that the set $\{\|U_n - a\| \leq \delta, \|V_n - b\| \leq \delta\} \cap \{g(U_n,V_n) - g(a,b)| > \varepsilon\}$ is empty, and has therefore zero probability—a fact used for the least inequality.

REMARK 2.5.1. *Proposition 2.5.2 extends to any fixed number, say k, of sequences* $U_n^{(i)} \xrightarrow{P} a_i, 1 \leq i \leq k$, *and a function* $g(u_1, u_2, \ldots, u_k)$ *of k variables which is continuous at* $(a_1, \ldots, a_k)$, *yielding:* $g(U_n^{(1)}, U_n^{(2)}, \ldots, U_n^{(k)}) \xrightarrow{P} g(a_1, a_2, \ldots, a_k)$. *The proof is entirely analogous. (Exercise 7.)*

COROLLARY 2.5.1. *If* $U_n \xrightarrow{P} a$, $V_n \xrightarrow{P} b$ *then (i)* $U_n + V_n \xrightarrow{P} a + b$, *(ii)* $U_n^T V_n \xrightarrow{P} a^T b$, *and (iii) assuming* $\|b\| \neq 0$, $U_n / \|V_n\| \xrightarrow{P} a/\|b\|$.

Proof. Use Proposition 2.5.2 with (i) $g(u,v) = u + v$, (ii) $g(u,v) = u^T v$ and (iii) $g(u,v) = u/\|v\|$.

2.5.1 Consistency of Sample Moments

EXAMPLE 2.5.1. consistency of the sample mean. *Let* $X_1, \ldots, X_n$ *be independent observations from an unknown distribution on* $\mathbb{R}^m$ *of which we assume a finite total variance* σ^2. *Since* $E\overline{X} = \mu$, *and* $var(\overline{X}) = \sigma^2/n \to 0$ *as* $n \to \infty$, *it follows that* $\overline{X}$ *is a consistent estimator of* μ.

EXAMPLE 2.5.2. *(consistency of the sample moments). Suppose a random sample X_1, ..., X_n is taken from a univariate distribution with a finite k-th moment, for some $k \geq 1$ (i.e., $E|X_j|^k < \infty$). Then it can be shown by the law of large numbers (LLN) that the sample moments $\widehat{m}_r$ are consistent estimators of population moments m_r for $r = 1,\ldots,k$,*

$$\widehat{m}_r = \frac{1}{n}\sum_{j=1}^{n} X_j^r \xrightarrow{a.s.} E(X^r) \quad = \quad m_r \qquad \textit{(the r-th 'population moment')}$$

$$r \quad = \quad 1,2,\ldots,k. \tag{2.43}$$

Note that $\widehat{m}_r$ is an unbiased estimator of m_r ($r = 1,2,\ldots,k$). Hence if $EX^{2k} \equiv m_{2k} < \infty$, then it follows from Proposition 2.5.1 that $\widehat{m}_r$ is a consistent estimator of m_r ($r = 1,2,\ldots,k$). Next consider the centered population moments $\mu_r = E(X - m_1)^r$, where $m_1 = \mu$ is the mean of the distribution (population). A natural estimator of μ_r is the (corresponding) centered r-th sample moment $\widehat{\mu}_r = \frac{1}{n}\sum_{j=1}^{n}(X_j - \overline{X})^r$. Note that by the binomial expansion,

$$\begin{aligned}\hat{\mu}_r &= \frac{1}{n}\sum_{j=1}^{n}\{X_j^r - \binom{r}{1}X_j^{r-1}\overline{X} + \cdots + (-1)^t\binom{r}{t}X_j^{r-t}\overline{X}^t + \\ &\qquad \cdots + (-1)^r\overline{X}^r\} \\ &= \frac{1}{n}\sum_{j=1}^{n}\sum_{t=0}^{r}(-1)^t\binom{r}{t}X_j^{r-t}\overline{X}^t = \sum_{t=0}^{r}(-1)^t\binom{r}{t}\overline{X}^t\,\widehat{m}_{r-t}\cdot \\ &= \sum_{t=0}^{r}(-1)^t\binom{r}{t}\widehat{m}_{r-t}\widehat{m}_1^t.\end{aligned} \tag{2.44}$$

It follows that the last sum converges in probability to

$$\sum_{t=0}^{r}(-1)^t\binom{r}{t}m_{r-t}m_1^t = E(X - m_1)^r, \tag{2.45}$$

provided $\widehat{m}_{r'} \xrightarrow{P} m_{r'}$ as $n \to \infty$ ($r' = 1,\ldots,r$). The latter (namely, (2.45)) is assured for all $r' = 1,\ldots,r$ if $EX^{2r} < \infty$ (by Proposition 2.5.1).

REMARK 2.5.2. *Example 2.5.2 extends to the case of a random sample X_1, ..., X_n from a multivariate distribution on $\mathbb{R}^m$ with a finite k-th moment, for some $k \geq 1$ (i.e., $E||X_j||^k < \infty$), since one can show by the LLN that the sample moments $\widehat{m}_{r_1\ldots r_m}, r_1 + \cdots + r_m = r$ are consistent estimators of population moments $m_{r_1\ldots r_m}, r_1 + \cdots + r_m = r$ for $r = 1,\ldots,k$.*

PROPOSITION 2.5.3. *Assume $X_1,\ldots,X_n$ are i.i.d.r.vec.'s from a multivariate distribution with finite mean μ and covariance matrix Σ. The sample covariance matrix estimator S_n in (2.34), that is also given by*

$$S_n = \frac{1}{n}\sum_{j=1}^{n}(X_j - \overline{X})(X_j - \overline{X})^T \tag{2.46}$$

is a consistent estimator of Σ.

The **proof** is left as an exercise 8.

2.6 The Multivariate Normal Distribution

REMARK 2.6.1. *The multivariate normal distribution play a central role in Large Sample Theory, the key to Nonparametric Data Analysis (see Bhattacharya and Rao (1976) [46] or Tsybakov (2009) [323]).*

1. A r.vec. $X : \Omega \to \mathbb{R}^m$ has a *multivariate normal* distribution if $\forall a \in \mathbb{R}^m$ the r.v. $a^T X$ has a normal distribution.

Assume X is as in §1 and $E(X) = \mu, Cov(X) = \Sigma$, (notation $X \tilde{-} N_m(\mu, \Sigma)$). The characteristic function of X is $\varphi_X(t) = e^{it^T\mu - \frac{1}{2}t^T\Sigma t}$.
If $X \tilde{-} N_m(\mu, \Sigma)$, where Σ is positive definite then the joint p.d.f. of X is given

$$\phi_X(x) = (2\pi)^{-p/2}|\Sigma|^{-1/2}exp(-\frac{1}{2}(x-\mu)^T\Sigma^{-1}(x-\mu)) \tag{2.47}$$

If $Z = (Z^1, ..., Z^m) \tilde{-} N_m(0, I_m)$, then the marginals $Z^1, \ldots, Z^m$ are independent. $X \tilde{-} N_m(\mu, \Sigma)$, where Σ is positive definite then $(X-\mu)^T\Sigma^{-1}(X-\mu)$ has a χ^2_m distribution.
If $X = (X^1...X^m)^T \sim N_m(\mu, \Sigma)$, then the joint characteristic function of X is given by

$$\Phi_X(t) = e^{it^T\mu - \frac{1}{2}t^T\Sigma t} \tag{2.48}$$

Assume $X \tilde{-} N_m(\mu, \Sigma)$, and L_1, L_2 are linear combinations in the components $X^1, ..., X^m$ of X, that is $L_1 = C_1^T X, L_2 = C_2^T X$. Then L_1, L_2 are independent (notation : $L_1 \sqcup L_2$) if $C_1^T \Sigma C_2 = 0$.

It follows that if $X_1, \ldots, X_n$ are i.i.d.r.vec.'c from a multivariate normal distribution, then $\bar{X}_n \sqcup S_n$.

Consider m i.i.d.r.vec.'s $(Y_1, ..., Y_m)$ from $\mathcal{N}_m(0, \Sigma)$, and the associated sampling matrix Y. Let $M = YY^T$ The probability distribution of M on the set of $p \times p$ symmetric matrices is called the *Wishart distribution* with scale matrix Σ and m degrees of freedom. Notation $M \sim W_m(\cdot|\Sigma)$.

PROPOSITION 2.6.1. *Assume $M \sim W_m(\cdot|\Sigma)$. Then (i) $E(M) = m\Sigma$. (ii) If B is a $p \times q$ matrix, then $B^T MB \sim W_m(\cdot|B^T\Sigma B)$. (iii) If $\Sigma > 0, a$ is a fixed vector, $a^T a > 0$, then $\frac{a^T Ma}{a^T \Sigma a} \sim \chi^2_m$. (iv) $M_1 \sim W_{m_1}(\cdot|\Sigma), M_2 \sim W_{m_2}(\cdot|\Sigma)$ are independent, then $M_1 + M_2 \sim W_{m_1+m_2}(\cdot|\Sigma)$.*

One reference for the following result is Mardia et al. (1972) [231].

THEOREM 2.6.1. *(Cochran's theorem) Assume C is a symmetric matrix and X is a sampling matrix from $\mathcal{N}_m(0, \Sigma)$. Then (i) XCX^T can be written as a weighted sum of rk(C) independent $W_1(\cdot|\Sigma)$ distributed random matrices. (ii) If in addition C is idempotent, then $XCX^T \sim W_{rk(C)}(\cdot|\Sigma)$. (iii) $nS_n = (n-1)S_u \sim W_{n-1}(\cdot|\Sigma)$.*

2.7 Convergence in Distribution

The main reference for this section is Ferguson (1996) [115]. In this and other sections of this chapter, we include proofs to insure that the key prerequisites are mathematically sound.

2.7.1 Convergence in Distribution (Weak Convergence)

The material in this section, including proofs, is from from Ferguson (1996) [115].

DEFINITION 2.7.1. *A sequence of random vectors $Y_n, n \in \mathbb{N}$, is convergent in distribution (or in law) to a random vector Y, if the sequence of joint c.d.f.'s $F_{Y_n}(y)$ converges to $F_Y(y)$ at all vectors y where F_Y is continuous. Notation $Y_n \to_d Y$, or $Y_n \to_d Y$.*

The following, including the proofs, is a combination of the Helly-Bray theorem with the continuity theorem , and is taken straight from Ferguson (1996) [115].

THEOREM 2.7.1. *Assume $X_n, n \in \mathbb{N}$, is a sequence of random vectors in $\mathbb{R}^m$. The following are equivalent:*
(a) $X_n \to_d X$.
(b) $Ef(X_n) \longrightarrow Ef(X) = \int_{\mathbb{R}^m} f dP_X$ as $n \to \infty$ for all real-valued continuous functions f on $\mathbb{R}^m$, with compact support.
(c) X_n converges weakly to X, that is

$$Ef(X_n) \longrightarrow Ef(X) \quad \text{as } n \to \infty \tag{2.49}$$

for all bounded real-valued continuous functions f on $\mathbb{R}^m$.
(d) $Ef(X_n) \longrightarrow Ef(X)$ as $n \to \infty$ for all measurable functions f on $\mathbb{R}^m$, such that X takes values on the set of points of continuity of f with probability 1.
(e) The sequence of characteristic functions of X_n converges to the characteristic function of X.

Proof of THEOREM 2.7.1. Obviously, $(d) \Rightarrow (c)$ and $(c) \Rightarrow (b)$. We will show $(d) \xrightarrow{Part(1)} (a) \xrightarrow{Part(2)} (b) \xrightarrow{Part(3)} (c) \xrightarrow{Part(4)} (d)$

(1) Proof of $(d) \xrightarrow{Part(1)} (a)$. Let x^0 be continuity point of F_X. Then $F_X(x^0) = Eg(X)$, where $g(x)$ is the indicator function,

$$g(x) = \begin{cases} 1 & , \text{if } x \leq x^0 \\ 0 & , \text{otherwise.} \end{cases}$$

The continuity set of g contains all points x except those such that $x \leq x^0$ with equality for at least one component. Because x^0 is a continuity point of F_X, we have $F_X(x^0 + \varepsilon 1) - F_X(x^0 - \varepsilon 1) \to 0$ as $\varepsilon \to 0$, which implies that the continuity set of g has probability 1 under the distribution of X. Hence, $F_{X_n}(x^0) \to F_X(x^0)$.

(2) Proof of $(a) \xrightarrow{Part(2)} (b)$. Let g be continuous and vanishing outside a compact set, C. Then g is uniformly continuous: For every $\varepsilon > 0$, there exists a number $\delta > 0$ such that $|x-y| < \delta$ implies $|g(x)-g(y)| < \varepsilon$.

Let $\varepsilon > 0$ and find such a $\delta > 0$. Slice C by finite sets of parallel hyperplanes at a distance at a distance of at most $\delta/\sqrt{m}$ apart, one set for each dimension, each hyperplane having probability zero under F_X (only countably many parallel planes can have positive mass). This cuts $\mathbb{R}^m$ into parallelepipeds of the form $(b,c] = \{x : b < x \leq c\} = \{x : b_i < x_i \leq c_i, \text{ for all } i = 1,\ldots,m.\}$ On any such parallelepipeds $|g(x)-g(c)| \leq \varepsilon$.
Thus, $|g(x) - \tilde{g}(x)| \leq \varepsilon$ for all x, where $\tilde{g}(x) = \sum_{all(b,c]} g(c)I_{(b,c]}(x)$. This is essentially a finite sum since g vanishes outside a compact set, and it may be rewritten as a finite sum of the form, $\tilde{g}(x) = \sum_i a_i I_{(-\infty,x_i]}(x)$ (see Ferguson(1996)[115]), with F_X continuous at each x_i. Thus, $X_n \xrightarrow{\mathcal{L}} X$ implies that

$$E\tilde{g}(X_n) = \sum_i a_i F_{X_n}(x) \to \sum_i a_i F_X(x) = E\tilde{g}(X).$$

Finally,

$$|Eg(X_n) - Eg(X)| \leq |Eg(X_n) - E\tilde{g}(X_n)| + |E\tilde{g}(X_n) - E\tilde{g}(X)| + |E\tilde{g}(X) - Eg(X)| \leq$$

$$\leq 2\varepsilon + |E\tilde{g}(X_n) - E\tilde{g}(X)| \to 2\varepsilon.$$

Since this is true for all $\varepsilon > 0, Eg(X_n) \to Eg(X)$.
(3) Proof of $(b) \Rightarrow Part(3)(c)$
Let g be continuous, $|g(x)| < A$ or all x, and $\varepsilon > 0$. By Chebysheff's inequality, find B such that $P\{|X| \geq B\} < \varepsilon/(2A)$. Find h continuous so that

$$h(x) = \begin{cases} 1 & \text{, if } |x| \leq B \\ 0 & \text{, if} |x| \geq B+1 \end{cases} \quad \text{and } 0 \leq h(x) \leq 1 \text{ for all } x$$

Then,

$$|Eg(X_n) - Eg(X)| \leq$$

$$\leq |Eg(X_n) - Eg(X_n)h(X_n)| + |Eg(X_n)h(X) - Eg(X)h(X_n)| + |Eg(X)h(X) - Eg(X)|$$

The middle term converges to 0 because $g \cdot h$ is continuous and vanishes outside a compact set. The first term is bounded by $\varepsilon/2$,
$|Eg(X_n) - Eg(X_n)h(X_n)| \leq E|g(X_n)||1-h(X_n)| \leq AE(1-h(X_n))$
$= A(1-Eh(X_n)) \to A(1-Eh(X)) \leq \varepsilon/2$,
and, similarly, the last term is bounded by $\varepsilon/2$. Therefore, $|Eg(X_n) - Eg(X)|$ is bounded by something that converges to ε.
Since this is true for all $\varepsilon > 0$, $\lim_{n\to\infty} |Eg(X_n) - Eg(X)| = 0$.

(4) To prove $(c) \xrightarrow{Part(4)} (d)$, we use the following lemma.

LEMMA 2.7.1. *Let g be bounded measurable with $P\{X \in C(g)\} = 1$. Then, for every $\varepsilon > 0$ there exist bounded continuous functions f and h such that $f \le g \le h$ and $E(h(X) - f(X)) < \varepsilon$.*

Proof of lemma. Let $f_k, h_k, k = 1, \ldots,$ be defined as follows :

$$f_k(x) = \inf_y(g(y) + k|x-y|), h_k(x) = \sup_y(g(y) - k|x-y|).$$

Obviously $f_1(x) \le \cdots \le f_k(x) \le g(x) \le g_k(x) \cdots \le g_1(x)$, and since that $|f_k(x') - f_k(x)| \le k|x' - x|, |h_k(x') - h_k(x)| \le k|x' - x|$, the sequences $f_k, g_k, k = 1, 2, \ldots$ are continuous. They are also bounded since g is bounded. Therefore their limits, $f_0 = \lim_{k\to\infty} f_k, h_0 = \lim_{k\to\infty} h_k$, satisfy the inequalities $f_0 \le g \le h_0$. At any point $x \in C(g), f_0(x) = g(x) = h_0(x)$. Indeed given $x \in C(g), \varepsilon > 0$, there is a $\delta > 0$, such that if $|y - x| < \delta$, then $|g(y) - g(x)| < \varepsilon$, and if B is a lower bound for g, and we select $k > \frac{g(x)-B}{\delta}$, then

$$f_0(x) \ge f_k(x) =$$

$$= \min(\inf_{|y-x|<\delta}(g(y) + k|x-y|), \inf_{|y-x|\ge\delta}(g(y) + k|x-y|)) \ge$$

$$\min(g(x) - \varepsilon, B + \delta(\frac{g(x) - B)}{\delta})) = g(x) - \varepsilon.$$

Therefore, for any $\varepsilon > 0, |f_0(x) - g(x)| < \varepsilon$, and similarly, for any $\varepsilon > 0, |h_0(x) - g(x)| < \varepsilon$, proving the claim. Given that $P\{X \in C(g)\} = 1$, it follows that $Ef_0(\mathbf{X}) = Eg(\mathbf{X}) = Eh_0(\mathbf{X})$. On the other hand, from the Monotone Convergence Theorem $Ef_k(\mathbf{X}) \nearrow Ef_0(\mathbf{X}), Eh_k(\mathbf{X}) \searrow Eh_0(\mathbf{X})$, therefore given $\varepsilon > 0$, there is a k such that $Eh_k(\mathbf{X}) - Eh_0(\mathbf{X}) < \varepsilon/2, Ef_0(\mathbf{X}) - Ef_k(\mathbf{X}) < \varepsilon/2$, thus $|Eh_k(\mathbf{X}) - Ef_k(\mathbf{X})| < \varepsilon$, done.

(Proof of $(c) \xrightarrow{Part(4)} (d)$ from here)
Let g be bounded measurable with $P\{X \in C(g)\} = 1$, let $\varepsilon > 0$, and find f and h as in the lemma. Then,
$Eg(X) - \varepsilon \le Ef(X) = \lim Ef(X_n) \le \liminf Eg(X_n)$
$\le \limsup Eg(X_n) \le \lim Eh(X_n) = Eh(X) \le Eg(X) + \varepsilon$. Let $\varepsilon \to 0$, and conclude $Eg(X) = \lim Eg(X_n)$.
Finally, the implication $(a) \Rightarrow (b)$ or (c) or (d) follows from the other implications that were proved.
We now show that $(a) \Leftrightarrow (e)$, that is $X_n \to_d X \Leftrightarrow \varphi_{X_n}(t) \to \varphi_X(t)$, for all $t \in \mathbb{R}^p$.

Proof. $(\Rightarrow)$ This follows immediately from Theorem 2.7.1, because $\exp\{ it^T\mathrm{X} = \cos t^T\mathrm{X} + i \sin t^T\mathrm{X} \}$ is bounded and continuous.
$(\Leftarrow)$ Let g be continuous with compact support. Then g is bounded, $|g(x)| \le B$ say, and uniformly continuous. Let $\varepsilon > 0$. Find $\delta > 0$ such that $|x - y| < \delta \Rightarrow |g(x) - g(y)| < \varepsilon$.
To show $Eg(X_n) \to Eg(X)$, let $Y_\sigma \sim \mathcal{N}_m(0, \sigma^2 I)$ be independent of the X_n and X. Then

$|Eg(X_n) - Eg(X)| \leq |Eg(X_n) - Eg(X_n + Y_\sigma)|$
$+|Eg(X_n + Y_\sigma) - Eg(X + Y_\sigma)|$
$+|Eg(X + Y_\sigma) - Eg(X)|$
The first term is

$$\leq E\{|g(X_n) - g(X_n + Y_\sigma)|I(|Y_\sigma| \leq \delta)\}$$
$$+E\{|g(X_n) - g(X_n + Y_\sigma)|I(|Y_\sigma|) > \delta)\}$$
$$\leq \varepsilon + 2BP\{|Y_\sigma| > \delta\} \leq 2\varepsilon$$

for σ sufficiently small. Similarly, the third term $\leq 2\varepsilon$.
It remains to show that

$$Eg(X_n + Y_\sigma) \to Eg(X + Y_\sigma)$$

The characteristic function of $\mathcal{N}_m(0, \alpha^2 I)$ is

$$\varphi(t) = [\frac{1}{\sqrt{2\pi}\alpha}]^m \int e^{it^T z - z^T z/(2\alpha^2)} dz = e^{-t^T t\alpha^2/2}$$

Using this with $\alpha = 1/\sigma$, and making the change of variables $u = x + y$ for y, we find

$Eg(X_n + Y_\sigma) = [\frac{1}{\sqrt{2\pi}\sigma}]^m \iint g(x+y) e^{-y^T y/(2\sigma^2)} dy dF_{X_n}(X)$
$= [\frac{1}{\sqrt{2\pi}\sigma}]^m \int g(u) e^{-(u-x)^T(u-x)/(2\sigma^2)} dF_{X_n}(X) du$
$= [\frac{1}{\sqrt{2\pi}\sigma}]^m \int g(u) \int [\frac{\sigma}{\sqrt{2\pi}}]^m \int e^{it^T(u-x) - \sigma^2 t^T t/2} dt dF_{X_n}(X) du$
$= [\frac{1}{2\pi}]^m \int g(u) \int e^{it^T u - \sigma^2 t^T t/2} \varphi_{X_n}(-t) dt du$
$\to [\frac{1}{2\pi}]^m \int g(u) \int e^{it^T u - \sigma^2 t^T t/2} \varphi_X(-t) dt du$
using the Lebesque Dominated Convergence Theorem
($|e^{it^T u} \varphi_{X_n}(-t)| \leq 1$ and g has compact support). Undoing the previous steps, we see that this last expression is equal to $Eg(X + Y_\sigma)$, done.

2.8 Limit Theorems

Most of this section follows from the lecture notes of Bhattacharya (1997) [31].

2.8.1 The Central Limit Theorem

The most important convergence theorem in law is the following theorem, abbreviated as *CLT*.

THEOREM 2.8.1 (Central Limit Theorem). *If Y_n is a sequence of i.i.d.r.vec.'s in $\mathbb{R}^m$ with common mean vector zero and a finite common covariance matrix Σ, then*

$$n^{-\frac{1}{2}}(Y_1 + \cdots + Y_n) \to_d N_m(0, \Sigma). \tag{2.50}$$

For a proof of the CLT, we consider the sequence $X_n = \sqrt{n}\bar{Y}_n = n^{-\frac{1}{2}}(Y_1 + \cdots + Y_n)$. Note that

$$\varphi_{X_n}(t) = E(\exp(in^{-\frac{1}{2}}t^T \sum_{j=1}^{n} Y_j)) = (E(\exp(in^{-\frac{1}{2}}t^T Y_1)))^n = (\varphi_{Y_1}(n^{-\frac{1}{2}}t))^n \quad (2.51)$$

Let $\sigma^2(t) = E((t^T X_1)^2)$. From equation (2.51) it follows that

$$\lim_{n\to\infty} \varphi_{X_n}(t) = \exp(-\frac{\sigma^2(t)}{2}). \quad (2.52)$$

The right hand side of equation (2.52) is $\varphi_U(1)$, where $U \sim \mathcal{N}(0, \sigma^2(t))$. Finally note that if $X \sim \mathcal{N}_m(0, \Sigma)$, then $t^T X \sim N(0, \sigma^2(t))$, and from Theorem 2.7.1 it follows that $X_n \xrightarrow{\mathcal{L}} X$, done.

We will sometimes use the alternate notation $\Phi_{0,\Sigma}$ for $N_m(0,\Sigma)$, and denote the corresponding joint c.d.f. by $\Phi_{0,\Sigma}(x)$, and, whenever Σ is positive definite, the joint p.d.f. by

$$\varphi_{0,\Sigma}(x) = \frac{1}{\sqrt{(2\pi)^m|\Sigma|}} e^{-\frac{1}{2}x^T\Sigma^{-1}x}.$$

The CLT says that if Σ is positive definite, then

$$\mathbb{P}(n^{-\frac{1}{2}}(Y_1 + \cdots + Y_n) \le x) \longrightarrow \Phi_{0,\Sigma}(x) = \frac{1}{\sqrt{(2\pi)^m|\Sigma|}} \int \cdots \int_{y \le x} e^{-\frac{1}{2}y^T\Sigma^{-1}y} dy \quad (2.53)$$

for all x. A useful extension of the multivariate CLT (see Cramer (1946) [76]) is

THEOREM 2.8.2. *(Multivariate Lindeberg-Feller-Levy CLT). Let $X_1, X_2, \ldots$ be independent r. vec.'c in $\mathbb{R}^m$, with the joint c.d.f.'s $F_1, F_2, \ldots,$ such that $EX_n = \mu_n$ and $Cov(X_n) = \Sigma_n$. Assume $\Sigma = \lim_{n\to\infty} \frac{1}{n}(\sum_{j=1}^{n} \Sigma_j)$ is a positive definite covariance matrix, and the following Lindeberg condition is satisfied:*

$$\forall \varepsilon > 0, \ \lim_{n\to\infty} \frac{1}{n} \sum_{k=1}^{n} \int_{\|x-\mu_k\| > \varepsilon\sqrt{n}} \|x - \mu_k\|^2 dF_k(x) = 0.$$

Then

$$\frac{1}{\sqrt{n}} \sum_{i=1}^{n} (X_i - \mu_i) \to_d N_m(0, \Sigma)$$

2.8.2 *Basic Large Sample Theory*

If $x \in \mathbb{R}^m$ and $\delta > 0$, the ball centered at x of radius δ is

$$B_x(\delta) = \{y \in \mathbb{R}^m, \|y - x\| \le \delta\}.$$

A number of key results in limit theory of statistics are summarized in the following

THEOREM 2.8.3 (generalized Slutsky theorem). *Suppose U_n, V_n, W_n $(n \geq 1)$ are three sequences of random vectors such that $U_n \xrightarrow{P} a$, $V_n \xrightarrow{P} b$, $W_n \to_d W$. Let $h(u,v,w)$ be a function valued in $\mathbb{R}^q$, which is continuous on a neighborhood $B_{\delta_1}(a) \times B_{\delta_2}(b) \times \mathbb{R}^m$ for some $\delta_1 > 0$, $\delta_2 > 0$. Then $h(U_n,V_n,W_n) \to_d h(a,b,W)$.*

Proof. We will first prove that $h(U_n,V_n,W_n) - h(a,b,W_n) \xrightarrow{P} 0$. The desired result would follow from this, using the continuity of $w \to h(a,b,w)$. Fix $\varepsilon > 0$ and $\theta > 0$, however small. In view of the convergence in distribution of W_n there exists $A = A(\theta)$ such that $\mathbb{P}(\|W_n\| > A) < \theta/3$ for all n (Exercise 3). In view of (uniform) continuity of h on the compact set $B_{\delta_1}(a) \times B_{\delta_2}(b) \times B_0(A)$, there exists $\delta = \delta(\varepsilon) > 0$ such that $\|h(u,v,w) - h(a,b,w)\| \leq \varepsilon$ for all (u,v,w) satisfying $\|u-a\| \leq \delta$, $\|v-b\| \leq \delta$ and $\|w\| \leq A$. Now since $U_n \xrightarrow{P} a$, $V_n \xrightarrow{P} b$, there exists a positive integer $n(\theta,\varepsilon)$ such that

$$\mathbb{P}(\|U_n - a\| > \delta) < \frac{\theta}{3}, \quad \mathbb{P}(\|V_n - b\| > \delta) < \frac{\theta}{3} \quad \forall\, n \geq n(\theta,\varepsilon). \tag{2.54}$$

Hence

$$\begin{aligned} & \mathbb{P}(\|h(U_n,V_n,W_n) - h(a,b,W_n)\| > \varepsilon) \\ \leq\; & \mathbb{P}(\|U_n - a\| > \delta) + \mathbb{P}(\|V_n - b\| > \delta) + \mathbb{P}(\|W_n\| > A) \qquad (2.55) \\ & + \mathbb{P}(\{\|U_n - a\| \leq \delta, \|V_n - b\| \leq \delta, \|W_n\| \\ \leq\; & A, \|h(U_n,V_n,W_n) - h(a,b,W_n)\| > \varepsilon\}) \leq 3\frac{\theta}{3} = \theta, \end{aligned}$$

since the set within curly brackets in the last term in (2.55) is empty and has, therefore, probability zero.

In particular we obtain the following corollary, the proof of which is left as an exercise (2).

COROLLARY 2.8.1. *(i). Suppose $Z_n \to_d Z$ and g is a continuous function. Then $g(Z_n) \to_d g(Z)$.*

(ii). Suppose $X_n \to_d X$ and $Y_n \xrightarrow{P} 0$. Then $X_n + Y_n \to_d X$.

The following simple result is widely used in asymptotic theory.

THEOREM 2.8.4. *Suppose W_n is a sequence of random vectors and $g(n)$ a sequence of constants, $g(n) \uparrow \infty$, such that $g(n)(W_n - \mu) \to_d V$. Then for every function H which is continuously differentiable in a neighborhood of μ, one has*

$$g(n)[H(W_n) - H(\mu)] \to_d H'(\mu)V. \tag{2.56}$$

Proof. By the mean value theorem there exists μ^* between μ and W_n such that the left side of (2.56) is

$$g(n)(W_n-\mu)H'(\mu^*) = g(n)(W_n-\mu)\{H'(\mu)+o_p(1)\} \qquad (2.57)$$

where $o_p(1) \to 0$ in probability. Note that this is a consequence of the fact that $W_n - \mu \xrightarrow{P} 0$ (since $g(n) \uparrow \infty$ and $g(n)(W_n-\mu) \to_d V$ (Exercise 10)). The convergence now follows. The most commonly used consequence of Theorem 2.8.4 is the following, called the delta method.

THEOREM 2.8.5. *(Cramer's delta method) Theorem 2.8.4 extends to the case where Z_j, $j \geq 1$, are i.i.d. k-dimensional random vectors with mean vector μ and covariance matrix $\Sigma = ((\sigma_{ij}))$, while H is vector-valued and continuously differentiable (as a function of k variables) in a neighborhood of μ. In this case*

$$\sqrt{n}[H(\overline{Z}) - H(\mu)] \to_d D_\mu H \cdot V = N_m(0, D_\mu H \Sigma D_\mu H^T), \qquad (2.58)$$

with $D_\mu H = (\partial H^j(z)/\partial z_i)_{i=1,\dots,k,j=1,\dots,p}|z=\mu$, and V is $N_k(0,\Sigma)$.

EXAMPLE 2.8.1 (asymptotic distribution of the studentized sample mean). *Let Y_n be a sequence of i.i.d. p-dimensional random vectors with a positive covariance matrix Σ. The studentized p-dimensional version of the t-statistic is given by*

$$\begin{aligned} T_n &= \sqrt{n}S_{u,n}^{-\frac{1}{2}}(\overline{Y}-\mu) = \sqrt{n-1}S_n^{-\frac{1}{2}}(\overline{Y}-\mu) \\ &= \sqrt{\frac{n-1}{n}}\tilde{T}_n, \qquad \tilde{T}_n = \sqrt{n}S_n^{-\frac{1}{2}}(\overline{Y}-\mu). \end{aligned} \qquad (2.59)$$

Here $EY_n = \mu$, $Cov(Y_n) = \Sigma > 0$ (finite). Also note, by the CLT and by Exercise 8 that

$$\sqrt{n}(\overline{Y}-\mu) \to_d N_m(0,\Sigma),$$

and

$$S_n \xrightarrow{P} \Sigma,$$

so that Slutsky's theorem applies to show that $T_n \to_d N_m(0, I_m)$.

EXAMPLE 2.8.2 (asymptotic distribution of Hotelling T^2). *For any random sample $X_1,\dots,X_n$ from a distribution on $\mathbb{R}^m$ with finite mean vector μ and covariance matrix Σ we consider the analogue of the Hotelling like statistic T_n^2 given by*

$$T_n^2 = n(\bar{X}-\mu)^T S_n^{-1}(\bar{X}-\mu). \qquad (2.60)$$

Note that $T_n^2 = \|T_n\|^2$, where T_n is the studentized sample mean vector in Example 2.8.1, and since the square norm is a continuous function on $\mathbb{R}^m$, from Example 2.8.1 and Corollary 2.8.1 we see that $T_n^2 \to_d \chi_m^2$.

2.9 Elementary Inference

In general, for any unknown distributions Q on $\mathbb{R}^m$, a confidence region (c.r.) of confidence level $1-\alpha$ for a parameter $\theta \in \Theta \subseteq \mathbb{R}^q$, is a set $C_\alpha(x_1,\ldots,x_n) \subseteq \mathbb{R}^q$, depending on random sample $x_1,\ldots,x_n$, such that

$$P(\theta \in C_\alpha(x_1,\ldots,x_n)) = 1-\alpha. \tag{2.61}$$

Often the parameter is a pair $\theta = (\theta_1,\theta_2)$, and we are interested in estimating only θ_1. In this case θ_2 is a *nuisance parameter*, and we may replace it by one of its consistent estimators $\hat{\theta}_2$ in the expression of a statistic with a known asymptotic distribution, leading to a formula for the confidence region of θ_1 that is independent of θ_2.

For example, given i.i.d.r.vec.s $X_1,\ldots,X_n \sim Q$, where the probability distribution Q has finite moments of order three, mean vector μ, positive definite covariance matrix $\Sigma > 0$, for n large enough, by the CLT we have:

$$P(n(\bar{X}_n-\mu)^T\Sigma^{-1}(\bar{X}_n-\mu) \leq \chi^2_{p,\alpha}) \approx 1-\alpha, \tag{2.62}$$

where the probability area of the upper tail above $\chi^2_{p,\alpha}$, under the density function of a χ^2_p-distributed is α. In this example if we consider the parameter $\theta = (\mu,\Sigma) \in \mathbb{R}^p \times Sym^+(p)$, the dimension $m = p+\frac{1}{2}p(p+1)$, and the nuisance parameter is the parameter $\theta_2 = \Sigma$. Since the sample covariance matrix is a consistent estimator we may plug $\hat{\theta}_2 = S_n$ in place of Σ in equation (2.62), and from Slutsky's theorem it follows that for n large enough, a confidence region $C_\alpha(x_1,\ldots,x_n)$ for μ at confidence level $1-\alpha$, is given by

$$C_\alpha(x_1,\ldots,x_n) = \{\mu \in \mathbb{R}^p, n(\bar{x}-\mu)^T S_n^{-1}(\bar{x}-\mu) \leq \chi^2_{p,\alpha}\}. \tag{2.63}$$

In general, since we do not know if the true distribution has positive definite covariance ($\Sigma \in Sym^+(p)$), the best one can do is to find an **approximate confidence region**, based on asymptotics, as above, or another approximation. Typically with $P(\theta \in C_\alpha(x_1,\ldots,x_n)) < 1-\alpha$. The difference between the *nominal coverage* $(1-\alpha)$ and the *true coverage* $(P(\theta \in C_\alpha(x_1,\ldots,x_n))$ is called *coverage error*.

Hypothesis testing. Nonparametric Approach. An χ^2 based test for $H_0 : \mu = \mu_0$ vs. $H_a : \mu \neq \mu_0$ can be obtained using the above confidence region. Indeed note that we fail to reject H_0 at level α if and only if $\mu_0 \in C_\alpha$. It follows that at level α we reject H_0 if

$$n(\mu_0-\bar{x})^T S_n^{-1}(\mu_0-\bar{x}) \geq \chi^2_{p,\alpha}, \tag{2.64}$$

2.9.1 Simultaneous Confidence Intervals for μ.

A variant of the following result is given with a proof in Johnson and Wichern (2008) [170], in the case of multivariate normal distributions.

THEOREM 2.9.1. *Assume* **X** *is a random sampling matrix from a multivariate distribution in* $\mathbb{R}^m$ *with mean vector* μ *and covariance matrix* Σ). *Then for any* $\alpha \in (0,1)$, *with probability* $1-\alpha$, *for any vector* $a \in \mathbb{R}^m, a^T\mu$ *is inside the intervals:*

$$L.B. = a^T\bar{x} - \sqrt{\chi^2_{p,\alpha} a^T S_n a}$$
$$U.B. = a^T\bar{x} + \sqrt{\chi^2_{p,\alpha} a^T S_n a} \tag{2.65}$$

As a corollary, if $n > p$, with probability of about $1-\alpha$, for $j = 1,\dots,p$, μ^j is inside the interval

$$\left(\bar{x}^j_n - \sqrt{\chi^2_{p,\alpha}} s_j, \bar{x}^j_n + \sqrt{\chi^2_{p,\alpha}} s_j\right). \tag{2.66}$$

These are simultaneous χ^2-based c.i.'s for the components of the mean vector for $n-p$ large. Simultaneous z-based c.i.'s for the components of the mean vector, for $n-p$ large, are given by the following theorem.

THEOREM 2.9.2. *(Bonferroni simultaneous confidence intervals) Assume* $x_1,\dots,x_n$ *is a random sample from a probability distribution* Q *on* $\mathbb{R}^p$, *with mean* $\mu = (\mu^1,\dots,\mu^p)^T$ *and covariance matrix* Σ, *with* $\Sigma > 0, n > p$. *Then for any* $\alpha \in (0,1)$, *with probability* at least $1-\alpha$, *for* $j = 1,\dots,p$, μ^j *is inside the interval*

$$\left(\bar{x}^j_n - z(\frac{\alpha}{2p})\frac{s_j}{\sqrt{n}}, \bar{x}^j_n + z(\frac{\alpha}{2p})\frac{s_j}{\sqrt{n}}\right), \tag{2.67}$$

where $z(\beta)$ *is the cutoff point of the upper tail enclosing a probability area* β *under the graph of the p.d.f. of a standard normally distributed r.v.*

Bonferroni simultaneous c.i.'s are based on the basic Bonferroni inequality:

$$P(\cup^p_{a=1} U_a) \le \sum_{a=1}^{p} P(U_a). \tag{2.68}$$

Comparison of the large sample simultaneous c.i.'s for the components of the mean vector ($n-p$ large) using these two methods is usually done by utilizing the z and χ^2-tables.

2.10 Comparison of Two Mean Vectors

Assume that we have i.i.d.r.vec.'s $X_{a,1},\dots,X_{a,n_a}, a = 1,2$ from two independent distributions in $\mathbb{R}^m$, $X_{a,1} \sim (\mu_a, \Sigma_a), a = 1,2$.

In the case of *matched pairs*, $n_1 = n_2 = n$, we simply assume that the difference vector $D = X_{1,1} - X_{2,1}$ has a mean μ_D and we are the null hypothesis

$H_0 : \mu_D = \mu_0$ vs. $H_1 : \mu_D \neq \mu_0$, and use the large sample rejection region at significance level α, level given in (2.64).

The case when the population covariance matrices are unequal can be answered in the large sample case, as follows.

For testing $H_0 : \mu_1 - \mu_2 = \delta_0$, assume the two samples are i.i.d.r.vec's $X_{a,j_a} \sim \mu_a, \Sigma_a, j_a = 1, \dots, n_a, a = 1, 2$ from two independent multivariate populations, of the total sample size $n = n_1 + n_2$, s.t. $\frac{n_1}{n} \to q \in (0,1)$, as $n \to \infty$, and $\frac{1}{q}\Sigma_1 + \frac{1}{1-q}\Sigma_2 > 0$, Then from the C.L.T and Slutsky's theorem 2.8.3, under H_0, it follows that

$$T^2 = (\overline{X_1} - \overline{X_2} - \delta_0)^T (\frac{1}{n_1} S_1 + \frac{1}{n_2} S_2)^{-1} (\overline{X_1} - \overline{X_2} - \delta_0) \to_d C \sim \chi^2_m, \qquad (2.69)$$

as $n \to \infty$, therefore we reject H_0 at level α if

$$(\overline{X_1} - \overline{X_2} - \delta_0)^T (\frac{1}{n_1} S_1 + \frac{1}{n_2} S_2)^{-1} (\overline{X_1} - \overline{X_2} - \delta_0) > \chi^2_m(\alpha). \qquad (2.70)$$

If the distributions are unknown and the samples are small, for testing, one may use the *nonparametric bootstrap*, a methodology that will be presented in Section 2.13.

2.11 Principal Components Analysis (P.C.A.)

Principal components (p.c.'s) are linear combinations in the given covariates of a multivariate random vector. The first principal component captures most of the total variance of this random vector, the second p.c. points in a direction perpendicular on the direction of the first p.c. that has a largest variance amongst all such covariates, etc. Principal components are used in *dimension reduction*.

Assume $X = (X^1, X^2, \dots, X^m) : \Omega \to \mathbb{R}^m$ is a random vector. A *simple linear combination* (s.l.c.) in X is a linear combination $L = l_1 X^1 + l_2 X^2 + \dots + l_m X^m = l^T X$, with $l^T l = 1$. We are interested in those s. l. c.'s of largest variance.

Assume $\lambda_i, i = 1, \dots, p$ are the eigenvalues of $\Sigma = Cov(X)$ in their decreasing order and $e_i, i = 1, \dots, p$ are corresponding norm one eigenvectors. Note that the eigenspaces corresponding to distinct eigenvalues are orthogonal, therefore w.l.o.g. we may assume that these vectors form an orthobasis of $\mathbb{R}^m$, therefore the matrix $\Gamma = (e_1 \dots e_m)$ is orthogonal.

If $\mu = E(X)$ is the mean vector, then the principal components of X are the covariates $Y^1, \dots, Y^m$ of $Y = \Gamma^T (X - \mu)$.

Large sample properties - the univariate case Note that if $X_1, X_2, \dots, X_n$ are i.i.d.r.v.'s from a distribution with finite fourth moments μ_4 and variance σ^2, then, from the delta method Theorem 2.8.5, it follows that for n large enough $\sqrt(n)(S^2 - \sigma^2) \to_d N(0, \mu_4 - \sigma^4)$, as shown in Ferguson (1996, p.46) [115].

A large sample hypothesis testing for the percentage of variability explained by the first P.C.'s is suggested as follows by Mardia et al. (1972, p. 232) [231]. If we assume that $\rho > 0$, we may not test $H_0 : \lambda_j = 0, j = k+1,..,p$, since in this case Σ would be singular. A reasonable hypothesis testing problem is that $\lambda_j = 0, j = k+1,..,p$ are small, that is

$$H_0 : \frac{\lambda_1 + \ldots + \lambda_k}{\lambda_1 + \ldots + \lambda_m} = \rho \text{ vs } H_1 : \frac{\lambda_1 + \ldots + \lambda_k}{\lambda_1 + \ldots + \lambda_m} > \rho.$$

If r the sample counterpart of ρ, then by the delta method, under H_0 the large sample distribution of r is normal with mean ρ and variance τ^2, where $\tau^2 = \frac{2Tr(\Sigma^2)}{nTr^2(\Sigma)}(\rho^2 - 2\alpha\rho + \alpha)$, where

$$\alpha = \frac{\lambda_1^2 + \ldots + \lambda_k^2}{\lambda_1^2 + \ldots + \lambda_m^2}.$$

For testing H_0 above, one may use Slutsky's theorem and a large sample z-test; one should studentize r by replacing τ^2 with its sample counterpart,

$$\widehat{\tau}^2 = \frac{2Tr(S_n^2)}{nTr^2(S_n)}(\rho^2 - 2\widehat{\alpha}\rho + \widehat{\alpha}), \tag{2.71}$$

where

$$\widehat{\alpha} = \frac{\widehat{\lambda}_1^2 + \cdots + \widehat{\lambda}_k^2}{\widehat{\lambda}_1^2 + \cdots + \widehat{\lambda}_m^2}. \tag{2.72}$$

2.12 Multidimensional Scaling

An important result in data visualization in higher numerical spaces, is known as multidimensional scaling(MDS) (see Schoenberg (1935) [300], Gower (1966) [130], Mardia et al. (1972, p. 397) [231]). Let $D = (d_{rs})_{r,s=1,\ldots,k}$ be a distance matrix between k points and consider A and B defined by $A = (a_{rs})$, $a_{rs} = -\frac{1}{2}\, d_{rs}^2$, $\forall r, \forall s = 1,\ldots k, B = \tilde{H}A\tilde{H}^T$, where $\tilde{H} = I_k - k^{-1}1_k 1_k^T$ is the centering matrix in $\mathbb{R}^k$. A version of Schoenberg's theorem suitable for our purposes is given below. Theorem 2.12.1 is an edited version of Theorem 14.2.1 in Mardia et al. (1972, pp. 397–398) [231].

THEOREM 2.12.1.(*a*) *If D is a matrix of Euclidean distances $d_{rs} = \|x^r - x^s\|$, $x^j \in \mathbb{R}^m$, and we set $X = (x^1 \ldots x^k)^T, k > p$, then B is given by*

$$b_{rs} = (x^r - \overline{x})^T(x^s - \overline{x}), r, s = 1, \ldots, k \tag{2.73}$$

In matrix form, $B = (\tilde{H}X)(\tilde{H}X)^T$.

(b) *Conversely, assume B is a $k \times k$ symmetric positive semidefinite matrix B of rank p, having zero row sums. Then B is a centered inner product matrix for a configuration X constructed as follows. Let $\lambda_1 \geq \cdots \geq \lambda_m$ denote the positive eigenvalues of B with corresponding eigenvectors $x_{(1)}, \ldots, x_{(p)}$ normalized by*

$$x_{(i)}^T x_{(i)} = \lambda_i. \tag{2.74}$$

Then the k points $P_r \in R^m$ with coordinates $x^r = (x_{r1}, \ldots, x_{rp})^T, r = 1, \ldots, k$, where x^r is the rth row of the $k \times p$ matrix $(x_{(1)} \ldots x_{(p)})$, have center $\bar{x} = 0$, and $B = \tilde{H} A \tilde{H}^T$ where $A = -\frac{1}{2}(\|x^r - x^s\|^2)_{r,s=1,\ldots,k}$.

2.13 Nonparametric Bootstrap and the Edgeworth Expansion

The main references for this section are Bhattacharya and Denker (1990)[36] and Bhattacharya et al. (2014) [40]. Here we describe the precise level of asymptotic accuracy of the nonparametric bootstrap procedure for estimating the true distribution of a class of commonly used statistics, when the common distribution of the observations has a density component. The technique for this analysis, and for all others which demonstrate the superiority of the bootstrap over normal approximation, involves the so-called *Edgeworth expansion* (see Bhattacharya and Ghosh (1978) [38]).

2.13.1 Edgeworth Expansions for Statistics

Let X_j $(1 \leq j \leq n)$ be i.i.d. random vectors in $\mathbb{R}^m$, and let f be an $\mathbb{R}^k$-valued Borel measurable on $\mathbb{R}^m$. Let $Z_j = f(X_j), 1 \leq j \leq n$, and assume a statistic T is given by

$$T(X_1, \ldots, X_n) = H(\bar{Z}),\ \bar{Z} := \frac{1}{n}\sum_{j=1}^{n} Z_j = \frac{1}{n}\sum_{j=1}^{n} f(X_j), \tag{2.75}$$

where, for simplicity H is assumed to be sufficiently smooth real-valued function defined in a neighborhood of the mean of Z_1,

$$\mu := EZ_j = E\bar{Z}. \tag{2.76}$$

Note that the theory in this section is also valid if H is vector valued (Bhattacharya and Denker (1990) [36]). Most classical statistics are of this form or may be approximated within the desired level of accuracy by such functions $H(\bar{Z})$. Since the asymptotic performance of only the pivotal bootstrap is of a higher order than that of the normal approximation, we will assume *studentization*, meaning that the variance of $H(Z_1)$ is 1 or

$$(gradH)(\mu)^T \Sigma (gradH)(\mu) = 1, \tag{2.77}$$

where Σ is the $k \times k$ covariance matrix of Z_1

$$\Sigma := CovZ_1. \tag{2.78}$$

Supposed that H is $s-1$ times continuously differentiable ($s \geq 2$) in a neighborhood of μ so that one has the Taylor expansion,

$$\begin{aligned} H(\bar{Z}) - H(\mu) &= (gradH)(\mu)^T(\mu)(\bar{Z}-\mu) \\ &\quad + \sum_{|\nu|=1}^{s-1} (D^\nu H)(\mu)\frac{(\bar{Z}-\mu)^\nu}{\nu!} + R(\bar{Z},\mu). \end{aligned} \tag{2.79}$$

Here $\nu = (\nu_1, \nu_2, \ldots, \nu_k)$ is a multi-index, i.e., $\nu \in (\mathbb{Z}^+)^k$, $|\nu| = \nu_1 + \nu_2 + \ldots + \nu_k$, $\nu! = \nu_1!\nu_2!\ldots\nu_k!, z^\nu = (z_1^{\nu_1} z_2^{\nu_2} \ldots z_k^{\nu_k})$ (for $z = (z_1, z_2, \ldots, z_k) \in \mathbb{R}^k$, $\nu \in (\mathbb{Z}^+)^k$), and $D^\nu = (\frac{\partial}{\partial z_1})^{\nu_1} \cdots (\frac{\partial}{\partial z_k})^{\nu_k}$. The remainder term $R(\bar{z},\mu)$ satisfies the relation

$$|R(\bar{z},\mu)| = o(\|\bar{z}-\mu\|^{s-1}) \quad as \quad \|\bar{z}-\mu\| \to 0. \tag{2.80}$$

The normalized statistic $W_n = \sqrt{n}(H(\bar{z}) - H(\mu))$ then may be expressed as

$$\begin{aligned} W_n &= \sqrt{n}(H(\bar{Z}) - H(\mu)) \\ &= (gradH)^T(\mu)\sqrt{n}(\bar{Z}) - H(\mu) + \sum_{|\nu|=2}^{s-1} (n^{-(|\nu|-1)/2}(D^\nu H)(\mu)\frac{(\sqrt{n}(\bar{Z}-\mu))^\nu}{\nu!} \\ &\quad + \sqrt{n}R(\bar{Z},\mu) \\ &= \widetilde{W_n} + \sqrt{n}R(\bar{Z},\mu) = \widetilde{W_n} + o_p(n^{-(s-2)/2}). \end{aligned} \tag{2.81}$$

Since $\widetilde{W}_n$ is $\sqrt{n}$-times a polynomial in $\bar{Z}-\mu$, one may calculate moments and cumulants of $\widetilde{W}_n$ in terms of those of $Z_j - \mu$ and the derivatives $(D^\nu H)(\mu)(1 \leq |\nu| \leq s-1)$. Let $\kappa_{r,n}$ denote the r-th cumulant $\widetilde{W}_n$ *omitting all terms of order* $o(n^{-(s-2)/2})$, $1 \leq r \leq s-2$. One may show that like the normalized mean $\sqrt{n}(\bar{Y}-\mu)$ of i.i.d. random variables with finite s moments, the r-th cumulant of $\widetilde{W}_n$ is of the order $o(n^{-(s-2)/2})$:

$$\begin{aligned} \kappa_{1,n} &= \sum_{j=1}^{s-2} n^{-j/2} b_{1j}, \\ \kappa_{2,n} &= \sigma^2 + \sum_{j=1}^{s-2} n^{-j/2} b_{2j}, (\sigma^2 = 1,\ by\ (2.77)) \\ \kappa_{r,n} &= n^{-(r-2)/2} b_r + \sum_{j=r-1}^{s-2} n^{-j/2} b_{rj} \ (3 \leq r \leq s). \end{aligned} \tag{2.82}$$

The of $\widetilde{W_n}$ may then be formally expressed as

$$\begin{aligned}\gamma_n(t) := Ee^{it\widetilde{W_n}} &\simeq exp\left\{-\frac{t^2}{2} + \sum_{r=1}^{s-2} n^{-r/2}\kappa_{r+2,n}\frac{(it)^{r+2}}{(r+2)!}\right\} \\ &\simeq exp\left\{-\frac{t^2}{2}\right\}\left[1 + \sum_{r=1}^{s-2} n^{-r/2} q_r(it)\right]\end{aligned} \tag{2.83}$$

where $q_r(it)$ is a polynomial in it whose coefficients involve the constants b_r, b_{rj} in equation (2.82). One may now apply the inverse of the Fourier transform on the right side of equation (2.83) to arrive at a "formal" Edgeworth density

$$\psi_{s-2,n}(x) = \left[1 + \sum_{r=1}^{s-2} n^{-r/2} q_r(-\frac{d}{dx})\right]\varphi(x), \tag{2.84}$$

where φ is the standard normal density on $\mathbb{R}^1$, and $q_r(-\frac{d}{dx})$ is obtained by formally replacing it by $-\frac{d}{dx}$ in $q_r(it)$. [Note: $(it)^j e^{-t^2/2}$ is the Fourier transform of the function $\left(-\frac{d}{dx}\right)\varphi(x)$]. A scheme roughly of this kind was proposed by F.Y. Edgeworth (1905)[93]. The theorem below provides a generally usable sufficient condition for the validity of the formal Edgeworth expansion.

THEOREM 2.13.1. *(Bhattacharya and Ghosh (1978) [38]). Assume*
(A_1) $E|f_i(X_1)|^s < \infty$, $1 \le i \le k$, for some integer $s \ge 2$,
(A_2) H is $(s-1)$-times continuously differentiable in a neighborhood of $\mu = EZ_1$, and
(A_3) Cramér's condition

$$\limsup|Ee^{i\xi^T Z_1}| < 1 \ \ (\xi \in \mathbb{R}^k) \tag{2.85}$$

is satisfied. Then

$$\sup_{B\in C}|\mathbb{P}(W_n \in B) - \int_B \psi_{s-2,n}(x)dx| = o(n^{-(s-2)/2}) \tag{2.86}$$

for every class C of Borel subets of $\mathbb{R}^1$ satisfying

$$\sup_{B\in C}\int_{(\partial B)^\varepsilon}\psi(x)dx = O(\varepsilon^a) \quad for\ some\ a > 0. \tag{2.87}$$

$$(\partial B)^\varepsilon := \{y : |y-x| < \varepsilon \quad for\ some\ x \in (\partial B)\}$$

REMARK 2.13.1. *Consider the following:*

1. *The class C of all intervals satisfies equation (2.87) with $a = 1$.*

2. *Since generally $p < k$, the distribution of Z_1 is singular w.r.t. Lebesgue measure in $\mathbb{R}^k$, even if X_1 has a density on $\mathbb{R}^m$. (In the case of r, $p = 2$ and $k = 5$). However, it may be shown that $Z_1 + \ldots + Z_k$ has a density component w.r.t. Lebesgue measure, if X_1 has a positive density component on a nonempty open subset of $\mathbb{R}^m$ (Bhattacharya and Ghosh (1978) [38]).*
3. *For estimating the distribution of $W_n = \sqrt{n}(H(\bar{Z}) - H(\mu))$ from the sample, one may substitute sample moments for the population moments in the Edgeworth density $\psi_{s-2,n}(x)$. Because the sampling errors involved in this substitution only $s = 3$ or 4 seem effective. This leads to the empirical Edgeworth approximation of the distribution on W_n to be discussed in the next subsection.*

2.13.2 *Bootstrap and the Empirical Edgeworth Expansion.*

Let $\hat{P}_n = \frac{1}{n}\sum_{j=1}^n \delta_{X_j}$ be the empirical distribution of $X_1, \ldots, X_n$ and let $X_1*, \ldots, X_n^*$ be n i.i.d. observations from $\hat{P}_n$, conditionally given $X_1, \ldots, X_n$. The i.i.d. r. vectors $X_1^*, \ldots, X_n^*$ can be thought of as a sample with replacement from $X_1, \ldots, X_n$, also known as a *bootstrap resample* from $X_1, \ldots, X_n$. One may formally calculate the Edgeworth expansion of the distribution of the bootstrapped statistic $W_n^* = \sqrt{n}(H(\bar{Z}^*) - H(\bar{Z}))$ under the empirical $\hat{P}_n^* = \frac{1}{n}\sum_{j=1}^n \delta_{X_j^*}$. One sees that this expansion $\psi_{s-2,n}^*$ is obtained by replacing population moments by sample moments in $\psi_{s-2,n}$. In other words, $\psi_{s-2,n}^*$, is precisely the so-called empirical *Edgeworth expansion* mentioned in Remark 2.13.1 number 3 above. For the sample mean $\bar{X}^*$ or $\bar{Z}^*$, this was proved to be a valid expansion by Babu and Singh (1984) [9], and extended to the present case of W_n^* in Bhattacharya (1987) [30]. See also Beran (1987) [16].

THEOREM 2.13.2. *(Babu and Singh theorem(1984) [9], Bhattacharya (1987) [30]). Under the hypothesis of Theorem 26.3.2, one has*

$$\sup_{B \in \mathcal{C}} |P^*(W_n^* \in B) - \int_B \psi_{s-2,n}^*(x)dx| = o_P(n^{-(s-2)/2}) \tag{2.88}$$

for every class $\mathcal{C}$ of Borel subset of $\mathbb{R}^1$ satisfying equation (2.87), where the subscript P in (2.88)*, means that the left hand side in that equation is a $o(n^{-(s-2)/2})$ with probability* 1.

It follows from Theorem 2.13.2 that the asymptotic comparison between the bootstrap estimate $P^*(W_n^* \in B)$ and the normal approximation $\int_B \varphi(x)dx$ (of $\mathbb{P}(W_n \in B)$) is the same as that between $\int_B \psi_{s-2,n}^*(x)dx$ and $\int_B \varphi(x)dx$ (up to order $o_P(n^{-(s-2)/2})$). Now, taking $s = 3$, one has

$$P^*(W_n^* \leq x) - \int_{-\infty}^{x} \psi_{1,n}^*(y)dy = o_P(n^{-\frac{1}{2}}) \quad , \tag{2.89}$$

where

$$\psi^*_{1,n}(y) = \varphi(y)\left(1 + n^{-\frac{1}{2}}\tilde{q}^*_1(y)\right). \tag{2.90}$$

Here $\tilde{q}_1(y)\varphi(y) = q_1\left(-\frac{d}{dy}\right)\varphi(y)$, with q_1 is as in Theorem 26.3.2, and $\tilde{q}^*_1(y)$ is obtained by replacing population moments by sample moments in $\tilde{q}_1(y)$. It turns out that q_1 is a polynomial of degree three, so that

$$\int_{-\infty}^{x} q_1\left(-\frac{d}{dy}\right)\varphi(y)dy = p_1(x)\varphi(x), \quad say, \tag{2.91}$$

where $p_1(x)$ is a polynomial of degree 2, as is $p^*_1(x)$. Hence

$$\int_{-\infty}^{x} \psi_{1,n}(y) = \int_{-\infty}^{x} \varphi(y)dy + n^{-\frac{1}{2}}p_1(x)\varphi(x),$$

$$\int_{-\infty}^{x} \psi^*_{1,n}(y) = \int_{-\infty}^{x} \varphi(y)dy + n^{-\frac{1}{2}}p^*_1(x)\varphi(x). \tag{2.92}$$

It follows from Theorem 26.3.2, 2.13.2, relations with equations (2.89), (2.92) that

$$\mathbb{P}(W_n \leq x) - \int_{-\infty}^{x} \varphi(y)dy = n^{-\frac{1}{2}}p_1(x)\varphi(x) + o_p(n^{-1}) \tag{2.93}$$

$$\mathbb{P}(W_n \leq x) - P^*(W^*_n \leq x) = n^{-\frac{1}{2}}(p_1(x) - p^*_1(x))\varphi(x) + o_p(n^{-1}). \tag{2.94}$$

Since the difference between a polynomial of sample moments (which are coefficients of $p^*_1(x)$) and that of the corresponding population moments is of the form $H(\bar{Z}) - H(\mu)$, one may show that

$$n^{\frac{1}{2}}(p_1(x) - p^*_1(x)) \rightarrow_d Y \sim \mathcal{N}(0, \sigma^2_b(x)), \tag{2.95}$$

where $\sigma^2_b(x)$ is a polynomial of degree four. Hence

$$n(\mathbb{P}(W_n \leq x) - P^*(W^*_n \leq x)) \rightarrow_d Y \sim \mathcal{N}(0, \sigma^2_b(x)\varphi^2(x)), \tag{2.96}$$

In other words,

COROLLARY 2.13.1. *The error of pivotal bootstrap approximation is*

$$\sup_x |\mathbb{P}(W_n \leq x) - P^*(W^*_n \leq x)| = O_p(n^{-1}), \tag{2.97}$$

whereas the error of the normal approximation is $O(n^{-\frac{1}{2}})$, as given by equation (2.93).

The proof of equation (2.97) actually is based on two relations,

$$\sup_x \left| \mathbb{P}(W_n \leq x) - \int_{-\infty}^{x} \psi_{1,n}^*(y)dy \right| = O_p(n^{-1}), \tag{2.98}$$

and

$$\sup_x \left| P^*(W_n^* \leq x) - \int_{-\infty}^{x} \psi_{1,n}^*(y)dy \right| = O_p(n^{-1}). \tag{2.99}$$

REMARK 2.13.2. *Consider the following:*

1. *The pivotal bootstrap is not only superior to the normal approximation (the error of the normal approximation is $O(n^{-\frac{1}{2}})$, while that of the bootstrap $O_P(n^{-1})$), it is superior to the one-term correction to the normal, namely the empirical Edgeworth, as shown by Bhattacharya and Qumsiyeh (1989) [45].*
2. *In the case when the statistics W_n is not pivotal, one can show (Bhattacharya and Denker (1990) [36], Part I) that errors of the bootstrap approximation and the normal approximation are of the same order.*
3. *Bootstrap is often used for nonpivotal statistics $\sqrt{n}(\hat{\theta}_n - \theta)$ to provide confidence intervals such as $[c_{\frac{\alpha}{2}}, c_{\frac{1-\alpha}{2}}]$, where*

$$\hat{c_p} := \sup\left\{c : P^*(\theta_n^* - \hat{\theta}_n \leq c) \leq p\right\}.$$

This is the so-called percentile method, *which bypasses the estimation of the standard error of $\hat{\theta}_n$ (see Efron (1982) [97], Efron (2003) [98], Efron and Tibshirani (1993) [99]).*

2.14 Nonparametric Function Estimation

For functions $F : \mathbb{R}^m \to \mathbb{R}$, denote by $\|\cdot\|_2$ and $\|\cdot\|_\infty$ the L^2 and L^∞ norms, respectively. Recall that

$$\|F\|_\infty = \sup_{x \in \mathbb{R}^m} |F(x)|, \tag{2.100}$$

and

$$\|F\|_2 = \int_{\mathbb{R}^m} |F(x)|^2 \lambda(dx), \tag{2.101}$$

where λ is the Lebesgue measure.

Let $X_1, X_2, \ldots, X_n$ be a random sample from a distribution P on $\mathbb{R}$. A consistent estimator of $P_{X_1} = P$ is the *empirical* $\widehat{P}_n = \frac{1}{n}\sum_{i=1}^n \delta_{X_i}$. Let $\widehat{F}_n(x) \equiv \widehat{P}_n((-\infty, n])$ be the c. d. f. of $\widehat{P}_n$ and $F(x)$ that of P. The *Glivenko–Cantelli theorem*(see Ferguson (1996) [115]) states that with probability 1

$$\left| \widehat{F}_n(x) - F(x) \right|_\infty \longrightarrow 0 \quad \text{as } n \to \infty. \tag{2.102}$$

Suppose now that F is absolutely continuous with a p.d.f. f. Since $\widehat{P}_n$ is discrete, one may estimate f using a *kernel density estimator*. That is a density of the random variable $\widehat{X}_n + hZ$ where $\widehat{X}_n$ has the distribution $\widehat{P}_n$ (conditionally, given $X_1, \ldots, X_n$) and Z is independent of $\widehat{X}_n$ and has a nice density, say, K, and h is a small positive number, called the *bandwidth* satisfying

$$h \equiv h_n \longrightarrow 0 \quad \text{as } n \to \infty. \tag{2.103}$$

Note that $\widehat{X}_n + hZ$ has the density

$$\widehat{f}_n(x) = \frac{1}{n}\sum_{i=1}^{n} K_h(x - X_i), \tag{2.104}$$

where K_h is the density of hZ, namely,

$$K_h(y) = \frac{1}{h}K\left(\frac{y}{h}\right). \tag{2.105}$$

To see that $\widehat{f}_n$ is the density of $\widehat{X}_n + hZ$, note that the latter has the distribution function

$$\begin{aligned} &\widetilde{\mathbb{P}}(\widehat{X}_n + hZ \leq x) = E[\widetilde{\mathbb{P}}(hZ \leq x - \widehat{X}_n | \widehat{X})] \\ = \; &\int_{-\infty}^{x-y} K_h(u) d\widehat{P}_n(y) = \frac{1}{n}\sum_{i=1}^{n}\int_{-\infty}^{x-X_i} K_h(u)du, \end{aligned} \tag{2.106}$$

where the superscript $\widetilde{}$ indicates that the probabilities are computed given $X_1, \ldots, X_n$. Differentiating (2.106) w.r.t. x one arrives at (2.104).

Alternatively, one may think of (2.104) as obtained by spreading out the point mass δ_{X_i} with a density centered at X_i and concentrating most of the density near X_i ($1 \leq i \leq n$). Now we show that the *bias* $E\widehat{f}_n(x) - f(x) \to 0$. For this write

$$\begin{aligned} E\widehat{f}_n(x) &= E\frac{1}{n}\sum_{i=1}^{n}\frac{1}{h}K\left(\frac{x - X_i}{h}\right) = E\frac{1}{h}K\left(\frac{x - X_i}{h}\right) \\ &= \frac{1}{h}\int_{-\infty}^{\infty} K\left(\frac{x-y}{h}\right) f(y)dy \\ &= \int_{-\infty}^{\infty} K(v) f(x - vh)dv \longrightarrow f(x) \text{ as } h \downarrow 0, \end{aligned} \tag{2.107}$$

if f is bounded and is continuous at x (by the Lebesgue Dominated Convergence Theorem). Also,

$$\begin{aligned}
\operatorname{var}(\widehat{f}_n(x)) &= \frac{1}{n}\operatorname{var} K_h(x - X_1) \qquad (2.108)\\
&= \frac{1}{n}\left\{\frac{1}{h^2} EK^2\left(\frac{x-X_1}{h}\right) - \frac{1}{n}\left(E\frac{1}{h}K\left(\frac{x-X_1}{h}\right)\right)^2\right\}\\
&= \frac{1}{nh}\int_{-\infty}^{\infty} K^2(v) f(x - vh)dv - \frac{1}{n}\left(\int_{-\infty}^{\infty} K(v) f(x - vh)dv\right)^2 \to 0
\end{aligned}$$

as $n \to \infty$,

if (i) f is bounded, (ii) f is continuous at x, (iii) $K^2(v)$ is integrable,

$$nh \longrightarrow \infty \quad \text{and} \quad h \longrightarrow 0 \qquad (\text{as } n \to \infty). \tag{2.109}$$

Thus, under the hypothesis (i)—(iii) and (2.109), one has (by (2.107) and (2.108))

$$E\left(\widehat{f}_n(x) - f(x)\right)^2 = \operatorname{var}\widehat{f}_n(x) + (\operatorname{Bias}\widehat{f}_n(x))^2 \longrightarrow 0 \quad \text{as } n \to \infty. \tag{2.110}$$

In other words, under the above assumptions, $\widehat{f}_n(x) \to f(x)$ *in probability* as $n \to \infty$.

Note that the convergences (2.107) and (2.108) do not really require boundedness of f on all of $\mathbb{R}$. For example, if one takes K to have a compact support then it is enough to require that f is continuous at x. We have proved that under mild assumptions the kernel estimator with *kernel K kernel estimator* is a *consistent estimator* of f at every point of continuity of f.

By choosing an appropriately smooth and symmetric kernel K one may make the error of approximation $\widehat{f}_n(x) - f(x)$ reasonably small.

A measure of the (squared) error of approximation is provided by the so-called , or *MISE* given by

$$\operatorname{MISE}(\widehat{f}_n) = \int_{\mathbb{R}} E[\widehat{f}_n(x) - f(x)]^2 dx = \int_{\mathbb{R}} [\operatorname{var}\widehat{f}_n(x) + (\operatorname{Bias}\widehat{f}_n(x)^2] dx. \tag{2.111}$$

Write

$$c_1 = \int v^2 K(v)dv, \quad c_2 = \int K^2(v)dv, \quad c_3 = \int (f''(x))^2 dx, \tag{2.112}$$

and assume c_1, c_2, c_3 are finite and that

$$\int K(v)dv = 1, \quad \int vK(v)dv = 0. \tag{2.113}$$

Now it follows from (2.107) that

$$\begin{aligned} E\widehat{f}_n(x) &= \int_{\mathbb{R}} K(v)\left[f(x) - vhf'(x) + \frac{v^2h^2}{2}f''(x) + o(h^2)\right]dv \\ &= f(x) + \frac{c_1h^2}{2}f''(x) + o(h^2), \\ (\text{Bias}\,\widehat{f}_n)(x) &= \frac{c_1h^2}{2}f''(x) + o(h^2), \end{aligned} \tag{2.114}$$

if f'' is continuous and bounded. Then

$$\int_{\mathbb{R}} (\text{Bias}\,\widehat{f}_n)^2(x)dx = \frac{c_3c_1^2h^4}{4} + o(h^4). \tag{2.115}$$

Next, by (2.108) and (2.109),

$$\text{var}\,\widehat{f}_n(x) = \frac{c_2f(x)}{nh} + o\left(\frac{1}{n}\right), \tag{2.116}$$

$$\int_{\mathbb{R}} \text{var}\,\widehat{f}_n(x)dx = \frac{c_2}{nh} + O\left(\frac{1}{n}\right), \text{ as } n \to \infty. \tag{2.117}$$

Hence

$$\text{MISE}(\widehat{f}_n) = \frac{c_1^2c_3}{4}h^4 + \frac{c_2}{nh} + o(h^4) + O\left(\frac{1}{n}\right). \tag{2.118}$$

Neglecting the two smaller order terms, the asymptotically optimal choice of the bandwidth h, for a given kernel K as above, is obtained by

$$h_n = \arg\min_h \left\{\frac{c_1^2c_3}{4}h^4 + \frac{c_2}{nh}\right\} = \left(\frac{c_2}{c_1^2c_3}\right)^{\frac{1}{5}} n^{-\frac{1}{5}}. \tag{2.119}$$

The corresponding asymptotically minimal MISE is

$$\text{MISE}\,\widehat{f}_n = \frac{c_4}{n^{4/5}} + o\left(n^{-4/5}\right) \quad c_4 := \frac{5}{4}\left(c_1^{2/5}c_2^{4/5}c_3^{1/5}\right). \tag{2.120}$$

We have arrived at the following result.

THEOREM 2.14.1. *Assume f'' is continuous and bounded. Then for any choice of a symmetric kernel K satisfying (2.113), and $0 < c_i < \infty$ $(i = 1, 2, 3)$, the asymptotically optimal bandwidth h is given by the extreme right side of (2.119), and the asymptotically minimal MISE is given by (2.120).*

From the expression (2.104) it follows that $\widehat{f}_n(x)$ is, for each n, a sum of i.i.d. random variables. By the Lindeberg CLT it now follows that, under the hypothesis of Theorem 2.14.1 one has (Exercise 12)

$$\frac{\widehat{f}_n(x) - E\widehat{f}_n(x)}{\sqrt{\text{var}\widehat{f}_n(x)}} \to_d Z \sim N(0,1), \quad \text{if } f(x) > 0. \tag{2.121}$$

Also check, using (2.114), (2.115) and (2.119), that

$$\frac{E\widehat{f}_n(x) - f(x)}{\sqrt{\operatorname{var}\widehat{f}_n(x)}} \longrightarrow c_3^{-\frac{1}{2}} \frac{f''(x)}{\sqrt{f(x)}} = \gamma, \quad \text{say, if } f(x) > 0. \tag{2.122}$$

Hence

$$\frac{\widehat{f}_n(x) - f(x)}{\sqrt{\operatorname{var}\widehat{f}_n(x)}} \xrightarrow{\mathcal{L}} N(\gamma, 1) \quad \text{if } f(x) > 0. \tag{2.123}$$

To remove the asymptotic bias γ, one may choose a slightly sub-optimal bandwidth $h_n = o(n^{-\frac{1}{5}})$ (Exercise 13). Since $\operatorname{var}\widehat{f}_n(x)$ involves $f''(x)$, for setting confidence regions for $f(x)$, one may resort to bootstrapping.

REMARK 2.14.1. *It has been shown by Epanechnikov (1969, pp. 153–158) [109]. that the constant c_4 in MISE is minimized (under the hypothesis of Theorem 2.14.1) by the kernel*

$$K(v) = \frac{3}{4\sqrt{5}}\left(1 - \frac{1}{5}v^2\right)\mathbf{1}_{\{|v| \le \sqrt{5}\}}. \tag{2.124}$$

However, the loss of efficiency is rather small if, instead of (2.124), one chooses any symmetric kernel with high concentration, such as the (standard) Normal density or the triangular density (Lehmann, E. L. (1998), pp. 415, 416) [216].

The asymptotic theory presented above has extensions to the multidimensional case. We provide a brief sketch of the arguments here, leaving the details to Exercise 14. Let f be a p.d.f. on $\mathbb{R}^m$, and let K be a symmetric kernel density with finite second moments $c_{1,i,j} = \int v_i v_j K(\mathbf{v})d\mathbf{v}$, and with $c_2 = \int K^2(\mathbf{v})d\mathbf{v}$. Then if the second partial derivatives $\frac{\partial f}{\partial x_i \partial x_j}$ are continuous and bounded, one uses the kernel estimate

$$\widehat{f}_n(\mathbf{x}) = \frac{1}{h^m}\sum_{i=1}^{n} K\left(\frac{\mathbf{x} - X_i}{h}\right), \tag{2.125}$$

based on i.i.d. observations $X_1, \ldots, X_n$ with p.d.f. f. Then

$$\begin{aligned}
E\widehat{f}_n(x) &= E\frac{1}{h^m}K\left(\frac{\mathbf{x} - X_1}{h}\right) = \frac{1}{h^m}\int_{\mathbb{R}^m} K\left(\frac{\mathbf{x} - \mathbf{v}}{h}\right) f(\mathbf{v})dv_1 \ldots dv_d \\
&= \int_{\mathbb{R}^m} K(\mathbf{u}) f(\mathbf{x} - h\mathbf{u}) d\mathbf{u} \\
&= \int_{\mathbb{R}^m} K(\mathbf{u})\left[f(\mathbf{x}) - h\mathbf{u}\cdot \operatorname{grad} f(\mathbf{x}) + \frac{h^2}{2}\sum u_i u_j \frac{\partial^2 f(\mathbf{x})}{\partial x_i \partial x_j} + o(h^2)\right] d\mathbf{u} \\
&= f(\mathbf{x}) + \frac{h^2}{2}\sum c_{1,i,j}\frac{\partial^2 f(\mathbf{x})}{\partial x_i \partial x_j} + o(h^2),
\end{aligned}$$

so that

$$\text{Bias}\,\widehat{f}_n(\mathbf{x}) = E\widehat{f}_n(\mathbf{x}) - f(\mathbf{x}) = \frac{h^2}{2}\sum c_{1,i,j}\frac{\partial^2 f(\mathbf{x})}{\partial x_i \partial x_j} + o(h^2). \qquad (2.126)$$

Also,

$$\begin{aligned} E\left(\frac{1}{h^m}K\left(\frac{\mathbf{x}-X_1}{h}\right)\right)^2 &= \frac{h^m}{h^{2p}}\int_{\mathbb{R}^m} K^2(\mathbf{u})\Bigg[f(\mathbf{x}) - h\mathbf{u}\cdot \text{grad} f(\mathbf{x}) \\ &\quad + \frac{h^2}{2}\sum u_i u_j \frac{\partial^2 f(\mathbf{x})}{\partial x_i \partial x_j} + o(h^2)\Bigg] d\mathbf{u} \\ &= \frac{1}{h^m}c_2 f(\mathbf{x}) + o\left(\frac{h^2}{h^m}\right), \\ &\quad \text{var}\left(\frac{1}{h^m}K\left(\frac{\mathbf{x}-X_1}{h}\right)\right) = \frac{1}{h^m}c_2 f(\mathbf{x}) + \\ &\quad \text{var}(\widehat{f}_n(\mathbf{x})) = \frac{1}{nh^m}c_2 f(\mathbf{x}) + o\left(\frac{1}{nh^m}\right) \quad (2.127) \end{aligned}$$

Hence

$$\begin{aligned} E(\widehat{f}_n(\mathbf{x}) - f(\mathbf{x}))^2 &= (\text{Bias}\,\widehat{f}_n(\mathbf{x}))^2 + \text{var}(\widehat{f}_n(\mathbf{x})) \\ &= \frac{1}{nh^m}c_2 f(\mathbf{x}) + \frac{h^4}{4}\left(\sum_{i,j} c_{1,i,j}\frac{\partial^2 f(\mathbf{x})}{\partial x_i \partial x_j}\right)^2 \\ &\quad + o(h^4) + o\left(\frac{1}{nh^m}\right). \end{aligned} \qquad (2.128)$$

Therefore,

$$\text{MISE}\,\widehat{f}_n = \int_{\mathbb{R}^m} E(\widehat{f}_n(\mathbf{x}) - f(\mathbf{x}))^2 d\mathbf{x} = \frac{c_2}{nh^m} + \frac{h^4}{4}\widetilde{c}_3 + o(h^4) + o\left(\frac{1}{nh^m}\right), \qquad (2.129)$$

where

$$\widetilde{c}_3 = \int_{\mathbb{R}^m}\left(\sum_{i,j} c_{1,i,j}\frac{\partial^2 f(\mathbf{x})}{\partial x_i \partial x_j}\right)^2 d\mathbf{x}.$$

As before, the asymptotically optimal bandwidth is given by

$$h_n = \arg\min_h \left\{\frac{h^4}{4}\widetilde{c}_3 + \frac{c_2}{nh^m}\right\} = \left(\frac{c_2 p}{\widetilde{c}_3}\right)^{\frac{1}{p+4}} n^{-\frac{1}{p+4}}, \qquad (2.130)$$

and the asymptotically minimal MISE is

$$\text{MISE}\,\widehat{f}_n = C_2^{\frac{4}{p+4}}\widetilde{C}_3^{\frac{p}{p+4}}\left(\frac{1}{4}p^{\frac{4}{p+4}} + p^{-\frac{p}{p+4}}\right) n^{-\frac{4}{p+4}} + o\left(n^{-\frac{4}{p+4}}\right). \qquad (2.131)$$

Multi-dimensional versions of (2.121)—(2.123) may now be derived (Exercise 14).

Unlike classical statistics, where the number of covariates (marginals) studied was relatively small, modern data analysis (including imaging and bioinformatics) deals with low sample size high dimensional (LSSHD) statistics, that is, the number p of variables is huge, while the sample size n is small, even by classical asymptotics, thus making statistical inference difficult. To address this challenge, statisticians are using various techniques including regularization and *dimension reduction*, as discussed in the section 2.15.

2.15 Data Analysis on Hilbert Spaces - an Introduction

Direct generalization of multivariate techniques to functional data analysis is not in general feasible; for the remainder of this chapter, some procedures for the one-and multi-sample problem will be modified, so as to become suitable for functional data (see Munk et al. (2008) [252]). For an extensive discussion of functional data see the monograph by Ramsey and Silverman (1997) [286]. In Munk et al. (2008)[252], the problem of identifying the projective shape of a planar curve was be considered as a practical application.

The union-intersection principle of Roy and Bose (1953) [293] provides us with a projection pursuit type technique to construct multivariate procedures from a family of univariate procedures. A case in point is Hotelling's (1931) [161] multivariate T^2-statistic that can be constructed from a family of univariate student statistics. It is easy to see that further extension to infinite dimensional Hilbert spaces along similar lines breaks down, in particular because the rank of the sample covariance operator cannot exceed the finite sample size and consequently cannot be injective, not even when the population covariance operator is one-to-one.

Several alternatives could be considered. One possibility is to project the data onto a Euclidean subspace of sufficiently high dimension and perform a Hotelling test with these finite dimensional data. This includes spectral-cut-off regularization of the inverse of the sample covariance operator as a special case. Another option is a Moore-Penrose type of regularization of this operator.

We will, however, consider another modification that seems more appropriate as it yields at once a more realistic hypothesis and a mathematically tractable procedure. That is, we will consider approximate equality of means instead of exact equality. Therefore, we will replace the usual hypothesis with a "neighborhood hypothesis".

This kind of modified hypothesis has a long history and has been developed in different situations. It has been, e.g., proposed by Hodges and Lehmann (1954) [159] for testing whether multinomial cell probabilities are approximately equal. Dette and Munk (1998) [84] extended this approach for the purpose of validating a model in a nonparametric regression framework. For methodological aspects and a more recent discussion, we refer to Goutis and Robert (1998) [129], Dette and Munk (2003) [85], and Liu and Lindsay (2005) [222]. The underlying idea is that the hypothesis is often formulated on the basis of theoretical considerations that will never cover reality completely. Hence in practice such a hypothesis will always be rejected if the sample size is large enough.

It is therefore more realistic to test a somewhat larger hypothesis that also includes parameters in a neighborhood of the original one. Also see Berger and Delampady (1987) [20], who employ the term "precise hypothesis" instead of "neighborhood hypothesis", whereas Liu and Lindsay (2005) [222] coined the phrase "tubular models". Dette and Munk (1998) [84] and Munk and Dette (1998) [251] considered L^2-neighborhood hypotheses in nonparametric regression models.

A further advantage is that neighborhood hypotheses often lead to simpler asymptotic analysis. This in turn makes it possible to interchange the role of a neighborhood hypothesis and its alternative without complicating the testing procedure. This is particularly relevant for goodness-of-fit type tests, where traditionally the choice of the null hypothesis is usually dictated by mathematical limitations rather than statistical considerations. Accepting a model after a goodness of fit test always leaves the statistician in an ambiguous situation as to whether the model has not been rejected for other reasons, e.g. because of lack of data, an inefficient goodness of fit test at hand, or because of the large variability of the data. In contrast, the approach we will soon describe allows to validate a hypotheses at a given level α, instead of accepting a model without any further evidence in favor of the model. In fact, this is equivalent to reporting on a confidence interval for a certain distance measure between models.

The section is organized as follows. In Subsection 2.15.1 we briefly review some basic concepts for Hilbert space valued random variables, and in Subsection 2.15.2 we briefly discuss the difficulties with studentization in infinite dimensional Hilbert spaces. Subsections 2.15.3 and 2.15.4 are devoted respectively to a suitably formulated version of the functional one-and multi-sample problem. The theory is applied to the recognition of the projective shape of a planar curve in Chapter 3.

2.15.1 Random Elements in Hilbert Spaces

Let $(\Omega, \mathcal{W}, \mathbb{P})$ be an underlying probability space, $\mathbb{H}$ a separable Hilbert space over the real numbers with inner product $\langle \bullet, \bullet \rangle$ and norm $|| \bullet ||$, and $\mathcal{B}_{\mathbb{H}}$ the σ-field generated by the open subsets of $\mathbb{H}$. A random element in $\mathbb{H}$ is a mapping $X : \Omega \to \mathbb{H}$ which is $(\mathcal{W}, \mathcal{B}_{\mathbb{H}})$ - measurable. Let us write $\mathbb{P}_X = P$ for the induced probability measure on $(\mathbb{H}, \mathcal{B}_{\mathbb{H}})$.

The probability distribution P is uniquely determined by its characteristic functional

$$\tilde{P}(x) = \mathbb{E}\, e^{i\langle x,X\rangle} = \int_{\mathbb{H}} e^{i\langle x,y\rangle} dP(y), x \in \mathbb{H}. \tag{2.132}$$

Assuming that

$$E\,||X||^2 < \infty, \tag{2.133}$$

the Riesz representation theorem ensures the existence of a vector $\mu \in \mathbb{H}$ and an operator $\Sigma : \mathbb{H} \to \mathbb{H}$, uniquely determined by the properties

$$\mathbb{E}\,\langle x, X\rangle = \langle x, \mu\rangle \,\forall\, x \in \mathbb{H}, \tag{2.134}$$

$$\mathbb{E}\,\langle x, X-\mu\rangle\,\langle y, X-\mu\rangle = \langle x, \Sigma\, y\rangle \,\forall\, x,y \in \mathbb{H}. \tag{2.135}$$

The operator Σ is linear, Hermitian, semi-definite positive; it has, moreover, finite trace and is consequently compact. Any operator with these properties will be referred to as a covariance operator, and any covariance operator is induced by some random element.

It follows from the Minlos–Sazanov theorem that for $\mu \in \mathbb{H}$ and $\Sigma : \mathbb{H} \to \mathbb{H}$ a covariance operator, the functional

$$\varphi(x) = e^{i\langle x,\mu\rangle - \frac{1}{2}\langle x,\Sigma x\rangle}, x \in \mathbb{H}, \tag{2.136}$$

is the characteristic functional of a probability measure on $\mathbb{H}$, which is called the Gaussian measure with parameters μ and Σ and will be denoted by $\mathcal{G}(\mu,\Sigma)$. The parameters represent respectively the mean and covariance operator of the distribution.

Let $\mathbb{H}^p$ be the real, separable Hilbert space of all p-tuples $x = (x_1,\ldots,x_p)^*, x_j$
$\in \mathbb{H}$ for $j = 1,\ldots,p$. The inner product in $\mathbb{H}^p$ is given by $\langle x,y\rangle_p = \sum_{j=1}^{p} \langle x_j, y_j\rangle$, for $x,y \in \mathbb{H}^p$.

2.15.2 *Why Studentization Breaks Down in a Hilbert Space*

Let $X_1,\ldots,X_n$ be independent copies of a random element X in $\mathbb{H}$ with

$$\mathbb{E}\,||X||^4 < \infty, \tag{2.137}$$

mean $\mu \in \mathbb{H}$, and covariance operator $\Sigma : \mathbb{H} \to \mathbb{H}$. Estimators of μ and Σ are respectively

$$\overline{X} = \frac{1}{n}\sum_{i=1}^{n} X_i,\; S = \frac{1}{n}\sum_{i=1}^{n}(X_i - \overline{X}) \otimes (X_i - \overline{X}), \tag{2.138}$$

where for $a,b \in \mathbb{H}$ the operator $a \otimes b : \mathbb{H} \to \mathbb{H}$ is defined by $(a \otimes b)(x) = \langle b,x\rangle\, a, x \in \mathbb{H}$.

Immediate extension of the union-intersection principle would suggest to use the Hotelling-type test statistic

$$T_n^2 = n \sup_{u \in \mathbb{H}:||u||=1} \frac{\langle \overline{X}, u \rangle^2}{\langle u, Su \rangle}, \tag{2.139}$$

for testing the classical hypothesis that $\mu = 0$. The studentization, however, now in general causes a problem since under the assumption that

$$\mathbb{P}\{X_1, \ldots, X_n \text{ are linearly independent}\} = 1, \tag{2.140}$$

it will be shown that

$$\mathbb{P}\{T_n^2 = \infty\} = 1, \tag{2.141}$$

even when Σ is supposed to be injective.

To prove (2.141), let us first observe that (2.140) entails that $\mathbb{P}\{\overline{X} \in$ linear span of $X_1 - \overline{X}, \ldots, X_n - \overline{X}\} = 0$. For if $\overline{X}$ were an element of the linear span there would exist scalars $\alpha_1, \ldots, \alpha_n$ such that $\overline{X} = \sum_{i=1}^n \alpha_i (X_i - \overline{X})$. Because of the linear independence of the X_i this means that the vector $\alpha = (\alpha_1, \ldots, \alpha_n)^* \in \mathbb{R}^n$ must satisfy

$$(I_n - \frac{1}{n} 1_n 1_n^*)\alpha = 1_n, \tag{2.142}$$

where I_n is the $n \times n$ identity matrix and 1_n a column of n numbers 1. This is impossible because the matrix on the left in (2.142) is the projection onto the orthogonal complement in $\mathbb{R}^n$ of the line through 1_n. Hence with probability 1 there exist $\overline{X}_1, \overline{X}_2$ such that $\overline{X} = \overline{X}_1 + \overline{X}_2$, and

$$\begin{cases} \overline{X}_1 \neq 0, \overline{X}_1 \perp X_i - \overline{X} :: \text{for } i = 1, \ldots, n, \\ \overline{X}_2 \in \text{linear span of } X_1 - \overline{X}, \ldots, X_n - \overline{X}. \end{cases} \tag{2.143}$$

Choosing $u = \overline{X}_1$ we have on the one hand that $\langle \overline{X}, \overline{X}_1 \rangle^2 = ||\overline{X}_1||^4 > 0$, and on the other hand we have $S\overline{X}_1 = n^{-1} \cdot \sum_{i=1}^n \langle X_i - \overline{X}, \overline{X}_1 \rangle (X_i - \overline{X}) = 0$, so that (2.141) follows.

A possible modification of this statistic is obtained by replacing S^{-1} with a regularized inverse of Moore-Penrose type and by considering

$$\begin{aligned} &\sup_{u \in \mathbb{H}:||u||=1} \frac{\langle \overline{X}, u \rangle^2}{\left\langle u, (\alpha I + S)^{-1} u \right\rangle} \\ &= \text{largest eigenvalue of } (\alpha I + S)^{-1/2} \left(\overline{X} \otimes \overline{X} \right) (\alpha I + S)^{-1/2}, \end{aligned}$$

where I is the identity operator. We conjecture that perturbation theory for compact operators in Hilbert spaces leads to the asymptotic distribution of

$(\alpha I + S)^{-1/2}$ and subsequently to the asymptotic distribution of this largest eigenvalue, in the same vein as this kind of result can be obtained for matrices. See, for instance, Watson (1983) [333] for sample covariance matrices and Ruymgaart and Yang (1997) [295] for functions of sample covariance matrices. Watson's (1982) result has been obtained for sample covariance operators on Hilbert spaces by Dauxois et al. (1982) [80]. As has been explained in the introduction, however, here we prefer to pursue the approach of modifying the hypothesis.

2.15.3 *The One-Sample Problem in a Hilbert Space*

Let $X_1, \ldots, X_n$ be as defined above and suppose we want to test hypotheses regarding μ. As has been argued above, we will change the usual hypothesis. This modified hypothesis may make more sense from an applied point of view and leads, moreover, to simpler asymptotics. To describe these hypotheses suppose that

$$M \subset \mathbb{H} \text{ is a linear subspace of dimension } m \in \mathbb{N}_0, \tag{2.144}$$

and let $\delta > 0$ be an arbitrary given number. Let us denote the orthogonal projection onto M by Π, and onto $M^{\perp}$ by $\Pi^{\perp}$. It is useful to observe that

$$\left\langle \Pi^{\perp}x, \Pi^{\perp}y \right\rangle = \left\langle x, \Pi^{\perp}y \right\rangle \; \forall\, x, y \in \mathbb{H}. \tag{2.145}$$

Furthermore let us introduce the functional

$$\varphi_M(x) = ||x - M||^2,\; x \in \mathbb{H}, \tag{2.146}$$

representing the squared distance of a point $x \in \mathbb{H}$ to M (finite dimensional subspaces are closed).

The neighborhood hypothesis to be tested is

$$\mathcal{H}_\delta : \mu \in M_\delta \cup B_\delta, \text{ for some } \delta > 0, \tag{2.147}$$

where $M_\delta = \{x \in \mathbb{H} : \varphi_M(x) < \delta^2\}$ and $B_\delta = \{x \in \mathbb{H} : \varphi_M(x) = \delta^2,$ $\langle \Pi^{\perp}x, \Sigma\Pi^{\perp}x\rangle > 0\}$. The alternative to (2.147) is

$$\mathcal{A}_\delta : \mu \in M_\delta^c \cap B_\delta^c. \tag{2.148}$$

The usual hypothesis would have been: $\mu \in M$. It should be noted that $\mathcal{H}_\delta$ contains $\{\varphi_M < \delta^2\}$ and that $\mathcal{A}_\delta$ contains $\{\varphi_M > \delta^2\}$. These are the important components of the hypotheses; the set B_δ is added to the null hypothesis by mathematical convenience, i.e. because the asymptotic power on that set is precisely α, as will be seen below.

For testing hypotheses like (2.147) see Dette and Munk (1998). These authors also observe that testing

$$\mathcal{H}'_\delta : \mu \in \left(M'_\delta\right)^c \cup B_\delta \text{ versus } \mathcal{A}'_\delta : \mu \in M'_\delta \cap B^c_\delta, \tag{2.149}$$

where $M'_\delta = \{x \in \mathbb{H} : \varphi_M(x) > \delta^2\}$, can be done in essentially the same manner; see also Remark 2.13.1. This may be very useful in practice. When, for instance, M is the subspace of all polynomials of degree at most $m-1$, it is more appropriate to test if one wants to establish that the mean value function is close to such a polynomial. In the traditional set-up interchanging null hypothesis and alternative would be virtually impossible due to mathematical difficulties, just as this is the case in the classical goodness-of-fit problems.

The reason that it is mathematically easier to deal with the present hypotheses is that the test statistic, which is based on

$$\varphi_M(\overline{X}) - \delta^2, \tag{2.150}$$

has a simple normal distribution in the limit for large sample sizes.

LEMMA 2.15.1. *We have*

$$\sqrt{n}\{\varphi_M(\overline{X}) - \varphi_M(\mu)\} \to_d \mathcal{N}(0, v^2), \text{ as } n \to \infty, \tag{2.151}$$

where

$$v^2 = 4\left\langle \Pi^\perp \mu, \Sigma\, \Pi^\perp \mu \right\rangle. \tag{2.152}$$

If $v^2 = 0$ the limiting distribution $\mathcal{N}(0,0)$ is to be interpreted as the distribution which is degenerate at 0.

Proof. The central limit theorem for $\mathbb{H}$-valued random variables yields the existence of a $\mathcal{G}(0,\Sigma)$ random element G, such that

$$\sqrt{n}(\overline{X} - \mu) \to_d G, \text{ as } n \to \infty, \tag{2.153}$$

in $(\mathbb{H}, \mathcal{B}_{\mathbb{H}})$. It is easy to see that $\varphi_M : \mathbb{H} \to \mathbb{R}$ is Fréchet differentiable at any $\mu \in \mathbb{H}$, tangentially to $\mathbb{H}$, with derivative the linear functional

$$2\left\langle \Pi^\perp \mu, h \right\rangle, h \in \mathbb{H}. \tag{2.154}$$

According to the functional delta method we may conclude

$$\sqrt{n}\{\varphi_M(\overline{X}) - \varphi_M(\mu)\} \to_d 2\left\langle \Pi^\perp \mu, G \right\rangle. \tag{2.155}$$

The random variable on the right in (2.155) is normal, because G is Gaussian, and clearly its mean is 0. Therefore its variance equals

$$\mathbb{E}\left\langle \Pi^\perp \mu, G \right\rangle \left\langle \Pi^\perp \mu, G \right\rangle = \left\langle \Pi^\perp \mu, \Sigma\, \Pi^\perp \mu \right\rangle, \tag{2.156}$$

according to the definition of Σ (cf.(2.135)).

LEMMA 2.15.2. *We have*

$$\hat{v}_n^2 = 4\left\langle \Pi^{\perp}\overline{X}, S\,\Pi^{\perp}\overline{X}\right\rangle \to_p v^2,\ as\ n \to \infty. \tag{2.157}$$

Proof. By simple algebra we find

$$\left\langle \Pi^{\perp}\overline{X}, S\,\Pi^{\perp}\overline{X}\right\rangle = \tag{2.158}$$

$$= \left\langle \Pi^{\perp}\overline{X}, \frac{1}{n}\sum_{i=1}^{n}\left\langle X_i - \overline{X}, \Pi^{\perp}\overline{X}\right\rangle (X_i - \overline{X}) \right\rangle =$$

$$= \frac{1}{n}\sum_{i=1}^{n}\left\langle X_i - \overline{X}, \Pi^{\perp}\overline{X}\right\rangle^2 =$$

$$= \frac{1}{n}\sum_{i=1}^{n}\{\left\langle X_i - \mu, \Pi^{\perp}\mu\right\rangle + \left\langle X_i - \mu, \Pi^{\perp}(\overline{X} - \mu)\right\rangle$$

$$+ \left\langle \mu - \overline{X}, \Pi^{\perp}\mu\right\rangle + \left\langle \mu - \overline{X}, \Pi^{\perp}(\overline{X} - \mu)\right\rangle\}^2.$$

According to the weak law of large numbers and the definition of covariance operator we have

$$\frac{1}{n}\sum_{i=1}^{n}\left\langle X_i - \mu, \Pi^{\perp}\mu\right\rangle^2 \xrightarrow{\mathrm{P}} \mathbb{E}\left\langle X - \mu, \Pi^{\perp}\mu\right\rangle^2 =$$

$$= \left\langle \Pi^{\perp}\mu, \Sigma\Pi^{\perp}\mu\right\rangle,\ \text{as}\ n \to \infty.$$

All the other terms tend to 0 in probability. As an example consider

$$\left\langle \mu - \overline{X}, \Pi^{\perp}\mu\right\rangle^2 \le ||\overline{X} - \mu||^2||\Pi^{\perp}\mu||^2 \xrightarrow{\mathrm{P}} 0,$$

as $n \to \infty$. The lemma follows from straightforward combination of the above.

For $0 < \alpha < 1$ let $\xi_{1-\alpha}$ denote the quantile of order $1 - \alpha$ of the standard normal distribution. Focusing on the testing problem (2.147), (2.148) let us decide to reject the null hypothesis when $\sqrt{n}\{\varphi_M(\overline{X}) - \delta^2\}/\hat{v} > \xi_{1-\alpha}$. The corresponding power function is then

$$\beta_n(\mu) = \mathbb{P}\{\sqrt{n}\{\varphi_M(\overline{X}) - \delta^2\}/\hat{v} > \xi_{1-\alpha}\}, \tag{2.159}$$

when $\mu \in \mathbb{H}$ is the true parameter.

THEOREM 2.15.1. *(asymptotics under the null hypothesis and fixed alternatives.) The power function in* (2.159) *satisfies*

$$\lim_{n\to\infty}\beta_n(\mu) = \begin{cases} 0, & \varphi_M(\mu) < \delta^2, \\ \alpha, & \varphi_M(\mu) = \delta^2, v^2 > 0, \\ 1, & \varphi_M(\mu) > \delta^2. \end{cases} \tag{2.160}$$

Hence the test has asymptotic size α, and is consistent against the alternatives $\mu : \varphi_M(\mu) > \delta^2$.

Proof. If $v^2 > 0$ it is immediate from Lemma's 1 and 2 that $\sqrt{n}\{\varphi_M(\overline{X}) - \delta^2\}/\hat{v} \longrightarrow_d \mathcal{N}(0,1)$. The result now follows in the usual way by observing that $\sqrt{n}\{\delta^2 - \varphi_M(\mu)\}$ tends to either ∞ (when $\varphi_M(\mu) < \delta^2$), to 0 (when $\varphi_M(\mu) = \delta^2$) or to $-\infty$ (when $\varphi_M(\mu) > \delta^2$). If $v^2 = 0$ we still have that $\sqrt{n}\{\varphi_M(\overline{X}) - \delta^2\}/\hat{v}$ tends in probability to ∞ (when $\varphi_M(\mu) < \delta^2$) or to $-\infty$ (when $\varphi_M(\mu) > \delta^2$).

To describe the sampling situation under local alternatives (including the null hypothesis) we assume now that

$$X_1, \ldots, X_n \text{ are i.i.d. } (\mu_{n,t}, \Sigma), \tag{2.161}$$

where Σ is as above and

$$\mu_{n,t} = \mu + \frac{t}{\sqrt{n}}\gamma, t \geq 0, \tag{2.162}$$

for some (cf. (2.147) and below)

$$\mu \in B_\delta, \gamma \in \mathbb{H} : \left\langle \mu, \Pi^\perp \gamma \right\rangle > 0. \tag{2.163}$$

Under these assumptions it follows that $\mu_{n,0} = \mu$ satisfies $\mathcal{H}_\delta$, and $\mu_{n,t}$ satisfies $\mathcal{A}_\delta$ for each $t > 0$. Let Φ denote the standard normal c.d.f.

THEOREM 2.15.2. *(asymptotic power). We have*

$$\lim_{n\to\infty} \beta_n(\mu_{n,t}) = 1 - \Phi\left(\xi_{1-\alpha} - 2t\frac{\langle \mu, \Pi^\perp \gamma\rangle}{v}\right), t > 0. \tag{2.164}$$

Proof. We may write $X_i = X_i' + (t/\sqrt{n})\gamma$, where the X_i' are i.i.d. (μ, Σ). It is easy to see from this representation that we still have

$$\hat{v}_n \longrightarrow_p v^2 > 0, \text{ as } n \to \infty \; \forall \, t > 0. \tag{2.165}$$

Exploiting once more the Fréchet differentiability of φ_M (see ((2.154))) we obtain

$$\sqrt{n}\frac{\varphi_M(\overline{X}) - \delta^2}{\hat{v}} = \tag{2.166}$$

$$= \sqrt{n}\frac{\varphi_M(\overline{X}') - \varphi(\mu)}{\hat{v}} + 2t\frac{\langle \Pi^\perp \mu, \Pi^\perp \gamma\rangle}{\hat{v}} + o_p(1) \rightarrow_d$$

$$\rightarrow_d \mathcal{N}(2t\frac{\langle \mu, \Pi^\perp \gamma\rangle}{v}, 1), \text{ as } n \to \infty,$$

and the result follows.

REMARK 2.15.1. *To corroborate the remark about interchanging null hypothesis and alternative made at the beginning of this section, just note that an asymptotic size α test for testing $\mathcal{H}'_\delta$ versus $\mathcal{A}'_\delta$ in* (2.149) *is obtained by rejecting $\mathcal{H}'_\delta$ when*

$$\sqrt{n}\frac{\varphi_M(\overline{X}) - \delta^2}{\hat{v}} < \xi_\alpha,\ \alpha \in (0,1). \tag{2.167}$$

This allows to assess the approximate validity of the model within the neighborhood δ. Of course, from (2.167) *we immediately get a confidence interval for δ as well.*

REMARK 2.15.2. *The expression in* (2.166) *remains valid for $t = 0$ or $\gamma = 0$. In either case the corresponding mean satisfies the null hypothesis assumption and the limit in* (2.166) *equals α.*

There is an objective, data-driven method to select the parameter δ, say, that determines the size of the neighborhood hypothesis. Given any level $\alpha \in (0,1)$ for the test, one might determine the smallest value $\hat{\delta}(\alpha)$ for which the neighborhood hypothesis is not rejected. It should be realized that modification of Hotelling's test will require a more or less arbitrary regularization parameter.

2.15.4 The Multi-Sample Problem in a Hilbert Space

Let $X_{j1},\dots,X_{jn_j}$ be i.i.d. with mean μ_j and covariance operator Σ_j, where $n_j \in \mathbb{N}$, s.t. $\sum_j n_j = n$, and let these random elements satisfy the moment condition in (2.137): all of this for $j = 1,\dots,p$. Moreover these p samples are supposed to be mutually independent, and their sample sizes satisfy

$$\begin{cases} \frac{n_j}{n} = \lambda_j + o\left(\frac{1}{\sqrt{n}}\right), \text{ as } n = n_1 + \cdots + n_p \to \infty, \\ \lambda_j \in (0,1), j = 1,\dots,p. \end{cases} \tag{2.168}$$

Let us define

$$\overline{X}_j = \frac{1}{n_j}\sum_{i=1}^{n_j} X_{ji},\ \overline{X} = \frac{1}{p}\sum_{j=1}^{p}\frac{n_j}{n}\overline{X}_j,\ j = 1,\dots,p. \tag{2.169}$$

Furthermore, let the functionals $\psi_n : \mathbb{H}^p \to \mathbb{R}$ be given by

$$\psi_n(x_1,\dots,x_p) = \sum_{j=1}^{p}\left|\left|\frac{n_j}{n}x_j - \overline{x}_n\right|\right|^2 \tag{2.170}$$

where $x_1,\dots,x_p \in \mathbb{H}$ and $\overline{x}_n = \frac{1}{p}\sum_{j=1}^{p}\frac{n_j}{n}x_j$. Defining $\psi : \mathbb{H}^p \to \mathbb{R}$ by

$$\psi(x_1,\dots,x_p) = \sum_{j=1}^{p}||\lambda_j x_j - \overline{x}||^2, \tag{2.171}$$

where $\overline{x} = \frac{1}{p}\sum_{j=1}^{p} \lambda_j x_j$, it is readily verified that

$$\sqrt{n}\{\psi_n(x_1,\ldots,x_p) - \psi(x_1,\ldots,x_p)\} \to 0, \text{ as } n \to \infty, \tag{2.172}$$

provided that condition (2.168) is fulfilled.

The neighborhood hypothesis in this model can be loosely formulated as "approximate equality of the means". More precisely the null hypothesis

$$\mathcal{H}_{p,\delta} : \mu = (\mu_1,\ldots,\mu_p)^* \in M_{p,\delta} \cup B_{p,\delta}, \tag{2.173}$$

where $M_{p,\delta} = \{x \in \mathbb{H}^p : \psi(x) < \delta^2\}$ and $B_{p,\delta} = \{x \in \mathbb{H}^p : \psi(x) = \delta^2$, $\sum_{j=1}^{p} \lambda_j \langle \lambda_j x_j - \overline{x}, \Sigma_j(\lambda_j x_j - \overline{x})\rangle > 0\}$, will be tested against the alternative

$$\mathcal{A}_{p,\delta} : \mu = (\mu_1,\ldots,\mu_p)^* \in M^c_{p,\delta} \cap B^c_{p,\delta}. \tag{2.174}$$

Let us introduce some further notation and set

$$\tau_p^2 = 4\sum_{j=1}^{p} \lambda_j \langle \lambda_j\mu_j - \overline{\mu}, (\lambda_j\mu_j - \overline{\mu})\rangle, \overline{\mu} = \frac{1}{p}\sum_{j=1}^{p} \lambda_j\mu_j. \tag{2.175}$$

Writing S_j for the sample covariance operator of the j-th sample (cf.(3.2)) the quantity in (2.175) will be estimated by

$$\hat{\tau}_{p,n}^2 = 4\sum_{j=1}^{p} \lambda_j \langle \lambda_j\overline{X}_j - \overline{X}, S_j(\lambda_j\overline{X}_j - \overline{X})\rangle. \tag{2.176}$$

THEOREM 2.15.3. *The test that rejects $\mathcal{H}_{p,\delta}$ for*

$$\sqrt{n}\frac{\psi_n(\overline{X}_1,\ldots,\overline{X}_p) - \psi_n(\mu_1,\ldots,\mu_p)}{\hat{\tau}_{p,n}} > \xi_{1-\alpha}, 0 < \alpha < 1, \tag{2.177}$$

has asymptotic size α, and is consistent against fixed alternatives $\mu = (\mu_1,\ldots,\mu_p)^*$ with $\psi(\mu) > \delta^2$.

Proof. Because the p samples are independent the central limit theorem in (2.153) yields

$$\sqrt{n}\begin{pmatrix} \overline{X}_1 - \mu_1 \\ \vdots \\ \overline{X}_m - \mu_m \end{pmatrix} \to_d \begin{pmatrix} G_1 \\ \vdots \\ G_m \end{pmatrix}, \tag{2.178}$$

where $G_1,\ldots,G_m$ are independent Gaussian random elements in $\mathbb{H}$, and

$$G_j =_d \mathcal{G}\left(0, \frac{1}{\lambda_j}\Sigma_j\right). \tag{2.179}$$

It follows from (2.172) that

$$\sqrt{n}[\psi_n(\overline{X}_1,\ldots,\overline{X}_m) - \psi_n(\mu_1,\ldots,\mu_m) \tag{2.180}$$

$$-\{\psi(\overline{X}_1,\ldots,\overline{X}_m)-\psi(\mu_1,\ldots,\mu_m)\}]=o_p(1).$$

Moreover, a slight modification of Lemma 2 yields that $\langle \overline{X}_j-\overline{X},S_j(\overline{X}_j-\overline{X})\rangle \xrightarrow{\mathrm{P}} \langle \mu_j-\overline{\mu},\Sigma_j(\mu_j-\overline{\mu})\rangle$ and hence

$$\hat{\tau}^2_{n,m} \xrightarrow{\mathrm{P}} \tau^2_m. \tag{2.181}$$

This means that the statistic on the left in (2.177) and the one obtained by replacing ψ_n with ψ in that expression will exhibit the same first order asymptotics. The proof will be continued with the latter, simpler version. A simple calculation shows that $\psi:\mathbb{H}^m\to\mathbb{R}$ is Fréchet differentiable at any $x\in\mathbb{H}^m$, tangentially to $\mathbb{H}^m$. Writing $\overline{h}=\frac{1}{p}\sum_{j=1}^m\lambda_jh_j$, for any $h_1,\ldots,h_m\in\mathbb{H}$, its derivative is equal to

$$2\sum_{j=1}^m\langle\lambda_jx_j-\overline{x},\lambda_jh_j-\overline{h}\rangle=2\sum_{j=1}^m\langle\lambda_jx_j-\overline{x},\lambda_jh_j\rangle. \tag{2.182}$$

Application of the delta method with the functional ψ in the basic result (2.178) yields

$$\sqrt{n}\{\psi(\overline{X}_1,\ldots,\overline{X}_m)-\psi(\mu_1,\ldots,\mu_m)\}\to_d 2\sum_{j=1}^m\langle\lambda_j\mu_j-\overline{\mu},\lambda_jG_j\rangle. \tag{2.183}$$

According to (2.179) we have

$$\lambda_jG_j=_d\mathcal{G}\left(0,\lambda_j\Sigma_j\right), \tag{2.184}$$

and because of the independence of the G_j it follows that

$$2\sum_{j=1}^p\langle\lambda_j\mu_j-\overline{\mu},\lambda_jG_j\rangle=_d\mathcal{N}(0,\tau_p^2), \tag{2.185}$$

where τ_p^2 is defined in (2.175). Exploiting the consistency in (2.181) the proof can be concluded in much the same way as that of Theorem 2.15.1. Just as in that theorem we need here that $\tau_p^2>0$ at the alternative considered in order to ensure consistency.

2.16 Exercises

Exercise 1. *(a) Prove that if $Y_n\xrightarrow{P}Y$, then Y_n converges in distribution to Y. (b) Prove that if Y_n converges in distribution to a constant c (i.e., the distribution of Y_n converges to the Dirac measure $\delta_{\{c\}}$), then $Y_n\xrightarrow{P}c$.*

[Hint: (a) Assume $Y_n\xrightarrow{P}Y$. Then

$$\begin{aligned}\mathbb{P}(Y_n\le t)&=\mathbb{P}(\{Y_n\le t\}\cap\{|Y_n-Y|\le\varepsilon\})+\mathbb{P}(\{Y_n\le t\}\cap\{|Y_n-Y|>\varepsilon\})\\&\le\mathbb{P}(Y\le t+\varepsilon)+\mathbb{P}(|Y_n-Y|>\varepsilon)\end{aligned}$$

First pick ε *small enough so that* $\mathbb{P}(X \le t+\varepsilon) \le \mathbb{P}(X \le t)+\eta/2$. *Then pick n large enough so that* $\mathbb{P}(|Y_n - Y| > \varepsilon) < \eta/2$. *Then, for n large enough, we have* $F_{Y_n}(t) \le F_Y(t)+\eta$. *Similarly, for n large enough, if* F_Y *is continuous at t, we have* $F_{Y_n}(t) \ge F_Y(t)-\eta$. *This shows that* $lim_{n\to\infty}F_{Y_n}(t) = F_Y(t)$ *at continuity points t of* F_Y.*]*

Exercise 2. *Prove Corollary 2.8.1.*

Exercise 3. *Suppose* P_n $(n \ge 1)$, *P are probability measures on* $(\mathbb{R}, \mathcal{B}^1)$ *such that* P_n *converges weakly to P. Show that* $\{P_n : n \ge 1\}$ *is* tight: for every $\varepsilon > 0$ there exists $A_\varepsilon > 0$ such that $P_n(\{x : |x| > A_\varepsilon\}) < \varepsilon$ for all *n*.
[Hint: (i) Find points of continuity $-B_\varepsilon$, C_ε *of the distribution function F of P such that* $F(-B_\varepsilon) < \varepsilon/3$, $F(C_\varepsilon) > 1-\varepsilon/3$.
(ii) Find N_ε *such that* $F_n(-B_\varepsilon) < \varepsilon/3$ *and* $F_n(C_\varepsilon) > 1-\varepsilon/3$ *for all* $n \ge N_\varepsilon$, *where* F_n *is the distribution function of* P_n. *Then* $P_n([-B_\varepsilon, C_\varepsilon]) > 1-\frac{2\varepsilon}{3}$ *for all* $n \ge N_\varepsilon$.
(iii) For $n = 1, \ldots, N_\varepsilon$, *find* $D_\varepsilon > 0$ *such that* $P_n([-D_\varepsilon, D_\varepsilon]) > 1-\varepsilon$ $(1 \le n \le N_\varepsilon)$.
(iv) Let $A_\varepsilon = \max B_\varepsilon, C_\varepsilon, D_\varepsilon$ *to get* $P_n(\{x : |x| > A_\varepsilon\}) < \varepsilon$ *for all n.]*

Exercise 4. *Assume* X_j, $j \ge 1$, *are i.i.d. real-valued, with* $EX_j = \mu$, $\mathrm{var}(X_j) = \sigma^2 > 0$, $EX_j^4 < \infty$. *Prove that*
(a) $\sqrt{n}(s^2 - \sigma^2) \to_d N(0, E(X_1-\mu)^4 - \sigma^4)$, *and*
(b) $\sqrt{n}(\frac{1}{s} - \frac{1}{\sigma}) \to_d N(0, [E(X_1-\mu)^4 - \sigma^4]\cdot[1/4\sigma^6])$.
(c) What is the asymptotic distribution of $\sqrt{n}(\log s - \log \sigma)$ *?*
[Hint: (a) Consider $U_j = X_j - \mu$, $j \ge 1$, $s^2 = (\frac{n}{n-1})\frac{1}{n}\sum_{j=1}^{1}(U_j - \overline{U})^2$
$= (\frac{n}{n-1})\left[\frac{1}{n}(\Sigma_{j=1}^n)U_j^2 - \overline{U}^2\right]$, *so that* $\sqrt{n}(s^2-\sigma^2) - \sqrt{n}(\frac{1}{n}\Sigma_{j=1}^n(U_j^2 - \sigma^2)) \xrightarrow{P} 0$.
(b) $\sqrt{n}(\frac{1}{s} - \frac{1}{\sigma}) - \sqrt{n}[H(\overline{z}) - H(\sigma^2)] \xrightarrow{P} 0$, *where* $z_j = U_j^2$, $EZ_j = \sigma^2$, $H(z) = z^{-1/2}$, $H(\overline{Z}) = (\frac{1}{n}\Sigma_{j=1}^n U_j^2)^{-1/2}$, $H(\sigma^2) = 1/\sigma$. *Apply Theorem 2.8.4.]*

Exercise 5. *(a) Let* X_n *have the discrete uniform distribution on* $\{0, \frac{1}{n}, \frac{2}{n}, \ldots, 1\}$ *(i.e.,* $\mathbb{P}(X_n = \frac{k}{n}) = \frac{1}{n+1}$ $(k = 0, 1, \ldots, n)$. *Show that* X_n *converges in distribution to the uniform distribution on* $[0,1]$ *(with constant density* 1*).*
(b) Use (a) to prove that $(1/n+1)\sum_{k=0}^n f(k/n) \to \int_0^1 f(x)dx$ *for every continuous function f on* $[0,1]$.
(c) Extend (b) to the case of all bounded measurable f on $[0,1]$ *with a finite set of discontinuities.*

Exercise 6. *Show that if a random variable* T_n *has a Student t distribution with n degrees of freedom then* $T_n \to_d Z$, *where* $Z \sim N(0,1)$.

Exercise 7. *Extend Proposition 2.5.2 to k sequences* $U_n^{(i)} \xrightarrow{P} a_i$ $(1 \le i \le k)$ *and a function g of k vector variables continuous at* $(a_1, a_2, \ldots, a_k)$, *as stated in Remark 2.5.1.*

Exercise 8. *Prove that the sample covariance matrix is a consistent estimator of the covariance matrix (see Proposition 2.5.3).*

Exercise 9. *Let X_n $(n \geq 1)$ be a sequence of i.i.d. random variables, and assume that the infimum and supremum of values of X_1 are m and M, respectively. That is, $\mathbb{P}(m \leq X_1 \leq M) = 1$, $\mathbb{P}(X_1 < a) > 0 \ \forall \ a > m$, $\mathbb{P}(X_1 > b) > 0$ $\forall \ b < M$ (Here X_n real-valued, but m and/or M may be infinite). Prove that $\max\{X_1,\dots,X_n\} \xrightarrow{P} M$ and $\min\{X_1,\dots,X_n\} \xrightarrow{P} m$.*

Exercise 10. *Prove that if Y_n converges in distribution to a constant c (i.e., the distribution of Y_n converges to the Dirac measure $\delta_{\{c\}}$), then $Y_n \xrightarrow{P} c$.*

Exercise 11. *Let P_n $(n \geq 1)$, P be probability measures on $(\mathbb{R}, \mathcal{B}^1)$ such that P_n converges weakly to P.*
(a) Give an example to show that $P_n(B)$ need not converge to $\mathbb{P}(B)$ for all Borel sets B.
(b) Give an example to show that the distribution function F_n of P_n may not converge to the distribution function F of P at every point x.

Exercise 12. *Let the hypothesis of Theorem 2.14.1 hold.*
(a) Derive (2.120) and (2.123). Show that
(b) $\{(\widehat{f}_n(x) - f(x))/\sqrt{\operatorname{var}\widehat{f}_n(x)} : x$ such that $f(x) > 0\}$ converges in law to a Gaussian process, and compute the mean and covariance function of this Gaussian process.

Exercise 13. *Assume that $h \equiv h_n = o(n^{-1/5})$ and (2.109) holds, and prove that with this bandwidth (2.123) holds with $\gamma = 0$, under the hypothesis of Theorem 2.14.1.*

Exercise 14. *Consider a pdf f on $\mathbb{R}^m$, having continuous and bounded second derivatives.*
(a) State and prove the analog of Theorem 2.14.1.
(b) Derive the multidimensional versions of (2.121)–(2.123).

Chapter 3

Geometry and Topology of Manifolds

3.1 Manifolds, Submanifolds, Embeddings, Lie Group actions

A *topological space* is a pair (M,τ), where M is a nonempty set, and τ is a set of parts of M, such that (i) M is in τ, (ii) for any finite set of parts in τ, their intersection is still in τ, and (iii) for any set of parts in τ, their union is in τ. A set in τ is said to be *open*, and its complement is said to be a *closed set* . A function $f:(M_1,\tau_1)\rightarrow(M_2,\tau_2)$ between two topological spaces is *continuous* if $\forall W\in\tau_2, f^{-1}(W)\in\tau_1$. If in addition f is one to one and its inverse is continuous, f is said to be a .

EXAMPLE 3.1.1. *A* metric space *is a pair (M,ρ), where M is a nonempty set, and*
$\rho: M\times M\rightarrow[0,\infty)$ is a function such that $\forall x_1,\forall x_2,\forall x_3\in M$, (i) if $\rho(x_1,x_2)=0$, then $x_1=x_2$, (ii) $\rho(x_1,x_2)=\rho(x_2,x_1)$(symmetry) and (iii) $\rho(x_1,x_2)\leq\rho(x_1,x_3)+\rho(x_3,x_2)$ (triangle inequality). The metric topology τ_ρ *of this metric space is the set of all subsets U of M, with the property that for any point $x\in U$, there is a positive number r such that the* open ball

$$B_r(x)=\{y\in M;\rho(y,x)<r\},\tag{3.1}$$

is a subset of U.

The *closure* $cl(D)$ of a subset D of a topological space (M,τ) is the intersection of all closed sets containing S. The subset D is said to be in (M,τ) if $cl(D)=M$.

EXAMPLE 3.1.2. *The set $\mathbb{Q}^m$ is a countable dense subset of the Euclidean space $(\mathbb{R}^m,\tau_0)$. Here τ_0 is the Euclidean metric topology.*

DEFINITION 3.1.1. *Given a subset N of a topological space (M,τ), the* induced topology $\tau|N$ *the set of all intersections $N\cap U, U\in\tau$.*

The inclusion map $\iota:(N,\tau|N)\rightarrow(M,\tau)$ is an example of a continuous function.

DEFINITION 3.1.2. *A relationship on a set M is a subset R of $M\times M$. The relationship R is reflexive if $\forall x\in M,(x,x)\in R$. The relationship R is symmetric if $\forall x,y$, if $(x,y)\in R$, then $(y,x)\in R$. The relationship R is transitive if $\forall x,y,z$, $(x,y)\in R$ and $(y,z)\in R$, then $(x,z)\in R$. An equivalence relationship on M is a relationship on M that is reflexive, symmetric and transitive.*

EXAMPLE 3.1.3. *Consider the relationship R_m on $\mathbb{R}^m$ defined as follows $(x,y)\in R_m$ if $x-y\in\mathbb{Z}^m$. It is elementary to show that R_m is an equivalence relationship.*

Given an equivalence relationship R on M and an element $x\in M$, the *equivalence class* of x is the set $[x]_R=\{y\in M,(y,x)\in R\}$. Note that two equivalence classes that do intersect are equal, therefore the set of equivalence classes give a partition of M..

DEFINITION 3.1.3. *Given an equivalence relationship R on M, the quotient set M/R is the set $M/R = \{[x]_R, x \in M\}$. If M has in addition a topology τ on it, the quotient topology τ/R on M/R is the set $V \subseteq M/R, \pi_R^{-1}(V) \in \tau$, where $\pi_R : M \to M/R$ is given by $\pi_R(x) = [x]_R$. and the pair $(M/R, \tau/R)$ is a quotient topological space.*

EXAMPLE 3.1.4. *Consider the numerical space $\mathbb{R}^m$ with the metric topology τ_0 associated with the Euclidean distance on $\mathbb{R}^m$. The quotient topological space $(\mathbb{R}^m/R_m, \tau_0/R_m)$ is called flat torus .*

A topological space is said to be *connected* if it can not be written as a union of two disjoint open sets. Connectivity is a topological property: if C is a connected subset of a topological space (M_1, τ_1) and $f : (M_1, \tau_1) \to (M_2, \tau_2)$ is a continuous function, then $f(C)$ is a connected subset in (M_2, τ_2). Note that the only connected sets of the Euclidean line are the *intervals*, which leads to the following

THEOREM 3.1.1. *(Generalized Darboux theorem) If (M, τ) is a connected topological space and $f : (M, \tau) \to \mathbb{R}$, is a continuous function, such that there are two values of f having opposite signs, then the equation $f(x) = 0$ has a solution in M.*

Given a topological space (M, τ), an *open neighborhood* of point on M is an open set that contains that point. The topological space (M, τ) is said to be *Hausdorff* if $\forall x_1, x_2$ in $M, x_1 \neq x_2$ there are disjoint open neighborhoods U_1 of x_1 and U_2 of x_2. The topological space (M, τ) is said to be *separable* if there is a dense countable subset D of M.

EXAMPLE 3.1.5. *The Euclidean space $\mathbb{E}^m = (\mathbb{R}^m, \tau_0)$ is both Hausdorff and separable.*

A metric space $\mathcal{M}$, is *paracompact* if from any open cover $(U_\alpha)_{\alpha \in A}$, of $\mathcal{M}$, one can extract a locally finite subcover. Given a paracompact space $\mathcal{M}$, and an open cover $(U_\alpha)_{\alpha \in A}$, *partition of the unity* , is an indexed set of functions $(\varphi_\alpha)_{\alpha \in A}$, for which (i) $\varphi_\alpha \geq 0$, (ii) $supp\varphi_\alpha \subset U_\alpha$, and (iii) $\sum_{\alpha \in A} \varphi_\alpha = 1$. Here $supp\varphi_\alpha$ is the closure of the set of points $p \in \mathcal{M}$, with $\varphi(p) > 0$.

EXAMPLE 3.1.6. *Any compact metric space is paracompact.*

Recall that a function of several real variables $f : U \subseteq \mathbb{R}^m \to \mathbb{R}^p$ defined on an open set U is said to be differentiable at a point x_0 if there exists a linear map $T : \mathbb{R}^m \to \mathbb{R}^p$ such that

$$\lim_{h \to 0} \frac{1}{||h||}(f(x_0 + h) - f(x_0) - T(h)) = 0. \tag{3.2}$$

We use the notation $d_{x_0}f$ for T; this is *the differential of f at x_0*. If the function is differentiable at x_0, then for all indices $j = 1,\dots,m$, the partial derivatives $\frac{\partial f}{\partial x^j}(x_0)$, exist, and if $e_j, j = 1,\dots,m$ is the standard basis of $\mathbb{R}^m$, then $\frac{\partial f}{\partial x^j}(x_0) = d_{x_0}f(e_j)$.

DEFINITION 3.1.4. *The function $f : U \subseteq \mathbb{R}^m \to \mathbb{R}^p$ is differentiable, if $\forall x \in U$, f is differentiable at the point x.*

DEFINITION 3.1.5. *A C^r diffeomorphism between two open sets in $\mathbb{R}^m$ is a differentiable one-to-one and onto map from the first set to the second set, with a differentiable inverse, that has continuous partial derivatives up the order r.*

3.1.1 Manifolds and Their Tangent Bundles

Assume $\mathcal{M}$ is a metric space. A *chart* on $\mathcal{M}$ is a *homeomorphism* of an open subset $U \subseteq \mathcal{M}$ onto an open subset of $\mathbb{R}^m$. An m dimensional chart $x : U \to x(U) \subseteq \mathbb{R}^m$ is typically defined by the pair (U,x), and if $\mathcal{M}$ is *connected* metric space, by the *invariance of the domain theorem* (see Brower (1912 [167],1913 [168]), any other chart (V,y) is also m-dimensional.

EXAMPLE 3.1.7. *If $F = (F_1,\dots,F_p) : \mathbb{R}^m \times \mathbb{R}^p \to \mathbb{R}^p$, is a C^1 differentiable function, and the matrix $(grad_y F_1 \dots grad_y F_p)$ has rank p at any point (x_0,y_0) with $F(x_0,y_0) = 0$, then, by the implicit function theorem, there is an open neighborhood V of such a point (x_0,y_0) in $\mathbb{R}^m \times \mathbb{R}^p$ such that the the projection $x : V \cap F^{-1}(0) \to x(V \cap F^{-1}(0))$ is a chart on $F^{-1}(0)$.*

DEFINITION 3.1.6. *A collection $\mathcal{A} = \{(U_\alpha,x_\alpha)\}$ of $\mathbb{R}^m$-valued charts on $(\mathcal{M},\rho)$ is called the atlas of class C^r if the following conditions are satisfied:*
(i) $\cup_\alpha U_\alpha = \mathcal{M}$.
(ii) Whenever $U_\alpha \cap U_\beta$ is not empty, then the map $x_\beta \circ x_\alpha^{-1} : x_\alpha(U_\alpha \cap U_\beta) \to x_\beta(U_\alpha \cap U_\beta)$ is a C^r diffeomorphism.

DEFINITION 3.1.7. *An m-dimensional manifold of class C^r is a triple $(\mathcal{M},\rho,A_\mathcal{M})$ where $(\mathcal{M},\rho)$ is a metric space and $A_\mathcal{M}$ is an $\mathbb{R}^m$-valued atlas of class C^r on $\mathcal{M}$.*

Note that since any manifold $\mathcal{M}$ is *locally compact*, every closed bounded subset of a complete manifold $\mathcal{M}$ is compact.

EXAMPLE 3.1.8. *Assume $F : \mathbb{R}^{m+p} \to \mathbb{R}^p$ is a differentiable function, such that for each point $x \in F^{-1}(0)$, the matrix $\left(\frac{\partial F_j}{\partial x_i}\right)_{i=\overline{1,m+p},j=\overline{1,p}}$ has rank p. Then from example 3.1.7 $\mathcal{M}_F = F^{-1}(0)$ is a differentiable manifold of dimension m with an atlas of charts given around a point $x \in F^{-1}(0)$, by a projection $x \to x_I$ where $I \subset \overline{1,m+p}$ is a set of indices, with $I \cup J = \overline{1,m+p}$, $I \cap J = \emptyset$ and rank $\left(\frac{\partial F_j}{\partial x_i}\right)_{i=\overline{1,m+p},j\in J} = p$.*

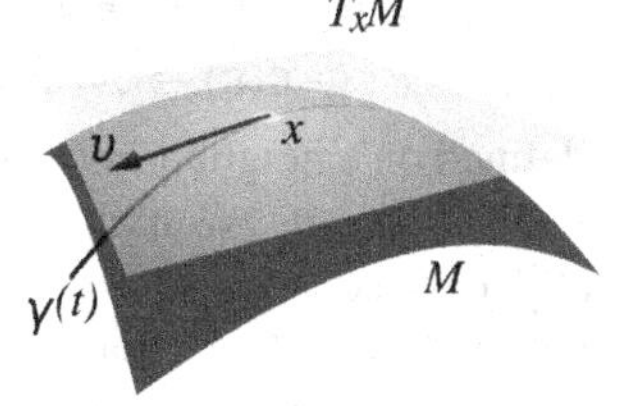

Figure 3.1: *Tangent plane to a surface at a given point.*

EXAMPLE 3.1.9. *A particular case of an m-dimensional manifold obtained as set of zeros of a differentiable function as in example 3.1.8 is the m-dimensional sphere or radius one,* $\mathcal{M} = S^m = \{x \in \mathbb{R}^{m+1} : \|x\| = 1\}$*, since* $S^m = F^{-1}(0)$*, where* $F((x^1,\ldots,x^{m+1})) = \sum_{a=1}^{m+1}(x^j)^2 - 1$ *and* $rank\left(\frac{\partial F_j}{\partial x_i}(x)\right)_{i=\overline{1,m+1},j\in J} = 1$*, for any point* $x \in S^m$.

DEFINITION 3.1.8. *Assume* $\mathcal{M},\mathcal{N}$ *are two manifolds of dimension m, respectively p. A function* $f : \mathcal{M} \to \mathcal{N}$ *is* differentiable function between manifolds, *if f is continuous and for any charts* $(U,x_U) \in \mathcal{A}_{\mathcal{M}}, (V,x_V) \in \mathcal{A}_{\mathcal{N}}$*, with* $V \cap f(\mathcal{M}) \neq \emptyset$*, the* local representative $f_{x_U,x_V} : x_U(U \cap f^{-1}(V)) \to \mathbb{R}^p$*, given by* $f_{x_U,x_V}(x) = x_V(f(x_U^{-1}(x)))$ *is a differentiable.*

A *manifold diffeomorphism* is an invertible differentiable function $f : \mathcal{M} \to \mathcal{N}$ that has a differentiable inverse; in this case $\mathcal{M}$ and $\mathcal{N}$ are said to be *diffeomorphic manifolds.*

If $(\mathcal{M},A_{\mathcal{M}}),(\mathcal{N},A_{\mathcal{N}})$ are manifolds, their *product* $\mathcal{M} \times \mathcal{N}$ has a natural structure of manifold with the atlas $A_{\mathcal{M}\times\mathcal{N}}$ of charts $(U \times V, x_U \times x_V)$ given by $x_U \times x_V(u,v) = (x_U(u),x_V(v))$. A differentiable *curve* on a manifold $\mathcal{M}$ is a differentiable function from an interval to $\mathcal{M}$. Two curves $c_a, a = 1,2$ defined on a neighborhood of $0 \in \mathbb{R}$ are *tangent* at p if $c_1(0) = c_2(0) = p$ and there is a chart (U,x) around p such that $(x \circ c_1)'(0) = (x \circ c_2)'(0)$.

DEFINITION 3.1.9. *The set of all curves tangent at p to the curve c is called* tangent vector v_p *at* $p = c(0)$*, and is labeled* $v_p = \frac{dc}{dt}(0)$.

If (U,x) *is a chart around p with* $x(p) = x_0$*, the tangent vector to the curve* $c_{p,x_0,i}(t) = x^{-1}(x_0 + te_i)$ *is labeled* $\frac{\partial}{\partial x^i}|_{p,x}$ *or* $\frac{\partial}{\partial x^i}|_p$*, if x is known.*

If $\mathcal{M}$ is a submanifold of $\mathbb{R}^N$, then two curves c_1,c_2 on $\mathcal{M}$ with $c_1(0) = c_2(0) = p$ have the same tangent vector in the sense of multivariable calculus, if $\frac{dc_1}{dt}(0) = \frac{dc_2}{dt}(0)$, therefore if $x : U \to \mathbb{R}^m$, is a chart around p, then $\frac{d(x \circ c_1)}{dt}(0) = \frac{d(x \circ c_2)}{dt}(0)$, showing that the definition 3.1.9 corresponds to the classical notion of tangent vector.

DEFINITION 3.1.10. *The $T_p\mathcal{M}$ at a point p of a manifold $\mathcal{M}$ is the set of all tangent vectors $v_p = \frac{dc}{dt}(0)$ to curves $c : (-\varepsilon, \varepsilon) \to \mathcal{M}$ with $c(0) = p$.*

Assume (U,x) is a chart defined on a neighborhood of p, with $x(p) = 0$, and c is a differentiable curve in $\mathcal{M}, c(0) = p$. We may write $c(t) = x^{-1}(x(c(t)))$, therefore any curve is locally defined around $p = c(0)$ by a curve $\gamma(t) = x(c(t)), \gamma : (-\varepsilon, \varepsilon) \to \mathbb{R}^m$, with $\gamma(0) = 0$. Note that the curves $t \to c(t)$ and $t \to x^{-1}(t\frac{d\gamma}{dt}|_{t=0})$ define the same tangent vector at p, therefore any tangent vector is uniquely determined by a vector $w \in \mathbb{R}^m$ and the tangent space $T_p\mathcal{M}$ inherits a structure m dimensional vector space from the $\mathbb{R}^m$, which is independent on the choice of the chart (U,x).

Let (U,x) be a chart around p with two curves defined around $0 \in \mathbb{R}$ $c_1(0) = c_2(0) = p$ and both c_1 and c_2 tangent at p. Then given another chart (V,y) around p we have;

$$\frac{d(y \circ c_1)}{dt}(0) = \left(\frac{\partial y^j}{\partial x^i}\bigg|_{c(o)} \right)_{i,j=1,\ldots,m} \frac{d(x \circ c_1)}{dt}(0) \tag{3.3}$$

Note that $x \circ c_1 : (-\varepsilon, \varepsilon) \to \mathbb{R}^m$ and similarily $y \circ c_1 : (-\varepsilon, \varepsilon) \to \mathbb{R}^m$, hence $x(p) = (x^1(c_1(0)), \cdots, x^m(c_1(0)))^T = x \circ c_1(0)$ and we also have that $y \circ c_1(t) = (y^1 \circ c_1(t), \cdots, y^m \circ c_1(t))^T$ and for $i = 1, \cdots, m$ we have

$$y^i(c(t)) = y^i \circ x^{-1}(x^1(c(t)), \cdots, x^m(c(t)))$$

$$\frac{d(y \circ c_1)}{dt}(0) = \left(\frac{d(y^1 \circ c_1)}{dt}(0), \cdots, \frac{d(y^m \circ c_1)}{dt}(0) \right)^T$$

where

$$\frac{d(y^1 \circ c_1)}{dt}(0) = \sum_{i=1}^{m} \frac{\partial(y^1 \circ x^{-1})}{\partial x^i}\bigg|_{x(c(0))} \frac{d(x^i \circ c_1)}{dt}(0)$$

$$\vdots \qquad \vdots$$

$$\frac{d(y^j \circ c_1)}{dt}(0) = \sum_{i=1}^{m} \frac{\partial(y^j \circ x^{-1})}{\partial x^i}\bigg|_{x(c(0))} \frac{d(x^i \circ c_1)}{dt}(0)$$

$$\vdots \qquad \vdots$$

$$\frac{d(y^m \circ c_1)}{dt}(0) = \sum_{i=1}^{m} \frac{\partial(y^m \circ x^{-1})}{\partial x^i}\bigg|_{x(c(0))} \frac{d(x^i \circ c_1)}{dt}(0)$$

Also note that; $\frac{\partial(y^j \circ x^{-1})}{\partial x^i}\Big|_{x(c(0))} \frac{d(x^i \circ c_1)}{dt}(0) = \frac{\partial y^j}{\partial x^i}\Big|_{c(0)} \frac{d(x^i \circ c_1)}{dt}(0)$

Furthermore, $p \in M, T_pM = \left\{ \frac{dc}{dt}(0) \middle| c : (-\varepsilon, \varepsilon) \to M, \ differentiable, c(0) = p \right\}$.

And we have that it is a vectors spaces as well. Let $v_{1,p}$ and $v_{2,p}$ be tangent vectors in T_pM we have that $v_{1,p}+v_{2,p} = \frac{dc}{dt}(0)$ such that

$$c(t) = x^{-1}\left(t\left[\left.\frac{d(x\circ c_1)}{dt}\right|_{x(p)} + \left.\frac{d(x\circ c_2)}{dt}\right|_{x(p)}\right] + x(p)\right)$$

The *tangent space of a manifold* $\mathcal{M}$, $T\mathcal{M}$, is the disjoint union of all tangent spaces to $\mathcal{M}$ at different points of $\mathcal{M}$, $T\mathcal{M} = \sqcup_{p\in\mathcal{M}} T_p\mathcal{M}$. The *tangent bundle* is the triple $(T\mathcal{M}, \Pi, \mathcal{M})$, where the $\Pi : T\mathcal{M} \to \mathcal{M}$ associates to each tangent vector its base point, $\Pi(v_p) = p$.

The tangent space $T\mathcal{M}$ has a natural structure of $2m$ dimensional manifold, for which the map Π is differentiable. Indeed if (U, x) is a chart on $\mathcal{M}$, then (TU, Tx) is a chart on $T\mathcal{M}$, where

$$TU = \{v \in T\mathcal{M}, \Pi(v) \in U\} \tag{3.4}$$

and

$$Tx(\frac{dc}{dt}(0)) = (x(c(0)), \frac{d(x\circ c)}{dt}(0)), \text{if } c(0) \in U. \tag{3.5}$$

A *vector field* V is a differentiable *section* of Π, that is $V : \mathcal{M} \to T\mathcal{M}$, is a differentiable function for which, $\Pi \circ V = Id_{\mathcal{M}}$. An equivalent definition is that a vector field on $\mathcal{M}$ is an assignment to every point of $\mathcal{M}$ of a tangent vector to $\mathcal{M}$ at that point, that is differentiable as a function of that point. That is, for each $p \in \mathcal{M}$, we have a tangent vector $V(p) \in T_p\mathcal{M}$. An example of a vector field on the sphere is suggested by the velocity of the wind at a given moment at different points on the surface of a planet.

Vector fields are generated by *one parameter groups of diffeomorphisms* (actions of the additive group $(\mathbb{R}, +)$ by diffeomorphishms of the manifold). The set of all differentiable vector fields on $\mathcal{M}$ is labeled $\Gamma(T\mathcal{M})$.

DEFINITION 3.1.11. *A vector field X on $\mathcal{M}$ can be also regarded as a differential operator $X : C^\infty(\mathcal{M}) \to C^\infty(\mathcal{M})$ on the algebra of smooth functions on $\mathcal{M}$. That is an operator with the properties (i) $X(f+g) = X(f) + X(g)$, (ii) $X(fg) = fX(g) + X(f)g$, and (iii) $X(1) = 0$.*
The Lie bracket $[X, Y]$ of two vector fields X, Y on $\mathcal{M}$ is defined as the differential operator on $C^\infty(\mathcal{M})$ given by

$$[X,Y](f) = X(Y(f)) - Y(X(f)). \tag{3.6}$$

3.1.2 Embeddings of Manifolds in Euclidean Spaces

Manifolds are abstractions. To visualize a manifold as a submanifold of an Euclidean space, one should *embed* it in a numerical space. If $f : \mathcal{M}_1 \to \mathcal{M}_2$

is a differentiable function between manifolds, its *tangent map* is the function $df : T\mathcal{M}_1 \to T\mathcal{M}_2$, given by

$$df(\frac{dc}{dt}|_{c(0)}) = \frac{d(f \circ c)}{dt}|_{f(c(0))}, \tag{3.7}$$

for all differentiable curves c defined on an interval containing $0 \in \mathbb{R}$. The *differential* of f at a point p is the restriction of the tangent map, regarded as a linear function $d_pf : T_p\mathcal{M}_1 \to T_{f(p)}\mathcal{M}_2$.

DEFINITION 3.1.12. *An* immersion *of a manifold $\mathcal{M}$ in an Euclidean space $\mathbb{R}^N$ is a differentiable $j : \mathcal{M} \to \mathbb{R}^N$, for which differential d_pj is a one-to-one function from $T_p\mathcal{M}$ to $\mathbb{R}^N$. An* embedding *of a manifold $\mathcal{M}$ into an Euclidean space $\mathbb{R}^N$ is a one-to-one immersion $j : \mathcal{M} \to \mathbb{R}^N$, for which j is a homeomorphism from $\mathcal{M}$ to $j(\mathcal{M})$, endowed with metric topology induced by the Euclidean distance.*

EXAMPLE 3.1.10. *The inclusion map of a flat torus $(S^1)^N$ in $(\mathbb{R}^2)^N$ is an embedding.*

PROPOSITION 3.1.1. *Any compact manifold $\mathcal{M}$ can be embedded into $\mathbb{R}^N$, for some large enough N.*

For a proof, assume $\mathcal{M}$ is m dimensional. W.l.o.g. we may consider a finite atlas $\mathcal{A}$ on $\mathcal{M}$, whose charts are $(U_a, \varphi_a), a = 1, \dots, N$. Assume $\eta_a, a = 1, \dots, N$ is a partition of unity subordinated to this open covering, and for $a = 1, \dots, N$, define $\psi_a : \mathcal{M} \to \mathbb{R}^m$,

$$\psi_a(p) = \begin{cases} \eta_a(p)\varphi_a(p) & \text{, if } p \in U_a \\ 0 & \text{, otherwise.} \end{cases}$$

It is easy to see that the function $F : \mathcal{M} \to \mathbb{R}^{(m+1)N}$ given by

$$F(p) = (\eta_1(p), \dots, \eta_N(p), \psi_1(p), \dots, \psi_N(p)), \tag{3.8}$$

is one to one. To prove that F is an embedding, given that $\mathcal{M}$ is compact, it suffices to show that F is an immersion. By the product rule, given $v \in T_p\mathcal{M}$,

$$d_pF(v) = (d_p\eta_1(v), \dots, d_p\eta_k(v), d_p\eta_1(v)\varphi_1(p) + \eta_1(p)d_p\varphi_1(v), \dots, \\ d_p\eta_N(v)\varphi_N(p) + \eta_N(p)d_p\varphi_N(v), \tag{3.9}$$

therefore if $d_pF(v) = 0$, it follows that $\forall a = 1, \dots, N, d_p\eta(v) = 0$, and since $\sum_a \eta_a(p) = 1$, there is an a such that $d_p\varphi_a(v) = 0$, and since $\varphi_a : U_a \to \mathbb{R}^m$ is a diffeomorphism, we get $v = 0$. Therefore F is an immersion, done.
The following result, due to Whitney (1944)[336], can not be improved if $m = 2^k, k \in N$,

THEOREM 3.1.2. *(Whitney's embedding theorem) Any smooth m dimensional manifold can be embedded into $\mathbb{R}^{2m}$.*

3.1.3 Lie Groups and Their Lie Algebras

A *Lie group* is a differentiable manifold $\mathcal{G}$ which has in addition an algebraic structure of *group*, for which the group multiplication $\circ : \mathcal{G} \times \mathcal{G} \to \mathcal{G}$, and the operation of taking the inverse $g \to g^{-1}$ are differentiable functions.

The group $GL(m, \mathbb{R})$ of invertible matrices, with the matrix multiplication is a Lie group. The *orthogonal group* $O(m)$ (see Exercise 16), and its connected component, the *special orthogonal group* $SO(m)$ of all $m \times m$ orthogonal matrices A with $Det(A) = 1$, are also Lie groups.

DEFINITION 3.1.13. *A one parameter subgroup of a Lie group $\mathcal{G}$ is differentiable group homomorphism γ from $(\mathbb{R}, +)$ to $\mathcal{G}$.*

EXAMPLE 3.1.11. *Assume Y is a $m \times m$ real matrix, and define $e^Y \in GL(p, \mathbb{R})$ given by*

$$e^Y = \exp(Y) = \sum_{k=0}^{\infty} \frac{1}{k!} Y^k. \tag{3.10}$$

Any one parameter subgroup $\gamma(t)$ of $GL(m, \mathbb{R})$, is given by $\gamma(t) = e^{tA}$, where $A \in \mathcal{M}(m, \mathbb{R})$.

REMARK 3.1.1. *In general, given a tangent vector u in the tangent space $T_{1_\mathcal{G}}\mathcal{G}$ at the identity element of a Lie group $\mathcal{G}$, there is a unique one parameter subgroup $\varphi_u : \mathbb{R} \to \mathcal{G}$, with $d_0\varphi_u(1) = u$. In view of example 3.1.11, one defines the* exponential map $\exp_\mathcal{G} : T_{1_\mathcal{G}}\mathcal{G} \to \mathcal{G}$, *as follows:*

$$\exp_\mathcal{G}(u) = \varphi_u(1). \tag{3.11}$$

The tangent space $\mathfrak{g} \doteq T_{1_\mathcal{G}}\mathcal{G}$ has an additional structure of Lie algebra, derived from the group operations. This structure measures the departure of the group structure from being commutative. The commutator of a pair of elements $x, y \in \mathcal{G}$, is the element $\{x, y\} = xyx^{-1}y^{-1}$. The commutator is the identity iff x and y commute. To measure the infinitesimal effect of departure from commutativity, one may consider the differentiable path $\varphi_{u,v} : \mathbb{R} \to \mathcal{G}$, associated with a pair of tangent vectors $u, v \in \mathfrak{g}$, given by

$$\varphi_{u,v}(t) = \{\exp_\mathcal{G}(tu), \exp_\mathcal{G}(tv)\}. \tag{3.12}$$

Note that $\varphi_{u,v}(t)$ is a constant map, whenever $\exp_\mathcal{G}(u)$ *and* $\exp_\mathcal{G}(v)$ *commute. Also note that since the Lie group exponential map is a local diffeomorphism around* $0 \in \mathfrak{g}$, *for t small enough, the local representative of $\varphi_{u,v}(t)$ in* (3.12) *has the form*

$$\exp_\mathcal{G}^{-1} \varphi_{u,v}(t) = t^2[u, v]_\mathfrak{g} + \omega(t), \tag{3.13}$$

where $\lim_{t\to 0} \frac{1}{t^2}\omega(t) = 0$.

DEFINITION 3.1.14. *The operation $b_\mathcal{G} : \mathfrak{g} \times \mathfrak{g} \to \mathfrak{g}$, given by*

$$b_\mathcal{G}(u, v) = [u, v]_\mathfrak{g}, \tag{3.14}$$

is called bracket operation *and the pair* $(\mathfrak{g},[u,v]_{\mathfrak{g}})$ *turns out to have the following properties*

- *linearity in the first variable:*

$$[au+bv,w]_{\mathfrak{g}} = a[u,w]_{\mathfrak{g}} + b[v,w]_{\mathfrak{g}}, \forall a,b \in \mathbb{R}, \forall u,v,w \in \mathfrak{g},$$

- *skew symmetry:*

$$[u,v]_{\mathfrak{g}} = [v,u]_{\mathfrak{g}}, \forall u,v, \in \mathfrak{g},$$

- *the Jacobi identity:*

$$[u,[v,w]]_{\mathfrak{g}} + [w,[u,v]]_{\mathfrak{g}} + [v,[w,u]]_{\mathfrak{g}} = 0, \forall u,v,w \in \mathfrak{g},$$

EXAMPLE 3.1.12. *A straightforward computation starting from the definition 3.1.14 and equation* (3.10) *shows that the Lie bracket of the Lie algebra* $\mathfrak{gl}(m,\mathbb{R})$ *of the general linear group* $GL(m,\mathbb{R})$ *is given by the commutator of the two matrices:*

$$[u,v] = uv - vu. \tag{3.15}$$

The Lie algebra $\mathfrak{k}$ *of a Lie subgroup K of* $GL(m,\mathbb{R})$*, is a linear subspace of* $\mathfrak{gl}(m,\mathbb{R})$ *and its Lie bracket operation is also given by* (3.15)*. It can be also obtained as the value the Lie bracket of the vector fields generated by the left translation from the two tangent vectors u and v. Finally, since any (finite dimensional) connected Lie group can be represented as a group of matrices, the Lie algebra operations are essentially all derived from* (3.15)*.*

For the rest of the chapter, we will assume that a differentiable function has a continuous differential.

DEFINITION 3.1.15. *A subset* $\mathcal{M}$ *of the Euclidean space* $\mathbb{R}^N, d_0$ *is a submanifold, if any point* $p_0 \in \mathcal{M}$*, there is a* $\mathbb{R}^C$*-valued differentiable function F defined on an open neighborhood* $U \subseteq \mathbb{R}^N$ *of* p_0*, such that* $M \cap U = F^{-1}(0)$ *and* $\forall p \in U \subseteq \mathbb{R}^N$ *the rank of the differential* dF_p *is C.*

Note that while often times the function F in definition 3.1.15 is globally defined on a neighborhood U of $\mathcal{M}$ (see, for instance, Example 3.1.9), in general, the function F is defined only locally.

Many manifolds in statistics do not arise as submanifolds of an Euclidean space though. They are *spaces of orbits* of *Lie group actions.*

Assume $\mathcal{K}$ is a Lie group with unit $1_{\mathcal{K}}$. Consider a function $\alpha : \mathcal{K} \times \mathcal{M} \to \mathcal{M}$, and for each element $k \in \mathcal{K}$, define the function $\alpha_k : \mathcal{M} \to \mathcal{M}$, given by

$$\alpha_k(p) = k \cdot p = \alpha((k,p)). \tag{3.16}$$

DEFINITION 3.1.16. α *is a differentiable left action on* $\mathcal{M}$ *if*

$$\begin{aligned} &\alpha \text{ is a differentiable function}, \\ &k \cdot (h \cdot p)) = (k \cdot h) \cdot p, \forall k,h \in \mathcal{G}, \forall p \in \mathcal{M}, \\ &1_{\mathcal{K}} \cdot p = p, \forall p \in \mathcal{M}. \end{aligned} \tag{3.17}$$

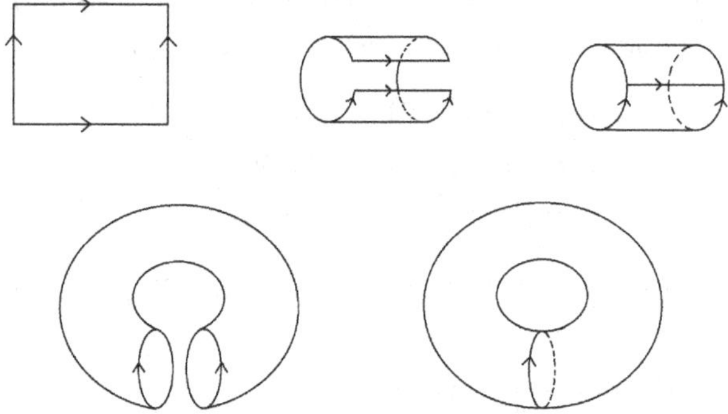

Figure 3.2: *A diffeomorphism between a flat torus and a torus in 3D.*

A simple example of group action is the *adjoint action* of $GL(m,\mathbb{R})$ on $\mathcal{M}_m(\mathbb{R})$, set of $m \times m$ matrices:

$$T \cdot A = TAT^{-1}. \tag{3.18}$$

DEFINITION 3.1.17. *If $\mathcal{K}$ acts on $\mathcal{M}$, the* orbit *of a point $p \in \mathcal{M}$ is the set $\mathcal{K}(p) = \{k \cdot p, g \in \mathcal{K}\}$ and the* isotropy group $\mathcal{K}_p = \{k \in \mathcal{K}, k \cdot p = p\}$.

The or *quotient space* of the group action (6.20) is the set of orbits

$$\mathcal{M}/\mathcal{G} = \{\mathcal{G}(p) | p \in \mathcal{M}\}. \tag{3.19}$$

EXAMPLE 3.1.13. *The action of the additive group $\mathbb{Z}^p$ on $\mathbb{R}^p$ is given by $\alpha(k,x) = k + x$. The quotient space $\mathbb{T}^p = \mathbb{R}^p/Z^p$ is called the p-dimensional flat torus. Note that $\mathbb{R}^2/Z^2$ can be identified with $S^1 \times S^1$, a surface in $\mathbb{R}^4$. This surface can be also obtained by "gluing" the opposite sides of the square $[0,1]^2$, via the identifications $(x,0) \sim (x,1)$ and $(0,y) \sim (1,y)$. Thus flat torus $S^1 \times S^1$ is diffeomorphic to a torus of revolution (think of the surface of a bagel), as shown in a number of steps in Figure 3.2.*

In another example, if we consider the action of the multiplicative group $\mathbb{R}^* = \mathbb{R}\backslash\{0\}$ on $\mathbb{R}^{m+1}$, given by scalar multiplication

$$\alpha(\lambda, x) = \lambda x \tag{3.20}$$

the quotient space is the m-dimensional *real projective space* $\mathbb{R}P^m$, set of all lines in $\mathbb{R}^{m+1}$ going through $0 \in \mathbb{R}^{m+1}$.

EXAMPLE 3.1.14. *To show that $\mathbb{R}P^2$ is a manifold, note that $\mathbb{R}P^2$ is a metric space with the distance between two lines in $\mathbb{R}^3$ given by the measure of their* *acute angle. If* $u \in \mathbb{R}^3 \setminus \{0\}, u = (u^1, u^2, u^3)$, *we label its orbit* $\mathbb{R}^*(u)$ *by* $[u] = [u^1 : u^2 : u^3]$. *Consider the affine open subsets* A_1, A_2, A_3 *given by*

$$\begin{aligned} A_1 &= \{[1 : x^2 : x^3], x^2, x^3 \in \mathbb{R}\}, \\ A_2 &= \{[x^1 : 1 : x^3], x^1, x^3 \in \mathbb{R}\}, \\ A_3 &= \{[x^1 : x^2 : 1], x^1, x^2 \in \mathbb{R}\}, \end{aligned} \tag{3.21}$$

and the charts $\{(A_i, x_i), i = 1,2,3\}$, *given by* $x_1 : A_1 \to \mathbb{R}^2, x_1([u]) = (\frac{u^2}{u^1}, \frac{u^3}{u^1}) = (x_1^2, x_1^3)$,
$x_2 : A_2 \to \mathbb{R}^2, x_2([u]) = (\frac{u^1}{u^2}, \frac{u^3}{u^2}) = (x_2^1, x_2^3)$,
and $x_3 : A_3 \to \mathbb{R}^2, x_3([u]) = (\frac{u^1}{u^3}, \frac{u^2}{u^3}) = (x_3^1, x_3^2)$,
It is obvious that $A_1 \cup A_2 \cup A_3 = \mathbb{R}P^2$.
Note that $A_1 \cap A_2 = \{[u] \in \mathbb{R}P^2, u^1 \neq 0, u^2 \neq 0$ *and if* $[u] \in A_1 \cap A_2$ *and* $x_1([u]) = (x_1^2, x_1^3)$, *and* $x_2([u]) = (x_2^1, x_2^3)$, *then since* $x_1^2 = \frac{u^2}{u^1}, x_1^3 = \frac{u^3}{u^1}, x_2^1 = \frac{u^1}{u^2}, x_2^3 = \frac{u^3}{u^2}$ *it follows that* $x_2^1 = \frac{1}{x_1^2}, x_2^3 = \frac{x_1^3}{x_1^2}$, *showing that* $x_2^{-1} \circ x_1$ *is analytic. The reader may check that* $x_3^{-1} \circ x_1, x_3^{-1} \circ x_2$ *and their inverses are also analytic functions, showing that* $\mathbb{R}P^2$ *is an analytic manifold. Similarly one can show that* $\forall m \in \mathbb{N}^*, \mathbb{R}P^m$ *is an m-dimensional analytic manifold.*

Similarly, if we consider the action of the multiplicative group $\mathbb{C}^* = \mathbb{C} \setminus \{0\}$ on $\mathbb{C}^{m+1} \setminus \{0\}$, given by scalar multiplication

$$\alpha(\lambda, z) = \lambda z \tag{3.22}$$

the quotient space is the m-dimensional *complex projective space* $\mathbb{C}P^m$, set of all complex lines in $\mathbb{C}^{m+1}$ going through $0 \in \mathbb{C}^{m+1}$. One can show that $\mathbb{C}P^m$ is a $2m$ dimensional analytic manifold, using similar transition maps to those in the case of $\mathbb{R}P^m$.
An m dimensional vector subspace of a real vector space V will be called *m–plane* in V. The *Grassmannian* $G_m(V)$ is the set of $m-planes$ in V. In particular $G_1(\mathbb{R}^m)$ is the $(m-1)$ dimensional real projective space, thus data analysis on real Grassmannians is an extension of axial data analysis.

PROPOSITION 3.1.2. *The Grassmannian* $G_m(\mathbb{R}^d)$ *is a* $m(d-m)$ *- dimensional manifold.*

Indeed, assuming that W is an m-dimensional linear subspace of $\mathbb{R}^d$, we decompose $\mathbb{R}^d = W \oplus W^\perp$. If $\tilde{W}$ is another such subspace, that is close to W, then $\tilde{W}$ is the graph of a linear map $T : W \to W^\perp$ and if we select two bases in W and $W^\perp$ such a linear transformation is uniquely determined by the matrix C associated with this linear map with respect to these two bases and we consider the local chart x_W defined by

$$x_W(\tilde{W}) = C, \tag{3.23}$$

taking values in $\mathcal{M}(m,d-m,\mathbb{R}) \sim \mathbb{R}^{m(d-m)}$. The differentiability of the transition maps is left to the reader (see Exercise 53).

Consider the set $\mathcal{R}^{(0)}_{m,k}(\subset \mathcal{R}_{m,k})$ of all $m\times k$ matrices S of full rank k and the left action α of the group of non singular $k\times k$ matrices on $\mathcal{R}^{(0)}_{m,k}$, given by $\alpha(X,L) = XL$, where $\mathrm{L} \in \mathcal{R}^{(0)}_{m,k}$. This group action defines an equivalence relation in $\mathcal{R}^{(0)}_{m,k}$. Two elements in $\mathcal{R}^{(0)}_{m,k}$ are considered equivalent if the column vectors of the two matrices span the same k - plane. Hence the equivalence classes of $\mathcal{R}^{(0)}_{m,k}$ are in one to one correspondence with the points of the Grassmannian $G_{k,m-k}$, and $G_k(\mathbb{R}^m)$ can be thought of as the quotient space

$$\mathcal{R}^{(0)}_{m,k}/\mathcal{R}^{(0)}_{k,k}.$$

Furthermore, the Grassmannian can be thought of as the quotient space $\mathcal{O}(m)/\mathcal{O}(k)\times\mathcal{O}(m-k)$ (see Exercise 54).

In general, a differentiable action of a Lie group $\mathcal{K}$ on a manifold $\mathcal{M}$ is a free action, if for all points $p\in\mathcal{M}$, the isotropy group $\mathcal{K}_p$ is the trivial group $\{1_{\mathcal{G}}\}$ and the quotient space $\mathcal{M}/\mathcal{G}$ is a $dim\mathcal{M}-dim\mathcal{G}$ dimensional manifold. If $\mathcal{H}$ is a *Lie subgroup* of a Lie group $\mathcal{K}$, the *left coset* of $k\in\mathcal{K}$ is $k\mathcal{H} = \{kh, h\in\mathcal{H}\}$, and the *quotient space* $\mathcal{K}/\mathcal{H}$ is the set $\mathcal{K}/\mathcal{H} = \{k\mathcal{H}, k\in\mathcal{K}\}$.

Let $Sym(m+1,\mathbb{R})$ be the space of $(m+1)\times(m+1)$ symmetric matrices.

EXAMPLE 3.1.15. *The Veronese-Whitney map* $j:\mathbb{R}P^m\to Sym(m+1,\mathbb{R})$, *given by*

$$j([x]) = xx^T, x^Tx = 1 \tag{3.24}$$

is an embedding of the real projective space in a Euclidean space of symmetric matrices.

DEFINITION 3.1.18. *Assume the Lie group $\mathcal{G}$ acts on both $\mathcal{M}$ and on $\mathbb{R}^N$. A $\mathcal{G}$-*equivariant *embedding j of $\mathcal{M}$ in $\mathbb{R}^N$ is an embedding j that is compatible with the two group actions, that is*

$$j(g\cdot p) = g\cdot j(p). \tag{3.25}$$

A manifold $\mathcal{M}$ is said to be a *homogeneous space* if there is a Lie group $\mathcal{K}$ acting transitively on $\mathcal{M}$. If $\mathcal{M}$ is a homogeneous space, and the transitive group $\mathcal{K}$ is known, a $\mathcal{K}$-equivariant embedding j of $\mathcal{M}$ is simply called *equivariant embedding*.

The *cotangent space* $T^*\mathcal{M}$ of a manifold $\mathcal{M}$ is the disjoint union of all cotangent spaces to $\mathcal{M}$ at different points of $\mathcal{M}$, $T\mathcal{M} = \sqcup_{p\in\mathcal{M}}T^*_p\mathcal{M}$. The *cotangent bundle* is the triple $(T^*\mathcal{M},\Pi^*,\mathcal{M})$, where the $\Pi^*: T^*\mathcal{M},\Pi,\mathcal{M}$ associates to each cotangent vector its base point, $\Pi(\omega_p) = p$.

The cotangent space $T^*\mathcal{M}$ has a natural structure of $2m$ dimensional manifold, for which the map Π^* is differentiable. A *Pfaff form* ω is a differentiable *section* of Π^*, that is $\omega:\mathcal{M}\to T^*\mathcal{M}$, is a differentiable function for which,

$\Pi^* \circ \omega = Id_{\mathcal{M}}$. An equivalent definition is that a Pfaff form on $\mathcal{M}$ is *an assignment* to every point of $\mathcal{M}$ of a linear function from $T_{\mathcal{M}}$ to $\mathcal{M}$ at that point, that is differentiable as a function of that point. That is, for each $p \in \mathcal{M}$, we have a cotangent vector $\omega(p) \in T_p^* \mathcal{M}$.

3.1.4 Hilbert Manifolds

In this section we assume that $\mathbb{H}$ is a complete separable over the reals. Any such space is isometric with the spacel_2 of sequences $x = (x_n)_{n\in\mathbb{N}}$ of real numbers for which the series $\sum_{n=0}^{\infty} x_n^2$ is convergent, with the scalar product $< x, y >= \sum_{n=0}^{\infty} x_n y_n$. A Hilbert space with the norm $||v|| =< v, v >$, induced by the scalar product becomes a Banach space. One can show that if f is differentiable at $x \in \mathbb{B}$, then L is unique. L is the differential of f at x, and it is labeled $d_x f$.

DEFINITION 3.1.19. *A function f defined on an open set U of a Banach space $(\mathbb{B}, ||\cdot||), f : U \to \mathbb{B}$, is differentiable at a point $x \in U$, if there is a bounded linear operator $L : \mathbb{B} \to \mathbb{B}$, such that if we set*

$$\omega_x(h) = f(x+h) - f(x) - L(h), \tag{3.26}$$

then

$$\lim_{h\to 0} \frac{||\omega_x(h)||}{||h||} = 0. \tag{3.27}$$

DEFINITION 3.1.20. *A chart on a separable metric space $\mathcal{M}, \rho$ is a one to one homeomorphism $\varphi_U : U \to \varphi(U)$ defined on a open subset of $\mathcal{M}$ to a Hilbert space $\mathbb{H}$. A is a separable metric space $\mathcal{M}$, that admits an open covering by domain of charts, such that the transition maps $\varphi_V \circ \varphi_U^{-1} : \varphi_V(U \cap V) \to \varphi_V(U \cap V)$ are differentiable.*

EXAMPLE 3.1.16. *The $P(\mathbb{H})$ of a Hilbert space $\mathbb{H}$, space of all one dimensional linear subspaces of $\mathbb{H}$, has a natural structure of Hilbert manifold modeled over $\mathbb{H}$. (see Exercise 31).*

3.2 Riemannian Structures, Curvature, Geodesics

Assume V is a vector space over $\mathbb{R}$ and let $SP(V)$ be the set of *all scalar products on V*. Recall that a scalar product on V is a symmetric bilinear form $B : V \times V \to \mathbb{R}$, such that $B(v, v) > 0, \forall v \in V, v \neq 0_V$.

If $\mathcal{M}$ is an m-dimensional manifold, let $SP\mathcal{M} = \bigsqcup_{u\in M} P(T_p\mathcal{M})$. One can show that $SP\mathcal{M}$ is an $(m + \frac{m(m+1)}{2})$ dimensional manifold, and the projection $\Pi : SP\mathcal{M} \to \mathcal{M}, \Pi(B_p) = p$ is differentiable. A *Riemannian structure* is a *differentiable section* of the projection Π, that is a differentiable function $g : \mathcal{M} \to SP\mathcal{M}$ such that $\Pi \circ g = Id_{\mathcal{M}}$, which is same as saying that at each point $p \in \mathcal{M}$ is given a bilinear form $g(p) : T_p\mathcal{M} \times T_p\mathcal{M} \to \mathbb{R}$ that is symmetric and positive definite, and $g(p)$ depends smoothly on $p \in \mathcal{M}$. A manifold $\mathcal{M}$ together with a Riemannian structure g is called a *Riemannian manifold.*

EXAMPLE 3.2.1. *(Spivak (1976) [316]) Using a partition of the unity, one can show that any paracompact manifold has a Riemannian structure.*

EXAMPLE 3.2.2. *A submanifold of the Euclidean space has a natural* Riemannian structure induced *by the Euclidean scalar product*.

Let $x = (x^1, \ldots, x^m)$ be a *system of local coordinates* associated with a chart φ defined on $U \subset \mathcal{M}$, and let $\frac{\partial}{\partial x^i}$, $i = 1, \ldots, m$ be the corresponding local frame of vector fields. Give the local coordinates system $\varphi : U \to \mathbb{R}^m$, one associates the matrix function ${}_\varphi g : \varphi(U) \to Sym^+(m, \mathbb{R}), {}_\varphi g = (g_{ij})_{i,j=1,\ldots m}$ with the Riemannian metric g, where

$$
{}_\varphi g(\varphi(p))_{ij} = g\left(\frac{\partial}{\partial x^i}|_p, \frac{\partial}{\partial x^j}|_p\right) \tag{3.28}
$$

Note that in equation (3.28), the metric g is an abbreviation for g_U, the induced metric on the U, regarded as an open submanifold of $\mathcal{M}$.

Conversely, if for a given atlas $\mathcal{A}$ on $\mathcal{M}$ we define the matrices

$$
({}_\varphi g_{ij})_{i,j=1,\ldots,m} : \varphi(U) \to Sym^+(m, \mathbb{R}) \tag{3.29}
$$

with the property that for any two of charts $(U, x_U), (V, y_V)$ and for any point $p \in U \cap V$, with $x = x_U(p), y = y_V(p)$,

$$
{}_{x_U} g_{ij}(x) = \sum_{a,b=1}^{m} \frac{\partial y^a}{\partial x^i} \frac{\partial y^b}{\partial x^j} \, {}_{x_V} g_{ab}(y), \forall i, j \in \overline{1, m} \tag{3.30}
$$

then the functions (3.29) define a Riemannian structure on $\mathcal{M}$.

Usually the subscript x_U in the notation ${}_{x_U} g$ is dropped, as the chart x_U is assumed to be known.

In the case of a submanifold $\mathcal{M} \subset \mathbb{R}^N$, the induced Riemannian structure g is given in terms with a *local parametrization* ϕ (inverse of the chart (U, x)) on $\mathcal{M}$, by the *first fundamental form*

$$
g_{ij} = \frac{\partial \phi}{\partial x^i} \cdot \frac{\partial \phi}{\partial x^j}. \tag{3.31}
$$

For example if we use spherical coordinates for a point on $\mathbb{S}^2$,

$$
x = \cos r \cos \theta, y = \cos r \sin \theta, z = \sin r \tag{3.32}
$$

the induced Riemannian structure on the manifold is given locally by the matrices

$$
g(r, \theta) = \begin{pmatrix} 1 & 0 \\ 0 & cos^2 r. \end{pmatrix} \tag{3.33}
$$

The classical local (covariant-contravariant) notation for the Riemannian structure (infinitesimal point varying squared Riemannian norm) is ds^2. We have

$$ds^2 = \sum_{i,j=1}^{m} g_{ij}(x)dx^i dx^j \tag{3.34}$$

Recall that a tensor is *covariant* in the upper indices and *contravariant* in the lower ones. This representation of ds^2 is that of a Riemannian structure, since the coefficients $g_{ij}(x)$ are not constant, depending also on x. For example, if we use spherical coordinates (3.32) for a point the induced Riemannian structure on the sphere is

$$(r,\theta) \to g_{r,\theta} = dr^2 + cos^2 r d\theta^2 \tag{3.35}$$

However, for different parameterizations, a locally flat Riemannian structure may also have nonconstant coefficients $g_{ij}(x)$.
For example the flat Riemannian structure given by the dot product on $\mathbb{R}^2$, in Cartesian coordinates

$$ds^2 = dx^2 + dy^2 \tag{3.36}$$

and in polar coordinates $r, \theta, x = r\cos\theta, y = r\sin\theta$, the same Riemannian structure has the form

$$ds^2 = dr^2 + r^2 d\theta^2. \tag{3.37}$$

The key invariant of a Riemannian manifold $(\mathcal{M}, g)$ was introduced by Riemann in his celebrated Habilitation thesis delivered in a lecture at the Georg-August University of Göttingen in 1854, as a measure for the deviation of $(\mathcal{M}, g)$ from being *flat* (locally isometric to an Euclidean space). See Spivak (1979) [316]. It is the *curvature*. In case of a surface (dim $\mathcal{M} = 2$), this can be reduced to a scalar function on $\mathcal{M}$, called the curvature function of $\mathcal{M}$, that coincides with the *Gaussian curvature* (see Gauss (1828) [123]) in case $\mathcal{M}$ is a submanifold of $\mathbb{R}^3$. This function is zero if and only if the Riemannian manifold is locally isometric to a Euclidean plane (is flat). In particular, the curvature function of a non-flat surface is constant only in the cases of famous *non-Euclidean* geometries: the *elliptic geometry* (of constant positive curvature), due to Riemann (essentially a real projective plane), and the *hyperbolic geometry* (curvature is constant negative), due to Bolyay, Lobachevsky and, possibly, to Gauss.

An intuitive approach to curvature is via *moving frames* (see Cartan (1946) [59]). An *orthoframe field* on a Riemanian manifold $\mathcal{M}$ of dim m is an ordered set of m vector fields on $\mathcal{M}$, which are orthonormal w.r.t. the Riemannian structure g on $\mathcal{M}$. Locally such an orthoframe field is given by a function,

$$\begin{aligned} &x \to (v_1(x), \ldots, v_m(x)), \text{where} \\ &g(v_i(x), v_j(x)) = \delta_{ij}, \forall i \leq j \end{aligned} \tag{3.38}$$

Cartan used instead *orthocoframe* fields. Such a coframe field is $(\omega^1, \ldots, \omega^m)$, where

$$g = (\omega^1)^2 + \cdots + (\omega^m)^2. \tag{3.39}$$

Cartan approached the geometry of manifolds using *differential forms* rather than vector fields.

Recall that if V is a real vector space, its dual space is $V^* = L(V, \mathbb{R})$. Assume $e_1, \ldots, e_m$ basis of V. Then the dual basis of V^* is $e^* = (e_1^*, \ldots, e_m^*)$ with

$$e_i^*(e_j) = \delta_j^i, \forall i, j = 1, \ldots, m. \tag{3.40}$$

Often the notation used is $e_i^* = e^i$.

Any co-vector $\theta \in V^*$, (linear form) can be written as $\theta = \sum_1^m a_i e_i^*$

DEFINITION 3.2.1. *Assume $p \in \mathcal{M}$ where $\mathcal{M}$ is a manifold. The cotangent space at the point p, $T_p^*\mathcal{M}$, is the dual of the tangent space $T_p\mathcal{M}$, $T_p^*\mathcal{M} = L(T_p\mathcal{M}, \mathbb{R})$, space of linear forms on $T_p\mathcal{M}$.*

The canonical basis of $T_p\mathcal{M}$ associated with a chart (U, X) around $p \in \mathcal{M}$ is $(\frac{\partial}{\partial x^i}|_p, i = 1, \ldots, m)$, therefore with the notation above $(e_i)_{i=1,\ldots,m} = (\frac{\partial}{\partial x^i})_{i=1,\ldots,m}$ Its dual basis e^i is labeled $e^i = dx^i|_p$. By definition $dx^i|_p(\frac{\partial}{\partial x^j}|_p) = \delta_j^i$. Another interpretation of dx^i is that of a differential of the function $x^i : U \to \mathbb{R}$.

A cotangent vector at the point $u \in \mathcal{M}$ can be decomposed w.r.t. the basis $(dx^i|_p)_{i=1,\ldots,m}$ in $T_p^*\mathcal{M}$ as $\theta_p = \sum_{i=1}^m a_i dx^i|_u$

The *cotangent space* $T^*\mathcal{M}$ of a manifold is the disjoint union of the cotangent spaces at various points of that manifold, $T^*\mathcal{M} = \bigsqcup_{u \in \mathcal{M}} T_p^*\mathcal{M}$.
$T^*\mathcal{M}$ is has a structure of manifold of dimension $2m$.

The projection $\pi^* : T\mathcal{M} \to \mathcal{M}, \pi^*(\theta) = p$ if $\theta \in T_p^*\mathcal{M}$ is differentiable as a function between manifolds.

A *1 - differential form* is a differentiable *section* $\theta : \mathcal{M} \to T^*\mathcal{M}$ of the projection Π^*, i.e. $\Pi^* \circ \theta = 1_{\mathcal{M}}$ $\theta(p) \in T_p^*\mathcal{M}$

The local representative is ${}_x\theta = \sum_{i=1}^m a_i(x) dx^i$, where $a_i(x), x = x(p)$ are differentiable functions.

With *Einstein's summation convention*, a convention that will be used for the rest of this chapter, $\theta = a_i dx^i$.

A *2-skew symmetric form*, is a bilinear form on a vector space V, with the property $b(v, w) = -b(w, v), \forall v, w \in V$.

The set $\bigwedge^2 V^*$ of all skew symmetric bilinear forms on a vector space V of dimension m has a vector space structure and $dim_{\mathbb{R}} \bigwedge^2 V^* = \frac{1}{2}m(m-1)$. In terms of the dual basis $(e^i, i = 1, \ldots, m)$ of a given basis $e_i, i = 1, \ldots, m)$ of V, a basis of $\bigwedge^2 V^*$ is $(e^i \wedge e^j, 1 \le i < j \le m)$ where

$$(e^i \wedge e^j)(e_k, e_l) = \delta_k^i \delta_l^j - \delta_l^i \delta_k^j, \forall l, k = 1, \ldots, m. \tag{3.41}$$

The space of *skew symmetric tensors* on $\mathcal{M}$ is $\bigwedge^2 T^*\mathcal{M} = \bigsqcup_{p \in \mathcal{M}} \bigwedge^2 (T_u^*\mathcal{M})$.

This space has a structure of differentiable manifold of dimension $\frac{1}{2}m(m+1)$. The projection $\bigwedge^2\pi : \bigwedge^2 T^*\mathcal{M} \to \mathcal{M}, \bigwedge^2\pi(b) = p$ if $b \in \bigwedge^2(T_p^*\mathcal{M})$ is differentiable, as a function between manifolds, and a section of this projection is called a *2-differential form.* If (U,x) is a chart around $p \in \mathcal{M}$, a basis for $\bigwedge^2(T_p\mathcal{M})$ is given by $(dx^i|_p \wedge dx^j|_p, 1 \leq i < j \leq m)$, therefore locally any 2-differential form ω on $\mathcal{M}$ can be written w.r.t. a chart (U,φ) as

$$\omega = \sum_{i<j} a_{ij}(x)dx^i \wedge dx^j = \frac{1}{2}\sum a_{ij}(x)dx^i \wedge dx^j, \tag{3.42}$$

where $a_{ij} = -a_{ji}$ are differentiable functions on $\varphi(U)$.

The set of all 2-differentiable forms on $\mathcal{M}$ is $\bigwedge^2\mathcal{M}$, and the set of all 1-differentiable forms on $\mathcal{M}$ is $\bigwedge^1\mathcal{M}$.

The *differential operator* $d : \bigwedge^1\mathcal{M} \to \bigwedge^2\mathcal{M}$ is defined by

$$d(\sum a_i dx^i) = \frac{1}{2}\sum_{i\neq j}(\frac{\partial a_i}{\partial x^j} - \frac{\partial a_j}{\partial x^i})dx^i \wedge dx^j \tag{3.43}$$

Note that

$$d(df) = 0, \tag{3.44}$$

for all differentiable functions $f : \mathcal{M} \to \mathbb{R}$, and by *Poincaré's lemma* and closed 1-differential form θ (i.e. with $d\theta = 0$) is locally *exact* (i.e. $\theta = df$.)

Assume the differential forms ω^j satisfy (3.39). Cartan introduced the *Levi–Civita connection forms* $(\omega^i_j)_{i,j=1,\ldots,m}$, defined by the following properties:

$$\omega^i_j + \omega^j_i = 0, \forall i,j \leq m,$$
$$d\omega^i + \sum_{j=1}^{m} \omega^i_j \wedge \omega_i = 0, \tag{3.45}$$

The *Cartan–curvature forms* of a Riemannian manifold are given by the matrix $\Omega = (\Omega^i_j)_{i,j=1,\ldots,m}$, having as entries differential forms on $\mathcal{M}$, as follows: $\Omega^i_j \in \bigwedge^2\mathcal{M}$

$$\Omega^i_j = d\omega^i_j + \sum_{k=1}^{m} \omega^i_k \wedge \omega^k_j$$
$$\Omega^i_j = \frac{1}{2}\Omega^i_{jlm}\omega^l \wedge \omega^m, \Omega^i_{jlm} + \Omega^i_{jml} = 0. \tag{3.46}$$

Note that since the connection forms satisfy to $\omega^j_i + \omega^i_j = 0$, the functions Ω^j_{iml}

satisfy to $\Omega^j_{iml} + \Omega^i_{jml} = 0$. The functions Ω^i_{jlm} define the *curvature* with respect to the given orthoframe field. In the case of a surface, if we set $\omega^1_2 = \omega$, equations (3.45) amount to

$$\begin{aligned} d\omega^1 &= -\omega \wedge \omega^2 \\ d\omega^2 &= \omega \wedge \omega^1 \end{aligned} \tag{3.47}$$

For a surface (2D manifold), its curvature function $k(x) = k = \Omega^1_{212}$ is given by the equation:

$$d\omega = k\omega^1 \wedge \omega^2, \tag{3.48}$$

Note that the wedge product is skew symmetric if x^1, x^2 are local coordinates on a surface, then

$$dx^2 \wedge dx^2 = 0, dx^1 \wedge dx^1 = 0, dx^1 \wedge dx^2 = -dx^2 \wedge dx^1 \neq 0. \tag{3.49}$$

Also, from definition (3.44), for any local coordinate x^i we have $ddx^j = 0$.

EXAMPLE 3.2.3. *We consider a surface of revolution, obtained by rotating a curve in a half plane, parameterized by arc length parameter u, around the boundary of this half plane. The location of a point $p(X,Y,X)$ on this surface can be parameterized as follows*

$$\begin{aligned} X &= x(u)\cos v \\ Y &= x(u)\sin v && (3.50) \\ Z &= y(u), && (3.51) \end{aligned}$$

where $x(u) > 0, y(u)$ are the coordinates on the curve, and $x'(u)^2 + y'(u)^2 = 1$. The nonzero components of the induced Riemannian metric are $g_{11}(u,u) = 1, g_{22}(u,v) = x^2(u)$, therefore, in local coordinates $g = du^2 + x^2(u)dv^2$. We may then select the orthocoframe $\omega^1 = du, \omega^2 = x(u)dv$, the Levi-Civita connection form has the expression

$$\omega = \lambda_1 du + \lambda_2 dv, \tag{3.52}$$

where $\lambda_i = \lambda_i(u,v)$. Then $d\omega^1 = ddu = 0 = -\omega \wedge (x(u)dv) = -(\lambda_1 du + \lambda_2 dv) \wedge (x(u)dv) = -\lambda_1 x(u) du \wedge dv$, and it follows that $\lambda_1 = 0$. From the second Levi-Civita connection equation we get

$$d\omega^2 = x'(u)du \wedge dv = \lambda_2 dv \wedge du = -\lambda_2 du \wedge dv, \tag{3.53}$$

therefore $\lambda_2 = -x'(u)$, and $\omega = -x'(u)dv$.

We may now compute the curvature function on the surface from equation (3.48)*. Using the differential of a product we get:*

$$d\omega = d(x'udv) = x''(u)du \wedge dv.$$

From the right hand side of (3.48) *we have*

$$k\omega^1 \wedge \omega^2 = kx(u)du \wedge dv.$$

From (3.48) *it follows that the curvature function is*

$$k(u,v) = -\frac{x''(u)}{x(u)}. \tag{3.54}$$

In particular, in a sphere of radius R in $\mathbb{R}^3$ can be regarded as a surface of revolution, obtained from planar curve with $x(u) = Rcos(\frac{u}{R})$, therefore the curvature is constant $k(u,v) = \frac{1}{R^2}$. From the example above, and from the following theorem it follows that a round sphere is not locally isometric to an Euclidean plane.

THEOREM 3.2.1. *(Riemann).* $(\mathcal{M}, g)$ *is locally Euclidean if and only if one of the following two sets of identities holds true (i)*$\Omega^i_{jkl} = 0, \forall i,j,k,l = 1,\ldots,N,$ *and, (ii)* $\Omega^i_j = 0, \forall i,j = 1,\ldots,N.$

A Riemannian structure on $\mathcal{M}$ is *not a* metric on $\mathcal{M}$. However, on a connected manifold $\mathcal{M}$, a Riemannian structure induces a on $\mathcal{M}$, given by

$$\rho_g(p,q) := \inf\left\{\int_0^1 \left[g\left(\frac{dc}{dt}, \frac{dc}{dt}\right)_{c(t)}\right]^{1/2} dt\right\}, \quad p,q \in \mathcal{M}, \tag{3.55}$$

where the infimum is taken over all rectifiable curves $c : [0,1] \to \mathcal{M}$ with $c(0) = p, c(1) = p$.

The length of a rectifiable curve $c : [a,b] \to \mathcal{M}$ is given by

$$L^b_a c = \int_a^b \sqrt{g(\frac{dc}{dt}, \frac{dc}{dt})}dt \tag{3.56}$$

and its *arc length* is the nondecreasing function $s : [a,b] \to \mathbb{R}^+$, given by

$$s(t) = \int_a^t \sqrt{g(\frac{dc}{du}, \frac{dc}{du})}du \tag{3.57}$$

If the curve c is injective, then the arc length is a one-to-one function, and can be used to re-parameterize it.

One can show that ρ_g given by

$$\rho_g(u_1,u_2) = \inf_{\gamma(\alpha)=u_1, \gamma(\beta)=u_2} L^\beta_\alpha \gamma \tag{3.58}$$

has the properties of a distance on $(\mathcal{M}, g)$; this is the *Riemannian distance*

(Chavel (1993, p. 19) [68]). It is also called the *geodesic distance* on $(\mathcal{M},g)$. Given two points p_1,p_2 on $\mathcal{M}$, we consider the space $\Omega(p_1,p_2)$, set of piecewise differentiable paths $c:[0,1]\to\mathcal{M}$. An important functional defined on $\Omega(p_1,p_2)$ is the the *(kinetic) energy functional* $E:\Omega(p_1,p_2)\to\mathbb{R}$, given by

$$E(\gamma)=\frac{1}{2}\int_0^1 g(\frac{d\gamma}{dt},\frac{d\gamma}{dt})dt. \tag{3.59}$$

DEFINITION 3.2.2. *A curve* $c\in\Omega(p_1,p_2)$ *is* geodesic *if it is a minimizer of the energy functional in* (3.59).

In the language of Physics, geodesics are trajectories of *particles* following the general *principle of least action*, where given a *Lagrangian (structure)* $\mathcal{L}: T\mathcal{M}\to\mathbb{R}^+$ on a manifold and its *associated action* $\mathcal{S}:\Omega(p_1,p_2)\to\mathbb{R}$:

$$\mathcal{S}(c)=\int_0^1 \mathcal{L}(\frac{dc}{dt})\mathrm{d}t. \tag{3.60}$$

one states that the free falling particles run along a curve c minimizing the action in (3.60).

Note that given any two points $p_1=c(s_1),p_2=c(s_2)$ on a geodesic c parameterized by its arc length s, since $\|\frac{dc}{ds}\|=1$,

$$(L_{s_1}^{s_2}c)^2=2E_{s_1}^{s_2}c=2(s_2-s_1)^2. \tag{3.61}$$

On the other, from the definition of the geodesic, and from the Cauchy–Schwartz inequality, it follows that for any piecewise differentiable curve γ, parameterized by arclength, we have

$$2E_{s_1}^{s_2}c\le 2E_{s_1}^{s_2}\gamma\le(L_{s_1}^{s_2}\gamma)^2. \tag{3.62}$$

From (3.61) and (3.62) it follows that a geodesic is locally a minimizer of the *length functional.*

To derive the ordinary differential equations of a geodesic in local coordinates, we consider a chart (U,φ) on $\mathcal{M}$. Note that if $c\in\Omega(p_1,p_2)$ is a geodesic, and $0<t_1<t_2<1$, are such that $c(t)\in U,\forall t,t_1\le t\le t_2$, then the restriction $c|[t_1,t_2]$ is a geodesic on U with the induced Riemannian structure $g|U$. Indeed for any rectifiable path $\lambda:[t_1,t_2]\to U,\lambda(t_a)=c(t_a),a=1,2$, one can define the path $\gamma:[0,1]\to\mathcal{M}$, that equals to λ on $[t_1,t_2]$ and with c outside $[t_1,t_2]$. Since the energy c is smaller than the energy of γ, it turns out that $E(c|[t_1,t_2])\le E(\lambda)$.

Let $x(t)=\varphi(c(t))$ be the local representative of the restriction of c to $[t_1,t_2]$, then $x(t)$ is a minimizer of the action

$$S(x)=\int_{t_1}^{t_2} L(x(t),\frac{dx}{dt})\mathrm{d}t, \tag{3.63}$$

where L, the local representative of the Lagrangian $\mathcal{L}$ with respect to the chart $(TU, T\varphi)$ in (3.4), (3.5) is given by

$$L(x^1,\ldots,x^m,y^1,\ldots,y^m) = \sum_{a,b=1}^{m} g_{ab}(x^1,\ldots,x^m)y^a y^b, \tag{3.64}$$

and $g_{ij} = \varphi g_{ij}$ are given in equation (3.28). The path $t \mapsto x(t), t_1 \leq t \leq t_2$ is a solution of the Euler-Lagrange equations

$$\frac{\partial L(x(t),x'(t))}{\partial x^a} - \frac{\mathrm{d}}{\mathrm{d}t}\frac{\partial L(x(t),x'(t))}{\partial y^a} = 0 \quad \text{for } a = 1,\ldots,m. \tag{3.65}$$

Let $(g^{ab})_{a,b=1,\ldots,m}$ be the inverse of $(g_{ab})_{a,b=1,\ldots,m}$. The *Christoffel symbols of the first kind* are defined by

$$[ab,c] = 1/2(\frac{\partial g_{ac}}{\partial x^b} + \frac{\partial g_{bc}}{\partial x^a} - \frac{\partial g_{ab}}{\partial x^c}). \tag{3.66}$$

and the *Christoffel symbols of the second kind* are given by

$$\Gamma_{ab}^c = g^{cr}[ab,r]. \tag{3.67}$$

In this notation, from the o.d.e. system (3.65), it is straightforward to obtain the following equivalent ODE system:

$$\frac{d^2x^k}{dt^2} + \Gamma_{ab}^k \frac{dx^a}{dt}\frac{dx^b}{dt} = 0, \forall k = 1,\ldots,m. \tag{3.68}$$

An equivalent definition of geodesics is via .

DEFINITION 3.2.3. *The* Levi-Civita connection *on a Riemannian manifold is an operator* $\nabla : \Gamma(T\mathcal{M}) \times \Gamma(T\mathcal{M}) \to \Gamma(T\mathcal{M})$, *with the properties*

$$\begin{aligned} \nabla(fX+Y,Z) &= f\nabla(X,Z) + \nabla(Y,Z) \\ \nabla(X,fY) &= f\nabla(X,Y) + X(f)Y \\ \nabla(X,Y) - \nabla(Y,X) &= [X,Y] \\ X(g(Y,Z)) &= g(\nabla(X,Y),Z) + g(Y,\nabla(X,Z)). \end{aligned} \tag{3.69}$$

The common notation for $\nabla(X,Y)$ *is* $\nabla_X Y$.

One can show that the Levi-Civita connection is unique and a geodesic on a Riemannian manifold is a curve that is autoparallel ($\nabla_{\frac{dc}{dt}(0)}\frac{dc}{dt}(0) = 0$). For curves on a submanifold $\mathcal{M}$ of $\mathbb{R}^N$, with the Riemannian structure induced by the scalar product in $\mathbb{R}^N$, we have the following characterization

THEOREM 3.2.2. *A curve* $c : (a,b) \to \mathcal{M}$ *is a geodesic on the if and only if* $\forall t \in (a,b), \frac{d^2c}{dt^2}(t) \perp T_{c(t)}\mathcal{M}$.

EXAMPLE 3.2.4. *A point on a geodesic on the round sphere, travels at constant speed on a great circle on the sphere. Indeed if* u, v *are vectors of norm one in* $\mathbb{R}^N, u^T v = 0$, *then* $c(t) = R(\cos \frac{t}{R} u + \sin \frac{t}{R} v)$, *then* $\forall t, \frac{d^2c}{dt^2}(t) \perp T_{c(t)}S^{N-1}$, *therefore c is the unique geodesic with* $c(0) = u, \frac{dc}{dt}(0) = v$.

Since a geodesic equation is an ODE system of second order, for any point $(u, v) \in T\mathcal{M}$, there is a unique geodesic $\gamma(t)$, defined on an interval $(-\varepsilon, \varepsilon)$ around $t = 0$ with $\gamma(0) = u, \frac{d\gamma}{dt}(0) = v$.

Since the equations (3.68) are homogeneous, if $t \to c(t)$ is a geodesic, then so is the path $t \to c(at)$, for any nonzero a. Therefore one can assume, by scaling the tangent vector at time $t = 0$ if necessary , that there is a neighborhood V of $0 \in T_p\mathcal{M}$ such that for any $v \in V$, the unique geodesic γ_v with $\gamma_v(0) = p, \dot{\gamma}(0) = v$ is defined at $t = 1$. The map $Exp_p : V \to \mathcal{M}$ given by

$$Exp_p v = \gamma_v(1) \tag{3.70}$$

If V is small enough, the *Riemannian exponential map* is a one-to-one from V to $Exp_p(V)$.

REMARK 3.2.1. *Positive curvature pulls geodesics closer, negative curvature sends geodesics apart, and zero curvature does not have any local effect on the angle made by two geodesics. See Figure 3.3.*

A Riemannian manifold is *geodesically complete*, if for any pair of its points p, q there is a minimal geodesic joining p and q. A useful result, given with a full proof in Chavel (2006, p. 26) [69] is the following

THEOREM 3.2.3. *(Hopf–Rinow) The following three statements are equivalent (i)* $(\mathcal{M}, \rho_g)$ *is geodesically complete, (ii)* Exp_p *is defined on the entire tangent space* $T_p\mathcal{M}$, *(iii)* $(\mathcal{M}, \rho_g)$ *is complete as a metric space.*

These three equivalent properties of completeness in theorem 3.2.3 are in turn equivalent to a third property: *all closed bounded subsets of* (M, ρ_g) *are compact* (Chavel (2006, p. 28) [69]).

If, moreover, $Exp_p : T_p\mathcal{M} \to \mathcal{M}$ is one to one, then the Riemannian manifold is diffeomorphic to $\mathbb{R}^m$.

Riemannian manifolds are curved, so that geodesics starting at a point p may meet for a second time in the cut locus of p. Technical details on cut locus and normal coordinates are as follows.

If the manifold is complete, its exponential map at p is defined on the entire tangent space $T_p\mathcal{M}$. An open set $U \subset \mathcal{M}$ is said to be a *normal neighborhood* of p $(\in U)$, if Exp_p is a diffeomorphism on a neighborhood V of the origin of $T_p\mathcal{M}$ onto U, with V such that $tv \in V$ for $0 \leq t \leq 1$, if $v \in V$. Suppose $U = Exp_p V$ is a normal neighborhood of p. Then $(x^1, x^2, ..., x^m)$ are said to be the *normal coordinates* of a point $q \in U$ w.r.t. a fixed orthobasis $(v_1, v_2, ..., v_m)$ of $T_p\mathcal{M}$ if

$$q = Exp_p(x^1 v_1 + x^2 v_2 + ... + x^d v_m). \tag{3.71}$$

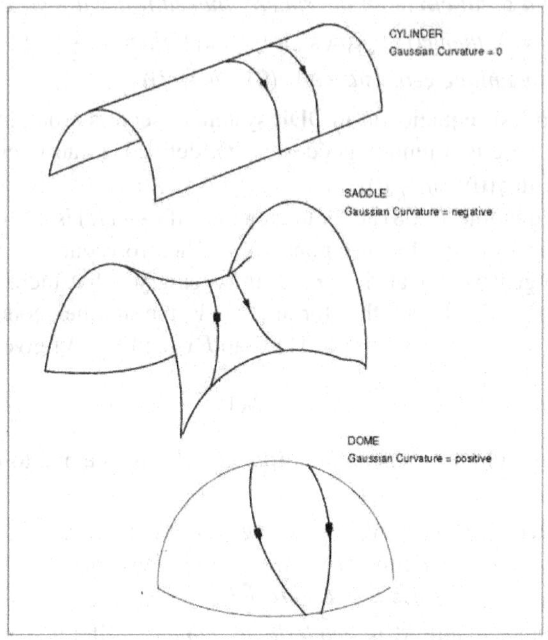

Figure 3.3: *Behavior of geodesics dictated by curvature.*

Let $v \in T_p\mathcal{M}$, be such that $g(v,v) = 1$. The set of numbers $s > 0$, such that the geodesic segment $\{Exp_p tv : 0 \leq t \leq s\}$ is minimizing is either $(0,\infty)$ or $(0,r(v)]$, where $r(v) > 0$. We will write $r(v) = \infty$ in the former case. If $r(v)$ is finite, then $Exp_q r(v)v$ is the *cut point of p in the direction v* (Kobayashi and Nomizu, (1996) [199], p.98). Let $S_p\mathcal{M} = \{v \in T_p\mathcal{M} : g_p(v,v) = 1\}$; then the largest open subset of $\mathcal{M}$ in which a normal coordinate system around p is defined is $Exp_p(V(p))$, where $V(p) = \{tv : 0 \leq t < r(v),\ v \in S_p\mathcal{M}\}$.

DEFINITION 3.2.4. *The cut locus of p is* $C(p) = Exp_p\{r(v)v : v \in S_p\mathcal{M},\ r(v)$ *finite*$\}$ *(Kobayashi and Nomizu (1996) [199], p. 100).*

Note that $C(p)$ has volume measure 0, and $\mathcal{M}$ is the disjoint union of $Exp_p(V(p))$ and $C(p)$. The *injectivity radius at the point p* is $r_p = \inf\{r(v) : v \in S_p\mathcal{M}\}$.

EXAMPLE 3.2.5. *For the m-dimensional sphere of radius one, with the Rie-*

mannian structure induced by the infinitesimal Euclidean metric on $\mathbb{R}^{m+1}$, *the exponential map is given by*

$$Exp_p(v) = \cos(\|v\|)p + \sin(\|v\|)\|v\|^{-1}v, \quad (v \in T_pS^m, v \neq 0). \tag{3.72}$$

Also, $V(p) = \{v \in T_pS^m : \|v\| < \pi\}$ and $C(p) = -p$.

We may now determine the exponential map when $\mathcal{M}$ is a real (complex) projective space $\mathbb{R}P^d$ ($\mathbb{C}P^{d/2}$ for d even) of constant (*constant holomorphic*) curvature. In this case $\mathcal{M}$ is a quotient of a round sphere S, and the projection map $\pi : S \to \mathcal{M}$ is a Riemannian submersion. If we denote by Exp the exponential map for both the sphere and projective space, we have $Exp_{\pi(p)}d\pi(v) = \pi(Exp_p(v))$. If $p \in S^d$, then since $\mathbb{R}P^d$ is a homogeneous space, for $[p] \in \mathbb{R}P^d$, we may assume without loss of generality that $p = (1,0,..,0)$. Then $C([p]) = \{[q] : q = (0, q^1, ..., q^d) \in S^d\} = \mathbb{R}P^{d-1}$ is the *projective hyperplane from infinity* of the point [p]. Similarly, we may assume that the point $[u] \in \mathbb{C}P^{d/2}$ is represented by $u = (1,0,..,0)$ and in this case $C([p])$ is $\mathbb{C}P^{d/2-1}$ the *complex projective hyperplane at infinity* of the point $[u]$.

The *sectional curvatures* associated with a orthocoframe $(\omega^i), i = \overline{1,m}$ are the functions $\Omega^i_{jij}, 1 \le i < j \le m$, defined in (3.46). A Riemannian manifold has *nonpositive curvature* if all the sectional curvatures are negative or zero.

For following result, see Helgason (1978) [147], p. 74.

THEOREM 3.2.4. *(Hadamard-Cartan) If* $(\mathcal{M}, g)$ *is a simply connected Riemannian manifold of nonpositive sectional curvature then for any point* $p \in \mathcal{M}$, *the cut locus* $C(p) = \emptyset$ *and* $Exp_p : T_p\mathcal{M} \to \mathcal{M}$ *is a distance expanding diffeomorphism.*

From theorem 3.2.4, we arrive to the following

DEFINITION 3.2.5. *A complete Riemannian manifold of nonpositive sectional curvature is called Hadamard manifold, or Cartan-Hadamard manifold*

An isometry τ of $(\mathcal{M}, g)$ is a diffeomorphism of $\mathcal{M}$ such that

$$g(\tau(q))(d\tau_q(v), d\tau_q(v)) = g(q)(v, v)$$

for all $v \in T_q\mathcal{M}$.

DEFINITION 3.2.6. *Isometries of a Riemannian manifold* $(\mathcal{M}, g)$ *form a group* $I_g\mathcal{M}$. *The isotropy group of the action of* $I_g\mathcal{M}$ *on* $\mathcal{M}$ *will be labeled* $H_p = I_g\mathcal{M}_p$. *If there is a point* $p \in \mathcal{M}$ *such that the orbit* $I_g\mathcal{M}(p) = \mathcal{M}$, *the manifold* $(\mathcal{M}, g)$ *is a Riemannian homogeneous space.*

Riemannian homogeneous spaces can be also described algebraically as follows (see Kobayashi and Nomizu (1969)[199], ch. 14). A Riemannian structure g on $\mathcal{M}$ is called $\mathcal{K}$*-invariant* if $\tilde{\alpha}_k$ is an isometry for all $k \in \mathcal{K}$; i.e. if

$$g_{\tilde{\alpha}_k(q)}(d(\tilde{\alpha}_k)_q(u), d(\tilde{\alpha}_k)_q(v)) = g_q(u, v), \ \forall\, q \in \mathcal{M}, \ u, v \in T_q\mathcal{M}. \tag{3.73}$$

(In other words, for all k,q as above, $d(\tilde{\alpha}_k)_q$ is a linear isometry from the inner-product space $(T_q\mathcal{M}, g_q)$ to the inner-product space $(T_{\tilde{\alpha}_k(q)}\mathcal{M}, g_{\tilde{\alpha}_k(q)})$.) A *Riemannian homogeneous space* for a Lie group $\mathcal{K}$ is a Riemannian manifold $(\mathcal{M}, g)$, where $\mathcal{M}$ is a homogeneous space for $\mathcal{K}$, and the Riemannian structure g is $\mathcal{K}$-invariant. The following is a standard result from the theory of Riemannian homogeneous spaces.

LEMMA 3.2.1. *Let $\tilde{\alpha}$ be a smooth, transitive action of a Lie group $\mathcal{K}$ on a manifold $\mathcal{M}$, let $q_0 \in \mathcal{M}$, and let $\mathcal{H}$ be the isotropy group at q_0. A scalar product g_{q_0} on $T_{q_0}\mathcal{M}$ can be extended to a $\mathcal{K}$-invariant Riemannian metric on $\mathcal{M}$ if and only if the inner product g_{q_0} is $\mathcal{H}$-invariant, in which case there is a unique $\mathcal{K}$-invariant extension.*

Partial proof. Necessity is obvious. To prove sufficiency, assume that for each $q \in \mathcal{M}$, choose an arbitrary $k_1 = k_1(q) \in \mathcal{C}_q$ (so $\tilde{\alpha}_{k_1}(q_0) = q$) and define an inner product g_q on $T_q\mathcal{M}$ by $g_q(u,v) = g_{q_0}(((d\tilde{\alpha}_{k_1})_{q_0})^{-1}(u), ((d\tilde{\alpha}_{k_1})_{q_0})^{-1}(v))$. Then clearly (3.73) is satisfied with $q = q_0$ and $k = k_1$. Furthermore g_{q_0} is the *only* inner product on $T_q\mathcal{M}$ for which this is true, establishing uniqueness. Then for any $k \in \mathcal{C}_q$, we have $k = k_1 h$ for some $h \in \mathcal{H}$, and using the group-action properties and the chain rule for maps between manifolds, we have

$$(d\tilde{\alpha}_k)_{q_0} = (d\tilde{\alpha}_{k_1})_{q_0} \circ (d\tilde{\alpha}_h)_{q_0}. \tag{3.74}$$

A straightforward computation shows that for $u, v \in T_{q_0}\mathcal{M}$ we have

$$g_{\tilde{\alpha}_k(q_0)}(((d\tilde{\alpha}_k)_{q_0})(u), ((d\tilde{\alpha}_k)_{q_0})(v)) = g_{q_0}(u,v).$$

Hence for all $k \in \mathcal{C}_q$, $(d\tilde{\alpha}_k)_{q_0}$ is a linear isometry from $(T_{q_0}\mathcal{M}, g_{q_0})$ to $(T_q\mathcal{M}, g_q)$, and $(d\tilde{\alpha}_{k^{-1}})_q = \left((d\tilde{\alpha}_k)_{q_0}\right)^{-1}$ is a linear isometry in the other direction. Now let $q_1, q_2 \in \mathcal{M}$ be arbitrary, and let $k \in \mathcal{K}$ be such that $\tilde{\alpha}(q_1) = q_2$. Transitivity implies that there exist $k_i \in \mathcal{C}_{q_i}$, $i = 1,2$, such that $k = k_2 k_1^{-1}$. Then $(d\tilde{\alpha}_k)_{q_1} = \left((d\tilde{\alpha}_{k_2})_{q_0}\right)^{-1} \circ (d\tilde{\alpha}_{k_1})_{q_0}$, a composition of linear isometries, hence a linear isometry.

It remains only to establish that the assignment $q \mapsto g(q)$ is smooth. The Lie theory technique behind this result would require additional definitions and a significant departure from the main path of this chapter. For details, we refer the reader to Helgason (1978) [147].

DEFINITION 3.2.7. *A complete Riemannian manifold $(\mathcal{M}, g)$ with the property that for each point $p \in M$ the geodesic symmetry $s_p : \mathcal{M} \to \mathcal{M}$ given by*

$$s_p(Exp_p(v)) = Exp_p(-v) \tag{3.75}$$

is an isometry and is said to be symmetric.

REMARK 3.2.2. *Any Riemannian symmetric space $\mathcal{M}$ is a Riemannian homogeneous space, since for any pair of points $p_1, p_2 \in \mathcal{M}$, if we consider the midpoint p of a minimal geodesic joining these two points (which exists by completeness), then the geodesic symmetry s_p takes p_1 to p_2. Conversely, if $(\mathcal{M}, g)$ is Riemannian homogeneous space, it is also complete, therefore to show that $(\mathcal{M}, g)$ is symmetric it suffices to prove that the geodesic symmetry at one point only, is an isometry.*

Assume $\mathcal{G}$ is a transitive group of isometries of the Riemannian homogeneous space $(\mathcal{M}, g)$ and $p_0 \in \mathcal{M}$ is a fixed point on $\mathcal{M}$. Let $\mathcal{H} = \mathcal{G}_{p_0}$ be the isotropy group at p_0. If the geodesic s_{p_0} is an isometry, then the automorphism $\sigma : \mathcal{G} \to \mathcal{G}$ given by

$$\sigma(g) = s_{p_0} g s_{p_0}^{-1} \tag{3.76}$$

is involutive, that is $\sigma^2 = Id_{\mathcal{G}}$. Therefore if $e = 1_{\mathcal{G}}$ is the unit element in $\mathcal{G}$, it turns our that the differential $s_e = d_e\sigma$ is a linear idempotent isomorphism of the Lie algebra $\mathfrak{g}$ of $\mathcal{G}$. Note that

$$(s_{p_0} h s_{p_0}^{-1})(exp_{p_0}(v)) = s_{p_0} h(exp_{p_0}(-v) =$$
$$= s_{p_0}(exp_{p_0}(-d_{p_0}h(v)) = exp_{p_0}(d_{p_0}h(v)) = h(exp_{p_0}(v)),$$

thus if $h \in \mathcal{H}$, then $\sigma(h) = h$. It follows that if $\mathfrak{h}$ is the Lie algebra of $\mathcal{H}$, and $x \in \mathfrak{h}$, then $s_e(x) = x$. Set $\mathfrak{m} = \{x \in \mathfrak{g}, s_e(x) = -x\}$. If $x \in \mathfrak{h}, y \in \mathfrak{m}$, then $s_e([x,y]) = [s_e(x), s_e(y)] = [x, -y] = -[x,y]$, thus $[x,y] \in \mathfrak{m}$. On the other hand, if $x \in \mathfrak{m}, y \in \mathfrak{m}$, then $s_e([x,y]) = [s_e(x), s_e(y)] = [-x, -y] = [x,y]$, thus $[x,y] \in \mathfrak{h}$, which proves the following

PROPOSITION 3.2.1. *(canonical decomposition). The Lie algebra $\mathfrak{g}$ admits the decomposition $\mathfrak{g} = \mathfrak{h} \oplus \mathfrak{m}$, where*

$$[\mathfrak{h}, \mathfrak{h}] \subseteq \mathfrak{h}, [\mathfrak{h}, \mathfrak{m}] \subseteq \mathfrak{m}, [\mathfrak{m}, \mathfrak{m}] \subseteq \mathfrak{h}. \tag{3.77}$$

DEFINITION 3.2.8. *Assume $\}$ is a Lie algebra. The center of $\mathfrak{g}$ is $\mathfrak{z} = \{x \in \mathfrak{g}, [x,y] = 0, \forall y \in \mathfrak{g}\}$. A pair $(\mathfrak{g}, s)$ is called an orthogonal symmetric algebra if s is an involutive Lie algebra automorphism of $\mathfrak{g}$, such that $\mathfrak{h} = \{x \in \mathfrak{g}, s(x) = x.\}$ is a compactly embedded Lie subalgebra of $\mathfrak{g}$ and $\mathfrak{z} \cap \mathfrak{h} = 0$.*

DEFINITION 3.2.9. *Assume $\mathcal{H}$ is a Lie subgroup of the Lie group $\mathcal{G}$, and σ is an involutive group automorphism of $\mathcal{G}$, such that $\mathcal{H} = \{g \in \mathcal{G}, \sigma(g) = g.\}$. The triple $(\mathcal{G}, \mathcal{H}, \sigma)$ is called a symmetric triple.*

From Kobayashi and Nomizu (1969, pp.231-232), one can show the following

THEOREM 3.2.5. *Assume $\mathcal{G}$ is a transitive group $\{\alpha_g, g \in \mathcal{G},\}$ of isometries of the Riemannian manifold $\mathcal{M}$, and assume $\mathcal{H}$ is the isotropy group of a given point $p_0 \in \mathcal{M}$, such that $(\mathcal{G}, \mathcal{H}, \sigma)$, is a symmetric triple. Let $\pi : \mathcal{G} \to \mathcal{M}$, be given by $\pi(g) = \alpha_g(p_0)$. Then, given $X \in \mathfrak{g}$, any geodesic starting at p_0, with* *the tangent vector $d_e\pi(X)$, is of the form $exp(tX)(p_0)$ where $exp(tX)$ is the one parameter subgroup of $\mathcal{G}$, with the tangent vector at e equal to $X \in \mathfrak{g}$.*

DEFINITION 3.2.10. *A is an embedding $j : \mathcal{M} \to \mathbb{R}^k$ of a Riemannian manifold $(\mathcal{M}, g)$ for which the geodesic distance ρ_g is the arc distance induced by j. A Riemannian embedding is said to be* at a point *p of $\mathcal{M}$, if every isometry of $j(\mathcal{M})$ that keeps $j(p)$ fixed is the restriction of an Euclidean isometry.* A two point homogeneous space *is a Riemannian manifold such that for each two pairs of its points (p,q), (p',q') with $\rho_g(p,q) = \rho_g(p',q')$, there is an isometry τ with $\tau(p) = p'$ and $\tau(q) = q'$.*

It is known that $\mathcal{M}$ is a two point homogeneous space if and only if, for each $p \in \mathcal{M}$, the isotropy group H_p is *transitive* on every *geodesic sphere* $S(p,r) := \{x \in \mathcal{M} : \rho_g(x,p) = r\}$, $r > 0$ (Chavel (1993) [68], p. 147). That is, given $q, q' \in S(p,r)$ there exists $h \in H_p$ such that $h \cdot q = h \cdot q'$. Finally, note that Kobayashi (1968) [196] gave a general construction of an isometric embedding of a compact symmetric space, which can be used to provide an equivariant embedding of any two point homogeneous space (including the Cayley plane) into an Euclidean space.

The volume of a Borelian set included in the domain U of a chart, is

$$Vol_g(B) = \int_B \sqrt{|g|}\lambda_m(dx), \tag{3.78}$$

where $|g|$ is the determinant of the matrix $(g_{ij})_{i,j=\overline{1,m}}$, and λ_m is the Lebesgue measure on $\mathbb{R}^m$. For example, the volume of a surface $\mathcal{M}$ (two-dimensional submanifold) in $\mathbb{R}^N$, can be obtained as follows. Let g_{ij} be the *first fundamental form* (induced Riemannian structure) defined by (3.30) and $|g|$. The area of a Borelian subset $B \subset U$ is given by the integral

$$\text{Area}_g(B) = \iint_B \sqrt{|g|}\, dx_1 dx_2 \tag{3.79}$$

The area of $\mathcal{M}$ is then given by

$$\text{Area}_g(\mathcal{M}) = \sum_\alpha \iint_{U_\alpha} \eta_\alpha \sqrt{|g_\alpha|}\, dx_1^\alpha dx_2^\alpha, \tag{3.80}$$

where $(U_\alpha)_{\alpha \in A}$ is an open cover with domain of charts on $\mathcal{M}$ and $(\eta_\alpha)_{\alpha \in A}$ is a partition of unity *subordinated* to $(U_\alpha)_{\alpha \in A}$.

3.3 The Laplace–Beltrami Operator

The space of differentiable functions on manifolds is a linear space, that is open and dense in a Hilbert space of functions. Therefore functional analysis on manifolds is not as much affected by curvature, and borrows most of its ideas from linear functional data analysis. Here, we assume that $\mathcal{M}$ is a compact connected orientable Riemannian manifold with the Riemannian structure g. The Riemannian structure locally is given by the metric tensor $(g_{ij}(x)), x \in U$ associated with a chart (U, φ_U) on $\mathcal{M}$. Let ρ_g denote the induced Riemannian distance. The Riemannian structure g induces a natural isomorphism $\Theta_g : T^*\mathcal{M} \to T\mathcal{M}$ between the tangent and cotangent bundle, given by

$$\Theta_g(v)(w) = g_p(v, w), \forall v, w \in T_p\mathcal{M}. \tag{3.81}$$

The *gradient* of a differentiable function f on $\mathcal{M}$, is the vector field $grad_g f$ given by

$$grad_g f = \Theta_g^{-1} df, \tag{3.82}$$

where df is the differential of f. The *divergence* of a vector field X w.r.t. g is given by

$$div_g X = Tr(Y \to \nabla_Y X), \tag{3.83}$$

where ∇ is the Levi–Civita connection.

Finally, the *Laplace–Beltrami operator* applied to a differentiable function f on $\mathcal{M}$, is $\Delta_g f$ given by

$$\Delta_g f = div_g grad_g f. \tag{3.84}$$

In Chavel (2006, p.150) [69] one shows that in terms of local coordinates, the expression of the Laplace–Beltrami operator is given by

$$\Delta_g = -\frac{1}{\sqrt{G(x)}} \sum_{j,k} \frac{\partial}{\partial x^j} (g(x)^{jk} \sqrt{G(x)} \frac{\partial}{\partial x^k}), \tag{3.85}$$

where $(g(x)^{ij})$ is the inverse of the matrix $(g(x)_{ij})$, and $G(x)$ is the determinant of this matrix. Note that since for any two functions f, h on $\mathcal{M}$, $< f, \Delta h >=< \Delta f, h >$, and $< \Delta f, f >= \|df\|^2$, it follows that the Laplace-Beltrami operator is *elliptic symmetric, positive semidefinite second order differential operator* on $C^\infty(\mathcal{M})$ (see Chavel (2006) [69], Chapter III), and its eigenfunctions are forming a complete orthonormal basis for $L^2(\mathcal{M})$, the space of square integrable functions on $\mathcal{M}$.

Let $\Delta \doteq \Delta_g : \mathcal{C}^\infty(\mathcal{M}) \to \mathcal{C}^\infty(\mathcal{M})$ Laplace-Beltrami operator on $(\mathcal{M}, g)$ given in (3.84). Let ϕ_λ and λ be an eigenfunction and the corresponding eigenvalue, respectively. Note that there are countably many eigenvalues, multiplicity included, and $\lambda_k \geq 0, k \in \mathbb{N}$ with no upper bound. This means that for each λ_k, a corresponding eigenfunction (which in general will occur with multiplicity) will be denoted by $\phi_{\lambda_k} = \phi_k, k \in \mathbb{N}$. Furthermore, we will use the convention that $\lambda_0 = 0$ with $\phi_0 = 1$ and that $\lambda_k \leq \lambda_{k+1}$ for $k \in \mathbb{N}$.

We extend the operator Δ to a linear operator on functions $f : \mathcal{M} \to \mathbb{C}$. Denote by $\|\cdot\|_2$ and $\|\cdot\|_\infty$ the L^2 and L^∞ norms on $\mathcal{M}$, respectively. Let $\mathcal{E}_k \subset L^2(\mathcal{M})$, $k \in \mathbb{N}$, denote the eigenspace associated with the eigenvalue λ_k,

$k \in \mathbb{N}$. The dimension of $\mathcal{E}_k$ will be denoted by $\dim \mathcal{E}_k < \infty$ for $k \in \mathbb{N}$. The multiplicity of eigenfunctions whose eigenvalues are less than a certain constant is determined by Weyl's formula (see Weyl (1912)[335]

$$\lim_{\Lambda \to \infty} \Lambda^{-\dim \mathcal{M}/2} \#\{\lambda_k | \lambda_k < \Lambda\} = \frac{\text{vol}_g(\mathcal{M})}{(2\sqrt{\pi})^{\dim \mathcal{M}} \Gamma(1 + \dim \mathcal{M}/2)}, \tag{3.86}$$

where # denotes the cardinality of a finite set, $\dim \mathcal{M}$ denotes the dimension of $\mathcal{M}$, $\text{vol}_g(\mathcal{M})$ denotes the volume of $(\mathcal{M}, g)$ and $\Gamma(\cdot)$ is the gamma function (see Minakshisundaram and Pleijel (1949) [244]). We note that if ϕ_k is a complex valued eigenfunction of Δ, then so is $\overline{\phi}_k$ where overbar denotes complex conjugation. Consequently, a real basis for $L^2(\mathcal{M})$, made of eigenfunctions of the Laplace-Beltrami operator can be chosen, and for $h \in L^2(\mathcal{M})$, the "components" $\hat{h}_k$ relative to this basis will be defined by

$$h = \sum_{k=0}^{\infty} \sum_{\mathcal{E}_k} \hat{h}_k \phi_k, \quad \text{where } \hat{h}_k = \int_{\mathcal{M}} h \overline{\phi}_k, \tag{3.87}$$

for all $k \in \mathbb{N}$. We note that summation over $\mathcal{E}_k$ means over all eigenfunctions ϕ_k in the eigenspace $\mathcal{E}_k$, $k \in \mathbb{N}$ and integration is defined by the usual partition of unity argument.

For concreteness, let us consider specific examples. The sphere $S^{m-1} \subset \mathbb{R}^m$ is the set of unit vectors in $p-$dimensional Euclidean space. In the case where $m = 3$, we note that any point in S^2 can almost surely be represented by

$$\omega = (\cos\varphi \sin\vartheta, \sin\varphi \sin\vartheta, \cos\vartheta)', \tag{3.88}$$

where $\varphi \in [0, 2\pi)$, $\vartheta \in [0, \pi)$, and superscript $'$ denotes transpose.

The orthogonal group $O(m)$ consists of the space of $m \times m$ real orthogonal matrices, however, this group is not connected. The connected component consisting of those real orthogonal matrices having determinant equal to unity, $SO(m)$, is called the special orthogonal group. Again in the case of $m = 3$, $SO(3)$ can be represented in the following way. Let

$$u(\varphi) = \begin{pmatrix} \cos\varphi & -\sin\varphi & 0 \\ \sin\varphi & \cos\varphi & 0 \\ 0 & 0 & 1 \end{pmatrix}, \quad a(\vartheta) = \begin{pmatrix} \cos\vartheta & 0 & \sin\vartheta \\ 0 & 1 & 0 \\ -\sin\vartheta & 0 & \cos\vartheta \end{pmatrix},$$

where $\varphi \in [0, 2\pi)$, $\vartheta \in [0, \pi)$. The well known Euler angle decomposition says, any element of $SO(3)$, can almost surely be uniquely written as:

$$g = u(\varphi_1) a(\vartheta) u(\varphi_2),$$

where $\varphi_1 \in [0, 2\pi)$, $\varphi_2 \in [0, 2\pi)$, $\vartheta \in [0, \pi)$.

Consider S^2, which can be identified with the quotient space $SO(3)/SO(2)$.

The following is an example of construction of orthonormal bases for this space. Let

$$\phi_{kq}(\omega) = \begin{cases} \sqrt{2}\sqrt{\dfrac{(2k+1)(k-q)!}{4\pi(k+q)!}}P_q^k(\cos\vartheta)\cos(q\varphi) & q = 1,\ldots,k \\ \sqrt{\dfrac{(2k+1)}{4\pi}}P_0^k(\cos\vartheta) & q = 0 \\ \sqrt{2}\sqrt{\dfrac{(2k+1)(k-|q|)!}{4\pi(k+|q|)!}}P_{|q|}^k(\cos\vartheta)\sin(|q|\varphi) & q = -1,\ldots,-k \end{cases} \tag{3.89}$$

where $\varphi \in [0,2\pi)$, $\vartheta \in [0,\pi)$, P_q^k are the Legendre functions, $-k \leq q \leq k$ and $k \in \mathbb{N}$. We note that we can think of (3.89) as the vector entries to the $2k+1$ vector

$$\phi_k(\omega) = \left(\phi_{kq}(\omega)\right) ,$$

$|q| \leq k$ and $k \in \mathbb{N}_0$. In this situation $\{\phi_{kq} : |q| \leq k, k \in \mathbb{N}_0\}$ are the eigenfunctions of the Laplace-Beltrami operator on S^2 with eigenvalues $\lambda_k = k(k+1)$, $k \in \mathbb{N}_0$ and hence form a complete orthonormal basis over $L^2(S^2)$.

In another example, in the case of $SO(3)$, let

$$D^\ell_{q_1q_2}(\varphi_1,\vartheta,\varphi_2) = \mathrm{e}^{-iq_1\varphi_1}d^\ell_{q_1q_2}(\cos\vartheta)\mathrm{e}^{-iq_2\varphi_2}, \tag{3.90}$$

where, $d^\ell_{q_1q_2}$ for $-\ell \leq q_1,q_2 \leq \ell$, $\ell = 0,1,\ldots$ are related to the Jacobi polynomials. Define the $(2\ell+1)\times(2\ell+1)$ matrix by

$$D^\ell(g) = \left(D^\ell_{q_1q_2}(g)\right), \tag{3.91}$$

where $-\ell \leq q_1,q_2 \leq \ell$, $\ell \geq 0$ and $g \in SO(3)$. We note that (3.91) constitute the collection of inequivalent irreducible representations of $SO(3)$. In this situation $\{\sqrt{2\ell+1}D^\ell_{q_1q_2} : |q_1|,|q_2| \leq k, k \in \mathbb{N}\}$ are the eigenfunctions of the Laplace-Beltrami operator on $SO(3)$ with eigenvalues $\lambda_\ell = \ell(\ell+1)/2$, $\ell \in \mathbb{N}$ and hence form a complete orthonormal basis over $L^2(SO(3))$. Here we used the extension of the Laplace–Beltrami operator to a linear space, extension of $C^\infty(\mathcal{M})$ over the field of complex numbers, a standard technique for studying the *spectrum of a symmetric operator* vis a self-adjoint extension.)

3.3.1 *Harmonic Analysis on Homogeneous Spaces*

Recall that a Riemannian homogeneous space admits a transitive group of isometries $\mathcal{K}$, acting transitively on $\mathcal{M}$. For every $p_0 \in \mathcal{M}$, let $\mathcal{K}_{p_0} = \{w \in \mathcal{K} : wp_0 = p_0\}$ denote the isotropy subgroup of p_0. As shown in Section 3.1, if

$\mathcal{M}$ is a homogeneous compact connected Riemannian manifold, then for every $p \in \mathcal{M}$, $\mathcal{K}_p$ is a closed subgroup of $\mathcal{K}$ and there exists a diffeomorphism of $\mathcal{K}/\mathcal{K}_p \simeq \mathcal{M}$. The classical example is the diffeomorphism of the 2–sphere S^2 with the quotient set of 3×3 rotation matrices modulo 2×2 rotation matrices $SO(3)/SO(2)$. A differentiable function $f : \mathcal{M} \to \mathbb{R}$ is called a *zonal function* with respect to $p_0 \in \mathcal{M}$ if it is constant on the orbit $\mathcal{K}_{p_0}p$ of a point $p \in \mathcal{M}$. Furthermore, we note that if $\mathcal{M}$ is homogeneous,

$$\sum_{\mathcal{E}_k} |\phi_k(p)|^2 = \dim \mathcal{E}_k \;\; \forall p \in \mathcal{M} \tag{3.92}$$

for $k = 0, 1, \ldots$ (see Giné (1975) [124]). If, in addition $\mathcal{M}$ is a two point homogeneous space, then for any $p, q \in \mathcal{M}$, with $\rho_g(p, p_0) = \rho_g(q, p_0)$, there exists a $\tau \in \mathcal{K}_{p_0}$ such that $\tau p = q$. Our primary example of harmonics on two point homogeneous space is the $m-1$ dimensional round hypersphere, $\mathbb{S}^{m-1}$, for $m \geq 3$. Indeed, for some $\omega = (\omega_1, \ldots, \omega_m)^t \in S^{m-1}$, the $m-1$ spherical coordinates can be represented by:

$$\begin{aligned} \omega_1 &= \sin\theta_{m-1} \cdots \sin\theta_2 \sin\theta_1 \\ \omega_2 &= \sin\theta_{m-1} \cdots \sin\theta_2 \cos\theta_1 \\ &\;\;\vdots \\ \omega_{m-1} &= \sin\theta_{m-1} \cos\theta_{m-2} \\ \omega_m &= \cos\theta_{m-1} \end{aligned} \tag{3.93}$$

where $\theta_1 \in [0, 2\pi)$ and $\theta_j \in [0, \pi)$ for $j = 2, \ldots, m-1$. The invariant measure is

$$d\omega = \frac{\Gamma(m/2)}{2\pi^{m/2}} \sin^{m-2}\theta_{m-1} \cdots \sin\theta_2 d\theta_1 \cdots d\theta_{m-1}.$$

Let $C_r^{\mu}(t)$, $t \in [-1, 1]$ be a polynomial of degree r determined by the power series

$$(1 - 2t\alpha + \alpha^2)^{-\mu} = \sum_{r=0}^{\infty} C_r^{\mu}(t)\alpha^r. \tag{3.94}$$

One notices that $C_r^{1/2}(t)$ are the classical Legendre polynomials. Thus for general μ, these polynomials are generalizations of the classical Legendre polynomials and are called the Gegenbauer (ultraspherical) polynomials.

Let $k = (k_1, k_2, \ldots, k_{m-2})$ and $\mathcal{K}_\ell = \{\ell \geq k_1 \geq k_2 \geq \cdots \geq k_{m-2} \geq 0\}$. The collection of eigenfunctions of Δ is usually written as

$$\left\{Y_k^{\ell,i} : k \in \mathcal{K}_\ell, \ell \geq 0, i = 1, 2\right\}, \tag{3.95}$$

where

$$\Delta Y_k^{\ell,i} = \lambda_\ell Y_k^{\ell,i}, \text{ and } \lambda_\ell = \ell(\ell + m - 2),$$

$\ell > 0$. Thus each $\ell \geq 0$, determines the eigenspace $\mathcal{E}_\ell$, where

$$\dim \mathcal{E}_\ell = \frac{(2\ell+m-2)(l+m-3)!}{\ell!(m-2)!}. \tag{3.96}$$

Collectively, (3.95) is called the spherical harmonics for $L^2(S^{m-1})$ and (3.95) forms a complete orthonormal basis (see M*ü*ller (1998) [250]).

3.3.2 Harmonics on Semi-Simple Lie Groups

If in addition the manifold has a group structure with the group action and inverse mapping being continuous hence is a Lie group, more refinements can be made. Let $\mathcal{K}$ be a compact connected semi-simple Lie group and fix once and for all, a maximal torus $\mathbb{T}$. Let $\mathfrak{g}$ and $\mathfrak{t}$ be the corresponding Lie algebras and denote by $\mathfrak{t}^*$ the dual space of $\mathfrak{t}$ possessing the Weyl group invariant inner product. Let $K \subset \mathfrak{t}^*$ be the fundamental Weyl chamber and denote by Φ, the set of real roots. Let $\Phi_+ = \{\alpha \in \Phi : \langle \alpha, \beta \rangle > 0, \beta \in K\}$ be the set of positive roots. Finally denote by $I^* \subset \mathfrak{t}^*$ the integral portion of $\mathfrak{t}^*$. Thus we can define $\bar{K} \cap I^*$, where the overbar in this case denotes set theoretic closure. Thus associated with $\mathcal{K}$ will be the dual objects $\widehat{\mathcal{K}} = \bar{K} \cap I^*$.

Consider an irreducible representation $U : \mathcal{K} \to \mathrm{Aut}(V)$, where V is some finite dimensional vector space and $\mathrm{Aut}(V)$ are the automorphisms of V. Then the collection of inequivalent irreducible representations of $\mathcal{K}$ can be enumerated as $\{U_\nu : \nu \in \widehat{\mathcal{K}}\}$ (see Br*ö*cker and tom Diek (1985) [52, p. 242]). The dimension of the *irreducible representations* are

$$d_\nu = \prod_{\alpha \in \Phi_+} \frac{\langle \alpha, \nu + \rho \rangle}{\langle \alpha, \rho \rangle}$$

for $\nu \in \widehat{\mathcal{K}}$ where $\rho = 2^{-1} \sum_{\alpha \in \Phi_+} \alpha$, the half sum of the positive roots and the inner product is induced by the Killing form.

The Killing form induces also a Riemannian structure on $\mathcal{K}$. Consequently, let Δ be the Laplace-Beltrami operator on $\mathcal{K}$. The components of the irreducible representations are the eigenfunctions of Δ so that

$$\left\{ \sqrt{d_\nu}\, U_\nu : \nu \in \widehat{\mathcal{K}} \right\} \tag{3.97}$$

is a complete orthonormal basis of $L^2(\mathcal{K})$. We note that the eigenvalues are

$$\lambda_\nu = \|\nu + \rho\|^2 - \|\rho\|^2 \tag{3.98}$$

for $\nu \in \widehat{\mathcal{K}}$ where the norm is with respect to the Killing form. The multiplicity of the eigenfunctions with respect to a fixed eigenvalue is therefore d_ν^2 for $\nu \in \widehat{\mathcal{K}}$. Weyl's formula (3.86) for $\mathcal{K}$ is thus

$$\lim_{T \to \infty} T^{-\dim \mathcal{K}/2} \sum_{\lambda_\nu < T} d^2(\nu + \rho) = \frac{\mathrm{vol}(\mathcal{K})}{(2\sqrt{\pi})^{\dim \mathcal{K}} \Gamma(1 + \dim \mathcal{K}/2)}. \tag{3.99}$$

See Minakshisundaram and Pleijel (1949) [244], Giné (1975) [124], or Hendriks (1990) [150].

The primary example of a Lie group is the space of $N \times N$ rotation matrices, $SO(N)$. For $N = 2k+1$ odd, let

$$\widehat{\mathcal{K}} = \left\{ j \in \mathbb{Z}^k : j_1 \geq j_2 \geq \cdots \geq j_k \geq 0 \right\}, \tag{3.100}$$

for $N = 2k$ even, let

$$\widehat{\mathcal{K}} = \left\{ j \in \mathbb{Z}^k : j_1 \geq j_2 \geq \cdots \geq |j_k| \geq 0 \right\}, \tag{3.101}$$

where $\mathbb{Z}$ denotes the set of all integers. One notices that in the even case, an extra set of indices come from the relation $|j_k|$. The particular form of $\widehat{\mathcal{K}}$, as classified by the *Dynkin diagrams*, for $SO(N)$ when $N = 2k+1$ reflects the B_k root structure $k \geq 2$, while $N = 2k$ reflects the D_k root structure $k \geq 3$ (see Bröcker and T. tom Diek (1985) [52]).
Consider Δ the Laplace-Beltrami operator on $SO(N)$. For $N = 2k+1$, the corresponding eigenvalue is

$$\lambda_j = j_1^2 + \cdots + j_k^2 + (2k-1)j_1 + (2k-3)j_2 + \cdots + j_k \tag{3.102}$$

while for $N = 2k$

$$\lambda_j = j_1^2 + \cdots + j_k^2 + (2k-2)j_1 + (2k-4)j_2 + \cdots + 2j_{k-1}. \tag{3.103}$$

Further details of the eigenstructure of $SO(N)$ are provided in Appendix B in Kim (1998) [187].

3.4 Topology of Manifolds

In this section we present a technical overview of homology as used in our procedures. For a detailed treatment see Greenberg and Harper (1981) [131], Milnor (1963) [242] and Spanier (1981) [313].

3.4.1 Background on Algebraic Topology

In algebraic topology, one associates to topological spaces and continuous functions between them, algebraic objects and morphisms between these objects. This association is *functorial* in the sense of *category theory* (see Mac Lane (1998, Ch I) [226]), therefore, as a result, two spaces having the same *homotopy type* one associates isomorphic objects.

Assume $(M, \tau_M), (N, \tau_N)$ are topological spaces and $C(M,N)$ is the space of continuous functions from (M, τ_M) to (N, τ_N). A *homotopy* between $f_0, f_1 \in C(M,N)$ is a continuous function $F : [0,1] \times M \to N$, such that $F(a,x) =$

$f_a(x), \forall a = 0, 1, \forall x \in M$. Notation $f_0 \cong f_1$ We say that (M, τ_M) and (N, τ_N) have the same *homotopy type* if there are continuous functions $f \in C(M,N), g \in C(N,M)$ such that $g \circ f \cong Id_M, f \circ g \cong Id_N$.

Homology is an algebraic functor for "counting holes in various dimensions" in topological spaces. Roughly speaking, the homology of X, denoted $H_*(X)$, is a sequence of commutative groups ($\mathbb{Z}$ - modules) $\{H_k(X) : k = 0, 1, 2, 3, \ldots\}$, where $H_k(X)$ is called the k-dimensional homology group of X. The rank of $H_k(X)$ (as a $\mathbb{Z}$-module), called the k-th Betti number of X, β_k, is a coarse measurement of the number of different holes in the space X that can be "sensed" by using subcomplexes of dimension k. The k-th Betti number of X can be also interpreted as the dimension β_k of the vector space $H_k(X, \mathbb{R})$, if real coefficients are used instead.

For example, if M is a topological space, β_0 is equal to the number of connected components of M. These are the types of features (holes) in M that can be detected by using points and edges– with this construction one is answering the question: are two points connected by a sequence of edges or not? The simplest basis for $H_0(M, \mathbb{R})$ consists of a choice of vertices in M, one in each *path-component* of M (the points p_0, p_1 on M are in the same path-component, if there is a continuous map $\gamma\colon [0,1] \to M$, with $\gamma(0) = p_0, \gamma(0) = p_1$). Likewise, the simplest basis for $H_1(M)$ consists of loops in M, each of which surrounding a $1D$ hole in M. For example, if M is a graph, then the space $H_1(M, \mathbb{R})$ is a vector space, whose dimension encodes the number of "independent" cycles in the graph.

3.4.2 Homology

There are numerous variants of homology: here we begin by using simplicial homology with integer coefficients. Given a set V, a *k-simplex* is an unordered subset $\{v_0, v_1, \ldots, v_k\}$ where $v_i \in V$ and $v_i \neq v_j$ for all $i \neq j$. The faces of this k-simplex consist of all $(k-1)$-simplices of the form $\{v_0, \ldots, v_{i-1}, v_{i+1}, \ldots, v_k\}$ for some $0 \leq i \leq k$. Geometrically, the k-simplex can be described as follows: given $k+1$ points in $\mathbb{R}^m$ ($m \geq k$), the k-simplex is a convex body bounded by the union of $(k-1)$ linear subspaces of $\mathbb{R}^m$ of defined by all possible collections of k points (chosen out of these $k+1$ points). Topologically, a k-simplex in (M, τ_M) is a continuous function from a geometric k-simplex to (M, τ_M).

A *simplicial complex* is a collection of simplices with vertices V, which is closed with respect to inclusion of faces. A *triangulated surface* gives a concrete example, where the vertices of the triangulation correspond to V. The orderings of the vertices correspond to an orientation. Any abstract simplicial complex on a (finite) set of points V has a *geometric realization* in some $\mathbb{R}^m$. Let X denote a simplicial complex.

Let X denote a simplicial complex. Define for each $k \geq 0$, the $\mathbb{Z}$-module $S_k(X)$ whose basis is the set of oriented k-simplices of X; that is, a k-

simplex $\{v_0,\dots,v_k\}$ together with an order type denoted $[v_0,\dots,v_k]$ where a change in orientation corresponds to a change in the sign of the coefficient: $[v_0,\dots,v_i,\dots,v_j,\dots,v_k] = -[v_0,\dots,v_j,\dots,v_i,\dots,v_k]$ if odd permutation is used. For k larger than the dimension of X, we set $S_k(X) = 0$. The boundary map is defined to be the linear transformation $\partial_k : S_k(X) \to S_{k-1}(X)$ which acts on basis elements $[v_0,\dots,v_k]$ via

$$\partial_k[v_0,\dots,v_k] := \sum_{i=0}^{k}(-1)^i[v_0,\dots,v_{i-1},v_{i+1},\dots,v_k]. \tag{3.104}$$

This gives rise to a : a sequence of $\mathbb{Z}$-modules and linear transformations

$$\cdots \stackrel{\partial_{k+2}}{\to} S_{k+1}(X) \stackrel{\partial_{k+1}}{\to} S_k(X) \stackrel{\partial_k}{\to} S_{k-1}(X) \cdots \stackrel{\partial_{k-1}}{\to} S_2(X) \stackrel{\partial_2}{\to} S_1(X) \stackrel{\partial_1}{\to} S_0(X) \tag{3.105}$$

Consider the following two submodules of $S_k(X)$: the cycles ($Z_k(X)$, subcomplexes without boundary) and the boundaries ($B_k(X)$, subcomplexes which are themselves boundaries) formally defined as:

- $k-$cycles: $Z_k(X) = \ker(\partial_k : S_k(X) \to S_{k-1}(X))$
- $k-$boundaries: $B_k(X) = \operatorname{im}(\partial_{k+1} : S_{k+1}(X) \to S_k(X))$

One may show that $\partial_k \circ \partial_{k+1} = 0$; that is, the boundary of a chain has empty boundary. It follows that $B_k(X)$ is a $\mathbb{Z}$-submodule of $Z_k(X)$. This has great implications. The k-cycles in X are the basic objects which count the presence of a "hole of dimension k" in X. But, certainly, many of the k-cycles in X are measuring the same hole; still other cycles do not really detect a hole at all – they bound a subcomplex of dimension $k+1$ in X. We say that two cycles ζ and η in $Z_k(X)$ are homologous if their difference is a boundary:

$$[\zeta] = [\eta] \;\leftrightarrow\; \zeta - \eta \in B_k(X).$$

The k-dimensional homology of X, denoted $H_k(X)$ is the quotient $\mathbb{Z}$ - module

$$H_k(X) := \frac{Z_k(X)}{B_k(X)}. \tag{3.106}$$

Specifically, an element of $H_k(X)$ is an equivalence class of homologous k-cycles. This inherits the structure of a $\mathbb{Z}$ -module in the natural way $[\zeta]+[\eta] = [\zeta+\eta]$ and $c[\zeta] = [c\zeta]$.

REMARK 3.4.1. *By arguments utilizing* barycentric subdivision, *one may show that the homology $H_*(M)$ is independent of the choice of the simplicial complex on M. For a simple example, the reader is encouraged to contemplate the "physical" meaning of $H_1(M)$. Elements of $H_1(M)$ are equivalence classes of (finite collections of) oriented cycles in the* 1*-skeleton of M, the equivalence relation being determined by the* 2*-skeleton of M.*

REMARK 3.4.2. *Note that singular homology is functorial. Let $f : X \to X'$ be a continuous simplicial map: f takes each k-simplex of X to a k'-simplex of X', where $k' \leq k$. Then, the map f induces a linear transformation $S_k(f) : S_k(X) \to S_k(X')$. One can show that $S_k(f)$ takes cycles to cycles and boundaries to boundaries; hence there is a well-defined linear transformation $H_k(f) : H_k(X) \to H_k(Y)$.*

For the category of topological spaces and continuous maps, *functoriality* means that if $f : M \to N, g : N \to P$ are continuous functions then $H_k(g \circ f) = H_k(g) \circ H_k(f)$.

An *exact sequence* is a sequence

$$\cdots \to H_k \xrightarrow{i_k} H_{k-1} \xrightarrow{i_k} H_1 \xrightarrow{\partial_k} H_0. \tag{3.107}$$

of groups $H_k, k \geq 0$, and homomorphisms j_k, such that the kernel of j_{k-1} equals the range of j_k. The exact sequence is short, if it contains at most three nonzero terms.

Given a subset $A \subset M$, of a topological space, with the induced topology, one may form the short exact sequence

$$0 \to S_*(A) \to S_*(M) \to S_*(M)/S_*(A) \to 0 \tag{3.108}$$

where $S_*(M)$ denotes the simplicial complex on M. The boundary map on $S_*(M)$ leaves $S_*(A)$ invariant and therefore induces a boundary map on the quotient complex. The corresponding homology, $H_k(M,A) = H_k(S_*(M)/S_*(A))$, is called *relative simplicial homology* of the pair (M,A), and is denoted by $H_k(M,A)$. Note that relative homology is given by the relative cycles, chains whose boundaries are chains on A, modulo the relative boundaries (chains that are homologous to a chain on A, i.e. chains that would be boundaries, modulo A again).

If (M,A), are as above and U is a subset of A, we say that *U can be excised* if the inclusion map of the pair $(M \backslash U, A \backslash U)$ into (M,A) induces an isomorphism on the relative homologies $H_k(M,A)$ to $H_k(M \backslash U, A \backslash U)$.

THEOREM 3.4.1. *(Excision theorem). If the closure of U is contained in the interior of A, then U can be excised.*

The *Eilenberg-Steenrod (E-S) axioms* (see Eilenberg and Steenrod (1952) [100] apply to a *homology theory,* sequence of functors $H_k, k \in \mathbb{N}$, from the category of pairs (M,A) of topological spaces to the category of abelian groups, together with a natural transformation $\partial : H_k(M,A) \to H_{k-1}(A, \emptyset)$ called the boundary map. These axioms are:

- *Homotopy axiom*: Homotopic maps induce the same map in homology. That is, if $g : (M,A) \to (N,B)$ is homotopic to $h : (M,A) \to (N,B)$, then their induced maps are the same.
- *Excision axiom*: If (M,A) is a pair and closure of U is contained in the interior of A, then U can be excised.

- *Dimension axiom*: Let $\diamond$ be the one-point space; then $H_k(\diamond) = 0 \forall k \neq 0$.
- *Additivity axiom:* If $M = \sqcup_\alpha M_\alpha$, the disjoint union of a family of topological spaces M_α, then $H_k(M) \cong \bigoplus_\alpha H_k(M_\alpha)$.
- *Exactness axiom*: Each pair (M,A) induces a long exact sequence in homology, via the inclusions $i : A \to M$ and $j : M \to (M,A)$, where M is the shorthand for $(M,)$:

$$\cdots \to H_k(A) \xrightarrow{i_k} H_k(M) \xrightarrow{j_k} H_k(M,A) \xrightarrow{\partial_k} H_{n-1}(A) \to \cdots . \tag{3.109}$$

The *coefficient group* of a homology theory is $H_0(\diamond)$. The following result is due to Milnor (1962) [241].

THEOREM 3.4.2. *Let $H_k, k \in \mathbb{N}$ be an additive homology theory on the category $\mathcal{W}$ consisting of all pairs (X,A) such that both X and A have the homotopy type of a CW-complex; and all continuous maps between such pairs, with coefficient group G. Then for each (X,A) in $\mathcal{W}$, there is a natural isomorphism between $H_k(X,A)$ and the k-th singular homology group of (X,A) with coefficients in G.*

We are now showing that the homotopy axiom holds true for the *singular cubic homology*. Assume that the coefficient group is the additive group of integers $\mathbb{Z}$. We consider the cube I^k, where $I = [0,1]$, and the set of *singular cubes* in a topological space (M,τ) : $I_k(M) = \{F : I^k \to M, F \text{continuous}\}$. Consider the chain complex $\mathcal{C}_k(M)$ generated by $I_k(M)$, and the corresponding boundary operators $\partial_k : C_k(M) \to C_{k-1}(M)$,

$$(\partial_k F)(t^1,\ldots,t^{k-1}) = \sum_{j=1}^{k}(-1)^j (F(t^1,\ldots,1_j,\ldots t^{k-1}) - F(t^1,\ldots,0_j,\ldots t^{k-1})). \tag{3.110}$$

The singular cubic homology groups are given by $H_k(M) = \frac{Ker\partial_k}{Im\partial_{k-1}}$.
Consider the *cubical complexes* $C_*(M)$ and $C_*(N)$. Let $f : M \to M$ be a continuous map: f takes each k-cube in X to a k-cube of Y. Then, the map f induces a linear transformation $f_k : C_k(M) \to C_k(N)$.

DEFINITION 3.4.1. *Given chain complexes $(A,\partial^A),(B,\partial^B)$ and chain complexes morphisms $f,g : A \to B$, a* chain homotopy *from f to g is a collection of maps $h_k : A_k \to B_{k+1}$ such that $f_k - g_k = \partial^B_{k+1} h_k + h_{k-1}\partial^A_k$, or simply*

$f-g=\partial^B h+h\partial^A$.

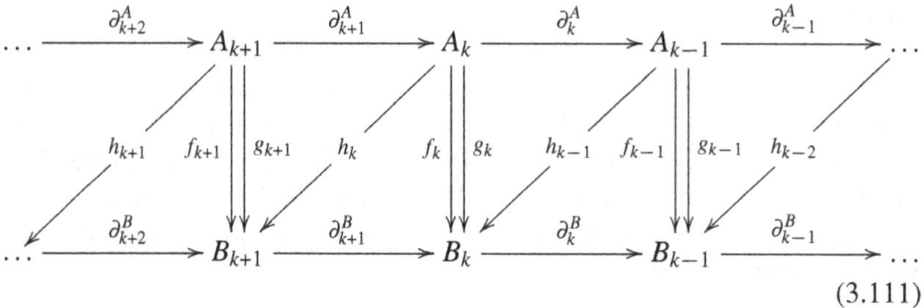

(3.111)

From definition 3.4.1, if we apply the identity $f-g=\partial^B h+h\partial^A$ to a boundary $\partial^A a$, the result is a boundary, an immediate proof of

PROPOSITION 3.4.1. *If h is a homotopy from f to g between chain complexes, then the induced maps in homology are equal, $f_*=g_*$.*

Note that if $H: I\times M\to N$ is a homotopy between two continuous maps $f,g: M\to N$, then the map $h_k: C_k(M)\to C_{k+1}(N)$ given by

$$h_k([F])=[H\circ(Id_I\times F)]. \tag{3.112}$$

is a chain homotopy from f_* to g_*, thus the induced maps in the cubic singular homology $H_k(f),H_k(g): H_k(C_*M)\to H_k(C_*N)$ are equal. Moreover the definition of the relative cubic homology of a pair (M,A) is similar with the simplicial counterpart. One may form the short exact sequence

$$0\to C_*(A)\to C_*(M)\to C_*(M)/C_*(A)\to 0 \tag{3.113}$$

The boundary map on $C_*(M)$ leaves $C_*(A)$ invariant and therefore induces a boundary map on the quotient complex. The corresponding homology, $H_k(M,A)=H_k(C_*(M)/C_*(A))$, is called *relative singular cubic homology* of the pair (M,A), and by abuse of notation, is also denoted by $H_k(M,A)$. From (3.113) and from the long exact sequences of homology (3.107) of the pairs (M,A), and (N,B), due to the fact that $H_k(f)=H_k(g)$, in the absolute homology, and a type of argument, one can show that $H_k(f)=H_k(g)$, in the relative homology, thus proving the Homotopy axiom, for the singular cubic homology. The other E-S axioms are easier to prove, and are left to the reader.

DEFINITION 3.4.2. *If $f: X\to X'$ is a morphism of cubical chain complexes, the sequence of homomorphism $H_k(f): H_k(X)\to H_k(X')$, $f_k([\zeta])=[C_k(f)(\zeta)], k=0,1,\dots$ are homology homomorphisms induced by f.*

A *homotopy equivalence* is a continuous function $f: M\to M$, for which there is a function $g: N\to M$, such that $g\circ f\cong Id_M, f\circ g\cong Id_N$.
Due to functoriality and the homotopy axiom, it follows that if $f: M\to N$ is a homotopy equivalence, then $H_k(f): H_k(M)\to H_k(N)$ induces an isomorphism

and, in particular, the ranks of the homology groups , called (Betti numbers) are equal, that is $rkH_k(M) = rkH_k(N)$. Therefore the Betti numbers are homotopy invariants.

Assume p_0 is a point on a topological space M, and $C(M, p_0) = \{f : \mathbb{S}^1 \to M, f(1) = p_0, f \text{continuous}\}$. A point base preserving homotopy in (M, p_0) is a homotopy $H : I \times \mathbb{S}^1 \to M, H(t,1) = p_0, \forall t \in I$ The *fundamental group* of a connected topological space M, $\pi_1(M)$, is set of all point base preserving homotopy classes of functions $f \in C(M, p_0)$. The fundamental group has a group structure can be obtained by extending the composition structure given by $[f] * [g] = [h]$, where

$$h(e^{2\pi t}) = \begin{cases} f(e^{4\pi t}), & \text{for } 0 \le t \le 0.5 \\ \\ g(e^{2\pi(2t-1)}), & \text{for} 0.5 \le t \le 1 \end{cases} \tag{3.114}$$

By way of example, $\pi_1(\mathbb{S}^1) = \mathbb{Z}$. Given a topological space M, its *kdimensional homotopy group*, $\pi_k(M)$, set of all homotopy classes of functions $f : \mathbb{S}^k \to M$, has a group structure. This group structure can be obtained by extending the composition structure $*$ of $\pi_1(M)$ to higher dimensions.

In the category of topological spaces and continuous functions, a map $\pi : E \to B$ *has the homotopy lifting property (h.l.p.) with respect to M* if: for any homotopy $f : I \times M \to B$ and for any map $\tilde{f}_0 : M \to E$ lifting $f_0 = f|_{\{0\}M \times M}$ (i.e., so that $f_0 = \pi \tilde{f}_0$), there exists a homotopy $\tilde{f} : I \times M \to E$ lifting f (i.e., so that $f = \pi \tilde{f}$) with $\tilde{f}_0 = \tilde{f}|_{M \times \{0\}}$. A is a map $f : E \to B$, that has the h.l.p. with respect to any topological space M. The *fiber* of this fibration (over a point b_0 in B) is $F = F_{b_0} = f^{-1}(\{b_0\})$, where, given such a fibration, there is a *long exact sequence of morphisms* :

$$\cdots \to \pi_k(F) \to \pi_k(E) \to \pi_k(B) \to \pi_{k-1}(F) \to \cdots \to \pi_0(B). \tag{3.115}$$

Note that the last two arrows are not morphisms; they are functions.

Two points u_1, u_2 on $S(\mathbf{H})$ represent the same point on $P(\mathbf{H})$ iff there is a complex number z of modulus 1, such that $u_2 = zu_1$. Thus $P(\mathbf{H})$ is a quotient $P(\mathbf{H}) = S(\mathbf{H})/\mathbb{S}^1$, and one may consider the *fibration*

$$\mathbb{S}^1 \to S(\mathbf{H}) \to P(\mathbf{H}). \tag{3.116}$$

From the exact sequence of the fibration (3.116), it follows that $\pi_2(P(\mathbf{H})) = \mathbb{Z}$ and all the other homotopy groups of $P(\mathbf{H})$ are trivial.

The homotopy group $\pi_k(M, p_0)$ can be also described in terms of continuous functions between pairs, as set of homotopy classes of continuous functions from $(I^k, \partial I^k) \to (M, p_0)$, therefore one may define the *Hurewicz map* $H_k : \pi_k(M) \to H_k(M)$ by associating to the homotopy class of $F : I^k \to M$, its corresponding singular cubic homology class.

THEOREM 3.4.3. *(Hurewicz) If M is (k - 1)-connected, the Hurewicz map is an isomorphism $\forall j \leq k$ when $k \geq 2$ and also the Hurewicz homomorphism is an* epimorphism *from $\pi_{k+1}(M)$ to $H_{k+1}(M)$ whenever M is $(k-1)$-connected, for $k \geq 2$.*

As an application of Theorem 3.4.3, one may compute the homotopy groups of $\mathbb{P}(H)$). See exercise (58)

3.4.3 Differential Topology

In Differential Topology one studies the homeomorphism type of topological manifolds, by using differentiable structures on them, as well as differentiable functions between them. One of the key tools, Morse theory, has been instrumental in classifying manifolds [309]. A classic reference is Milnor (1963) [242].

For some smooth function on a d-dimensional manifold $f : \mathcal{M} \to \mathbb{R}$, consider a point $p \in \mathcal{M}$ where the differential $d_p f$ vanishes. In local coordinates we have $\partial f_x/\partial x^1(x(p)) = 0, \dots, \partial f_x/\partial x^d(x(p)) = 0$, where $f_x = f \circ x^{-1}$. Then the point p is called a *critical point* , and the value $f(p)$ is called a *critical value*. A critical point $p \in \mathcal{M}$ is called *non-degenerate* if the Hessian matrix $(\frac{\partial^2 f_x}{\partial x^i \partial x^j}(x(p)))_{i,j=1,\dots,d}$ is nonsingular. A function that has only non-degenerate critical points is aid to be a *Morse function.*

Since the Hessian matrix at a critical point is non-degenerate, it has a mixture of positive and negative eigenvalues. The number η of negative eigenvalues of the Hessian at a critical point called the *Morse index* at that point. In Milnor(1963)[242] one shows that

PROPOSITION 3.4.2. *(Morse Lemma) Given a critical point $p \in \mathcal{M}$ of f with index η, there is a chart (U,x) around p, $x = (x^1,\dots,x^d)$ so that*

$$f(x(q)) = f(0) - x^1(q)^2 - \cdots - x^\eta(q)^2 + x^{\eta+1}(q)^2 + \cdots + x^d(q)^2$$

for all $q \in U$.

DEFINITION 3.4.3. *Let $\mathbb{D}^k$ be the unit disk in $\mathbb{R}^k$, whose boundary $\partial\mathbb{D}^k$ is the unit sphere $\mathbb{S}^{k-1}$. Assume $\varphi : \mathbb{S}^{k-1} \to M$ is a continuous function. Take the disjoint union $M \sqcup \mathbb{D}^k = (\{0\} \times M) \cup (\{1\} \times \mathbb{D}^k)$ and consider the equivalence relationship, where $(0, \varphi(z)) \equiv (1, z), \forall z \in \mathbb{S}^{k-1}$. The quotient space under $\equiv$ with the quotient topology is the space $M \sqcup_\varphi e^k$, obtained by attaching to M a k-dimensional cell along φ.*

Based on this result, one is able to show that at a critical point $p \in \mathcal{M}$ of a Morse function, with $f(p) = a$ say, that the sublevel set $\mathcal{M}_{f \leq a}$ has the same homotopy type as that of the sublevel set $\mathcal{M}_{f \leq a-\varepsilon}$ (for some small $\varepsilon >$ 0) with an η-dimensional cell attached to it. In fact, for a compact $\mathcal{M}$, its homotopy type is that of a cell complex with one η-dimensional cell for each critical point of index η. This cell complex is known as a *CW complex*, if the cells are attached in the order of their dimension. The homology of a CW complex Y, can be obtained from the description of the cell attachment maps (see Greenberg and Harper (1981) [131]), with one generator of $H_q(Y)$ for each q-dimensional cell that does not bound a $q+1$-dimensional cell, therefore given a Morse function on a manifold $\mathcal{M}$, one may derive its homology $H_*(M)$.

As an illustration, due to Bubenik et al. (2010)[53], that will be used in Section 26.1.1, let us consider a real valued function f that is a mixture of two bump functions on the disk of radius 10 in $\mathbb{R}^2$, see Figure 3.4.

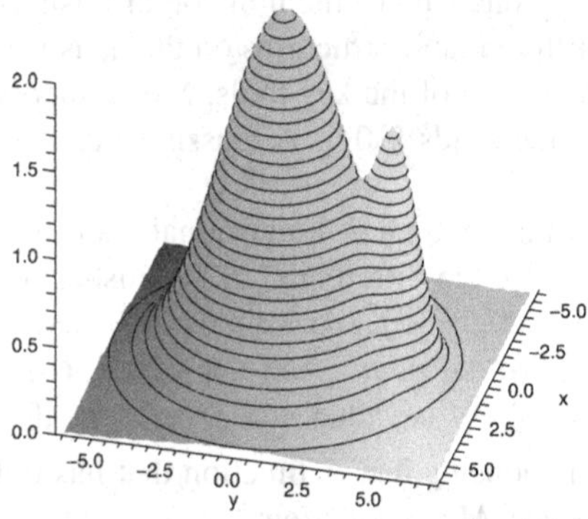

Figure 3.4 *A mixture of two bump functions and various contours below which are the sublevel sets. (Source: Bubenik et al. (2010), Figure 2.1. Reproduced by permission of American Mathematical Society).*

In this example, the maximum of f equals 2, so $\mathcal{M}_{f\leq 2} = \mathcal{M}$. This sublevel set is the disk and therefore has no interesting topology since the disk is contractible. In contrast, consider the sublevel sets when $r = 1$, 1.2, and 1.5 (see Figures 3.5, 3.6, and 3.7).

In these cases, the sublevel sets $\mathcal{M}_{f\leq r}$ have non-trivial topology, namely one, two and one hole(s) respectively, each of whose boundaries is one-dimensional. This topology is detected algebraically by the first integral homology group $H_1(\mathcal{M}_{f\leq r})$ which will be referred to as the homology of degree 1 at level r. This group enumerates the topologically distinct cycles in the sublevel set. In the first and third cases, for each integer $z \in \mathbb{Z}$, there is a cycle which wraps around the hole z times. We have $H_1(\mathcal{M}_{f\leq r}) = \mathbb{Z}$. In the second case, we have two generating non-trivial cycles and so $H_1(\mathcal{M}_{f\leq r}) = \mathbb{Z} \oplus \mathbb{Z}$.

By way of another example let us consider the d-dimensional sphere $\mathbb{S}^d = \{x \in \mathbb{R}^{d+1}, \|x\| = 1\}$ and the height function $h : \mathbb{S}^d \to \mathbb{R}$, with $h(x^1,\ldots,x^{d+1}) = x^{d+1}$. One may show that h is a Morse function, and using the indices at the two critical points, one may derive the homology $H_*(S^d)$. See exercise 56. The famous set of Morse inequalities (see Milnor (1963) [242]) states that if β_k is

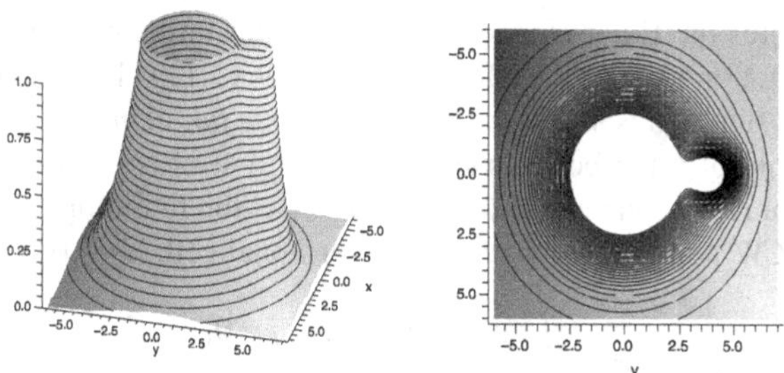

Figure 3.5 *The sublevel set at r = 1 has one hole. (Source: Bubenik et al. (2010), Figure 2.2. Reproduced by permission of American Mathematical Society).*

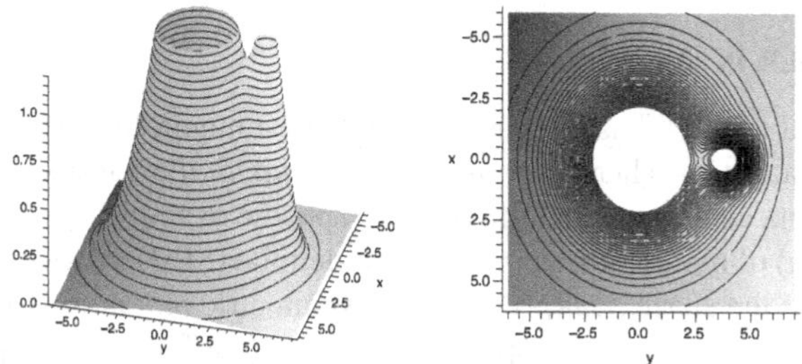

Figure 3.6 *The sublevel set at r = 1.2 has two holes. (Source: Bubenik et al. (2010), Figure 2.3. Reproduced by permission of American Mathematical Society).*

the $k-$th Betti number and m_k is the number of critical points of index k, then

$$\begin{aligned}
\beta_0 &\leq m_0 \\
\beta_1 - \beta_0 &\leq m_1 - m_0 \\
\beta_2 - \beta_1 + \beta_0 &\leq m_2 - m_1 + m_0 \qquad (3.117) \\
&\cdots \\
\chi(M) = \sum_{k=0}^{d} (-1)^k \beta_k &= \sum_{k=0}^{d} (-1)^k m_k
\end{aligned}$$

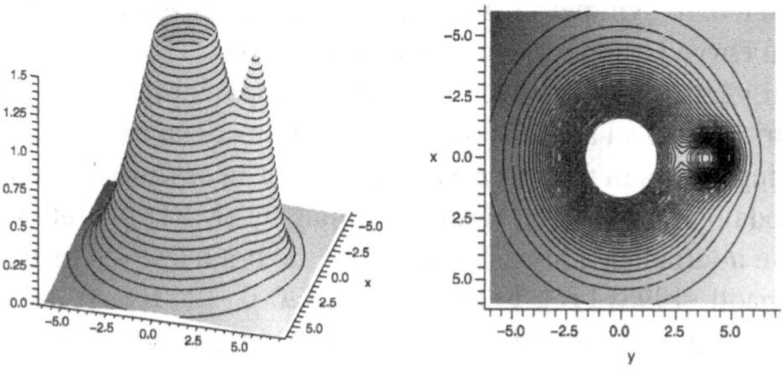

Figure 3.7 *The sublevel set at r = 1.5 has one hole. (Source: Bubenik et al. (2010), Figure 2.4. Reproduced by permission of American Mathematical Society).*

where χ denotes the Euler characteristic. Note that in the case of $\mathbb{C}P^d$, with the Morse function (3.166) the Morse inequalities (3.117) become equalities. Such a Morse function is said to be a *perfect Morse function*.

An example of a perfect Morse function is give in Nicolaescu (2011, p. 64)[254]. As a result, one can show that $\mathbb{C}P^d$, has the homotopy type of a $2d$ dimensional CW complex with one $2j$-dimensional cell, $\forall j = 0, \ldots, d$, therefore

$$H_k(\mathbb{C}P^d) = \begin{cases} \mathbb{Z}, & \text{for } k = 2j,\ j = 0, \ldots, d \\ 0, & \text{otherwise} \end{cases} \tag{3.118}$$

For details see exercise 57.

3.5 Manifolds in Statistics

The numerical space $\mathbb{R}^m$ is the basic example of a manifold arising as an object space in statistics. Therefore, *multivariate data analysis* is a key particular example of data analysis on manifolds. As a specific example, we consider the space $Sym^+(m)$ of $m \times m$ symmetric positive-definite (SPD) matrices, a convex open subset of the space $Sym(m)$ of all real symmetric $m \times m$ matrices. When endowed with the metric g_0 inherited from $Sym(m) \cong \mathbb{R}^{\frac{m(m+1)}{2}}$, $g_0(u,v) = Tr(uv)$, the space $Sym^+(m)$ is flat, and therefore provides an example in which the extrinsic sample mean has a simple closed-form expression. However, with this metric the analysis may require to keep track of the boundary of $Sym^+(m)$ in $Sym(m)$, since the space $(Sym^+(m), g_0)$ is incomplete.

As an alternative, Arsigny et al. (2006) [5] and Schwartzman (2006) [301] consider a $GL^+(m)$-invariant metric on $Sym^+(m)$ associated with the left action of α^+ given in equation (3.163) which is a transitive Lie group action. The isotropy group of α^+ is $SO(m)$. According to (3.73), given $H \in GL^+(m)$, with $\alpha^+(H, I_m) = S$, and $\forall U, V \in Sym(m) = T_{I_m}Sym^+(m)$ this *canonical metric* g_{can} is given by

$$g_{\text{can}}(S)(d_{I_m}\alpha^+_H(U), d_{I_m}\alpha^+_H(V)) = g_0(U, V), \tag{3.119}$$

which is independent of H (since the isotropy group of α^+ is $SO(m)$, which leaves invariant the metric g_0). $Sym^+(m)$, with respect to which $Sym^+(m)$ is a Cartan-Hadamard symmetric space. The space $Sym^+(m)$ with this metric is complete but curved, suggesting an intrinsic analysis.

In recent years, there has been a rapid development in the application of nonparametric statistical analysis on manifolds to medical imaging. In particular, data taking values in the space $Sym^+(3)$ appear in *diffusion tensor imaging* (DTI), a modality of magnetic resonance imaging (MRI) that allows visualization of the internal anatomical structure of the brain's white matter (Basser and C. Pierpaoli (1996) [14], LeBihan et al. (2001) [213]). At each point in the brain, the local pattern of diffusion of the water molecules at that point is described by a *diffusion tensor* (DT), a 3×3 SPD matrix. A DTI image is a 3D rectangular array that contains at every voxel (volume pixel) a 3×3 SPD matrix that is an estimate of the true DT at the center of that voxel. (Thus DTI differs from most medical-imaging techniques in that, at each point, what the collected data are used to estimate is a matrix rather than a scalar quantity.) At each voxel, the estimated DT is constructed from measurements of the diffusion coefficient in at least six directions in three-dimensional space. The eigenvalues of the DT measure diffusivity, an indicator of the type of tissue and its health, while the eigenvectors relate to the spatial orientation of the underlying neural fibers. The first papers on nonparametric inference for diffusion tensors are due to Osborne (2012) [257] and Osborne et al. (2013) [258].

Curves and surfaces naturally occurring in the real world provide another important class of manifolds with carrying data on them. In addition these elementary examples, some special manifolds or more generally, some orbifolds, play a key role in statistical data analysis.

REMARK 3.5.1. *Unlike vector spaces, or spaces of diffusion tensors, object spaces that are encountered in object data analysis, such as spaces of directions, spaces of axes, and shape spaces, which are described in the following subsections have the distinct property of being compact. For this reasons, in addition to usual extensions of the notion of mean vector, this property leads to new parameters and statistics associated with random objects, such as the* antimeans, *which are specific for data analysis on manifolds (see Patrangenaru et.al(2014a)[270]). Such new developments are to be the discussed in current and future research in Object Data Analysis (see Patrangenaru et. al(2015a)[271]).*

3.5.1 Spaces of Directions, Axial Spaces and Spaces of Frames

A simple example is the *unit circle* $\mathbb{S}^1 = \{z \in \mathbb{C}, \bar{z}z = 1\}$, which may be regarded as a sample space for periodic events, or for directions at a given location in a plane.

Statistics on a higher dimensional *sphere* $\mathbb{S}^m = \{\mathbf{x} \in \mathbb{R}^{m+1}, \mathbf{x}^\mathrm{T}\mathbf{x} = 1\}$ have also been studied by statisticians. Important contributions in this area is due to Watson (1983) [333]. *Directional data analysis* includes data analysis on spheres, on special orthogonal groups $SO(m)$, which are particular cases of *Stiefel manifolds* or on real projective spaces $\mathbb{R}P^m, m \geq 2$, also known as *axial data analysis*.

In vector - cardiography (Downs, 1972) [88] and in astronomy (Prentice, 1986, 1989 [283, 284]) one comes across observations on *Stiefel manifolds* $V_{3,2}$. The Stiefel manifolds $V_{m,r}$ is the set of orthonormal r-frames in $\mathbb{R}^m$. In can be shown that $V_{m,r}$ are homogeneous spaces that can be identified with the quotient space $O(m)/O(m-r)$. Hendriks and Landsman (1998) [154] studied asymptotic distributions of extrinsic means on Stiefel manifolds.

3.5.2 $\mathcal{G}$-Shape Spaces

An important class of object spaces having a manifold structure or a *manifold stratification* that is relevant in *landmark based high level image analysis*, is the class of *$\mathcal{G}$-shape spaces* associated with a transitive action of a Lie group $\mathcal{G}$ on the space of configurations of k labeled landmarks in $\mathbb{R}^m$ (see Patrangenaru and Patrangenaru (2004) [275]). To be more specific, given an action $\alpha : \mathcal{G} \times \mathbb{R}^m \to \mathbb{R}^m$, we consider the associated *diagonal action* $\alpha_k : \mathcal{G} \times (\mathbb{R}^m)^k \to (\mathbb{R}^m)^k$, given by:

$$\alpha_k(g,(x^1,\dots,x^k)) = (\alpha(g,x^1),\dots,(\alpha(g,x^k)). \tag{3.120}$$

The orbits of (3.120) are called *$\mathcal{G}$-shapes*. The quotient $(\mathbb{R}^m)^k/\mathcal{G}$, space of such orbits has a structure of orbifold with singularities. An obvious singularity is the orbit of $\mathbf{0_k} = (0,\dots,0) \in \mathbb{R}^k$. Even if this orbit is removed, singularities might still be present. A subspace of $(\mathbb{R}^m)^k/\mathcal{G}$, made of $\mathcal{G}$-shapes of generic configurations, will be called *$\mathcal{G}$-shape space*. Examples of $\mathcal{G}$-shape spaces that have a manifold structure are given subsequently in this chapter.

3.5.3 Kendall Shape Spaces

In this subsection, $\mathcal{G}$ is the group direct similarities of $\mathbb{R}^m$. A *similarity* is a function $f : \mathbb{R}^m \to \mathbb{R}^m$, that uniformly dilates distances, that is, for which there is $k > 0$, such that $\|f(x) - f(y)\| = k\|x - y\|, \forall x,y \in \mathbb{R}^m$. One can show (see Exercise 49) that a similarity is of the form

$$f(x) = Ax + b, A^T A = cI_m, c > 0. \tag{3.121}$$

A *direct similarity* is given by (3.121), where A has a positive determinant. Direct similarities form under composition, the *group of direct similarities*. The sample spaces considered here are due to David G. Kendall (1984) [177], who called them *shape spaces of k-ads in* $\mathbb{R}^m$, and used the notation Σ^k_m; technically, in the language of $\mathcal{G}$-shapes, these are spaces of *direct similarity shapes* of k-ads. A k-ad in m dimensions is an ordered set of points in $\mathbb{R}^m$. A comprehensive account of Kendall's shape spaces in arbitrary dimensions can be found in Kendall et al. (1999) [179].

The orbit of a k-ad under the action of the group of direct similarities, is called a direct similarity shape. Here we consider only k-ads in which the k points are not all equal. The *Kendall shape*(direct similarity shape) can be also obtained as follows. First one removes translation by centering the k-ad $\mathbf{x} = (x^1,\dots,x^k)$ to

$$\xi = (x^1 - \bar{x},\dots,x^k - \bar{x}). \tag{3.122}$$

Note that the set of all centered k-ads lie in a vector subspace L^m_k in $(R^m)^k$ of dimension $mk - m$, $L^m_k = \{\xi = (\xi^1,\dots,\xi^k) \in (\mathbb{R}^m)^k : \xi^1 + \dots + \xi^k = 0\}$, and,

$L_k^{m*} = L_k^m \backslash \{0\}$. If the size is not relevant, its effect is removed by scaling ξ to unit size as

$$\mathbf{u} = \frac{\xi}{|\xi|}. \tag{3.123}$$

Since size is not relevant in this case, its effect is removed by scaling ξ to unit size as $\mathbf{u} = \frac{\xi}{|\xi|}$. The quantity $\mathbf{u}$, often called *preshape*, lives on $S(L_k^{m*})$, the unit sphere in L_k^m called the *preshape sphere.* Note that $S(L_k^{m*}) \sim S^{mk-m-1}$, the unit sphere centered at the origin in $\mathbb{R}^{mk-m}$. Finally, the Kendall shape $[\mathbf{x}]$ of this k-ad is the orbit, of $\mathbf{u} = (u^1, \ldots, u^k)$ under all rotations around 0_m in $\mathbb{R}^m$ (Dryden and Mardia (1998, pp. 56-57)) [91]. That is,

$$[\mathbf{x}] = \{A\mathbf{u} = (Au^1, \ldots, Au^k) : A \in SO(m)\}. \tag{3.124}$$

Thus the *Kendall shape space*, set of direct similarity shapes of k-ads in m dimensions is the compact orbifold $\Sigma_m^k = S(L_k^{m*})/SO(m) \sim S^{mk-m-1}/SO(m)$.

For $m > 2$ (and in particular $m = 3$), $\Sigma_m^k = S(L_k^{m*})/SO(m)$ have singularities, for the same reasons as $S\Sigma_m^k$ does, as described above.

In the case $m = 2$, the Kendall shape space Σ_2^k has a particularly nice manifold structure though. Indeed, if we regarding each k-ad in the plane as an ordered set of k complex numbers $\mathbf{z} = (z^1, \ldots, z^k)$, the centered k-ad

$$\begin{aligned} \zeta &= (\zeta^1, \ldots, \zeta^k) \\ \zeta^j &= z^j - \bar{z}, \forall j = 1, \ldots, k, \end{aligned} \tag{3.125}$$

has the same direct similarity shape as $\mathbf{z} = (z^1, \ldots, z^k)$, and lies in the complex hyperplane

$$L_k = \{\zeta \in \mathbb{C}^k : \zeta^1 + \cdots + \zeta^k = 0\} \sim \mathbb{C}^{k-1}. \tag{3.126}$$

It we consider a second k-ad $\mathbf{z}' = (z'^1, \ldots, z'^k)$, having the same Kendall shape as the k-ad $\mathbf{z} = (z^1, \ldots, z^k)$, they differ by a translation, followed by a rotation and a scaling. That is

$$z'^j = \rho e^{i\theta} z^j + b, \forall j = 1, \ldots, k, \rho > 0, b \in \mathbb{C}, \theta. \tag{3.127}$$

From (3.127), the centered k-ad $\zeta' = (\zeta'^1, \ldots, \zeta'^k), \zeta'^j = z'^j - \overline{z'}, \forall j = 1, \ldots, k$ satisfies the following:

$$\zeta'^j = \rho e^{i\theta} \zeta^j, \forall j = 1, \ldots, k, \rho > 0. \tag{3.128}$$

If in (3.128) we set $\lambda = \rho e^{i\theta}$, then $\lambda \in \mathbb{C}^* = \mathbb{C}\backslash\{0\}$, and from this equation, we see that the two k ads have the same Kendall shape, if and only if

$$\zeta' = \lambda \zeta. \tag{3.129}$$

Note that ζ', ζ are nonzero complex vectors in L_k. The k-ads $\mathbf{z}, \mathbf{z}'$ *have the same Kendall shape if and only if the centered k-ads* ζ', ζ *differ by a complex multiple, which is same as saying* $[\zeta'] = [\zeta] \in \mathbb{P}(L_k)$. Therefore we obtained the following

THEOREM 3.5.1. *The Kendall planar shape space Σ_2^k can be identified with the complex projective space $\mathbb{P}(L_k)$. Moreover, since L_k has complex dimension $k-1$, $P(L_k) \simeq P(\mathbb{C}^{k-1}) = \mathbb{C}P^{k-2}$, therefore the Kendall planar shape analysis is data analysis on the complex projective space $\mathbb{C}P^{k-2}$.*

3.5.4 Planar Size-and-Shape Manifolds

In this example $m = 2$ and $\mathcal{G}$ is the group *roto-translations* of the Euclidean plane. Consider k-ads, $k > 2$, ordered set of k points in the numerical plane, not all the same. A manifold of interest in shape analysis is the *planar size-and-shape space* $S\Sigma_2^k$ (see Dryden and Mardia (1998, p. 57) [91]). Two planar k-ads have the same size-and shape if they differ by a *direct isometry of the Euclidean plane*. Translate a k-ad $\mathbf{x}$ by $-\bar{\mathbf{x}}$ (i.e. center $\mathbf{x}$) to get a k-ad $\xi = (\xi^1, \ldots, \xi^k) \in L_k^2$ given by (3.137). Two k-ads $\mathbf{x}, \mathbf{x}'$ have the same direct similarity size-and-shape if there is a $w \in \mathbb{S}^1 = \{\zeta \in \mathbb{C}, |\zeta| = 1\}$, such that $\xi' = w\xi$, where $\xi, \xi' \in L_k^2$ are the centered $\mathbf{x}, \mathbf{x}'$ respectively. Let $[\mathbf{x}]_S, [\mathbf{x}]$ and $r(\mathbf{x})$ be the size-and-shape, shape, and size of $\mathbf{x}$, respectively, where $r^2(\mathbf{x}) =: \sum_{i=1}^k |\xi^i|^2$. Define the one-to-one map $\Psi : S\Sigma_2^k \to (0,\infty) \times \Sigma_2^k$ given by

$$\Psi([\mathbf{x}]_S) = (r(\mathbf{x}), [\mathbf{x}]), \tag{3.130}$$

where Σ_2^k was introduced in subsection 3.5.3. Since an Euclidean similarity is obtained from an isometry, followed by a scaling it follows that

THEOREM 3.5.2. *The planar size-and-shape space $S\Sigma_2^k$ can be identified with $(0,\infty) \times \Sigma_2^k$, a differentiable manifold of dimension $2k-3$.*

In the representation (3.130), the second coordinate on the right side is the planar similarity shape of a k-ad, thus allowing a natural mathematical modeling for the change in shape with size growth.

3.5.5 Size-and-Shape Manifolds in Higher Dimensions

In this example the group $\mathcal{G}$ is the full group $\mathcal{E}_m$ of isometries of the m dimensional Euclidian space. The *size-and-reflection-shape* $[\mathbf{x}]_{RS}$ of the k-ad $\mathbf{x}$ in $\mathbb{R}^m$ is the $\mathcal{E}_n$-orbit of $\mathbf{x}$, under the action $\alpha(A,b),\mathbf{x}) = (Ax^1 + b, \ldots, Ax^k + b), A \in O(m), b \in \mathbb{R}^m$, of the $\mathcal{E}_n$ on the set of all centered k-ads. The terminology (size-and-reflection-shape) is due to Dryden and Mardia (1998, p. 57)) [91]. The set of all size-and-reflection-shapes of centered k-ads *in general position* ξ, i.e. k-ads for which $\{\xi_1, \ldots, \xi_k\}$ spans $\mathbb{R}^m$, is the *size-and-reflection-shape space* $SR\Sigma_{m,0}^k$. This space is a manifold, since the action of an orthogonal matrix on $\mathbb{R}^m$ is uniquely determined by its action on a basis of $\mathbb{R}^m$, and a centered k-ad in general position includes such a basis. Recall that, by the fundamental theorem of Euclidean geometry, *any isometry of the Euclidean space is a linear transformation* of $\mathbb{R}^m$ of the form

$$x' = Bx + b, B^T B = I_m, \tag{3.131}$$

therefore, if one centers the k-ad $\mathbf{x}$ to $\xi = (x^1 - \overline{x}, \ldots, x^k - \overline{x}) \in L_k^m$, then $[\mathbf{x}]_{RS} = [\xi]_{RS}$, and if a k-ad $\mathbf{x}'$, having the same reflection-shape as $\mathbf{x}$, differs from $\mathbf{x}$ by the isometry (3.131), then

$$\overline{x'} = B\overline{x} + b. \tag{3.132}$$

Therefore $\xi' = B\xi$, where $\xi' \in L_k^m$ is obtained by centering the k-ad $\mathbf{x}'$. Thus, if we set $L_{k,m,0} = \{\xi \in L_k^m, \text{rk}\xi = m\}$, the manifold $S\Sigma_{m,0}^k$ can be represented as a quotient $L_{k,m,0}/O(m)$ and the manifold dimension of $S\Sigma_{m,0}^k$ is $km - \frac{m(m+1)}{2}$.

3.5.6 Size-and-Reflection Shape Manifolds in Higher Dimensions

In higher dimensions, it is easier, and equally important, to study shapes of configurations with respect to the full group of isometries, as opposed to studying $SO(m)$-shapes only. The *size-and-reflection shape* of the k-ad $\mathbf{x}$ is orbit of the centered k-ad ξ under the action of the orthogonal group, defined for *the set* $\mathcal{C}_{k,m,0}$ *of all k-ads* $\mathbf{x}$ *in general position,* i.e., k-ads such that the linear span of $\{\xi^j, j = 1, \ldots, k\}$ (of the centered $\mathbf{x}$) is $\mathbb{R}^m$:

$$[\mathbf{x}]_{RS} = \{A\xi : A \in O(m)\}. \tag{3.133}$$

A k-ad is said to be if it is not contained in any hyperplane of $\mathbb{R}^m$; in particular for such a k-ad, $k > m$. The set of all size-and-reflection shapes of k-ads in general position in $\mathbb{R}^m$, is the *size-and-reflection-shape manifold* $SR\Sigma_{m,0}^k$ which is the set

$$SR\Sigma_{m,0}^k = \{[\mathbf{x}]_R, \mathbf{x} \text{ in general position }\} = \{[\mathbf{x}]_{\mathbf{RS}}, \text{rk}\mathbf{x} = \mathbf{m}.\}, \tag{3.134}$$

where rk$\mathbf{x}$ is the rank of $\mathbf{x}$. The manifold dimension, codimension of the $O(m)$ -orbits in $(\mathbb{R}^m)^k$, is dim $SR\Sigma_{m,0}^k = km - \frac{m(m+1)}{2}$.

3.5.7 Linear Shape Spaces and Affine Shape Spaces as Grassmannians

This subsection is concerned with a description of object spaces of *linear shape spaces* and *reflection-affine shapes.* Here, the group $\mathcal{G}$ considered in Subsection 3.5.2 is the *general linear group* $GL(m, \mathbb{R})$, in the case of linear shapes, and the *affine group* in m dimensions, $Aff(m)$, in the case of reflection-affine shapes.

The *linear shape space* as an object space was first considered in connection with analysis of 2D gel electrophoresis images (see [235]), and the *affine shape space,* was first considered for data analysis via nonparametric statistical analysis of landmark based affine shape data in which each observation $\mathbf{x} = (x_1, \ldots, x_k)$ is a k-ad, $k > m+1$ in m dimensions, where the points $x^j, j = 1, \ldots, k$ represent k locations on a scene, such as landmarks on a remote planar scene observed in aerial or satellite images (see Patrangenaru and Mardia (2013) [274]). An *affine transformation* $\gamma = \gamma_{A,b}$ of $\mathbb{R}^m$ is given by

$$\gamma(x) = y = Ax + b, A \in GL(m, \mathbb{R}), b \in \mathbb{R}^m. \tag{3.135}$$

Let $(x_1, x_2, ..., x_k)$ be a k-ad of points in $\mathbb{R}^m$. The affine group $Aff(m)$, is the group of transformations given in equation (3.135). The action α of $Aff(m)$ on $\mathbb{R}^m$ induces a diagonal action $\alpha_k : Aff(m) \times (\mathbb{R}^m)^k \to (\mathbb{R}^m)^k$, given by

$$\alpha_k((A,b),(x_1, \ldots, x_k)) = (\gamma_{A,b}(x_1), \ldots, \gamma_{A,b}(x_k)), A \in GL(m, \mathbb{R}), b \in \mathbb{R}^m. \tag{3.136}$$

To be specific, we consider only *k-ads in general position.* The *reflection-affine shape space* $A\Sigma_m^k$ of such k-ads, is the space of orbits of this action. The reflection-affine shape of the k-ad $\mathbf{x}$ is labeled $a\tilde{\sigma}(\mathbf{x})$, and the reflection-affine shape space is the space of $A\Sigma_m^k \cong (\mathbb{R}^m)^k / Aff(m)$, of all such orbits.
To identify topologically this space of objects, we first remove translation (the vector b in equation (3.135)), by *centering the k-ad* $\mathbf{x} = (x_1, \ldots, x_k)$ to

$$\begin{aligned} \xi &= (\xi_1, \ldots, \xi_k) \\ \xi_j &= x_j - \bar{x}, \forall j = 1, \ldots, k. \end{aligned} \tag{3.137}$$

Note that the set of centered k-ads lies in the vector subspace L_k^m in $(R^m)^k$ of dimension $mk - m, L_k^m = \{\xi = (\xi_1, \ldots, \xi_k) \in (\mathbb{R}^m)^k : \xi_1 + \cdots + \xi_k = 0\}$, and, $L_k^{m*} = L_k^m \backslash \{0\}$. $k \geq m$ of set $(\mathbb{R}^m)^k$. The group of affine transformations acts diagonally on left on $(\mathbb{R}^m)^k$. The action (3.136) induces an action β_k of $GL(m,)$ on L_k^m, given by

$$\beta_k((A,(\xi_1, \ldots, \xi_k)) = (A\xi_1, \ldots, A\xi_k), A \in GL(m, \mathbb{R}), (\xi_1, \ldots, \xi_k) \in L_k^m, w \tag{3.138}$$

whose orbit set is in a one to one correspondence with the orbit set of the action (3.136). Indeed, the *reflection-affine shape space* can be regarded as the space of orbits of the action (3.138) on L_k^m. Since $k - 1 > m$, the centered k-ads (at $0 \in \mathbb{R}^m$) $(\xi_1, \ldots, \xi_k)$ and $(\eta_1, \ldots, \eta_k)$ (of $\mathbf{x}$ and $\mathbf{y}$ respectively), are in the same orbit of (3.136), iff the k-ads $\mathbf{x}$ and $\mathbf{y}$ have the same reflection-affine shape.

DEFINITION 3.5.1. *A configuration is in general position if the affine subspace spanned by* $x_1, \ldots, x_k$ *is* $\mathbb{R}^m$.

If we consider only k-ads in general position, then ξ and η are on the same orbit of β_k iff their columns span the same m-dimensional space. Therefore a reflection-affine shape in L_k^1 of a k-ad in general position, can be represented as point on $G_m(L_k^1)$. We arrived at the following

THEOREM 3.5.3. *If $k > m+1$ then the reflection-affine shape space of k-ads in general position in $\mathbb{R}^m$ is in a one-to-one correspondence with the space of m-dimensional linear subspaces of L^1_k.*

COROLLARY 3.5.1. *The reflection affine shape space of k-ads in general position in $\mathbb{R}^m$, with $k > m+1$, is in a one-to-one correspondence with the Grassmann manifold $G_m(\mathbb{R}^{k-1})$.*

This result, due to Sparr (1998) [315], is given here the a new proof (see Patrangenaru and Mardia[274]). The *reflection-affine shape space $A\Sigma^k_m$* of k-ads in m-dimensions can be stratified as follows: $A\Sigma^k_m = (A\Sigma^k_m)_m \cup (A\Sigma^k_m)_{m-1} \cup \cdots \cup (A\Sigma^k_m)_0$, where $(A\Sigma^k_m)_m$ is the space of reflection-affine shapes if the configuration is in general position in $\mathbb{R}^m$, and $(A\Sigma^k_m)_r$ is the space of reflection-affine shapes if the configuration spans a subspace of dimension $r, \forall r = 0,\dots,m$. We will focus on $(A\Sigma^k_m)_m$. If the space spanned by the columns of η^T is the same as the space spanned by the columns of ξ^T, then $\mathbf{y}$ and $\mathbf{x}$ have the same reflection-affine shape.

THEOREM 3.5.4. *(Sparr) If $k > m+1$ then the reflection-affine shape space of k-ads in general position in $\mathbb{R}^m$ is in a one-to-one correspondence with the space of m-dimensional linear subspaces of L^1_k.*

Recalling the stratification of the reflection-affine shape space, in view of Theorem 3.5.4, $A\Sigma^k_m$ can be viewed as a union of Grassmann manifolds: $A\Sigma^k_m = G_m(L^1_k) \cup G_{m-1}(L^1_k) \cup \dots G_1(L^1_k)$.

A *direct affine transformation* of $\mathbb{R}^m$ is given by

$$y = Ax + b, A \in GL(m,\mathbb{R}), b \in \mathbb{R}, det(A) > 0. \tag{3.139}$$

The *direct affine shape $a\sigma(\mathbf{x})$* of the k-ad $\mathbf{x}$, is similarly defined as the orbit of $\mathbf{x}$ under the *diagonal* action α_k of the *group of direct affine transformations $\mathcal{A}ff_+(m)$* on $(\mathbb{R}^m)^k$, which is induced by the action of $\mathcal{A}ff_+(m)$ given by (3.139) on each of the $\mathbb{R}^m$ factors of $(\mathbb{R}^m)^k$. In other words, $(x_1,\dots,x_k)$ and $(y_1,\dots,y_k)$ have the same affine shape if $(y_1,\dots,y_k) = \alpha_k((A,b),(x_1,\dots,x_k), det(A) > 0)$. Since to each k-ad $(x_1,\dots,x_k)$, we associate the k-ad $(\xi_1,\dots,\xi_k)$ which is centered at $0 \in \mathbb{R}^m$ ($\bar{\xi}_k = 0$), and the k-ads $(x_1,\dots,x_k)$ and $(\xi_1,\dots,\xi_k)$ have the same affine shape, it follows that two k-ads $\mathbf{x}$ and $\mathbf{y}$ have the same reflection-affine shape iff their corresponding centered k-ads $y_j = Ax_j + b$, $\forall j = 1,\dots,k, det(A) > 0$. Summing up over j results in $\bar{y}_k = A\bar{x}_k + b, det(A) > 0$. Therefore, the k-ads $\mathbf{x}$ and $\mathbf{y}$ have the same direct affine shape iff the center k-ads ξ, η, given by $x_j - \bar{x}_k = \xi_j$ and $y_j - \bar{y}_k = \eta_j$ differ by an orientation preserving linear transformation, $\eta_j = A\xi_j, \forall j = 1,\dots,k, det(A) > 0$.

This filters out translation, thus α_k induces an action β_k of $GL_+(m,\mathbb{R})$ on L^m_k, given by $\beta_k(A,\xi) = A\xi = \eta$.

The *direct affine shape space $A_+\Sigma^k_m$* of k-ads in m-dimensions can be stratified as follows: $A_+\Sigma^k_m = (A_+\Sigma^k_m)_m \cup (A_+\Sigma^k_m)_{m-1} \cup \cdots \cup (A_+\Sigma^k_m)_0$, where $(A_+\Sigma^k_m)_m$ is the space of direct affine shapes if the configuration is in general position in

$\mathbb{R}^m$, and $(A_+\Sigma_m^k)_r$ is the space of direct affine shapes if the configuration spans a subspace of dimension $r, \forall r = 0, \ldots, m-1$. We will focus on $(A_+\Sigma_m^k)_m$. If the space spanned by the columns of η^T is the same as the space spanned by the columns of ξ^T, then y and x have the same direct affine shape.

THEOREM 3.5.5. *The direct affine shape space of k-ads in general position in* $\mathbb{R}^m, k > m+1$ *is in a one-to-one correspondence with the Grasmann manifold* $G_m^o(L_k^1)$, *space of oriented m-dimensional linear subspaces* L_k^1.

Similarly, for $k > m$, the orbits of the diagonal action of the general linear group $\mathcal{G} = GL(m,\mathbb{R})$ on k-ads in $\mathbb{R}^m$ form the orbifold $L\Sigma_m^k$, *linear shape space* that admits a *stratification*

$$L\Sigma_m^k = L\Sigma_m^{k,m} \cup \cdots \cup L\Sigma_m^{k,1},$$

where $L\Sigma_m^{k,r}$ is the space of k-ads $x_1, \ldots, x_k$ such that the sampling matrix $\mathbf{x} = (x_1 \ldots x_k)$ has rank r. For fixed r the rows of the sampling matrix span an r-plane in $\mathbb{R}^k$, therefore we have the following result

THEOREM 3.5.1. *For* $r = 1, \ldots, m$ *the linear space* $L\Sigma_m^{k,r}$ *is in a one-to-one correspondence with* $G_r(\mathbb{R}^k)$.

In particular the top stratum $L\Sigma_m^{k,m}$, of linear shapes of k-ads in general position, form a Grassmann manifold $G_m(\mathbb{R}^k)$.

3.5.8 Projective Shape Spaces

A d - dimensional *projective subspace* of $\mathbb{R}P^m$ is a projective space $P(V)$, where V is a $(d+1)$-dimensional vector subspace of $\mathbb{R}^{m+1}$. A codimension one projective subspace of $\mathbb{R}P^m$ is also called a *hyperplane*. The *linear span* of a subset D of $\mathbb{R}P^m$ is the smallest projective subspace of $\mathbb{R}P^m$ containing D. We say that k points in $\mathbb{R}P^m$ are in *general position* if their linear span is $\mathbb{R}P^m$. If k points in $\mathbb{R}P^m$ are in general position, then $k \geq m+2$.

The numerical space $\mathbb{R}^m$ can be *embedded* in $\mathbb{R}P^m$, preserving collinearity. An example of such an *affine embedding* is

$$h((u^1, \ldots, u^m)) = [u^1 : \ \ldots \ : u^m : 1] = [\tilde{u}], \tag{3.140}$$

where $\tilde{u} = (u^1, \ldots, u^m, 1)^T$, and in general, an affine embedding is given for any $A \in Gl(m+1,\mathbb{R})$, by $h_A(u) = [A\tilde{u}]$. The complement of the range the embedding h in (3.140) is the hyperplane $\mathbb{R}P^{m-1}$, set of points $[x^1 : \cdots : x^m : 0] \in \mathbb{R}P^m$.

Conversely, the *inhomogeneous (affine) coordinates* $(u^1, \ldots, u^m)$ of a point $p = [x^1 : x^2 : \cdots : x^{m+1}] \in \mathbb{R}P^m \backslash \mathbb{R}P^{m-1}$ are given by

$$u^j = \frac{x^j}{x^{m+1}}, \forall j = 1, \ldots, m. \tag{3.141}$$

Consider now the linear transformation from $\mathbb{R}^{m+1}$ to $\mathbb{R}^{m'+1}$ defined by the matrix $B \in M(m+1, m'+1; \mathbb{R})$ and its kernel $K = \{x \in \mathbb{R}^{m'+1}, Bx = 0\}$. The *projective map* $\beta : \mathbb{R}P^{m'} \backslash P(K) \to \mathbb{R}P^m$, *associated with* B is defined by $\beta([x]) = [Bx]$. In particular, a *projective transformation* β of $\mathbb{R}P^m$ is the projective map associated with a nonsingular matrix $B \in GL(m+1, \mathbb{R})$ and its action on $\mathbb{R}P^m$:

$$\beta([x^1 : \cdots : x^{m+1}]) = [B(x^1, \ldots, x^{m+1})^T]. \tag{3.142}$$

In affine coordinates (inverse of the affine embedding (3.140)), the projective transformation (3.142) is given by $v = f(u)$, with

$$v^j = \frac{a^j_{m+1} + \sum_{i=1}^m a^j_i u^i}{a^{m+1}_{m+1} + \sum_{i=1}^m a^{m+1}_i u^i}, \forall j = 1, \ldots, m \tag{3.143}$$

where $\det B = \det((a^j_i)_{i,j=1,\ldots,m+1}) \neq 0$. An affine transformation of $\mathbb{R}^m, v = Au + b, A \in GL(m, \mathbb{R}), b \in \mathbb{R}^m$, can be regarded as a particular case of projective transformation α, associated with the matrix $B \in GL(m+1, \mathbb{R})$, where

$$B = \begin{pmatrix} A & b \\ 0^T_m & 1 \end{pmatrix}. \tag{3.144}$$

A *projective frame* in an m dimensional projective space (or *projective basis* in the computer vision literature, see e.g. Hartley (1993) [142]) is an ordered set of $m+2$ projective points in general position. An example of projective frame in $\mathbb{R}P^m$ is the *standard projective frame* is $([e_1], \ldots, [e_{m+1}], [e_1 + \cdots + e_{m+1}])$.

In projective shape analysis it is preferable to employ coordinates invariant with respect to the group PGL(m) of projective transformations. A projective transformation takes a projective frame to a projective frame, and its action on $\mathbb{R}P^m$ is determined by its action on a projective frame, therefore if we define the *projective coordinate(s)* of a point $p \in \mathbb{R}P^m$ w.r.t. a projective frame $\pi = (p_1, \ldots, p_{m+2})$ as being given by

$$p^\pi = \beta^{-1}(p), \tag{3.145}$$

where $\beta \in PGL(m)$ is a projective transformation taking the standard projective frame to π, these coordinates have automatically the invariance property.

REMARK 3.5.2. *Assume* $u, u_1, \ldots, u_{m+2}$ *are points in* $\mathbb{R}^m$, *such that* $\pi = ([\tilde{u}_1], \ldots, [\tilde{u}_{m+2}])$ *is a projective frame. If we consider the* $(m+1) \times (m+1)$ *matrix* $U_m = [\tilde{u}^T_1, \ldots, \tilde{u}^T_{m+1}]$, *the projective coordinates of* $p = [\tilde{u}]$ *w.r.t.* π *are given by*

$$p^\pi = [y^1(u) : \cdots : y^{m+1}(u)], \tag{3.146}$$

where

$$v(u) = U_m^{-1} \tilde{u}^T \tag{3.147}$$

and

$$y^j(u) = \frac{v^j(u)}{v^j(u_{m+2})}, \forall j = 1, \ldots, m+1. \tag{3.148}$$

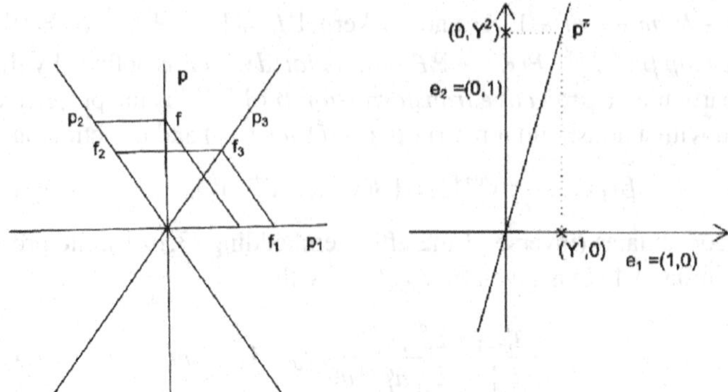

Figure 3.8 *Projective coordinates for m = 1. Projective frame $\pi = (p_1, p_2, p_3)$ and a projective point p (left) and projective coordinates of p with respect to π (right).*

This representation of projective coordinates is displayed in Figure 3.8 for m = 1. In this figure on the left one constructs the coordinates y^1, y^2 of f with respect to the frame $\pi = ([f_1], [f_2], [f_3])$ and on the right hand side one shows the corresponding projective point $[y^1 : y^2]$. Note that in our notation, the superscripts are reserved for the components of a point whereas the subscripts are for the labels of points. The projective coordinate(s) of x are given by the point $[z^1(x) : \cdots : z^{m+1}(x)] \in \mathbb{R}P^m$.

DEFINITION 3.5.2. *Two configurations of points in $\mathbb{R}^m$ have the same projective shape if they differ by a projective transformation of $\mathbb{R}^m$.*

Unlike similarities or affine transformations, projective transformations of $\mathbb{R}^m$ do not have a group structure under composition (the domain of definition of the composition of two such maps is smaller than the maximal domain of a projective transformation in $\mathbb{R}^m$), therefore rather than considering the projective shapes of configurations in $\mathbb{R}^m$, we consider projective shapes of configurations in $\mathbb{R}P^m$. A *projective shape* of a k-ad (configuration of k landmarks or labeled points) is the orbit of that k-ad under projective transformations with respect to the diagonal action

$$\alpha_k(h; p_1, \ldots, p_k) = (\alpha_h(p_1), \ldots, \alpha_h(p_k)). \tag{3.149}$$

Since the action (3.149) of $h \in PGL(m)$ on $[x] \in \mathbb{R}P^m$, when expressed in inhomogeneous coordinates (3.141) reduces to (3.143), if two configurations Γ_1, Γ_2 of points in $\mathbb{R}^m$ have the same projective shape, then $h(P_1), h(P_2)$ have the same projective shape in $\mathbb{R}P^m$ (h is the affine embedding given by (3.140)).

REMARK 3.5.3. *Patrangenaru (2001) [268] considered the set $G(k,m)$ of k-ads $(p_1,...,p_k), k > m+2$ for which $\pi = (p_1,...,p_{m+2})$ is a projective frame. PGL(m) acts simply transitively on $G(k,m)$ and the* projective shape space $P\Sigma_m^k$, *is the quotient* $G(k,m)/PGL(m)$. *Using the projective coordinates* $(p_{m+3}^{\pi},\dots,p_k^{\pi})$ *given by* (3.145) *one can show that* $P\Sigma_m^k$ *is a manifold diffeomorphic with* $(\mathbb{R}P^m)^{k-m-2}$.

The projective frame representation has two useful features: firstly, the projective shape space has a manifold structure, thus allowing to use an asymptotic theory for means on manifolds in Bhattacharya and Patrangenaru (2003, 2005) [42, 43], an secondly, it can be extended to infinite dimensional , such as projective shapes of curves, as shown in Munk et al. (2008) [252]. This approach has the advantage of being inductive in the sense that each new landmark of a configuration adds an extra marginal axial coordinate, thus allowing to detect its overall contribution to the variability of the configuration as well as correlation to the other landmarks.

In general, the manifolds considered in this book are homogeneous spaces. The first results in density estimation on homogeneous manifolds are due to Beran (1968) [19].

REMARK 3.5.4. *A new, label independent methodology to projective shape analysis of k-ads in general position, was developed by Kent and Mardia (2012) [185] (see also Kent and Mardia (2006) [184]). However, their approach was not yet used in practical applications beyond dimension* 1, *possibly due to the fact that it requires nonlinear approximations to the matrix solution of the equation in A*

$$A = \frac{m}{k}\sum_{i=1}^{k}\frac{x_i x_i^T}{x_i^T A^{-1} x_i} \tag{3.150}$$

in terms of a k-ad of points in $\mathbb{R}P^m$ *given in their spherical representation , the key step in Kent and Mardia(2012)[185] description of a projective shape.*

3.6 Exercises

Exercise 15. *The following are manifolds: (i) Any open set in* $\mathbb{R}^m$ *is an m-dimensional manifold. (ii) The unit sphere in* $\mathbb{R}^m$ *is an (m-1)-dimensional manifold. (iii) The torus* $\mathcal{T}^N = (\mathbb{S}^1)^N$ *is an N-dimensional manifold.*

Exercise 16. *Show that the orthogonal group* $O(m)$ *is a submanifold of dimension* $\frac{m(m-1)}{2}$ *of* $\mathcal{M}_m(\mathbb{R})$, *the Euclidean space of* $m \times m$ *matrices.*
Hint (Victor Pambuccian)). $O(m) = \{A \in \mathcal{M}_m(\mathbb{R}), AA^T = I_m\}$, *Note that* $F(A) = AA^T - I_m = 0$ *is symmetric, and the dimension of the space of symmetric matrices is* $dim(Sym(m,\mathbb{R}) = m + (m-1) + \cdots + 1 = \frac{m(m+1)}{2}$
$O(m)$ *is the set of zeros of the function F, therefore it suffices to show that F is differentiable, and for any matrix* $A \in O(m)$, *the differential* $d_A F : M_m(\mathbb{R}) \to Sym(m,\mathbb{R})$ *is onto. Since*

$$F(A+H) - F(A) = AH^T + HA^T + HH^T, \tag{3.151}$$

and since the function $L : M_m(\mathbb{R}) \to Sym(m,\mathbb{R})$,

$$L(H) = AH^T + HA^T \tag{3.152}$$

is linear and the function $\omega : M_m(\mathbb{R}) \to Sym(m,\mathbb{R})$,

$$\omega(H) = HH^T \tag{3.153}$$

has the property that $lim_{H\to 0}\frac{\omega(H)}{H} = lim_{H\to 0}H^T = 0$, *we have*

$$F(A+H) = F(A) + L(H) + \omega(H),\ lim_{H\to 0}\frac{\omega(H)}{H} = 0 \tag{3.154}$$

Thus F is differentiable at A and the differential at A is $d_A F = L$ *that is*

$$d_A F(H) = AH^T + HA^T. \tag{3.155}$$

$d_A F$ *is onto for any* $A \in O(m)$, *if the equation* $d_A F(X) = Y$ *has a solution* $X \in M_m(\mathbb{R}), \forall Y \in Sym(m,\mathbb{R})$. *Take* $X = \frac{1}{2}YA$, *then* $d_A F(X) = AX^T + XA^T = A(\frac{1}{2}YA)^T + \frac{1}{2}YAA^T = \frac{1}{2}AA^T Y^T + \frac{1}{2}Y = \frac{1}{2}Y^T + \frac{1}{2}Y = Y$ *is symmetric* $\frac{1}{2}Y + \frac{1}{2}Y = Y$.

Exercise 17. *Show that the projective space* $\mathbb{C}P^d$ *is an analytic manifold of real dimension 2d.*

Exercise 18. *Show that if* $p > 1$ *is an integer, the function* $z \to z^p$ *is an immersion from* S^1 *to* $\mathbb{R}^2$, *which is not an embedding.*

Exercise 19. *Show that if* $\mathcal{M}$ *is an m-dimensional manifold, then its tangent space* $T\mathcal{M}$ *is a 2m dimensional manifold. Also show that the projection map* $\Pi : T\mathcal{M} \to \mathcal{M}$, *given by* $\Pi(\frac{dc}{dt}(0) = c(0)$, *is a differentiable map .*

Exercise 20. *Note that even if an immersion* $j : M \to \mathbb{R}^d$ *is one-to-one, its range j(M) may not be a manifold. Give such an example of immersion* $j : (0,1) \to \mathbb{R}^2$.

Exercise 21. *(i) Show that* $O(m+1)$ *acts transitively on* $\mathbb{R}^m$ *via* $A \cdot [x] = [Ax]$. *(ii) Show that the restriction of the action* (13.35) *to* $O(m) \times Sym^+(m+1)$, *is an action of the orthogonal group on the set of positive semi-definite symmetric matrices. (iii) Show that Veronese-Whitney map* (12.3) *in* (12.3) *is an equivariant embedding.*

Exercise 22. *Show that the projective plane* $\mathbb{R}P^2$ *is a surface that can not be embedded in* $\mathbb{R}^3$.

Exercise 23. *Show that the surface of a bagel-like solid of revolution, obtained by rotation of a circle or radius* $R = 1$ *around a line in the plane of that circle at distance 2 from its center, is a submanifold of the Euclidean space of codimension one (a surface). This is a torus of revolution.*

Solution. *We have the following parametrization of the torus, as a surface of revolution:*

$$\begin{aligned} x &= (2+\cos\theta)\cos\phi \\ y &= (2+\cos\theta)\sin\phi \\ z &= \sin\theta, \theta, \phi \in \mathbb{R}. \end{aligned} \tag{3.156}$$

We eliminate the parameters θ, ϕ *in* (3.156) *and obtain the implicit equation*

$$(2-\sqrt{x^2+y^2})^2+z^2=1, \tag{3.157}$$

which, after removing the square root is equivalent to the implicit equation for a point $u=(x,y,z)$ *on the torus:*

$$F(x,y,z)=(x^2+y^2+z^2+3)^2-16(x^2+y^2)=0 \tag{3.158}$$

Note that the gradient $\nabla_u F=0$ *only if* $u=0$ *or* u *satisfies to* $x^2+y^2=5, z=0$. *Such a point* u *does not lie on the torus, therefore for any point* u *on the torus,* $d_uF \neq 0$, *thus* $\mathrm{rank} d_uF=1$ *and the torus of revolution, set* $F^{-1}(0)$ *is a surface in* $\mathbb{R}^3$ *(submanifold of codimension 1 of* $\mathbb{R}^3$*).*

Exercise 24. *Show that the m dimensional torus in 3.1.13 can be embedded as a hypersurface in an Euclidean space.*

Hint. *For* $p=2$ *see exercise 22. Then use induction. For details, see Patrangenaru (1984)[263].*

Exercise 25. The Lie algebra of a Lie group. *Recall that the Lie algebra of a Lie group is the tangent space at the identity, of the group, together with an internal operation generated by the commutator operation in the group. Assume* $\mathfrak{g}$ *is the Lie algebra of the Lie group G. If* $g(t), h(t)$ *are curves in G with* $g(0)=h(0)=1_G$ *whose tangent vectors at* 0 *are respectively* $v, w \in \mathfrak{g}$, *then the Lie algebra operation* $[v,w]$ *is defined as the tangent vector at* 0 *of the curve*

$$c(t)=g(\sqrt{t})h(\sqrt{t})g^{-1}(\sqrt{t})h^{-1}(\sqrt{t}) \tag{3.159}$$

Exercise 26. *Show that the Lie algebra bracket operation is anticommutative.*

Exercise 27. *We showed that in* $\mathfrak{gl}(m)$, *Lie algebra of the general linear group* $GL(m,\mathbb{R})$ *the bracket operation is given by* $[v,w]=vw-wv$.

Exercise 28. *Show that for any linear Lie group (Lie subgroup of* $GL(m,\mathbb{R})$*), the Lie algebra bracket is the same as the bracket in* $\mathfrak{gl}(m)$.

Exercise 29. *A classical result clams that the exponential map* $A \to e^A = I + A + \frac{1}{2!}A^2 +$ *gives a local diffeomorphism from the Lie algebra of a Lie group to a neighborhood of the identity of that group.*

Exercise 30. *Show that if* $A \in \mathfrak{so}(m)$ *then* $e^A \in SO(m)$.

Exercise 31. *A homogeneous space $\mathcal{M}$ can be represented as a quotient K/H of two groups. Indeed, because a transitive group action implies that there is only one group orbit, $\mathcal{M}$ is isomorphic to the quotient space G/H where H is the isotropy group G_u. The choice of $u \in \mathcal{M}$ does not affect the isomorphism type of G/G_u because all of the isotropy groups are conjugate.*

Exercise 32. *Show that the sphere $\mathbb{S}^m$ is the quotient space $O(m+1)/O(m)$.*

Exercise 33. *Show that the projective space in example 3.1.16 is a Hilbert manifold.*
Hint. *Define the distance between two vector lines as their angle, and, given a line $\mathbb{L} \subset \mathbb{H}$, show a neighborhood U_L of L can be mapped, via a homeomorphism φ_L onto an open neighborhood of the orthocomplement $\mathbb{L}^\perp$ by using the decomposition $\mathbb{H} = \mathbb{L} \oplus \mathbb{L}^\perp$. Then for two perpendicular lines L_1 and L_2, show that the transition maps $\varphi_{L_1} \circ \varphi_{L_2}^{-1}$ are differentiable as maps between open subsets in $\mathbb{L}_1^\perp$, respectively in $\mathbb{L}_2^\perp$. Use a countable orthobasis of $\mathbb{H}$ and the lines $L_n, n \in \mathbb{N}$ generated by the vectors in this orthobasis to cover $P(\mathbb{H})$ with the open sets $U_{L_n}, n \in \mathbb{N}$. Finally use the fact that for any line $\mathbb{L}^\perp$ and $\mathbb{H}$ are isometric as Hilbert spaces.*

Exercise 34. *What is the surface area of the torus in Exercise 20?*

Exercise 35. *Let $g = g_x$ and $h = g_y$ be the local matrices associated with the Riemannian structure g on M. Show that*

$$g_{ij}(x(u)) = \frac{\partial y^a}{\partial x^i}(x(u))\frac{\partial y^b}{\partial x^j}(x(u))h_{ab}(y(u)) \tag{3.160}$$

Exercise 36. The Riemannian structure on the unit sphere. *Consider the unit sphere $S^m \in \mathbb{R}^{m+1}$. Let $x = x(u)$ be the inverse of the stereographic projection from the North pole, $u : \mathbb{R}^m \to S^m$. Show that*

$$u(x) = \left(\frac{2x}{\|x\|^2+1}, \frac{\|x\|^2-1}{\|x\|^2+1} \right) \in S^m \subset \mathbb{R}^{m+1}. \tag{3.161}$$

and compute the induced Riemannian structure ${}_xg_{ij}$ with respect to these coordinates, and the infinitesimal squared geodesic distance ds^2 in equation (3.34).

Exercise 37. *Show that the unit sphere S^m has constant sectional curvature.*

Exercise 38. *Find all the surfaces of rotation of constant positive curvature in $\mathbb{R}^3$, and show that the only surfaces of rotations of positive curvature that are complete are the round spheres.*
Hint. *Integrate the differential equation* (3.54)*, in the case $k(u,v) = \frac{1}{R^2}$.*

Exercise 39. Poincaré model for hyperbolic geometry. *An example of an abstract Riemannian structure is*

$$ds^2 = \frac{dx^2 + dy^2}{(1-x^2-y^2)^2} \tag{3.162}$$

defined on a unit open disk in plane. This is the Poincaré hyperbolic disk. *Show that the Poincaré disk is a Riemannian manifold of constant curvature* $k = -1$.

Exercise 40. *Show that the great circles are the geodesics on the sphere.*

Exercise 41. *Find the geodesics of the Poincaré disk in Exercise 24, and show that the exponential map is defined on the entire tangent plane to the Poincaré disk at any point on this Riemannian manifold, thus the Poincaré disk is complete as a Riemannian manifold.*

Exercise 42. *Find two perpendicular geodesics passing through the point* $u = (1,0,0)$ *of the torus in Exercise 2.*

Exercise 43. *Show that the set of isometries of a Riemannian manifold* $I_g\mathcal{M}$. *Show that the isotropy group* H_p *in 3.2.6 is a subgroup of* $I_g\mathcal{M}$
Show that the group $O(m+1)$ *acts on* S^m *with the Riemannian metric induced by* $\mathbb{R}^{m+1}$ *as a group of isometries, via the matrix multiplication* $A \cdot p = Ap$. *What is the isotropy group of this action at the North Pole* $p = (0,\ldots,0,1)$?

Exercise 44. *Show that the actions* α *given by*

$$\alpha : GL(m) \times Sym^{+}(m) \to Sym^{+}(m), \alpha(H,S) = HSH^{T}, \tag{3.163}$$

and its restriction and $\alpha^{+} : GL^{+}(m) \times Sym^{+}(m) \to Sym^{+}(m)$ *are transitive Lie group actions on* $Sym^{+}(m)$.

Exercise 45. *Usually, in directional statistics, one regards an axial distribution as one corresponding to an* S^m *valued random variable* X *such that* X *and* $-X$ *have the same distribution (Watson (1983) [333], Ch. 5, Fisher et al. (1996) [116], Beran and Fisher (1998)[18]). Show that this is the same as giving a probability distribution on* $\mathbb{R}P^m$.

Exercise 46. *(i) Show that* $\mathbb{R}P^m$ *is a quotient of the action by isometries of the multiplicative group* $-1, 1$ *on* S^m *via* $(-1) \cdot x = -x$.
(ii) Show that $\mathbb{R}P^m$ *carries a Riemannian structure of constant sectional curvature 1.*
(iii) Show that if $p = [1:0:\ldots,0]$, *the cut locus* $C(p) = \mathbb{R}P^{m-1}$ *is the projective hyperplane from infinity of the point* p.

Exercise 47. *(i) Show that the Veronese–Whitney map defined in* (3.1.15) *is an* $SO(m)$ *- equivariant embedding of* $\mathbb{R}P^m$ *into the space* $(S(m+1,\mathbb{R})$ *of real symmetric matrices, where the action of* $SO(m)$ *on* $(S(m+1,\mathbb{R})$ *is induced by the adjoint action* (13.35).
(iii) Show that the Euclidean distance d_0 *on* $(S(m+1,\mathbb{R})$ *is given by*

$$d_0(A,B) = Tr((A-B)(A-B)^t) = Tr((A-B)^2), \tag{3.164}$$

(iv) Show that the embedding in (i) is an isometric embedding, where $\mathbb{R}P^m$ *carries a Riemannian structure in exercise 2.(ii) and* $(S(m+1,\mathbb{R})$ *has the Euclidean distance in* (3.164).

Exercise 48. *Show that the projective space $\mathbb{R}R^m$ is the quotient space $O(m+1)/(O(m)\times O(1))$.*

Exercise 49. *Show that a similarity of $\mathbb{R}^m$ is given by an equation of the form* (3.121).
Hint. *Assume $k>0$, such that $\forall x,y, \|f(x)-f(y)\| = k\|x-y\|$. Define the function $g:\mathbb{R}^m\to\mathbb{R}^m, g(x)=\frac{1}{k}f(x)$. Note that g is an isometry, and by the Fundamental theorem of geometry, it is a composition of at most $m+1$ reflections in hyperplanes. Then show that a reflection is an affine map, and use induction, to prove the claim.*

Exercise 50. *Consider the centering matrix $\mathbf{H_k}=\mathbf{I_k}-\frac{1}{\mathbf{k}}\mathbf{1_k}\mathbf{1_k^T}$.*
(i) Show that $\mathbf{H_k^2}=\mathbf{H_k}\mathbf{H_k^T}=\mathbf{H_k^T}=\mathbf{H_k}$, and rank$(\mathbf{H_k})=k-1$. (ii) Show that the range of the linear endomorphism h of $\mathbb{C}^k$, given by $h(\mathbf{z})=\mathbf{H_k}\mathbf{z}$ is the subspace L_k in (3.126)*. (iii) If we regard $\mathbb{C}^{k-1}$ as a the subspace $\{(\mathbf{z},\mathbf{0}),\mathbf{z}\in\mathbb{C}^{\mathbf{k-1}}\}$ of $\mathbb{C}^k$, show that the restriction $h_1=h|\mathbb{C}^{k-1}$ is an isomorphism of vector spaces, and the inverse of h_1 is given by $h_1^{-1}(\zeta)=H^T\zeta$, where H is the matrix of the first $k-1$ columns of $\mathbf{H_k}$. (iv) The linear isomorphism h_1^{-1} induces a bijection from the projective space $P(L_k)$ to $\mathbb{C}P^{k-2}$. (v) If $\mathbf{z}\in\mathbb{C}^k=\mathcal{M}(1,k;\mathbb{C})$ is a k-ad, then the map ψ given by $\psi(\sigma(\mathbf{z}))=[\mathbf{H^T}\mathbf{H_k}\mathbf{z}]$ is a bijection from Kendall's shape space Σ_2^k to $\mathbb{C}P^{k-2}$.*

Exercise 51. *Let $(S(k-1,\mathbb{C})$ be the set of $k-1$ by $k-1$ self adjoint matrices. Define the (complex) Veronese Whitney map $j:\mathbb{C}^{k-2}\to(S(k-1,\mathbb{C})$ by*

$$j([\zeta])=\zeta\zeta^*, \tag{3.165}$$

(i) Show that j is an embedding.(ii) Show that the unitary group $U(k-1)$ of $(k-1)\times(k-1)$ complex matrices A with $A^*A=I$ acts on $\mathbb{C}P^{k-2}$ via $A\cdot[\zeta]=[A\zeta]$, and on $(S(k-1,\mathbb{C})$ via the adjoint action (13.35)*. (ii) Show that the embedding* (3.165) *is $U(k-1)$-equivariant.*

Exercise 52. *Show that the complex projective space $\mathbb{C}P^m$ can be represented as the quotient space $U(m+1)/(U(m)\times U(1))$.*

Exercise 53. *Show that the Grassmann manifold $G_m(\mathbb{R}^d)$, has a structure of $m(d-m)$ dimensional differentiable manifold.*

Exercise 54. *Show that Grassmann manifold of m-dimensional subspaces in $\mathbb{R}^{m+r}$ can be represented as $O(m+r)/(O(m)\times O(r))$.*

Exercise 55. *Let $x=(1,2,\ldots,m+1)\in\mathbb{R}^{m+1}$ be the point. Find the representation as a point $\mathbb{R}P^m$, of the projective shape of the $m+2$-ad $([e_1],\ldots,[e_{m+1}],[e_1+\cdots+e_{m+1}],[x])$ with respect to the standard projective frame $([e_1],\ldots,[e_{m+1}],[e_1+\cdots+e_{m+1}]$.*

Exercise 56. *Show that the height function $h:\mathbb{S}^d\to\mathbb{R}$, with $h(x^1,\ldots,x^{d+1})=x^{d+1}$ is a Morse function on the unit sphere. Derive the homology $H_*(S^d)$.*

Exercise 57. *Show that the function* $f : \mathbb{C}P^d \to \mathbb{R}$:

$$f([z^0, \ldots, z^d] = \frac{\sum_{j=0}^{d} j|z_j|^2}{\sum_{j=0}^{d} |z_j|^2}. \tag{3.166}$$

is a perfect Morse function, and prove that the homology groups $H_*(\mathbb{C}P^d)$ *are given by* (3.118).

Hint. *Recall that the sets* $A_k, k = 0, \ldots, d, A_k = \{[z] \in \mathbb{C}P^d, z^k \neq 0\}$ *give an atlas on* $\mathbb{C}P^d$, *and for* $k = 0, \ldots, d$, *the affine coordinates* $v^j = v^j(k) = \frac{z^j}{z^k}$ *are defined for* $[z] \in A_k$. *an open set in* $\mathbb{C}^d$. *Show that* $f|A_k$ *has only one critical point* $p_k \in A_k$, *with* $v^j(k) = 0, j = 0, \ldots, d, j \neq k$, *and* p_k *is nondegenerate to the index* $\eta(p_k) = 2k$.

Exercise 58. *The homotopy groups of the projective space of a separable complex Hilbert space* $\mathbb{P}(H)$ *are given by*

$$\pi_k(\mathbb{P}(H)) = \begin{cases} \mathbb{Z}, & \text{for } k = 2, \\ 0, & \text{otherwise} \end{cases} \tag{3.167}$$

Chapter 4

Consistency of Fréchet Moments on Manifolds

4.1 Introduction

In a visionary paper, Fréchet (1948) introduced the notion of *random element* in a metric space, and noted that those new concepts and results "are setting the bases of a new Chapter, that we believe will insure a great future of Probability theory" as they have important implications in both Statistics and Experimental Sciences. In that paper, Fréchet defined the mean of a probability measure on a separable metric space $(\mathcal{M},\rho)$, and hinted at a proof of the consistency of the sample mean for a random sample from such a distribution. *Consistency*, this important feature required of an estimator, was given a complicated first proof by Ziezold (1977) [345]. Essentially Ziezold showed that the *Fréchet sample mean set* is a consistent estimator of the *Fréchet mean set* if $(\mathcal{M},\rho)$ is compact, and according to Huckemann (2011)[164], Ziezold's proof goes through without any difficulty in the case that (M,ρ) is complete as well. Bhattacharya and Patrangenaru (2003)[42] gave a new proof of the consistency of the Fréchet sample mean on a complete separable metric space, which includes the case of a *random object (r.o.)* on a complete space with a manifold stratification. They also introduced the *Fréchet total variance* of a r.o. in a 2002 paper [41], and showed that this parameter is consistently estimated by the *Fréchet total sample variance* if $(\mathcal{M},\rho)$ is complete. Given a probability measure Q on a manifold $\mathcal{M}$, the Fréchet moments of Q were called by Patrangenaru (1998)[266] *extrinsic moments*, if ρ is induced by the *chord distance* via an embedding of $\mathcal{M}$ into an Euclidean space, respectively *intrinsic moments*, if

ρ is the Riemannian distance induced by a Riemannian structure g on $\mathcal{M}$. A generic consistency result for *extrinsic sample means on submanifolds*, due to Hendricks (1990a) [151], was extended by Patrangenaru (1998) [266] and Bhattacharya and Patrangenaru (2003) [42] to extrinsic sample means on *j-nonfocal distributions* on a manifold $\mathcal{M}$, with an embedding j

4.2 Fréchet Means and Cartan Means

In this section we consider a $\mathcal{M}$*-valued random object (r.o.)* X, i.e. a measurable map on a probability space $(\Omega, \mathcal{A}, Pr)$ into $(\mathcal{M}, \mathcal{B}_{\mathcal{M}})$, where $\mathcal{B}_{\mathcal{M}}$ denotes the *Borel sigma-algebra generated by open subsets of* $\mathcal{M}$. To each r.o. X, we associate a probability measure $Q = P_X$ on $\mathcal{B}_{\mathcal{M}}$ given by $Q(B) = Pr(X^{-1}(B))$, therefore from now, when we are referring to probability measures or distributions on $\mathcal{M}$, we mean probability measures defined on $\mathcal{B}_{\mathcal{M}}$. The aim of this section is to develop nonparametric statistical inference procedures for *measures of location and spread of distributions on arbitrary complete manifolds.*

In statistics one mainly considers two types of distances on a manifold $\mathcal{M}$: a geodesic distance (arc distance), that is the Riemannian distance ρ_g associated with a Riemannian structure g on $\mathcal{M}$ (see Chapter 3) and a chord distance, the distance ρ_j induced by the Euclidean distance on $\mathbb{R}^N$ via an embedding $j : \mathcal{M} \to \mathbb{R}^N$, that is given by

$$\rho_j(p,q) = \|j(q) - j(p)\|_0^2. \tag{4.1}$$

An *intrinsic data analysis* on a manifold is a statistical analysis of a probability measure, using a Riemannian distance based statistics, while an *extrinsic data analysis* is a statistical analysis based on a chord distance based statistics. If for a given probability measure Q on a metric space $(\mathcal{M}, \rho)$, the function ${}_p\rho$: $\mathcal{M} \to \mathbb{R}, {}_p\rho(q) = \rho(p,q)$, is in $\mathcal{L}_Q^r(\mathcal{M})$, we define the *r-th Fréchet function* on $\mathcal{M}$ that associates to each point $p \in \mathcal{M}$ its expected r-th power of the distance to a random point on $\mathcal{M}$, $\mathcal{F}_r(p) = \int \rho^r(p,x)Q(dx)$. In statistical analysis on manifolds, the most studied Fréchet function, is $\mathcal{F}_2$, therefore for the rest of this chapter $\mathcal{F} = \mathcal{F}_2$:

$$\mathcal{F}(p) = \int \rho^2(p,x)Q(dx), \tag{4.2}$$

DEFINITION 4.2.1. *If Q is a probability measure on $\mathcal{M}$, the total Fréchet variance of Q, $t\Sigma_{\mathcal{F}}(Q)$, is the the infimum of the value of the Fréchet function; this is the lowest value of $\mathcal{F}$ on $\mathcal{M}$ if $(\mathcal{M}, \rho)$ is complete. If $\rho = \rho_g$ is a geodesic distance on $\mathcal{M}$ the total Fréchet variance is called the total intrinsic variance $t\Sigma_{g,I}(Q)$. If ρ is the chord distance induced on $\mathcal{M}$ via an embedding $j : \mathcal{M} \to \mathbb{R}^N$ the total Fréchet variance is called the total extrinsic variance $t\Sigma_{j,E}(Q)$. These indices of spread are also labeled $t\Sigma_I$ and $t\Sigma_E$, if we consider a given Riemannian structure g or embedding j, and study only one distribution Q.*

For the rest of this chapter we will assume that for any distance ρ considered, $(\mathcal{M},\rho)$ is complete as a metric space.

EXAMPLE 4.2.1. *The terminology in definition 4.2.1 is motivated by the classical notion of total variance of a random vector in multivariate analysis. For a given multivariate distribution Q of a random vector X on $\mathbb{R}^N$, having finite second moments, the intrinsic total variance of X is the mean and the total variance of X, $\sum_{i=1}^N \sigma_i^2$, sum of variances σ_i of the marginal distributions X_i of X.*

DEFINITION 4.2.2. *Let Q be a probability measure on $\mathcal{M}$ with a distance ρ. The set of minimizers of $\mathcal{F}$ in* (4.2) *is called the* Fréchet mean set. *If $\mathcal{F}$ has a unique minimizer, this is called the Fréchet mean of Q, and is labeled $\mu_{\mathcal{F}}(Q)$. If (M,g) is a Riemannian manifold, the Fréchet mean (set) with respect to arc distance ρ_g is called the* intrinsic mean (set) *of Q. If the intrinsic mean exists, it is labeled $\mu_{g,I}(Q)$, or $\mu_g(Q)$ or μ_g, or μ_I. The intrinsic mean set is labeled $I(g,Q)$ or $I(Q)$. If $j:\mathcal{M}\to\mathbb{R}^N$ is an embedding, the Fréchet mean (set) with respect to the induced chord distance is called the* extrinsic mean (set) *of Q. If the extrinsic mean exists, it is labeled $\mu_{j,E}(Q)$, or $\mu_E(Q)$ or μ_j, or μ_E.*

EXAMPLE 4.2.2. *The uniform distribution on a compact Riemannian manifold, is a probability measure whose p.d.f. relative to the volume measure* (25.6) *is given by $f(p) = 1/vol(\mathcal{M}), \forall p \in \mathcal{M}$. The intrinsic mean set of the uniform distribution on a Riemannian homogeneous space $\mathcal{M}$ is the entire space $\mathcal{M}$.*

In general, the Fréchet mean set was introduced and first studied by Ziezold (1977) [345]. The points in the intrinsic mean set bear different names in the literature. Kobayashi and Nomizu (1996) [199], Karcher (1977) [174] and Emery and Mokobodzki (1991) [108] call them *centers of mass* or *barycenters*, while Kendall (1990) [180] and Le (1998) [210] use the name of *Karcher means* and Pennec and Ayache (1998) [280] call them *Fréchet expectations*. In fact the **earliest considerations on sample means on Riemannian manifolds are due to Cartan** (1927)[57]; it would most appropriate to call a minimizer of the second order Fréchet function associated with an empirical distribution for a Riemanian distance, a *Cartan mean*. Therefore, often an intrinsic sample mean will be called a Cartan mean.

REMARK 4.2.1. *The set of parameters of a finite dimensional probability distributions model was equipped by C. R. Rao (1945) [287] with a Riemannian metric, called the Fisher information metric. This led to subsequent differential geometric developments with applications to second order efficiency by Efron (1975) [95] and Amari (1985) [3]. In this context, mean values defined by Oller and Corcuera (1995) [256] turn out to be critical points of $\mathcal{F}$, while the intrinsic means defined here are minimizers of $\mathcal{F}$. This explains, for example, why in Oller and Corcuera (1995) [256] the von Mises distribution on S^d is found to have two mean values, while in fact there is only one intrinsic mean.*

If Q is a probability measure on a complete Riemannian manifold $(\mathcal{M},g)$, let $C(q)$ be the cut-locus of a point $q \in \mathcal{M}$. We set $C(Q) = \cup_{q \in I(Q)} C(q)$. For a point $q \in \mathcal{M}$, with $Q(C(q)) = 0$, we denote by $\lambda_Q = \lambda_{Q,q}$ the *image measure of* Q *under* Exp_q^{-1} on $\mathcal{M} \backslash C(q)$.

THEOREM 4.2.1. *Assume $(\mathcal{M}, g)$ is a complete connected Riemannian manifold. Let $I(Q)$ be the intrinsic mean set of Q and set $C(Q) = \cup_{q \in I(Q)} C(q)$. (a) If there is a point p on $\mathcal{M}$ such that $\mathcal{F}(p)$ is finite, then the intrinsic mean set is a nonempty compact set. (b) If $q \in I(Q)$ and $Q(C(Q)) = 0$, then*

$$\int_{V(q)} v \lambda_Q(dv) = 0. \tag{4.3}$$

(c) Suppose $(\mathcal{M}, g)$ has nonpositive curvature, and $\mathcal{M}$ is simply connected. Then every probability measure Q on $\mathcal{M}$ has an intrinsic mean, provided $\mathcal{F}(p)$ is finite for some p.

Proof. (a) It follows from the triangle inequality (for ρ_g) that if $\mathcal{F}(p)$ is finite for some p, then $\mathcal{F}$ is finite and continuous on $\mathcal{M}$. To show that a minimizer exists, let l denote the infimum of $\mathcal{F}$ and let $p_n \in \mathcal{M}$ be such that $\mathcal{F}(p_n) \to l$ as $n \to \infty$. By the triangle and the Schwartz inequalities, and by integration w.r.t. Q, one has

$$\begin{aligned} \rho_g^2(p_n, p_1) &\leq 2\rho_g^2(p_n, x) + 2\rho_g^2(x, p_1), \forall\, x \in \mathcal{M}, \\ \rho_g^2(p_n, p_1) &\leq 2(\mathcal{F}(p_n) + \mathcal{F}(p_1)). \end{aligned} \tag{4.4}$$

Hence since $F(p_n)$ $(n \geq 1)$ is a bounded sequence, so is p_n $(n \geq 1)$. By completeness of $\mathcal{M}$, p_n has a subsequence converging to some point p^*. Then $\mathcal{F}(p^*) = l$, so that p^* is a minimizer. Also the inequalities (4.4) applied this time to p^* and an arbitrary minimizer μ, show that $\rho_g^2(\mu, p^*) \leq 4l$. In other words, the set of minimizers is bounded. It is also a closed set, since its complement is clearly open, proving compactness of the intrinsic mean set.

To prove (b), note that $Exp_q(V(q))$ has Q-probability 1. Consider an arbitrary point x in $Exp_q(V(q))$; then with probability one there is a unique geodesic, say $\gamma_{x,\mu}$ joining x and μ with $\gamma_{x,\mu}(0) = x$, $\gamma_{x,\mu}(1) = \mu$. Also let $\mu_v(t)$ be the geodesic starting at μ $(\mu_v(0) = \mu)$ with tangent vector v $((d\mu_v(t)/dt)(0) = v)$. Let $\alpha_{v,x}$ be the angle made by the vectors tangent to these geodesics at μ. Then (see Helgason (1978)[147], p. 77, and Oller and Corcuera (1995)[256], Proposition 2.10)

$$d_\mu \mathcal{F}(v) = 2 \int \rho_g(x, \mu) \|v\| \cos(\alpha_{v,x}) Q(dx). \tag{4.5}$$

Select a point $q \in I(Q)$ and write the integral in (4.5) in normal coordinates on $T_q\mathcal{M}$. If $\mu \in I(Q)$, then μ is a critical point of $\mathcal{F}$. Then we select $\mu = q$, and evaluate the right hand side of (4.5) at $v = v_i = \frac{\partial}{\partial x^i}$. Note that given that Exp_q is a radial isometry, the right hand side of (4.5) in this case is $2\int x^i \lambda_Q(dx)$, where x^i are the normal coordinates of an arbitrary point of $Exp_q(V(q))$. Then in such coordinates, (4.5) becomes (4.3)

For part (c) of the theorem, we adapt the proof of Kobayashi and Nomizu (1996)[199], Theorem 9.1, to our situation as follows. By part (a) there is a point q in the intrinsic mean set. By theorem 3.2.4, since $\mathcal{M}$ is simply connected and complete, $C(q) = \emptyset$, and we define a map G on $\mathcal{M}$ by

$$G(p) = \int_{\mathcal{M}} \|Exp_q^{-1}(p) - v\|^2 \lambda_Q(dv). \tag{4.6}$$

Since on a simply connected manifold of nonpositive curvature Exp_q is expanding, we have $G(p) \leq \mathcal{F}(p)$. On the other hand by part (b), $G(p) = G(q) + \|Exp_q^{-1}(p)\|^2$ and, since Exp_q is a radial isometry, $\mathcal{F}(q) = G(q)$. Therefore, q is in fact the unique minimizer of $\mathcal{F}$. If $\mathcal{M}$ has points with positive curvature or is not simply connected, the existence of the intrinsic mean is not granted in general. If $\mathcal{M}$ is flat a sufficient condition for the existence of the intrinsic mean is that the support of Q is contained in a geodesically convex open normal neighborhood of $\mathcal{M}$ and $\mathcal{F}(p)$ is finite for some p. In general, if the infimum of the injectivity radii is a positive number $r(\mathcal{M})$ and the *scalar curvature* of $(\mathcal{M}, g)$ is bounded from above by $(\pi/r(\mathcal{M}))^2$, and if the support of Q is contained in a closed geodesic ball $\overline{B}_\rho$ of radius $\rho = r(\mathcal{M})/4$, then the intrinsic mean exists. To see this, note that, when restricted to the closed geodesic ball $\overline{B}_{2\rho}$, $\mathcal{F}$ has a unique minimum at some point in $\overline{B}_\rho$ (see Karcher (1977), Theorem 1.2). Clearly, this minimum value is no more than ρ^2. On the other hand, if $p \in (\overline{B}_{2\rho})^c$, then $\mathcal{F}(p) \geq \rho_g^2(p, \overline{B}_\rho) > \rho^2$. This proves the uniqueness of the minimum of $\mathcal{F}$ in $\mathcal{M}$, when the support of Q is contained in $\overline{B}_\rho$. Necessary and sufficient conditions for the existence of the intrinsic mean of absolutely continuous radially distributed probability measures on $\mathbb{C}P^{d/2}$ are given in Le (1998) and in Kendall et al. (1999).

DEFINITION 4.2.3. *If $C(q)$ has Q-measure zero, for some $q \in \mathcal{M}$, an intrinsic moment w.r.t. a given set of normal coordinates of an arbitrary order $s = (s^1, \ldots, s^d) \in Z_+^d$ can be defined by*

$$\int x^s \lambda_Q(dx), where x^s = (x^1)^{s^1} \ldots (x^d)^{s^d} \tag{4.7}$$

if the integral in (4.7) *is finite.*

REMARK 4.2.2. *Recall that the first considerations on intrinsic sample means are due to Élie Cartan in his 1927 edition of his* Geometry of Riemannian Spaces *(see Cartan (1946) [59] for the second edition). For this historical reason, if the intrinsic sample mean would carry a person's name, that should be* Cartan mean.

A general constructive algorithm for the Cartan mean, due to Groisser (2004)[133], is given here. The idea is to express the Cartan mean as a fixed point of a one parameter group of diffeomorphisms generated by a vector field

X on $\mathcal{M}$ (see Chapter 3 for a definition), or equivalently as the unique zero of X in a certain convex neighborhood.

DEFINITION 4.2.4. *Let $(A,d_A),(B,d_B)$ be metric spaces and let $k \in [0,1)$. We call a map $F: A \to B$ a contraction with constant k if $d_B(F(x),F(y)) \leq d_A(x,y), \forall x,y \in A$.*

The *Groisser algorithm* is proved using a version of the following Contracting Mapping Theorem:

THEOREM 4.2.2. *Let $B = B_\rho(p_0)$ be an open ball in a metric space (A,d_A), with $(closure(B),d_A)$ complete. Suppose that $B \subset U \subset A$, that $F: U \to A$ is a contraction with constant k, and that $d(p_0,F(p_0)) < (1-\rho)k$. Then F preserves B and has a unique fixed point p. Furthermore $p \in B$ and $\lim_{n\to\infty} F_n(q) = p, \forall q \in B$.*

The maps we use arise from certain vector fields, perhaps defined only locally, on a Riemannian manifold. To describe these maps, let ∇ be the Levi-Civita connection on a Riemannian manifold $(\mathcal{M},g)$, not assumed complete. If X is a C^1 vector field defined on some open set $V \subset \mathcal{M}$, then at each point $p \in V$ we can view the covariant derivative $\nabla_{.}X : T_p\mathcal{M} \to T_p\mathcal{M}$, given by $v \to \nabla_v X$ as a linear endomorphism of $T_p\mathcal{M}$. Call X *nondegenerate* on a subset $U \subset V$ if this endomorphism $(\nabla_v X)_p$ is invertible for all $p \in U$. When referring to bounds on $((\nabla_v X)_p)^{-1}$ and other linear transformations, throughout this paper we use the operator norm: $\|T\| = sup_{\|v=1\|}\|T(v)\|$. A C^1 vector field X defined on an open set in $\mathcal{M}$ and nondegenerate on a subset U defines a map $\Phi_X : U \to M$ by

$$\Phi_X(p) = Exp_p(((\nabla_v X)_p)^{-1}X_p, \tag{4.8}$$

assuming that $Exp_p(-((\nabla_v X)_p)^{-1}X_p)$ is defined for all $p \in U$. (In this section we use both X_p and $X(p)$ to denote the value of a vector field X at a point p.) Note that zeroes of X are fixed-points of Φ_X, and if X is not too large pointwise then the converse is true as well. Groisser (2004)[133] proved the following:

THEOREM 4.2.3. *Let $(\mathcal{M},g)$ be a Riemannian manifold and let $U \subset \mathcal{M}$ be open. Given $\varepsilon > 0, k_1 > 0, k_2 > 0$, let $X_{\varepsilon,k_1,k_2}(U)$ denote the set of nondegenerate vector fields X on U satisfying the following conditions pointwise on U: (i) $\|X\| \leq \varepsilon$ (ii)$\|(\nabla_v X)^{-1}\| \leq k_1$, and (iii) $\nabla\nabla X\| \leq k_2$. If both εk_1 and $k_2 k_1^{-1}$ are sufficiently small, and $X \in X_{\varepsilon,k_1,k_2}(U)$, then $\Phi_X : U \to \mathcal{M}$ is a contraction, where the distance function on U is the geodesic distance on $\mathcal{M}$. If U is a ball $B = B_\rho(p_0)$, and if ρ is sufficiently small and ε, k_1, k_2 are as above, then there exists a positive $\varepsilon_1 \leq \varepsilon$ such that if $\|X(p_0) \leq \varepsilon_1, \|$ then Φ_X preserves B and hence has a unique fixed point p in B; the point p is also the unique zero of X in B. For all q in some possibly smaller open ball centered at p_0, the iterates $(\Phi_X)n(q)$ converge to p.*

Now let $q_1,\ldots,q_n$ be a finite set of a geodesic convex subset $U \subset \mathcal{M}$, and let the vector field Y on U defined by

$$Y(p) = \frac{1}{n}\sum_{i=1}^{n} Log_p(q_i) \in T_p\mathcal{M}, \tag{4.9}$$

where Log_p is the inverse of the Riemannian exponential map on $T_p\mathcal{M}$. This inverse is globally defined if $(\mathcal{M}, g)$ is a Hadamard-Cartan manifold (see Chapter 3).

For the structure of probability measures which are invariant under a group of isometries one has the following simple result.

PROPOSITION 4.2.1. *Suppose K is a group of isometries of $(\mathcal{M}, g)$ which leaves the measure Q invariant. Then the intrinsic mean set is left invariant by K. In this case Q induces a quotient measure on the space of orbits $\mathcal{M}/K$ and the mean set of Q is a union of orbits.*

Proof. Since $\rho_g(p,q) = \rho_g(\tau(p), \tau(q))$ for all $p, q \in \mathcal{M}$, if Q is invariant under τ then one has $\mathcal{F}(\tau(p)) = \mathcal{F}(p)$ (See (4.2)). In particular, this is true when p is a minimizer of $\mathcal{F}$ and $\tau \in K$. The claim follows from these observations.

4.2.1 Consistency of Fréchet Sample Means

DEFINITION 4.2.5. *Let $X_1, \dots, X_n$ be independent random variables with a common distribution Q on a metric space $(\mathcal{M}, \rho)$, and consider their empirical distribution $\hat{Q}_n = \frac{1}{n}\sum_{k=1}^{n} \delta_{X_k}$. The Fréchet sample mean (set) is the Fréchet mean (set) of $\hat{Q}_n$, i.e. the (set of) minimizer(s) $\hat{p}$ of $p \to \frac{1}{n}\sum_{j=1}^{n} \rho^2(X_j, p)$. If $(\mathcal{M}, g)$ is a Riemannian manifold, then the Fréchet sample mean (set) of $\hat{Q}_n$ for the distance $\rho = \rho_g$ is called the Cartan mean (set). If $j : \mathcal{M} \to \mathbb{R}^N$ is an embedding, the Fréchet sample mean (set) of $\hat{Q}_n$ for the chord distance on $\mathcal{M}$ is called the extrinsic sample mean (set).*

REMARK 4.2.3. *The lack of a closed formula for $\overline{x}_I$ is one of the main drawbacks of intrinsic analysis on a nonflat Riemannian manifold. Even for any distribution Q on a simply connected manifold of nonpositive curvature, for which according to theorem 4.2.1, the intrinsic mean exists if $\mathcal{F} \in \mathcal{L}^2(Q)$, there is no exact formula for $\overline{x}_I$. The computation of the intrinsic mean set of a probability measure on a nonflat manifold $\mathcal{M}$ often involves nonstandard numerical algorithms, even if $\mathcal{M}$ has a Riemannian metric of maximum degree of mobility.*

The following result, due to Bhattacharya and Patrangenaru (2003) [42], establishes the strong consistency of the Fréchet sample mean as an estimator of the Fréchet mean of the underlying distribution.

THEOREM 4.2.4. *Let Q be a probability measure on a metric space $(\mathcal{M}, \rho)$ such that every closed bounded subset of $\mathcal{M}$ is compact. Assume $\mathcal{F}$ is finite on $\mathcal{M}$. (a) Then, given any $\varepsilon > 0$, there exist a P–null set N and $n(\omega) < \infty$*

$\forall\, \omega \in N^c$ such that the Fréchet (sample) mean set of $\hat{Q}_n = \hat{Q}_{n,\omega}$ is contained in the ε-neighborhood of the Fréchet mean set of Q for all $n \geq n(\omega)$. (b) If the Fréchet mean of Q exists then every measurable choice from the Fréchet (sample) mean set of $\hat{Q}_n$ is a strongly consistent estimator of the Frećéhet mean of Q.

Proof. (a) We will first prove that for every compact subset K of $\mathcal{M}$ one has

$$\sup_{p\in K} |\mathcal{F}_{n,\omega}(p) - \mathcal{F}(p)| \longrightarrow 0 \quad \text{a.s. as } n \to \infty,$$
$$\mathcal{F}_{n,\omega}(p) := \int \rho^2(x,p)\hat{Q}_{n,\omega}(dx) \equiv \tfrac{1}{n}\textstyle\sum_{j=1}^n \rho^2(X_j,p). \tag{4.10}$$

To prove (4.10) first observe that for a given $p_0 \in K$ one has, in view of the strong law of large numbers (SLLN) applied to $n^{-1}\sum_{j=1}^n \rho(X_j, p_0)$,

$$\begin{aligned} &\sup_{p\in K} \frac{1}{n}\sum_{j=1}^n \rho(X_j,p) \\ &\leq \frac{1}{n}\sum_{j=1}^n \rho(X_j,p_0) + \sup_{p\in K}\rho(p,p_0) \\ &\leq \int \rho(x,p_0)Q(dx) + 1 + \operatorname{diam} K = A, \text{say}, \end{aligned} \tag{4.11}$$

which holds for all $n \geq n_1(\omega)$, where $n_1(\omega) < \infty$ outside a P-null set N_1. Fix $\varepsilon' > 0$. From (4.11) one obtains, using the inequality $|\rho^2(X_j,p) - \rho^2(X_j,p')| \leq \{\rho(X_j,p) + \rho(X_j,p')\}\rho(p,p')$, the bound

$$\sup_{\{p,p'\in K:\rho(p,p')<\delta_1\}} |\mathcal{F}_{n,\omega}(p) - \mathcal{F}_{n,\omega}(p')| \leq 2A\delta_1 = \varepsilon'/3 \quad \forall n \geq n_1(\omega)(\omega \notin N_1), \tag{4.12}$$

where $\delta_1 := A/6\varepsilon'$. For the next step in the proof of (4.10), let $\delta_2 > 0$ be such that $|\mathcal{F}(p) - \mathcal{F}(p')| < \varepsilon'/3$ if $p, p' \in K$ and $\rho(p,p') < \delta_2$. Let $\delta = \min\{\delta_1, \delta_2\}$, and $\{q_1, q_2, \ldots, q_r\}$ a δ-net in K, i.e., $\forall\, p \in K$ there exists $q(p) \in \{q_1, \ldots, q_r\}$ such that $\rho(p, q(p)) < \delta$. By the SLLN, there exists a P-null set N_2 and $n_2(\omega) < \infty\ \forall\, \omega \notin N_2$ such that

$$\max_{i=1,2,\ldots,r} |\mathcal{F}_{n,\omega}(q_i) - \mathcal{F}(q_i)| < \varepsilon'/3 \quad \forall n \geq n_2(\omega) \quad (\omega \notin N_2). \tag{4.13}$$

Note that by (4.12), (4.13), and the fact that $|\mathcal{F}(q(p)| - \mathcal{F}(p)| < \varepsilon'/3\ \forall\, p \in K$, one has

$$\sup_{p\in K}|\mathcal{F}_{n,\omega}(p)-\mathcal{F}(p)|\leq$$
$$\sup_{p\in K}|\mathcal{F}_{n,\omega}(p)-\mathcal{F}_{n,\omega}(q(p))|+\sup_{p\in K}|\mathcal{F}_{n,\omega}(q(p))-\mathcal{F}(q(p))|$$
$$+\sup_{p\in K}|\mathcal{F}(q(p)-F(p)|$$
$$\varepsilon'/3+\varepsilon'/3+\varepsilon'/3=\varepsilon' \quad \forall\, n\geq n(\omega):=\max\{n_1(\omega),n_2(\omega)\}, \tag{4.14}$$

outside the P-null set $N_3 = N_1 \cup N_2$. This proves (4.14).

To complete the proof of (a), fix $\varepsilon > 0$. Let C be the (compact) Fréchet mean set of Q, $\ell := \min\{\mathcal{F}(p) : p \in C\}$. Write $C^\varepsilon := \{p : \rho(p,C) < \varepsilon\}$. It is enough to show that there exist $\theta(\varepsilon) > 0$ and $n(\omega) < \infty$ $\forall\,\omega$ outside a P-null set N such that

$$\begin{aligned}&\mathcal{F}_{n,\omega}(p) \qquad\qquad \leq \ell+\theta(\varepsilon)/2 \quad \forall\, p\in C\\ &\mathcal{F}_{n,\omega}(p) \quad \geq \ell+\theta(\varepsilon) \quad \forall\, p\in\mathcal{M}\backslash C^{\varepsilon}, \quad \forall\, n\geq n(\omega) \quad (\omega\notin N).\end{aligned} \tag{4.15}$$

For (4.15) implies that $\min\{\mathcal{F}_{n,\omega}(p) : p \in \mathcal{M}\}$ is not attained in $\mathcal{M}\backslash C^\varepsilon$ and, therefore, the Fréchet mean set of $\hat{Q}_{n,\omega}$ is contained in C^ε, provided $n \geq n(\omega)$ $(\omega \notin N)$. To prove (4.15) we will first show that *there exists a compact set D containing C and $n_3(\omega) < \infty$ outside a P-null set N_3 such that both $\mathcal{F}(p)$ and $\mathcal{F}_{n,\omega}(p)$ are greater than $\ell + 1$ $\forall\, p \in \mathcal{M}\backslash D$, for all $n \geq n_3(\omega)$ $(\omega \notin N_3)$.* If $\mathcal{M}$ is compact then this is trivially true, by taking $\mathcal{M} = D$. So assume $\mathcal{M}$ is noncompact. Fix $p_0 \in C$ and use the inequality $\rho(x,q) \geq |\rho(q,p_0) - \rho(x,p_0)|$ to get

$$\int \rho^2(x,q)Q(dx) \geq \int \{\rho^2(q,p_0)+\rho^2(x,p_0)-2\rho(q,p_0)\rho(x,p_0)\}Q(dx),$$

or,

$$\mathcal{F}(q)\geq \rho^2(q,p_0)+\mathcal{F}(p_0)-2\rho(q,p_0)F^{\frac{1}{2}}(p_0). \tag{4.16}$$

Similarly, using $\hat{Q}_{n,\omega}$ in place of Q,

$$\mathcal{F}_{n,\omega}(q)\geq \rho^2(q,p_0)+\mathcal{F}_{n,\omega}(p_0)-2\rho(q,p_0)F^{\frac{1}{2}}_{n,\omega}(p_0). \tag{4.17}$$

Since $\mathcal{M}$ is unbounded, one may take q at a sufficiently large distance Δ from C such that, by (4.16), $\mathcal{F}(q) > \ell+1$ on $\mathcal{M}\backslash D$, where $D := \{q : \rho(q,C) \leq \Delta\}$. Since $\mathcal{F}_{n,\omega}(p_0) \to \mathcal{F}(p_0)$ a.s., by (4.17) one may find a P-null set N_3 and $n_3(\omega) < \infty$ such that $\mathcal{F}_{n,\omega}(q) > \ell + 1$ on $\mathcal{M}\backslash D$ $\forall\, n \geq n_3(\omega)$ $(\omega \notin N_3)$. This proves the italicized statement above.

Finally, let $D_\varepsilon := \{p \in D : \rho(p,C) \geq \varepsilon\}$. Then D_ε is compact and $\ell_\varepsilon := \min\{\mathcal{F}(p) : p \in D_\varepsilon\} > \ell$, so that there exists $\theta = \theta(\varepsilon)$, $0 < \theta(\varepsilon) < 1$, such that $\ell_\varepsilon > \ell + 2\theta$. Now apply (4.14) with $K = D$ to find $n_4(\omega) < \infty$ outside a P-null set N_4 such that $\forall n \geq n_4(\omega)$ one has (i) $\mathcal{F}_{n,\omega}(p) \leq \ell + \theta/2\ \forall p \in C$ and (ii) $\mathcal{F}_{n,\omega}(p) > \ell + \theta\ \forall p \in D_\varepsilon$. Since $\mathcal{F}_{n,\omega}(p) > \ell + 1$ on $\mathcal{M}\backslash D\ \forall n \geq n_3(\omega)$ $(\forall \omega \notin N_3)$, one has $\mathcal{F}_{n,\omega}(p) > \ell + \theta\ \forall p \in D_\varepsilon \cup (\mathcal{M}\backslash D) = \mathcal{M}\backslash C^\varepsilon$ if $n \geq n(\omega) :=$ $\max\{n_3(\omega), n_4(\omega)\}$ for $\omega \notin N$, where $N = N_3 \cup N_4$. This proves (4.15), and the proof of part (a) is complete.

Part (b) is an immediate consequence of part (a).

REMARK 4.2.4. *A theorem of Ziezold (1977) [345] for general separable (pseudo) metric spaces implies the conclusion of part (b) of Theorem 4.2.4, but was formulated in a less transparent way. In metric spaces such that all closed bounded subsets are compact, the present theorem provides (i) strong consistency for Fréchet sample means and (ii) uniform convergence to the Fréchet mean of Q of arbitrary measurable selections from the sample mean set. This applies to both intrinsic and extrinsic means of Q and $\hat{Q}_n$ on manifolds. It is also a significant fact, that on non-compact spaces the proper topological condition ensuring that the uniqueness of the Fréchet mean implies strong consistency of its sample counterpart, a fact first pointed out in Bhattacharya and Patrangenaru (2003) [42].*

Under the hypothesis of Theorem 4.2.4(a), the Hausdorff distance between the Cartan mean set and the intrinsic mean set does not in general go to 0, as the following example shows.

EXAMPLE 4.2.3. *Consider n independent random variables $X_1, \ldots, X_n$ with the same distribution on S^1, that is absolutely continuous w.r.t. the uniform distribution. Then with probability one, we may assume that for $i \neq j$, $X_i \neq X_j$. Assume $X_j = e^{i\theta_j}$, and let $X_j^* = e^{i\theta_j} = -X_j$, where the arguments θ_j^* are in the increasing order of their indices. $\mathcal{F}(e^{i\theta})$ is periodic with period 2π and is a piecewise quadratic function; on each interval $[\theta_j^*, \theta_{j+1}^*]$, $\mathcal{F}(e^{i\theta}) = \sum_{k=1}^n (2\pi\varepsilon_{k,j} + (-1)^{\varepsilon_{k,j}}(\theta - \theta_k))^2$ where $\varepsilon_{k,j} \in \{0,1\}$. Therefore, the points of local minima have the form $\frac{1}{n}\sum_{k=1}^n(\theta_k + 2\pi\varepsilon_{j,k}(-1)^{\varepsilon_{j,k}})$ and each local minimum value $m_j = m_j(\theta_1, \theta_2, \ldots, \theta_n)$ is a quadratic form in $\theta_1, \ldots, \theta_n$. Since $\varepsilon_{k,j} \in \{0,1\}$, there are at most 2^n such possible distinct quadratic polynomials. Given that the each of the variables θ_j is continuous, the probability that there is a fixed pair of indices $i \neq j$, such that $m_i(\theta_1, \ldots, \theta_n) = m_j(\theta_1, \ldots, \theta_n)$ is zero. This shows that, with probability one, all the local minima are distinct and the Cartan mean exists. On the other hand, the intrinsic mean set of the uniform measure on the circle is the whole circle, proving that in this case the Hausdorff distance between the Cartan mean and the intrinsic mean set is π with probability one.*

Let $X_1, \ldots, X_n$ be independent random variables with a common distribution Q. The *intrinsic total sample variance* $t\Sigma_i(\hat{Q}_n)$ is the intrinsic total variance of the empirical $\hat{Q}_n$.

DEFINITION 4.2.6. *Let Q be a probability measure on $\mathcal{M}$ with a distance ρ. If $\mathcal{F}_1$ in (4.2) has a unique minimizer, this is called the Fréchet median of Q, and is labeled $\nu_{\mathcal{F}}(Q)$. If (M,g) is a Riemannian manifold, the Fréchet median (set) with respect to arc distance ρ_g is called the intrinsic median (set) of Q.*

REMARK 4.2.5. *The Fréchet function $\mathcal{F}_1$, and related asymptotic results have been also obtained by Hendricks and Landsman (2007) [155]. Note that the square root of a distance function is a distance function as well. Therefore the median set of a probability measure Q w.r.t. the distance ρ is the mean set of the same probability measure relative to the distance $\sqrt{\rho}$. Thus all the asymptotic results for Fréchet sample median estimator can be derived in terms from the corresponding results for Fréchet sample means.*

THEOREM 4.2.5. *Let Q be a probability measure on $(\mathcal{M},g)$ with finite $\mathcal{F}$. Then the sample intrinsic total variance $t\Sigma_I(\hat{Q}_n)$ is a strongly consistent estimator of the population intrinsic total variance $t\Sigma_I$*

Proof. In the proof of Theorem 4.2.4, one showed that $\frac{1}{n}\sum_{j=1}^{n}\rho_g^2(X_j,p)$ converges a.s. to $\mathcal{F}(p)$ uniformly for all p belonging to the compact intrinsic mean set $I(Q)$. Thus outside a P -null set N_1, $\forall\delta>0$ there exists $n_\delta^{(1)}(\omega)$ with the property

$$\sup_{p\in I(Q)}|\frac{1}{n}\sum_{j=1}^{n}\rho_g^2(X_j(\omega),p)-t\Sigma_I(Q)|<\delta,\forall n\geq n_\delta^{(1)}(\omega),\omega\notin N_1 \tag{4.18}$$

Let $\overline{Y}_{n,I}$ be a measurable selection, for each n, from the Cartan mean set. Fix $\delta>0$. Again, from Theorem (4.2.4), with probability one, for all sufficiently large $n, I(Q)$ is contained in the δ-neighborhood of $I(Q)$. Then, there exists $p_\delta(\omega)\in I(Q)$ and $n_\delta^{(2)}(\omega)$ such that

$$\rho_g(\overline{Y}_{n,i}(\omega),p_\delta(\omega))<\delta,\forall n\geq n_\delta^{(2)}(\omega),\omega\notin N_2, \tag{4.19}$$

where N_2 is a P-null set independent of δ. Hence $\forall\omega\notin N_1\cup N_2$ and $\forall n\geq max\{n_\delta^{(1)}(\omega),n_\delta^{(2)}(\omega)\}$, one has

$$\begin{aligned}
&|\frac{1}{n}\sum_{j=1}^{n}\rho_g^2(X_j(\omega),\overline{Y}_{n,i}(\omega))-t\Sigma_i(Q)|\leq|\frac{1}{n}\sum_{j=1}^{n}\rho_g^2(X_j(\omega),p_\delta(\omega))-t\Sigma_i(Q)|+\\
&\quad+|\frac{1}{n}\sum_{j=1}^{n}\rho_g^2(X_j(\omega),\overline{Y}_{n,i}(\omega))-\frac{1}{n}\sum_{j=1}^{n}\rho_g^2(X_j(\omega),p_\delta(\omega))|<\\
&<\delta+\frac{1}{n}\sum_{j=1}^{n}\{\rho_g(X_j(\omega),\overline{Y}_{n,i}(\omega))+\rho_g(Y_{n,j}(\omega),p_\delta(\omega))\}\cdot\rho_g(\overline{Y}_{n,i}(\omega),p_\delta(\omega))<\\
&<\delta+\delta\{\frac{2}{n}\sum_{j=1}^{n}\rho_g(X_j(\omega),p_\delta(\omega))+\delta\}\leq\delta+\delta^2+\delta\cdot\frac{2}{n}\sum_{j=1}^{n}\rho_g(X_j(\omega),p_\delta(\omega))\leq\\
&\leq\delta+\delta^2+2\delta(t\Sigma_i(Q)+\delta)^{1/2},
\end{aligned} \tag{4.20}$$

which proves the desired result.

REMARK 4.2.6. *The main drawback of intrinsic data analysis is that there are no known necessary and sufficient conditions for a probability measure Q on a complete Riemannian manifold (M,g) for the existence of $\mu_{g,I}(Q)$. Important results in this direction, including a algorithm for $\mu_{g,I}(Q)$ for a fairly concentrated probability measure Q are due to Groisser (2004)[133].*

Fortunately, this situation is not encountered with extrinsic means. Here we consider a general metric space $\mathcal{M}$, that is embedded in a numerical space as a closed subset, with the distance on $\mathcal{M}$ induced by the Euclidean distance in the numerical space, where $\mathcal{M}$ is embedded. If $j : \mathcal{M} \to \mathbb{R}^N$ is such an embedding, so that $\mathcal{M}$ with the induced chord distance is complete as a metric space, it follows that $j(M)$ is a *closed* subset (or a closed submanifold in case $\mathcal{M}$ is a manifold) of $\mathbb{R}^N$.

DEFINITION 4.2.7. *Assume ρ_0 is the Euclidean distance in $\mathbb{R}^N$. A point x of $\mathbb{R}^N$ such that there is a unique point p in M for which $\rho_0(x, j(M)) = \rho_0(x, j(p))$ is called j-nonfocal. A point which is not j-nonfocal is said to be j-focal.*

For example, the only focal point of S^m in $\mathbb{R}^{m+1}$ is the origin.
A probability measure Q on $\mathcal{M}$ induces a probability measure $j(Q)$ on $\mathbb{R}^N$.

DEFINITION 4.2.8. *A probability measure Q on $\mathcal{M}$ is said to be j-nonfocal if the mean μ of $j(Q)$ is a j-nonfocal point.*

If x is a j-nonfocal point, its *projection* on $j(\mathcal{M})$ is the unique point $y = P_j(x) \in j(\mathcal{M})$ with $\rho_0(x, j(M)) = \rho_0(x, y)$.

THEOREM 4.2.6. *If μ is the mean of $j(Q)$ in $\mathbb{R}^N$. Then (a) the extrinsic mean set is the set of all points $p \in \mathcal{M}$, with $\rho_0(\mu, j(p)) = \rho_0(\mu, j(\mathcal{M}))$ and (b) If $\mu_{j,E}(Q)$ exists then μ exists and is j-nonfocal and $\mu_{j,E}(Q) = j^{-1}(P_j(\mu))$.*

Proof. (a) If $p, q \in \mathcal{M}$, then $\| j(p) - j(q)\|^2 = \| j(p) - \mu\|^2 + 2\langle j(p) - \mu, \mu - j(q)\rangle + \|\mu - j(q)\|^2$ and if we integrate this identity over $\mathcal{M}$ w.r.t. Q, given that $\int_{\mathcal{M}} xQ(dx) = \int_{\mathbb{R}^k} xj(Q)(dx) = \mu$, we get

$$\mathcal{F}(p) = \| j(p) - \mu\|^2 + \int_{\mathcal{M}} \|\mu - x\|^2 j(Q)(dx). \tag{4.21}$$

In particular for any points $p, q \in \mathcal{M}$, $\mathcal{F}(p) - \mathcal{F}(q) = \rho_0^2(\mu, j(p)) - \rho_0^2(\mu, j(q))$ and (a) follows by selecting $j(q)$ to be a minimizer of $\mathcal{F}$. (b) If $\mu_{j,E}(Q)$ exists then μ exists and from part (a) it follows that the distance from an arbitrary point on $j(\mathcal{M})$ to μ has the unique minimizer $j(\mu_{j,E}(Q))$, that is μ is nonfocal and since $\rho_0(\mu_{j,E}(Q), \mu) = \rho_0(\mu, j(\mathcal{M}))$, that is $j(\mu_{j,E}(Q)) = P_j(\mu)$.
For the next theorem, to keep notations simple we will assume that $\mathcal{M}$ is a submanifold of $\mathbb{R}^N$ (the embedding j is the inclusion map). The result can be rephrased by assuming instead that j is an arbitrary embedding, since $j(\mathcal{M})$ is a submanifold of $\mathbb{R}^N$.

THEOREM 4.2.7. *The set of focal points of a submanifold $\mathcal{M}$ of $\mathbb{R}^N$ that has no flat points with the induced Riemannian structure, is a closed subset of $\mathbb{R}^N$ of Lebesgue measure* 0.

Proof. A point p is nonfocal, with $\rho_0(p,\mathcal{M}) = r$, if and only if the (hyper)sphere $S(p,r)$ of radius r centered at p has a unique point x in common with $\mathcal{M}$. In this case the interior of the ball $B(p,r)$ is included in $\mathbb{R}^N \backslash \mathcal{M}$ and $T_x\mathcal{M} \subseteq T_xS(p,r)$; x is the point of absolute minimum of the function L_p defined on $\mathcal{M}$ by $L_p(y) = \rho_0^2(p,y)$. Let $u = (u^1,...,u^d)$ be coordinates of points $y = y(u)$ on $\mathcal{M}$, with $y(0) = x$. In Milnor (1963) [242], p. 36, it is shown that x is a degenerate critical point of L_p if and only if p is a focus. Moreover from the computations in Milnor (1963) [242], p. 35, it follows that if $K_1, K_2,...,K_s$ are the nonzero principal curvatures of $\mathcal{M}$ at the point x, and $|t| < \min\{|K_1|^{-1}, |K_2|^{-1},...,|K_s|^{-1}\}$ for any unit vector v in $\nu_x\mathcal{M}$, the normal space at the point x, the matrix

$$((\partial y/\partial u^i(0)) \cdot (\partial y/\partial u^j(0)) - tv \cdot (\partial^2 y/\partial u^i \partial u^j(0)))$$

is positive definite. In particular since $r < \min\{|K_1|^{-1}, |K_2|^{-1},...,|K_s|^{-1}\}$, the matrix

$$((\partial y/\partial u^i(0)) \cdot (\partial y/\partial u^j(0)) - (p - y(0)) \cdot (\partial^2 y/\partial u^i \partial u^j(0)))$$

is positive definite. There is a neighborhood N of p and an open neighborhood U of 0 such that for any $u \in U$ and $q \in N$ the matrix of the second partial derivatives of $L_q(y(u))$, namely,

$$((\partial y/\partial u^i(u)) \cdot (\partial y/\partial u^j(u)) - (q - y(u)) \cdot (\partial^2 y/\partial u^i \partial u^j(u))),$$

is positive definite. Since the manifold topology of $\mathcal{M}$ coincides with the induced topology, one may assume that there is a ball $B(x,\varepsilon)$, such that $y(U) = \mathcal{M} \cap B(x,\varepsilon)$. Let ε be as small as necessary. Since x is the only common point of $\mathcal{M}$ and $S(p,r)$, and the set $\mathcal{M} \backslash \text{ Int } B(x,\varepsilon)$ is closed, there is a number δ, $r > \delta > 0$, such that $\rho_0(p, \mathcal{M} \backslash \text{ Int } B(x,\varepsilon)) = r + \delta$. Let $q \in \text{Int } B(p, \delta/2)$, and $z \in \mathcal{M} \backslash \text{ Int } B(x,\varepsilon)$. Then $\rho_0(q,z) > |\rho_0(q,p) - \rho_0(p,z)| > r + \delta - \delta/2 > \rho_0(q,x)$. It follows that $\rho_0(q,\mathcal{M}) = \rho_0(q, \mathcal{M} \cap \text{Int}B(x,\varepsilon))$. If $y \in \text{Int } B(x,\varepsilon) \backslash \{x\}$ is such that $\rho_0^2(q,y) = \rho_0^2(q,\mathcal{M})$ it follows by the positive definiteness of the displayed matrix above that y is an isolated point of minimum of L_q, proving that the set of nonfocal points is open.

Let $G(\infty)$ be the set of foci of $\mathcal{M}$, and let G be the set of focal points. It is known (Milnor (1963) [242], p. 33) that $G(\infty)$ has Lebesgue measure zero. If x is a point on $\mathcal{M}$, we define $G(x)$ to be the set of all points f in $\mathbb{R}^N$ such that there is at least another point x' $(\neq x)$ on $\mathcal{M}$ with $\rho_0(x,f) = \rho_0(x',f) = \rho_0(f,\mathcal{M})$. Another description of $G(x)$ is as the set of all centers f of spheres of $\mathbb{R}^N$ that are tangent to $\mathcal{M}$ at least at two points, one of which is x, and whose interiors are disjoint from $\mathcal{M}$. The tangent space $T_x\mathcal{M}$ is included in the tangent space at x to such a sphere. Therefore the normal line at x to such a sphere is included in the normal space $\nu_x\mathcal{M}$, which means that a point in $G(x)$ is in $\nu_x\mathcal{M}$. We show that on each ray starting at x in $\nu_x\mathcal{M}$ (x is the zero element, if $\nu_x\mathcal{M}$ is regarded as a vector space) there is at most one point in $G(x)$. Indeed if f_1, f_2 are two distinct points on such a ray starting at x, assume f_1 is closer to x than f_2. Let x', x'' be such that $\rho_0(x', f_1) = \rho_0(x, f_1) = \rho_0(f_1, \mathcal{M})$, $\rho_0(x'', f_2) = \rho_0(x, f_2) = \rho_0(f_2, \mathcal{M})$. Then x' is a point of $\mathcal{M}$ in the interior of $S(f_2, \rho_0(f_2, \mathcal{M}))$, a contradiction. Given that $G(x)$ intersects the radii coming out of x in $\nu_x\mathcal{M}$ at most at one point, the Lebesgue measure of $G(x)$ in $\nu_x\mathcal{M}$ is zero.

Let $N\mathcal{M}$ be the disjoint union of $\nu_x\mathcal{M}$, $x \in \mathcal{M}$. $N\mathcal{M}$ is the normal bundle of $\mathcal{M}$ and it is a manifold of dimension k. We define the map $N : N\mathcal{M} \to \mathbb{R}^k$ by $N((x, \nu_x) = x + \nu_x$. One may show (Milnor (1963)[242]) that the critical values of N are the foci of $\mathcal{M}$. Therefore if $f = N((x, \nu_x))$ is a focal point that is not a focus, f is a regular value of N. Thus, if λ represents the Lebesgue volume form in $\mathbb{R}^k$, then $N^*\lambda$ is a volume form on $N^{-1}(\mathbb{R}^k \backslash G(\infty))$, and the Lebesgue measure of $G \backslash G(\infty)$ is

$$\lambda(G \backslash G(\infty)) \leq \int_{N\mathcal{M} \backslash G(\infty)} N^*\lambda.$$

If we apply Fubini's theorem integrating over the base $\mathcal{M}$ the integral in each fiber (normal space $\nu_x\mathcal{M}$)) , we see that the integrand in $\nu_x\mathcal{M}$ is a volume form that is a multiple $C(x)$ of the Lebesgue measure in $\nu_x\mathcal{M}$. Therefore

$$\int_{N\mathcal{M} \backslash G(\infty)} N^*\lambda = \int_{\mathcal{M}} C(x) \left(\int_{G(x)} \lambda_x(dv) \right) \mathrm{vol}_{\mathcal{M}}(dx),$$

which is zero since

$$\int_{G(x)} \lambda_x(dv) = \lambda_x(G(x)) = 0.$$

The following theorem links the intrinsic mean of Q on a Riemannian manifold with its extrinsic mean under an embedding which is equivariant at a point p (see definition 3.2.10). Note that for every h in the isotropy group H_p defined in 3.2.6 the *differential* dh maps $T_p\mathcal{M}$ into itself.

THEOREM 4.2.8. *Let $j : \mathcal{M} \to \mathbb{R}^k$ be a Riemannian embedding which is equivariant at p. Assume that $0 \equiv (p, 0)$ is the only fixed point of $T_p\mathcal{M}$ under the family of maps $\{dh : h \in H_p\}$. Assume also that Q is a probability measure on $\mathcal{M}$ which is invariant under the isotropy group H_p, and $\mu(j(Q))$ is finite and nonfocal. (a) Then either $\mu_E(Q) = p$ or $\mu_E(Q) \in C(p)$, the cut locus of p. The same holds for the intrinsic mean $\mu_g(Q)$ if it exists. (b) If, in addition to the hypothesis above, $\mathcal{M}$ is a compact two point homogeneous space other than the round sphere and $\mu_g(Q)$ exists, then $\mu_g(Q) = \mu_E(Q) = p$.*

Proof. (a) The mean $\mu(j(Q))$ of Q, regarded as a measure on the ambient Euclidean space, is invariant under each Euclidean isometry $\hat{h}$, say, which extends $h \in H_p$. For Q, as a measure on the Euclidean space, is invariant under $\hat{h}$, $\forall h \in H_p$, due to the equivariant embedding at p and of the invariance of Q on $\mathcal{M}$ under H_p. It now follows that $\mu_E(Q)$ is invariant under H_p. Suppose now that $\mu_E(Q) \neq p$. We will show that in that case $\mu_E(Q) \in C(p)$. If this is not so then $\mu_E(Q) \in Exp_p(V(p))$. Then there exists a unique minimizing geodesic joining p and $\mu_E(Q)$. Because of uniqueness this geodesic, say γ, is left invariant by the isometries $h \in H_p$. Then $\dot{\gamma}(0)$ is invariant under $dh \, \forall \, h \in H_p$, contradicting the hypothesis that 0 is the only invariant vector in $T_p\mathcal{M}$ under $\{dh : h \in H_p\}$.

Suppose next $\mu_g(Q)$ exists. Then $\mu_g(Q)$ is invariant under H_p, since $\mathcal{F}(y) = \mathcal{F}(hy) \forall h \in H_p$ due to the invariance of Q under H_p. The same argument as above now shows that either $\mu_g(Q) = p$ or $\mu_g(Q) \in C(p)$.

(b) It follows from a classification theorem due to Wang (1952) that beside the round spheres, there are only four types of two point homogeneous spaces, namely, the real projective spaces, complex projective spaces, quaternionic projective spaces and the Cayley projective planes (see also (Helgason (1978) [147], p. 535)). It is known from Warner (1965)[332] that for any point $p \in \mathcal{M}, C(p)$ is a strong deformation retract of $\mathcal{M}\backslash\{p\}$, and in particular $C(p)$ has the homotopy type of $\mathcal{M}\backslash\{p\}$. On the other hand, if $\mathcal{M}$ is one of the two point homogeneous spaces other than a sphere given by Wang's classification, then the cohomology of $\mathcal{M}\backslash\{p\}$ is not trivial. This shows that in this case $\mathcal{M}\backslash\{p\}$ is not homotopically trivial, and therefore $C(p)$ is also not homotopically trivial. This implies that if $\mathcal{M}$ is not a sphere, $C(p)$ has at least two points q, q'. Moreover since the isotropy group H_p is transitive on the geodesic sphere $S(p, \rho_g(p,q))$, we may assume that $\rho_g(p,q) = \rho_g(p,q') = r$. Hence if $\mu_E(Q) \in C(p)$ there exists $q' \in C(p)\backslash\{\mu_E(Q)\}$ such that $\rho_g(p, \mu_E(Q)) = \rho_g(p,q')$. By the transitivity of H_p on $S(p,r)$ there exists $h \in H_p$ such that $h(\mu_E(Q)) = q'$, contradicting the invariance of $\mu_E(Q)$. By (a), $\mu_E(Q) = p$.

The same argument applies to $\mu_g(Q)$ if it exists.

EXAMPLE 4.2.4. *Let Q be a probability measure on a sphere, with $\mu(j(Q)) \neq 0$, such that the group leaving Q invariant is the stabilizer of a given point p. Then $\mu_E(Q)$ is either p or the antipodal point of p on the sphere. The same is true of $\mu_g(Q)$ if it exists. Such examples of probability distributions are given in Watson (1983) [333], p. 136, and Fisher (1993), including the von Mises distributions. Another example of an invariant distribution is given by the Dimroth–Watson distribution on the real projective plane $\mathbb{R}P^2$, whose Radon–Nykodim derivative at the point $[x]$ w.r.t. the volume measure of a constant curvature Riemannian structure on $\mathbb{R}P^2$ is proportional to* $\exp[k(p \cdot x)^2]$, *$x \in S^2$, and is $O(2)$ invariant. A general $O(2)$ invariant measure with a density on $\mathbb{R}P^2$ has the Radon–Nykodim derivative w.r.t. the volume form at the* *point* $[x]$ *proportional to* $f((p \cdot x)^2)$, $x \in S^2$, *where* f *is a density of an absolutely continuous positive measure on a finite interval. An example of equivariant embedding of* $\mathbb{R}P^2$ *furnished with a Riemannian structure with constant curvature into the space of symmetric matrices* $S(3,\mathbb{R})$ *is provided by the Veronese-like map* $j[u] = uu^T$. *The Euclidean distance* ρ_0 *on* $S(3,\mathbb{R})$ *is given by* $\rho_0^2(A,B) := Tr((A-B)(A-B))$. *As such if* u,v *are in* S^2, $\rho_0^2(j[u], j[v]) = Tr(uu^T - vv^T)(uu^T - vv^T) = Tr(uu^T uu^T - 2uu^T vv^T + vv^T vv^T) = 2(1-(u \cdot v)^2)$. *The fact that the embedding is equivariant follows from the action of isometries of* $O(3)$ *on* $S(3,\mathbb{R})$, *by simultaneous left and right multiplication. For details see exercises 3 and 4 for at the end of this section.*

DEFINITION 4.2.9. *Consider an embedding* $j : \mathcal{M} \to \mathbb{R}^N$. *Assume* $(x_1, ..., x_n)$ *is a sample from a j-nonfocal probability measure* Q *on* $\mathcal{M}$, *and the function* $p \to \sum_{r=1}^{n} \|j(p) - j(x_r)\|^2$ *has a unique minimizer on* $\mathcal{M}$; *this minimizer is the extrinsic sample mean.*

From Theorem 4.2.6 the *extrinsic sample mean* is given by

$$\bar{x}_E := j^{-1}\left(P_j(\overline{j(x)})\right) \tag{4.22}$$

THEOREM 4.2.9. *Assume* Q *is a nonfocal probability measure on the manifold* $\mathcal{M}$ *and* $X = \{X_1, \ldots, X_n\}$ *are i.i.d.r.o.'s from* Q. *(a) If the sample mean* $\overline{X}$ *is a j-nonfocal point then the extrinsic sample mean is given by* $j^{-1}(P_j(\overline{X}))$. *(b)* $\overline{X}_E$ *is a strongly consistent estimator of* $\mu_{j,E}(Q)$.

Proof. (a) If $\overline{X}$ is a j-nonfocal point then by Theorem 4.2.6, applied to the empirical $\hat{Q}_n$, the image via j of the extrinsic sample mean estimator is the projection of the extrinsic sample mean estimator, $j(\overline{X}_{j,E}) = P_j(j(\overline{X}))$.

(b) By the SLLN, $\overline{X}$ converges to the mean μ of $j(Q)$ almost surely. Since F^c is open, by theorem 4.2.7, and the projection P_j from F^c to $j(\mathcal{M})$ is continuous, $j^{-1}(P_j(j(\overline{X})))$ converges to $\mu_E(Q)$ almost surely.

REMARK 4.2.7. *In particular from Theorem 4.2.6 , if* Q *is focal, the extrinsic mean set has at least two points. Therefore by Theorem 4.2.4 the extrinsic sample mean set may have more than one point, and a sequence of selections from the extrinsic sample mean set may not have a limit.*

COROLLARY 4.2.1. *Assume* $\mathcal{M}$, Q *and the equivariant embedding* j *are as in Theorem 4.2.7(b). Then the extrinsic sample mean is a strongly consistent estimator of the intrinsic mean of* Q.

4.3 Exercises

Exercise 59. *Show that set* F *of Veronese–Whitney focal points of* $\mathbb{R}P^m$ *in* $S(m+1,\mathbb{R})$ *is contained in the set of matrices in* $S_+(m+1,\mathbb{R})$ *and consists of those matrices whose largest eigenvalues are of multiplicity at least 2. The projection* $P_j : S_+(m+1,\mathbb{R}) \backslash F \to j(\mathbb{R}P^m)$ *associates to each nonnegative definite* *symmetric matrix A with a largest eigenvalue of multiplicity one, the matrix $j([m])$ where m is the eigenvector of A. corresponding to that eigenvalue.*

Hint $S_+(m+1,\mathbb{R})$ is convex, so the mean under Q of matrices in $S_+(m+1,\mathbb{R})$ is a matrix in $S_+(m+1,\mathbb{R})$. Therefore, we are interested in determining only the projection of a semipositive matrix on $\mathcal{M}$. If A is in $S(N,\mathbb{R})$ and T is an orthogonal matrix, then $d_0(A,M) = d_0(T(A),M)$. Given A in $S_+(m+1,\mathbb{R})$ there is T in $O(m+1)$ such that $T(A) = \mathrm{diag}(\eta_a)_{a=1,\dots,m+1} = D$, and the entries of D are all nonnegative, in increasing order. Let $x = (x^a)$ be a unit vector in $\mathbb{R}^{m+1}$. After elementary computations we get

$$\rho_0^2(D,j([x])) = 1 + \sum \eta_a^2 - 2\sum \eta_a (x^a)^2 \geq \rho_0^2(D,j([e_{m+1}])), \tag{4.23}$$

where e_{m+1} is the eigenvector of D of length one corresponding to the highest eigenvalue. Note that if η_{m+1} has multiplicity two or more, then for any $t \in [0,2\pi]$, and for any unit vector $x = (x^a) \in \mathbb{R}^{m+1}$, we have

$$\rho_0^2(D,j[x]) \geq \rho_0^2(D,j[\cos t e_m + \sin t e_{m+1}]) = \rho_0^2(D,j[e_{m+1}]), \tag{4.24}$$

and D is focal.

If η_{m+1} is simple, i.e. has a multiplicity of one, then $\rho_0^2(D,j[x]) \geq \rho_0^2(D,j[e_{m+1}])$ and the equality holds only if $[x] = [e_{m+1}]$. In this last case D is a nonfocal and $P_{\mathcal{M}}(D) = j([e_{m+1}])$.

Exercise 60. *Assume $Y = [X]$, where $X^TX = 1$ is a random variable on $\mathbb{R}P^m$, Show that a probability distribution P_Y on $\mathbb{R}P^m$ is Veronese–Whitney nonfocal if the highest eigenvalue of $E[XX^T]$ is simple and in this case the extrinsic mean of Y is $\mu_E = [m]$, where m is a unit eigenvector of $E[X_1X_1^T]$ corresponding to the highest eigenvalue.*

Exercise 61. *Under $[x_1],\dots,[x_n]$ is a sample on $\mathbb{R}P^m$. Show that the Veronese–Whitney sample mean $[\overline{x}]_E$ exists if and only if the highest eigenvalue of $K_n = \frac{1}{n}\sum_{n=1}^{n} x_r x_r^T$ is simple, and in this case $\overline{[x]}_E = [m]$, where m is a unit eigenvector of K_n.*

Exercise 62. *Show that the Veronese–Whitney sample mean $[\overline{x}]_E$ in exercise 61 is also the MLE for the mean of a Bingham distribution (see Kent (1992) [181]), whose density function is proportional to* $\exp(x^TAx)$, *where A is a symmetric matrix, and for the mean of the Dimroth–Watson distribution, whose density function at $[x]$ is proportional to* $\exp(k(\mu\cdot x)^2)$, *where k is a constant.*

Exercise 63. *(a) Show that the intrinsic mean axis of the Dimroth–Watson distribution is the same as the Veronese-Whitney mean.*
(b) If $[X] = \{[X_1],\dots,[X_n]\}$ are i.i.d.r.v.'s from a and that the Veronese–Whitney sample mean $[\overline{X}]_E$ of a Dimroth–Watson distribution with $\mu \neq 0$, show that $[\overline{X}]_E$ is a consistent estimator of the intrinsic mean of that distribution.
(c) If the largest eigenvalue of the symmetric matrix A is simple, show that the extrinsic sample mean or of the Bingham distribution in exercise 63 is a consistent estimator of the extrinsic mean of that distribution.

Exercise 64. *Show that set F of Veronese–Whitney focal points of $\mathbb{C}P^{k-2}$ in $S(k-1,\mathbb{C})$ is contained in the set of nonnegative definite self adjoint matrices $S_+(k-1,\mathbb{C})$, and consists of those matrices whose largest eigenvalues are of multiplicity at least 2. The projection $P_j : S_+(k-1,\mathbb{C})\backslash F \to j(\mathbb{R}P^{k-2})$ associates to each nonnegative definite self adjoint matrix A with a largest eigenvalue of multiplicity one, the matrix $j([m])$ where m is the eigenvector of A. corresponding to that eigenvalue.*

Exercise 65. *(Bhattacharya and Patrangenaru (2003) [42])Assume $Y = [Z]$, where $Z^*Z = 1$ is a random variable on $\mathbb{C}P^{k-2}$, Show that a probability distribution P_Y on $\mathbb{C}P^{k-2}$ is Veronese–Whitney nonfocal if the highest eigenvalue of $E[ZZ^*]$ is simple and in this case the extrinsic mean of Y is $\mu_E = [m]$, where m is an eigenvector of $E[ZZ^*]$ corresponding to the highest eigenvalue, $m^*m = 1$.*

Exercise 66. *(Patrangenaru (1998) [266]) Under $[\zeta_1],\ldots,[\zeta_n]$ is a sample on $\mathbb{C}P^{k-2}$. Show that the Veronese–Whitney sample mean $[\overline{\zeta}]_E$ exists if and only if the highest eigenvalue of $K_n = \frac{1}{n}\sum_{n=1}^{n}\zeta_r\zeta_r^*$ is simple, and in this case $\overline{[\zeta]}_E = [m]$, where m is a unit eigenvector of K_n.*

Exercise 67. *Show that the Veronese–Whitney sample mean $[\overline{x}]_E$ in exercise 8 is also the MLE for the mean of a complex Bingham distribution (Kent, 1994), whose Radon–Nikodym derivative w.r.t. the uniform distribution has the expression*

$$f_A(z) = C(A)^{-1}\exp(z^*Az), \qquad z \in \mathbb{C}S^{k-2} \tag{4.25}$$

where A is a Hermitian matrix.

Exercise 68. *The Dryden–Mardia distribution on $\mathbb{C}P^{k-2}$ is induced by a $\mathbb{C}^{k-1}$ valued random variable Z which has a multivariate normal distribution with mean ν and covariance matrix $\sigma^2 I_{2k-2}$. The variable X on $\mathbb{C}P^{k-2}$ corresponding to Z is $X = [Z]$. Assume j is the Veronese–Whitney embedding of $\mathbb{C}P^{k-2}$ and let $Q = P_X$. (i) Show that*

$$\mu(j(Q)) = \alpha I_{k-1} + \beta\nu\nu^*, \tag{4.26}$$

where $\alpha > 0$, $\alpha+\beta > 0$. (ii) Show that the Dryden–Mardia distribution is Veronese–Whitney nonfocal, and the extrinsic sample mean is a consistent estimator of the extrinsic mean.

Hint. For (i) see Kent and Mardia (1997) [183], for (ii) select the orthogonal coordinates with the first axis along ν, we notice that as a matrix, $E(j(X))$ is conjugate with a diagonal matrix whose diagonal entries are all α except for the entry $\alpha+\beta$ in the upper left corner, showing that $E(j(X))$ is nonfocal for j. By Theorem 4.2.9 the extrinsic mean of the Dryden–Mardia distribution exists and the extrinsic sample mean is a consistent estimator of the extrinsic mean.

Exercise 69. *When is the extrinsic sample mean of the complex Bingham distribution in* (4.25) *inconsistent?*

Solution. *Assume λ_A is the largest eigenvalue of A and let V_A be the eigenspace* *corresponding to* λ_A. *Then the extrinsic mean set of the complex Bingham distribution for planar shapes is the set* $\{[\mu] | \mu \in V_A \backslash 0\}$. *The extrinsic mean exists only if* V_A *has dimension one over* $\mathbb{C}$. *Therefore if* $\dim_{\mathbb{C}} V_A \geq 2$, *even if the extrinsic sample mean (Procrustes estimate) exists, it is inconsistent by Theorem 4.2.9.*

Chapter 5

Nonparametric Distributions for Fréchet Sample Means

5.1 Introduction

Recall that a Fréchet mean of a probability measure Q on a complete metric space (M, ρ) is the unique minimizer of the function $\mathcal{F}(x) = \int \rho^2(x,y)Q(dy)$ (see Fréchet (1948) [121]). A Fréchet mean associated with a geodesic distance ρ_g determined by the Riemannian structure g on a complete manifold $\mathcal{M}$ is said to be an intrinsic mean, $\mu_g(Q)$. It is known that if Q is sufficiently concentrated then $\mu_g(Q)$ exists (see Theorem 5.4.2 (a) below). Also recall that the extrinsic mean $\mu_E(Q) = \mu_{j,E}(Q)$ of a probability measure Q on a complete manifold $\mathcal{M}$ relative to an embedding $j : M \to \mathbb{R}^k$ is the Fréchet mean associated with the restriction to $j(\mathcal{M})$ of the Euclidian distance in $\mathbb{R}^k$. In Chapter 4, it was shown that such an extrinsic mean of Q exists if the ordinary mean of $j(Q)$ is a nonfocal point of $j(\mathcal{M})$, i.e., if there is a *unique* point x_0 on $j(\mathcal{M})$ having the smallest distance from the mean of $j(Q)$; in this case $\mu_{j,E}(Q) = j^{-1}(x_0)$.

DEFINITION 5.1.1. *A probability distribution Q has small flat support on a Riemannian manifold $\mathcal{M}$ if $supp(Q)$ is included in a flat geodesic ball.*

In general is is easier to compute an extrinsic mean, except for the case when Q has small flat support. It may be pointed out that if Q is highly concentrated, as may be the case with medical imaging data (see the example in Bhattacharya and Patrangenaru (2003) [42]), the Cartan and extrinsic means are virtually indistinguishable.

We now provide a summary of the main results in this chapter. The first part of Section 5.2 is devoted to nonparametric inference for the total intrinsic variance of a probability measure Q on a manifold $\mathcal{M}$. In Theorem 5.2.1, it is shown that for i.i.d. observations, the sample total intrinsic variance is asymptotically normal around the total intrinsic variance with an explicit formula of the covariance matrix; additional results are given for the nonparametric bootstrap counterparts (see Corollary 5.4.3), and for asymptotic distributions of the sample total intrinsic variance for a distribution with small flat support. Section 5.2 is also focused on asymptotic distributions of Fréchet mean sets, and Fréchet means, assuming the Fréchet population mean is unique. In 5.4.1 it is shown that under some rather general assumptions, the image of the Fréchet mean under a chart φ that contains the Fréchet mean in its domain, is asymptotically normally distributed around the image of the Fréchet mean of Q. This leads to the asymptotic distribution theory of the intrinsic sample mean on a Riemannian manifold $\mathcal{M}$ (Theorems 5.4.2, 5.4.3). In Corollaries 2.3 and 2.4, bootstrap confidence regions are derived for the Fréchet mean, with or without a pivot.

Section 5.4 is devoted to asymptotics of extrinsic sample means. The original results due to Hendriks and Landsman (1998) [154] and Patrangenaru (1998) [266] were independent of each other and used different estimators for the extrinsic mean. In this chapter, we present an extension of Patrangenaru's approach. Extrinsic means are commonly used in directional, axial, and shape statistics. In the particular case of directional data analysis, i.e. when $M = S^{m-1}$ is the unit sphere in $\mathbb{R}^m$, Fisher et al. (1996) [116] provided an approach for inference using computationally efficient bootstrapping which gets around the problem of increased dimensionality associated with the embedding of the manifold $\mathcal{M}$ in a higher dimensional Euclidean space. In Corollary 5.5.2 confidence regions (c.r.) are derived for the extrinsic mean $\mu_{j,E}(Q)$ Nonparametric bootstrap methods on abstract manifolds are also derived in this section (Theorem 5.5.3, Proposition 5.5.3).

If one assumes that Q has a nonzero absolutely continuous component with respect to the volume measure on $\mathcal{M}$, then from some results in Section 2.13 (see also Bhattacharya and Ghosh (1978) [38], Babu and Singh (1984) [9], Beran (1987) [16], Hall (1988, 1997) [136, 137]), one derives bootstrap based confidence regions for $\mu_E(Q)$ with coverage error $O_p(n^{-2})$ (Theorem 5.5.1) (see Section 2.13, Bhattacharya and Qumsiyeh (1989) [45], Bhattacharya and Denker (1990) [36]). One may also use the nonpivotal bootstrap to construct confidence regions based on the percentile method of Hall (1997) [137] for general Q with a coverage error no more than $O_p(n^{-m/(m+1)})$, where m is the dimension of the manifold (see Remark 5.4.4 and Proposition 5.5.3). For a definition of the coverage error, see Section 2.13. This is particularly useful in those cases where the asymptotic dispersion matrix is difficult to compute.

5.2 Asymptotic and Bootstrap Distributions of Fréchet Total Variances on Manifolds

The main focus of this section is the derivation of asymptotic distributions of total intrinsic sample variance, intrinsic and extrinsic sample means, and of confidence regions based on these distributions. We provide classical CLT-based confidence regions and tests based on them, as well as those based on Efron's bootstrap (Efron (1982) [97]).

The total Fréchet variance explains the variability of a random object in its entirety. A random variable X on $\mathcal{M}$ is constant if and only if $t\Sigma_{\mathcal{F}}(P_X) = 0$. In particular, given a Riemannian structure g on $\mathcal{M}$, it is useful to estimate $t\Sigma_g(Q)$ with confidence. Let $X_1,\dots,X_n$ be independent identically distributed random objects (i.i.d.r.o.'s) with a common distribution Q on $\mathcal{M}$, for which μ_g exists. Consider the random variable

$$T_n = n^{-1}\sum_{k=1}^{n}\rho_g^2(\mu_g, X_k). \tag{5.1}$$

THEOREM 5.2.1. *Let $X_1,\dots,X_n$ be independent identically distributed random objects (i.i.d.r.o.'s) with a common distribution $Q = P_{X_1}$ on $\mathcal{M}$, for which μ_g exists, and assume $Q(C(\mu_g)) = 0$ and the intrinsic moments (defined in Chapter 4) of order four or less about μ_g are finite, then $n^{1/2}(T_n - t\Sigma_g)$ converges in law to $\mathcal{N}(0, E[\rho_g^4(\mu_g, X_1)] - (t\Sigma_g)^2)$.*

Proof. We define the $T_{\mu_g}\mathcal{M}$-valued vector valued random variables V_k such that $Exp_{\mu_g}V_k = X_k$. V_k are i.i.d.r.v.'s with the probability density $\lambda_Q = \lambda_{Q,\mu_g}$. The mean and variance of $\|V_1\|$ can be estimated in terms of the first four order moments of λ_Q. Since V_1 has mean zero, it turns out that $E(\|V_1\|^2) = Tr\Sigma(\lambda_Q)$ and $Var(\|V_1\|^2) = E(\|V_1\|^4) - (E(\|V_1\|^2)^2 = E((\sum_{j=1}^{n}(V_1{}^j)^2)^2) - (Tr\Sigma(\lambda_Q))^2$. As such, $W_n = n^{1/2}(n^{-1}\sum_{k=1}^{n}\|V_k\|^2) - (Tr\Sigma(\lambda_Q))^2$ is asymptotically $N(0, E((\sum_{j=1}^{n}(V_1{}^j)^2)^2) - (Tr\Sigma(\lambda_Q))^2)$. Since Exp_{μ_g} is a radial isometry, $d(\mu_g, X_k) = \|V_k\|$ and the asymptotic distribution is $N(0, E[\rho_g^4(mu_g, X)] - (t\Sigma_g)^2)$.

REMARK 5.2.1. *Since $n^{1/2}(T_n - t\Sigma_g)$ converges weakly to $N(0, E[d^4(\mu_g, X_1)] - (t\Sigma_g)^2)$, it follows that T_n converges in probability to $t\Sigma_g$, thus $t\Sigma_g(\hat{Q}_n) - t\Sigma_g$ and $T_n - t\Sigma_g$ are asymptotically equivalent. However, this falls short of saying that $n^{1/2}(t\Sigma_g(\hat{Q}_n) - t\Sigma_g)$ and $n^{1/2}(T_n - t\Sigma_g)$ are asymptotically equivalent, which together with Theorem 5.2.1 would suffice to yield percentile confidence regions for $t\Sigma_g$.*

A useful weaker result of Theorem 5.2.1 is the following:

THEOREM 5.2.2. *If Q has small flat support on $\mathcal{M}$ and has finite moments up to the fourth order, then $n^{1/2}(t\hat{\Sigma}_{g,n} - t\Sigma_g)$ converges in law to a random vector with a $\mathcal{N}(0, Var(\rho_g{}^2(X_1, \mu_g)))$ distribution.*

Proof. Given the small support assumption, let $W_1, \ldots, W_n$ be i.i.d.r.$\vec{v}$'s in $\mathbb{R}^m$ representing the i.i.d.r.o.'s $X_1, \ldots, X_n$ with the distribution $Q = P_{X_1}$. Assume $E(W_1) = \mu$.

$$t\hat{\Sigma}_{g,n} - t\Sigma_g = Tr(\frac{1}{n}\sum_{i=1}^{n}(W_i - \overline{W})(W_i - \overline{W})^T - TrE((W_1 - \mu)(W_1 - \mu)^T) =$$

$$= \frac{1}{n}\sum_{i=1}^{n} Tr((W_i - \overline{W})(W_i - \overline{W})^T) - TrE((W_1 - \mu)(W_1 - \mu)^T) =$$

$$= \frac{1}{n}\sum_{i=1}^{n} Tr((W_i - \mu)(W_i - \mu)^T) - TrE((W_1 - \mu)(W_1 - \mu)^T) =$$

$$= \frac{1}{n}\sum_{i=1}^{n} \|W_i - \mu\|^2 - E(\|W_1 - \mu\|^2). \quad (5.2)$$

From the C.L.T. (applied here for a random sample from the probability distribution of distribution of $\|W_1 - \mu\|^2$), it follows that

$$\sqrt{n}(t\hat{\Sigma}_{g,n} - t\Sigma_g) = \sqrt{n}(\overline{\|W - \mu\|^2} - E(\|W_1 - \mu\|^2) \to_d Y, \quad (5.3)$$

where $Y \sim \mathcal{N}(0, Var(\|W_1 - \mu\|^2) = \mathcal{N}(0, Var(d_g{}^2(X, \mu_g)))$, done ■

Under the assumptions of Theorem 5.2.2 if we set $S^2 = \frac{1}{n}\sum_{i=1}^{n}(\rho_g{}^2(X_i, \overline{X}_g) - \overline{\rho_g{}^2(X, \overline{X}_g)})^2$, then we obtain

COROLLARY 5.2.1. *If Q has small flat support on $\mathcal{M}$ then*

$$n^{\frac{1}{2}}(\frac{(t\hat{\Sigma}_{g,n} - t\Sigma_g)}{S}) \quad (5.4)$$

converges in distribution to a random variable with a standard normal distribution.

Further, we get

COROLLARY 5.2.2. *A $100(1-\alpha)\%$ large sample symmetric confidence interval for $t\Sigma_g$ is given by*

$$(t\hat{\Sigma}_{g,n} - z_{\frac{\alpha}{2}}\frac{S}{\sqrt{n}}, t\hat{\Sigma}_{g,n} + z_{\frac{\alpha}{2}}\frac{S}{\sqrt{n}}) \quad (5.5)$$

.

Note that for any affine coordinate U we have

LEMMA 5.2.1. *Assume $U_1, \ldots, U_n$ are i.i.d.r.v.'s from a probability distribution Q with finite mean μ, variance σ^2, third and fourth moments about zero*

$\mu_{3,0}, \mu_{4,0}$, *and let* $\overline{U^a} = \frac{1}{n}\sum_{i=1}^n U_i^a$ *be the sample estimator of* $\mu_{a,0}$. *Then for n large enough,* $\sqrt{n}(\hat{\sigma}^2 - \sigma^2) \to_m W$ *where* $\hat{\sigma}^2 = \overline{U^2} - \overline{U}^2$ *and*

$$W \sim \mathcal{N}(0, \sigma^2(6\mu^2 - \sigma^2) + 3\mu^4 - 4\mu_{3,0}\mu + \mu_{4,0}), \tag{5.6}$$

therefore if we assume $\sigma^2(6\mu^2 - \sigma^2) + 3\mu^4 - 4\mu_{3,0}\mu + \mu_{4,0} > 0$. and studentize we obtain the following:

PROPOSITION 5.2.1. *Under the hypothesis of Lemma 5.2.1 if we set*

$$T = \frac{\sqrt{n}(\hat{\sigma}^2 - \sigma^2)}{(\widehat{\sigma^2}(6\overline{U}^2 - \widehat{\sigma^2}) + 3\overline{U}^4 - 4\overline{U^3}\overline{U} + \overline{U^4})^{\frac{1}{2}}}, \tag{5.7}$$

then T has asymptotically a standard normal distribution.

COROLLARY 5.2.3. *A large sample* $100(1-\alpha)\%$ *confidence interval for* σ^2 *is given by*

$$(\widehat{\sigma^2} - \frac{z_\beta}{\sqrt{n}}(\widehat{\sigma^2}(6\overline{u}^2 - \widehat{\sigma^2}) + 3\overline{u}^4 - 4\overline{u^3}\overline{u} + \overline{u^4})^{\frac{1}{2}},$$
$$\widehat{\sigma^2} + \frac{z_\gamma}{\sqrt{n}}(\widehat{\sigma^2}(6\overline{u}^2 - \widehat{\sigma^2}) + 3\overline{u}^4 - 4\overline{u^3}\overline{u} + \overline{u^4})^{\frac{1}{2}}), \tag{5.8}$$

where $\beta + \gamma = \alpha$.

If Q has small support, the Cartan mean estimator $\overline{X}_{n,g}$ exists and is unique. Let $V_k{}^{(n)} \in T_{\overline{X}_{n,g}}M$ be such that $X_k = Exp_{\overline{X}_{n,g}}(V_k{}^{(n)})$. The *intrinsic sample covariance estimator* is

$$\hat{\Sigma}_I =: \frac{1}{n}(\sum_{j=1}^n (V_j{}^{(n)} - \overline{V^{(n)}})(V_j{}^{(n)} - \overline{V^{(n)}})^t \tag{5.9}$$

where $\overline{V^{(n)}}$ is the sample mean of $V_j{}^{(n)}, j = 1, \ldots, n$

We have the studentized asymptotic result:

THEOREM 5.2.3. *If* $X_1, \ldots, X_n$ *are i.i.d.r.v.'s with a common distribution Q of small flat support on* $\mathcal{M}$, *then*

$$T_{n,g} := (n\hat{\Sigma}_I)^{-1/2} Exp_{\mu_g}{}^{-1}(\overline{X}_{n,g})$$

converges in law to $N_m(0, I_m)$.

Assume g_x is the Riemannian structure at $x \in M$. If Q has small flat support, since $X_1, \ldots, X_n$ are i.i.d.r.o.'s with a common distribution Q for which $\overline{X}_{n,g}$ exists, we get

COROLLARY 5.2.4. *Under the hypothesis of Theorem 5.2.3,*

$$T = g_{\mu_g}(T_{n,g}, T_{n,g})$$

converges in law to a χ^2_m *distribution.*

The statistic $T = T(Q, \mu_g)$ in Corollary 5.2.4 is *asymptotically pivotal*, since it converges in law to a distribution independent of unknown parameters. One may use then the standard bootstrap procedure to get a statistic $T^*(Q^*, X^*|X)$ from T, by random sampling $\{X_r^*\}_{r=1,\ldots,n}$ with repetition from the empirical $\hat{Q}_n$, conditionally given $\{X_r\}_{r=1,\ldots,n}$. Let $\hat{\Sigma}_g{}^*$ be the statistic obtained from $\hat{\Sigma}_g$ by substituting X_r^* for X_r when $r = 1,\ldots,n$. Then $T^*(Q^*, X^*|X)$ is obtained from $T(Q, \mu_g)$ by substituting $X_1^*,\ldots,X_n^*$ for $X_1,\ldots,X_n$, $\overline{X}_{n,g}$ for μ_g and $\hat{\Sigma}_g{}^*$ for $\hat{\Sigma}_g$. Using the standard bootstrap techniques via Edgeworth expansions as shown in [36], or in Hall (1997) [137], we get

THEOREM 5.2.4. *If Q is a probability distribution with small support which has a nonzero absolutely continuous component w.r.t. the volume measure of the flat Riemannian manifold $\mathcal{M}$, then the distribution of the statistic T given in Corollary 5.2.4 can be approximated uniformly by the bootstrap distribution of $T^*(Q^*, X^*|X)$ with an error $0_p(n^{-1})$.*

Theorem 5.2.4 yields an effective computational method for confidence regions of nominal coverage $(1-\alpha)$ based on Efron's method (see Efron(1982) citeEfron:1982) with a coverage error $O_p(n^{-1})$ (see, e.g. Bhattacharya and Denker (1990) [36]). A $(1-\alpha)100\%$ bootstrap confidence region for μ_g with an error $0_p(n^{-1})$ can be obtained as follows: let P^* denote the conditional distribution of $X_1^*,\ldots,X_n^*$ under the empirical. We set

$$c_n(1-\alpha) := sup\{c|P^*(T^*(Q^*, X^*|X) \leq c) \leq 1-\alpha\}$$

and obtain the following

COROLLARY 5.2.5. *A $(1-\alpha)100\%$ bootstrap confidence region for μ_g is given by*

$$\{\mu : T(Q, \mu) \leq c_n(1-\alpha)\} \tag{5.10}$$

Such a confidence region is approximately the $Exp_{\overline{X}_{n,g}}$- image of an ellipsoid in $T_{\overline{X}_{n,g}}M$. Note that once the coordinates in the regular geodesic neighboring ball are given, simultaneous confidence intervals can also be derived in terms of these coordinates.

An index of spread of a probability measure Q on a Riemannian manifold $\mathcal{M}$, which coincides with the total variance when $\mathcal{M}$ is an Euclidian space can be defined as follows:

DEFINITION 5.2.1. *Assume Q is the probability distribution of the $\mathcal{M}$-valued variate X and $x \in M$ is an element in the intrinsic mean set of Q. The intrinsic total variance of Q (intrinsic second total moment about μ_g) is $t\Sigma_g = E(d_g{}^2(X,x))$. In particular if μ_g exists, $t\Sigma_g = E(d_g{}^2(X,\mu_g))$. Higher intrinsic total moments about μ_g can be similarly defined.*

In [41] one shows that if $X_1,\ldots,X_n$ are i.i.d.r.v.'s with a common distribution

Q, the *total intrinsic sample variance* $t\hat{\Sigma}_{g,n}$, which is the intrinsic total variance of $\hat{Q}_n$, is a strongly consistent estimator of $t\Sigma_g$. There, one defines T_n as :

$$T_n = n^{-1} \sum_{j=1}^{n} d_g^2(\mu_g, X_k)$$

and shows that if μ_g exists, the intrinsic moments of order four about μ_g are finite, and the cut-locus of μ_g has Q-measure zero, then $n^{1/2}(T_n - t\Sigma_g)$ converges in law to $N(0, E[d^4(\mu_g, X_1)] - (t\Sigma_g)^2)$.

For the rest of the section we will prove a similar result and give confidence intervals for $t\Sigma_g$. If A is a matrix, we label by $\Delta(A)$ the vector of diagonal entries of A, and by $S(A)$ the sum of all entries of A. We assume in addition that Q has small support and to studentize this result, we need the following extension to the general multivariate case of Theorem 8 in [113] .

LEMMA 5.2.2. *Assume* $V_1, \ldots, V_n$ *are i.i.d.r.v.'s in* $\mathbf{R^p}$ *with finite moments of order four. For* $j, k = \overline{1, p}$*, let* $\mu_{4;j,k} = E((V_{1,j} - \mu_j)^2 (V_{1,k} - \mu_k)^2)$ *If* Σ_4 *is the matrix* $(\mu_{4;j,k})$*, then* $n^{1/2}(\Delta(S) - \Delta(\Sigma))$ *converges in law to* $N(0, \Sigma_4)$

Assume $X_1, \ldots, X_n$ are real i.i.d.r.v.'s with the common distribution Q with small support included in s flat topological disc U of the manifold $\mathcal{M}$. Select a point $q \in supp(Q)$, and let V_k be the T_qM valued random vectors, such that $X_k = Exp_q(V_k)$. Under the small support assumption, the total intrinsic sample covariance estimator is

$$t\hat{\Sigma}_{g,n} =: \frac{1}{n}\Big(\sum_{j=1}^{n} \|V_j - \overline{V}\|^2\Big) = TrS_n = \sum_{j=1}^{n} s_{jj} \tag{5.11}$$

where S is the sample covariance of $V_1, \ldots, V_n$. On the other hand, again by the small support assumption, it follows that

$$t\Sigma_g = Tr(Cov(V_1)) = \sum_{j=1}^{n} \sigma_{jj}. \tag{5.12}$$

Note that the formulas are independent of q, since the *local holonomy group* of U is trivial. Let Σ_4 be the matrix in the Lemma 5.2.2 corresponding to V_1.

THEOREM 5.2.5. *If* Q *has small flat support on* $\mathcal{M}$ *then* $n^{1/2}(t\hat{\Sigma}_{g,n} - t\Sigma_g)$ *converges in law to* $N(0, S(\Sigma_4))$.

For a proof, it suffices to use the delta method for the function $g : \mathbf{R}^p \to \mathbf{R}$, given by $g(x^1, \ldots, x^p) = x^1 + \ldots + x^p$.
We studentize the previous theorem, by selecting a matrix estimator $\hat{\Sigma}(V)_4$ of Σ_4 corresponding to the i.i.d.r.v.'s $V_1, \ldots, V_n$

COROLLARY 5.2.6. *If Q has small flat support on $\mathcal{M}$ then*

$$n^{\frac{1}{2}}\left(\frac{(t\hat{\Sigma}_{g,n} - t\Sigma_g)}{S(\hat{\Sigma}(V)_4)^{1/2}}\right) \tag{5.13}$$

converges in law to $N(0,1)$.

and get

COROLLARY 5.2.7. *A $100(1-\alpha)\%$ large sample symmetric confidence interval(c.i.) for $t\Sigma_g$ is given by*

$$(t\hat{\Sigma}_{g,n} - S(\hat{\Sigma}(V)_4)^{1/2})z_{\frac{\alpha}{2}}, t\hat{\Sigma}_{g,n} + S(\hat{\Sigma}(V)_4)^{1/2})z_{\frac{\alpha}{2}}) \tag{5.14}$$

.

5.3 Elementary CLT for Extrinsic Means

We consider now asymptotic distributions of Fréchet means. We start with an elementary method for constructing confidence regions for extrinsic means $\mu_E(Q)$ on closed submanifolds $\mathbb{R}^N$. The embedding j is the inclusion map of $\mathcal{M}$ in $\mathbb{R}^N$, and the projection P_j is the projection $P_{\mathcal{M}}$ on $\mathcal{M}$. Let H be the projection on the affine subspace $\mu_E(Q) + T_{\mu_E}\mathcal{M}$. We would like to determine the asymptotic distribution of $H(\overline{X})$. While $H(\overline{X})$ is not the same as $P_{\mathcal{M}}(\overline{X})$, its asymptotic distribution is easier to compute. For large samples the extrinsic sample mean is close to the extrinsic mean and, therefore, $H(\overline{X})$ and $P_{\mathcal{M}}(\mu)$ will be close to each other. When $\mathcal{M}$ is a linear variety of $\mathbb{R}^N$, the two maps coincide. Thus for concentrated data the delta method for H gives a good estimate of the distribution of the extrinsic sample mean. Assume that around $P_{\mathcal{M}}(\mu)$ the implicit equations of $\mathcal{M}$ are $F^1(x) = ... = F^c(x) = 0$, where $F^1, ..., F^c$ are functionally independent. Then $\overline{X} - H(\overline{X})$ is in $\nu_{P_{\mathcal{M}}(\mu)}\mathcal{M}$, the orthocomplement of $T_{P_{\mathcal{M}}(\mu)}\mathcal{M}$; thus it is a linear combination of the gradients $grad_{P_{\mathcal{M}}(\mu)}F^1, ..., grad_{P_{\mathcal{M}}(\mu)}F^c$. We need to evaluate the differential of the map H at μ, in terms of $F^1, ..., F^c$. Set $\nu_\alpha = \|grad_{P_{\mathcal{M}}(\mu)}F^\alpha\|^{-1} grad_{P_{\mathcal{M}}(\mu)}F^\alpha$ $(\alpha = 1, ..., c)$, and

$$(h^{\alpha\beta}(\mu))_{\alpha,\beta=1,...,c} = ((h_{\alpha\beta}(\mu))_{\alpha,\beta=1,...,c})^{-1}, h_{\alpha\beta}(\mu) = \nu_\alpha \cdot \nu_\beta \tag{5.15}$$

Then $x - H(x) = \sum_\beta t^\beta(x,\mu)\nu_\beta$ where $t^\beta(x,\mu) = \sum_\alpha h^{\alpha\beta}(\mu)(x - P_{\mathcal{M}}(\mu)) \cdot \nu_\alpha$. Therefore,

$$H(x) = x + h^{\alpha\beta}(\mu)((P_{\mathcal{M}}(\mu) - x) \cdot \nu_\alpha)\nu_\beta, d_\mu H(v) = v - \sum_\alpha h^{\alpha\beta}(\mu)(v \cdot \nu_\alpha)\nu_\beta, \tag{5.16}$$

that is

$$G_i^j = \frac{\partial H^j}{\partial x^i}(\mu) = \delta_{ij} - \sum_{\alpha,\beta} h^{\alpha\beta}(\mu)\nu_\alpha^i \nu_\beta^j \tag{5.17}$$

where $\delta_{ij} = 1$ or 0 according as $i = j$ or $i \neq j$. By the delta method we arrive at

THEOREM 5.3.1. *Let $\{X_k\}_{k=1,\dots,n}$ be a random sample from a nonfocal distribution Q on the submanifold $\mathcal{M}$, given in a neighborhood of $\mu_E(Q)$ by the equations $F^1(x) = \cdots = F^c(x) = 0$. Assume Q has mean μ and covariance matrix Σ as a distribution in the ambient numerical space. If G is the matrix given by (5.17), then $n^{\frac{1}{2}}(H(\overline{X}) - P_{\mathcal{M}}(\mu))$ converges weakly to $\mathcal{N}_{N-c}(0, G\Sigma G^t)$ in the tangent space of $\mathcal{M}$ at the extrinsic mean $\mu_E(Q) = P_{\mathcal{M}}(\mu)$ of Q.*

Since the sample counterpart of

$$\Gamma = G\Sigma G^t \tag{5.18}$$

may be difficult to compute and one may use Efron's percentile bootstrap (see Efron (1982) [97]) to obtain a confidence region for $\mu_E(Q)$.

COROLLARY 5.3.1. *Under the hypothesis of Theorem 5.3.1, one may construct an asymptotic $(1-\alpha)$-confidence region for $\mu_E(Q) = P_{\mathcal{M}}\mu$, using the bootstrapped statistic $n^{\frac{1}{2}}(H(\overline{X}^*) - \overline{H}(\overline{X}))$. Here $\overline{H}(\overline{X})$ is the projection of $\overline{X}$ on the affine subspace $\overline{X}_E + T_{\overline{X}_E}\mathcal{M}$, and $\overline{X}^*$ is the mean of a random sample with repetition of size n from the empirical $\hat{Q}_n$ considered as a probability measure on the Euclidean space in which $\mathcal{M}$ is embedded.*

REMARK 5.3.1. *Suppose $\mathcal{F}$ is finite and Q is nonfocal. By Theorem 4.2.7, there exists $\delta > 0$ such that $\overline{X}$ is nonfocal, if $\|\overline{X} - \mu\| < \delta$. Since $P(\|\overline{X} - \mu\| \geq \delta) = O(n^{-1})$, one may define $\overline{X}_E$ to be any measurable selection from the sample extrinsic mean set, if $\overline{X}$ is focal. Theorem 5.3.1 and a corresponding version of Corollary 5.3.1 hold for this $\overline{X}_E$.*

5.4 A Central Limit Theorem for Fréchet Means and Bootstrapping.

Given $q \in M$, the *exponential map* $Exp_q : U \to M$ is defined on an open neighborhood U of $0 \in T_qM$ by the correspondence $v \to \gamma_v(1)$, where $\gamma_v(t)$ is the unique geodesic satisfying $\gamma(0) = q, \dot{\gamma}(0) = v$, provided $\gamma(t)$ extends at least to $t = 1$. Thus if (M, g) is geodesically complete or, equivalently, (M, ρ_g) is complete as a metric space, then Exp_q is defined on all of T_qM. In this chapter, unless otherwise specified, all Riemannian manifolds are assumed to be *complete.*

Note that if $\gamma(0) = p$ and $\gamma(t)$ is a geodesic, it is generally not true that the geodesic distance between p and $q = \gamma(t_1)$, say, is minimized by $\gamma(t), 0 \leq t \leq t_1$ (consider e.g. the great circles on the sphere S^2 as geodesics). Let $t_0 = t_0(p)$ be the supremum of all $t_1 > 0$ for which this minimization holds. If $t_0 < \infty$, then $\gamma(t_0)$ is the *cut point of p along γ*. The *cut locus* $C(p)$ of p is the union of all cut points of p along all geodesics γ starting at p (For example, $C(p) = \{-p\}$ on S^2).

In this section we deal with both intrinsic and extrinsic means. Hence we will often consider a general distance ρ on a complete differentiable manifold $\mathcal{M}$. In particular, (M,ρ) with the metric topology is complete as a metric space. We consider only those probability measures Q on $\mathcal{M}$ for which the Fréchet mean $\mu_{\mathcal{F}} =: \mu_{\mathcal{F}}(Q)$ exists.

REMARK 5.4.1. *(Bhattacharya and Lin(2013)[39]) The additional assumption in Bhattacharya and Patrangenaru (2005)[43] on the existence of a chart (U,ϕ) such that $Q(U) = 1$, or Remark 5.5. in Bhattacharya and Bhattacharya(2012)[27] are in fact unnecessary, due to the consistency of the Fréchet mean (see also Bhattacharya and Patrangenaru(2014)[44]).*

Consider a chart (U,φ) on M, with $\mu_{\mathcal{F}} \in U$. If $X_i (i = 1,\dots,n)$ are i.i.d. with common distribution Q and defined on a probability space $(\Omega,\mathcal{A},P)$, let $\mu_{n,\mathcal{F}}$ be a measurable selection form the Fréchet mean set (w.r.t. ρ) of the empirical $\hat{Q}_n = \frac{1}{n}\sum_{i=1}^{n}\delta_{X_i}$. Let $\mu_n = \phi(\mu_{n,\mathcal{F}})$ be a measurable selection from the Fréchet mean set of $\hat{Q}_n = \frac{1}{n}\sum_{i=1}^{n}\delta_{X_i}$. Assume ρ^2 is twice differentiable as a function on $M \times M$. It follows that the Fréchet function in equation (4.2) is twice differentiable on $\mathcal{M}$. We consider the local representative F^{ϕ} of the Fréchet function. Since μ is the point of minimum of

$$F^{\phi}(\theta) := \int_M \rho^2(\phi^{-1}(\theta),q)Q(dq) = E(\rho^2(\phi^{-1}(\theta),X), \tag{5.19}$$

where X is a random object on $\mathcal{M}$ with $P_X = Q$. Similarly μ_n is a local minimum of

$$F_n^{\phi}(\theta) := \int_M \rho^2(\phi^{-1}(\theta),q\hat{Q}_n(dq) = \frac{1}{n}\sum_{i=1}^{n}\rho^2(\phi^{-1}(\theta),X_i). \tag{5.20}$$

Write the Euclidean gradient of $\theta \to \frac{1}{2}\rho^2(\phi^{-1}(\theta),q)$ as

$$\Psi(u;\theta) = \frac{1}{2}grad_{\theta}(\rho^2(\phi^{-1}(\theta),q)) = (\frac{1}{2}\frac{\partial}{\partial\theta^r}(\rho^2(\phi^{-1}(\theta),q)))_{r=1}^{m} = (\Psi^r(q;\theta))_{r=1}^{m}. \tag{5.21}$$

For the $\frac{1}{2}$ scalar multiple, see Schwartzman (2014)[302]. One has the Taylor expansion

$$0 = \frac{1}{\sqrt{n}}\sum_{i=1}^{n}\Psi^r(X_i;\mu_n) = \frac{1}{\sqrt{n}}\sum_{i=1}^{n}\Psi^r(X_i;\mu) +$$

$$+\frac{1}{n}\sum_{i=1}^{n}\sum_{r'=1}^{m} D_{r'}\Psi^r(X_i;\mu))\sqrt{n}(\mu_n^{r'} - \mu^{r'}) + R_n^r, (1 \le r \le d) \tag{5.22}$$

where

$$R_n^r = \sum_{r'=1}^{m}\sqrt{n}(\mu_n^{r'} - \mu^{r'})\frac{1}{n}\sum_{i=1}^{n}\{D_{r'}\Psi^r(X_i;\theta_n) - D_{r'}\Psi^r(X_i;\mu)\}, \tag{5.23}$$

and θ_n lies on the line segment joining μ and μ_n (for sufficiently large n). We will assume

$$E|\Psi(X_1;\mu)|^2 < \infty,$$
$$E|D_{r'}\Psi^r(X_1;\mu)|^2 < \infty(\forall r,r'). \tag{5.24}$$

To show that R_n^r is negligible, write

$$u^{r,r'}(x,\varepsilon) := \sup_{\{\theta:\|\theta-\mu\|\leq\varepsilon\}} |D_{r'}\Psi^r(x;\theta) - D_{r'}\Psi^r(x;\mu)|$$

and assume

$$\delta^{r,r'}(c) := Eu^{r,r'}(X_i,c) \to 0, as \ \ c\downarrow 0, (1 \leq r,r' \leq m). \tag{5.25}$$

One may then rewrite (5.22) in vectorial form as

$$0 = \frac{1}{\sqrt{n}}\sum_{i=1}^{n}\Psi(X_i;\mu) + (\Lambda+\delta_n)\sqrt{n}(\mu_n-\mu) \tag{5.26}$$

where

$$\Lambda = E((D_{r'}\Psi^r(X_i;\mu)))_{r,r'=1}^m \tag{5.27}$$

and $\delta_n \to 0$ in probability as $n \to \infty$, if $\mu_n \to \mu$ in probability. If, finally, we assume Λ is nonsingular then (5.26) leads to the equation

$$\sqrt{n}(\mu_n-\mu) = \Lambda^{-1}(\frac{1}{\sqrt{n}}\sum_{i=1}^{n}\Psi(X_i;\mu)) + \delta_n' \tag{5.28}$$

where δ_n' goes to zero in probability as $n \to \infty$. We have then arrived at the following theorem, due to Bhattacharya and Patrangenaru (2005)[43].

THEOREM 5.4.1. *(CLT for Fréchet sample means.) Let Q be a probability measure on a differentiable manifold $\mathcal{M}$ endowed with a metric ρ such that every closed and bounded set of (M,ρ) is compact. Assume (i) the Fréchet mean $\mu_{\mathcal{F}}$ exists, (ii) the map $\theta \to (\rho^\phi)^2(\theta,u)$ is twice continuously differentiable on $\phi(U)$, (iii) the integrability conditions* (5.24) *hold as well as the relation* (5.25), *and (iv) Λ, defined by* (5.27), *is nonsingular. Then (a) every measurable selection μ_n from the (sample) Fréchet mean set of $\hat{Q}_n^\phi = \frac{1}{n}\sum_{i=1}^n \delta_{\tilde{X}_i}$ is a consistent estimator of μ, and (b)*

$$\sqrt{n}(\varphi(\mu_n) - \varphi(\mu) \to_d \mathcal{N}_m(0,\Lambda^{-1}\Sigma(\Lambda^T)^{-1}), \tag{5.29}$$

where Σ is the covariance matrix of $\Psi(X_1;\mu)$.

The *proof* given here is from Bhattacharya and Patrangenaru (2005)[43]. Part (a) follows from Theorem 2.3 in Bhattacharya and Patrangenaru (2003)[43]. The proof of part (b) is as outlined above, and it may also be derived from standard proofs of the CLT for M-estimators (see e.g. Huber (1980)[163], pp. 132-134). ■

As an immediate corollary one obtains

COROLLARY 5.4.1. *Let $(\mathcal{M},g)$ be a Riemannian manifold and let $\rho = \rho_g$ be the geodesic distance. Let Q be a probability measure on $\mathcal{M}$. Assume that (i) the intrinsic mean $\mu_g = \mu_{\mathcal{F}}$ exists, (ii) the map $\rho^2 : \mathcal{M} \times \mathcal{M} \to \mathbb{R}$ is twice continuously differentiable and Λ, defined by (5.27), is nonsingular. Then the conclusions of Theorem 5.4.1 hold for the Cartan mean $\mu_{n,g}$ of $\hat{Q}_n = \frac{1}{n}\sum_{i=1}^{n}\delta_{X_i}$, with $\mu = \phi(\mu_g)$.*

We now prove one of the main results of this section.

THEOREM 5.4.2. *(CLT for Cartan means.) Let (M,g) be a Riemannian manifold and let $\rho = \rho_g$ be the geodesic distance. Let Q be a probability measure on $\mathcal{M}$ whose support is contained in a closed geodesic ball $\overline{B}_r \equiv \overline{B}_r(x_0)$ with center x_0 and radius r which is disjoint form the cut locus $C(x_0)$. Assume $r < \frac{\pi}{4K}$, where K^2 is the supremum of sectional curvatures in $\overline{B}_r$ if this supremum is positive, or zero if this supremum is nonpositive. Then (a) the intrinsic mean μ_g (of Q) exists, and (b) the conclusion of Theorem 2.1 holds for the image $\mu_n = \phi(\mu_{n,g})$ of the Cartan mean $\mu_{n,g}$ of $\hat{Q}_n = \frac{1}{n}\sum_{i=1}^{n}\delta_{X_i}$, under the inverse ϕ of the exponential map, $\phi = (Exp_{x_0})^{-1}$.*

Proof. (a) It is known that under the given assumptions, there is a *local minimum* μ_g, say, of the Fréchet function F, which belongs to B_r and that this minimum is also the *unique* minimum in $\overline{B}_{2r}$ (Karcher (1977) [174], Kendall (1990) [180], Theorem 7.3), and Le (2001) [211]). We now show that μ_g is actually the unique *global minimum* of $\mathcal{F}$. Let $p \in (\overline{B}_{2r})^c$. Then $\rho(p,x) > r, \forall x \in \overline{B}_r$. Hence

$$\mathcal{F}(p) = \int_{\overline{B}_r} \rho^2(p,x)Q(dx) > \int_{\overline{B}_r} r^2 Q(dx) = r^2. \tag{5.30}$$

On the other hand,

$$\mathcal{F}(\mu_g) \le \mathcal{F}(x_0) = \int_{\overline{B}_r} \rho^2(x_0,x)Q(dx) \le r^2 \tag{5.31}$$

proving $\mathcal{F}(p) > \mathcal{F}(\mu_g)$.
(b) In view of Corollary 5.4.1, we only need to show that the Hessian matrix $\Lambda \equiv \Lambda(\mu)$ of $F \circ \phi^{-1}$ at $\mu := \phi(\mu_g)$ is nonsingular, where $\phi = Exp_{x_0}^{-1}$. Now according to Karcher (1977) [174], Theorem 1.2, for every geodesic curve $\gamma(t)$ in $B_r, t \in (c,d)$ for some $c < 0, d > 0$,

$$\frac{\rho^2}{dt^2}\mathcal{F}(\gamma(t)) > 0 \quad (c < t < d). \tag{5.32}$$

Let $\psi = Exp_{\mu_g}$ denote the exponential map at μ_g, and let $\gamma(t)$ be the unique geodesic with $\gamma(0) = \mu_g$ and $\dot{\gamma}(0) = v$, so that $\gamma(t) = \psi(tv)$. Here we identify the tangent space $T_{\mu_g}M$ with $\mathbb{R}^m$. Applying (2.16) to this geodesic (at $t = 0$), and writing $G = F \circ \psi$, one has

$$\frac{\rho^2}{dt^2}\mathcal{F}(\psi(tv))|_{t=0} = \sum v^i v^j (D_i D_j G)(0) > 0, \forall v \neq 0, \tag{5.33}$$

i.e., the Hessian of G is positive definite at $0 \in \mathbb{R}^m$. If $x_0 = \mu_g$, this completes the proof of (b).

Next let $x_0 \neq \mu_g$. Now $F \circ \phi^{-1} = G \circ (\psi^{-1} \circ \phi^{-1})$ on a domain that includes $\mu = \phi(\mu_g) \equiv (Exp_{x_0})^{-1}(\mu_g)$. Write $\psi^{-1} \circ \phi^{-1} = f$. Then in a neighborhood of μ,

$$\frac{\partial^2 (G \circ f)}{\partial u^r \partial u^{r'}}(u) = \sum_{j,j'} (D_j D_{j'} G)(f(u)) \frac{\partial f^j}{\partial u^r}(u) \frac{\partial f^{j'}}{\partial u^{r'}}(u) + \\ + \sum_j (D_j G)(f(u)) \frac{\partial^2 f^j}{\partial u^r \partial u^{r'}}(u). \tag{5.34}$$

The second sum in (5.33) vanishes at $u = \mu$, since $(D_j G)(f(\mu)) = (D_j G)(0) = 0$ as $f(\mu) = \psi^{-1}\phi^{-1}(\mu) = \psi^{-1}(\mu_g) = 0$ is a local minimum of G. Also f is a diffeomorphism in a neighborhood of μ. Hence , writing $\Lambda_{r,r'}(\mu)$ as the (r,r') element of $\Lambda(\mu)$,

$$\Lambda_{r,r'}(\mu) = \frac{\partial^2 (F \circ \phi^{-1})}{\partial u^r \partial u^{r'}}(\mu) = \sum_{j,j'} (D_j D_{j'} G)(0) \frac{\partial f^j}{\partial u^r}(\mu) \frac{\partial f^{j'}}{\partial u^{r'}}(\mu).$$

This shows, along with (5.32), that $\Lambda = \Lambda(\mu)$ is positive definite ■

REMARK 5.4.2. *If the supremum of the sectional curvatures (of a complete manifold $\mathcal{M}$) is nonpositive, and the support of Q is contained in $\overline{B}_r$ then the hypotheses of Theorem 5.4.2 are satisfied, and the conclusions (a), (b) hold. One may apply this even with $r = \infty$.*

REMARK 5.4.3. *The assumptions in Theorem 5.4.2 on the support of Q for the existence of μ_g is too restrictive for general applications. But without additional structures it can not be entirely dispensed with, as is easily shown by letting Q be the uniform distribution on the equator of S^2. For the complex projective space $\mathbb{C}P^{\frac{m}{2}}$, m-even, necessary and sufficient conditions for the existence of the intrinsic mean μ_g of an absolutely continuous (w.r.t. the volume measure) Q with radially symmetric density are given in Le(1998)[210], Kendall et al. (1999) [179].*

It may be pointed out that it is the assumption of some symmetry, i.e., the invariance of Q under a group of isometries, that often causes the intrinsic mean set to contain more than one element (see, e.g. Bhattacharya and Patrangenaru (2003)[42], Proposition 2.2.). The next result is therefore, expected to be more generally applicable then Theorem 5.4.2.

THEOREM 5.4.3. *(CLT for Cartan means.) Let Q be absolutely continuous w.r.t. the volume measure on an m dimensional Riemannian manifold (M, g). Assume that (i) μ_g exists, (ii) the integrability conditions (2.9) hold, (iii) the*

Hessian matrix Λ of $F \circ \phi^{-1}$ at $\mu = \phi(\mu_g)$ is nonsingular and (iv) the covariance matrix Σ of $\Psi(\tilde{X}_i; \mu)$ is nonsingular. Then

$$\sqrt{n}(\mu_n - \mu) \to_d \mathcal{N}_m(0, \Gamma), \tag{5.35}$$

where $\Gamma = \Lambda^{-1}\Sigma(\Lambda^T)^{-1}$.

This theorem follows from Theorem 5.4.1 and Remark 5.4.2.
By replacing Λ and Σ by their sample counterparts $\hat{\Lambda}$ and $\hat{\Sigma}$, respectively, and assuming that Σ is nonsingular, one obtains the following:

$$n[\varphi(\mu_n) - \varphi(\mu)]^T(\hat{\Lambda}\hat{\Sigma}^{-1}\hat{\Lambda}^T)[\varphi(\mu_n) - \varphi(\mu)] \xrightarrow{\mathcal{L}} \chi^2_m, \tag{5.36}$$

where χ^2_m is the chi-square distribution with m degrees of freedom. From (5.36) one obtains a confidence region for ν (and of $\mu = \varphi^{-1}(\nu)$).

In order to obtain a confidence region for $\mu_{\mathcal{F}}$ using the CLT in Theorem 5.4.1 in the traditional manner, one needs to estimate the covariance matrix $\Gamma = \Lambda^{-1}\Sigma(\Lambda^T)^{-1}$. For this one may use proper estimates of Λ and Σ, namely,

$$\hat{\Lambda}(\theta) := \frac{1}{n}\sum_{i=1}^{n}(Grad\Psi)(\tilde{X}_i, \mu_n), \quad \hat{\Sigma} = Cov\hat{Q}_n^{\phi},$$
$$\hat{\Gamma} := \hat{\Lambda}^{-1}\hat{\Sigma}(\hat{\Lambda}^t)^{-1}, \quad \hat{\Gamma}^{-1} = \hat{\Lambda}^t\hat{\Sigma}^{-1}\hat{\Lambda} \tag{5.37}$$

The following corollary is now immediate. Let $\chi^2_{m,1-\alpha}$ denote the $(1-\alpha)$-th quantile of the chi-square distribution with m degrees of freedom.

COROLLARY 5.4.2. *Under the hypothesis of Theorem 5.4.2, if Σ is nonsingular, a confidence region for $\mu_{\mathcal{F}}$ of asymptotic level $1-\alpha$ is given by $U_{n,\alpha} := \phi^{-1}(D_{n,\alpha})$, where $D_{n,\alpha} = \{\nu \in \phi(U) : n(\mu_n - \nu)^t\hat{\Gamma}^{-1}(\mu_n - \nu) \le \chi^2_{m,1-\alpha}\}$.*

We now turn to the problem of bootstrapping a confidence region for $\mu_{\mathcal{F}}$. Let $X^*_{i,n}$ be i.i.d. with common distribution $\hat{Q}_n$ (conditionally, given $\{X_i : 1 \le i \le n\}$). Write $\tilde{X}^*_{i,n} = \phi(X^*_{i,n}), 1 \le i \le n$ and let μ^*_n be a measurable selection from the Fréchet mean set of $\hat{Q}^{*,\phi}_n := \frac{1}{n}\sum_{i=1}^n \delta_{\tilde{X}^*_{i,n}}$. Let $E^*_{n,\alpha}$ be a subset of $\phi(U)$, such that $P^*(\mu^*_n - \mu_n \in E^*_{n,\alpha}) \to 1-\alpha$ in probability, where P^* denotes the probability under $\hat{Q}_n$.

COROLLARY 5.4.3. *In addition to the hypothesis of Theorem 5.4.1, assume C is nonsingular. Then $\phi^{-1}(\{(\mu_n - E^*_{n,\alpha}) \cap \phi(U)\})$ is a confidence region for $\mu_{\mathcal{F}}$ of asymptotic level $(1-\alpha)$.*

Proof. One may write equation (5.19) and relation (5.19), with μ and μ_n replaced by μ_n and μ^*_n, respectively, also replacing $\tilde{X}_i$ by $\tilde{X}^*_i$ in (5.22). To show that a new version of (2.11) holds with similar replacements (also replacing Λ by $\hat{\Lambda}$), with a δ^*_n (in place of δ_n) going to zero in probability, one may apply Chebyshev's inequality with a first order absolute moment under $\hat{Q}_n$, proving that $\hat{\Lambda}^* - \hat{\Lambda}$ goes to zero in probability. Here $\hat{\Lambda}^* = \frac{1}{n}\sum_{i=1}^{n}(Grad\Psi)(\tilde{X}_i^*;\mu_n^*)$. One then arrives at the desired version of (5.19) , replacing $\mu_n, \mu, \Lambda, \tilde{X}_i$ by $\mu_n^*, \mu_n, \hat{\Lambda}, \tilde{X}_i^*$, respectively, and with the remainder (corresponding to $\delta_n^{'}$) going to zero in probability. ■

REMARK 5.4.4. *In Corollary 5.4.3, we have considered the so called percentile bootstrap of Hall (1997) [137] (also see Efron (1982) [97]), which does not require the computation of the standard error $\hat{\Lambda}$. For this as well as for the CLT based confidence region given by Corollary 5.4.3, one can show that the coverage error is no more than $O_p(n^{-m/(m+1)})$ or $O(n^{-m/(m+1)})$, as the case may be (Bhattacharya and Chan (1996) [35]). One may also use the bootstrap distribution of the pivotal statistic $n(\mu_n-\mu)^T\hat{\Gamma}^{-1}(\mu_n-\mu)$ to find $c_{n,\alpha}^*$ such that*

$$P^*(n(\mu_n^*-\mu_n)^T\hat{\Gamma}^{*-1}(\mu_n^*-\mu_n) \leq c_{n,\alpha}^*) \simeq 1-\alpha, \tag{5.38}$$

to find the confidence region

$$D_{n,\alpha}^* = \{v \in \phi(U) : n(\mu_n - v)^T\hat{\Gamma}^{-1}(\mu_n - v) \leq c_{n,\alpha}^*\}. \tag{5.39}$$

In particular, if Q has a nonzero absolutely continuous component w.r.t. the volume measure on M, then so does Q^ϕ w.r.t. the Lebesgue measure on $\phi(U)$) (see do Carmo (1992) [87], p.44). Then assuming (a) $c_{n,\alpha}^*$ is such that the P^*-probability in (2.23) equals $1-\alpha+O_p(n^{-2})$ and (b) some additional smoothness and integrability conditions of the third derivatives of Ψ, one can show that the coverage error (i.e. the difference between $1-\alpha$ and $P(\mu \in D_{n,\alpha}^*)$ is $O_p(n^{-2})$ (see Bhattacharya and Ghosh (1978) [38], Chandra and Ghosh (1979) [62], Hall (1988) [136], (1997) [137] and Bhattacharya and Denker (1990) [36]). It follows that the coverage error of the confidence region $\phi^{-1}(D_{n,\alpha}^* \cap \phi(U))$ for $\mu_{\mathcal{F}}$ is also $O(n^{-2})$. We state one such result precisely.

COROLLARY 5.4.4. *(Bootstrapping the Cartan mean) Suppose the hypothesis of Theorem 5.4.3 holds. Then*

$$\sup_{r>0}|P^*(n(\mu_n^*-\mu_n)^T\hat{\Gamma}^{*-1}(\mu_n^*-\mu_n) \leq r) -$$
$$-P(n(\mu_n-\mu)^T\hat{\Gamma}^{-1}(\mu_n-\mu) \leq r)| = O_p(n^{-2}),$$

and the coverage error of the pivotal bootstrap confidence region is $= O_p(n^{-2})$.

REMARK 5.4.5. *The assumption of absolute continuity of the probability measure Q is reasonable for most applications. Indeed this is assumed in most parametric models in directional and shape analysis (see, e.g. Watson (1983) [333], Dryden and Mardia (1998) [91]).*

REMARK 5.4.6. *The results of this section may be extended to the two-sample problem, or to paired samples, in a fairly straightforward manner. For example,* *in the case of paired observations $(X_i, Y_i), i = 1, \cdots, n$, let X_i have (marginal) distribution Q, and intrinsic mean μ_g, and let Q_2 and ν_g be the corresponding quantities for Y_i. Let $\phi = Exp_{x_0}^{-1}$ for some x_0, and let μ, ν and μ_n, ν_n be the images under ϕ of the intrinsic population and sample means. Then one arrives at the following*

$$\sqrt{n}(\mu_n - \mu) - \sqrt{n}(\nu_n - \nu) \to_d \mathcal{N}_m(0, \Gamma), \tag{5.40}$$

where Γ is the covariance matrix of $\Lambda_1^{-1}\Psi(\tilde{X}_i; \mu) - \Lambda_2^{-1}\Psi(\tilde{Y}_i; \nu)$. Here Λ_i is the Hessian matrix of $F \circ \phi^{-1}$ for $Q_i (i = 1, 2)$. Assume Γ is nonsingular. Then a CLT-based confidence region for $\gamma := \mu - \nu$ is given in terms of $\gamma_n := \mu_n - \nu_n$ by $\{v \in \mathbb{R}^m : n(\gamma_n - v)\hat{\Gamma}^{-1}(\gamma_n - v) \leq \chi^2_{m,1-\alpha}\}$. Alternatively, one may use a bootstrap estimate of the distribution of $\sqrt{n}(\gamma_n - \gamma)$ to derive a confidence region.

5.5 The CLT for Extrinsic Sample Means and Confidence Regions for the Extrinsic Mean

From Theorem 5.4.1 one may derive a CLT for extrinsic sample means similar to Corollary 5.4.2. In this section, however, we use another approach which, for extrinsic means, is simpler to apply and generally less restrictive.

Recall that the extrinsic mean $\mu_{j,E}(Q)$ of a nonfocal probability measure Q on a manifold $\mathcal{M}$ w.r.t. an embedding $j : M \to \mathbb{R}^N$, when it exists, is given by $\mu_{j,E}(Q) = j^{-1}(P_j(\mu))$, where μ is the mean of $j(Q)$ and P_j is the projection on $j(\mathcal{M})$ (see Bhattacharya and Patrangenaru (2003)[42], Proposition 3.1, for example). Often the extrinsic mean will be denoted by $\mu_E(Q)$, or simply μ_E, when j and Q are fixed in a particular context. To insure the existence of the extrinsic mean set, in this section we will assume that $j(\mathcal{M})$ is closed in $\mathbb{R}^N$.

Assume $(X_1, ..., X_n)$ are i.i.d. $\mathcal{M}$-valued random objects whose common probability distribution is Q, and let $\overline{X}_E := \mu_E(\hat{Q}_n)$ be the *extrinsic sample mean*. Here $\hat{Q}_n = \frac{1}{n}\sum_{j=1}^n \delta_{X_j}$ is the empirical distribution. Two different CLT's for the extrinsic sample mean are currently available. The first one, concerns a distribution on a *submanifold* $\mathcal{M}$ of $\mathbb{R}^N$ (with j the inclusion map) and was derived by Hendriks and Landsman (1996, 1998)[152],[154]. Independently, Patrangenaru (1998)[266] derived a CLT for distribution on an embedded manifold, using a different estimator. Differentiable manifolds that are not apriori submanifolds of $\mathbb{R}^N$ arise in new areas of data analysis such as in shape analysis, in high level image analysis, or in signal and image processing (see e.g. Dryden and Mardia (1998) [91], Kendall (1984 [177], 1995 [178]), Kendall et al. (1999) [179], Goodall (1991) [125], Srivastava and Klassen (2001) [318], Mardia and Patrangenaru (2001a) [232]). These manifolds, known under the names of shape spaces and projective shape spaces, are quotient spaces of submanifolds of $\mathbb{R}^N$ (spaces of orbits of actions of Lie groups), rather than submanifolds of $\mathbb{R}^N$. Our approach is a generalization of the adapted frame method of Patrangenaru (1998)[266] to closed embeddings in $\mathbb{R}^N$. This method leads to an appropriate dimension reduction in the CLT and, thereby, reduces computational intensity. This method extends the results of Fisher et al. (1996) [116] who considered the case $M = S^m$.

Assume j is an embedding of an m dimensional manifold $\mathcal{M}$ such that $j(\mathcal{M})$ is closed in $\mathbb{R}^N$, and Q is a j-nonfocal probability measure on $\mathcal{M}$ such that $j(Q)$ has finite moments of order two (or of sufficiently high order as needed). Let μ and Σ be, respectively, the mean and covariance matrix of $j(Q)$ regarded as a probability measure on $\mathbb{R}^N$. Let $\mathcal{F}$ be the set of focal points of $j(\mathcal{M})$, and let $P_j : \mathcal{F}^c \to j(\mathcal{M})$ be the projection on $j(\mathcal{M})$. P_j is differentiable at μ and has the differentiability class of $j(\mathcal{M})$ around any nonfocal point. In order to evaluate the differential $d_\mu P_j$ we consider a special orthonormal frame field that will ease the computations. Assume $p \to (f_1(p), \dots, f_m(p))$ is a local frame field on an open subset of $\mathcal{M}$ such that, for each $p \in M$, $(d_p j(f_1(p)), \dots, d_p j(f_m(p)))$ are orthonormal vectors in $\mathbb{R}^N$. A local frame field $(e_1(y), e_2(y), \dots, e_N(y))$ defined on an open neighborhood $U \subseteq \mathbb{R}^N$ is *adapted to the embedding* j if it is an orthonormal frame field and

$$e_r(j(p)) = d_p j(f_r(p), \forall r = 1, \dots, m, \forall p \in j^{-1}(U). \tag{5.41}$$

Let $e_1, e_2, \dots, e_N$ be the canonical basis of $\mathbb{R}^N$ and assume $(e_1(y), e_2(y), \dots, e_N(y))$ is an adapted frame field around $P_j(\mu) = j(\mu_E)$. Then $d_\mu P_j(e_b) \in T_{P_j(\mu)} j(\mathcal{M})$ is a linear combination of $e_1(P_j(\mu)), e_2(P_j(\mu)), \dots, e_m(P_j(\mu))$:

$$d_\mu P_j(e_b) = \sum_{a=1}^{m} \left(d_\mu P_j(e_b)\right) \cdot e_a\left(P_j(\mu)\right) e_a\left(P_j(\mu)\right), \forall b = 1, \dots, N. \tag{5.42}$$

By the delta method, $n^{\frac{1}{2}}(P_j(\overline{j(X)}) - P_j(\mu))$ converges weakly to a random vector V having a $\mathcal{N}_N(0, \Sigma_\mu)$ distribution. Here $\overline{j(X)} = \frac{1}{n} \sum_{i=1}^n j(X_i)$ and

$$\Sigma_\mu = \left[\sum_{a=1}^{m} d_\mu P_j(e_b) \cdot e_a(P_j(\mu)) e_a(P_j(\mu))\right]_{b=1,\dots,N} \Sigma \left[\sum_{a=1}^{m} d_\mu P_j(e_b) \cdot e_a(P_j(\mu)) e_a(P_j(\mu))\right]_{b=1,\dots,N}^T, \tag{5.43}$$

where Σ is the covariance matrix of $j(X_1)$ w.r.t. the canonical basis $e_1, \dots, e_N$. The asymptotic distribution $\mathcal{N}_N(0, \Sigma_\mu)$ is degenerate and can be regarded as a distribution on $T_{P_j(\mu)} j(\mathcal{M})$, since the range of $d_\mu P_j$ is $T_{P_j(\mu)} j(\mathcal{M})$. Note that

$$d_\mu P_j(e_b) \cdot e_a(P_j(\mu)) = 0, \quad \text{for} \quad a = m+1, \dots, N.$$

REMARK 5.5.1. *An asymptotic distribution of the extrinsic sample mean can be obtained as a particular case of Theorem 5.4.1. The covariance matrix in* *that theorem depends both on the way the manifold is embedded and on the chart used. We provide below an alternate CLT, which applies to an arbitrary embedding, leading to pivots and are independent of the chart used.*

The tangential component $tan(v)$ of $v \in \mathbb{R}^N$ w.r.t. the basis $e_a(P_j(\mu)) \in T_{P_j(\mu)}j(\mathcal{M}), a = 1, \dots, m$ is given by

$$tan(v) = (e_1(P_j(\mu))^T v \dots e_m(P_j(\mu))^T v)^T. \tag{5.44}$$

Then the random vector $(d_{\mu_E} j)^{-1}(tan(P_j(\overline{j(X)}) - P_j(\mu)))$ has the following covariance matrix w.r.t. the basis $f_1(\mu_E), \cdots, f_m(\mu_E)$:

$$\Sigma_{j,E} = (e_a(P_j(\mu))^T \Sigma_\mu e_b(P_j(\mu)))_{1 \le a,b \le m} =$$

$$\left[\sum d_\mu P_j(e_b) \cdot e_a(P_j(\mu))\right]_{a=1,\dots,m} \Sigma \left[\sum d_\mu P_j(e_b) \cdot e_a(P_j(\mu))\right]^T_{a=1,\dots,m}. \tag{5.45}$$

DEFINITION 5.5.1. *The matrix $\Sigma_{j,E}$ given by* (5.45) *is the* extrinsic covariance matrix *of the j-nonfocal distribution Q (of X_1) w.r.t. the basis$f_1(\mu_E), \dots, f_m(\mu_E)$.*

When j is fixed in a specific context, the subscript j in $\Sigma_{j,E}$ may be omitted. If, in addition, rank $\Sigma_\mu = m$, Σ_E is invertible and we define the *j-standardized mean vector*

$$\overline{Z}_{j,n} =: n^{\frac{1}{2}} \Sigma_E{}^{-\frac{1}{2}} (\overline{X}^1{}_j \dots \overline{X}^m{}_j)^T. \tag{5.46}$$

PROPOSITION 5.5.1. *Assume $\{X_r\}_{r=1,\dots,n}$ are i.i.d.r.o.'s from the j-nonfocal distribution Q, with finite mean $\mu = E(j(X_1))$, and assume the extrinsic covariance matrix $\Sigma_{j,E}$ of Q is finite. Let $(e_1(y), e_2(y), \dots, e_N(y))$ be an orthonormal frame field adapted to j. Then (a) the extrinsic sample mean $\overline{X}_{j,E}$ has asymptotically a normal distribution in the tangent space to $\mathcal{M}$ at $\mu_{j,E}(Q)$ with mean* 0 *and covariance matrix $n^{-1}\Sigma_{j,E}$, and (b) if $\Sigma_{j,E}$ is nonsingular, the j-standardized mean vector $\overline{Z}_{j,n}$ given in* (5.46) *converges weakly to a $\mathcal{N}_m(0_m, I_m)$-distributed random vector.*

As a particular case of Proposition 5.5.1, when j is the inclusion map of a submanifold of $\mathbb{R}^N$, we get the following result for nonfocal distributions on an arbitrary closed submanifold $\mathcal{M}$ of $\mathbb{R}^N$:

COROLLARY 5.5.1. *Assume $M \subseteq \mathbb{R}^N$ is a closed m dimensional submanifold of $\mathbb{R}^N$. Let $\{X_r\}_{r=1,\dots,n}$ be i.i.d.r.o.'s from the nonfocal distribution Q on $\mathcal{M}$ with the finite mean vector $\mu = E(X_1)$ and covariance matrix Σ. Let $(e_1(y), e_2(y), \dots, e_N(y))$ be an orthonormal frame field adapted to $\mathcal{M}$. Let $\Sigma_E :=$ $\Sigma_{j,E}$, where $j : M \to \mathbb{R}^N$ is the inclusion map. Then (a) $n^{\frac{1}{2}} tan(j(\overline{X}_E) - j(\mu_E))$ converges weakly to $\mathcal{N}_m(0_m, \Sigma_E)$, and (b) if Σ induces a nonsingular bilinear form on $T_{j(\mu_E)}j(\mathcal{M})$, then $\|\overline{Z}_{j,n}\|^2$ in* (5.46) *converges weakly to the chi-square distribution χ^2_m.*

REMARK 5.5.2. *The CLT for extrinsic sample means as stated in Proposition 5.5.1 or Corollary 5.5.1 can not be used to construct confidence regions for extrinsic means, since the population extrinsic covariance matrix is unknown. In order to find a consistent estimator of* $\Sigma_{j,E}$, *note that* $\overline{j(X)}$ *is a consistent estimator of* μ, $d_{\overline{j(X)}}P_j$ *converges in probability to* $d_\mu P_j$, *and* $e_a(P_j(\overline{j(X)}))$ *converges in probability to* $e_a(P_j(\mu))$ *and, further,*

$$S_{j,n} = n^{-1}\sum(j(X_r) - \overline{j(X)})(j(X_r) - \overline{j(X)})^T$$

is a consistent estimator of Σ. *It follows that*

$$\left[\sum_{a=1}^{m} d_{\overline{j(X)}}P_j(e_b)\cdot e_a(P_j(\overline{j(X)}))e_a(P_j(\overline{j(X)}))\right] S_{j,n}$$
$$\left[\sum_{a=1}^{m} d_{\overline{j(X)}}P_j(e_b)\cdot e_a(P_j(\overline{j(X)}))e_a(P_j(\overline{j(X)}))\right]^T \tag{5.47}$$

is a consistent estimator of Σ_μ, *and* $tan_{P_j(\overline{j(X)})}v$ *is a consistent estimator of* $tan(v)$.

If we take the components of the bilinear form associated with the matrix (5.47) w.r.t. $e_1(P_j(\overline{j(X)})), e_2(P_j(\overline{j(X)})), ..., e_m(P_j(\overline{j(X)}))$, we get a consistent estimator of $\Sigma_{j,E}$, called the *the sample extrinsic covariance matrix*, given by

$$S_{j,E,n} = \left[\left[\sum d_{\overline{j(X)}}P_j(e_b)\cdot e_a(P_j(\overline{j(X)}))\right]_{a=1,...,m}\right]\cdot S_{j,n}$$
$$\left[\left[\sum d_{\overline{j(X)}}P_j(e_b)\cdot e_a(P_j(\overline{j(X)}))\right]_{a=1,...,m}\right]^T, \tag{5.48}$$

and obtain the following results:

THEOREM 5.5.1. *(Adapted frame CLT) Assume* $j : M \to \mathbb{R}^N$ *is a closed embedding of* $\mathcal{M}$ *in* $\mathbb{R}^N$. *Let* $\{X_r\}_{r=1,...,n}$ *be a random sample from the j-nonfocal distribution Q, and let* $\mu = E(j(X_1))$ *and assume* $j(X_1)$ *has finite second order moments and the extrinsic covariance matrix* $\Sigma_{j,E}$ *of* X_1 *is nonsingular. Let* $(e_1(y), e_2(y), \ldots, e_N(y))$ *be an orthonormal frame field adapted to j. If the* sample extrinsic covariance matrix $S_{j,E,n}$ *is given by* (5.48), *then for n large enough* $S_{j,E,n}$ *is nonsingular (with probability converging to one) and (a) the statistic*

$$n^{\frac{1}{2}}S_{j,E,n}^{-\frac{1}{2}}tan(P_j(\overline{j(X)}) - P_j(\mu)) \tag{5.49}$$

converges weakly to $N_m(0_m, I_m)$, *so that the following*

$$n\|S_{j,E,n}^{-\frac{1}{2}}tan(P_j(\overline{j(X)}) - P_j(\mu))\|^2 \tag{5.50}$$

converges weakly to χ^2_m, *and (b) the statistic*

$$n^{\frac{1}{2}} S^{-\frac{1}{2}}_{j,E,n} tan_{P_j(\overline{j(X)})}(P_j(\overline{j(X)}) - P_j(\mu)) \tag{5.51}$$

converges weakly to $N_m(0_m, I_m)$, *so that*

$$n \| S^{-\frac{1}{2}}_{j,E,n} tan_{P_j(\overline{j(X)})}(P_j(\overline{j(X)}) - P_j(\mu)) \|^2 \tag{5.52}$$

converges weakly to χ^2_m.

COROLLARY 5.5.2. *Under the hypothesis of Theorem 5.5.1, a confidence region for* μ_E *of asymptotic level* $1-\alpha$ *is given by (a)* $C_{n,\alpha} := j^{-1}(U_{n,\alpha})$, *where* $U_{n,\alpha} = \{\nu \in j(\mathcal{M}) : n\|S^{-\frac{1}{2}}_{j,E,n} tan_{\nu}(P_j(\overline{j(X)}) - \nu)\|^2 \leq \chi^2_{m,1-\alpha}\}$, *or by (b)* $D_{n,\alpha} := j^{-1}(V_{n,\alpha})$, *where* $V_{n,\alpha} = \{\nu \in j(\mathcal{M}) : n\|S^{-\frac{1}{2}}_{j,E,n} tan_{P_j(\overline{j(X)})}(P_j(\overline{j(X)}) - \nu)\|^2 \leq \chi^2_{m,1-\alpha}\}$.

REMARK 5.5.3. *To simplify the notation,* $\forall a = 1, \dots n$, *let* $Y_a = j(X_a)$. *Then, if* $\bar{Y}_n$ *is the sample mean vector of the i.i.d.r.vec.'s* $Y_1, \dots, Y_n$,

$$\sqrt{n}[P_j(\bar{Y}_n) - P_j(\mu)] = \sqrt{n}[(DP_j)_\mu(\bar{Y}_n - \mu)] + o_P(1) \to_d \mathcal{N}_N(0, \Sigma_\mu) \ as n \to \infty, \tag{5.53}$$

where, from (5.43), *we have*

$$\Sigma_\mu = (DP_j)_\mu \Sigma (DP_j)^T_\mu. \tag{5.54}$$

In equation (5.54), $(DP_j)_y$ *is the* $N \times N$ *Jacobian matrix of* P_j *at* y.

Given that $P_j : U \subset \mathbb{R}^N \to j(\mathcal{M})$ *is a differentiable function between manifolds (here* U *is an open subset of the set of nonfocal points), it follows that the differential* $d_\mu P_j$ *maps* $\mathbb{R}^N$ *to* $T_{P_j(\mu)}(j(\mathcal{M}))$, *regarded as a subspace of* $T_{P_j(\mu)}(\mathbb{R}^N) = \mathbb{R}^N$. *For this reason it was convenient to express the values of this differential on the canonical basis of* $T_{P_j(\mu)}(\mathbb{R}^N) = \mathbb{R}^N$ *as linear combinations of a basis in of* $T_{P_j(\mu)}(j(\mathcal{M}))$. *Therefore, in* (5.43), *we selected a convenient local section of the bundle of orthonormal frames of* $\mathcal{M}$, *say* $y \to (e_1(y), e_2(y), \dots, e_m(y))$ *of the subspace* $T_y(j(\mathcal{M})) \subset T_y(\mathbb{R}^N) = \mathbb{R}^N$ *(for* y *in a neighborhood of* $P_j(\mu)$ *in* $j(\mathcal{M})$*). The following proposition then follows from* (5.53).

PROPOSITION 5.5.2. *Assume* Q *is* j *nonfocal, and* $E\|Y_1\|^2 < \infty$. *Let* B *denote the* $m \times N$ *matrix whose rows* $e_i^T(P_j(\mu)), i = 1, \dots, m$, *form an orthonormal basis of* $T_{P_j(\mu)}(j(\mathcal{M}))$ *regarded as a subspace of* R^N. *Then*

$$\sqrt{n}[B(P_j(\bar{Y}_n) - P_j(\mu))] \to_d \mathcal{N}_m(0_m, \Sigma_E) as \ n \to \infty. \tag{5.55}$$

Also, writing B_n *for the* $m \times N$ *matrix whose rows are* $e_i^T(P_j(\bar{Y}_n)), i = 1, \dots, m$, *one has*

$$\sqrt{n}[B_n(P_j(\bar{Y}_n) - P_j(\mu))] \to_d \mathcal{N}_m(0_m, \Sigma_E) \ as \ n \to \infty. \tag{5.56}$$

Proof. *Equation* (5.55) *follows from Proposition 5.5.1, while* (5.56) *follows from* (5.55) *and a Slutsky type argument (see Theorem 2.8.3).*

REMARK 5.5.4. *The* extrinsic covariance matrix *of Q w.r.t. the orthogonal basis $e_1(P_j(\mu)),\dots,e_m(P_j(\mu))$ is $\Sigma_E = B\Sigma_\mu B^T$. Note that* (5.55) *expresses the asymptotic multivariate normal distribution $N_m(0_m, n^{-1}\Sigma_E)$ of the coordinates $(V_{1n},\dots,V_{mn})^T$ of the random vector $tan(P_j(\bar{Y}_n)-P_j(\mu))$ with respect to the basis $e_a^T(P_j(\mu)), a=1,\dots,m$. Under broad conditions, the extrinsic covariance matrix Σ_E is a nonsingular $m\times m$ matrix. For this it is sufficient to require that the distribution of $tanP_j(Y_1)$ does not have support on a $(m-1)$- dimensional subspace of $T_{P_j(\mu)}(j(\mathcal{M}))$.*

The following is a consequence of Theorem 5.5.1:

THEOREM 5.5.2. *Let the hypothesis of Theorem 5.5.1 hold. Then (a) one has the following asymptotic of the* extrinsic Hotelling statistic*:*

$$\sqrt{n}[B_n(P_j(\bar{Y}_n)-P_j(\mu))] \to_d V \sim \mathcal{N}_m(0,\Sigma_E) \text{ as } n\to\infty, \text{ and}$$
$$n[B_n(P_j(\bar{Y}_n)-P_j(\mu))]^T\hat{(}S_{j,E})^{-1}[B_n(P_j(\bar{Y}_n)-P_j(\mu))] \to_d T \sim \chi^2_m, \quad \text{as } n\to\infty. \tag{5.57}$$

Here $S_{j,E}$, is the sample extrinsic covariance matrix *given by*

$$S_{j,E} = B_n(DP)_{\bar{Y}_n}\hat{\Sigma}(DP)^T_{\bar{Y}_n}B_n^T, \tag{5.58}$$

$\hat{\Sigma}$ being the sample covariance matrix of $Y_i, i=1,\dots,n$. Also, (b) a confidence region for $P_j(\mu)$ of asymptotic level $1-\alpha$ is given by

$$C_{1-\alpha} = \{\nu\in j(\mathcal{M}), n[B_n(P_j(\bar{Y}_n)-\nu)]^T {S_{j,E}}^{-1}[B_n(P_j(\bar{Y}_n)-\nu)] \le \le \chi^2_m(1-\alpha). \tag{5.59}$$

Theorem 5.5.1 and Corollary 5.5.2 involve pivotal statistics. The advantages of using pivotal statistics in bootstrapping for confidence regions are well known (See, e.g., Babu and Singh (1984) [9], Beran (1987) [16], Hall (1988) [136], (1992) [137], Bhattacharya and Qumsiyeh (1989) [45] and Bhattacharya and Denker (1990) [36]).

At this point we recall the steps that one takes to obtain a bootstrapped statistic from a pivotal statistic. If $\{X_r\}_{r=1,\dots,n}$ is a random sample from the unknown distribution Q, and $\{X_r^*\}_{r=1,\dots,n}$ is a random sample from the empirical $\hat{Q}_n$, conditionally given $\{X_r\}_{r=1,\dots,n}$, then the statistic

$$T(X,Q) = n\|S_{j,E,n}^{-\frac{1}{2}}tan(P_j(\overline{j(X)})-P_j(\mu))\|^2$$

given in Theorem 5.5.1.a has the bootstrap analog

$$T(X^*,\hat{Q}_n) = n\|{S^*_{j,E,n}}^{-\frac{1}{2}}tan_{P_j(\overline{j(X)}))}(P_j(\overline{j(X^*)})-P_j(\overline{j(X)}))\|^2.$$

Here $S^*_{j,E,n}$ is obtained from $S_{j,E,n}$ substituting $X_1^*, \ldots, X_n^*$ for $X_1, \ldots X_n$, and $T(X^*, \hat{Q}_n)$ is obtained from $T(X,Q)$ by substituting $X_1^*, \ldots, X_n^*$ for $X_1, \ldots, X_n$, $\overline{j(X)}$) for μ and $S^*_{j,E,n}$ for $S_{j,E,n}$.

The same procedure can be used for the vector valued statistic

$$V(X,Q) = n^{\frac{1}{2}} S_{j,E,n}^{-\frac{1}{2}} tan(P_j(\overline{j(X)}) - P_j(\mu)),$$

and as a result we get the bootstrapped statistic

$$V^*(X^*, \hat{Q}_n) = n^{\frac{1}{2}} {S^*_{j,E,n}}^{-\frac{1}{2}} tan_{P_j(\overline{j(X)}))}(P_j(\overline{j(X^*)}) - P_j(\overline{j(X)})).$$

For the rest of this section, we will assume that $j(Q)$, when viewed as a measure on the ambient space $\mathbb{R}^N$, *has finite moments of sufficiently high order*. If $\mathcal{M}$ is compact then this is automatic. In the noncompact case finiteness of moments of order twelve, along with an assumption of a nonzero absolutely continuous component, is sufficient to ensure an Edgeworth expansion up to order $O(n^{-2})$ of the pivotal statistic $V(X,Q)$ (See Bhattacharya and Ghosh (1978) [38], Bhattacharya and Denker (1990) [36], Hall (1988) [136], Fisher et al. (1996) [116] and Chandra and Ghosh (1979) [62]). We then obtain the following results:

THEOREM 5.5.3. *Let $\{X_r\}_{r=1,\ldots,n}$ be a random sample from the j-nonfocal distribution Q which has a nonzero absolutely continuous component w.r.t. the volume measure on $\mathcal{M}$ induced by j. Let $\mu = E(j(X_1))$ and assume the covariance matrix Σ of $j(X_1)$ is defined and the extrinsic covariance matrix $\Sigma_{j,E}$ is nonsingular and let $p \to (e_1(p), e_2(p), \ldots, e_N(p))$ be an orthonormal frame field adapted to j. Then the distribution function of*

$$n\|S_{j,E,n}^{-\frac{1}{2}} tan(P_j(\overline{j(X)}) - P_j(\mu))\|^2$$

can be approximated by the bootstrap extrinsic Hotelling distribution of

$$n\|{S^*_{j,E,n}}^{-\frac{1}{2}} tan_{P_j(\overline{j(X)})}(P_j(\overline{j(X^*)}) - P_j(\overline{j(X)}))\|^2$$

with a coverage error $0_p(n^{-2})$.

One may also use nonpivotal bootstrap confidence regions, especially when $S_{j,E,n}$ is difficult to compute. The result in this case is the following (see Bhattacharya and Chan (1996)[35]).

PROPOSITION 5.5.3. *Under the hypothesis of Proposition* 3.1, *the distribution function of $n\|tan(P_j(\overline{j(X)}) - P_j(\mu))\|^2$ can be approximated uniformly by the bootstrap distribution of*

$$n\|tan_{P_j(\overline{j(X)})}(P_j(\overline{j(X^*)}) - P_j(\overline{j(X)}))\|^2$$

to provide a confidence region for μ_E with a coverage error no more than $O_p(n^{-\frac{m}{m+1}})$.

REMARK 5.5.5. *Note that Corollary 5.5.1(b) provides a computationally simpler scheme than Corollary 5.5.1(a) for large sample confidence regions; but for bootstrap confidence regions Theorem 5.5.3, which is the bootstrap analog of Corollary 5.5.1(a), yields a simpler method. The corresponding* $100(1-\alpha)\%$ *confidence region is* $C^*_{n,\alpha} := j^{-1}(U^*_{n,\alpha})$ *with* $U^*_{n,\alpha}$ *given by*

$$U^*_{n,\alpha} = \{\mu \in j(\mathcal{M}) : n\|S_{j,E,n}^{-\frac{1}{2}} tan(P_j(\overline{j(X)}) - P_j(\mu))\|^2 \leq c^*_{1-\alpha}\}, \tag{5.60}$$

where $c^*_{1-\alpha}$ *is the upper* $100(1-\alpha)\%$ *point of the values*

$$n\|S^*_{j,E,n}{}^{-\frac{1}{2}} tan_{P_j(\overline{j(X)})}(P_j(\overline{j(X^*)}) - P_j(\overline{j(X)}))\|^2 \tag{5.61}$$

among the bootstrap resamples. One could also use the bootstrap analog of the confidence region given in Corollary 5.5.1(b) for which the confidence region is $D^*_{n,\alpha} := j^{-1}(V^*_{n,\alpha})$ *with* $V^*_{n,\alpha}$ *given by*

$$V^*_{n,\alpha} = \{\mu \in j(\mathcal{M}) : n\|S_{j,E,n}^{-\frac{1}{2}} tan_{P_j(\overline{j(X)})}(P_j(\overline{j(X)}) - P_j(\mu))\|^2 \leq d^*_{1-\alpha}\}, \tag{5.62}$$

where $d^*_{1-\alpha}$ *is the upper* $100(1-\alpha)\%$ *point of the values*

$$n\|S^*_{j,E,n}{}^{-\frac{1}{2}} tan_{P_j(\overline{j(X^*)})}(P_j(\overline{j(X^*)}) - P_j(\overline{j(X)}))\|^2 \tag{5.63}$$

among the bootstrap resamples. The region given by (5.62) *and* (5.63) *has coverage error* $O_p(n^{-2})$.

5.6 Exercises

Exercise 70. *If* Q *is rotationally symmetric on* S^m *(such as the von Mises distribution), then the intrinsic mean set of* Q *is a union of parallel* $m-1$ *dimensional spheres or poles of the axis of rotation, since the space of orbits is 1-dimensional. Let* $SO(m)$ *be the special orthogonal group (or group of rotations). The* $SO(m)$ *invariant measures on* S^m *depend on one function of one real variable, as shown in Watson (1983)[333], Section 4.2.*

Exercise 71. *Let* $M = S^m$, *be the unit sphere centered at the origin of* $\mathbf{R}^{m+1}$.
a. Show that a probability measure Q *is nonfocal on* S^m, *if* $\mu = \int_{\mathbf{R}^{m+1}} xQ(dx) \neq 0$.
b. Show that if Q *is nonfocal, then the extrinsic mean is given by* $\mu_E = \frac{\mu}{\|\mu\|}$.
c.Show that H *defined in* (5.16) *is given by*

$$H(\overline{X}) - \mu = \overline{X} - \mu_Q - \{(\overline{X} - \mu_Q) \cdot \mu\}\mu. \tag{5.64}$$

Exercise 72. *Use Exercise 71, to show that :*
a. $\sqrt{n}(H(\overline{X}) - \mu)$ *converges in distribution to a m-dimensional normal distribution supported by the tangent space* $T_\mu S^m$ *identified with the linear subspace*

$\{v \in \mathbf{R}^{m+1} : v^T \mu = 0\}$.

b. As a measure on $\mathbf{R}^{m+1}$, *the distribution in part a. has mean* 0 *and covariance matrix* Γ *in* (13.37) *given by* $\Gamma := \Sigma + (\mu^T \Sigma \mu)\mu\mu^T - 2\Sigma(\mu\mu^T)$, *where* Σ *is the covariance matrix of* Q *viewed as a measure on* $\mathbb{R}^{m+1}$.

Chapter 6

Nonparametric Inference for Two Samples on Manifolds

6.1 Introduction

Hendriks and Landsman (1998) [154] developed asymptotic test statistics for the equality of two extrinsic means on a submanifold $\mathcal{M}$ of the Euclidean space $\mathbb{R}^D$, based on independent samples drawn from the two distributions. Using a general local orthonormal frame field methodology, developed in Patrangenaru (1998) [266], and in Bhattacharya and Patrangenaru (2005) [43], one may find an alternate two sample test statistic for the equality of two extrinsic means. Such an approach is due to A. Bhattacharya (2008) [25], who tested the equality of mean extrinsic objects by taking the pooled extrinsic sample mean as an estimator of the hypothesized extrinsic mean in two correlated pairs (or not) of objects on a manifold embedded in the Euclidean space; note that this test statistic can not be extended to matched pairs, since it is derived from the a test for the equality of two extrinsic means, by pooling without pairing the observations measured for each individual.

Nonparametric tests for the difference between mean vectors in matched pairs in $\mathbb{R}^m$ are built using standard asymptotic distributions for the sample mean difference of the matched pairs, or for the difference between sample means in the unmatched case, and are leading to χ^2_m-tests; the key algebraic property used to derive such asymptotics, is the additive group structure of $\mathbb{R}^m$, under the operation of vector sum. Similarly, comparison of mean axes (Beran and Fisher (1998) [18]), or of extrinsic mean planar projective shapes (see Mardia and Patrangenaru (2005) [233]), as well as estimation of 3D motion in computer vision (eg. Tuzel et al. (2005) [324]), leads to natural inference problems on groups of rotations, as additional examples of matched pairs analysis on Lie groups.

However, in most situations, the dimension of the Lie group $\mathcal{G}$ of transformations is higher than the dimension of the sampling manifold $\mathcal{M}$ on which $\mathcal{G}$ is acting. Ideally, the dimensions of transitive Lie group $\mathcal{G}$ and of $\mathcal{M}$ should be the same. This is the case if the sampling manifold $\mathcal{M}$ itself has a Lie group structure. In Section 6.3 we define the change C in a matched pair of random objects X, Y on a Lie group $(\mathcal{G}, \circ)$ by $C = X^{-1} \circ Y$. This allows us to reduce a two-sample testing question for means in paired data on $\mathcal{G}$, to a one-sample mean change test on $\mathcal{G}$. These are applied the nonparametric inference for testing for a one-sample extrinsic mean on a manifold developed in Section 6.3. Section 6.6 is dedicated to developing a nonparametric methodology for comparing the extrinsic means of two populations on a manifold.

6.2 Two-sample Test for Total Extrinsic Variances

In this short section, we present a large samples test for comparing the total intrinsic variances in two independent populations, due to A. Bhattacharya (2008) [25]. Here we will use the asymptotic distribution of the sample total extrinsic variance to construct nonparametric tests to compare two probability distributions Q_1 and Q_2 on an embedded manifold $\mathcal{M}$.

Let $X_1, \dots, X_{n_1}$ and $Y_1, \dots, Y_{n_2}$ be two iid samples from Q_1 and Q_2, respectively, that are mutually independent. Let $\mu_{i,E}$ and V_i denote the extrinsic means and total variances of $Q_i, i = 1, 2$, respectively. Similarly denote by $\hat{\mu}_{i,E}$ and $\hat{V}_i$ the sample extrinsic means and total variances. We want to test the hypothesis $H_0 : V_1 = V_2 + \delta$. From fact that the samples are independent, and the asymptotic distribution of the total sample extrinsic variance if $n = n_1 + n_2, \frac{n_a}{n} \to p_a, a = 1, 2$ as $n \to \infty$, we get that, under H_0,

$$\sqrt{n}(\hat{V}_1 - \hat{V}_2) \to_d Y, Y \sim \mathcal{N}(\delta, \frac{\sigma_1^2}{p_1} + \frac{\sigma_2^2}{p_2}), \tag{6.1}$$

$$\frac{\hat{V}_1 - \hat{V}_2 - \delta}{\sqrt{\frac{s_1^2}{n_1} + \frac{s_2^2}{n_2}}} \to_d Z, Z \sim \mathcal{N}(0, 1), \tag{6.2}$$

where $\sigma_1^2 = Var[d^2(X_1, \mu_{1,E})], \sigma_2^2 = Var[d^2(X_2, \mu_{2,E})]$ and s_1^2, s_2^2 are their sample estimates. Hence to test if H_0 is true, we can use the test statistic

$$T = \frac{\hat{V}_1 - \hat{V}_2 - \delta}{\sqrt{\frac{s_1^2}{n_1} + \frac{s_2^2}{n_2}}}. \tag{6.3}$$

For a test of asymptotic size α, we reject H_0 if $|T| > z_{\frac{\alpha}{2}}$.

REMARK 6.2.1. *One can also construct a bootstrap confidence interval for* $V_1 - V_2 - \delta$ *and use that to test* $H_0 : V_1 - V_2 = \delta$, *for example by taking for* $a = 1,2$ *bootstrap resamples* $X^*_{a,i_a}, i_a = 1,\dots,n_a$, *from each of the two samples* $X_{a,i_a}, i_a = 1,\dots,n_a$, *and compute the pivotal bootstrap statistics*

$$T^* = \frac{\hat{V}_1^* - \hat{V}_2^* - \delta}{\sqrt{\frac{s_1^{2*}}{n_1} + \frac{s_2^{2*}}{n_2}}}, \tag{6.4}$$

and, for a large number N of resamples, compute the upper and lower $\frac{\alpha}{2}$ *the bootstrap percentiles* $T^*_{1-\frac{\alpha}{2}}, T^*_{\frac{\alpha}{2}}$. *The hypothesis is rejected at level* α *if* 0 *is not in the interval* $(T^*_{1-\frac{\alpha}{2}}, T^*_{\frac{\alpha}{2}})$.

6.3 Bhattacharya's Two-Sample Test for Extrinsic Means of Independent Populations on a Manifold

We now turn to two-sample tests for extrinsic means of distributions on an arbitrary m dimensional manifold M given in Bhattacharya and Patrangenaru (2014) [44]). There is a large sample test for equality of two extrinsic means, originally due to A. Bhattacharya (2008) [25] (see also Bhattacharya and Bhattacharya (2012) [27], p.42). Let $X_{ak_a} : k_a = 1,\dots,n_a, a = 1,2$ be two independent random samples drawn from distributions $Q_a, a = 1,2$ on M, and let j be an embedding of M into $\mathbb{R}^N$. Denote by μ_a the mean of the induced probability $Q_a \circ j^{-1}$ and Σ_a its covariance matrix ($a = 1,2$). Then the extrinsic mean of Q_a is $\mu_{a,j} = j^{-1}(P_j(\mu_a))$, assuming Q_a is nonfocal. Write $Y_{ak_a} = j(X_{ak_a}) k_a = 1,\dots,n_a, a = 1,2$ and let $\bar{Y}_a, a = 1,2$ be the corresponding sample means. Assuming finite second moments of $Y_{a1}, a = 1,2$, which is automatic if M is compact, one has, by Theorem 5.5.2

$$\sqrt{n_a} B_a[P_j(\bar{Y}_a) - P_j(\mu_a)] \to_d \mathcal{N}_m(0, \Sigma_{a,j}), a = 1,2, \tag{6.5}$$

where $\Sigma_{a,j}$ is the extrinsic covariance matrix of Q_a, and B_a are the same as in Proposition 5.5.2 and Theorem 5.5.2, but with Q replaced by Q_a (a=1,2). That is, $B_a(y)$ is the $m \times N$ matrix of an orthonormal basis (frame) of $T_y(j(M)) \subset T_y(\mathbb{R}^N) = \mathbb{R}^N$ for y in a neighborhood of $P_j(\mu_a)$, and $B_a = B_a(P_j(\mu_a))$. Similarly, $C_a \doteq \Sigma_{\mu_a,a} = (DP_j)_{\mu_a} \Sigma_a (DP_j)^T_{\mu_a}, (a = 1,2)$. The null hypothesis $H_0 : \mu_{1,j} = \mu_{2,j}$, say, is equivalent to $H_0 : P_j(\mu_1) = P_j(\mu_2) = \pi$, say. Then, under the null hypothesis, letting $B = B(\pi)$, one has $B_1 = B_2 = B$, and

$$[B(\frac{1}{n_1}C_1 + \frac{1}{n_2}C_2)B^T]^{-1/2}B[P_j(\bar{Y}_1) - P_j(\bar{Y}_2)] \to_d \mathcal{N}_m(0_m, I_m),$$
$$\text{as } n_1 \to \infty, n_2 \to \infty. \tag{6.6}$$

For statistical inference one estimates C_a by

$$\hat{C}_a = (DP_j)_{\bar{Y}_a}\hat{\Sigma}_a(DP_j)^T_{\bar{Y}_a} \tag{6.7}$$

where $\hat{\Sigma}_a$ is the sample covariance matrix of sample $a(a = 1,2)$. Also B is replaced by $\hat{B} = B(\hat{\pi})$ where $\hat{\pi}$ is a sample estimate of π. Under H_0, both $P_j(\bar{Y}_1)$ and $P_j(\bar{Y}_2)$ are consistent estimates of π, so we take a "pooled estimate"

$$\hat{\pi} = P_j(\frac{1}{n_1+n_2}(n_1P_j(\bar{Y}_1) + n_2P_j(\bar{Y}_2))). \tag{6.8}$$

We, therefore, have the following result due to Bhattacharya and Bhattacharya (2012)[27].

THEOREM 6.3.1. *Assume the extrinsic sample covariance matrix $\hat{\Sigma}_{a,j}$ is nonsingular for $a = 1,2$. Then, under $H_0 : \mu_{1,j} = \mu_{2,j}$, one has:*

$$(\hat{B}[P_j(\bar{Y}_1) - P_j(\bar{Y}_2)])^T[\hat{B}(\frac{1}{n_1}\hat{C}_1 + \frac{1}{n_2}\hat{C}_2)\hat{B}^T]^{-1}(\hat{B}[P_j(\bar{Y}_1) - P_j(\bar{Y}_2)])$$
$$\to_d \chi^2_m, \tag{6.9}$$

as $n_1 \to \infty, n_2 \to \infty$ □

For the two-sample intrinsic test, given a Riemannian structure g on $\mathcal{M}$, let $\mu_{1,g}, \mu_{2,g}$ denote the intrinsic means of Q_1 and Q_2, assumed to exist, and consider $H_0 : \mu_{1,g} = \mu_{2,g}$. Denoting by $\hat{\mu}_{1,g}, \hat{\mu}_{2,g}$ the sample intrinsic means, (5.29) implies that, under H_0,

$$(\frac{1}{n_1}\Lambda_1^{-1} + \frac{1}{n_2}\Lambda_2^{-1})^{-1/2}[\varphi_p(\hat{\mu}_{1,g}) - \varphi_p(\hat{\mu}_{2,g})] \to_d \mathcal{N}_m(0, I_m),$$
$$\text{as } n_1 \to \infty, n_2 \to \infty, \tag{6.10}$$

where $\varphi_p = Exp_p^{-1}$ for some convenient p in $\mathcal{M}$, and Λ_a, Σ_a are as in Theorem 5.4.3 with Q_a in place of $Q(a = 1,2)$. One simple choice for p is the pooled estimate lying on the distance minimizing geodesic connecting $\mu_{1,g}, \mu_{2,g}$ at a distance $\frac{n_2}{n}\rho_g(\mu_{1,g}, \mu_{2,g})$ from $\mu_{1,g}$. With this choice we write $\hat{\varphi}$ for φ_p. Let $\hat{\Sigma}_a$ and $\hat{\Lambda}_a$ be the same as $\hat{\Sigma}, \hat{\Lambda}$ in Theorem 5.4.3 and Corollary 5.4.2, but using the a-th sample and with $\hat{\varphi}$ for $\varphi(a = 1,2)$. We thus arrive at the following

THEOREM 6.3.2. *The test to reject* $H_0 : \mu_{1,g} = \mu_{2,g}$ *iff*

$$[\hat{\varphi}(\hat{\mu}_{1,g}) - \hat{\varphi}(\hat{\mu}_{2,g})]^T \tilde{\Sigma}_p^{-1} [\hat{\varphi}(\hat{\mu}_{1,g}) - \hat{\varphi}(\hat{\mu}_{2,g})] > \chi^2_m(1-\alpha) \qquad (6.11)$$

has asymptotic level of significance α, *where*

$$\tilde{\Sigma}_p = [(\frac{1}{n_1}\hat{\Lambda}_1^{-1}\hat{\Sigma}_1\Lambda_1^{-1} + \frac{1}{n_2}\hat{\Lambda}_2^{-1}\hat{\Sigma}_2\Lambda_2^{-1})]$$

REMARK 6.3.1. *In an effort to address the problem of testing for the equality of extrinsic means in matched pairs on arbitrary manifolds, in Subsection 4.6.2 of Bhattacharya and Bhattacharya (2012) [27], one considers the more general case of correlated r.o.'s. The matching there does not seem to play a role in the methodology developed, since the test statistic is obtained by first independently projecting the sample means* $j(\bar{X}_1)$ *and* $j(\bar{X}_2)$. *Such a choice ignores the matching, given that applying a permutation in the sample indices of the observations in the second sample say, does not affect the value of the statistic* $\bar{Y}_2$, *and ultimately of the test statistic, so essentially only the sample means are paired, not the values* $X_{a,1}, X_{a,2}$ *of individual observations* $a = 1,\dots,n$. *Note that while in the multivariate case, the mean of the difference is the difference of the means, for an arbitrary manifold valued data, individual differences have no meaning, and by pairing* $j(\bar{X}_1)$ *and* $j(\bar{X}_2)$ *one ignores the within samples matching. While the extension of a to sample problem for correlated r.o.'s will be addressed in a series of exercises at the end of this chapter, in the next section, we will take a different approach of extending the classical methodology for hypothesis testing for matched pairs, that also allows for null hypotheses when the means are not necessarily equal.*

6.4 Test for Mean Change in Matched Pairs on Lie Groups

In some particular cases when there is a Lie group $\mathcal{G}$ acting on the manifold $\mathcal{M}$, such that for each pair of points $p_1, p_2 \in \mathcal{M}$, one may find a unique element $g = g(p_1, p_2) \in \mathcal{G}$, with $g \cdot p_1 = p_2$, given matched samples $(X_i, Y_i), i = 1,\dots,n$ from two distributions on $\mathcal{M}$, one may associate a random sample $G_i = g(X_i, Y_i), i = 1,\dots,n$ on $\mathcal{G}$, thus moving the matched sample analysis on $\mathcal{G}$. A general example of this kind is provided by the left action or by the adjoint action of a Lie group $\mathcal{G}$ on itself. However, this approach may be extended to the more general case, when the existence of the element unique $g = g(p_1, p_2) \in \mathcal{G}$, with $g \cdot p_1 = p_2$, is insured only for pairs (p_1, p_2) of points that are not too far apart. This approach was used for example in Beran and Fisher (1998) [18] with $\mathcal{M} = \mathbb{R}P^2$, and $\mathcal{G} = SO(3)$, the group of rotations in $\mathbb{R}^3$, and in Mardia and Patrangenaru (2005) [233] with $\mathcal{M} = P\Sigma_2^k$, and $\mathcal{G} = SO(3)^{k-4}$, thus reducing the two-sample test for extrinsic mean multiple axes to a one-sample test for extrinsic mean multiple rotations. The drawback in Mardia and

Patrangenaru (2005) [233] is that such an analysis leads to an increase in dimensionality, forcing the extrinsic sample covariance matrices to be degenerated, when the sample sizes are small, an impediment in computing pivotal statistics, that are preferred for a lower coverage error.

Recall that the case of paired random vectors X, Y in $\mathbb{R}^m$, comparing their mean vectors, is usually done in terms of the difference vector $D = Y - X$ to eliminate much of the influence of extraneous unit to unit variation (Johnson and Wichern (2007 [170], p. 274)), without increasing the dimensionality. Similarly, for matched pairs on the unit circle $Z_1, Z_2 \in S^1$ (matched circular data), rather than testing for the equality $\mu_{j,1} = \mu_{j,2}$ of the two extrinsic means, one may test if the extrinsic mean of $\bar{Z}_1 Z_2$ is the unit of S^1. In both these cases, one deals with matched paired data on a commutative Lie group. In this section we extend these technique to paired random objects on an arbitrary *Lie group*, that is not necessarily commutative.

Assume X and Y are paired *r.o.'s on a Lie group*$(\mathcal{G}, \circ)$ (see Section 3.1). The *change from X to Y* was defined by Crane and Patrangenaru (2011) [78] as the r. o. $C =: X^{-1} \circ Y$. We say that *there is no mean change* from X to Y if the mean change from X to Y is the identity of the group $\mathcal{G}$, that is the null hypothesis is

$$H_0 : \mu_C = 1_{\mathcal{G}}, \tag{6.12}$$

where μ_C is the extrinsic mean of C with respect to an embedding of $\mathcal{G}$ in $\mathbb{R}^D$. The particular case, for testing H_0 in (6.12), on the group $(\mathbb{R}P^3)^q$ is analyzed in detail in Chapter 24.

In general, to compute the p-value for the hypothesis 6.12 one may use the asymptotic χ^2_m distribution in Chapter 5, leading to the following result.

THEOREM 6.4.1. *Assume $J : \mathcal{G} \to \mathbb{R}^k$ is a closed embedding of the m-dimensional Lie group $\mathcal{G}$ in $\mathbb{R}^k$. Let $\{X_r, Y_r\}_{r=1,\dots,n}$ be a random sample of matched pairs for which the change $C_1 = X_1^{-1} \circ Y_1$ has a the j-nonfocal distribution, and let $\mu = E(j(C_1))$ and assume $j(C_1)$ has finite second order moments and the extrinsic covariance matrix $\Sigma_{j,E}$ of C_1 is nonsingular. Let $(e_1(p), e_2(p), \dots, e_k(p))$ be an orthonormal frame field adapted to j. If $G(J,C)_n$ is the extrinsic sample mean covariance matrix, since for n large enough $G(J,C)_n$ is nonsingular (with probability converging to one) then the p-value associated with H_0 is given by*

$$p = Pr(T > n \| G(J,C)_n^{-\frac{1}{2}} tan(P_j(\overline{j(C)}) - P_j(j(1_{\mathcal{G}})))\|^2), \tag{6.13}$$

where T has a χ^2_m distribution

Given the i.i.d.'s matched pairs $(X_i, Y_i) \in \mathcal{G}^2, i = 1, \dots, n$, and the corresponding changes $C_i = X_i^{-1} Y_i \in \mathcal{G}, i = 1, \dots, n$, it is known (see Bhattacharya and Patrangenaru (2005) [43]) that

$$\sqrt{n} tan(j(\bar{C}) - j(\mu_j)) \to_d \mathcal{N}_m(0, \Sigma_j), \tag{6.14}$$

where Σ_j is the extrinsic covariance matrix of C, and $tan(v)$ is the tangential component in $T_{\mu_{a,j}} j(\mathcal{G})$ of a vector $v \in \mathbb{R}^N$ with respect to the decomposition $\mathbb{R}^N = T_{\mu_{a,j}} j(\mathcal{G}) \oplus (T_{\mu_{a,j}} j(\mathcal{G}))^{\perp}$.
Let $S_{j,n}$ be the sample extrinsic covariance matrix, obtained from the i.i.d.r.o.'s $\{X_r\}_{r=1,\dots,n}$ from the unknown distribution Q. At this point we recall from Bhattacharya and Patrangenaru (2005) [43] the steps that one takes to obtain a bootstrapped statistic from a pivotal statistic. If $\{X_r^*\}_{r=1,\dots,n}$ is a random sample from the empirical $\hat{Q}_n$, conditionally given $\{X_r\}_{r=1,\dots,n}$, then the studentized vector valued statistic

$$V(X,Q) = n^{\frac{1}{2}} S_{j,n}^{-\frac{1}{2}} tan(P(\overline{j(X)}) - P(\mu)) \tag{6.15}$$

leads to the bootstrapped statistic

$$V^*(X^*, \hat{Q}_n) = n^{\frac{1}{2}} {S_{j,n}^*}^{-\frac{1}{2}} tan_{P(\overline{j(X)}))}(P_j(\overline{j(X^*)}) - P_j(\overline{j(X)})). \tag{6.16}$$

Here $S_{j,n}^*$ is obtained from $S_{j,n}$ substituting $X_1^*, \dots\dots, X_n^*$ for $X_1, \dots., X_n$, and $T(X^*, \hat{Q}_n)$ is obtained from $T(X,Q)$ by substituting $X_1^*, \dots\dots, X_n^*$ for $X_1, \dots., X_n$, $\overline{j(X)}$ for μ and $S_{j,n}^*$ for $S_{j,n}$.

COROLLARY 6.4.1. *A* $(1-\alpha)100\%$ *bootstrap confidence region for* μ_j *is* $C_{n,\alpha}^* := j^{-1}(U_{n,\alpha}^*)$ *with* $U_{n,\alpha}^*$ *given by*

$$U_{n,\alpha}^* = \{\mu \in j(\mathcal{G}) : n\|S_{j,n}^{-\frac{1}{2}} tan(P(\overline{j(X)}) - P(\mu))\|^2 \leq c_{1-\alpha}^*\}, \tag{6.17}$$

where $c_{1-\alpha}^*$ *is the upper* $100(1-\alpha)\%$ *point of the values*

$$n\|{S_{j,n}^*}^{-\frac{1}{2}} tan_{P(\overline{j(X)})}(P(\overline{j(X^*)}) - P(\overline{j(X)}))\|^2 \tag{6.18}$$

among the bootstrap resamples.

6.5 Two-Sample Tests for Extrinsic Means on a Manifold that Admit a Simply Transitive Lie Group Action

Recall that given an embedding $j : \mathcal{M} \to \mathbb{R}^N$, we consider the *chord distance* on $\mathcal{M}$, given by $d(x_1,x_2) = d_0(j(x_1), j(x_2))$, where d_0 is the Euclidean distance in $\mathbb{R}^N$. If $X_{ak_a} : k_a = 1, \dots, n_a, a = 1,2$ are i.i.d.r.o.'s drawn from distributions $Q_a, a = 1,2$ on a $\mathcal{M}$, if we denote by μ_a the mean of the induced probability $Q_a \circ j^{-1}$ and by Σ_a its covariance matrix ($a = 1,2$), then the extrinsic mean of Q_a is $\mu_{a,j} = j^{-1}(P(\mu_a))$, assuming Q_a is j-nonfocal, and the extrinsic sample mean is $\bar{X}_{a,j} = j^{-1}(P(\bar{Y}_a))$. Here, again, P is the projection from $\mathbb{R}^N$ to $j(\mathcal{M})$, and is we write $Y_{ak_a} = j(X_{ak_a}) k_a = 1, \dots, n_a, a = 1,2$ then $\bar{Y}_a, a = 1,2$ is the corresponding sample mean. Assuming finite second moments of $Y_{a1}, a = 1,2$, which is automatic if $\mathcal{M}$ is compact, from (6.14) we have

$$\sqrt{n_a} tan(j(\bar{X}_{a,j}) - j(\mu_{a,j})) \to_d \mathcal{N}_m(0, \Sigma_{a,j}), a = 1,2, \tag{6.19}$$

where $\Sigma_{a,j}$ is the extrinsic covariance matrix of Q_a, and $tan(v)$ is the tangential component in $T_{\mu_{a,j}}j(\mathcal{M})$ of a vector $v \in \mathbb{R}^N$ with respect to the decomposition $\mathbb{R}^N = T_{\mu_{a,j}}j(\mathcal{M}) \oplus (T_{\mu_{a,j}}j(\mathcal{M}))^{\perp}$.

DEFINITION 6.5.1. *An action of a Lie group $\mathcal{G}$ on a manifold $\mathcal{M}$, is a differentiable function $\alpha : \mathcal{G} \times \mathcal{M} \to \mathcal{M}$, such that*

$$\alpha(1_{\mathcal{G}}, x) = x, \forall x \in \mathcal{M},$$
$$\alpha(g, \alpha(h,x)) = \alpha(g \odot h, x), \forall g \in \mathcal{G}, \forall h \in \mathcal{G}, \forall x \in \mathcal{M}. \tag{6.20}$$

$\mathcal{M}$ has a simply transitive Lie group of isometries $\mathcal{G}$, if there is an Lie group action $\alpha : \mathcal{G} \times \mathcal{M} \to \mathcal{M}$ by isometries with the property that given $x \in \mathcal{M}$, for any object $y \in \mathcal{M}$, there is a unique $g \in \mathcal{G}$ such that $\alpha(g,x) = y$.

A two-sample hypothesis testing problem for extrinsic means on the embedded manifold $\mathcal{M}$ that admits a simply transitive Lie group of isometries $\mathcal{G}$, can be formulated as follows:

$$H_0 : \mu_{2,j} = \alpha(\delta, \mu_{1,j})$$
$$\text{versus}$$
$$H_1 : \mu_{2,j} \neq \alpha(\delta, \mu_{1,j}). \tag{6.21}$$

Given a fixed object $x \in \mathcal{M}$, the mapping $\alpha^x : \mathcal{G} \to \mathcal{M}, \alpha^x(g) = \alpha(g,x)$ is bijective, therefore the hypothesis problem (6.21) is equivalent to the following hypothesis testing problem on the Lie group $\mathcal{G}$:

$$(1)\ H_0 : (\alpha^{\mu_{1,j}})^{-1}(\mu_{2,j}) = \delta,$$
$$\text{versus}$$
$$H_1 : (\alpha^{\mu_{1,j}})^{-1}(\mu_{2,j}) \neq \delta \tag{6.22}$$

Let $H : \mathcal{M}^2 \to \mathcal{G}$, defined by

$$H(x_1, x_2) = (\alpha^{x_1})^{-1}(x_2). \tag{6.23}$$

THEOREM 6.5.1. *Assume $X_{a,k_a}, k_a = 1, \ldots, n_a$ are identically independent distributed random objects (i.i.d.r.o.'s) from the independent probability measures $Q_a, a = 1,2$ with finite extrinsic moments of order $s, s \leq 4$ on the m dimensional manifold $\mathcal{M}$ on which the Lie group $\mathcal{G}$ acts simply transitively. Let $n = n_1 + n_2$ and assume $\lim_{n\to\infty} \frac{n_1}{n} \to \pi \in (0,1)$. Let $\varphi : \mathfrak{g} \to \mathcal{G}$ and L_δ be respectively, a chart with $\varphi(1_{\mathcal{G}}) = 0_{\mathfrak{g}}$, and the left translation by $\delta \in \mathcal{G}$. Then under H_0,*
(i.) The sequence of random vectors

$$\sqrt{n}(\varphi \circ L_\delta^{-1}(H(\bar{X}_{n_1,j}, \bar{X}_{n_2,j}))) \tag{6.24}$$

converges weakly to $\mathcal{N}_m(0_m, \Sigma_j)$, for some covariance matrix Σ_j that depends

linearly on the extrinsic covariance matrices $\Sigma_{a,j}$ *of* $Q_a, a = 1,2$.
(ii.) If (i.) holds and Σ_j *is positive definite, then the sequence*

$$n(\varphi \circ L_{\bar{\delta}}^{-1}(H(\bar{X}_{n_1,j}, \bar{X}_{n_2,j})))^T \Sigma_j^{-1} (\varphi \circ L_{\bar{\delta}}^{-1}(H(\bar{X}_{n_1,j}, \bar{X}_{n_2,j}))) \tag{6.25}$$

converges weakly to χ^2_m *distribution.*

The following result is a direct consequence of Cramer's delta method, applied to functions between embedded manifolds.

LEMMA 6.5.1. *Assume* $F : \mathcal{M}_1 \to \mathcal{M}_2$ *is a differentiable function between manifolds. For* $a = 1,2$, *assume dim* $\mathcal{M}_a = m_a$, *and* $j_a : \mathcal{M}_a \to \mathbb{R}^{N_a}$ *is an embedding. Let* X_n *be a sequence of r.o.'s on* M_1 *such that*

$$\sqrt{n} tan_{j_1(\nu)}(j_1(X_n) - j_1(\nu)) \to_d \mathcal{N}_{m_1}(0, \Sigma), \tag{6.26}$$

then

$$\sqrt{n} tan_{j_2(F(\nu))}(j_2(F(X_n)) - j_2(F(\nu))) \to_d \mathcal{N}_{m_2}(0, d_\nu F \Sigma (d_\nu F)^T), \tag{6.27}$$

where $d_\nu F : T_\nu \mathcal{M}_1 \to T_{F(\nu)} \mathcal{M}_2$ *is the differential of* F *at* μ.

By the inverse function theorem, the mapping $H : \mathcal{M} \times \mathcal{M} \to \mathcal{G}$ is continuous. Given that, according to Bhattacharya and Patrangenaru (2003) [42], for $a = 1,2$, the extrinsic sample mean $\bar{X}_{n_a,j}$ is a consistent estimator of $\mu_{a,j}$, for $a = 1,2$, by the continuity theorem (Billingsley (1995) [48], p.334) a consistent estimator for $(\alpha^{\mu_{1,j}})^{-1}(\mu_{2,j})$ is $H(\bar{X}_{n_1,j}, \bar{X}_{n_2,j})$. From Bhattacharya and Patrangenaru [43], for $a = 1,2$,

$$\sqrt{n_a} tan_{\mu_{a,j}}(j(\bar{X}_{n_a,j}) - j(\mu_{a,j})) \to_d N_m(0_m, \Sigma_{a,j}), \tag{6.28}$$

and, since $\frac{n_a}{n} \to \pi$, it follows that

$$\sqrt{n} tan_{(\mu_{1,j}, \mu_{2,j})}(j^2((\bar{X}_{n_1,j}, \bar{X}_{n_2,j})) - j^2((\mu_{1,j}, \mu_{2,j}))) \to_d N_{2m}(0_{2m}, \Sigma), \tag{6.29}$$

where

$$\Sigma = \begin{pmatrix} \frac{1}{\pi}\Sigma_{1,j} & 0 \\ 0 & \frac{1}{1-\pi}\Sigma_{2,j} \end{pmatrix}. \tag{6.30}$$

We apply Lemma 6.5.1 to the function $F : \mathcal{M}^2 \to \mathcal{G}$, given by $F = L_{\delta^{-1}} \circ H$, and select a convenient chart φ, and obtain equation (6.24). Theorem 6.5.1ii. is an immediate consequence of part (i.) plus a weak continuity argument (Billingsley (1995) [48], p.334) ■

COROLLARY 6.5.1. *(Two sample test on a homogeneous space) For* $a = 1,2$, *assume* $X_{a,j_a}, j_a = 1,\ldots,n_a$, *are i.i.d.r.o.'s from independent populations on the* m *dimensional embedded manifold* $\mathcal{M}$ *on which the Lie group* $\mathcal{G}$ *acts simply transitively. Let* $n = n_1 + n_2$, *and assume* $\lim_{n\to\infty} \frac{n_1}{n} \to \pi \in (0,1)$. *Assume* Σ_j *is* *positive definite and $\hat{\Sigma}_j$ is a consistent estimator for Σ_j. The asymptotic p-value for the hypothesis testing problem 6.21 is given by $p = P(T \geq T_\delta^2)$ where*

$$T_\delta^2 = n((\varphi \circ L_\delta^{-1}(H(\bar{X}_{n_1,j}, \bar{X}_{n_2,j})))^T (\hat{\Sigma}_j)^{-1} (\varphi \circ L_\delta^{-1}(H(\bar{X}_{n_1,j}, \bar{X}_{n_2,j}))), \quad (6.31)$$

and T has a χ_m^2 distribution.

If the distributions are unknown and the samples are small, an alternative approach is to use Efron's nonparametric bootstrap (see Efron (1982) [97]). If $\max(n_1, n_2) \leq \frac{m}{2}$, the sample mean $\hat{\Sigma}_j$ in Corollary 8.4.1 does not have an inverse, and pivotal nonparametric bootstrap methodology can not be applied. In this case one may use a nonpivotal bootstrap methodology for the two-sample problem H_0 (see Bhattacharya and Ghosh (1978) [38], Hall and Hart (1990) [138], Fisher et al. (1996) [116] or Hall (1997) [137]).

REMARK 6.5.1. *Beran and Fisher (1998) [18] were the first to use group actions in hypothesis testing problems, a technique later used in Mardia and Patrangenaru (2005)[233]. The drawback in Mardia and Patrangenaru (2005)[233] was that their analysis led to an increase in dimensionality, forcing the extrinsic covariance matrices to be degenerated. In this section we consider in particular case when $\mathcal{M}$ itself has a Lie group structure $\odot$, and the group action is by left translations: $\alpha : \mathcal{M}^2 \to \mathcal{M}, \alpha(g,x) = g \odot x$. Given two objects x, y the* change from *x to y is $c = x^{-1} \odot y$. Given two random objects X, Y, from Theorem 6.5.1 we may estimate the change from the extrinsic mean of X to the extrinsic mean of Y. Note that this change of means is the mean change defined in Section 6.3 only if X, Y are matched pairs on a commutative Lie group $(\mathcal{G}, \odot)$.*

REMARK 6.5.2. *The methodology of simply transitive groups on manifolds for two sample hypothesis testing problem, was first used by Osborne [257] in the context of intrinsic means. That case parallels the case of extrinsic means, with some specific adjustments as shown in Chapter 8.*

6.6 Nonparametric Bootstrap for Two-Sample Tests

The primary referencees for this section are Bhattacharya and Patrangenaru (2014) [44] and Osborne et al.(2013) [258]. The topic here concerns the use of Efron's bootstrap (Efron (1979) [96]) for the approximation of true coverage probabilities for confidence regions for the Fréchet means and for p-values of two-sample tests for equality of Fréchet means. That the bootstrap outperforms the traditional approximation of the distribution of the standardized sample mean by the CLT when the underlying distribution is absolutely continuous was first proved by Singh (1981) [307]. Extensions to studentized, or pivoted, (multivariate) sample means were obtained by Babu and Singh (1984) [9], who derived asymptotic expansions of the distribution.

Indeed, such an expansion holds for more general statistics which are smooth functionals of sample means or which admit a stochastic expansion by a polynomial of sample means (Bhattacharya (1987) [30]). This is derived from refinements of the multivariate CLT by asymptotic expansions (Bhattacharya (1977) [29], Bhattacharya and Ghosh (1978) [38], Chandra and Ghosh (1979) [62]). As a consequence, the chi-square approximation of the distributions of statistics appearing in (5.36), for example, has an error of the order $O(n^{-1})$, whereas the bootstrap approximation of these distributions has an error $O(n^{-2})$ (See, e.g., Beran (1988) [17], Hall (1987) [135], Bhattacharya and Qumsiyeh (1989) [45], Hall (1997) [137]). Therefore, bootstrapping the statistics such as appearing in (5.36) would lead to smaller coverage errors than the classical chi-square approximations. In the two-sample case with sample sizes m and n, the error of approximation by bootstrap of the distributions of asymptotically chisquare statistics is $O(N^{-2})$ where $N = \min\{m,n\}$. In the present context of generally high-dimensional manifolds, often the bootstrap approximation of the covariance is singular, which sometimes makes the bootstrap approximation of the distributions either not feasible, or subject to further errors in case of rather arbitrary augmentations.

THEOREM 6.6.1. *(Osborne et al. (2013) [258]). Under the hypotheses of Theorem 6.5.1i., assume in addition, that for $a = 1,2$ the support of the distribution of $X_{a,1}$ and the extrinsic mean $\mu_{a,j}$ are included in the domain of the chart φ and $\varphi(X_{a,1})$ has an absolutely continuous component and finite moments of sufficiently high order. Then the joint distribution of*

$$V = \sqrt{n}(\varphi \circ L_{\delta}^{-1}(H(\bar{X}_{n_1,j},\bar{X}_{n_2,j}))) \tag{6.32}$$

can be approximated by the bootstrap joint distribution of

$$V^* = \sqrt{n}(\varphi \circ L_{\delta}^{-1}(H(\bar{X}^*_{n_1,j},\bar{X}^*_{n_2,j})) \tag{6.33}$$

with an error $O_p(n^{-\frac{1}{2}})$, where, for $a = 1,2$, $\bar{X}^*_{n_a,j}$ are the extrinsic means of the bootstrap resamples $X^*_{a,k_a}, k_a = 1,\dots,n_a$, given $X_{a,k_a}, k_a = 1,\dots,n_a$.

We briefly describe now bootstrapping procedures for two independent samples, or match pairs, testing problems on manifolds. Consider, for example, the test $H_0 : P_j(\mu_1) = P_j(\mu_2)$ for the equality of extrinsic means of two distributions on M. Let $\theta = P_j(\mu_1) - P_j(\mu_2)$, and write $T(X|\theta)$ for the statistic on the left of (6.9), but with $P_j(\bar{Y}_1) - P_j(\bar{Y}_2)$ replaced by $P_j(\bar{Y}_1) - P_j(\bar{Y}_2) - \theta$. Whatever the true θ is, the asymptotic distribution of $T(X|\theta)$ is chi-square with m degrees of freedom. Hence a bootstrap-based test of asymptotic level α is to reject H_0 iff $T(X|0) > c^*_{1-\alpha}$, where $c^*_{1-\alpha}$, is the $(1-\alpha)$th quantile of the bootstrapped values $T(X|\bar{Y}_1) - P_j(\bar{Y}_2)$. It follows that the p-value of the test is given by: p-value = proportion of bootstrapped values $T(X^*|P_j(\bar{Y}_1) - P_j(\bar{Y}_2)$ which exceed the observed value of $T(X|0)$. Similar bootstrapping procedures apply to testing the equality of sample intrinsic means, and to match pair problems.

As pointed out above, for a manifold of a high dimension m, the bootstrap version of the covariance matrix, as appears, for example, within square brackets [ ] is often singular if the sample size is not very large. Suppose the classical chi-square approximation is not considered reliable enough, especially if the p-value that it provides is only marginally small. Then, instead, one may consider the very conservative Bonferroni procedure(see 2.68) of looking individually at each of the d contrasts provided, for example, by a P.C.A. (see Section 2.11) of the sample covariance matrix, and compute d p-values for the m tests for the m population contrasts, using the bootstrap approximations of the m distributions of sample contrasts. The estimated p-value is then set at d-times the minimum of these p-values. It should be pointed out that one should not just pick only the principal components which account for the largest variability. If the variability due to a particular contrast is small, while its true population mean value is nonzero, then the corresponding test will bring out the difference with high probability, i.e., with a small p-value (See Bhattacharya and Bhattacharya (2012) [27], p. 16).

6.7 Exercises

Exercise 73. *(a) Show that if j is an embedding of the m dimensional manifold $\mathcal{M}$ into $\mathbb{R}^N$, then $\tilde{j}(x,y) = (j(x), j(y))$ is an embedding of $\mathcal{M} \times \mathcal{M}$ into $\mathbb{R}^N \times \mathbb{R}^N$. (b) Show that if $Q_a, a = 1,2$ are j-nonfocal probability measures on $\mathcal{M}$, then their product measure Q is a $\tilde{j}$ nonfocal probability measure on $\mathcal{M} \times \mathcal{M}$.*

Exercise 74. *Consider the probability measures $Q_a, a = 1,2$ as in exercise 73. Let $\mu_{j,a}$ be the extrinsic means of $Q_a, (a = 1,2)$. Once again, we are interested in testing $H_0 : \mu_{j,1} = \mu_{j,2} = \mu_j$, say. (a) Show that the extrinsic mean of Q is $\mu_{\tilde{j}} = (\mu_{j,1}, \mu_{j,2})$. (b) If for $a = 1,2, \bar{Y}_a$ is the sample mean of $Y_{a,k} = j(X_{a,k}), k = 1,\ldots,n$ on $\mathbb{R}^N$, with $E(Y_{a1}) = \mu_a$, then the extrinsic sample mean is $\mu_{\tilde{j}} = (j^{-1}(P_j(\mu_1)), j^{-1}(P_j(\mu_2))$.*

Exercise 75. *Consider the manifold $\mathcal{M}$ as in exercise 73. If P_j is the projection operator on the set of all nonfocal points in $\mathbb{R}^N$ into $j(M)$, then the projection on $\tilde{j}(M)$ is $\tilde{P}_j = P_j \times P_j : \mathbb{R}^N \times \mathbb{R}^N \to j(M) \times j(M)$.*

Exercise 76. *Consider the probability measures $Q_a, a = 1,2$ as in exercise 74. Let B denote the $m \times N$ matrix of m orthonormal basis vectors of $T_j(\mu_j)(j(M)) \subset T_j(\mu_j)(\mathbb{R}^N)$. Under H_0, show that*

$$n^{1/2} B([P_j(\bar{Y}_1) - P_j(\bar{Y}_2)]) \to_d \mathcal{N}_m(0, \Sigma_{j,1} + \Sigma_{j,2} - \Sigma_{j,12} - \Sigma_{j,21}), \text{ as } n \to \infty. \tag{6.34}$$

On the right, $\Sigma_{j,a}$ is the extrinsic covariance matrix of X_{a1}, while $\Sigma_{j,ab} = B(DP_j)_{\mu_a} Cov(Y_{a1}, Y_{b1})(DP_j)_{\mu_b} B^T, a \neq b$.

Exercise 77. *Same assumptions as in exercise 76. From* (6.34) *derive the following result by the usual Slutsky type argument: The test which rejects H_0*

iff

$$n\hat{B}([P_j(\bar{Y}_1) - P_j(\bar{Y}_2)])^T[\hat{\Sigma}_{j,1} + \hat{\Sigma}_{j,2} - \hat{\Sigma}_{j,12} - \hat{\Sigma}_{j,21}]\hat{B}([P_j(\bar{Y}_1) - P_j(\bar{Y}_2)])$$
$$\geq \chi^2_m(1-\alpha) \tag{6.35}$$

has the asymptotic confidence coefficient $1-\alpha$. *Here* $\hat{B}$ *is the* $m \times N$ *matrix whose rows are orthonormal basis vectors of* $T_{j(\hat{\mu}_j)}(j(M))$, *with* $j(\hat{\mu}_j) = P_j(\frac{1}{2}[P_j(\bar{Y}_1) + P_j(\bar{Y}_2)])$. *Also,* $\hat{C}_a, \hat{\Sigma}_a$ *are as in* (6.9)

Chapter 7

Function Estimation on Manifolds

7.1 Introduction

This results in this chapter, the first part is mainly due to our friend **Peter T. Kim**, a top expert in the area of function estimation on Riemannian manifolds. We thank him for his generous introduction to this fascinating area.

In addition to estimation and inference on means, a common problem in statistics is that of nonparametric density and function estimation. In the case where the sample space is Euclidean, this problem has been extensively studied, dating back to an extension to the classical histogram in Rosenblatt (1956) [292] and a kernel density estimator in Parzen (1962) [262]. As with many other areas of statistics on manifolds, density estimation was first extended to the circle and sphere. Methodologies of this sort are described in Mardia (1975) [229], Watson (1983) [333], and Jupp and Mardia (1989) [173]. Subsequently, a number of papers extended these ideas to density and function estimation on compact manifolds. Amongst these are Ruymgaart [294], Mair and Ruymgaart [227] and Rooij and Ruymgaart (1991) [328].

In this chapter, we consider two approaches to function estimation on compact manifolds. The first one, which follows from Kim and Koo (2005) [190], utilizes the Laplace–Beltrami operator to perform statistical inverse estimation

. The second, due to Pelletier (2005) [278], extends the notion of kernel estimators to Riemannian manifolds.

7.2 Statistical Inverse Estimation

Following the Euclidean version (see Cavalier and Tsybakov (2002) [61], Koo (1993) [200]), we will now define a statistical inverse problem on $\mathcal{M}$. Consider a regression function f, of the response variable Y, on the measurement variable $X \in \mathcal{M}$, so that $E(Y|X) = f(X)$, where X is uniformly distributed on $\mathcal{M}$ and E denotes conditional, or, unconditional expectation. In particular, consider the signal plus noise model

$$Y = f(X) + \varepsilon, \tag{7.1}$$

where $E\varepsilon = 0$ and $E\varepsilon^2 = \sigma^2 > 0$. Let T be a (possibly unbounded) operator on $L^2(\mathcal{M})$.

A *statistical inverse problem* on $\mathcal{M}$, is an attempt to statistically estimate $T(f)$, where the statistical aspect comes from not knowing $T(f)$, but by observing a random sample of size n, $\{(X_1,Y_1),\ldots,(X_n,Y_n)\}$, coming from the joint distribution of (X,Y). With this information we can form an empirical version of the Fourier coefficients from the data,

$$\hat{f}_k^n = \frac{1}{n}\sum_{j=1}^{n} Y_j \overline{\phi_k}(X_j) \tag{7.2}$$

and define an appropriate nonparametric regression estimator of f by

$$f_\Lambda^n(x) = \sum_{\lambda_k \leq \Lambda} \hat{f}_k^n \phi_k(x), \quad x \in \mathcal{M}. \tag{7.3}$$

This allows us to define a statistical estimator $T(f_\Lambda^n)$ of $T(f)$ with the objective then to understand convergence properties for large samples.

The difficulty of recovering $T(f)$ largely boils down to whether the operator T is bounded or unbounded. In the former situation, as long as f_Λ^n is a consistent estimator of f, then $T(f_\Lambda^n)$ will consistently estimate $T(f)$. This problem can be deduced from the Fourier series-based estimator in Hendricks (1990) [150]. The unbounded situation, which is otherwise known as an ill-posed problem, is obviously more difficult but usually of more practical relevance. Thus the degree of ill-posedness (unboundedness) dictates the manner in which $T(f)$ can be recovered by $T(f_\Lambda^n)$ as $n \to \infty$ (see Rooij and Ruymgaart (1991) [328], Wahba (1977) [330]). It is indeed with respect to ill-posed estimation that we will proceed.

Consider an unknown probability distribution $\mathbb{P}_f$ depending on the regression function f in (7.1), belonging to some function class $\mathcal{F} \subset L^2(\mathcal{M})$ and suppose $\{b_n\}$ is some sequence of positive numbers. Let $T : L^2(\mathcal{M}) \to L^2(\mathcal{M})$ be a (possibly unbounded) linear operator. This sequence is called a *lower bound* for $T(f)$ if

$$\lim_{c\to 0}\liminf_{n}\inf_{T^n}\sup_{f\in\mathcal{F}}\mathbb{P}_f\left(\|T^n - T(f)\|_2 \geq cb_n\right) = 1, \tag{7.4}$$

where the infimum is over all possible estimators T^n based on $\{(X_1,Y_1),\ldots,(X_n,Y_n)\}$. Alternatively, the sequence in question is said to be an *upper bound* for $T(f)$ if there is a sequence of estimators $\{T^n\}$ such that

$$\lim_{c\to\infty}\limsup_{n}\sup_{f\in\mathcal{F}}\mathbb{P}_f\left(\|T^n - T(f)\|_2 \geq cb_n\right) = 0. \tag{7.5}$$

The sequence of numbers $\{b_n\}$ is called the optimal rate of convergence for $T(f)$ if it is both a lower bound and an upper bound with the associated estimators $\{T^n, n\geq 1\}$, being called *asymptotically optimal* (see Stone (1980) [321]). We will use the following notation. For two positive (random) sequences $\{a_n\}$ and $\{b_n\}$, let $a_n \ll b_n$ ($\ll_{\mathbb{P}}$) mean $a_n = O(b_n)$ ($O_{\mathbb{P}}$) as $n\to\infty$. If $a_n \ll b_n$ ($\ll_{\mathbb{P}}$) and $b_n \ll a_n$ ($\ll_{\mathbb{P}}$) then denote this by $a_n \asymp b_n$ ($\asymp_{\mathbb{P}}$). Let E denote the expectation.

The main results will be presented in terms of appropriate function subspaces of $L^2(\mathcal{M})$. First, on the space $C^\infty(\mathcal{M})$ of infinitely continuous differentiable functions on $\mathcal{M}$, consider the so-called Sobolev norm $\|\cdot\|_{\mathcal{H}_s}$ of order s defined in the following way. For any function $h = \sum_k\sum_{\mathcal{E}_k}\hat{h}_k\phi_k$, let

$$\|h\|^2_{\mathcal{H}_s} = |\hat{h}_0|^2 + \sum_{k\geq 1}\sum_{\mathcal{E}_k}\lambda_k^s|\hat{h}_k|^2. \tag{7.6}$$

One can verify that (7.6) is indeed a norm. Denote by $\mathcal{H}_s(\mathcal{M})\subset L^2(\mathcal{M})$ the (vector-space) completion of $C^\infty(\mathcal{M})$ with respect to (7.6). This will be called the *Sobolev space of order s*. For some fixed constant $Q>0$, define

$$\mathcal{H}_s(\mathcal{M},Q) = \left\{h\in\mathcal{H}_s(\mathcal{M}) : \|h\|^2_{\mathcal{H}_s}\leq Q\right\}. \tag{7.7}$$

We will need some regularity conditions. In particular, let $\mathbb{P}_1$ and $\mathbb{P}_2$ denote probability distributions having densities p_1 and p_2, respectively, where p_1 and p_2 are defined on $\mathcal{M}$. The Kullback–Leibler information divergence $\mathcal{D}(\mathbb{P}_1\|\mathbb{P}_2)$ is defined by

$$\mathcal{D}(\mathbb{P}_1\|\mathbb{P}_2) = \int p_1\log\left(\frac{p_1}{p_2}\right). \tag{7.8}$$

Conditional on $X=x$, the response variable Y has a probability distribution of the form

$$\mathbb{P}_{f(x)}(dy) = p(y|x,f(x))dy. \tag{7.9}$$

The Gaussian distribution has the form of

$$p(y|x,f(x)) = \{2\pi\sigma^2(x)\}^{-1/2}\exp\left\{-\frac{|y-f(x)|^2}{2\sigma^2(x)}\right\}. \tag{7.10}$$

We have the following conditions:

A1) There is a positive constant $0 < C_{\text{KL}} < \infty$ such that

$$\mathcal{D}(\mathbb{P}_{f_1(x)} \| \mathbb{P}_{f_2(x)}) \leq C_{\text{KL}} |f_1(x) - f_2(x)|^2.$$

A2) There is a positive constant $0 < C_V < \infty$ such that

$$\text{var}(Y|X = x) = E\left(|Y - f(x)|^2 \middle| X = x\right) \leq C_V \text{ for } x \in \mathcal{M}.$$

A3) X is uniformly distributed on $\mathcal{M}$.

THEOREM 7.2.1. *(Koo and Kim (2005)) Suppose* A1), A2), A3) *hold and T induces the operator $T_k : \mathcal{E}_k \to \mathcal{E}_k$, $k \geq 0$, such that*

$$c_0 \lambda_k^\beta \leq \inf_{\|v\|_2 = 1} \|T_k(v)\|_2^2 \leq \sup_{\|v\|_2 = 1} \|T_k(v)\|_2^2 \leq c_1 \lambda_k^\beta \tag{7.11}$$

for some $\beta \geq 0$ and $c_0, c_1 > 0$. If $f \in \mathcal{H}_{s+\beta}(\mathcal{M}, Q)$ with $s > \dim \mathcal{M}/2$ and $\beta \geq 0$, then $n^{-s/(2s+2\beta+\dim \mathcal{M})}$ is the optimal rate of convergence.

In this theorem, the degree of ill-posedness (7.11) is polynomial in the eigenvalues. Thus the optimal rate of recovery of the signal is inverse to the sample size n. As a side benefit, Theorem 7.2.1 provides us with optimal rate of recovery of the regression function f when $\beta = 0$.

COROLLARY 7.2.1. *Suppose* A1), A2), A3) *hold and $f \in \mathcal{H}_s(\mathcal{M}, Q)$ with $s > \dim \mathcal{M}/2$. If $\Lambda \asymp n^{1/(2s+\dim \mathcal{M})}$, then f_Λ^n achieves the optimal rate of convergence $n^{-s/(2s+\dim \mathcal{M})}$ for f.*

Another norm to consider is the following. Again, for any function $h = \sum_k \sum_{\mathcal{E}_k} \hat{h}_k \phi_k$, let

$$\|h\|_{\mathcal{A}_{s,\beta}}^2 = |\hat{h}_0|^2 + \sum_{k \geq 1} \sum_{\mathcal{E}_k} \lambda_k^s \exp\{\lambda_k^\beta\} |\hat{h}_k|^2, \tag{7.12}$$

for $s \geq 0$ and $\beta \geq 0$. One can verify that (7.12) is also a norm. For some fixed constant $Q > 0$, let $\mathcal{A}_{s,\beta}(\mathcal{M}, Q)$ denote the smoothness class of functions $h \in \mathcal{A}_{s,\beta}(\mathcal{M})$ which satisfy

$$\|h\|_{\mathcal{A}_{s,\beta}}^2 \leq Q.$$

We will call $\mathcal{A}_{s,\beta}(\mathcal{M}) \subset L^2(\mathcal{M})$, the class of analytic functions of order β.

THEOREM 7.2.2. *(Koo and Kim(2005)) Suppose* A1), A2), A3) *hold and T induces the operator $T_k : \mathcal{E}_k \to \mathcal{E}_k$ for $k \geq 0$ such that*

$$c_0 \exp(\lambda_k^\beta) \leq \inf_{\|v\|_2 = 1} \|T_k(v)\|_2^2 \leq \sup_{\|v\|_2 = 1} \|T_k(v)\|_2^2 \leq c_1 \exp(\lambda_k^\beta) \tag{7.13}$$

for some $\beta > 0$ *and* $c_0, c_1 > 0$. *If* $f \in \mathcal{A}_{s,\beta}(\mathcal{M}, Q)$ *with* $s \geq \dim \mathcal{M}/2$ *and* $\beta > 0$, *then*

$$(\log n)^{-s/(2\beta)}$$

is the optimal rate of convergence.

Notice that in this situation, where the degree of ill-posedness is exponential (7.13), the optimal rate of recovery of $T(f)$ is logarithmic and hence much slower than in the polynomial case (7.11). Indeed, we have that the more ill-posed the problem, the slower will be the optimal rate of recovery.

Consider the following remarks about these results. First, this approach is formulated in terms of the inverse operator where the eigenfunctions are contained in the orthonormal basis of the Laplace–Beltrami operator. An alternative approach, is to state the problem in terms of the direct operator, which in our notation would be T^{-1}, and use the orthonormal basis of the direct operator. The two approaches are equivalent and therefore the choice is of personal preference. The direct approach is presented in Rooij and Ruymgaart (1991) [328], Rooij and Ruymgaart (2001) [329], Ruymgaart (1993) [296].

A second remark is that one can also define minimaxity in L^2-risk (modulo a constant) as follows. If there exists a positive constant C_L such that

$$\inf_{T^n} \sup_{f \in \mathcal{F}} E_f \|T^n - T(f)\|_2 \geq C_L b_n,$$

then b_n is a lower bound in L^2-risk. Here the infimum is also taken over all possible estimators $\{T^n\}$ of $T(f)$. On the other hand, if there exists a positive constant C_U and a sequence of estimators T^n such that

$$\sup_{f \in \mathcal{F}} E_f \|T^n - T(f)\|_2 \leq C_U b_n,$$

then b_n is an upper bound in L^2-risk. One can easily modify the method of proof as in Yatracos (1988) [341], in order to prove minimaxity in L^2-risk. Consequently, the optimal rates of convergence in Theorem 7.2.1 and Theorem 7.2.2 are L^2-minimax rates of convergence as well.

Third, this theory holds intact for density estimation, though the theory has been presented in terms of regression. The only modification would be that the random sample would be based on $X_1, \ldots, X_n$ with (7.2) replaced by

$$\hat{f}_k^n = \frac{1}{n} \sum_{j=1}^{n} \overline{\phi_k}(X_j), \tag{7.14}$$

for $k \in \mathbb{N}$.

Finally, let us point out that as far as $L^2(\mathcal{M})$−rates of convergence are concerned, they only depend on $\dim \mathcal{M}$ and that these results are the same for an Euclidean space of the same dimension. However, it is expected that the geometry of $\mathcal{M}$ affects the constants of the sharp bound, i.e., constant plus rate of convergence.

7.2.1 Applications and Examples

In most cases it turns out that the operator T is diagonalizable by $\{\phi_k : k \in \mathbb{N}\}$. This means that there is a sequence $\{t_k : k \in \mathbb{N}\}$ such that $T_k = t_k \mathbf{I}_{\dim \mathcal{E}_k}$, for $k \in \mathbb{N}$ where $\mathbf{I}_{\dim \mathcal{E}_k} : \mathcal{E}_k \to \mathcal{E}_k$ denotes the identity operator. Thus,

$$T(f) = \sum_{k=0}^{\infty} \sum_{\mathcal{E}_k} t_k \hat{f}_k \phi_k. \tag{7.15}$$

In this case the estimator in question would appear as

$$T(f_\Lambda^n) = \sum_{\lambda_k \le \Lambda} t_k \hat{f}_k^n \phi_k \tag{7.16}$$

for $x \in \mathcal{M}$.

Let q be a polynomial of finite degree $\deg q < \infty$. Consider the case of $T = q(\Delta)$. In this situation,

$$\|T_k(v)\|_2^2 = q(\lambda_k)\|v\|_2^2,$$

for all $v \in \mathcal{E}_k$, $k = 0, 1, 2, \ldots$. Thus

$$T(f)(x) = \sum_{\lambda_k} q(\lambda_k) \hat{f}_k \phi_k(x),$$

for $x \in \mathcal{M}$ and

$$c_0 \lambda_k^{\deg q} \le \inf_{\|v\|_2=1} \|T_k(v)\|_2^2 \le \sup_{\|v\|_2=1} \|T_k(v)\|_2^2 \le c_1 \lambda_k^{\deg q}.$$

Thus the nonparametric estimator would be

$$T(f_\Lambda^n)(x) = \sum_{\lambda_k \le \Lambda} q(\lambda_k) \hat{f}_k^n \phi_k(x)$$

for $x \in \mathcal{M}$ and the optimal rate of recovery of $T(f)$ would be $n^{-s/(2s+2\deg q + \dim \mathcal{M})}$ as $n \to \infty$ for $f \in \mathcal{H}_{s+\deg q}(\mathcal{M}, Q)$, $s > \dim \mathcal{M}/2$ and $\deg q > 0$.

Now suppose the operator $T = \exp\{q(\Delta)\}$. In this situation,

$$\|T_k(v)\|_2^2 = \exp\{q(\lambda_k)\}\|v\|_2^2,$$

for all $v \in \mathcal{E}_k$, $k = 0, 1, 2, \ldots$. Thus

$$T(f)(x) = \sum_{\lambda_k} \exp\{q(\lambda_k)\} \hat{f}_k \phi_k(x),$$

for $x \in \mathcal{M}$ and

$$c_0 \exp(\lambda_k^{\deg q}) \le \inf_{\|v\|_2=1} \|T_k(v)\|_2^2 \le \sup_{\|v\|_2=1} \|T_k(v)\|_2^2 \le c_1 \exp(\lambda_k^{\deg q}).$$

Thus the nonparametric estimator would be

$$T(f_\Lambda^n)(x) = \sum_{\lambda_k \leq \Lambda} \exp\{q(\lambda_k)\} \hat{f}_k^n \phi_k(x),$$

for $x \in \mathcal{M}$ and the optimal rate of recovery of $T(f)$ would be $(\log n)^{-s/(2\deg q)}$ as $n \to \infty$ if $f \in \mathcal{A}_{s,\deg q}(\mathcal{M}, Q)$, $s > \dim \mathcal{M}/2$ and $\deg q > 0$.

For a particular example, consider an extension of Section 7.2 of Mair and F.H. Ruymgaart (1996) [227] to $\mathcal{M}$. Let $u : \mathcal{M} \times [0, L] \to \mathbb{R}$, for some $L > 0$ that satisfies the "heat" equation on $\mathcal{M}$

$$\left(\Delta + \partial_t\right) u = 0, \tag{7.17}$$

with initial condition $r(x) = u(x, 0)$, for $x \in \mathcal{M}$.

The problem is to determine the initial state r from the data $\{(X_1, Y_1), \ldots, (X_n, Y_n)\}$ at some fixed point in time $t = t_0 > 0$ satisfying

$$Y_j = u(X_j, t_0) + \varepsilon_j \quad \text{for} \quad j = 1, \ldots, n, \tag{7.18}$$

where it is assumed that the random errors ε_j are independent (with mean 0 and finite variance) of the uniformly distributed design points $X_j \in \mathcal{M}$, $j = 1, \ldots, n$.

At $t = t_0$, the solution to (7.17) is of the form

$$u(x, t_0) = \sum_{k \in \mathbb{N}} \sum_{\mathcal{E}_k} \exp(-t_0 \lambda_k) \hat{r}_k \phi_k(x), \ x \in \mathcal{M} \,. \tag{7.19}$$

An unbiased estimator of $\hat{r}_k$ is therefore

$$\hat{r}_k^n = \exp(t_0 \lambda_k) \frac{1}{n} \sum_{j=1}^{n} Y_j \phi_k(Z_j) \quad \text{for} \quad k \in \mathbb{N}.$$

Consequently, an estimator for r is

$$r_\Lambda^n(x) = \sum_{\lambda_k < \Lambda} \hat{r}_k^n \phi_k(x), \quad x \in \mathcal{M} \ .$$

One can see that the bound on the operator is that of Theorem 7.2.2, hence for $u(\cdot, t_0) \in \mathcal{A}_{s,1}(\mathcal{M}, Q)$, the optimal rate of recovery of $r(\cdot) = u(\cdot, 0)$ is $(\log n)^{-s/2}$.

In the situation where $\mathcal{M} = S^2$,

$$\hat{r}_{kq}^n = \exp(t_0 k(k+1)) \frac{1}{n} \sum_{j=1}^{n} Y_j \overline{\phi}_{kq}(Z_j) \quad \text{for} \quad , |q| \leq k, \ k \in \mathbb{N}.$$

and therefore,

$$r_\Lambda^n(x) = \sum_{k(k+1) < \Lambda} \sum_{q=-k}^{k} \hat{r}_k^n \phi_{kq}(x), \quad x \in S^2 \ .$$

For $\mathcal{M} = SO(3)$,

$$\hat{r}^{\ell,n}_{q_1q_2} = \exp(t_0\ell(\ell+1)/2)\frac{1}{n}\sum_{j=1}^{n} Y_j\overline{D}^{\ell}_{q_1q_2}(Z_j) \text{ for } ,|q_1|,|q_2| \leq \ell,\ \ell \in \mathbb{N}.$$

and therefore,

$$r^n_\Lambda(x) = \sum_{\ell(\ell+1)/2<\Lambda} \sum_{q_1=-\ell}^{\ell} \sum_{q_2=-\ell}^{\ell} (2\ell+1)\hat{r}^{\ell,n}_{q_1q_2} D^{\ell}_{q_1q_2}(x),\ x \in SO(3)\ .$$

7.3 Proofs of Main Results

We will prove Theorem 7.2.1 and Theorem 7.2.2 by first finding upper bounds. Following this we will establish lower bounds for the smoothness classes in question and demonstrate that the upper and lower bounds match so that the resulting bounds are optimal.

The approach of Healy and Kim (1996) [145] will be used for calculating the upper bounds, while the approach of Koo (1993) [200], Koo and Chung (1998) [201] and Kim and Koo (2002) [189], will be used to find the lower bounds.

We will let $C_1, C_2, \ldots$ denote positive constants independent of the sample size n.

7.3.1 Upper Bounds

LEMMA 7.3.1. *If $h \in \mathcal{H}_s(\mathcal{M}, Q)$ with $s > \dim M/2$, then*

$$||h||_\infty \leq C(\mathcal{M}, Q, s),$$

where $C(\mathcal{M}, Q, s)$ is a constant depending only on $\mathcal{M}$, Q and s.

Proof. Write $h = \sum_k \sum_{\mathcal{E}_k} \hat{h}_k \phi_k$. Observe that

$$\begin{aligned} |h(x)|^2 &\leq \Big(\sum_k \sum_{\mathcal{E}_k} \lambda_k^s |\hat{h}_k|^2\Big)\Big(\sum_k \sum_{\mathcal{E}_k} \lambda_k^{-s} |\phi_k(x)|^2\Big) \\ &\leq QZ(x,s), \end{aligned}$$

where $Z(x, \tilde{s})$ is the zeta function of Δ defined by

$$Z(x,\tilde{s}) = \sum_k \sum_{\mathcal{E}_k} \lambda_k^{-\tilde{s}} |\phi_k(x)|^2, \quad x \in \mathcal{M}, \tilde{s} \in \mathbb{C}.$$

It is known that $Z(x, \tilde{s})$ is a continuous function of x for fixed $\mathrm{Re}(\tilde{s}) > \dim \mathcal{M}/2$ (see Minakshisundaram and Pleijel (1949) [244]). Since $\mathcal{M}$ is compact,

$$\sup_{x\in\mathcal{M}} Z(x,s) \leq C(\mathcal{M}, s)$$

for a constant $C(\mathcal{M},Q,s)<\infty$. □

LEMMA 7.3.2. *Suppose* $f\in\mathcal{H}_{s+\beta}(\mathcal{M},Q)$ *with* $s>\dim M/2$. *If* T *satisfies (7.11) with* $\beta\geq 0$, *Then*

$$E\|T(f_\Lambda^n)-ET(f_\Lambda^n)\|_2^2\ll\frac{\Lambda^{\beta+\dim M/2}}{n}.$$

Proof. Observe that

$$\mathrm{var}(\hat{f}_k^n)=\frac{1}{n}\mathrm{var}(\phi_k(X))\leq\frac{1}{n}E|\phi_k(X)Y|^2\leq\frac{1}{n}\{\|f\|_\infty^2+C_V\}\|\phi_k\|_2^2. \quad (7.20)$$

From Lemma 7.3.2, $\|f\|_\infty\leq C(\mathcal{M},Q,s)$. By the definition of T_k, (3.86) and (7.20),

$$\begin{aligned}E\|T(f_\Lambda^n)-ET(f_\Lambda^n)\|_2^2 &= E\sum_{\lambda_k\leq\Lambda}\left\|T_k\left(\sum_{\mathcal{E}_k}(\hat{f}_k^n-\hat{f}_k)\phi_k\right)\right\|_2^2\\ &\ll \sum_{\lambda_k\leq\Lambda}\lambda_k^\beta\sum_{\mathcal{E}_k}\mathrm{var}(\hat{f}_k^n)\\ &\ll \Lambda^\beta n^{-1}\{C(\mathcal{M},Q,s)^2+C_V\}\#\{\lambda_k\leq\Lambda\}\\ &\ll n^{-1}\Lambda^{\beta+\dim M//2}.\end{aligned}$$

We note that Weyl's formula (3.86) is used to get the final line. □

We note that $\mathcal{A}_{s,\beta}(\mathcal{M},Q)\subset\mathcal{H}_{s+\beta}(\mathcal{M},Q)\subset\mathcal{H}_s(\mathcal{M},Q)$ for $\beta\geq 0$, hence Lemma 7.3.1 applies to analytic functions. Through a slight modification to the proof of Lemma 7.3.2, one has the following.

LEMMA 7.3.3. *Suppose* $f\in\mathcal{A}_{s,\beta}(\mathcal{M},Q)$ *with* $s>\dim M/2$ *and* $\beta\geq 0$. *If* T *satisfies (7.13) with* $\beta\geq 0$, *Then*

$$E\|T(f_\Lambda^n)-ET(f_\Lambda^n)\|_2^2\ll\frac{\Lambda^{s+\dim M/2}e^{\Lambda^\beta}}{n}.$$

LEMMA 7.3.4. *Suppose that* $f\in\mathcal{H}_{s+\beta}(\mathcal{M},Q)$ *with* $s>\dim\mathcal{M}/2$. *If* T *satisfies (7.11) with* $\beta\geq 0$, *then*

$$\|T(f)-ET(f_\Lambda^n)\|_2^2\ll\Lambda^{-s}.$$

Proof. Note that

$$\begin{aligned}\|T(f)-ET(f_\Lambda^n)\|_2^2 &= \left\|\sum_{\lambda_k>\Lambda}T_k\left(\sum_{\mathcal{E}_k}\hat{f}_k\phi_k\right)\right\|_2^2\\ &\ll \sum_{\lambda_k>\Lambda}\lambda_k^\beta\sum_{\mathcal{E}_k}|\hat{f}_k|^2\\ &\leq \sum_{\lambda_k>\Lambda}|\hat{f}_k|^2\lambda_k^{s+\beta}\lambda_k^{-s}\\ &\leq Q\Lambda^{-s}. \qquad \square\end{aligned}$$

In fact, again with only minor modifications, the same can be said about analytic functions.

LEMMA 7.3.5. *Suppose that $f \in \mathcal{A}_{s,\beta}(\mathcal{M},Q)$ with $\beta \geq 0$ and $s > \dim\mathcal{M}/2$. If T satisfies (7.13) with $\beta \geq 0$, then*

$$\|T(f) - ET(f_\Lambda^n)\|_2^2 \ll \Lambda^{-s}.$$

By putting together Lemma 7.3.2, Lemma 7.3.3, Lemma 7.3.4, and Lemma 7.3.5, upper bound estimates can be established as

$$\|T(f_\Lambda^n) - T(f)\|_2^2 \ll_P \begin{cases} n^{-1}\Lambda^{\beta+\dim M/2} + \Lambda^{-s} & \text{if } f \in \mathcal{H}_{s+\beta}(\mathcal{M},Q) \\ n^{-1}\Lambda^{\beta+\dim M/2}\, e^{\Lambda^{\beta}} + \Lambda^{-s} & \text{if } f \in \mathcal{A}_{s,\beta}(\mathcal{M},Q) \end{cases}$$

as $n \to \infty$. Thus the upper bound rates for Sobolev and analytic functions are

$$n^{-s/(2s+2\beta+\dim\mathcal{M})} \quad \text{and} \quad (\log n)^{-s/(2\beta)},$$

respectively.

7.3.2 Lower Bounds

To show that the upper bound rates are optimal rates, we calculate lower bound rates of convergence and show that these are the same as the upper bounds. In calculating the lower bounds we follow the popular approach:

- specify a subproblem;
- use Fano's lemma to calculate the difficulty of the subproblem.

7.3.2.1 Polynomial Ill-Posedness

Let N_n be a positive integer depending on n and define

$$V_n = \left\{(k,l) : l = 1,\ldots,\dim\mathcal{E}_k,\ k = N_n,\ldots,bN_n\right\},$$

where $b > 0$ can be chosen such that

$$C_1 N_n^{\dim M/2} \leq \#V_n \leq C_2 N_n^{\dim M/2}$$

for n sufficiently large using (3.86).

Let $\tau = \tau(n) = \{\tau_k^l : (k,l) \in V_n\}$, where $\tau_k^l \in \{0,1\}$, and consider the function

$$f_\tau = C_3 N_n^{-(s+\beta)/2-\dim M/4} \sum_{V_n}\sum_{\mathcal{E}_k} \tau_k^l \phi_k^l, \tag{7.21}$$

where C_3 is a positive constant to be chosen below. Finally, let

$$\mathcal{F}_n = \left\{f_\tau : \tau \in \{0,1\}^{\#V_n}\right\}.$$

Under the assumption that $s > \dim M/2$, we have the following lemma.

LEMMA 7.3.6. *For n sufficiently large, $\mathcal{F}_n \subset \mathcal{H}_{s+\beta}(\mathcal{M}, Q)$.*

Proof. Applying the Sobolev norm to (7.21), we get

$$\begin{aligned} \|f_\tau\|^2_{\mathcal{H}_s} &= C_3^2 N_n^{-(s+\beta)-\dim\mathcal{M}/2} \sum_{V_n}\sum_{\mathcal{E}_k} (\tau_k^l)^2 \lambda_k^{s+\beta} \\ &\le C_3^2 b^{s+\beta} N_n^{-\dim M/2} \#\{N_n \le \lambda_k \le bN_n\} \\ &\le C_3^2 b^{s+\beta} C_2. \end{aligned}$$

again Weyl's formula (3.86) is used in the last line. Choose C_3 so that $C_3^2 b^{s+\beta} C_2 \le Q$ and Consequently, we have shown that $f_\tau \in \mathcal{H}_{s+\beta}(\mathcal{M}, Q)$. □

Now let $f, g \in \mathcal{F}_n$ with $f \ne g$. Set $\delta = f - g$ and $\delta = \sum_{V_n} \delta_k$ with $\delta_k \in \mathcal{E}_k$. Observe that

$$\begin{aligned} \|T(\delta)\|_2^2 &= \sum_{V_n} \|T_k(\delta_k)\|_2^2 \\ &\ge C_4 \sum_{V_n} \lambda_k^\beta \|\delta_k\|_2^2 \\ &\ge C_5 N_n^{-s-\dim M/2}. \end{aligned} \tag{7.22}$$

It follows from (7.22) and Lemma 3.1 of Koo(1993) [200] that there exists an $\mathcal{F}_n^* \subset \mathcal{F}_n$ such that for all $f_1, f_2 \in \mathcal{F}_n^*$ with $f_1 \ne f_2$,

$$\|T(f_1) - T(f_2)\|_2 \ge C_6 N_n^{-s/2} \text{ and } \log(\#\mathcal{F}_n^* - 1) \ge C_7 N_n^{\dim M/2}. \tag{7.23}$$

Observe that

$$E\mathcal{D}(\mathbb{P}_{f_1(X)} \| \mathbb{P}_{f_2(X)}) \le C_{\mathrm{KL}} \|f_1 - f_2\|_2^2 \ll N_n^{-(s+\beta)} \quad \text{for any } f_1, f_2 \in \mathcal{F}_n^*. \tag{7.24}$$

By Fano's lemma (see Birgé (1983) [49], Koo (1993) [200], Yatrakos (1988) [341]), if T^n is any estimator of $T(f)$, then

$$\begin{aligned} &\sup_{f \in \mathcal{H}_{s+\beta}(\mathcal{M},Q)} \mathbb{P}_f(\|T^n - T(f)\|_2 \ge cN_n^{-s/2} | X_1, \ldots, X_n) \\ \ge\ &\sup_{f \in \mathcal{F}_n^*} \mathbb{P}_f(\|T^n - T(f)\|_2 > cN_n^{-s/2} | X_1, \ldots, X_n) \\ \ge\ & 1 - \frac{\sum_{j=1}^n \max_{f_1, f_2 \in \mathcal{F}_n^*} \mathcal{D}(\mathbb{P}_{f_1(X_j)} \| \mathbb{P}_{f_2(X_j)}) + \log 2}{\log(\#\mathcal{F}_n^* - 1)}. \end{aligned} \tag{7.25}$$

Choose $N_n \asymp n^{1/(s+\beta+\dim\mathcal{M}/2)}$. By A1, (7.23), (7.24) and (7.25), we have

$$\lim_{c \to 0} \liminf_n \inf_{T^n} \sup_{f \in \mathcal{H}_{s+\beta}(\mathcal{M},Q)} \mathbb{P}_f(\|T^n - T(f)\|_2 > cn^{-s/(2s+2\beta+\dim\mathcal{M})}) = 1.$$

7.3.2.2 Exponential Ill-Posedness

Let

$$f_\tau = C_8 N_n^{-s/2-\dim M/4} e^{-N_n^\beta/2} \sum_{N_n \leq \lambda_k \leq bN_n} \sum_{\mathcal{E}_k} \tau_k^l \phi_k^l$$

and

$$\mathcal{F}_n = \{f_\tau : \tau \in \{0,1\}^{\#V_n}\}.$$

Under the assumption that $s > \dim M/2$, we have the following lemma which can be proved by the argument for Lemma 7.3.6.

LEMMA 7.3.7. *For n sufficiently large,* $\mathcal{F}_n \subset \mathcal{A}_{s,\beta}(\mathcal{M}, Q)$.

Now let $f_1, f_2 \in \mathcal{F}_n$ with $f_1 \neq f_2$. Set $\delta = f_1 - f_2$ and $\delta = \sum_{V_n} \delta_k$ with $\delta_k \in \mathcal{E}_k$. Observe that

$$\begin{aligned} \|T(\delta)\|_2^2 &= \sum_{V_n} \|T_k(\delta_k)\|_2^2 \\ &\geq C_4 \sum_{V_n} e^{\lambda_k^\beta} \|\delta_k\|_2^2 \\ &\geq C_5 N_n^{-s-\dim M/2}. \end{aligned} \tag{7.26}$$

It follows from (7.26) and Lemma 3.1 of Koo (1993) [200] that there exists an $\mathcal{F}_n^* \subset \mathcal{F}_n$ such that for all $f_1, f_2 \in \mathcal{F}_n$ with $f_1 \neq f_2$,

$$\|T(f_1) - T(f_2)\|_2 \geq C_9 N_n^{-s/2} \text{ and } \log(\#\mathcal{F}_n^* - 1) \geq C_{10} N_n^{\dim M/2}. \tag{7.27}$$

Observe that

$$E\mathcal{D}(\mathbb{P}_{f_1(X)} \| \mathbb{P}_{f_2(X)}) \leq C_{\mathrm{KL}} \|f_1 - f_2\|_2^2 \ll N_n^{-s} e^{-N_n^\beta} \text{ for any } f, g \in \mathcal{F}_n^*. \tag{7.28}$$

Choose $N_n \asymp (\log n)^{1/\beta}$. By A1, (7.25), (7.27) (7.28) and , we have

$$\lim_{c\to 0} \liminf_n \inf_{T^n} \sup_{f \in \mathcal{A}_{s,\beta}(\mathcal{M},Q)} \mathbb{P}_f(\|T^n - T(f)\|_2 > c(\log n)^{-s/(2\beta)}) = 1.$$

7.4 Kernel Density Estimation

As with the previous methodology, let $(\mathcal{M}, g)$ be a compact Riemannian manifold without boundary having dimension d. Furthermore, assume that $(\mathcal{M}, \rho_g)$ is a complete metric space, where ρ_g is the Riemannian distance. Let X be a random object on $\mathcal{M}$ whose image measure is absolutely continuous with respect to the Riemannian volume measure v_g, resulting in an almost surely continuous density f on $\mathcal{M}$.

Now, let $K : \mathbb{R}_+ \to \mathbb{R}$ be a function generating a kernel density on $\mathbb{R}^m$ in the sense of Section 2.14. K is a nonnegative continuous function, with the following properties:

- $\int_{\mathbb{R}^m} K(\|x\|)\lambda_m(dx) = 1$,
- $\int_{\mathbb{R}^m} K(\|x\|)x\lambda_m(dx) = 0$,
- $\int_{\mathbb{R}^m} \|x\|^2 K(\|x\|)\lambda_m(dx) < \infty$,
- $\operatorname{supp} K = [0;1]$,
- $\sup K(x) = K(0)$,

where λ_m is the Lebesgue measure on $\mathbb{R}^m$. K is thus an isotropic kernel on $\mathbb{R}^m$ supported by the closed unit ball.

Let p and q denote two points on $\mathcal{M}$. The volume density function $\theta_p(q)$ is defined as

$$\theta_p(q) = \frac{\mu_{Exp_p^* g}}{\lambda_p}\left(Exp_p^{-1}(q)\right).$$

That is, $\theta_p(q)$ is the quotient of the canonical measure associated with the Riemannian structure $Exp_p^* g$ on $T_p(\mathcal{M})$ over the Lebesgue measure or the Euclidean structure g_p on $T_p(\mathcal{M})$. $\theta_p(q) = \theta_q(p)$ for p and q within a normal coordinate neighborhood on $\mathcal{M}$, defined in Section 3.2, as shown in Willmore (1993)[338].

DEFINITION 7.4.1. *Let $X_1, X_2, \ldots, X_n$ be i.i.d. random objects on $\mathcal{M}$ with density f. A Pelletier density estimator of f is the mapping $f_{n,K} : \mathcal{M} \to \mathbb{R}$ that associates the value*

$$f_{n,K}(p) = \frac{1}{n}\sum_{i=1}^{n} \frac{1}{r^m}\frac{1}{\theta_{X_i}(p)} K\left(\frac{\rho_g(p, X_i)}{r}\right) \tag{7.29}$$

for every $p \in \mathcal{M}$, where r denotes the bandwidth.

This Pelletier density estimator is a valid probability density on $\mathcal{M}$ that is consistent with the usual Euclidean kernel density estimators in (2.125) when $(\mathcal{M}, g) = (\mathbb{R}^m, g_0)$ is the Euclidean space, and $K(\|\cdot\|)$ is an isotropic kernel on $\mathbb{R}^m$. Indeed, the estimator $f_{n,K}$ in (25.1) is nonnegative. Furthermore, let $p_1, \ldots, p_n$ be a realization of $X_1, \ldots, X_n$. Then

$$\begin{aligned}
\int_{\mathcal{M}} f_{n,K}(p) vol_g(dp) &= \int_{\mathcal{M}} \frac{1}{n}\sum_{i=1}^{n} \frac{1}{r^m}\frac{1}{\theta_{p_i}(p)} K\left(\frac{\rho_g(p, p_i)}{r}\right) vol_g(dp) = \\
&= \int_{\mathcal{M}} \frac{1}{r^m}\frac{1}{\theta_{p_1}(p)} K\left(\frac{\rho_g(p, p_1)}{r}\right) vol_g(dp) = \\
&= \int_{\mathcal{B}_r(p_1)} \frac{1}{r^m}\frac{1}{\theta_{p_1}(p)} K\left(\frac{\rho_g(p, p_1)}{r}\right) vol_g(dp) = \\
&= \int_{\mathcal{B}_r(0)} \frac{1}{r^m} K\left(\frac{\|x\|}{r}\right) \lambda_m(dx) = 1,
\end{aligned} \tag{7.30}$$

thus $f_{n,K}$ is a probability density. Note that like in the Euclidean case, if the bandwidth is small enough, the kernels involved in Pelletier density estimators are centered on the observations, in the sense that the observations are intrinsic means for the distributions representing those kernels (see Pelletier (2005) [278]).

The key result is that under broad assumptions ($r \leq r_0$, where $0 < r_0 < inj_g(\mathcal{M})$ and $inj_g(\mathcal{M})$ is the injectivity radius of $(\mathcal{M}, g)$), the Pelletier density estimator is a consistent estimator of f :

THEOREM 7.4.1. *Let f be a twice differentiable probability density on $\mathcal{M}$ with a bounded second covariant derivative and $f_{n,K}$ be the estimator defined in* (25.1)*. If $r < r_0$, then there exists a constant C_f such that*

$$E_f \|f_{n,K} - f\|^2_{L^2(\mathcal{M})} \leq C_f \left(\frac{1}{nr^m} + r^4 \right)$$

Furthermore, for $r \sim n^{-1/(m+1)}$, $E_f \|f_{n,K} - f\|^2_{L^2(\mathcal{M})} = O\left(n^{-4/(m+4)}\right)$.

For a proof, we consider the usual decomposition of $E_f(\|f_{n,K} - f\|^2)$ (see Van der Vaart (1998) [327], or (2.110)):

$$E_f \|f_{n,K} - f\|^2_{L^2(\mathcal{M})} = \int_{\mathcal{M}} (E_f(f_{n,K}(p)) - f(p))^2 vol_g(dp) + \int_{\mathcal{M}} Var_f(f_{n,K}(p)) vol_g(dp), \tag{7.31}$$

and then, we find upper bounds for the right hand terms in (7.31). These are stated in the next two lemmas.

LEMMA 7.4.1. *Assume f is a probability density on $\mathcal{M}$ and $f_{n,K}$ is the Pelletier estimator, such that the conditions in Theorem 7.4.1 are satisfied. Then there is a constant C_b such that the biased term has the upper bound*

$$\int_{\mathcal{M}} (E_f(f_{n,K}(p)) - f(p))^2 vol_g(dp) \leq C_b r^4. \tag{7.32}$$

For a **proof** we first note that the pointwise bias $B(p) = E_f(f_{n,K}(p)) - f(p)$ can be written as follows:

$$\begin{aligned} B(p) &= \int_{B_r(p)} \frac{1}{r^m} K\left(\frac{\rho_g(p,q)}{r}\right) \left(\frac{f(q)}{\theta_q(p)} - \frac{f(p)}{\theta_p(q)}\right) vol_g(dp) = \\ &= \int_{B_r(p)} \frac{1}{r^m} K\left(\frac{\rho_g(p,q)}{r}\right) \frac{1}{\theta_p(q)} (f(q) - f(p)) vol_g(dp). \end{aligned} \tag{7.33}$$

If we consider the local representative of f in normal coordinates (3.71), relative to an orthogonal basis in $T_p\mathcal{M}$, a representative that for simplicity we label by f as well, then from the differentiability assumptions

$$f(x) = f(0) + \sum_{j=1}^{m} c_j x^j + R_2(p,x), \tag{7.34}$$

where $|R_2(p,x)| \leq C_r\|x\|^2, \forall x, \|x\| \leq r$. From (7.30) and (7.34), and given that $\int_{\mathbb{R}^m} K(\|x\|x\lambda_m(dx) = 0$, it follows that

$$|B(p)| = |\int_{B_r(0)} \frac{1}{r^m} K\left(\frac{\|x\|}{r}\right) R_2(p,x)\lambda_m(dx)| \leq$$

$$\leq \int_{B_r(0)} \frac{1}{r^m} K\left(\frac{\|x\|}{r}\right) C_r\|x\|^2\lambda_m(dx) = r^2 \int_{B_1(0)} K(\|u\|)C_r\|u\|^2\lambda_m(du), \quad (7.35)$$

therefore

$$B^2(p) \leq Dr^4 \quad (7.36)$$

where D is a positive constant, and, given that the volume of the compact manifold $\mathcal{M}$ is finite the inequality in (7.32) follows by integrating the inequality (7.36) over $\mathcal{M}$ with respect to the volume measure associated with the Riemannian structure g.

LEMMA 7.4.2. *Assume f is a probability density on $\mathcal{M}$ and $f_{n,K}$ is the Pelletier estimator, such that the conditions in Theorem 7.4.1 are satisfied. Then there is a constant C_v such that*

$$\int_{\mathcal{M}} Var_f(f_{n,K}(p))vol_g(dp) \leq \frac{C_v}{nr^m}. \quad (7.37)$$

For a **proof** we first note that the pointwise variance is given by

$$Var_f(f_{n,K}(p)) = Var_f(\frac{1}{n}\sum_{i=1}^{n} \frac{1}{r^m} \frac{1}{\theta_{X_i}(p)} K\left(\frac{\rho_g(p,X_i)}{r}\right)) =$$

$$= \frac{1}{nr^{2m}} Var_f(\frac{1}{\theta_{X_1}(p)} K\left(\frac{\rho_g(p,X_1)}{r}\right)) \leq$$

$$\leq \frac{1}{nr^{2m}} E_f(\frac{1}{\theta_{X_1}(p)} K\left(\frac{\rho_g(p,X_1)}{r}\right))^2 =$$

$$= \frac{1}{nr^{2m}} \int_{\mathcal{M}} \frac{1}{\theta_q^2(p)} K^2\left(\frac{\rho_g(p,q)}{r}\right) vol_g(dq). \quad (7.38)$$

Note that since $\mathcal{M}$ is compact, $\mathcal{M}^2$ is compact as well. Therefore there is a positice constant C_1, s.t. $\forall (p,q) \in \mathcal{M}, \frac{1}{\theta_q(p)} \leq C_1$. From (7.38), we get

$$\begin{aligned}
&\int_{\mathcal{M}} Var_f(f_{n,K}(p)) vol_g(dp) \leq \\
&\leq \frac{C_1}{nr^{2m}} \mathrm{vol}_g(\mathcal{M}) \int_{\mathcal{M}} K^2\left(\frac{\rho_g(p,q)}{r}\right) vol_g(dq) = \\
&= \frac{C_1}{nr^{2m}} \mathrm{vol}_g(\mathcal{M}) K^2(0) \int_{B_r(p)} \frac{1}{\theta_q(p)} vol_g(dq) = \\
&= \frac{C_1}{nr^{2m}} \mathrm{vol}_g(\mathcal{M}) K^2(0) \int_{B_r(0)} \frac{1}{\theta_q(p)} \lambda_m(dx) = \\
&= \frac{C_1 vol B_0(r)}{nr^{2m}} \mathrm{vol}_g(\mathcal{M}) K^2(0) = \frac{C_v}{nr^m},
\end{aligned} \tag{7.39}$$

done ■

Theorem 7.4.1 now follows directly from (7.31) and Lemmas 7.4.1, 7.4.2.

REMARK 7.4.1. *In a recent paper, Kim and Park (2013) [192] focus on a kernel function defined on the tangent space $T_p\mathcal{M}$ at a point on an m dimensional Riemannian manifold, in terms of the exponential map. They establish that the formula of the asymptotic behavior of the bias and the mean square error contains geometric quantities expressed by terms of polynomials in the component of the curvature tensor and its covariant derivatives. It is natural to replace $X_i - p$ in* (2.125) *by $Exp_p(X_i)$, which enabled Kim and Park (2013) to define the kernel density estimator of a probability density function f in terms of a smoothing parameter $h > 0$ as follows:*

$$\widehat{f}_n(p) = \frac{1}{h^m C_h} \sum_{i=1}^{n} K\left(\frac{1}{h} Exp_p(X_i)\right), \tag{7.40}$$

where $\int_{T_p\mathcal{M}} K(v)\lambda_m(dv) = 1$ and C_h is a positive constant such that

$$h^m C_h = \int_{\mathcal{M}} K\left(\frac{1}{h} Exp_p(x)\right) vol_g(dp). \tag{7.41}$$

The integral on the right hand side of (7.4.1) *is independent of x, (see Kim and Park (2013) [192]). Moreover Kim and Park (op.cit.) show that the estimator* (7.40) *is consistent and establish upper bound for its convergence in $L^2(\mathcal{M})$.*

REMARK 7.4.2. *Bhattacharya and Dunson (2010) [28] developed a nonparametric Bayesian method for density estimation of a r.o. on a separable metric space, and in particular on a manifold (see also Chapter 12 in Bhattacharya and Bhattacharya (2012) [27]). They combined a family of Pelletier density estimators in definition 7.4.1 (see Pelletier (2005) [278]) with a prior kernel. Thus, a class of mixture models for nonparametric density estimators* *on compact manifolds was introduced, and the consistency of these estimators was studied as well. The performance of their proposed Bayesian density estimator was evaluated through a stimulation study using Dirichlet process mixtures of complex Watson kernels (Dryden and Mardia (1998) [91], p. 114) and for density estimation based on Kendall planar shape gorilla skulls data (Dryden and Mardia (1998) [91], p. 317).*

Part II

Asymptotic Theory and Nonparametric Bootstrap on Special Manifolds

Chapter 8

Nonparametric Statistics on Homogeneous Hadamard Manifolds

8.1 Introduction

This idea of analyzing data on non-Euclidean sample spaces gained traction as computational power increased, allowing for this new direction to be implemented practically. Non-Euclidean data analysis includes directional data (Watson (1983) [333]), direct similarity shape (Kendall (1984) [177]), tectonic plates (Chang (1988) [63]), certain Lie groups (Kim (2000) [188]), Stiefel manifolds (Hendricks and Landsman (1998) [154]), projective shape manifolds (Patrangenaru (2001) [268]), and affine shape manifolds (Patrangenaru and Mardia (2003) [274]). A common features of all these sample spaces is that they are all *homogeneous spaces*. Given a homogeneous space, it is a statistician's choice to select an appropriate homogeneous Riemannian structure on the sample space, that in her or his view would best address the data analysis in question. While the analysis of directional and shape data that dominated object data analysis in its initial phase are on compact, more recently, in brain imaging , proteomics, and astronomy, homogeneous spaces of noncompact type arose as well (see Dryden et al. (2009) [89] and Bandulasiri et al. (2009) [10]).

Statistical analysis on general homogeneous spaces was first considered in the context of density estimation with the ground breaking paper by R. J. Beran (1968) [19], and in the context of function estimation via harmonic analysis on Lie groups , which are also homogeneous spaces, in Kim (1998, 2000) [187, 188], Koo and Kim (2008, 2008a) [202, 203]). For two-sample tests for means on Lie groups we refer to Chapter 6; the results in Section 6.4 are made more explicit here and extended somewhat.

The purpose of this chapter is to present a nonparametric methodology for analysis of data on a noncompact Riemannian space of nonconstant curvature that admits a simply transitive group of isometries. These methods follow from Osborne et al. (2013) [258]. The chapter is thus organized as follows. Section 8.2 reviews the general concepts of homogeneous spaces, homogeneous metric spaces, Riemannian homogeneous spaces, and the simply transitive group of isometries. General definitions and results concerning intrinsic means on Hadamard manifolds, including on Hadamard manifolds that are homogeneous spaces, are also summarized in Section 8.3. In Sections 8.2 and 8.3, we recall the construction of an invariant Riemannian metric (see Kobayashi and Nomizu (1963) [199], p.154, or Chapter 3). Here we focus on the case of such a metric on $Sym^+(m)$, that was coined by Schwartzman (2006) [301] as the *generalized Frobenius metric*. Section 8.4 presents a two-sample test statistic for intrinsic means on a Hadamard manifold that admits a simply transitive Lie group of isometries. The asymptotic distribution of this statistic is also presented, along with a result from nonparametric bootstrap that can be applied to this two-sample problem when sample sizes are small.

8.2 Considerations for Two-Sample Tests on Riemannian Homogeneous Spaces

A minimal requirement to enable comparisons of distributions is the homogeneity of the sample space that contains the data. To begin with, we consider a metric space without a differentiable structure. It is assumed that any two observations (points) on such a sample space $(\mathcal{M},\rho)$ can be brought into coincidence by a distance-preserving automorphism of this sample space. Formally, this property amounts to the transitivity of the left action of the isometry group on the sample space $(\mathcal{M},\rho)$.

DEFINITION 8.2.1. *If $(\mathcal{K},\cdot)$ is a group with the identity element $e\in\mathcal{K}$, and $\mathcal{M}$ a set, then*

- *a map $\alpha:\mathcal{K}\times\mathcal{M}\to\mathcal{M}$ is called a* left action of $\mathcal{K}$ on $\mathcal{M}$ *if the following two properties are satisfied:*

$$\begin{aligned} \alpha_{k\cdot h} &= \alpha_k\circ\alpha_h \quad \forall k,h\in\mathcal{K}, & (8.1)\\ \textit{and}\quad \alpha_e &= \textit{identity map}\ \ \mathcal{M}\to\mathcal{M}, & (8.2)\end{aligned}$$

  and for all $k\in\mathcal{K}$, we define $\alpha_k:\mathcal{M}\to\mathcal{M}$, by $\alpha_k(x):=\alpha(k,x)$.
- *Given a left-action $\alpha:\mathcal{K}\times\mathcal{M}\to\mathcal{M}$, for $x\in\mathcal{M}$, the* isotropy group at *x, is the subgroup $\mathcal{K}_x=\{k\in\mathcal{K}:\alpha_k(x)=x\}\subset\mathcal{K}$, and the* orbit *of x is the set $\mathcal{K}(x)=\{\alpha_k(x),k\in\mathcal{K}\}$.*

- *The left action* α *is called* transitive , *if for all* $x_1, x_2 \in \mathcal{M}$, *there exists a* $k \in \mathcal{K}$ *such that* $\alpha_k(x_1) = x_2$. *If, the action* α *is transitive, then the set* $\mathcal{M}$ *is called* $\mathcal{K}$ - homogeneous space, *or simply a* homogeneous space.

This algebraic definition 8.2.1 may be considered too abstract for the purposes of statistical analysis because we are primarily interested in analyzing distributions on metric spaces $(\mathcal{M}, \rho)$, rather than an abstract set. For this reason, we will only consider $\mathcal{K}$ - homogeneous spaces, relative to a subgroup $\mathcal{K}$ of the group of isometries $I(\mathcal{M}, \rho)$, where the group structure is given by the composition of functions. In such a situation, the action $\alpha : \mathcal{K} \times \mathcal{M} \to \mathcal{M}$ is $\alpha(f,x)) = f(x)$. Such a homogeneous space will be called a *homogeneous metric space*. As discussed in Section 8.1, Euclidean spaces, spheres, , projective spaces, groups of rotations, and Stiefel manifolds endowed with extrinsic distances are all homogeneous metric spaces. For illustrative purposes, we will focus on a particular non-compact homogeneous space, that being the set $Sym^+(p)$ of $p \times p$ positive definite matrices due to its utility for certain applications. $Sym^+(2)$ is the sample space for CMB data and $Sym^+(3)$ is that for DTI data .

PROPOSITION 8.2.1. *If* $GL^+(p,\mathbb{R})$ *is the group of matrices of positive determinant and let* $\alpha : GL^+(p,\mathbb{R}) \times Sym^+(p) \to Sym^+(p)$ *be given by*

$$\alpha(k,x) := \alpha_k(x) = kxk^T, \tag{8.3}$$

then α *is a transitive left action of* $GL^+(p,\mathbb{R})$ *on* $Sym^+(p)$, *and the isotropy group at* I_p *is the special orthogonal group* $SO(p)$.

Proof. If $x(1), x(2) \in Sym^+(p)$ and for $a = 1,2, \lambda_1(a) \geq \dots \lambda_p(a) > 0$ are the eigenvalues of $x(a)$ in their decreasing order, and $x_1(a), \dots, x_p(a)$ eigenvectors of $x(a)$ with $x(a)x_i(a) = \lambda_i(a)x_i(a), \forall i = 1, \dots p, x_j(a)^T x_i(a) = \delta_{ji}, \forall i,j = 1,\dots,p$, then by the SVD $x(a) = \sum_{i=1}^p \lambda_i(a)x_i(a)x_i(a)^T$. If we define $r(a) = \sum_{i=1}^p \sqrt{\lambda_i(a)} x_i(a)x_i(a)^T$, then $r(a) \in Sym^+(p)$, and, $r(a)^2 = x(a)$. If we set $k = r(2)r(1)^{-1} \in GL^+(p,\mathbb{R})$, then $\alpha_k(x(1)) = x(2)$. Hence α is transitive. ■

Given a sample space $\mathcal{M}$, in order to perform two sample tests for data $\mathcal{M}$, one may first endow $\mathcal{M}$ with a structure of $\mathcal{K}$ - homogeneous space, and, next find a distance ρ , such that $(\mathcal{M}, \rho)$ becomes a $\mathcal{K}$ - homogeneous metric space. There are two general methods of endowing $\mathcal{M}$ with a distance ρ, so that α becomes an action of $\mathcal{K}$ by isometries of $(\mathcal{M}, \rho)$. From now on, we will assume in addition that $\mathcal{M}$ is a manifold, $\mathcal{K}$ is a Lie group and the left action $\alpha : \mathcal{K} \times \mathcal{M} \to \mathcal{M}$ is smooth.

DEFINITION 8.2.2. *An isometric embedding of the metric space* $(\mathcal{M}, \rho)$ *is an injective map* $J : \mathcal{M} \to \mathbb{R}^D$ *such that* $\|J(x) - J(y)\| = \rho(x,y), \forall x,y \in \mathcal{M}$. *If in addition* $(\mathcal{M}, \rho)$ *is a* $\mathcal{K}$*-homogeneous metric space for the left action* α : $\mathcal{K} \times \mathcal{M} \to \mathcal{M}$. *such an embedding is called isometric equivariant embedding.*

Given a Riemannian metric g on $\mathcal{M}$, an isometric embedding of the Riemannian manifold $(\mathcal{M},g)$ is an embedding $J:\mathcal{M}\to\mathbb{R}^D$, such that

$$g_x(\xi,\eta)=d_xJ(\xi)^Td_xJ(\eta),\forall x\in\mathcal{M},\forall\xi,\eta\in T_x\mathcal{M}.$$

The existence of an isometric embedding of a Riemannian manifold in $\mathbb{R}^D$ given by Nash (1956) [253] does not provide explicitly the equations of such an embedding. The existence of an isometric embedding of a Riemannian homogeneous space is highly non-trivial (see Moore (1976) [249]). In the absence of an equivariant embedding of a homogeneous $\mathcal{K}$-space, one has to look for $\mathcal{K}$-invariant Riemannian structures on $\mathcal{M}$.

REMARK 8.2.1. *Isometric equivariant embeddings were previously used in statistics for two sample tests for means. The inclusion of the round sphere S^{N-1} as a subset of $\mathbb{R}^N$ is an example of isometric equivariant embedding that was used for directional data two sample tests (see Beran and Fisher (1998) [18]). The map $J:\mathbb{R}P^{N-1}\to Sym(N), J([x])=\frac{xx^T}{x^Tx}$, is an isometric equivariant embedding that was used for paired two sample tests for mean projective shapes by Crane and Patrangenaru (2011) [78]. In these examples the sample spaces are compact. It is more difficult to find an isometric equivariant embedding of a noncompact homogeneous space. For example, for the set $\mathcal{H}=\{x=(x_1,\ldots,x_N,x_{N+1})\in\mathbb{R}^{N+1},-\sum_{j=1}^N x_j^2+x_{N+1}^2=1\}$, with the distance $\rho(x,y)=-\sum_{j=1}^N(x_j-y_j)^2+(x_{N+1}-y_{N+1})^2$, there is no explicit isometric embedding into an Euclidean space. The sample space $Sym^+(p)$ is also noncompact, so we will consider another method to provide this space with a structure of homogeneous metric space.*

From Lemma 3.2.1 in Chapter 3, we know that for any homogeneous space $\mathcal{M}$ with a transitive group $\mathcal{K}$, whose isotropy group at q_0 is $\mathcal{H}$, a scalar product g_{q_0} on $T_{q_0}\mathcal{M}$ can be extended to a $\mathcal{K}$-invariant Riemannian metric on $\mathcal{M}$ if and only if the inner product g_{q_0} is $\mathcal{H}$-invariant, in which case such a $\mathcal{K}$-invariant extension is unique.

Recall that if U is an open set in $\mathbb{R}^D$, then the tangent space at a point $x\in U$ is $T_xU=\{x\}\times U$, and the tangent bundle of U is $TU=\cup_{x\in U}T_xU=U\times\mathbb{R}^D$. In particular, if $x\in Sym^+(p)$ and $T_xSym^+(p)=\{x\}\times Sym(p)$ then the tangent bundle $TSym^+(p)$ is $Sym^+(p)\times Sym(p)$. Bearing this in mind, given a tangent vector $(x,u)\in TSym^+(p)$, and referring to the action in equation (8.3), we get

$$(d_x\alpha_k)((x,u))=(kxk^T,kuk^T),\forall(x,u)\in TSym^+(p). \tag{8.4}$$

The *Frobenius* scalar product on $Sym(p)$ is given by $u\cdot_F v=Tr(uv)$.

DEFINITION 8.2.3. *The generalized Frobenius metric on $Sym^+(p)$ is the $GL^+(p,\mathbb{R})$-invariant metric g_F associated with the scalar product g_{I_p} on $T_{I_p}Sym^+(p)$ given by the Frobenius scalar product:*

$$g_{I_p}((I_p,u),(I_p,v))=:u\cdot_F v=Tr(uv). \tag{8.5}$$

Élie Cartan introduced following notion of simply transitive groups (see Cartan (1946, pp.275-293) [59]).

DEFINITION 8.2.4. *An action α of a Lie group $\mathcal{G}$ on a manifold $\mathcal{M}$ is simply transitive if one of the equivalent statements (i) or(ii) holds true: (i) for any two points $x_1, x_2 \in \mathcal{M}, x_1 \neq x_2$, there is a unique $k \in \mathcal{G}$, such that $\alpha_k(x_1) = x_2$, and, (ii) if $x \in \mathcal{M}$, and $\alpha^x : \mathcal{G} \to \mathcal{M}$ is given by $\alpha^x(k) = \alpha(k,x)$, then α^x is bijective.*

The proof of the equivalence of (i) and (ii) in Definition 8.2.4 is elementary.

REMARK 8.2.2. *Beran and Fisher (1998) [18] and Mardia and Patrangenaru (2005) [233] reduced a two sample problem for mean location on a certain homogeneous space (a sphere, or a projective space $\mathbb{R}P^m$) to a one sample problem on a Lie group acting on that homogeneous space ($SO(3)$, respectively $SO(m+1)$). This procedure is local, and has the drawback that it increases the dimensionality of the manifold where the data analysis is performed, since the dimension of a transitive Lie group $\mathcal{K}$ acting transitively on a manifold $\mathcal{M}$, is at least higher than the dimension of $\mathcal{M}$. Therefore, wherever the data is on a $\mathcal{K}$-homogeneous space, it is useful to check on the existence of a subgroup $\mathcal{G}$ of $\mathcal{K}$ that is simply transitive with the induced action $\alpha_{|\mathcal{G}\times\mathcal{M}}$. In this case, the dim $\mathcal{G}$ = dim $\mathcal{M}$, and the procedure does not increase the dimensionality.*

LEMMA 8.2.1. *A left action α of a group $\mathcal{G}$ on $\mathcal{M}$ is transitive, if and only if there is a point $x_0 \in \mathcal{M}$ such that the orbit $\mathcal{G}(x_0) = \mathcal{M}$.*

For **proof** of Lemma 8.2.1, assume $x_1, x_2 \in \mathcal{M}$ are arbitrary. From the hypothesis there exist $t_1, t_2 \in \mathcal{G}$, such that $x_1 = \alpha_{t_1}(x_0)$ and $x_2 = \alpha_{t_2}(x_0)$. From the properties of the action α, we obtain $x_0 = \alpha_{t_1^{-1}}(x_1)$ therefore $\alpha_{t_2}(\alpha_{t_1^{-1}}(x_1)) = x_2$, which can be also written as $\alpha_{t_2 t_1^{-1}}(x_1) = x_2$; that is, there exists a $t = t_2 {t_1}^{-1} \in \mathcal{G}$, such that $\alpha_t(x_1) = x_2$. This proves the transitivity of the group action α of $\mathcal{G}$ on $\mathcal{M}$.

PROPOSITION 8.2.2. *The group $T^+(p,\mathbb{R})$ of lower triangular $p \times p$, matrices with positive diagonal entries, with the restriction of the action 8.3 to $T^+(p,\mathbb{R}) \times Sym^+(p)$, is transitive on $Sym^+(p)$.*

The *proof* is immediate, since by the *Cholesky decomposition* of a positive definite covariance matrix Σ, there exist a *unique* matrix $t \in T^+(p,\mathbb{R})$ such that $\Sigma = tt^T$. We will set $t = c(\Sigma)$. This is same as saying that, given the action α in (8.3), for any matrix $\Sigma \in Sym^+(p)$, $\exists t \in T^+(p,\mathbb{R})$ such that $\alpha_t x(I_p) = \Sigma$. This shows that the orbit $T^+(p,\mathbb{R})(I_p) = Sym^+(p)$, and the result follows from Lemma 8.2.1.

8.3 Intrinsic Means on Hadamard Manifolds

Recall from Chapter 4 some general results for the intrinsic mean of a random object X.

DEFINITION 8.3.1. *Assume $(\Omega, \mathcal{A}, P)$ is a probability space, $(\mathcal{M}, g)$ is a* *complete Riemannian manifold, and $\mathcal{B}_\mathcal{M}$ is the Borel σ-filed generated by open subsets of $\mathcal{M}$. A random object (r.o.) is a mapping $X : \Omega \to \mathcal{M}$, that is $(\mathcal{A}, \mathcal{B}_\mathcal{M})$-measurable.*

In Chapter 3 we considered, in this context, the intrinsic mean set for X. Let $\rho = d_g$ be the geodesic distance on $\mathcal{M}$. Following Patrangenaru (1998) [266], we introduced the Fréchet function $\mathcal{F} : \mathcal{M} \to \mathbb{R}$, given in (4.2). Assuming $\mathcal{F}(a)$ is finite for some $a \in \mathcal{M}$, Patrangenaru (2001) [268] called the minimum of $\mathcal{F}$ on $\mathcal{M}$ to be *total intrinsic variance* $t\Sigma_g$ of X with respect to the Riemannian structure g. When $\mathcal{F}$ has a unique minimizer, that minimizer was called the *intrinsic mean (with respect to the Riemannian structure g)* of X and was labeled μ_g. Using Theorem 9.1 in Kobayashi and Nomizu (1969, p.109) [198], Patrangenaru (1998) [266] noted that μ_g exists if $\mathcal{M}$ is simply connected Hadamard manifold (see Definition 3.2.5).

Also recall that the *Cartan mean (set)* of a finite set $\{x_1, \dots, x_n\}$ of points on $\mathcal{M}$ is the intrinsic mean (set) of the empirical distribution $\hat{Q}_n = \frac{1}{n}\sum_{i=1}^n \delta_{x_i}$. In 1928, Cartan ([58]) showed that the baricenter (Cartan mean) of a finite set of points on a Hadamard manifold exists (see Cartan (1946) [59], p. 354), which in practice means that the sample mean vector from Multivariate Analysis naturally extends to a unique Cartan mean of a sample of points on a Hadamard manifold. Note that the Cartan mean set may have more than one point on a positively curved Riemannian manifold.

THEOREM 8.3.1. *The space $(Sym^+(p), g_F)$ a Hadamard manifold, with the Riemannian exponential mapping at I_p, given by $Exp_{I_p}(w) = \exp(w)$, $\forall w \in T_{I_p}Sym^+(p) = Sym(p)$, where $\exp(w) = \sum_{r=0}^\infty \frac{1}{r!} w^r$.*

Proof. From Proposition 8.2.2, the group $T^+(p, \mathbb{R})$ is a noncommutative simply transitive group of isometries of $Sym^+(p)$. Moreover, the mapping $t \to h(t) = tt^T = \alpha(t, I_p)$, endows $T^+(p, \mathbb{R})$ with the left invariant Riemannian structure $h^* g_F$, thus $(T^+(p, \mathbb{R}), h^* g_F)$ is a Riemannian homogeneous space in itself. Therefore, $\forall t \in T^+(p, \mathbb{R})$, $t \to t^{-1}$ there is an isometry of $(T^+(p, \mathbb{R}), h^* g_F)$, which is same as saying that for any geodesic γ, with $\gamma(0) = I_p$, the *geodesic reflection* on $(T^+(p, \mathbb{R}), h^* g_F)$, at the point I_p, given by $\gamma(s) \to \gamma(-s)$, is an isometry of $(T^+(p, \mathbb{R}), h^* g_F)$. Thus $(T^+(p, \mathbb{R}), h^* g_F)$ is a *symmetric space in the sense of Cartan* at the point I_p. This turns the space $(T^+(p, \mathbb{R}), h^* g_F)$, into a globally symmetric space in the sense of Cartan, given that for each point $t \in T^+(p, \mathbb{R})$, the left translation $L_t(u) = tu$ is also an isometry of $(T^+(p, \mathbb{R}), h^* g_F)$, that takes geodesics to geodesics, thus forcing the geodesic reflection at the point $t \in T^+(p, \mathbb{R})$ to be as well an isometry of $(T^+(p, \mathbb{R}), h^* g_F)$. This is a noncompact symmetric space, since $T^+(p, \mathbb{R})$ is an open subset in a linear space of matrices. From Theorem 3.1. in Helgason (1962) [146] stating that all the sectional curvatures of a symmetric space of noncompact type are non-positive and since an isometry between two Riemannian manifolds preserves the sectional curvature at corresponding points, and $h : (T^+(p,\mathbb{R}), h^* g_F) \to (Sym^+(p), g_F)$ is an isometry, it follows that $(Sym^+(p), g_F)$ a Hadamard manifold.

For the second claim, note that a geodesic γ on a Lie group $\mathcal{G}$ with a left invariant Riemannian metric, with $\gamma(0) = 1_{\mathcal{G}}$ is a *one parameter subgroup* (Helgason (1978)[147], p.94) of $\mathcal{G}$ and it is elementary to show that the one parameter subgroups of $GL(p,\mathbb{R})$ are given by $\gamma(s) = \exp sv, \forall s \in \mathbb{R}$, for some $v \in M(p,\mathbb{R})$. In particular, since $\mathcal{G} = T^+(p,\mathbb{R})$ is a Lie subgroup of $GL(p,\mathbb{R})$, the one parameter subgroups of $T^+(p,\mathbb{R})$ are of the form $s \to \exp(sv), \forall s \in \mathbb{R}$, for some $v \in \mathfrak{g}$, where $\mathfrak{g}$ is the set of lower triangular matrices, Lie algebra of $\mathcal{G}$, therefore the geodesics γ, of $(T^+(p,\mathbb{R}), h^* g_F)$, with $\gamma(0) = I_p$, are given by $\gamma(s) = \exp(sv)$, $\forall s \in \mathbb{R}$, for some $v \in \mathfrak{g}$. Finally, since h is an isometry, with $h(I_p) = I_p$, if $s \to \lambda(s)$ is a geodesic on $(Sym^+(p), g_F)$, with $\lambda(0) = I_p$, then $\gamma(s) = h^{-1}(\lambda(s))$ is a geodesic on $(T^+(p,\mathbb{R}), h^* g_F)$, with $\gamma(0) = I_p$. Thus $\gamma(s) = \exp(sv), \forall s \in \mathbb{R}$, for some $v \in \mathfrak{g}$, and $\lambda(s) = h(\gamma(s)) = \gamma(s)\gamma(s)^T = \exp(sv)(\exp(sv))^T = \exp(sw)$, where $w \in Sym(p)$. □

Given that two distinct points x, y on a Hadamard manifold $(\mathcal{M}, g)$, they can be joined by a unique geodesic

$$\gamma_{x,y} : \mathbb{R} \to \mathcal{M}, \gamma_{x,y}(0) = x, \gamma_{x,y}(d_g(x,y)) = y, \tag{8.6}$$

and an explicit formula of the geodesic $\gamma_{x,y}$ in equation (8.6) is very useful for computing the geodesic distance $d_g(x,y)$. We are in particular interested in this formula $d_{g_F}(x,y)$, given $x, y \in Sym^+(p)$, which is a direct consequence of theorem 8.3.1.

COROLLARY 8.3.1. *The geodesic distance $d_{g_F}(x,y)$ on $Sym^+(p)$ is given by*

$$d^2_{g_F}(y,x) = Tr\left((\log(c(x)^{-1} x (c(x)^{-1})^T))^2\right), \tag{8.7}$$

where $\log : Sym^+(p) \to Sym(p)$ is the inverse of the matrix exponential map $\exp : Sym(p) \to Sym^+(p)$.

Proof. Recall that $Exp_x : T_x Sym^+(p) \to Sym^+(p)$ is the Riemannian exponential at the point $x \in Sym^+(p)$, $Exp_x(w) = \gamma_{x,w}(1)$, where $\gamma_{x,w}$ it the geodesic on $Sym^+(p)$ with $\gamma_{x,w}(0) = x$, $\frac{d\gamma_{x,w}}{ds}(0) = w \in Sym(p)$. Here, for any point $x \in Sym^+(p)$, we identify $(x,w) \in T_x Sym^+(p) = \{x\} \times Sym(p)$ with $w \in Sym(p)$. Note that if $x \in Sym^+(p)$ then the $\alpha_{c(x)} : Sym^+(p) \to Sym^+(p)$, is an isometry taking geodesics centered at I_p to geodesics centered at x and since $\alpha_{c(x)}(z) = c(x) z c(x)^T, d_z \alpha_{c(x)}(v) = c(x) v c(x)^T, \forall v \in Sym(p)$. In addition, it follows that

$$\begin{aligned} Exp_x(c(x) v c(x)^T) &= \gamma_{x, c(x) v c(x)^T}(1) = \gamma_{x, d_{I_p} \alpha_{c(x)}(v)}(1) = \\ &= \gamma_{x, d_{I_p} \alpha_{c(x)}(v)}(1) = \gamma_{\alpha_{c(x)}(I_p), d_{I_p} \alpha_{c(x)}(v)}(1) = \\ &= \alpha_{c(x)}(\gamma_{I_p, v}(1)) = c(x) Exp_{I_p}(v) c(x)^T, \end{aligned} \tag{8.8}$$

and from Theorem 8.3.1 and equation (8.8) we get

$$Exp_x(c(x)vc(x)^T) = c(x)\exp(v)c(x)^T. \tag{8.9}$$

Given $w \in T_x Sym^+(p) \simeq Sym(p)$, if we solve for v in the equation $c(x)vc(x)^T = w$, we get $v = c(x)^{-1}w(c(x)^{-1})^T$, therefore, from equation (8.9) we see that

$$Exp_x(w) = c(x)\exp(c(x)^{-1}w(c(x)^{-1})^T)c(x)^T. \tag{8.10}$$

Assume $y = Exp_x(w)$, then $d^2_{g_F}(y,x) = g_{F_x}(w,w)$. One can easily show that

$$g_{F_x}(w,w) = Tr((x^{-1}w)^2). \tag{8.11}$$

We solve equation $y = Exp_x(w)$ for w by using equation (8.10), to obtain

$$w = c(x)\log(c(x)^{-1}y(c(x)^{-1})^T)c(x)^T, \tag{8.12}$$

where log is the inverse of the mapping $\exp : Sym(p) \to Sym^+(p)$, so that $\log(\Sigma)$ can be obtained as follows : assume $\Sigma = A\Lambda A^T$, where Λ is the diagonal matrix $\text{diag}(\lambda_1,\dots,\lambda_p), \lambda_j > 0, \forall j = 1,\dots,p$, and $A \in O(p)$, then

$$\log(\Sigma) = A\ \text{diag}(ln(\lambda_1),\dots,ln(\lambda_p))A^T. \tag{8.13}$$

If we plug $w = w(x,y)$ into the expression on the right hand side of equation (8.11), we get $d^2_{g_F}(x,y) = Tr((x^{-1}c(x)\log(c(x)^{-1}y(c(x)^{-1})^T)c(x)^T)^2)$. Given that $x = c(x)c(x)^T$, we have $x^{-1} = (c(x)^T)^{-1}c(x)^{-1}$, therefore

$$d^2_{g_F}(y,x) = Tr((c(x)^T)^{-1}\log(c(x)^{-1}y(c(x)^{-1})^T)c(x)^T)^2), \tag{8.14}$$

or, in a more compact form

$$d^2_{g_F}(y,x) = Tr((\log(c(x)^{-1}y(c(x)^{-1})^T))^2). \tag{8.15}$$

□

The intrinsic mean of a random object X on $Sym^+(p)$ with respect to the generalized Frobenius distance g_F, which is the minimizer of the function $\mathcal{F} : Sym^+(p) \to [0,\infty)$, will be called the *generalized Frobenius mean* and is labeled μ_F, and the intrinsic covariance matrix will be called the *generalized Frobenius covariance matrix* will be labeled Σ_F. As a consequence of equation (8.15) we obtain the following

COROLLARY 8.3.2. *(a) The generalized Frobenius mean of X is given by*

$$\mu_F = \underset{y \in Sym^+(p)}{\arg\min}\ E[Tr((\log(c(X)^{-1}y(c(X)^{-1})^T))^2)]. \tag{8.16}$$

The generalized Frobenius mean of a sample $x_1,\dots,x_n$ of points on $Sym^+(p)$ is given by

$$\mu_F = \underset{y \in Sym^+(p)}{\arg\min} \sum_{i=1}^{n}(Tr((\log(c(x_i)^{-1}y(c(x_i)^{-1})^T))^2)). \tag{8.17}$$

In order to compute the generalized Frobenius sample mean in (8.17), one may use fast algorithms, such as the ones given in Groisser (2004) [133] and also suggested independently by Pennec et al. (2006) [281].

8.4 Two-Sample Tests for Intrinsic Means on Homogeneous Hadamard Manifolds

Assume that the Hadamard homogeneous space $(\mathcal{M}, d_g)$ admits a simply transitive group of isometries $\mathcal{G}$. Under theis assumption, a two sample problem for intrinsic means on $\mathcal{M}$ can be transferred to a one sample problem on the Lie group $\mathcal{G}$, as follows. Note that the left action $\alpha : \mathcal{G} \times \mathcal{M} \to \mathcal{M}$ is simply transitive, which means that in the isotropy group $\mathcal{G}_x$, there is a trivial subgroup consisting in the identity element $1_{\mathcal{G}}$. More generally, given a fixed object $x \in \mathcal{M}$, the mapping $\alpha^x : \mathcal{G} \to \mathcal{M}$ is bijective. Therefore, the two sample mean hypothesis test for linear data $H_0 : \mu_1 = \mu_2 + \delta$ vs. $H_1 : \mu_1 \neq \mu_2 + \delta$ can be formulated in our more general setting for intrinsic means as follows:

$$\begin{aligned}(1)\ H_0 &: \mu_{1,g} = \alpha(\delta, \mu_{2,g}) \\ &\text{versus} \\ H_1 &: \mu_{1,g} \neq \alpha(\delta, \mu_{2,g}).\end{aligned} \tag{8.18}$$

(8.18) can be translated to a hypothesis testing problem on the Lie group $\mathcal{G}$, as follows:

$$\begin{aligned}(1)\ H_0 &: (\alpha^{\mu_{1,g}})^{-1}(\mu_{2,g}) = \delta, \\ &\text{versus} \\ H_1 &: (\alpha^{\mu_{1,g}})^{-1}(\mu_{2,g}) \neq \delta\end{aligned} \tag{8.19}$$

Let $H : \mathcal{M}^2 \to \mathcal{G}$, defined by

$$H(x_1, x_2) = (\alpha^{x_1})^{-1}(x_2). \tag{8.20}$$

THEOREM 8.4.1. *Assume $X_{a,j_a}, j_a = 1, \dots, n_a$ are identically independent distributed random objects (i.i.d.r.o.'s) from the independent probability measures $Q_a, a = 1,2$ with finite intrinsic moments of order $s, s \leq 4$ on the m dimensional Hadamard manifold $\mathcal{M}$ on which the Lie group $\mathcal{G}$ acts simply transitively. Let $n = n_1 + n_2$ and assume $\lim_{n\to\infty} \frac{n_1}{n} \to p \in (0,1)$. Let $\exp : \mathfrak{g} \to \mathcal{G}$ and L_δ be, respectively, the Lie group exponential map and the left translation by $\delta \in \mathcal{G}$. Then under H_0,*
i. The sequence of random vectors

$$\sqrt{n}((L_\delta \circ \exp)^{-1}(H(\bar{X}_{n_1,g}, \bar{X}_{n_2,g})) \tag{8.21}$$

converges weakly to $\mathcal{N}_m(0_m, \Sigma_g)$ for some covariance matrix Σ_g that depends linearly on the intrinsic covariance matrices $\Sigma_{a,g}$ of $Q_a, a = 1,2$.
ii. If (i.) holds and Σ_g is positive definite, then the sequence

$$n((L_\delta \circ \exp)^{-1}(H(\bar{X}_{n_1,g}, \bar{X}_{n_2,g}))^T \Sigma_g^{-1}((L_\delta \circ \exp)^{-1}(H(\bar{X}_{n_1,g}, \bar{X}_{n_2,g})) \tag{8.22}$$

converges weakly to the χ^2_m distribution.

By the multivariable inverse function theorem, the mapping $H : \mathcal{M} \times \mathcal{M} \to \mathcal{H}$ is continuous. Given that, according to Bhattacharya and Patrangenaru (2003) [42], for $a = 1,2$, the Cartan mean $\bar{X}_{n_a,g}$ is a consistent estimator of $\mu_{a,g}$, for $a = 1,2$. By the continuity theorem (Billingsley (1995) [48], p.334), a consistent estimator for $(\alpha^{\mu_{1,g}})^{-1}(\mu_{2,g})$ is $H(\bar{X}_{n_1,g}, \bar{X}_{n_2,g})$. From Bhattacharya and Patrangenaru (2005) [43], under the null hypothesis, for $a = 1,2$, the asymptotic distribution of $\sqrt{n_a}Exp^{-1}_{\mu_{a,g}}(\bar{X}_{n_1,g})$ is multivariate normal $\mathcal{N}_m(0_m, \Sigma_{a,g})$, where $\Sigma_{a,g}$ is the intrinsic covariance matrix of Q_a (the *intrinsic covariance matrix* is the covariance matrix in the CLT for Cartan means in Bhattacharya and Patrangenaru (2005) [43]). Since $\lim_{n\to\infty} \frac{n_1}{n} \to p \in (0,1)$, from Cramér's delta method it follows that asymptotically $\sqrt{n}((L_\delta \circ \exp)^{-1}(H(\bar{X}_{n_1,g}, \bar{X}_{n_2,g}))$ has $\mathcal{N}_m(0_m, \Sigma_g)$, where Σ_g depends on the intrinsic covariance matrices of the two populations $X_{a,1}, a = 1,2$ as described in Theorem 8.4.1(i). Theorem 8.4.1(ii). is an immediate consequence of part (i) and is a weak continuity argument (Billingsley (1995) [48], p.334).

COROLLARY 8.4.1. *For $a = 1,2$, assume $x_{a,j_a}, j_a = 1,\dots,n_a$, are random samples from independent populations on the m dimensional Hadamard manifold $\mathcal{M}$ on which the Lie group $\mathcal{G}$ acts simply transitively. Let $n = n_1 + n_2$, and assume $\lim_{n\to\infty} \frac{n_1}{n} \to p \in (0,1)$. Assume Σ_g is positive definite and $\hat{\Sigma}_g$ is a consistent estimator for Σ_g. The asymptotic p-value for the hypothesis testing problem* (8.18) *is given by $p = P(T \geq T_\delta^2)$ where*

$$T_\delta^2 = n((L_\delta \circ \exp)^{-1}(H(\bar{x}_{n_1,g}, \bar{x}_{n_2,g}))^T(\hat{\Sigma}_g)^{-1}((L_\delta \circ \exp)^{-1}(H(\bar{x}_{n_1,g}, \bar{x}_{n_2,g})), \tag{8.23}$$

and T has a χ^2_m distribution.

Often in practice, the sample sizes are too small to obtain reliable statistical results using parametric models. If the distributions are unknown and the samples are small, an alternative approach is for one to use Efron's nonparametric bootstrap (see Efron (1979) [96]). If $n_a \leq m, \forall a = 1,2$, the estimator $\hat{\Sigma}_g$ in Corollary 8.4.1 does not have an inverse, and pivotal nonparametric bootstrap methodology can not be applied. In this case, one may use nonpivotal bootstrap for the two-sample problem H_0, following Bhattacharya and Ghosh (1978) [38], Babu and Singh (1984) [9], Hall and Hart (1990) [138], Fisher et al. (1996) [116], Hall (1997) [137] and others.

THEOREM 8.4.2. *Under the hypotheses of Theorem 6.5.1i., assume in addition, that for $a = 1,2$ the distribution of* $\mathrm{Exp}^{-1}_{\mu_{g,a}} X_{a,1}$ *has an absolutely continuous component, and finite moments of sufficiently high order. Then the joint distribution of*

$$V = \sqrt{n}(\exp)^{-1}(H(\bar{X}_{n_1,g}, \bar{X}_{n_2,g}))$$

can be approximated by the bootstrap joint distribution of

$$V^* = \sqrt{n}(\exp)^{-1}(H(\bar{X}^*_{n_1,g}, \bar{X}^*_{n_2,g})) \tag{8.24}$$

with an error $O_p(n^{-\frac{1}{2}})$, *where, for* $a = 1,2$, $\bar{X}^*_{n_a,g}$ *are the Cartan means of the bootstrap resamples* X^*_{a,j_a}, $j_a = 1,\ldots,n_a$, *given* X_{a,j_a}, $j_a = 1,\ldots,n_a$.

Chapter 9

Analysis on Stiefel Manifolds

9.1 Stiefel Manifolds

This chapter presents the results from Hendriks and Landsman (1998) [154]. Let $Mat_{p,r}$ be the vector space of real $p \times r, r \leq p$ matrices, with the inner product $(A,B) = TrA^TB$, so that $\|A-B\|^2 = Tr(A-B)^T(A-B)$. This corresponds to the Euclidean inner product with respect to the identification of $Mat_{p,r}$ with $\mathbb{R}^{pr}$ by putting the matrix entries in one column of size pr. Let $Sym(r)$ be the vector space of symmetric $r \times r$ matrices, and Sym_r^+ the subset of semi positive-definite symmetric $r \times r$ matrices. Let E_r denote the $r \times r$ unit matrix. Then $\mathcal{M} = V_{p,r} \subset Mat_{p,r}$ denotes the Stiefel manifold $\mathcal{M} = V_{p,r} = \{V \in Mat_{p,r}, V^TV = E_r \in Sym(r)\}$ whose dimension is $m = pr - \frac{r(r+1)}{2}$. $\mathcal{M}$ can be considered as a compact submanifold of the vector space $Mat_{p,r}$ given by the equation

$$f(X) = X^TX = E_r, \tag{9.1}$$

where f is considered as a mapping between vector spaces with inner product $f : Mat_{p,r} \to Sym(r)$. Therefore, a point in the Stiefel manifold is a $p \times r$ matrix with orthogonal columns of Euclidean norm 1. As a particular case $V_{k,1}$ corresponds to the sphere S^{k-1}, and $V_{p,p}$ equals the orthogonal group $O(p)$ of $p \times p$-matrices X such that $X^TX = E_p$.

Recall that the Stiefel manifold $\mathcal{M} = V_{p,r}$ is a homogeneous space with respect to the Lie group $O(p)$ where the action is defined as the restriction of the action by matrix multiplication

$$O(p) \times Mat_{p,r} \ni (g,X) \mapsto gX \in Mat_{p,r}. \tag{9.2}$$

As a matter of fact, $\mathcal{M}$ admits the larger symmetry group $O(p) \times O(r)$, where the action is defined by restriction of the action

$$(O(p)\times O(r))\times Mat_{p,r}\ni(g,h,X)\mapsto gXh^T\in Mat_{p,r}. \tag{9.3}$$

It is clear that these symmetries are in fact isometries. Namely, it can be shown that $\|g(x-y)h^T\|^2=\|(x-y)\|^2$. If $U\in Mat_{p,r}$, then the elements X of $\mathcal{M}$ for which $L_U(X)=\|U-X\|^2=(U-X,U-X)$ is minimal are characterized by the condition that $U=Xc$, where $c\in Sym_r^+$ is the (semi) positive-definite symmetric square root of U^TU. X is uniquely defined if c is non singular, thus if U is of rank r. In that case, X is a non degenerate critical point of the function $L_U:\mathcal{M}\to\mathbb{R}$. If U is of rank less than r, uniqueness of X fails as there is a 1-1 correspondence between isometric (injective) mappings Y: Ker U →(Im $U)^{\perp}$ and solutions V to the equation $U=Vc$, given by $Y=V|_{\mathrm{ker}U}$. As such, the cut-locus $\mathcal{C}$ consists exactly of the $p\times r$-matrices U that are not of maximal rank, r. The nearest-point mapping $\Phi:Mat_{p,r}\setminus\mathcal{C}\to\mathcal{M}$ will be given by

$$\Phi(U)=U(U^TU)^{-1/2},U\notin\mathcal{C} \tag{9.4}$$

Notice that the linear mapping $Df|_X:Mat_{p,r}\to Sym(r)$ is given by $Df|_X(H)=H^TX+X^TH$ and that its transpose (or rather adjoint with respect to the inner products) $(Df|_X)^T:Sym(r)\to Mat_{p,r}$ is given by $(Df|_X)^T(c)=2Xc$. Thus, for $\mu\in\mathcal{M}$, we have $Df|_\mu(Df|_\mu)^T(c)=4c$.

The orthogonal projection on the tangent space $T_\mu(\mathcal{M})$ will be given by

$$tan_\mu(U)=U-(Df|_\mu)^T(Df|_\mu(Df|_\mu)^T)^{-1}Df|_\mu(U)=U-\frac{1}{2}\mu[\mu^TU+U^T\mu].$$

For $\mu\in\mathcal{M}$, a normal vector $n_\mu\in N_\mu(\mathcal{M})\subset Mat_{p,r}$ must lie in the range of $(Df|_\mu)^T$, so that $n_\mu=(Df|_\mu)^T(s)=2\mu s$ for some $s\in Sym(r)$. Thus, $2s=\mu^Tn_\mu$ and μ^Tn_μ is symmetric and $n_\mu=\mu[\mu^Tn_\mu]$. Since $T_\mu(\mathcal{M})$ is precisely the orthogonal complement, it consists of the matrices $W\in Mat_{p,r}$ for which μ^TW is skew-symmetric. To calculate the Weingarten mapping at the point μ in the normal direction n_μ, notice that n_μ is extended to a field of normal vectors on $\mathcal{M}$ by $n_\alpha=\alpha[\mu^Tn_\mu],\alpha\in\mathcal{M}$. And the Weingarten mapping is defined as the mapping $A_{n_\mu}:T_\mu(\mathcal{M})T\to T_\mu(\mathcal{M})T$ with $A_{n_\mu}(W)=-tan_\mu(D_W(n))=-tan_\mu(W\mu^Tn_\mu)$. Therefore for $W\in T_\mu(\mathcal{M})$,

$$\begin{aligned}A_{n_\mu}(W)&=-W\mu^Tn_\mu+\tfrac{1}{2}\mu[\mu^TW\mu^Tn_\mu+n_\mu^T\mu W^T\mu]\\&=\tfrac{1}{2}[E_p-\mu\mu^T]Wn_\mu^T\mu-\tfrac{1}{2}[Wn_\mu^T-n_\mu W^T]\mu\\&=-W\mu^Tn_\mu-\tfrac{1}{2}\mu W^Tn_\mu+\tfrac{1}{2}n_\mu W^T\mu\end{aligned} \tag{9.5}$$

REMARK 9.1.1. *Let $\mathcal{M}=O(p)$ be the set of orthogonal matrices. It coincides with the Stiefel manifold $V_{p,p}$. Then $\mu\mu^T=E_p$ and An_μ may be simplified* *slightly (cf. Hendriks and Landsman (1996a) [152], Example ii)*

$$A_{n_\mu}(W) = -\frac{1}{2}[W_n^T - n_\mu W^T]\mu = -\frac{1}{2}W\mu^T n_\mu - \frac{1}{2}n_\mu \mu^T W, W \in T_\mu(\mathcal{M}).$$

REMARK 9.1.2. *Let $\mathcal{M} = S^{k-1}$ be the unit sphere in $\mathbb{R}^k$. It is a special case of Stiefel manifold $V_{k,1}$. In this case $\mathcal{C} = \{0\}$, $U = \mathbb{R}^k \setminus \{0\}$ and $\Phi(a) = a/\|a\|$ for $a \in U$. Let $a \in U$ be the population mean then $\mu = a/\|a\|$, $\mu_n = \overline{X}_n/\|\overline{X}_n\|$, $tan_{\mu_n} = E_k - \mu_n\mu_n^T$, and $G_n = \|\overline{X}_n\| tan_{\mu_n} + \mu_n\mu_n^T$. See Hendriks and Landsman (1996a, 1996b) [152, 153].*

9.2 Special Orthogonal Groups $M = SO(p)$

The discussion differs of $SO(p)$ from the one on Stiefel manifolds only in the determination of the cut-locus and a necessary modification to the nearest-point mapping Φ.

Let $U \in Mat_{p,p}$, then the elements X of $\mathcal{M}$ for which $L_U(X) = \|U - X\|^2 = (U - X, U - X)$ is stationary are characterized by the condition that $U = Xc$ where $c \in Sym_p$ is a symmetric square root of $U^T U$. In order to have minimal distance, we need the square root with the largest trace. This means that if $det(U) > 0$, we can use the positive-definite root. If $det(U) < 0$, we must allow a negative root of the smallest eigenvalue of UTU, which will give rise to non uniqueness if the smallest eigenvalue has multiplicity greater than 1. In the same vein, if $det(U) = 0$ and the eigenvalue 0 has multiplicity greater than 1 then X is not unique. However, if 0 is a simple eigenvalue of $U^T U$, the condition that $det(X) = 1$ determines a single solution. Moreover one can prove that if $det(U) > 0$ or that if both $det(U) \leq 0$ and the lowest eigenvalue of $U^T U$ is simple, then the minimum X is non degenerate. Therefore, the cut-locus $\mathcal{C}$ is the set of $p \times p$ matrices U such that $det(U) \leq 0$ and the lowest eigenvalue of $U^T U$ is not simple.

9.3 Intrinsic Analysis on Spheres

For the d-dimensional unit sphere , $M = S^d = \{p \in \mathbb{R}^{d+1} : \|p\| = 1\}$, is the Stiefel manifold $V_{d=1,1}$. S^d is equipped with the Riemannian metric induced by the Euclidean metric on $\mathbb{R}^{d+1}$, the exponential map at a given point $p \in S^d$ is defined on the tangent space T_pM and is given, according to Bhattacharya and Patrangenaru (2005) [43], by

$$Exp_p(v) = \cos(\|v\|)p + \sin(\|v\|)\|v\|^{-1}v, \quad (v \in T_pS^d, v \neq 0). \tag{9.6}$$

If $x \in S^d, x \neq -p$, then there is a unique vector $u \in T_pM$ such that $x = Exp_pu$, and we will label this vector by $u = Log_px$. Since $T_pS^d = \{v \in \mathbb{R}^{d+1}, v \cdot p = 0\}$, it follows that

$$Log_p x = (1-(p\cdot x)^2)^{-\frac{1}{2}} arccos(p\cdot x)(x-(p\cdot x)p) \tag{9.7}$$

In particular, for $d = 2$, we consider the orthobasis $e_1(p), e_2(p) \in T_pS^2$, where $p = (p_1, p_2, p_3)^t \in S^2\backslash\{N,S\}(N = (0,0,1), S = (0,0,-1))$:

$$e_1(p) = ((p_1)^2+(p_2)^2)^{-\frac{1}{2}}(-p_2, p_1, 0)^t \tag{9.8}$$
$$e_2(p) = (-((p_1)^2+(p_2)^2)^{-\frac{1}{2}} p_1p_3, -(x^2+y^2)^{-\frac{1}{2}} p_2p_3, ((p_1)^2+(p_2)^2)^{\frac{1}{2}})^t$$

The *logarithmic coordinates* of the point $x = (x_1, x_2, x_3)^T$ are given in this case by

$$\begin{aligned} u^1(p) &= e_1(p)\cdot Log_p x \\ u^2(p) &= e_2(p)\cdot Log_p x \end{aligned} \tag{9.9}$$

For computations, one may use $a \cdot b = a^t b$.

In the case of the sphere S^2 from Example 12.1, it follows that if we consider an arbitrary data point $u = (u^1, u^2)$, and a second point $\theta = Log_p\lambda = (\theta^1, \theta^2)$, and evaluate the matrix of second order partial derivatives w.r.t. θ^1, θ^2 of

$$G(u,\theta) = arccos^2(cos||u|| + \frac{sin||u||}{||u||}(u^1\theta^1 + u^2\theta^2) - \frac{1}{2}||\theta||^2 cos||u||), \tag{9.10}$$

then

$$\frac{\partial^2 G}{\partial\theta^r\partial\theta^s}(u;0) = \frac{4u^ru^s}{||u||^2}(1-\frac{||u||}{tan||u||}) + \frac{4\delta_{rs}||u||}{tan||u||} \tag{9.11}$$

where δ_{rs} is the symbol of Kronecker and $||u||^2 = (u^1)^2 + (u^2)^2$. The matrix $\hat{\Lambda} = (\lambda_{rr'})_{r,r'=1,2}$ has the entries

$$\lambda_{rr'} = \frac{1}{n}\sum_{i=1}^{n}\frac{\partial^2 G}{\partial\theta^r\partial\theta^{r'}}(u_i;0). \tag{9.12}$$

Assume $\hat{C}$ is the sample covariance matrix of $u_j, j = 1,...,n$; A large sample confidence region for the intrinsic mean is given by Corollary 5.2.with $\mu_n = 0$.

In the case of a hypersphere in $\mathbb{R}^k$, $j(x) = x$ and $P_j = P_M$. We evaluate the statistic $||\overline{Z}_{j,n}||^2 = n||\Sigma_{j,E}{}^{-\frac{1}{2}} tan(P_M(\overline{X}) - P_M(\mu)))||^2$. The projection map is $P_M(x) = x/||x||$. P_M has the following property: if $v = cx$ then $d_xP_M(v) = 0$; on the other hand if the restriction of d_xP_M to the orthocomplement of $\mathbb{R}x$ is a conformal map, that is if $v\cdot x = 0$, then $d_xP_M(v) = ||x||^{-1}v$. In particular, if we select the coordinate system such that $x = ||x||e_k$, then one may take $e_a(P_M(x)) = e_a$, and we get

$$d_xP_M(e_b)\cdot e_a(P_M(x)) = ||x||^{-1}\delta_{ab}, \quad \forall a,b = 1,...,k-1, d_xP_M(e_k) = 0.$$

Since $e_k(P_M(\mu))$ points in the direction of μ, $d_\mu P_M(e_b) \cdot \mu = 0 \ \forall\, b = 1,...,k-1$, and we get

$$\Sigma_E = \|\mu\|^{-2} E([\mathbf{X} \cdot e_a(\mu/\|\mu\|)]_{a=1,...,k-1}[\mathbf{X} \cdot e_a(\mu/\|\mu\|)]^t_{a=1,...,k-1}) \quad (9.13)$$

which is the matrix G in formula (A.1) in Fisher et al (1996) [116].

EXAMPLE 9.3.1. *In the case of the sphere, one may approximate the asymptotic distribution in theorem 5.3.1 using the so the bootstrap distribution in Corollary 5.3.1 as follows: Consider a random sample with repetition* $v_1^*, ..., v_n^*$ *from a sample of unit normals* $v_1, ..., v_n$ *, and project the mean* $\overline{v}^*$ *of* $v_1^*, ..., v_n^*$ *on the tangent space to the sphere at* $\overline{v}$ *. If the number of such resamples is large (in the order of thousands), then the distribution of projections of the points* $\overline{v}^*$ *on the tangent plane at , is approximately the same as the distribution of* $H(\overline{v})$ *in* $T_\mu S^2$*. The probability error is of order* $0_P(n^{-1/2})$

Hendriks and Landsman (1998)[154] consider simulations of observations on the Stiefel manifold $V_{3,2}$ from the distributions with densities $f_{\mu,\kappa}$, with respect to the uniform distribution on $V_{3,2}$, given by

$$\begin{aligned} f_{\mu,\kappa}(X) &= C_\kappa exp(-\kappa Tr(X-\mu)^T(X-\mu)) = \\ &= C'_\kappa exp(2\kappa\mu^T X), \mu \in V_{3,2}, \kappa > 0, \end{aligned} \quad (9.14)$$

C_κ and C'_κ being a normalizing constant. The parameters μ and κ are the mean location and the concentration parameter, respectively. This model is a submodel of the von Mises–Fisher model (Khatri and Mardia (1977) [186]). They were interested in the performance of the χ^2-type procedures in the one-sample and two-sample cases, and showed that in spite of the seeming complexity in the formulation of the procedures everything can be implemented easily and performs well.

The method for generating uniform variates on Stiefel manifolds is to characterize it as the distribution which is invariant under the left action by $O(p)$. This property is shared by the multivariate standard normal distributions on $Mat_{3,2}$ and preserved by the mapping Φ (cf. Chikuse (1990) [70], James (1954) [169], Watson (1983b) [333]).

REMARK 9.3.1. *We note that the asymptotic results in this chapter seem to indicate that a more appropriate model for distributions on spheres* $S^d = V_{d+1,1}$*, that takes into account an arbitrary extrinsic covariance matrix could be modeled by distributions with densities* $f_{\mu,\Sigma}$*, with respect to the uniform distribution, given by*

$$f_{\mu,\Sigma}(X) = C_\Sigma exp(-(X-\mu)^T \Sigma^{-1}(X-\mu)), \mu \in V_{d+1,1}, \Sigma > 0, \quad (9.15)$$

Chapter 10

Asymptotic Distributions on Projective Spaces

Projective shape analysis is one of the areas where data analysis on a real projective shape plays a crucial role. In Chapter 3, we showed that real projective spaces are sample spaces for axial data, and complex projective space are sample spaces for direct similarity planar shape data (see Kendall (1984) [177]).

10.1 Asymptotic Distributions of the Sample Total Intrinsic Variance for Distributions with Small Support on Real Projective Spaces

Projective shape analysis is concerned with the properties of configurations of points, as they are seen in a central projection by external observers, in what can be considered a simplified analysis of vision in the absence of occlusions. A simple model for *monocular vision* is given by real projective geometry, the geometry of the projective plane $\mathbb{R}P^2$.

We recall from Chapter 3 that $\mathbb{R}P^m$ is the set of vector lines in $\mathbb{R}^{m+1}$, that is if $X = (X^1,...,X^{m+1}) \in \mathbb{R}^{m+1}\backslash 0$, the projective point $[X] = [X^1 : ... : X^{m+1}] \in \mathbb{R}P^m$ is

$$[X] = \{\lambda X, \lambda \in \mathbb{R}\backslash 0\} \tag{10.1}$$

The coordinates $(X^1,...,X^{m+1})$ of $[X]$ defined in (10.1) are called *homogeneous coordinates* and are determined up to a multiplicative constant, opposed to *affine coordinates* $(x^1,...,x^m)$ of $[X]$, obtained by "dehomogenizing" one variable, which are unique. The affine coordinates of $[X]=[X^1:\cdots:X^{m+1}]\in\mathbb{R}P^m$ w.r.t. the last variable are

$$x^j=\frac{X^j}{X^{m+1}} \tag{10.2}$$

A linear variety $\mathcal{V}$ of dimension d is given by

$$\mathcal{V}=\{[X],X\in V\backslash\mathbf{0}\}, \tag{10.3}$$

where V is a $(d+1)$- dimensional subspace of $\mathbb{R}^{m+1}$.

In projective shape analysis, we associate a random object to a (finite) random configuration of observed points. Recall from Section 3.5, that $m+2$ labeled points in $\mathbb{R}P^m$ form a *projective frame* if any subset of $m+1$ of these points are not included in a linear variety of dimension $m-1$. For $k>m+1$, we denote by $G(k,m)$ the set of ordered systems of k projective points $(p_1,\ldots,p_k)$, such that $(p_1,...,p_{m+1})$ are in general position. The group of projective transformations acts on $G(k,m)$ as in (3.142). The orbit of the k-ad $(p_1,...,p_k)$ via this action is a *projective shape* of this k-ad.

In Section 3.5, we showed that $P\Sigma_m^k$, the space of projective shapes of such k-ads, called the *projective shape space*, is a manifold diffeomorphic to $(\mathbb{R}P^m)^{k-m-2}$. Accordingly, inference for projective shapes is based on a statistical analysis on this product manifold and can be regarded as inference for multivariate axial analysis.

We defined the *projective coordinate* of an observation $(p_1,...,p_k)$ in general position, with respect to a projective frame. We associate with the j-th projective point $p_j=[x_j{}^t:1]$ in $\mathbb{R}P^m$, the $(m+1)\times(m+1)$ matrix $U_m=[\tilde{x}_1,...,\tilde{x}_{m+1}]$.

Recall that the projective coordinate of the configuration of Euclidean points $(x_1,\ldots,x_k)$ is given by

$$(p_j^\pi, j=m+3,...,k)$$

where

$$p_j^\pi=[\frac{z^1(x_j)}{\|z(x_j)\|}:\ldots:\frac{z^{m+1}(x_j)}{\|z(x_j)\|}],$$

$$z^j(x)=\frac{v^j(x)}{v^j(x_{m+2})}, for j=1,...,m+1$$

and

$$(v^1(x),...,v^{m+1}(x))=U_m^{-1}\tilde{x}.$$

The projective coordinates p_j^π and the *projective invariants* $(\iota_{sj}),s=1,\ldots,m,j=m+3,...,k$ of a projective shape due to Goodall and Mardia (1999) [127] determine each other; the relations between the two representations of a projective shape are :

$$p_j^\pi=[\iota_{1j}:\ldots:\iota_{mj}:1], \tag{10.4}$$

showing that these invariants are locally the affine coordinates of $(p_j^\pi, j = m+3,...,k)$ whenever the last homogeneous coordinate of each of the $p_j^{\pi'} s$ is not zero.

In planar image analysis data, the landmarks are often pictured sitting on an almost flat surface. In such a situation or when one takes into account only the small errors in data collection, there is little variability, which makes it appropriate to regard such observations as coming from a probability distributions of small flat support on $(\mathbb{R}P^2)^{k-4}$

In this section, we therefore confine ourselves only to probability measures Q with a *small flat support* on $P\Sigma_2^k$. Since any flat metric is locally isometric to an Euclidean metric, we consider a metric on $P\Sigma_2^k$ which is Euclidean in a neighborhood of the support of our distribution. In the invariant coordinates on the support of Q, and in particular if x_1,x_2 are two observations from Q with

$$x_r = ([\iota_{r,1j} : \iota_{r,2j} : 1])_{j=4,\dots,k}, \tag{10.5}$$

we may consider the distance

$$d^2(x_1,x_2) = \sum_{j=4}^{k}\sum_{s=1}^{2}(\iota_{1,sj} - \iota_{2,sj})^2. \tag{10.6}$$

In this case, if we use the invariant representation for random configurations of k landmarks, the intrinsic sample means and total sample variance are ordinary sample means and total sample variances , for which classical large sample and nonparametric bootstrap theory can be used. This approach requires a less computational complexity than for extrinsic means .

Assume $X_1,...,X_n$ are i.i.d.r.v.'s with a common distribution Q of small flat support on $P\Sigma_m^k$. For each $r, r = 1,...,n$, assume X_r has the invariant representation

$$X_r = (p_{r,j}^\pi), j = m+3,...,k, \tag{10.7}$$

where

$$p_{r,j}^\pi = [\iota_{r,1j} : \iota_{r,2j} : 1]. \tag{10.8}$$

Then the intrinsic sample mean $\overline{X}_{n,I}$ is given in invariant representation by

$$\overline{X}_{n,g} = (p_j^\pi), j = 4,\dots,k, \tag{10.9}$$

where

$$p_j^\pi = [\overline{\iota_{.,1j}} : \overline{\iota_{.,2j}} : 1]. \tag{10.10}$$

Patrangenaru (1999) [267] showed that if an external observer travels in 3D and records images of a configuration of points with with an ideal pinhole camera, then, with respect to a projective frame of presumably marked coplanar points, the recorded configuration is planar if and only if the projective coordinates are independent of the position of the observer. In the basic case, we assume $m = 2$ and $k = m+3 = 5$. The corresponding hypothesis testing problem is

$$H_0 : t\Sigma_g = 0 \ vs. \ H_1 : t\Sigma_g \neq 0, \tag{10.11}$$

where $t\Sigma_g$ is the *intrinsic total variance* of a probability measure Q with small flat support on $P\Sigma_2^5$.

Note that $P\Sigma_2^5$ is diffeomorphic to $\mathbb{R}P^2$ and, in invariant representation, the random variable associated with Q is

$$X = [\iota_{,1} : \iota_{,2} : 1]. \tag{10.12}$$

In this case, the intrinsic mean, and intrinsic total variance of X are respectively the mean and total variance of the random vector $\iota = (\iota_{,1}, \iota_{,2})^T$ and if $X_1, \ldots, X_n$ are i.i.d.r.v.'s the common distribution Q, and we similarly associate with X_r the random vector $\iota_r = (\iota_{r,1}, \iota_{r,2})^T$, we have

$$\overline{X}_{n,I} = \overline{\iota}_n = (\overline{\iota}_{,1} \overline{\iota}_{,2}) \tag{10.13}$$

and

$$t\hat{\Sigma}_{I,n} = n^{-1} \sum_{j=1}^{2} \sum_{r=1}^{n} (\iota_{r,j} - \overline{\iota_{,j}})^2. \tag{10.14}$$

The value of the $\hat{\Sigma}(V)_4$ given in Theorem 5.2.5, is in this case given by

$$\hat{\Sigma}(V)_4 = n^{-1} \sum_{i,j=1}^{m} \sum_{r=1}^{n} (\iota_{r,i} - \overline{\iota_{,i}})^2 (\iota_{r,j} - \overline{\iota_{,j}})^2. \tag{10.15}$$

As a consequence of Corollary 5.2.5, we obtain the following result.

THEOREM 10.1.1. *Assume $x_1, \ldots, x_n$ is a large random sample from a probability distribution with small flat support on $P\Sigma_2^5$. Then we reject the null hypothesis in equation 10.11 at level α if*

$$t\hat{\Sigma}_{I,n} > (S(\hat{\Sigma}(V)_4)^{1/2}) z_{\frac{\alpha}{2}}. \tag{10.16}$$

Indeed, from Corollary 5.2.5, it follows that a $100(1-\alpha)\%$ symmetric large sample confidence interval for $t\Sigma_g$ is given by

$$(t\hat{\Sigma}_{i,n} - (S(\hat{\Sigma}(V)_4)^{1/2}) z_{\frac{\alpha}{2}}, t\hat{\Sigma}_{i,n} + (S(\hat{\Sigma}(V)_4)^{1/2}) z_{\frac{\alpha}{2}}) \tag{10.17}$$

and the condition in the theorem says that 0 is not in this confidence region.

10.2 Asymptotic Distributions of VW-Means on Real Projective Spaces

In this section, we focus on the asymptotic distribution of sample means in axial data analysis and planar shape analysis.

The axial space is the $(N-1)$ dimensional real projective space $M = \mathbb{R}P^{N-1}$, which can be identified with the sphere $S^{N-1} = \{x \in \mathbb{R}^N \,|\, \|x\|^2 = 1\}$ with antipodal points identified (see e.g., Mardia and Jupp (2000) [230]). If $[x] = \{x, -x\} \in \mathbb{R}P^{N-1}$, $\|x\| = 1$, the tangent space at $[x]$ can be described as

$$T_{[x]}\mathbb{R}P^{N-1} = \{([x], v), v \in \mathbb{R}^N | v^t x = 0\}. \qquad (10.18)$$

We consider here the general situation when the distribution on $\mathbb{R}P^{N-1}$ may not be concentrated. Note that for N odd, $\mathbb{R}P^{N-1}$ cannot be embedded in $\mathbb{R}^N$, since for any embedding of $\mathbb{R}P^{N-1}$ in $\mathbb{R}^k$ with N odd, the first Stiefel–Whitney class of the normal bundle is not zero (Milnor and Stasheff (1974) [243], pp. 51).

The *Veronese–Whitney (VW)* embedding is defined for arbitrary N by the formula

$$j([x]) = xx^t, \|x\| = 1. \qquad (10.19)$$

The embedding j maps $\mathbb{R}P^{N-1}$ into a $\frac{1}{2}N(N+1) - 1$ dimensional Euclidean hypersphere in the space $S(N, \mathbb{R})$ of real $N \times N$ symmetric matrices, where the Euclidean distance d_0 between two symmetric matrices is

$$d_0(A, B) = Tr((A - B)^2).$$

This embedding, which was used by Watson (1983) [334], is preferred over other embeddings in Euclidean spaces because it is *equivariant* (see Kent (1992) [181]). This means that the special orthogonal group $SO(N)$ of orthogonal matrices with determinant +1 acts as a group of isometries on $\mathbb{R}P^{N-1}$ with the metric of constant positive curvature. It also acts on the left on $S_+(N, \mathbb{R})$, the set of nonnegative definite symmetric matrices with real coefficients, by $T \cdot A = TAT^t$. Also, $j(T \cdot [x]) = T \cdot j([x]) \forall T \in SO(N), \forall [x] \in \mathbb{R}P^{N-1}$.

Note that $j(\mathbb{R}P^{N-1})$ is the set of all nonnegative definite matrices in $S(N, \mathbb{R})$ of rank one and trace one. The following result (see Patrangenaru (1998) [266]), based on the geometry of real quadratic forms, is left to the reader as a useful exercise.

PROPOSITION 10.2.1. *(Patrangenaru (1998)) (a) The set $\mathcal{F}$ of the VW focal points of $j(\mathbb{R}P^{N-1})$ in $S_+(N, \mathbb{R})$ is the set of matrices in $S_+(N, \mathbb{R})$ whose largest eigenvalues are of multiplicity at least* 2. *(b) If j is the VW embedding in* (10.19)*, the projection $P_j : S_+(N, \mathbb{R}) \backslash \mathcal{F} \to j(\mathbb{R}P^{N-1})$ assigns to each nonnegative definite symmetric matrix A with a largest eigenvalue of multiplicity one, the matrix $j([m])$, where $m(\|m\| = 1)$ is an eigenvector of A corresponding to its largest eigenvalue.*

The following result of Prentice (1984) [282] is also needed in the sequel.

PROPOSITION 10.2.2. *(Prentice (1984) [282]) Assume* $[X_r]$, $\|X_r\| = 1$, $r = 1,...,n$ *is a random sample from a j-nonfocal, probability measure* Q *on* $\mathbb{R}P^{N-1}$. *Then the sample (VW-)extrinsic covariance matrix* $S_{j,E}$ *is given by*

$$S_{j,E\,ab} = n^{-1}(\eta_N - \eta_a)^{-1}(\eta_N - \eta_b)^{-1}\sum_r (m_a \cdot X_r)(m_b \cdot X_r)(m \cdot X_r)^2, \quad (10.20)$$

where $\eta_a, a = 1,...,N$, *are eigenvalues of* $K := n^{-1}\sum_{r=1}^{n} X_r X_r^t$ *in increasing order and* $m_a, a = 1,...,N$, *are corresponding linearly independent unit eigenvectors.*

Here we give a proof of formula (10.20), based on the equivariance of j, to prepare the reader for a similar but more complicated formula of the analogous estimator given later for $\mathbb{C}P^{k-2}$.

Since the map j is equivariant, w.l.o.g. one may assume that $j(\overline{X}_E) = P_j(\overline{j(X)})$ is a diagonal matrix, $\overline{X}_E = [m_N] = [e_N]$ and the other unit eigenvectors of $\overline{j(X)} = D$ are $m_a = e_a, \forall a = 1,...,N-1$. We evaluate $d_D P_j$. Based on this description of $T_{[x]}\mathbb{R}P^{N-1}$, one can select in $T_{P_j(D)}j(\mathbb{R}P^{N-1})$ the orthonormal frame $e_a(P_j(D)) = d_{[e_N]}j(e_a)$. Note that $S(N,\mathbb{R})$ has the orthobasis $F_a^b, b \leq a$ where, for $a < b$, the matrix F_a^b has all entries zero except for those in the positions $(a,b),(b,a)$ that are equal to $2^{-\frac{1}{2}}$; also $F_a^a = j([e_a])$. A straightforward computation shows that if $\eta_a, a = 1,...,N$, are the eigenvalues of D in their increasing order, then $d_D P_j(F_a^b) = 0, \forall b \leq a < N$ and $d_D P_j(F_a^N) = (\eta_N - \eta_a)^{-1} e_a(P_j(D))$; from this equation it follows that, if $\overline{j(X)}$ is a diagonal matrix D then the entry $S_{j,E\,ab}$ is given by

$$S_{j,E\,ab} = n^{-1}(\eta_N - \eta_a)^{-1}(\eta_N - \eta_b)^{-1}\sum_r X_r^a X_r^b (X_r^N)^2. \quad (10.21)$$

Taking $\overline{j(X)}$ to be a diagonal matrix and $m_a = e_a$ formula (10.20) follows. Note that $\mu_{E,j} = [\nu_N]$, where $(\nu_a), a = 1,...,N$ are unit eigenvectors of $E(XX^t) = E(j(Q))$ corresponding to eigenvalues in their increasing order. Let $T([\nu]) = n\|S_{j,E}{}^{-\frac{1}{2}} tan(P_j(\overline{j(X)}) - P_j(E(j(Q))))\|^2$ be the statistic given by (5.50).

We can now derive the following theorem as a special case of Theorem 5.5.1 (a).

THEOREM 10.2.1. *Assume* j *is the Veronese–Whitney embedding of* $\mathbb{R}P^{N-1}$ *and* $\{[X_r], \|X_r\| = 1, r = 1,...,n\}$ *is a random sample from a j-nonfocal probability measure* Q *on* $\mathbb{R}P^{N-1}$ *that has a nondegenerate j-extrinsic variance. Then* $T([\nu])$ *is given by*

$$T([\nu]) = n\nu^t[(\nu_a)_{a=1,...,N-1}]S_{j,E}{}^{-1}[(\nu_a)_{a=1,...,N-1}]^t\nu, \quad (10.22)$$

and, asymptotically, $T([\nu])$ *has a* χ^2_{N-1} *distribution.*

Proof. Since j is an isometric embedding and the tangent space $T_{[\nu_N]}\mathbb{R}P^{N-1}$ has the orthobasis $\nu_1,...,\nu_{N-1}$, if we select the first elements of the adapted moving frame in Theorem 5.5.1 to be $e_a(P_j(\nu_{E,j}) = (d_{[\nu_N]}j)(\nu_a)$

then the $a-th$ tangential component of $P_j(\overline{j(X)}) - P_j(\nu)$ w.r.t. this basis of $T_{P_j(E(j(Q)))}j(\mathbb{R}P^{N-1})$ equals up to a sign the a-th component of $m - v_N$ w.r.t. the orthobasis $v_1, ..., v_{N-1}$ in $T_{[v_N]}\mathbb{R}R^{N-1}$, namely $v_a^t m$. The result follows now from Theorem 5.5.1(a), $\square$.

REMARK 10.2.1. *If we apply Theorem 5.5.1(b) to the embedding j, we obtain a similar theorem due to Fisher, Hall, Jing and Wood (1996)[116], where $T([\nu])$ is replaced by $T([m])$.*

10.3 Asymptotic Distributions of VW-Extrinsic Means of Complex Projective Spaces

Similar asymptotic results can be obtained for the large sample distribution of Procrustes means of planar shapes , as we discuss below. From Theorem 3.5.1 in Chapter 3, recall that the planar shape space $M = \Sigma_2^k$ of an ordered set of k points in $\mathbb{C}$, at least two of which are distinct, can be identified in different ways with the complex projective space $\mathbb{C}P^{k-2}$ (see, for example, Kendall (1984) [177], Bhattacharya and Patrangenaru (2003) [42]).

Here we regard $\mathbb{C}P^{k-2}$ as a set of equivalence classes $\mathbb{C}P^{k-2} = S^{2k-3}/S^1$, where S^{2k-3} is the space of complex vectors in $\mathbb{C}^{k-1}$ of norm 1, and the equivalence relation on S^{2k-3} is by multiplication with scalars in S^1 (complex numbers of modulus 1). A complex vector $z = (z^1, z^2, ..., z^{k-1})$ of norm 1 corresponding to a given configuration of k landmarks, with the identification described in Bhattacharya and Patrangenaru (2003) [42], can be displayed in the Euclidean plane (complex line) with the superscripts as labels. If, in addition, r is the largest superscript such that $z^r \neq 0$ then we may assume that $z^r > 0$. Using this representative of the projective point $[z]$ we obtain a unique graphical representation of $[z]$, which will be called the *spherical representation.*

The *Veronese–Whitney* (or simply *VW*) embedding of $\mathbb{C}P^{k-2}$ in the space of hermitian matrices $S(k-1, \mathbb{C})$ given in this case by $j([z]) = zz^*$ where, if z is considered as a column vector, z^* is the adjoint of z, i.e., the conjugate of the transpose of z. The Euclidean distance in the space of hermitian matrices $S(k-1, \mathbb{C})$ is $d_0^2(A,B) = Tr((A-B)(A-B)^*) = Tr((A-B)^2)$.

Kendall (1984) [177] showed that the Riemannian metric induced on $j(\mathbb{C}P^{k-2})$ by d_0 is a metric of constant holomorphic curvature. The associated Riemannian distance is known as the *Kendall distance* and the full group of isometries on $\mathbb{C}P^{k-2}$ with the Kendall distance is isomorphic to the special unitary group $SU(k-1)$ of all $(k-1) \times (k-1)$ complex matrices A with $A^*A = I$ and $\det(A) = 1$.

A random variable $X = [Z], \|Z\| = 1$, valued in $\mathbb{C}P^{k-2}$ is j-nonfocal if the highest eigenvalue of $E[ZZ^*]$ is simple, and then the extrinsic mean of X is $\mu_{j,E} = [\nu]$, where $\nu \in \mathbb{C}^{k-1}, \|\nu\| = 1$, is an eigenvector corresponding to this eigenvalue (see Bhattacharya and Patrangenaru (2003) [42]). The extrinsic sample mean $\overline{[z]}_{j,E}$ of a random sample $[z_r] = [(z_r^1, \ldots, z_r^{k-1})], \|z_r\| = 1, r =$

$1,\ldots,n$, from such a nonfocal distribution exists with probability converging to 1 as $n \to \infty$, and is the same as that given by

$$\overline{[z]}_{j,E} = [m], \tag{10.23}$$

where m is a highest unit eigenvector of

$$K := n^{-1}\sum_{r=1}^{n} z_r z_r^*. \tag{10.24}$$

This means that $\overline{[z]}_{j,E}$ is the full Procrustes estimate for parametric families such as Dryden–Mardia distributions or complex Bingham distributions for planar shapes (see Kent (1992 [181],(1994 [182])) . For this reason, $\mu_{j,E} = [m]$ will be called the *Procrustes* mean of Q.

PROPOSITION 10.3.1. *Assume $X_r = [Z_r], \|Z_r\| = 1, r = 1,...,n$, is a random sample from a j-nonfocal probability measure Q with a nondegenerate j-extrinsic covariance matrix on $\mathbb{C}P^{k-2}$. Then the j-extrinsic sample covariance matrix $S_{j,E}$ as a complex matrix has the entries*

$$S_{j,E\,ab} =$$
$$n^{-1}(\eta_{k-1}-\eta_a)^{-1}(\eta_{k-1}-\eta_b)^{-1}\sum_{r=1}^{n}(m_a\cdot Z_r)(m_b\cdot Z_r)^*|m_{k-1}\cdot Z_r|^2. \tag{10.25}$$

The proof is similar to that given for Proposition 10.2.2 and is based on the equivariance of the VW map j, w.r.t. the actions of $SU(k-1)$ on $\mathbb{C}P^{k-2}$ and on the set $S_+(k-1,\mathbb{C})$ of nonnegative semidefinite self adjoint $(k-1)$ by $(k-1)$ complex matrices (see Bhattacharya and Patrangenaru (2003) [42]). Without loss of generality we may assume that K in (10.24) is given by $K = \operatorname{diag}\{\eta_a\}_{a=1,\ldots,k-1}$ and the largest eigenvalue of K is a simple root of the characteristic polynomial over $\mathbb{C}$, with $m_{k-1} = e_{k-1}$ as a corresponding complex eigenvector of norm 1. The eigenvectors over $\mathbb{R}$ corresponding to the smaller eigenvalues are given by $m_a = e_a, m'_a = ie_a, a = 1,\ldots,k-2$, and yield an orthobasis for $T_{[m_{k-1}]}j(\mathbb{C}P^{k-2})$. For any $z \in S^{2k-1}$ which is orthogonal to m_{k-1} in $\mathbb{C}^{k-1}$ w.r.t. the real scalar product, we define the path $\gamma_z(t) = [\cos t m_{k-1} + \sin t z]$. Then $T_{P_j(K)}j(\mathbb{C}P^{k-2})$ is generated by the vectors tangent to such paths $\gamma_z(t)$, at $t = 0$. Such a vector, as a matrix in $S(k-1,\mathbb{C})$, has the form $zm_{k-1}^* + m_{k-1}z^*$. In particular, since the eigenvectors of K are orthogonal w.r.t. the complex scalar product, one may take $z = m_a$, $a = 1,\ldots,k-2$, or $z = im_a$, $a = 1,\ldots,k-2$, and thus get an orthobasis in $T_{P_j(K)}j(M)$. When we norm these vectors to have unit lengths we obtain the orthonormal frame

$$e_a(P_j(K)) = d_{[m_{k-1}]}j(m_a) = 2^{-\frac{1}{2}}(m_a m_{k-1}^* + m_{k-1}m_a^*),$$

$$e'_a(P_j(K)) = d_{[m_{k-1}]}j(im_a) = i2^{-\frac{1}{2}}(m_a m_{k-1}^* - m_{k-1}m_a^*).$$

Since the map j is equivariant we may assume that K is diagonal. In this case $m_a = e_a$, $e_a(P_j(K)) = 2^{-\frac{1}{2}}E_a^{k-1}$ and $e'_a(P_j(K)) = 2^{-\frac{1}{2}}F_a^{k-1}$ where E_a^b has all entries zero except for those in the positions (a,b) and (b,a) that are equal to 1, and F_a^b is a matrix with all entries zero except for those in the positions (a,b) and (b,a) that are equal to i, respectively $-i$. Just as in the real case, a straightforward computation shows that $d_K P_j(E_a^b) = d_K P_j(F_a^b) = 0, \forall a \leq b < k-1$ and

$$d_K P_j(E_a^{k-1}) = (\eta_{k-1} - \eta_a)^{-1} e_a(P_j(K)), \, d_K P_j(F_a^{k-1}) = (\eta_{k-1} - \eta_a)^{-1} e'_a(P_j(K)).$$

We evaluate the extrinsic sample covariance matrix $S_{j,E}$ given in (3.8) using the real scalar product in $S(k-1,\mathbb{C})$, namely, $U \cdot V = Re\,Tr(UV^*)$. Note that

$$d_K P_j(E_b^{k-1}) \cdot e_a(P_j(K)) = (\eta_{k-1} - \eta_a)^{-1}\delta_{ba}, \quad d_K P_j(E_b^{k-1}) \cdot e'_a(P_j(K)) = 0$$

and

$$d_K P_j(F_b^{k-1}) \cdot e'_a(P_j(K))^t = (\eta_{k-1} - \eta_a)^{-1}\delta_{ba}, \quad d_K P_j(F_b^{k-1}) \cdot e_a(P_j(K)) = 0.$$

Thus we may regard $S_{j,E}$ as a complex matrix noting that in this case we get

$$S_{j,E\,ab} = n^{-1}(\eta_{k-1} - \eta_a)^{-1}(\eta_{k-1} - \eta_b)^{-1}\sum_{r=1}^{n}(e_a \cdot Z_r)(e_b \cdot Z_r)^*|e_{k-1} \cdot Z_r|^2, \tag{10.26}$$

thus proving formula (10.25) when K is diagonal. The general case follows by equivariance.

We now consider the statistic

$$T(\overline{(X)}_E, \mu_E) = n\|S_{j,E}{}^{-\frac{1}{2}}tan(P_j(\overline{j(X)}) - P_j(\mu_E))\|^2$$

given in (5.49) in the present context of random variables valued in complex projective spaces to get

THEOREM 10.3.1. *Let $X_r = [Z_r]$, $\|Z_r\| = 1$, $r = 1,\dots,n$, be a random sample from a Veronese-nonfocal probability measure Q on $\mathbb{C}P^{k-2}$. Then the quantity (5.50) is given by :*

$$T([m],[\nu]) = n[(m \cdot \nu_a)_{a=1,\dots,k-2}]S_{j,E}{}^{-1}[(m \cdot \nu_a)_{a=1,\dots,k-2}]^* \tag{10.27}$$

and asymptotically $T([m],[\nu])$ has a χ^2_{2k-4} distribution.

Proof. The tangent space $T_{[\nu_{k-1}]}\mathbb{C}P^{k-2}$ has the orthobasis specified by $\nu_1,\dots,\nu_{k-2},\nu_1^*,\dots,\nu_{k-2}^*$. Note that since j is an isometric embedding, we may select the first elements of the adapted moving frame in Corollary 5.5.1 to be $e_a(P_j(\mu)) = (d_{[\nu_{k-1}]}j)(\nu_a)$, followed by $e_a^*(P_j(\mu)) = (d_{[\nu_{k-1}]}j)(\nu_a^*)$. Then the $a-th$ tangential component of $P_j(\overline{j(X)}) - P_j(\mu)$ w.r.t. this basis of $T_{P_j(\mu)}j(\mathbb{C}P^{k-2})$ equals up to a sign the component of $m - \nu_{k-1}$ w.r.t. the orthobasis $\nu_1,\dots,\nu_{k-2}$

in $T_{[v_{k-1}]}\mathbb{C}P^{k-2}$, which is $v_a^t m$; and the $a^* - th$ tangential components are given by $v_a^{*t} m$, and together (in complex multiplication) they yield the complex vector $[(m \cdot v_a)_{a=1,\ldots,k-2}]$. The claim follows from this and formula (10.20) as a particular case of Corollary 5.5.1 □

We may derive from this the following large sample confidence regions:

COROLLARY 10.3.1. *Assume $X_r = [Z_r]$, $\|Z_r\| = 1$, $r = 1, \ldots, n$, is a random sample from a j-nonfocal, probability measure Q on $\mathbb{C}P^{k-2}$. An asymptotic $(1-\alpha)$-confidence region for $\mu_E^j(Q) = [v]$, is given by $R_\alpha(\mathbf{X}) = \{[v] : T([m],[v]) \leq \chi^2_{2k-4,\alpha}\}$, where $T([m],[v])$ is given in (5.4.9). If Q has a nonzero-absolutely continuous component w.r.t. the volume measure on $\mathbb{C}P^{k-2}$, then the coverage error of $R_\alpha(\mathbf{X})$ is of order $O(n^{-1})$.*

For small samples the coverage error could be quite large, and a bootstrap analogue of Theorem 10.3.1 is preferable.

THEOREM 10.3.2. *Let j be the VW embedding of $\mathbb{C}P^{k-2}$, and let $X_r = [Z_r]$, $\|Z_r\| = 1$, $r = 1, \ldots, n$ be a random sample from a j–nonfocal distribution Q on $\mathbb{C}P^{k-2}$, having a nonzero absolutely continuous component w.r.t. the volume measure on $\mathbb{C}P^{k-2}$. Assume in addition that the restriction of the covariance matrix of $j(Q)$ to $T_{[v]}j(\mathbb{C}P^{k-2})$ is nondegenerate. Let $\mu_E(Q) = [v]$ be the extrinsic mean of Q. For a resample $\{Z_r^*\}_{r=1,\ldots,n}$ from the sample consider the matrix $K^* := n^{-1}\sum Z_r^* Z_r^{*^*}$. Let $(\eta_a^*)_{a=1,\ldots,k-1}$ be the eigenvalues of K^* in their increasing order, and let $(m_a^*)_{a=1,\ldots,k-1}$ be the corresponding unit complex eigenvectors. Let $S_{j,E}^{*}{}^*$ be the matrix obtained from $S_{j,E}$ by substituting all the entries with $*$–entries. Then the bootstrap distribution function of*

$$T([m]^*,[m]) := n[(m_{k-1}^* \cdot m_a^*)_{a=1,\ldots,k-2}]S_{j,E}^{*}{}^*)^{-1}[(m_{k-1} \cdot m_a^*)_{a=1,\ldots,k-2}]^*$$

approximates the true distribution function of $T([m],[v])$ given in Theorem 10.3.1 with an error of order $O_p(n^{-2})$.

REMARK 10.3.1. *For distributions that are reasonably concentrated one may determine a non-pivotal bootstrap confidence region using Corollary 5.5.1(a).*

The chart used here features affine coordinates in $\mathbb{C}P^{k-2}$. Recall that the complex space $\mathbb{C}^{k-2}$ can be embedded in $\mathbb{C}P^{k-2}$ preserving collinearity. Such a standard affine *embedding , missing only a hyperplane at infinity, is* $(z^1, \ldots, z^{k-2}) \to [z^1 : \ \ldots \ : z^{k-1} : 1]$.

This leads to the notion of *affine* coordinates of a point

$$p = [z^1 : \ \ldots \ : z^m : z^{k-1}], z^{k-1} \neq 0,$$

to be defined as

$$(w^1, w^2, \ldots, w^{k-2}) = \left(\frac{z^1}{z^{k-1}}, \ldots, \frac{z^{k-2}}{z^{k-1}}\right).$$

To simplify the notation the simultaneous confidence intervals used in the next section, can be expressed in terms of simultaneous *complex confidence intervals*. If $z = x + iy, w = u + iv, x < u, y < v$ then we define the complex interval $(z, w) = \{c = a + ib | a \in (x, u), b \in (y, v)\}$.

10.4 Nonparametric Estimation and Testing for the Projective Shape of a Finite Configuration

In general, if $f : M_1 \to M_2$ is a differentiable function defined from the manifold M_1 to the manifold M_2 and $x \in M_1$, the differential of the function f at x is labeled $D_x f$. Assume $J : M \to \mathbb{R}^N$ is an embedding of the d dimensional complete manifold M (Spivak (1979) [316]). Recall from Chapter 4 that the extrinsic mean μ_J of a $J-$nonfocal random object (r.o.) Y on M by

$$\mu_J =: J^{-1}(P_J(\mu)), \tag{10.28}$$

where $\mu = E(J(Y))$ is the mean vector of $J(Y)$ and $P_J : \mathcal{F}^c \to J(M)$ is the ortho-projection on $J(M)$ defined on the complement of the set $\mathcal{F}$ of focal points of $J(M)$ (see Chapter 4). Also recall that the extrinsic covariance matrix of Y with respect to a local frame field $y \to (f_1(y), \dots, f_d(y))$ for which $(D_y J(f_1(y)), \dots, D_y J(f_d(y)))$ are orthonormal vectors in $\mathbb{R}^N$ was defined in Bhattacharya and Patrangenaru (2005) [43].

If Σ is the covariance matrix of $J(Y)$ regarded as a random vector on $\mathbb{R}^N$, then P_J is differentiable at μ. In order to evaluate the differential $D_\mu P_J$ one considers a special orthonormal frame field to ease the computations. A local ortho-frame field $(e_1(p), e_2(p), \dots, e_N(p))$ defined on an open neighborhood $U \subseteq \mathbb{R}^N$ of $P_J(M)$ is *adapted to the embedding J* if $\forall y \in J^{-1}(U), (e_r(J(y)) = D_y J(f_r(y)), r = 1, \dots, d$.

Let $e_1, e_2, \dots, e_N$ be the canonical basis of $\mathbb{R}^N$ and assume $(e_1(p), e_2(p), \dots, e_k(p))$ is an adapted frame field around $P_J(\mu) = J(\mu_J)$, and let Σ_E be the extrinsic covariance matrix of Y with respect to $(f_1(\mu_J), ..., f_d(\mu_J))$, given by (5.45).

From Subsection 3.5.8, $P\Sigma_m^k$ is homeomorphic to $M = (\mathbb{R}P^m)^q, q = k - m - 2$ and since RP^m, as a particular case of a Grassmann manifold, can be equivariantly embedded in the space $S(m+1)$ of $(m+1) \times (m+1)$ symmetric matrices (see Dimitric (1996) [86]) via $j : \mathbb{R}P^m \to S(m+1)$,

$$j([x]) = xx^T, x^t x = 1. \tag{10.29}$$

Mardia and Patrangenaru (2005) [233] considered the resulting equivariant embedding of the projective shape space $P\Sigma_m^k$:

$$J = j_k : P\Sigma_m^k = (\mathbb{R}P^m)^q \to (S(m+1))^q$$

defined by

$$j_k([x_1], ..., [x_q]) = (j([x_1]), ..., j([x_q])), \tag{10.30}$$

where $x_s \in \mathbb{R}^{m+1}, x_s^T x_s = 1, \forall s = 1, ..., q$.

REMARK 10.4.1. *The embedding j_k in* (10.30) *yields the fastest known computational algorithms in projective shape analysis. Basic axial statistics related to Watson's method of moments such as the sample mean axis (Watson(1983) [333]) and extrinsic sample covariance matrix (Prentice (1984) [282]) can be expressed in terms of $j_{m+3} = j$.*

A random projective shape Y of a k-ad in $\mathbb{R}P^m$ is given in axial representation by the multivariate random axes

$$(Y^1,\dots,Y^q), Y^s = [X^s], (X^s)^T X^s = 1, \forall s = 1,\dots,q = k-m-2. \qquad (10.31)$$

From Bhattacharya and Patrangenaru (2003) [42] or Mardia and Patrangenaru (2005) [233] it follows that in this representation, the extrinsic mean projective shape of $(Y^1,\dots,Y^q)$ exists if $\forall s = 1,\dots,q$, the largest eigenvalue of $E(X^s(X^s)^T)$ is simple. In this case μ_{j_k} is given by

$$\mu_{j_k} = ([\gamma_1(m+1)],\dots,[\gamma_q(m+1)]) \qquad (10.32)$$

where $\lambda_s(a)$ and $\gamma_s(a), a = 1,\dots,m+1$ are the eigenvalues in increasing order and the corresponding unit eigenvector of $E(X^s(X^s)^T)$.

If $Y_r, r = 1,\dots,n$ are i.i.d.r.o.'s (independent identically distributed random objects) from a population of projective shapes (in its multi-axial representation), for which the mean shape μ_{j_k} exists, from a general consistency theorem for extrinsic means on manifolds in Bhattacharya and Patrangenaru (2003) [42] it follows that the extrinsic sample mean $[\overline{Y}]_{j_k,n}$ is a strongly consistent estimator of μ_{j_k}. In the multivariate axial representation

$$Y_r = ([X_r^1],\dots,[X_r^q]), (X_r^s)^T X_r^s = 1; s = 1,\dots,q. \qquad (10.33)$$

Let J_s be the random symmetric matrix given by

$$J_s = n^{-1}\Sigma_{r=1}^n X_r^s (X_r^s)^T \quad , \quad s = 1,\dots,q, \qquad (10.34)$$

and let $d_s(a)$ and $g_s(a)$ be the eigenvalues in increasing order and the corresponding unit eigenvector of J_s, $a = 1,\dots,m+1$. Then the sample mean projective shape in its multi-axial representation is given by

$$\overline{Y}_{j_k,n} = ([g_1(m+1)],\dots,[g_q(m+1)]). \qquad (10.35)$$

REMARK 10.4.2. *Some of the results in this section can be found without a proof in Mardia and Patrangenaru (2005) [233]. Their proofs are given in Patrangenaru et al. (2010) [272], and for completeness we are giving them below as well.*

If a is a positive integer, $\overline{1,a}$ is the set of indices from 1 to a. To determine the extrinsic covariance matrix Σ_E of (10.31), we note that the vectors

$$f_{(s,a)} = (0,\dots,0,\gamma_s(a),0,\dots,0), \qquad (10.36)$$

with the only nonzero term in position s, $s \in \overline{1,q}$, $a \in \overline{1,m}$ yielding a basis in the tangent space at the extrinsic mean $T_{\mu_{j_k}}(\mathbb{R}P^m)^q$, that is orthonormal with respect to the scalar product induced by the embedding j_k. The vectors $e_{(s,a)}, \forall s \in \overline{1,q}, \forall a \in \overline{1,m}$ defined as follows :

$$e_{(s,a)} =: D_{\mu_{j_k}} j_k(f_{(s,a)}). \tag{10.37}$$

form an orthobasis of $T_{j_k(\mu_{j_k})}(\mathbb{R}P^m)^q$. We complete this orthobasis to an orthobasis of q-tuples of matrices $(e_i)_{i\in\mathcal{I}}$ for $(S(m+1))^q$, that is indexed by the set $\mathcal{I}$, the first indices of which are the pairs $(s,a), s = 1,\dots,q; a = 1,\dots,m$ in their lexicographic order.

Let E_a^b be the $(m+1)\times(m+1)$ matrix with all entries zero, except for an entry 1 in the position (a,b). The standard basis of $S(m+1)$ is given by $e_a^b = E_a^b + E_b^a, 1 \leq a \leq b \leq m+1$. For each $s = 1,...,q$, the vector $({}_s e_a^b) = (0_{m+1},...,0_{m+1}, e_a^b, 0_{m+1},...,0_{m+1})$ has all the components zero matrices $0_{m+1} \in S(m+1)$, except for the s-th component, which is the matrix e_a^b of the standard basis of $S(m+1,\mathbb{R})$; the vectors ${}_s e_a^b, s = 1,\dots,q, 1 \leq a \leq b \leq m+1$ listed in the lexicographic order of their indices (s,a,b) give a basis of $S(m+1)^q$.

Let Σ be the covariance matrix of $j_k(Y^1,\dots,Y^q)$ regarded as a random vector in $(S(m+1))^q$, with respect to this standard basis, and let $P =: P_{j_k} : (S(m+1))^q \to j_k((\mathbb{R}P^m)^q)$ be the projection on $j_k((\mathbb{R}P^m)^q)$. From (5.45) it follows that the extrinsic covariance matrix of $(Y^1,\dots,Y^q)$ with respect to the basis (10.36) of $T_{\mu_{j_k}}(\mathbb{R}P^m)^q$ is given by

$$\Sigma_E = \left[e_{(s,a)}(P(\mu)) \cdot D_\mu P({}_r e_a^b)\right]_{(s=1,\dots,q),(a=1,\dots,m)} \cdot \Sigma \\ \cdot \left[e_{(s,a)}(P(\mu)) \cdot D_\mu P({}_r e_a^b)\right]^T_{(s=1,\dots,q),(a=1,\dots,m)}. \tag{10.38}$$

Assume $Y_1,\dots,Y_n$ are i.i.d.r.o.'s from a j_k-nonfocal probability measure on $(\mathbb{R}P^m)^q$ and μ_{j_k} in (10.32) is the extrinsic mean of Y_1.

PROPOSITION 10.4.1. *In the case of the VW embedding $j = j_k$, the sample extrinsic covariance matrix estimator $S_{j_k,E,n}$ in (5.48) is given by the $(mq)\times(mq)$ symmetric matrix $S_{j,E,n}$, with the entries in pairs of indices $(s,a), s = 1,\dots,q; a = 1,\dots,m$, in their lexicographic order given by*

$$S_{j,E,n_{(s,a),(t,b)}} = n^{-1}(d_s(m+1) - d_s(a))^{-1}(d_t(m+1) - d_t(b))^{-1} \cdot \\ \cdot \sum_{r=1}^{n} (g_s(a)^T X_r^s)(g_t(b)^T X_r^t)(g_s(m+1)^T X_r^s)(g_t(m+1)^T X_r^t). \tag{10.39}$$

The proof of Proposition 10.4.1, based on the equivariance of the VW map j_k and similar to that given for Proposition 10.2.2, is left to the reader.

From Chapter 5, it follows that $S_{j_k,E,n}$ is a strongly consistent estimator of the population extrinsic covariance matrix in (10.38). In preparation for an asymptotic distribution of $\overline{Y}_{j_k,n}$ we set

$$D_s = (g_s(1)\dots g_s(m)) \in \mathcal{M}(m+1,m;\mathbb{R}), s = 1,\dots,q. \tag{10.40}$$

If $\mu = ([\gamma_1],\dots,[\gamma_q])$, where $\gamma_s \in \mathbb{R}^{m+1}, \gamma_s^T\gamma_s = 1$, for $s = 1,\dots,q$, we define a Hotelling's T^2 type-statistic

$$T(\overline{Y}_{j_k,n};\mu) = n(\gamma_1^T D_1,\dots,\gamma_q^T D_q) S_{j,E,n}{}^{-1}(\gamma_1^T D_1,\dots,\gamma_q^T D_q)^T. \tag{10.41}$$

THEOREM 10.4.1. *Assume $(Y_r)_{r=1,\dots,n}$ are i.i.d.r.o.'s on $(\mathbb{R}P^m)^q$, and Y_1 is j_k-nonfocal, with $\Sigma_E > 0$. Let $\lambda_s(a)$ and $\gamma_s(a)$ be the eigenvalues in increasing order and corresponding unit eigenvectors of $E[X_1^a(X_1^a)^T]$. If $\lambda_s(1) > 0$, for $s = 1,\dots,q$, then $T(\overline{Y}_{j_k,n};\mu_{j_k})$ converges weakly to χ^2_{mq}.*

If Y_1 is a j_k-nonfocal population on $(\mathbb{R}P^m)^q$, since $(\mathbb{R}P^m)^q$ is compact, it follows that $j_k(Y_1)$ has finite moments of sufficiently high order. According to Bhattacharya and Ghosh (1978) [38], this along with an assumption of a nonzero absolutely continuous component, suffices to ensure an Edgeworth expansion up to order $O(n^{-2})$ of the pivotal statistic $T(\overline{Y}_{j_k,n};\mu_{j_k})$, and implicitly the bootstrap approximation of this statistic.

COROLLARY 10.4.1. *Let $Y_r = ([X_r^1],\dots,[X_r^q]), X_{st}^T X_{st} = 1, s = 1,\dots q, r = 1,\dots,n$, be i.i.d.r.o.'s from a j_k-nonfocal distribution on $(\mathbb{R}P^m)^q$ which has a nonzero absolutely continuous component, and with $\Sigma_E > 0$. For a random resample with repetition $(Y_1^*,\dots,Y_n^*)$ from $(Y_1,\dots,Y_n)$, consider the eigenvalues of $\frac{1}{n}\sum_{r=1}^n X_{rs}^* X_{rs}^{*T}$ in increasing order and corresponding unit eigenvectors $d_s^*(a)$ and $g_s^*(a), a = 1,\dots,m+1$. Let $S_{j,E,n}{}^*$ be the matrix obtained from G_n, by substituting all the entries with $*-$entries. Then, from Section 2.13, the bootstrap distribution of the statistic*

$$T(\overline{Y}^*_{j_k};\overline{Y}_{j_k}) = n(g_1(m+1)^T D_1^*,\dots,g_q(m+1)^T D_q^*)\, S_{j,E,n}{}^{*-1} \\ (g_1(m+1)^T D_1^*,\dots,g_q(m+1)^T D_q^*)^T \tag{10.42}$$

approximates the true distribution of $T(\overline{Y}_{j_k};\mu_{j_k})$ given by (10.41), with an error of order $0_p(n^{-2})$.

REMARK 10.4.3. *The above corollary can be also found in Mardia and Patrangenaru (2005) [233]. Patrangenaru et al.(2010) [272]) noted that the condition $\Sigma_E > 0$ is missing from the corresponding result in Mardia and Patrangenaru (2005) [233] though, as well as in their Theorem 4.1.*

Theorem 10.4.1 and Corollary 10.4.1 are useful in estimation and testing for mean projective shapes. From Theorem 10.4.1 we derive large sample confidence region for μ_{j_k}.

COROLLARY 10.4.2. *Assume $(Y_r)_{r=1,\dots,n}$ are i.i.d.r.o.'s from a j_k-nonfocal* *probability distribution on* $(\mathbb{R}P^m)^q$, *and* $\Sigma_E > 0$. *An asymptotic* $(1-\alpha)$-*confidence region for* $\mu_{j_k} = [v]$ *is given by* $R_\alpha(\mathbf{Y}) = \{[v] : T(\overline{Y}_{j_k,n};[v]) \leq \chi^2_{mq,\alpha}\}$, *where* $T([\overline{Y}_{j_k},[v])$ *is given in* (10.41). *If the probability measure of* Y_1 *has a nonzero-absolutely continuous component w.r.t. the volume measure on* $(\mathbb{R}P^m)^q$, *then the coverage error of* $R_\alpha(\mathbf{Y})$ *is of order* $O(n^{-1})$.

For small samples the coverage error could be quite large, and the bootstrap analogue in Corollary 10.4.1 is preferable. Consider for example the one sample testing problem for mean projective shapes:

$$H_0 : \mu_{j_k} = \mu_0 \text{ vs. } H_1 : \mu_{j_k} \neq \mu_0. \tag{10.43}$$

COROLLARY 10.4.3. *The large sample p-value for the testing problem* (10.43) *is* $p = Pr(T > T(\overline{Y}_{j_k,n};\mu_0))$, *where* $T(\overline{Y}_{j_k,n};\mu)$ *is given by* (10.41).

In the small sample case, problem (10.43) can be answered based on Corollary 10.4.1 to obtain the following $100(1-\alpha)\%$ bootstrap confidence region for μ_{j_k} :

COROLLARY 10.4.4. *Under the hypotheses of Corollary 10.4.1, The corresponding* $100(1-\alpha)\%$ *confidence region for* μ_{j_k} *is*

$$C^*_{n,\alpha} := j_k^{-1}(U^*_{n,\alpha}) \tag{10.44}$$

with $U^*_{n,\alpha}$ *given by*

$$U^*_{n,\alpha} = \{\mu \in j_k((\mathbb{R}P^m)^q) : T(\overline{y}_{j_k,n};\mu) \leq c^*_{1-\alpha}\}, \tag{10.45}$$

where $c^*_{1-\alpha}$ *is the upper* $100(1-\alpha)\%$ *point of the values of* $T(\overline{Y}^*_{j_k};\overline{Y}_{j_k})$ *given by* (10.42). *The region given by* (10.44)–(10.45) *has coverage error* $O_p(n^{-2})$.

For the one sample hypothesis testing problem for mean projective shapes 10.43, if $\Sigma_{j_k,E,n}$ is singular and all the marginal axial distributions have positive definite extrinsic covariance matrices, one may use simultaneous confidence ellipsoids to estimate μ_{j_k}. Assume $(Y_r)_{r=1,\ldots,n}$ are i.i.d.r.o.'s from a j_k-nonfocal probability distribution on $(\mathbb{R}P^m)^q$. For each $s = 1,\ldots,q$ let $\Sigma_{j_k,s}$ be the extrinsic covariance matrix of Y_1^s, and let $\overline{Y}^s_{j_k,n}$ and $S_{j_k,s,n}$ be the extrinsic sample mean and the extrinsic sample covariance matrix of the s-th marginal axial and the probability measure of Y_1^s has a nonzero-absolutely continuous component w.r.t. the volume measure on $\mathbb{R}P^m$. For $s = 1,\ldots,q$ and for $[\gamma_s] \in \mathbb{R}P^m$, $\gamma_s^T\gamma_s = 1$, we consider the statistics:

$$T_s = T_s(\overline{Y}^s_{j_k,n},[\gamma_s]) = n\gamma^T D_s S_{j_k,E,n}{}^{-1} D_s^T \gamma \tag{10.46}$$

and the corresponding bootstrap distributions

$$T_s^* = T_s(\overline{Y}^{s*}_j, \overline{Y}^s_{j,n}) = ng_s(m+1)^T D_s^* S_{j_k,s,n}{}^{*-1} D_s^{*T} g_s(m+1). \tag{10.47}$$

Since by Theorem 10.4.1 T_s has asymptotically a χ^2_m distribution, we obtain the following

COROLLARY 10.4.5. *For $s = 1,\dots,q$ let $c^*_{s,1-\beta}$ be the upper $100(1-\beta)\%$ point of the values of T^*_s given by* (10.47). *We set*

$$C^*_{s,n,\beta} := j_k^{-1}(U^*_{s,n,\beta}) \tag{10.48}$$

where

$$U^*_{s,n,\beta} = \{\mu \in \mathbb{R}P^m : T_s(\bar{y}^s_{j,n};\mu) \le c^*_{s,1-\beta}\}. \tag{10.49}$$

Then

$$R^*_{n,\alpha} = \cap_{s=1}^q C^*_{s,n,\frac{\alpha}{q}} \tag{10.50}$$

with $C^*_{s,n,\beta}, U^*_{s,n,\beta}$ given by (10.48)-(10.49) *is a region of approximately at least* $100(1-\alpha)\%$ *confidence for μ_{j_k}. The coverage error is of order $O_p(n^{-2})$.*

REMARK 10.4.4. *If $\Sigma_{j_k,E,n}$ is singular one may construct confidence regions for μ_{j_k} using nonpivotal bootstrap.*

10.5 Two-Sample Tests for Means for Projective Shapes

We consider the problem of testing for the equality of the Fréchet means of two populations of projective shapes. We will assume that Q_1, Q_2 are probability measures on $P\Sigma^k_m$, and $\mu_{\mathcal{F}}(Q_i) = \mu_i$ exists, for $i = 1,2$. We treat separately, the cases $m = 1$ and $m > 1, m \neq 3$. Recall that $q = m-k-2$.

If $m = 1$, we assume in addition that there is a point $p \in P\Sigma^k_1 = (S^1)^q$ such that for $i = 1,2$, the support of Q_i and $\mu_i = \mu_I(Q_i)$ are contained in the complement of the cut locus of p. We identify the tangent space $T_p(S^1)^q$ with $\mathbb{R}^q$ and use the formula (3.70) in Chapter 3 for the exponential map at p, and set

$$v_i = Exp_p^{-1}(\mu_a), a = 1,2. \tag{10.51}$$

If $(X_{a,r_a}), a = 1,\dots,n_a; a = 1,2$, are random samples from Q_a, then we can obtain corresponding $(U_{a,r_a}), a = 1,\dots,n_a$, as a random sample from a multivariate distribution on $\mathbb{R}^q$ of mean v_a. Assume that $n_1 - k + 3$ and $n_2 - k + 3$ are large enough. A large sample χ^2 confidence region can be determined using standard multivariate methods (see (2.69)) : a $100(1-\alpha)\%$ confidence hyper-ellipsoid for $v_1 - v_2$ is given by

$$[\bar{U}_1 - \bar{U}_2 - (v_1 - v_2)]^T[\frac{1}{n_1}S_1 + \frac{1}{n_2}S_2]^{-1}[\bar{U}_1 - \bar{U}_2 - (v_1 - v_2)] \le \chi^2_{k-3}(\alpha). \tag{10.52}$$

Here we used the fact that $P\Sigma^k_1 = (S^1)^{k-3}$ has a Riemannian structure of a locally flat torus, therefore, if the two population means are not too far apart, which is the case when a two sample test for means is necessary, the restriction of the confidence region for the "difference of intrinsic means" is independent of the selection of the point p. Moreover, if the two samples are not large enough, one can use instead a bootstrap confidence region, by using nonparametric bootstrap (see Section 2.13).

For $m > 1$, we compare VW means (for a definition of the VW-embedding, see Subsection 3.5.8). Using the spherical representation, we reduce the problem to axial statistics. Since for $b = 1,2$, $\mu_b \in (\mathbb{R}P^m)^q$, we set in spherical representation $\mu_b = (\mu_{b1}, \dots, \mu_{b,q})$. Without loss of generality, we may assume that for each j, the angle between $\mu_{1,j}$ and $\mu_{2,j}$ are $\frac{\pi}{2}$ or less, and we consider the unique rotation $r_{m,j} \in SO(m+1)$, such that $r_{mj}(\mu_{1j}) = \mu_{2j}$ and the restriction of $r_{m,j}$ to the orthocomplement of the plane determined by $\mu_{1,j}$ and $\mu_{2,j}$ in $\mathbb{R}^{m+1}$ is the identity.

The equality $\mu_1 = \mu_2$ is equivalent to $r_{m,j} = I_{m+1}, j = 1, \dots, q$ where I_{m+1} is an $(m+1) \times (m+1)$ identity matrix. Assume $(X_{1,r})_{r=1,\dots,n_1}$, $(X_{2,s})_{s=1,\dots,n_2}$ are random samples from Q_1, Q_2 respectively. A consistent estimator of the Lie group valued random variable $r_m = (r_{m,j}, j = 1, \dots, q)$ is $\rho_m = (\rho_{m,j}, j = 1, \dots, q)$, where for each $j = 1, \dots, q$; $\rho_{m,j} \in SO(m+1)$ is the unique rotation defined as above, that brings the extrinsic sample means (mean directions) in coincidence, that is, superimposes $\bar{X}_{1,j,E}$ onto $\bar{X}_{2,j,E}$.

A particular case of practical interest is when $m = 2$. In this case, we will consider only the subcase $k = 5$, for which we give an application in the next section. In this case, we test the equality $r_2 = I_3$ in $SO(3)$. If $R \in SO(3), R \neq I_3$, let V_1, V_2, V_3 be an orthogonal basis of $\mathbb{R}^3$, such that $T(V_3) = V_3$. If we set $H(T) = [(V_1 \cdot T(V_1)), (V_1 \times T(V_1))^T)]$, then the map $H : SO(3) \longrightarrow \mathbb{R}P^3$ is a well defined diffeomorphism from $SO(3)$ to the axial space $\mathbb{R}P^3$. Modulo the diffeomorphism H, the equality $\mu_1 = \mu_2$ amounts to $H(r_2) = [1:0:0:0]$. The distribution of the resulting consistent estimator $H(\rho_2)$ of $H(r_2)$ is essentially given in Beran & Fisher (1998) [18], Theorem 2.1). Assume neither n_1 nor n_2 is small compared with $n = n_1 + n_2$. Let $G(r)$ be the affine coordinates of $H(r)$; if $H(r) = [H_0(r) : H_1(r) : H_2(r) : H_3(r)]$, then $G(r) = (G_1(r), G_2(r), G_3(r))$, with $G_a(r) = H_a(r)/H_0(r), a = 1,2,3$. Using Equation (5.13) of Beran & Fisher (1998, p.489) [18], it turns out that $n^{1/2}(\rho_2 - r_2)$ has a trivariate Gaussian distribution and is independent of n. Then by Cramér's delta method (see Ferguson (1996) [115], p.45, or Chapter 2), under the null hypothesis, it follows that if there are two constants c_1, c_2, such that for $b = 1,2$, $n_b/n \to c_b$, then $n^{1/2}\{G(\rho_2) - G(r_2)\}$ has a trivariate Gaussian distribution which is independent of n. Consequently, from Chapter 2, if one considers the resamples under the empirical $n^{1/2}\{G(\rho_2{}^*) - G(\rho_2)\}$, by nonpivotal bootstrap, this distribution will have asymptotically the same distribution as that of $n^{\frac{1}{2}}\{G(\rho_2) - G(r_2)\}$.

10.5.1 A Lie Group Structure on the Manifold of 3D Projective Shapes of Configurations Containing a Projective Frame

Note that, as shown in Crane and Patrangenaru [78] and Buibas et al. (2012) [55], unlike in other dimensions, the projective shape manifold $P\Sigma_3^k, k \geq 5$, has a Lie group structure , derived from the quaternion multiplication. Recall that if a real number x is identified with $(0,0,0,x) \in \mathbb{R}^4$, and if we label the quadruples $(1,0,0,0),(0,1,0,0)$, respectively $(0,0,1,0)$ by $\overrightarrow{i}, \overrightarrow{j}$, respectively $\overrightarrow{k}$, then the multiplication table given by

$\odot$	$\overrightarrow{i}$	$\overrightarrow{j}$	$\overrightarrow{k}$
$\overrightarrow{i}$	-1	$\overrightarrow{k}$	$-\overrightarrow{j}$
$\overrightarrow{j}$	$-\overrightarrow{k}$	-1	$\overrightarrow{i}$
$\overrightarrow{k}$	$\overrightarrow{j}$	$-\overrightarrow{i}$	-1

where $a \odot b$ product of a on the first column with b on the top row, is listed on the row of a and column of b, extends by linearity to a multiplication $\odot$ of $\mathbb{R}^4$. Note that $(\mathbb{R}^4,+,\odot)$ has a structure of a noncommutative field, the field of *quaternions*, usually labeled by $\mathbb{H}$. Note that if $h,h' \in \mathbb{H}$, then $\|h \odot h'\| = \|h\|\|h'\|$, and the three dimensional sphere inherits a group structure, the *group of quaternions of norm one*.

Moreover, since $\mathbb{R}P^3$ is the quotient $S^3/x \sim -x$

$$[x] \odot [y] =: [x \odot y], \tag{10.53}$$

is a well defined *Lie group* operator on $\mathbb{R}P^3$, called the group of *p-quaternions*. Note that if $h = t + x\overrightarrow{i} + y\overrightarrow{j} + z\overrightarrow{k}$, its *conjugate* is $\bar{h} = t - x\overrightarrow{i} - y\overrightarrow{j} - z\overrightarrow{k}$, and the inverse of h is given by

$$h^{-1} = \|h\|^{-2}\bar{h}, \tag{10.54}$$

As shown in Section 3.5.8, as a manifold, $P\Sigma_3^k$ is diffeomorphic with $(\mathbb{R}P^3)^q$, where $q = k-5$. With this identification, $P\Sigma_3^k \sim (\mathbb{R}P^3)^q$ inherits a Lie group structure from the group structure p-quaternions $\mathbb{R}P^3$ with the multiplication given by

$$\begin{aligned}([h_1],\ldots,[h_q]) \odot ([h'_1],\ldots,[h'_q]) &:= ([h_1] \odot [h'_1],\ldots,[h_q] \odot [h'_q]) = \\ &= ([h_1 \odot h'_1],\ldots,[h_q \odot h'_q]).\end{aligned} \tag{10.55}$$

The identity element is given by

$$1_{(\mathbb{R}P^3)^q} = ([0:0:0:1],\ldots,[0:0:0:1]), \tag{10.56}$$

and given a point $\mathbf{h} = ([h_1],\ldots,[h_q]) \in (\mathbb{R}P^3)^q$, from (10.54), its inverse is $\mathbf{h}^{-1} = \bar{\mathbf{h}} = ([\bar{h}_1],\ldots,[\bar{h}_q])$.

10.5.2 Nonparametric Bootstrap Tests for VW Mean 3D Projective Shape Change

Given two elements x,y in a Lie group $(\mathcal{G},\circ)$, we define the *change* c *from* x *to* y by $c = x^{-1} \circ y$. Similarly, following Section 6.4, given a (matched) pair of random objects X, Y on a Lie group $(\mathcal{G}, \circ)$ we define the *change C from X to Y* by $C = X^{-1} \circ Y$. This allows us to transfer a two sample problem for means of matched pairs on $\mathcal{G}$, to a one sample problem for the mean change on $\mathcal{G}$.

In particular, given $(\mathbb{R}P^3)^q$, has a Lie group structure given by (10.55), and, again, since from Section 3.5.8, $P\Sigma_3^k$ is homeomorphic to $M = (\mathbb{R}P^3)^q, q = k-5$, we define the *3D projective shape change* between matched pairs $\mathbf{H_1}, \mathbf{H_2}$ of projective shapes, as the change from $\mathbf{H_1}$ to $\mathbf{H_2}$, in terms of the induced Lie group structure on $P\Sigma_3^k$. Then, testing the existence of mean 3D projective shape change $\mathbf{Y}$ from $\mathbf{H_1}$ to $\mathbf{H_2}$ is transferred to the hypothesis testing problem

$$H_0 : \mu_{j_k} = \mathbf{1}_{(\mathbb{R}\mathbf{P}^3)^\mathbf{q}} \text{ vs. } H_1 : \mu_{j_k} \neq \mathbf{1}_{(\mathbb{R}\mathbf{P}^3)^\mathbf{q}}, \tag{10.57}$$

where μ_{j_k} is the VW-mean of $\mathbf{Y}$ (certainly other Fréchet mean may be considered, but we prefer the VW-mean, for which necessary and sufficient conditions of existence are known (see Section 10.4)). The explicit expression of the identity element $1_{(\mathbb{R}P^3)^q}$ in (10.57) is given in (10.56).

Once the hypothesis testing problem 10.57 was set, we apply the Corollaries 10.4.5, 10.43 from Section 10.4 to 3D mean projective shapes.

Assume $(\mathbf{H_{1,r}}, \mathbf{H_{2,r}})_{\mathbf{r=1,\dots,n}}$ are i.i.d.r.o.'s from paired distributions on $(\mathbb{R}P^3)^q$, such that $\mathbf{Y_1} = \bar{\mathbf{H}}_{\mathbf{1,1}}\mathbf{H_{2,1}}$ has a j_k-nonfocal probability distribution on $(\mathbb{R}P^3)^q$. Testing the hypothesis 10.57 in the case $m = 3$, at level α, amounts to finding a $1-\alpha$ confidence region for μ_{j_k} given by Corollary 10.4.5, and, if the sample is small and the extrinsic sample covariance matrix is degenerate, checking if $\mathbf{1}_{(\mathbb{R}\mathbf{P}^3)^\mathbf{q}}$ is in a $1-\alpha$ confidence region, amounts to finding the upper $\frac{\alpha}{q}$ cutoffs for the bootstrap distributions of the test statistics $T_s^*, s = 1, \dots, k-5$, and checking if the values of T_s, for $\mu_{j_k} = 1_{(\mathbb{R}P^3)^q}$ are all in the corresponding confidence intervals. That is

REMARK 10.5.1. *For $m = 3, s = 1, \dots, q = k-5$ let $c^*_{s,1-\beta}$ be the upper $100(1-\beta)\%$ point of the values of T_s^* given by* (10.47). *We set*

$$C^*_{s,n,\beta} := j^{-1}(U^*_{s,n,\beta}) \tag{10.58}$$

where

$$U^*_{s,n,\beta} = \{\mu_s \in \mathbb{R}P^3 : T_s(\overline{Y}^s_{j,n}; \mu_s) \leq c^*_{s,1-\beta}\}. \tag{10.59}$$

Then

$$R^*_{n,\alpha} = \prod_{s=1}^{q} C^*_{s,n,\frac{\alpha}{q}}, \tag{10.60}$$

with $C^*_{s,n,\beta}, U^*_{s,n,\beta}$ given by (10.58)–(10.59), *is a region of approximately at least $100(1-\alpha)\%$ confidence for μ_{j_k}. Then we fail to reject at level α the hypothesis that there is a nontrivial mean change in the 3D projective shapes $\mathbf{H_1}, \mathbf{H_2}$ if $\mathbf{1}_{(\mathbb{R}\mathbf{P}^3)^\mathbf{q}} \in \mathbf{R}^*_{\mathbf{n},\alpha}$.*

We consider now the case of a *two sample test for VW mean projective shapes based on independent samples,* using the general tests developed in Section 10.4. Here $\mathcal{M} = \mathcal{G} = (\mathbb{R}P)^q, \delta = 1_{(\mathbb{R}P^3)^q}, q = k-5$, and

$$\varphi([x_1],\dots,[x_q]) = (\varphi_{m+1}([x_1]),\dots,\varphi_{m+1}([x_q])). \tag{10.61}$$

In our examples we considered the group action $\alpha : (\mathbb{R}P)^q \times (\mathbb{R}P)^q \to (\mathbb{R}P)^q$ given by the multiplication (10.55):

$$\alpha(\mathbf{h},\mathbf{k}) = \mathbf{h} \odot \mathbf{k}. \tag{10.62}$$

Therefore the hypothesis testing $H_0 : \mu_{1,j_k} = \mu_{2,j_k}$ on the Lie group $((\mathbb{R}P)^q, \odot)$ is equivalent to the testing problem

$$H_0 : \mu_{1,j_k}^{-1} \odot \mu_{2,j_k} = 1_{(\mathbb{R}P^3)^q}. \tag{10.63}$$

From Section 2.13, if the sample sizes n_1, n_2 are small, it suffices to compute the bootstrap distribution of

$$D^* = \varphi(H(\bar{X}^*_{n_1,j_k}, \bar{X}^*_{n_2,j_k})), \tag{10.64}$$

where $H(\mathbf{h},\mathbf{k}) = \mathbf{h} \odot \mathbf{k}$ and φ is given by (10.61).

REMARK 10.5.2. *Given that* $\varphi(1_{(\mathbb{R}P^3)^q}) = \mathbf{0} \in (\mathbb{R}^3)^q$, *testing the hypothesis* (10.63) *at level* α *is equivalent to testing if* $\mathbf{0}$ *is inside a* $100(1-\alpha)\%$ *bootstrap confidence region for* $\varphi(\mu)$. *Since the group multiplication in* $((\mathbb{R}P^3)^q, \odot)$ *is a product of projective quaternion multiplications* (10.53), *one may use simultaneous bootstrap confidence intervals, based on the q affine marginal bootstrap distributions* $(D_1^*,\dots,D_q^*) = D^*$ *in* (10.64). *From the Bonferroni inequality* 2.68, *for each* $j = 1,\dots,q$, *we obtain a* $100(1-\frac{\alpha}{q})\%$ *confidence region* C_j^*, *that can be visualized as a* 3D *box, product of three* $100(1-\frac{\alpha}{3q})\%$ *simultaneous confidence intervals.*

Chapter 11

Nonparametric Statistics on Hilbert Manifolds

Chapters 4 through 6 provided general theory for adapting inferential methodology for multivariate data to the analysis of data lying on finite-dimensional manifolds. However, for data arising on infinite-dimensional Hilbert manifolds, methodology for inference must be adapted instead from techniques developed for data analysis on Hilbert spaces, as described in Chapter 2. As far as infinite dimensional data analysis is concerned, in this chapter our scope is limited to considering only the extension of the neighborhood hypothesis methodology on Hilbert spaces in Section 2.15, to the case of Hilbert manifolds. Applications of this material to shape analysis of planar contours can be found in Chapter 18. The primary reference for this chapter is Ellingson et al. (2013) [104].

11.1 Introduction

A number of statistical methodologies have been developed for the analysis of data lying on Hilbert spaces for the purpose of studying functional data. Some multivariate methods, such as PCA, have useful extensions in functional data analysis (Loève [223]). For dense functional data, the asymptotics of the resulting eigenvalues and eigenvectors were studied by Dauxois et al. (1982) [80]. Even for sparse functional data, these methods have been proved useful (see Hall et al. (2006) [139], Müller et al.(2006) [239]) and have multiple applications.

There are, nevertheless, techniques defined for multivariate analysis that often fail to be directly generalizable to infinite-dimensional data, especially when such data is nonlinear. New methodologies have been developed to account for these high dimensional problems, with many of these presented in a standard text by Ramsay and Silverman (2005) and given by references therein. For high dimensional inference on Hilbert spaces, Munk and Dette (1998) [251] utilized the concept of neighborhood hypotheses for performing tests on nonparametric regression models. Following from this approach, Munk et al. (2008) [252] developed one-sample and multi-sample tests for population means, as discussed in Chapter 2.

However, these methods do not account for estimation of means on infinitely dimensional curved spaces, such as Hilbert manifolds. In order to properly analyze such data, these methods must be further generalized and modified. A key example in which such data arises is in the statistical analysis of direct similarity shapes of planar contours (see [101]), which will be discussed in detail in Chapter 19.

This chapter will focus on the theory for analysis of data on general Hilbert manifolds. Section 12.1 will describe Hilbert manifolds. The second section will describe extrinsic analysis on such spaces and the last section will adapt the neighborhood hypothesis of Munk et al. (2008) [252] to this setting.

11.2 Hilbert Manifolds

In this section, we assume that **H** is a *separable, infinite dimensional Hilbert space* over the reals. Any such space is isometric with l_2, the space of sequences $x = (x_n)_{n\in\mathbb{N}}$ of reals for which the series $\sum_{n=0}^{\infty} x_n^2$ is convergent, with the scalar product $< x,y >= \sum_{n=0}^{\infty} x_n y_n$. A Hilbert space with the norm $\|v\| = \sqrt{< v,v >}$, induced by the scalar product, becomes a Banach space . Differentiability can be defined with respect to this norm.

DEFINITION 11.2.1. *A function f defined on an open set U of a Hilbert space* **H** *is Fréchet differentiable at a point $x \in U$, if there is a linear operator $T : \mathbf{H} \to \mathbf{H}$, such that if we set*

$$\omega_x(h) = f(x+h) - f(x) - T(h), \tag{11.1}$$

then

$$\lim_{h\to 0} \frac{\|\omega_x(h)\|}{\|h\|} = 0. \tag{11.2}$$

Since T in Definition 11.2.1 is unique, it is called the *differential* of f at x and is also denoted by $d_x f$.

DEFINITION 11.2.2. *A chart on a separable metric space $(\mathcal{M},\rho)$ is a one to one homeomorphism $\varphi : U \to \varphi(U)$ defined on an open subset U of $\mathcal{M}$ to a Hilbert space* **H**. *A Hilbert manifold is a separable metric space $\mathcal{M}$, that admits an open covering by domain of charts, such that the transition maps $\varphi_V \circ \varphi_U^{-1} : \varphi_U(U\cap V) \to \varphi_V(U\cap V)$ are differentiable.*

As an example of a Hilbert manifold, we can consider the projective space $P(\mathbf{H})$ of a Hilbert space $\mathbf{H}$, which is the space of all one dimensional linear subspaces of $\mathbf{H}$. The following provides details for why this is a Hilbert manifold.

EXAMPLE 11.2.1. *First, define the distance between two vector lines as their angle. Now, given a line $\mathbb{L} \subset \mathbf{H}$, a neighborhood $U_{\mathbb{L}}$ of $\mathbb{L}$ can be mapped via a homeomorphism $\varphi_{\mathbb{L}}$ onto an open neighborhood of the orthocomplement $\mathbb{L}^{\perp}$ by using the decomposition $\mathbf{H} = \mathbb{L} \oplus \mathbb{L}^{\perp}$. Then for two perpendicular lines $\mathbb{L}_1$ and $\mathbb{L}_2$, it is easy to show that the transition maps $\varphi_{\mathbb{L}_1} \circ \varphi_{\mathbb{L}_2}^{-1}$ are differentiable as maps between open subsets in $\mathbb{L}_1^{\perp}$ and $\mathbb{L}_2^{\perp}$, respectively. A countable orthobasis of $\mathbf{H}$ and the lines $\mathbb{L}_n$, $n \in \mathbb{N}$, which is generated by the vectors in this orthobasis, is used to cover $P(\mathbf{H})$ with the open sets $U_{\mathbb{L}_n}, n \in \mathbb{N}$. Finally, use the fact that for any line $\mathbb{L}$, $\mathbb{L}^{\perp}$ and $\mathbf{H}$ are isometric as Hilbert spaces.*

Note that the line $\mathbb{L}$ spanned by a nonzero vector $\gamma \in \mathbf{H}$ is usually denoted by $[\gamma]$ when regarded as a projective point on $P(\mathbf{H})$.

Similarly, one may consider complex Hilbert manifolds, modeled on Hilbert spaces over $\mathbb{C}$. A vector space over $\mathbb{C}$ can be regarded as a vector space over the reals by restricting the scalars to $\mathbb{R}$. Therefore, any complex Hilbert manifold automatically inherits a structure of real Hilbert manifold.

11.3 Extrinsic Analysis of Means on Hilbert Manifolds

Since a Hilbert manifold does not have a linear structure, standard methods for data analysis on Hilbert spaces, such as those described in Chapter 2 cannot directly be applied. To account for this nonlinearity, one may instead perform extrinsic analysis, in a manner similar to what is described in Chapter 4, by embedding this manifold in a Hilbert space.

DEFINITION 11.3.1. *An embedding of a Hilbert manifold $\mathcal{M}$ in a Hilbert space $\mathbb{H}$ is a one-to-one differentiable function $j : \mathcal{M} \to \mathbb{H}$, such that for each $x \in \mathcal{M}$, the differential $d_x j$ is one to one, and the range $j(\mathcal{M})$ is a closed subset of $\mathbb{H}$ and the topology of $\mathcal{M}$ is induced via j by the topology of $\mathbb{H}$.*

We now return to the previous example of $P(\mathbf{H})$ and consider a useful embedding of this space.

EXAMPLE 11.3.1. *We embed $P(\mathbf{H})$ in $\mathcal{L}_{HS} = \mathbf{H} \otimes \mathbf{H}$, the space of Hilbert–Schmidt operators of $\mathbf{H}$ into itself, via the Veronese–Whitney (VW) embedding j given by*

$$j([\gamma]) = \frac{1}{\|\gamma\|^2} \gamma \otimes \gamma. \tag{11.3}$$

If $\|\gamma\| = 1$, this definition can be reformulated as

$$j([\gamma]) = \gamma \otimes \gamma. \tag{11.4}$$

Given that $\gamma^*(\beta) =< \beta, \gamma >$ *equation* (11.4) *is equivalent to*

$$j([\gamma])(\beta) =< \beta, \gamma > \gamma. \tag{11.5}$$

The range of this embedding is the submanifold $\mathcal{M}_1$ *of rank one Hilbert–Schmidt operators of* $\mathbf{H}$.

To define a location parameter for probability distributions on a Hilbert manifold, the concept of extrinsic means from Bhattacharya and Patrangenaru (2003, 2005)) is extended to the infinite dimensional case as follows:

DEFINITION 11.3.2. *If* $j : \mathcal{M} \to \mathbb{H}$ *is an embedding of a Hilbert manifold in a Hilbert space, the chord distance* ρ *on* $\mathcal{M}$ *is given by* $\rho(x,y) = \|j(x) - j(y)\|$, *and given a random object X on* $\mathcal{M}$, *the associated Fréchet function is*

$$\mathcal{F}_j(x) = E(\|j(X) - j(x)\|^2). \tag{11.6}$$

The set of all minimizers of $\mathcal{F}_j$ *is called the extrinsic mean set of* X. *If the extrinsic mean set has one element only, then that element is called the extrinsic mean and is labeled* $\mu_{E,j}$ *or simply* μ_E.

The following proposition provides some additional details about extrinsic means on Hilbert manifolds.

PROPOSITION 11.3.1. *Consider a random object X on* $\mathcal{M}$ *that has an extrinsic mean set. Then (i)* $j(X)$ *has a mean vector* μ *and (ii) the extrinsic mean set is the set of all points* $x \in \mathcal{M}$, *such that* $j(x)$ *is at minimum distance from* μ. *(iii) In particular,* μ_E *exists if there is a unique point on* $j(\mathbf{M})$ *at minimum distance from* μ, *the projection* $P_j(\mu)$ *of* μ *on* $j(\mathbf{M})$, *and in this case* $\mu_E = j^{-1}(P_j(\mu))$.

Proof. Let $Y = j(X)$. Note that the Hilbert space $\mathbb{H}$ is complete as a metric space, therefore $\inf_{y\in\mathbb{H}} E(\|Y - y\|^2) = \min_{y\in\mathbb{H}} E(\|Y - y\|^2) \leq \min_{y\in j(\mathcal{M})} E(\|Y - y\|^2)$. From our assumption, it follows that $\inf_{y\in\mathbb{H}} E(\|Y - y\|^2) = \min_{y\in\mathbb{H}} E(\|Y - y\|^2)$ is finite, which proves (i). To prove (ii), assume that ν is a point in the extrinsic mean set, and that x is an arbitrary point on $\in \mathcal{M}$. Since $E(\|j(\nu) - Y\|^2) \leq E(\|j(x) - Y\|^2)$ and $j(\nu) - \mu$ and $j(x) - \mu$ are constant vectors, it follows that

$$\|j(\nu) - \mu\|^2 \leq \|j(x) - \mu\|^2 + 2E(< j(x) - j(\nu), \mu - Y >). \tag{11.7}$$

It is obvious that the expected value on the extreme right-hand side of equation (11.7) is zero. □

We now consider a property that is critical for having a well-defined, unique extrinsic mean, which will address item (iii) in Proposition 11.3.1.

DEFINITION 11.3.3. *A random object X on a Hilbert manifold* $\mathcal{M}$ *embedded in a Hilbert space is j-nonfocal if there is a unique point p on* $j(\mathcal{M})$ *at minimum distance from* $E(j(X))$.

To illustrate this idea, consider the following example involving a unit Hilbert sphere.

EXAMPLE 11.3.2. *The unit sphere $S(\mathbb{H}) = \{x \in \mathbb{H}, \|x\| = 1\}$ is a Hilbert manifold embedded in $\mathbb{H}$ via the inclusion map $j : S(\mathbb{H}) \to \mathbb{H}, j(x) = x$. A random object X on $S(\mathbb{H})$ of mean μ is j-nonfocal, if $\mu \neq 0$.*

Using this property, one may give an explicit formula for the extrinsic mean for a random object fon $P(\mathbf{H})$ with respect to the VW embedding, which we will call the VW mean.

PROPOSITION 11.3.2. *Assume $X = [\Gamma]$ is a random object in $P(\mathbf{H})$. Then the VW mean of X exists if and only if $E(\frac{1}{\|\Gamma\|^2}\Gamma \otimes \Gamma)$ has a simple largest eigenvalue, in which case, the distribution is j-nonfocal. In this case the VW mean is $\mu_E = [\gamma]$, where γ is an eigenvector for this eigenvalue.*

Proof. We select an arbitrary point $[\gamma] \in P(\mathbf{H}), \|\gamma\| = 1$. The spectral decomposition of $\Lambda = E(\frac{1}{\|\Gamma\|}\Gamma)$ is $\Lambda = \sum_{k=1}^{\infty} \delta_k^2 E_k, \delta_1 \geq \delta_2 \geq \dots$ where for all $k \geq 1, E_k = e_k \otimes e_k, \|e_k\| = 1$, therefore if $\gamma = \sum_{k=1}^{\infty} x_k e_k, \sum_{k=1}^{\infty} x_k^2 < \infty$, then $\|j(\gamma) - \mu\|^2 = \|\gamma \otimes \gamma\|^2 + \sum_{k=1}^{\infty} \delta_k^2 - 2 < \Lambda, \gamma \otimes \gamma >$. To minimize this distance it suffices to maximize the projection of the unit vector $\gamma \otimes \gamma$ on Λ. If $\delta_1 = \delta_2$ there are the vectors $\gamma_1 = e_1$ and $\gamma_2 = e_2$ are both maximizing this projection, therefore there is a unique point $j([\gamma])$ at minimum distance from Λ if and only if $\delta_1 > \delta_2$. □

A definition of a covariance parameter is also needed in order to define asymptotics and perform inference on an extrinsic mean. The following result is a straightforward extension of the corresponding finite dimensional result in Bhattacharya and Patrangenaru (2005) that was discussed in previous chapters. The tangential component $tan(v)$ of $v \in \mathbb{H}$ w.r.t. the orthobasis $e_a(P_j(\mu)) \in T_{P_j(\mu)} j(M), a = 1, 2, \cdots, \infty$ is given by

$$tan(v) = \sum_{a=1}^{\infty} (e_a(P_j(\mu)) \cdot v) e_a(P_j(\mu)). \tag{11.8}$$

Then given the j-nonfocal random object X, extrinsic mean μ_E, and covariance operator of $\tilde{\Sigma} = cov(j(X))$, if $f_a(\mu_E) = d_{\mu_E}^{-1}(e_a(P_j(\mu))), \forall a = 1, 2, \dots,$ then X has *extrinsic covariance operator* represented w.r.t. the basis $f_1(\mu_E), \cdots$ by the infinite matrix $\Sigma_{j,E}$:

$$\Sigma_{j,E} =$$

$$\left[\sum d_\mu P_j(e_b) \cdot e_a(P_j(\mu))\right]_{a=1,\dots} \tilde{\Sigma} \left[\sum d_\mu P_j(e_b) \cdot e_a(P_j(\mu))\right]_{a=1,\dots}^T. \tag{11.9}$$

With extrinsic parameters of location and covariance now defined, the asymptotic distribution of the extrinsic mean can be shown as in the final dimensional case (see Bhattacharya and Patrangenaru (2005) [43]):

PROPOSITION 11.3.3. *Assume $X_1,\ldots,X_n$ are i.i.d. random objects (r.o.'s) from a j-nonfocal distribution on a Hilbert manifold $\mathcal{M}$, for a given embedding $j : \mathcal{M} \to \mathbf{H}$ in a Hilbert space $\mathbf{H}$ with extrinsic mean μ_E and extrinsic covariance operator Σ. Then, with probability one, for n large enough, the extrinsic sample mean $\bar{X}_{n,E}$ is well defined. If we decompose $j(\bar{X}_{n,E}) - j(\mu_E)$ with respect to the scalar product into a tangential component in $T_{j(\mu_E)}j(\mathcal{M})$ and a normal component $N_{j(\mu_E)}j(\mathcal{M})$, then*

$$\sqrt{n}(tan(j(\bar{X}_{n,E}) - j(\mu_E))) \to_d \mathcal{G}, \tag{11.10}$$

where $\mathcal{G}$ has a centered Gaussian distribution in $T_{j(\mu_E)}j(\mathcal{M})$ with extrinsic covariance operator $\Sigma_{j,E}$.

11.4 A One-Sample Test of the Neighborhood Hypothesis

Following from a neighborhood method in the context of regression by Munk and Dette (1998) [251], Munk et al. (2008) [252] developed tests for means of random objects on Hilbert spaces. We now consider the adaptation of this methodology from Ellingson et al. (2013) [104] for tests for extrinsic means.

Assume Σ_j is the extrinsic covariance operator of a random object X on the Hilbert manifold $\mathcal{M}$ with respect to the embedding $j : \mathcal{M} \to \mathbb{H}$. Let $\mathbf{M}_0$ be a compact submanifold of $\mathcal{M}$. Let $\varphi_0 : \mathcal{M} \to \mathbb{R}$ be the function

$$\varphi_0(p) = \min_{p_0 \in \mathbf{M}_0} \|j(p) - j(p_0)\|^2, \tag{11.11}$$

and let $\mathbf{M}_0^\delta, \mathbb{B}_0^\delta$ be given respectively by

$$\begin{aligned} \mathbb{M}_0^\delta &= \{p \in \mathcal{M}, \varphi_0(p) \leq \delta^2\}, \\ \mathbb{B}_0^\delta &= \{p \in \mathcal{M}, \varphi_0(p) = \delta^2, \}. \end{aligned} \tag{11.12}$$

Since φ_0 is Fréchet differentiable and all small enough $\delta > 0$ are regular values of φ_0, it follows that $\mathbf{B}_0^\delta$ is a Hilbert submanifold of codimension one in $\mathcal{M}$. Let v_p be the normal space at a points $p \in \mathbf{B}_0^\delta$, orthocomplement of the tangent space to $\mathbb{B}_0^\delta$ at p. We define $\mathbb{B}_0^{\delta,X}$

$$\mathbb{B}_0^{\delta,X} = \{p \in \mathbf{B}_0, \Sigma_j|_{v_p} \text{is positive definite}\}. \tag{11.13}$$

DEFINITION 11.4.1. *The neighborhood hypothesis consists of the following two alternative hypotheses:*

$$\begin{aligned} H_0 &: \mu_E \in M_0^\delta \cup B_0^{\delta,X}, \\ H_0 &: \mu_E \in (M_0^\delta)^c \cap (B_0^{\delta,X})^c. \end{aligned} \tag{11.14}$$

Here, we consider neighborhood hypothesis testing for the particular situation in which the submanifold $\mathbf{M}_0$ consists of a point m_0 on $\mathcal{M}$. We set $\varphi_0 = \varphi_{m_0}$, and since $T_{m_0}\{m_0\} = 0$ we will prove the following result.

THEOREM 11.4.1. *If* $M_0 = \{m_0\}$, *the test statistic for the hypotheses specified in (11.14) has an asymptotically standard normal distribution and is given by:*

$$T_n = \sqrt{n}\{\varphi_{m_0}(\hat{\mu}_E) - \delta^2\}/s_n, \tag{11.15}$$

where

$$s_n^2 = 4\langle\hat{v}, S_{E,n}\hat{v}\rangle \tag{11.16}$$

and

$$S_{E,n} = \frac{1}{n}\sum_{i=1}^{n}(\tan_{\hat{\mu}} d_{\overline{j(X)}_n} P_j(j(X_i) - \overline{j(X)}_n)) \otimes$$
$$\otimes(\tan_{\hat{\mu}} d_{\overline{j(X)}_n} P_j(j(X_i) - \overline{j(X)}_n)) \tag{11.17}$$

is the extrinsic sample covariance operator for $\{X_i\}_{i=1}^n$, *and*

$$\hat{v} = (d_{\hat{\mu}_{E,n}} j)^{-1}\widehat{tan}_{j(\hat{\mu}_{E,n})}(j(m_0) - j(\hat{\mu}_{E,n})). \tag{11.18}$$

For the proof of this result, we first need the following useful extension of Cramer's delta method. The proof of this result is left to the reader.

THEOREM 11.4.2. *For* $a = 1,2$ *consider an embedding* $j_a : \mathcal{M}_a \to \mathbf{H}_a$ *of a Hilbert manifold* $\mathcal{M}_a$ *in a Hilbert space* $\mathbf{H}_a$. *Assume* $X_1,\dots,X_n$ *are i.i.d. r.o.'s from a* j_1*-nonfocal distribution on* $\mathcal{M}_1$ *for a given embedding* $j_1 : \mathcal{M}_1 \to \mathbf{H}_1$ *in a Hilbert space* $\mathbf{H}_1$, *with extrinsic mean* μ_E *and extrinsic covariance operator* Σ. *Let* $\varphi : \mathcal{M}_1 \to \mathcal{M}_2$ *be a differentiable function, such that* $\varphi(X_1)$ *is a* j_2*-nonfocal r.o. on* $\mathcal{M}_2$. *Then*

$$\sqrt{n}tan_{j_2(\varphi(\mu_E))}(j_2(\varphi(\bar{X}_{n,E})) - j_2(\varphi(\mu_E))) \to_d Y, \tag{11.19}$$

where $Y \sim \mathcal{N}(0, d_{\mu_E}\varphi^*\Sigma d_{\mu_E}\varphi)$. *Here* $d_{\mu_E}\varphi^*$ *is the adjoint operator of* $d_{\mu_E}\varphi$.

With this result, we now turn back to prove Theorem 11.4.1

Proof. The function φ_0 given in equation (11.11) defined on $\mathcal{M}$ can be written as a composite function $\varphi_0 = \Phi_A \circ j$, where $\Phi_A(x) = \|x - A\|^2$ is differentiable on $\mathbb{H}\backslash\{A\}$, with the differential at x given by $d_x\Phi_A(y) = 2 < y, x - A >$. Since $j(\mathcal{M})$ is a submanifold of $\mathbb{H}$, the restriction ϕ_A of $\Phi_A(x)$ to $j(\mathcal{M})$ is a differentiable function, with the differential

$$d_p\phi_A(y) = 2 < y, p - A >, \forall y \in T_p j(\mathcal{M}). \tag{11.20}$$

Note that $\varphi_{m_0}(p) = \phi_{j(m_0)}(j(p))$, therefore, given that the differential $d_p j$ is a vector space isomorphism, we obtain

$$d_p\varphi_{m_0}(u) = 2 < d_p j(u), j(p) - j(m_0) >, \forall u \in T_p\mathcal{M}, \tag{11.21}$$

and in particular

$$d_{\mu_E}\varphi_{m_0}(u) = 2 < d_{\mu_E}j(u), j(\mu_E) - j(m_0) >, \forall u \in T_{\mu_E}\mathcal{M}, \tag{11.22}$$

that is

$$d_{\mu_E}\varphi_{m_0} = 2d_{\mu_E}j \otimes tan(j(\mu_E) - j(m_0)). \tag{11.23}$$

Since the null hypothesis (11.14) is accepted as long as $\varphi_{m_0}(\mu_E) < \delta^2$, we derive the asymptotic distribution of $\varphi_{m_0}(\hat{\mu}_E)$ under $\varphi_{m_0}(\mu_E) = \delta^2$. From Theorem 11.4.2, it follows that

$$\sqrt{n}(\varphi_{m_0}(\hat{\mu}_E) - \varphi_{m_0}(\mu_E)) \to_d Y, \tag{11.24}$$

where $Y \sim \mathcal{N}(0, (d_{\mu_E}\varphi_{m_0})^*\Sigma_E d_{\mu_E}\varphi_{m_0})$, we see that the random variable

$$Z_n = \frac{\sqrt{n}(\varphi_{m_0}(\hat{\mu}_E) - \varphi_{m_0}(\mu_E))}{\sqrt{(d_{\mu_E}\varphi_{m_0})^*\Sigma_E d_{\mu_E}\varphi_{m_0}}} \tag{11.25}$$

has asymptotically a standard normal distribution. From equation (11.23), if we set

$$\begin{aligned} \nu = (d_{\mu_E}j)^{-1}tan(j(\mu_E) - j(m_0)), \\ \sigma^2 = 4 < \nu, \Sigma_E \nu >, \end{aligned} \tag{11.26}$$

then

$$Z_n = \frac{\sqrt{n}(\varphi_{m_0}(\hat{\mu}_E) - \delta^2)}{\sigma} \tag{11.27}$$

Finally we notice that $\hat{\nu}$ in (11.18) is a consistent estimator of ν in (11.26). Therefore s_n^2 in equation (11.16) is a consistent estimator of σ^2 in equation (11.26) and, from Slutsky's theorem, it follows that the test statistic T_n in equation (11.15) has asymptotically a $\mathcal{N}(0,1)$ distribution. □

Chapter 12

Analysis on Spaces of Congruences of k-ads

This chapter is concerned with nonparametric statistical analysis of landmark-based shape data, for which the effects of magnification and differences are taken into account. Our objective is to analyze extrinsic mean reflection-size-and-shapes associated with a random object on the reflection-size-and-shape manifold $SR\Sigma_{m,0}^k$ introduced in equation (3.134) in Section 3.5.5.

12.1 Introduction

It is a main focus of this chapter to construct *equivariant embeddings* of reflection-size-and-shape manifolds into vector spaces of matrices using an important representation of Schoenberg (1935) [300] that relates an *Euclidean distance matrix* of squared inter-point distances of a set of k points in $\mathbb{R}^m$ and the $k \times k$ positive semidefinite matrix of the inner products of the corresponding centered points. It is then appropriate to apply the general nonparametric theory in Chapter 5. A first approach to size-and-shape analysis using an Euclidean distance matrix, which differs from this one, is due to Lele (1993) [217]. Here, we introduce the Schoenberg embedding of the size-and-reflection-shape manifold $SR\Sigma_{m,0}^k$. The extrinsic sample mean size-and-reflection-shape under this embedding turns out to be the sample estimate $[MDS_m(W)]_{RS}$ in Dryden and

Mardia (1998, p. 281). We also provide the main ingredients for asymptotic nonparametric inference based on *Schoenberg means* on $SR\Sigma^k_{m,0}$. The necessary and sufficient condition for a probability measure Q on $SR\Sigma^k_{m,0}$ to be Schoenberg-nonfocal is that, in their decreasing order, the eigenvalues of rank m and $m+1$ of the mean matrix of the push forward distribution are distinct.

A second result of significance is the identification of the space $S\Sigma^k_2$ of size-and-shapes $[\mathbf{x}]_S$ of planar k-ads $\mathbf{x}$, with a noncompact manifold – the direct product of Σ^k_2 and $(0,\infty)$, where

$$[\mathbf{x}]_S = \{w\xi : w \in \mathbb{C}, |w| = 1\}. \tag{12.1}$$

In Section 12.2, an equivariant embeddeding ϕ of the planar *size-and-shape manifold* $S\Sigma^k_2$, extending the Veronese–Whitney embedding of the Kendall's shape space in Bhattacharya and Patrangenaru (2003) [42], is given. We show that the extrinsic mean size-and-shape of a probability measure Q on $S\Sigma^k_2$ exists (i.e. Q is ϕ-nonfocal) if and only if the largest eigenvalue λ_k of the push forward probability measure $\phi(Q) = Q\circ\phi^{-1}$ is simple, and in this case the extrinsic mean size and shape is $[\lambda_k\mathbf{u_0}]_S$, where $\mathbf{u_0}$ is a unit eigenvector of the mean matrix $\tilde{\mu}$ of $Q\circ\phi^{-1}$. This provides a proper venue for a future study of *allometry* - the dependence of shape on size, in various contexts.

12.2 Equivariant Embeddings of $S\Sigma^k_2$ and $RS\Sigma^k_{m,0}$

Recall from Chapter 10 the complex representation of the Kendall shape $\sigma(\mathbf{x})$ of a planar k-ad $\mathbf{x}$. If the scaling effect is removed by scaling ξ to unit size as

$$\mathbf{u} = \frac{\xi}{|\xi|}, \tag{12.2}$$

the transformed quantity $\mathbf{u}$ is called a *preshape.* The *quadratic Veronese–Whitney embedding* of Σ^k_2 into $S(k,\mathbb{C})$, the linear space of *selfadjoint complex matrices of order* k, (or simply the *Veronese–Whitney map*) is $j : \Sigma^k_2 \to S(k,\mathbb{C})$, where with $\mathbf{u}$ representing the preshape as in (13.1),

$$j([\mathbf{x}]) = \mathbf{u}\mathbf{u}^*, \mathbf{u}^*\mathbf{u} = 1. \tag{12.3}$$

Extend (12.3) to an embedding ϕ of the product model of $S\Sigma^k_2$ in Theorem 3.5.2 into $\mathbb{C}^{k^2}$ (regarded as the set of all $k\times k$ complex matrices) given by

$$\phi([\mathbf{x}]_S) = r\mathbf{u}\mathbf{u}^*, r > 0, \mathbf{u} \in L_k, \mathbf{u}^*\mathbf{u} = 1. \tag{12.4}$$

Note that the range of ϕ in (12.4) is a closed noncompact submanifold of $S(k,\mathbb{C})$.

The distance ρ on $S\Sigma^k_2$ is the Euclidean distance inherited from the embedding ϕ.

The embedding ϕ (as well as j) is $\mathcal{G}$-equivariant, where $\mathcal{G}$ is isomorphic to the group $SU(k-1)$ of $(k-1)\times(k-1)$ unitary matrices with determinant 1 (See Kendall et al. (1999), Bhattacharya and Patrangenaru (2003) [42]).

In higher dimensions, using an approach based on well known result MDS result due to Schoenberg (1935) [300] (see also Mardia et al. (1972, p. 397) [231]), presented in Bandulasiri and Patrangenaru (2005) [11] introduced the *Schoenberg embedding* of reflection shapes in higher dimensions (see Chapter 13). Let $S(k,\mathbb{R})$ denote the set of all $k\times k$ real symmetric matrices. The embedding considered here, of the size-and-reflection-shape manifold $J : SR\Sigma_{m,0}^k \to S(k,\mathbb{R})$, is given by

$$J([\xi]_{RS}) = \xi^T\xi. \tag{12.5}$$

The following result will be used to derive formulas for extrinsic parameters and for their estimators.

THEOREM 12.2.1. *The range of the Schoenberg embedding of $SR\Sigma_{m,0}^k$ is the subset $SM_{k,m}$ of $k\times k$ positive semidefinite symmetric matrices A with $rkA = m$, $A\mathbf{1_k} = \mathbf{0}$.*

Consider a $(k-1)\times k$ matrix H, whose rows are all of unit length, orthogonal to each other and orthogonal to the row vector $\mathbf{1_k^T}$.

PROPOSITION 12.2.1. *Let M_k be the space of $k\times k$ symmetric matrices A with $A\mathbf{1_k} = \mathbf{0}$. The map ϕ from M_k to $S(k-1,\mathbb{R})$, given by $\phi(A) = \mathbf{H}A\mathbf{H}^T$ is an isometry. In addition, $Tr(\phi(A)) = Tr(A)$.*

Proof. Since ϕ in Proposition 12.2.1 is a linear map, it suffices to show that $\|\phi(A)\| = \|A\|$. Here we consider the Euclidean norm of a matrix M given by $\|M\|^2 = Tr(MM^T)$. The claims are easily verified from the relations $\mathbf{HH}^T = \mathbf{I}_{k-1}, \mathbf{HH} = \mathbf{I}_k - \frac{1}{k}\mathbf{1}_k\mathbf{1}_k^T$, and the fact that for any matrices A,B, $Tr(AB) = Tr(BA)$, whenever both products make sense ■

We may then define an embedding ψ of the size-and-reflection-shape manifold as $\psi : SR\Sigma_{m,0}^k \to S(k-1,\mathbb{R})$, given by

$$\psi([\xi]_R) = \mathbf{H}\xi^{\mathbf{T}}\xi\mathbf{H}^{\mathbf{T}}. \tag{12.6}$$

From Proposition 12.2.1, it follows that the Schoenberg embedding and the embedding ψ induce the same distance on $SR\Sigma_{m,0}^k$.

REMARK 12.2.1. *The range of ψ is the set of $(k-1)\times(k-1)$ symmetric matrices of rank m. Note that for $k = m+1$, the range is the open convex subset $Sym^+(k-1,\mathbb{R}) \subset Sym(k-1,\mathbb{R})$ of positive definite symmetric matrices and the induced distance on $SR\Sigma_{m,0}^k$ is an Euclidean distance.*

REMARK 12.2.2. *Let $O(k)$ act on $SR\Sigma_{m,0}^k$ as $([\xi]_{RS}, A) \to [\xi A]_{RS}, A \in O(k)$. Then the embedding (12.5) is $O(k)$-equivariant. This action is not free. But in view of Proposition 12.2.1, the Schoenberg emebedding can be "tightened" to an $O(k-1)-$ equivariant embedding in $S(k-1,\mathbb{R})$.*

12.3 Extrinsic Means and Their Estimators

We recall from Chapter 5 that the *extrinsic mean* $\mu_{J,E}(Q)$ of a *nonfocal* probability measure Q on a manifold M w.r.t. an embedding $J : M \to \mathbb{R}^N$, when it exists, is the point from which the expected squared (induced Euclidean) distance under Q is minimum. It is given by $\mu_{J,E}(Q) = J^{-1}(P_J(\mu))$, where μ is the usual mean of $J(Q)$ as a probability measure on $\mathbb{R}^N$ and P_J is its projection on $J(M)$ (see Chapter 5, or Proposition 3.1 [42]). When the embedding J is given, and the projection $P_j(\mu)$ is unique, one often identifies $\mu_{j,E}$ with its image $P_j(\mu)$, and refer to the later as the extrinsic mean. The term "nonfocal Q" means that the projection (the minimizer of the distance from μ to points in $J(M)$) is unique. Often the extrinsic mean will be denoted by $\mu_E(Q)$, or simply μ_E, when J and Q are fixed in a particular context.

Assume $(X_1,\ldots,X_n)$ are i.i.d. M-valued random objects whose common probability measure is Q, and let $\overline{X}_E := \mu_{J,E}(\hat{Q}_n) = \mu_E(\hat{Q}_n)$ be the *extrinsic sample mean*. Here $\hat{Q}_n = \frac{1}{n}\sum_{j=1}^{n}\delta_{X_j}$ is the empirical distribution.

In this section we will consider extrinsic means associated with embeddings of the two types of manifolds described in this chapter.

12.3.1 *Mean Planar Shape and Mean Planar Size-and-Shape*

We first describe the extrinsic mean of a probability measure Q on Σ_2^k with respect to the Veronese–Whitney map j given in (12.3).

The squared distance in the space $S(k,\mathbb{C})$ of self adjoint matrices is $d_0^2(A,B) = Tr((A-B)(A-B)^*) = Tr((A-B)^2)$.

A probability measure Q on Σ_2^k may be viewed as a distribution of a random shape $[\mathbf{U}]$, where $\mathbf{U}$ is a random preshape. This probability measure Q is J-nonfocal if the largest eigenvalue of $E(\mathbf{U}\mathbf{U}^*)$ is simple, and in this case $\mu_{J,E}(Q) = [\mu]$, where $\mu \in S(L_k^2)$ is a unit eigenvector corresponding to this largest eigenvalue (see Bhattacharya and Patrangenaru (2003) [42]).

The extrinsic sample mean direct similarity planar shape $\overline{[\mathbf{X}]}_{J,E}$ of a random sample $[\mathbf{X_r}]$ with preshapes $\mathbf{U_r} = [U_r^1 : \cdots : U_r^k]$, $1_k^T\mathbf{U_r} = 0$, $\|\mathbf{U_r}\| = 1, r = 1,\ldots,n$, from such a nonfocal distribution exists with probability converging to 1 as $n \to \infty$ (see Bhattacharya and Patrangenaru (2003) [42]) and is given by

$$\overline{[\mathbf{X}]}_{J,E} = [\mathbf{U}], \tag{12.7}$$

where $\mathbf{U}$ is a unit eigenvector in $S(L_k^2)$ corresponding to the largest eigenvalue of

$$K := n^{-1}\sum_{r=1}^{n}\mathbf{U_r}\mathbf{U_r^*}. \tag{12.8}$$

This means that $\overline{[\mathbf{X}]}_{J,E}$ is given by a formula that is similar to the one for the *full Procrustes estimate* of the mean shape in parametric families such as

Dryden–Mardia type distributions or complex Bingham type distributions for planar shapes on Σ_2^k (Kent (1992) [181], (1994) [182]) . For this reason, the Veronese-Whitney extrinsic sample mean shape may be called the *Procrustes* mean estimate.

To compute the extrinsic mean of probability measures on $S\Sigma_2^k$, we proceed exactly as in the case of Σ_2^k, but under the additional assumption that the image of Q under ϕ, namely $Q \circ \phi^{-1}$, regarded as a probability measure on $\mathbb{C}^{k^2}(\approx \mathbb{R}^{2k^2})$ has finite second moments. With this assumption, let $\tilde{\mu}$ be the mean ($k \times k$ matrix) of $\phi(Q) = Q \circ \phi^{-1}$. Then $\tilde{\mu}$ is a Hermitian matrix, which is positive semidefinite. There exists a complex orthogonal matrix T such that $T\tilde{\mu}T^* = D = diag(\lambda_1, \dots, \lambda_k)$, where $0 \le \lambda_1 \le \dots \le \lambda_k$ are the eigenvalues of $\tilde{\mu}$. Then if $\mathbf{v} = \sqrt{r}T\mathbf{u}$, the squared distance between $\tilde{\mu}$ and an element $\phi([\mathbf{x}]_S)$ of $\phi(S\Sigma_2^k)$ is given by

$$
\begin{aligned}
Trace(\tilde{\mu} - r\mathbf{u}\mathbf{u}^*)^2 &= Trace(D - \mathbf{v}\mathbf{v}^*)^2 = \\
&= \sum_j \lambda_j^2 + \sum_j |v_j|^4 - 2\sum_j \lambda_j |v_j|^2 + \sum_{j \neq j'} |v_j \bar{v}'_j|^2 = \\
= \sum_j \lambda_j^2 + \sum_j |v_j|^4 - 2\sum_j \lambda_j |v_j|^2 &+ \sum_j |v_j|^2 \sum_j |v'_j|^2 - \sum_j |v_j|^4 = \\
&= \sum_j \lambda_j^2 - 2\sum_j \lambda_j |v_j|^2 + r^2(\mathbf{x})
\end{aligned}
\tag{12.9}
$$

noting that $|\mathbf{v}|^2 = r(\mathbf{x})$. We first minimize (12.9) for a given size $r = r(\mathbf{z})$. Clearly this is achieved by letting $\mathbf{v} = \sqrt{r}e_k$ (or $\sqrt{r}e^{i\theta}e_k$ for some θ), where e_k has 1 as its last (k-th) coordinate, and zeros elsewhere. Then $\mathbf{u} = \frac{1}{\sqrt{r}}T^*v$ is an eigenvector of $\tilde{\mu}$ in the eigenspace of the largest eigenvalue λ_k. With this choice (12.9) becomes

$$
\sum_j \lambda_j^2 - 2r\lambda_k + r^2 \tag{12.10}
$$

The minimum of (12.10) over all $r > 0$ is attained with $r = \lambda_k$. Hence the minimum of the (12.9) is achieved by an element $\mathbf{u} = \mathbf{u}_0$ where $\mathbf{u}_0$ is a unit vector in the eigenspace of λ_k, i.e., by the element $\phi([\mathbf{x}]_S) = \phi([\lambda_k \mathbf{u}_0]_S)$ of $\phi(S\Sigma_2^k)$. If λ_k is a simple eigenvalue of $\tilde{\mu}$, then this minimizer is *unique*. In this case the extrinsic mean μ_E of Q is the size-and-shape of $\lambda_k \mathbf{u}_0$. Hence the consistency theorem in Bhattacharya and Patrangenaru (2003)[42] applies in this case. The size of the mean μ_E is λ_k.

12.3.2 Extrinsic Mean Size-and-Reflection-Shapes

We consider now a random k-ad in general position $\mathbf{X}$, centered as $\mathbf{X_0} = (X^1 - \overline{X}, \dots, X^k - \overline{X}) \in (\mathbb{R}^m)^k \simeq M(m,k;\mathbb{R})$. Set

$$C = E(\mathbf{X_0}^T \mathbf{X_0}). \tag{12.11}$$

Obviously, $C1_k = 0, C \geq 0$.

The extrinsic mean size-and-reflection-shape of $[\mathbf{X}]_{RS}$ exists if $Tr(C - \xi^T\xi)^2$ has a unique solution $\xi \in M(m,k;\mathbb{R})$ up to an orthogonal transformation, with

$$\xi 1_k = 0, \mathrm{rk}\xi = m \tag{12.12}$$

That is the same as saying that given C, ξ is a *classical solution* in $\mathbb{R}^m$ to the MDS problem, as given in Mardia et al. (1972, p. 408) [231] in terms of the first largest m eigenvalues of C. Assume the eigenvalues of C in their decreasing order are $\lambda_1 \geq \dots \geq \lambda_k$. The classical solution of the MDS problem is *unique* (up to an orthogonal transformation) if $\lambda_m > \lambda_{m+1}$ and ξ^T can be taken as the matrix

$$V = (v_1 v_2 \dots v_m), \tag{12.13}$$

whose columns are orthogonal eigenvectors of C corresponding to the largest eigenvalues $\lambda_1 \geq \dots \geq \lambda_m$ of C, with

$$v_j^T v_j = \lambda_j, \forall j = 1, \dots, m. \tag{12.14}$$

Since the eigenvectors $v_1, \dots, v_m$ are linearly independent, $\mathrm{rk}\xi = m$. If v is an eigenvector of C for the eigenvalue $\lambda > 0$, since $C1_k = 0$ it follows that $v1_k = 0$. Therefore the classical solution ξ derived from the eigenvectors (12.13) satisfies (12.12) . In conclusion we have:

THEOREM 12.3.1. *Assume $C = \sum_{i=1}^k \lambda_i e_i e_i^T$ is the spectral decomposition of $C = E(\mathbf{X_0^T X_0})$, then the extrinsic mean μ_E size-and-reflection-shape exists if and only if $\lambda_m > \lambda_{m+1}$ and if this is the case, $\mu_E = [\xi]_{RS}$ where ξ^T can be taken as the matrix* (12.13) *satisfying* (12.14).

From theorem 12.3.1, it follows that given k-ads in general position in $\mathbb{R}^m$, $\{\mathbf{x}_1, \dots, \mathbf{x}_n\}, \mathbf{x}_j = (x_j^1, \dots, x_j^k), j = 1, \dots, n$, their extrinsic sample mean size-and-reflection-shape is $\overline{[\mathbf{x}]}_E = [\hat{\xi}]_{RS}$, where $\hat{\xi}$ is the classical solution in $\mathbb{R}^m$ to the MDS problem for the matrix

$$\hat{C} = \frac{1}{n} \sum_{j=1}^n \xi_j^T \ \xi_j. \tag{12.15}$$

Here ξ_j is the matrix obtained from $\mathbf{x}_j$ after centering , assuming $\hat{\lambda}_m > \hat{\lambda}_{m+1}$. Here $\hat{\lambda}_1 \geq \dots \geq \hat{\lambda}_k$ are the eigenvalues of $\hat{C}$. Indeed, the configuration of the sample mean is given by the eigenvectors corresponding to the m largest eigenvalues of $\hat{C}$.

REMARK 12.3.1. *From Remark 12.2.1, in the case $k = m+1$, the projection P_ψ is the identity map, therefore any distribution Q is ψ-nonfocal and $\psi(\mu_{E,\psi})$ is the mean μ of $\psi(Q)$.*

12.4 Asymptotic Distribution of Extrinsic Sample Mean Size-and-Reflection-Shapes

The reflection shape manifold $R\Sigma^k_{m,0}$ is the set of size-and-reflection-shapes in general position of size one, submanifold of $SR\Sigma^k_{m,0}$. Recall the embedding in (12.6), namely,

$$\psi([\xi]_{RS}) = HJ([\xi]_{RS})H^T = H\xi^T\xi H^T, \tag{12.16}$$

From Proposition 12.2.1 and with the notation in Theorem 12.2.1, it follows that $\phi(SM_{k,m})$ which is also the range of ψ in (12.16) is the set $\tilde{N}_{m,k}$ of $(k-1)\times(k-1)$ positive semidefinite symmetric matrices of rank m, and the restriction of ϕ to $SM_{m,k}$ is an isometry from $(SM_{m,k}, d_{k,0})$ to $(\tilde{N}_{m,k}, d_{k-1,0})$ where $d_{r,0}$ is the restriction of the Euclidean distance on the space of $r\times r$ symmetric matrices.

REMARK 12.4.1. *If, for $\eta = \xi H$, we set $\widetilde{s\sigma}(\eta) = [\xi]_{RS}$, the embedding ψ is equivariant with respect to the group actions of $O(k-1)$ on $SR\Sigma^k_{m,0}$ and on $S(k-1,\mathbb{R})$:*

$$\begin{aligned}\alpha(\widetilde{s\sigma}(\eta), A) &= \widetilde{s\sigma}(\eta A)\\ \beta(S,A) &= ASA^T.\end{aligned} \tag{12.17}$$

In this section we will derive the asymptotics for the extrinsic sample mean size-and-reflection-shape for samples from a ψ-nonfocal distributions Q on $SR\Sigma^k_{m,0}$. For this purpose we will use the general results for extrinsic means on a manifold in Bhattacharya and Patrangenaru (2005) (Theorem 3.1. and its corollaries) [11]. The tangent space to $\tilde{N}_{m,k}$ at $\psi([\xi]_{RS})$ is the range of the differential of ψ at $[\xi]_{RS}$. If we set $\eta = \xi H^T$, then $rank(\eta) = m$ and $\psi([\xi]_{RS}) = \eta^T\eta$, and

$$T_{\psi(\widetilde{s\sigma}(\eta))}\tilde{N}_{m,k} = T_{\eta^T\eta}\tilde{N}_{m,k} = \{v \in S(k-1,\mathbb{R}) : v = y^T\eta + \eta^T y, y \in \mathcal{M}(k-1,m,\mathbb{R})\} \tag{12.18}$$

Since $\eta\eta^T$ has an inverse, for any $y \in \mathcal{M}(k-1,m,\mathbb{R})$ we define the symmetric matrix $S = \frac{1}{2}(\eta^T(\eta\eta^T)^{-1}y + y^T(\eta\eta^T)^{-1}\eta)$ and obtain the following representation of the tangent space in (12.18):

$$T_{\eta^T\eta}\tilde{N}_{m,k} = \{v \in S(k-1,\mathbb{R}), v = S\eta\eta^T + \eta^T\eta S, S \in S(k-1,\mathbb{R})\}. \tag{12.19}$$

Given the equivariance of ψ, if $\eta^T\eta = A\Lambda A^T$, with $A \in O(k-1)$ and Λ the diagonal matrix with diagonal elements $\lambda_1, \dots \lambda_m, 0, \dots, 0$, then

$$T_{\eta^T\eta}\tilde{N}_{m,k} = AT_\Lambda N_{m,k}A^T. \tag{12.20}$$

From (12.19), it follows that the tangent space at Λ is given by

$$T_\Lambda \tilde{N}_{m,k} = \{v \in S(k-1:\mathbb{R}), v = \begin{pmatrix} V_m & W \\ W^T & \mathbf{0} \end{pmatrix},$$
$$V_m \in S(m,\mathbb{R}), W \in \mathcal{M}(k-1-m,m;\mathbb{R})\}. \tag{12.21}$$

A standard orthonormal basis in $S(k-1,\mathbb{R})$ given in Bhattacharya and Patrangenaru (2003) [42] is the basis

$$\tilde{E} = (E_1^1, \dots, E_{k-1}^{k-1}, 2^{-\frac{1}{2}}(E_i^j + E_j^i), 1 \le i < j \le k-1), \tag{12.22}$$

where E_i^j has all entries zero, except for the entry in the ith row and j^{th} column, which equals 1. From (13.9) it follows that $T_\Lambda \tilde{N}_{m,k}$ is spanned by the orthobasis

$$e(\Lambda) = (E_1^1, \dots, E_m^m, 2^{-\frac{1}{2}}(E_i^j + E_j^i), 1 \le i < j \le m \text{ or } 1 \le i \le m < j \le k-1). \tag{12.23}$$

Since, for any $A \in O(k-1)$, the map $v \to AvA^T$ is an isometry of $S(k-1,\mathbb{R})$, an orthobasis in the space $T_{\eta^T\eta}\tilde{N}_{m,k}$ is given then by

$$e(\eta^T\eta) = (E_1^1(\eta), \dots, E_m^m(\eta), 2^{-\frac{1}{2}}(E_i^j(\eta) + E_j^i(\eta)), \tag{12.24}$$

for $1 \le i < j \le m$ or $1 \le i \le m < j \le k-1$, where $E_i^j(\eta) = AE_i^jA^T$.

REMARK 12.4.2. *Since the asymptotic results are often presented in vector notation, it will be useful to order the orthobasis* (12.22) *in the following noncanonical way*

$$E = (E_1^1, \dots, E_m^m, 2^{-\frac{1}{2}}(E_i^j + E_j^i), 1 \le i < j \le m \text{ or } 1 \le i \le m < j \le k-1,$$
$$E_{m+1}^{m+1}, \dots, E_{k-1}^{k-1}, 2^{-\frac{1}{2}}(E_i^j + E_j^i), m+1 \le i < j \le k-1). \tag{12.25}$$

The extrinsic mean $\mu_E = \mu_{\psi,E}(Q)$ of a Schoenberg nonfocal probability measure Q on $SR\Sigma_{m,0}^k$ is given by $\mu_E = \psi^{-1}(P_\psi(\mu))$, where μ is the mean of $\psi(Q)$ in $S(k-1,\mathbb{R})$ and P_ψ is the projection on $\tilde{N}_{m,k}$. Following Bhattacharya and Patrangenaru (2005) [43], the extrinsic covariance operator $\Sigma_E = \Sigma_{\psi,E}$ is the restriction of the self-adjoint linear operator $d_\mu P_\psi \Sigma d_\mu P_\psi^T$ to $T_{P_\psi(\mu)}\tilde{N}_{m,k}$. The extrinsic covariance matrix is the matrix associated to Σ_E with respect to an orthobasis $e_1(P_\psi(\mu)), \dots, e_d(P_\psi(\mu))$ of $T_{P_\psi(\mu)}\tilde{N}_{m,k}, d = \frac{m}{2}(2k-m-1)$.

LEMMA 12.4.1. *Assume the mean μ of $\psi(Q)$ is a diagonal matrix Λ. The differential of the projection P_ψ at Λ with respect to the ordered orthobasis* (12.25) *is given by*

$$d_\Lambda P_\psi(E_i^i) = \begin{cases} E_i^i & i \le m \\ 0 & i > m \end{cases}, \quad d_\Lambda P_\psi(E_j^l + E_l^j) = \begin{cases} E_j^l + E_l^j & j < l \le m \\ \frac{\lambda_j}{\lambda_j - \lambda_l}(E_j^l + E_l^j) & j \le m < l \\ 0 & m < j < l. \end{cases} \tag{12.26}$$

Given the equivariance of the embedding ψ, from Lemma 12.4.1 we obtain the following

PROPOSITION 12.4.1. *If the spectral decomposition of the mean μ of $\psi(Q)$ is $\mu = \sum_{i=1}^{k-1} \lambda_i \tilde{e}_i \tilde{e}_i^t$, with $\lambda_1 \geq \cdots \geq \lambda_m > \lambda_{m+1} \geq \cdots \geq \lambda_{k-1}$, then*
(i) the tangent space $T_{\psi(\mu_E)} N_{m,k} = T_1 \oplus T_2$, where T_1 has the orthobasis

$$(\tilde{e}_1\tilde{e}_1^T, \ldots, \tilde{e}_m\tilde{e}_m^T, 2^{-\frac{1}{2}}(\tilde{e}_i\tilde{e}_j^T + \tilde{e}_j\tilde{e}_i^T), 1 \leq i < j \leq m)), \tag{12.27}$$

and T_2 has the orthobasis

$$(2^{-\frac{1}{2}}(\tilde{e}_j\tilde{e}_l^T + \tilde{e}_l\tilde{e}_j^T), 1 \leq j \leq m < l \leq k-1). \tag{12.28}$$

(ii) Let N be the orthocomplement of $T_{\psi(\mu_E)} N_{m,k}$ Then if

$$\begin{aligned} d_\mu P_\psi | T_1 &= Id_{T_1}, \\ d_\mu P_\psi(\tilde{e}_j\tilde{e}_l^T + \tilde{e}_l\tilde{e}_j^T) &= \frac{\lambda_j}{\lambda_j - \lambda_l}(\tilde{e}_j\tilde{e}_l^T + \tilde{e}_l\tilde{e}_j^T), \forall (j,l), 1 \leq j \leq m < l \leq k-1, \\ d_\mu P_\psi | N &= 0. \end{aligned} \tag{12.29}$$

An orthobasis of N in Proposition 13.4.2 is

$$2^{-\frac{1}{2}}(\tilde{e}_j\tilde{e}_l^T + \tilde{e}_l\tilde{e}_j^T), m < j < l \leq k-1). \tag{12.30}$$

The two orthobases (12.27), (13.53) and (13.55) yield an orthobasis $\tilde{\mathbf{e}}$ of $S(k-1, \mathbb{R})$. From proposition 13.4.2 it follows that the matrix D associated with the differential $d_\mu P_\psi$ relative to the orthobasis $\tilde{\mathbf{e}}$ is diagonal:

$$D = \begin{pmatrix} I_{\frac{m(m-1)}{2}} & \mathbf{0} & \mathbf{0} \\ \mathbf{0} & \Delta_{m(k-m-1)} & \mathbf{0} \\ \mathbf{0} & \mathbf{0} & \mathbf{0} \end{pmatrix}, \tag{12.31}$$

where

$$\Delta_{m(k-m-1)} = \begin{pmatrix} \frac{\lambda_1}{\lambda_1 - \lambda_{m+1}} & \cdots & 0 \\ \cdots & \cdots & \cdots \\ 0 & \cdots & \frac{\lambda_m}{\lambda_m - \lambda_{k-1}} \end{pmatrix}. \tag{12.32}$$

The space of symmetric matrices$S(k-1, \mathbb{R})$ regarded as its own tangent space at μ splits in three orthogonal subspaces

$$S(k-1, \mathbb{R}) = T_1 \oplus T_2 \oplus N, \tag{12.33}$$

leading to a decomposition of the covariance matrix Σ of $\psi(Q)$, with respect to the orthobasis of $S(k-1,\mathbb{R})$ obtained by augmenting the orthobasis (13.47) by an orthobasis of N, as follows:

$$\Sigma = \begin{pmatrix} \Sigma_{11} & \Sigma_{12} & \Sigma_{13} \\ \Sigma_{12}^T & \Sigma_{22} & \Sigma_{23} \\ \Sigma_{13}^T & \Sigma_{23}^T & \Sigma_{33} \end{pmatrix}. \tag{12.34}$$

If we change the coordinates in $\mathbb{R}^{k-1}$ by selecting an orthobasis $\tilde{\mathbf{e}}$, the eigenvectors $\tilde{e}_1,\dots,\tilde{e}_{k-1}$ of μ, in such a coordinate system, the mean is a diagonal matrix Λ and the matrix $\Sigma_\mu = D\Sigma D^T$, defined in Bhattacharya and Patrangenaru (2005) [43] is

$$\Sigma_\mu = \begin{pmatrix} \Sigma_{11} & \Sigma_{12}\Delta & 0 \\ \Delta\Sigma_{12}^T & \Delta\Sigma_{22}\Delta & 0 \\ 0 & 0 & 0 \end{pmatrix}, \tag{12.35}$$

and the extrinsic covariance matrix Σ_E defined in Bhattacharya and Patrangenaru (2005)[43], with respect to the basis $d_{\mu_\psi}^{-1}(e(\Lambda))$, with $e(\Lambda)$ as defined in (13.47), is

$$\Sigma_E = \begin{pmatrix} \Sigma_{11} & \Sigma_{12}\Delta \\ \Delta\Sigma_{12}^T & \Delta\Sigma_{22}\Delta \end{pmatrix}, \tag{12.36}$$

We assume now that $Y_1,\dots,Y_n$ are independent identically distributed random reflection objects from a ψ-nonfocal probability distribution Q on $S\Sigma_{m,0}^k$, with $\lambda_m > \lambda_{m+1}$ and let $\widetilde{s\sigma}(\eta)$ be the mean of $\psi(Q)$ and Σ the covariance matrix of $\psi(Q)$ with respect to the orthobasis $\tilde{\mathbf{e}}$ defined above. Let $\vec{W}$ be the vectorized form of a matrix $W \in S(k-1,\mathbb{R})$ with respect to the basis $\tilde{V}$. Assume $tan\vec{W}$ denote the component of $\vec{W}$ tangent to $\vec{\tilde{N}}_{m,k}$ at $\vec{\psi}(\mu_\psi)$.

THEOREM 12.4.1. *(a) The random vector $n^{\frac{1}{2}} tan(\vec{\psi}(\overline{Y}_E) - \vec{\psi}(\mu_E))$ converges weakly to a random vector having a $N(0,\Sigma_E)$ distribution, where Σ_E is given in* (13.61).
(b) If Σ_E is nonsingular, then $n\ tan(\vec{\psi}(\overline{Y}_E) - \vec{\psi}(\mu_E))^T \Sigma_E^{-1} tan(\vec{\psi}(\overline{Y}_E) - \vec{\psi}(\mu_E))^T$ converges weakly to a $\chi^2_{km-\frac{m(m+1)}{2}}$ distribution.

From Theorem 13.4.2 we obtain the following result:

COROLLARY 12.4.1. *Let G be a normally distributed matrix in $S(k-1,\mathbb{R})$, weak limit of $n^{\frac{1}{2}}(\overline{Y}-\mu)$. Assume the spectral decomposition of μ is $\mu = V\Lambda V^T$. Set $G^V = V^T G V = (g_{jl}^V)$ and $\tilde{G}^V = (\tilde{g}_{jl}^V)$ be determined by*

$$\tilde{g}_{jl}^V = \begin{cases} g_{jl}^V & 1 \le j \le l \le m \\ \frac{\lambda_j}{\lambda_j - \lambda_l} g_{jl}^V & 1 \le j \le m < l \le m-1 \\ 0 & m < j \le l \le k-1 \end{cases} \tag{12.37}$$

Then $n^{\frac{1}{2}}(\psi(\overline{Y}_E) - \psi(\mu_E))$ converges in distribution to the normally distributed random matrix $VG^V V^T$.

From Theorem 13.4.2 it follows that the extrinsic mean size-and-reflection-shape can be easily estimated using non-pivotal bootstrap . Assume $\{\mathbf{x}_1,\ldots,\mathbf{x}_n\}$ is a random sample of configurations $\mathbf{x}_j = (x_j^1,\ldots,x_j^k), j = 1,\ldots,n$. We resample at random and with repetition N times from this sample, where N is a reasonably large number, say $N \geq 500$. For each such resample $\mathbf{x}_1^*,\ldots,\mathbf{x}_n^*$ we compute the extrinsic sample mean $\overline{[\mathbf{x}]_{RS_E}}^*$. We then use a local parametrization of $SR\Sigma_{m,0}^k$ and find $(1-\alpha)100\%$ Bonferroni simultaneous confidence intervals for the corresponding $km - \frac{m(m+1)}{2}$ local coordinates.

Tests for the equality of two extrinsic mean size-and-shapes can be derived from the general theory for two sample tests for extrinsic means on manifolds developed by A. Bhattacharya (2008) [25].

REMARK 12.4.3. *For extrinsic mean size-and-reflection-shapes, $k = m+1$, from Remark 12.2.1, it follows that the space $SR\Sigma_{m,0}^{m+1}$ is isometric to a convex open subset of an Euclidean space of dimension $\frac{m(m+1)}{2}$ and, in view of Remark 12.3.1 this isometry carries the extrinsic means to ordinary means in this Euclidean space, and inference for means on $SR\Sigma_{m,0}^{m+1}$ follows from multivariate analysis. In particular, for $k = m+1$, at level α, a nonpivotal bootstrap test for the matched paired hypothesis $H_0 : \mu_{1,E} = \mu_{2,E}$ can be obtained as follows. Given matched pair samples $[\mathbf{x}_{1,\mathbf{i}}]_{RS}, [\mathbf{x}_{2,\mathbf{i}}]_{RS}, i = 1,\ldots,n$, consider $100(1-\alpha)$ simultaneous confidence intervals for the mean difference matrix $\psi(\mu_{1,E}) - \psi(\mu_{2,E})$ obtained from the bootstrap distribution of $\overline{\psi([\mathbf{x}_1]_{RS}) - \psi([\mathbf{x}_2]_{RS})}$, and reject H_0 if at least one of these intervals does not contain* 0.

12.5 Mean Size-and-Shape of Protein Binding Sites

To illustrate a practical situation in which the Schoenberg mean size-and-shape is useful, we consider binding sites of proteins, as discussed in Chapter 1. It is commonly hypothesized in bioinformatics literature that the structure and function of binding sites of proteins are in some way related. A first step towards exploring this problem involves finding common atoms of groups of binding sites and computing the Schoenberg sample mean size-and-shape of these structures.

To demonstrate, we will now present a few examples. First, we consider 4 protein binding sites, as in Ellingson (2011) [101]. These sites, obtained from the RCSB Protein Data Bank and shown in Figure 12.1, are found in the proteins 1phf, 1phg, 2cpp, and 1m85 and bind to the ligand heme. The matching of atoms can be performed using a number of algorithms, but for this example, we utilize TIPSA, a method presented in Ellingson and Zhang (2012) [107]. The sample extrinsic mean size-and-shape is shown in 12.2. As a second example, we return to the two sets of binding sites presented in Figures 1.8 and 1.9 in Chapter 1. As in the previous example, the atoms common to each sample of

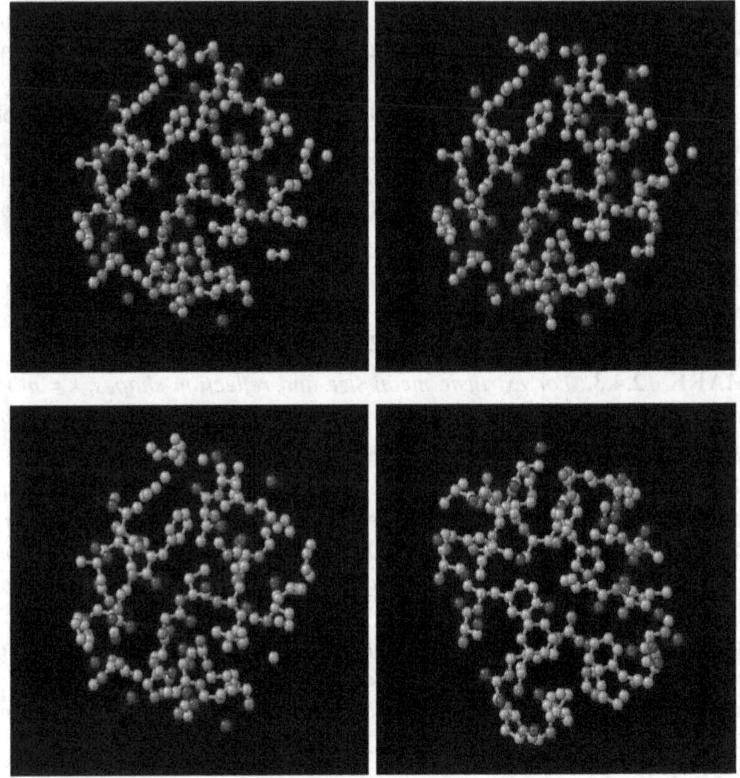

Figure 12.1 *Binding sites that bind to heme for proteins 1phf, 1phg, 2cpp, and 1m85. (Source: Bhattacharya et al.(2012), Figure 7. Reproduced by permission of John Wiley & Sons LTD).*

binding sites were obtained. Figure 12.3 shows the Schoenberg sample mean size-and-shape calculated for each of these samples.

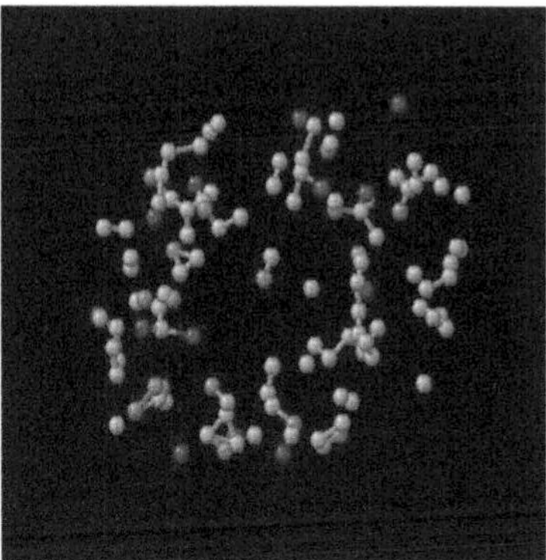

Figure 12.2 *The Schoenberg extrinsic mean reflection size-and-shape of the atoms common to the binding sites. (Source: Bhattacharya et al.(2012), Figure 7. Reproduced by permission of John Wiley & Sons LTD).*

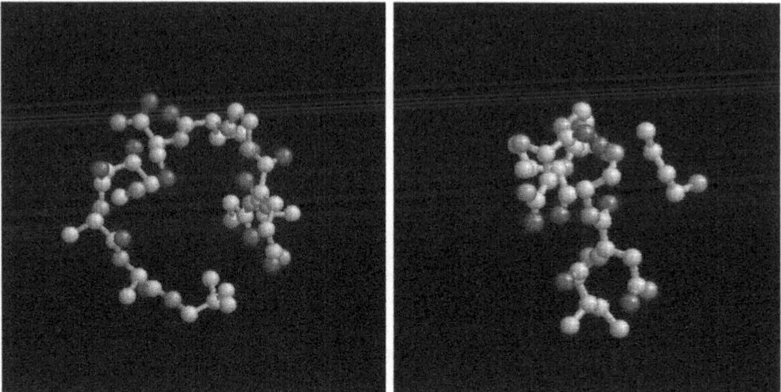

Figure 12.3 *The Schoenberg extrinsic mean reflection size-and-shape for samples of serine proteinase and acid proteinase, binding sites, as shown in Figures 1.8 and 1.9.*

Chapter 13

Similarity Shape Analysis

13.1 Introduction

In this chapter, we are concerned with a landmark based nonparametric analysis of similarity shape data. For landmark based shape data, one considers a k-ad $\mathbf{x} = (x^1, \ldots, x^k) \in (\mathbb{R}^m)^k$, which consists of k labeled points in $\mathbb{R}^m$ that represent coordinates of *landmarks* (see Chapter 1). Each landmark represents a location on an object having a certain significance, which may have a geometric or anatomical meaning. The landmark coordinates, themselves, are of no interest in shape analysis, as they depend on the position of the object relative to the device, such as a digital camera, that recorded the data.

The physical principle behind the data acquisition process is key in the selection of the transformation by which the coordinates of two landmark configurations are identified. For example, if the pixel coordinates of landmarks in a frontal digital camera image of a still scene are recorded, the coordinates obtained from two such photographs taken at the same distance differ by a *congruence*, and one may consider the *size-and-shape* of the k-ad, $s\sigma(x)$, as discussed in the previous chapter, which is defined as the orbit of ξ under the action of $SO(m)$, where $\xi = (\xi^1, \ldots, \xi^k)$ represents the centered configuration, given in (3.122).

On the other hand, if the landmark coordinates differ by a composition of a rotation, translation and scaling, one considers the direct similarity shape $[\mathbf{x}]$ of a k-ad in $\mathbb{R}^m$, as defined in (3.124). In the latter case, one "discards" the effects of magnification and differences that may arise because of variation in size or in equipment used or due to the manner in which images are taken and digitally recorded. In the language of group actions developed in Chapter 3, the transformations in this case would belong to the group of direct isometries, if size is taken into account, or the one generated by direct isometries and scaling, as proposed in a pioneering paper by Kendall (1984) [177] for measuring shapes, if the size is ignored. A general concept of $\mathcal{G}$*-shape*, associated with the action of a Lie group $\mathcal{G}$ on $\mathbb{R}^m$ is due to Patrangenaru and Patrangenaru (2004) [275]. Once the type of $\mathcal{G}$-shape data was identified, the sample space is known, and the main focus is the choice of a distance on it.

In this chapter, we consider only $\mathcal{G}$-shapes, where $\mathcal{G} = SO(m)$ or $\mathcal{G} = O(m)$. For reasons that will be evident in Chapter 22, our preference is for *extrinsic distances*. Our main focus here is to to construct (in Section 13.2) *equivariant embeddings* of such $O(m)$-spaces into vector spaces of matrices, using an important representation of Schoenberg (1935) [300] relating a matrix of squared Euclidean inter-point distances of a set of k points in $\mathbb{R}^m$ and the $k \times k$ positive semidefinite matrix of the inner products of the corresponding centered points, which is for size-and-shapes. It is then appropriate to apply the general nonparametric theory from Chapter 5. Recall from Section 3.5 that the reflection shape space is the manifold $\Sigma^k_{m,0}$ of reflection shapes of k-ads for which $\{\xi_1, \ldots, \xi_k\}$ spans $\mathbb{R}^m$. The manifold approach to reflection-shape analysis, including Schoenberg embeddings and connections to MDS in Section 2.12 was initiated by Bandulasiri and Patrangenaru (2005) [11].

REMARK 13.1.1. *Note that the mean reflection-shape, introduced by Dryden et al. (2008) [90], which in Bandulasiri et al. (2009) [10] was called* MDS-mean reflection shape, *is not an extrinsic mean, as pointed out in Section 5 of A. Bhattacharya (2009)[25]. The first correct formula for the Schoenberg reflection-shape is in fact the one given in Bandulasiri et al. (2009) [12], and, independently, in A. Bhattacharya (2009) [24].*

Given Remark 13.1.1, the manifold based reflection-shape analysis approach in this chapter has limited connection with the MDS based mean reflection-shape estimators in Dryden and Mardia (1998 [91], p. 281).

In Section 13.3, we introduce the Schoenberg embedding of the reflection-shape manifold $\Sigma^k_{m,0}$. The extrinsic sample mean reflection-shape under this embedding turns out to be the sample estimate $[MDS_m(W)]_{RS}$ in Dryden and Mardia (1998 [91], p. 281). However, the population mean defined by the latter (see Dryden and Mardia (1998 [91], pp. 88, 279), in the parametric or semiparametric setting, is in general different from the extrinsic mean described here, unless the underlying distribution has additional structure (such as isotropy, in the case $m = 2$). It now follows from the general results on the consistency of the extrinsic sample mean as an estimator of the extrinsic mean (see Chapter 4) that the estimate in Dryden and Mardia (1998) [91] is a consistent estimate of their target population parameter only for certain special classes of distributions.

Section 13.3 provides the main ingredients for asymptotic nonparametric inference based on *Schoenberg means* on $\Sigma_{m,0}^k$. The necessary and sufficient condition for a probability measure Q on $\Sigma_{m,0}^k$ to be Schoenberg-nonfocal is that, in their decreasing order, the eigenvalues of rank m and $m+1$ of the mean matrix of the push forward distribution are distinct. The extrinsic sample mean differs from the MDS sample mean estimate in Dryden and Mardia (1998) [91] or in Kent (1994) [182] (in the case $m = 2$).

13.2 Equivariant Embeddings of Σ_2^k and $R\Sigma_{m,0}^k$

Recall that in the complex representation of a Kendall shape $\sigma(\mathbf{x})$ of a planar k-ad $\mathbf{x}$, if the scaling effect is removed by scaling ξ to unit size as

$$\mathbf{u} = \frac{\xi}{|\xi|}, \tag{13.1}$$

the transformed quantity $\mathbf{u}$ is called a *preshape*. The *quadratic Veronese–Whitney embedding* of Σ_2^k into $S(k,\mathbb{C})$, the linear space of *selfadjoint complex matrices of order k*, (or simply the *Veronese–Whitney map*) is $j : \Sigma_2^k \to S(k,\mathbb{C})$, where with $\mathbf{u}$ representing the preshape as in (13.1),

$$j([\mathbf{x}]) = \mathbf{u}\mathbf{u}^*, \mathbf{u}^*\mathbf{u} = 1. \tag{13.2}$$

We now consider the case $m \geq 3$. Since$\xi \in L_k^m$, one may consider the map $J : R\Sigma_{m,0}^k \to L_k \subset Sym(k,\mathbb{R})$, where L_k is the linear subspace $L_k = \{A \in Sym(k,\mathbb{R}), A\mathbf{1}_k = 0\}$, where $\mathbf{1_k}$ is the $k \times 1$ vector $(1\ldots1)^T$, is given by:

$$A = J([\mathbf{x}]_R) = \mathbf{u}^{\mathbf{T}}\mathbf{u}, \tag{13.3}$$

where $\mathbf{u} = \xi/\|\xi\|$ and $\xi = (x^1 - \overline{x}, \ldots, x^k - \overline{x}) \in (\mathbb{R}^m)^k$ identified with $\mathcal{M}(m,k;\mathbb{R})$. This is the *Schoenberg embedding* .

REMARK 13.2.1. *The Schoenberg in* (13.3) *can be extended to the orbifold* $R\Sigma_{m,+}^k$ *of similarity shapes of ordered configurations of k points* $\mathbb{R}^m$ *that are not all the same. One may select an orthobasis of* L_k *(with the Euclidean norm induced from* $Sym(k,\mathbb{R})$*), and obtain the embedding* ψ *of this orbifold in* $Sym(k-1,\mathbb{R})$*, considered by Dryden et al. (2008) [90]. The theory developed in this chapter is valid only on the regular part of this orbifold, and given that the singular part has measure zero, this situation is fairly general.*

Since J is differentiable, to show that J is an embedding it suffices to show that J and its derivative are both one to one. If $J([\mathbf{x}]_R) = J([\mathbf{x}']_R)$, from (13.3) the Euclidean distances between corresponding landmarks of the scaled configurations are equal

$$\|u^i - u^j\| = \|u'^i - u'^j\|, \quad \forall i,j = 1,\dots,k. \tag{13.4}$$

Moreover since $\sum_{i=1}^k u^i = \sum_{i=1}^k u'^i = 0$, by the fundamental theorem of Euclidean geometry, there is a matrix $T \in O(m)$ such that $u'^i = Tu^i, \forall i = 1,\dots,k$. If we set $B = \frac{\|\xi'\|}{\|\xi\|}T$, $b = \overline{x}' - B\overline{x}$, it follows that $x'^i = Bx^i + b, \forall i = 1,\dots,k$ with $B^T B = cI_m$, and from (3.121), we see that $[\mathbf{x}]_R = [\mathbf{x}']_R$. Thus J is one to one. The proof of the injectivity of the derivative of J is left to the studious reader.

THEOREM 13.2.1. *The range of the Schoenberg embedding of $R\Sigma^k_{m,0}$ is the subset $M_{k,p}$ of positive semidefinite matrices of trace 1 in L_k.*

Since L_k isometric to $Sym(k-1,\mathbb{R})$, rather than using this embedding, one may embed $R\Sigma^k_{m,0}$ directly in $Sym(k-1,\mathbb{R})$, via ψ given by

$$\psi([\mathbf{x}]_R) = \mathbf{H}J([\mathbf{x}]_R)\mathbf{H}^T, \tag{13.5}$$

where $\mathbf{H}$ is a $(k-1)\times k$ matrix, with $\mathbf{H^T H = I_{k-1}}$.

COROLLARY 13.2.1. *The range of ψ is the set $\tilde{M}_{k,p}$ of rank m and trace 1 matrices in $Sym(k-1,\mathbb{R})$.*

The tangent space to $\tilde{M}_{m,k}$ at $\psi([\xi]_R)$ is the range of the differential of ψ at $[\xi]_R$. If we set $\eta = \xi H^T$, then $rank(\eta) = m$ and $\psi([\xi]_R) = \eta^T\eta$, and

$$\begin{aligned} T_{\psi(\tilde{\sigma}(\eta))}\tilde{M}_{m,k} &= T_{\eta^T\eta}\tilde{M}_{m,k} \\ &= \{v \in Sym(k-1,\mathbb{R}) : Trv = 0, v = y^T\eta + \eta^T y, y \in \mathcal{M}(k-1,m,\mathbb{R})\} \end{aligned} \tag{13.6}$$

Since $\eta\eta^T$ has an inverse, for any $y \in \mathcal{M}(k-1,m,\mathbb{R})$ we define the symmetric matrix $S = \frac{1}{2}(\eta^T(\eta\eta^T)^{-1}y + y^T(\eta\eta^T)^{-1}\eta)$ and obtain the following representation of the tangent space in (13.6):

$$T_{\eta^T\eta}\tilde{M}_{m,k} = \{v \in Sym(k-1,\mathbb{R}), Trv = 0, v = S\eta\eta^T + \eta^T\eta S, S \in S(k-1,\mathbb{R})\}. \tag{13.7}$$

Given the equivariance of ψ, if $\eta^T\eta = A\Lambda A^T$, with $A \in O(k-1)$ and Λ the diagonal matrix with diagonal elements $\lambda_1,\dots\lambda_m,0,\dots,0,\sum\lambda_j = 0$, then

$$T_{\eta^T\eta}\tilde{M}_{m,k} = AT_\Lambda M_{m,k}A^T. \tag{13.8}$$

From (13.7) it follows that the tangent space at Λ is given by

$$\begin{aligned} T_\Lambda\tilde{M}_{m,k} = \{v \in Sym(k-1:\mathbb{R}), v = \begin{pmatrix} V_m & W \\ W^T & \mathbf{0} \end{pmatrix}, V_m \in S(m,\mathbb{R}), \\ TrV_m = 0, W \in \mathcal{M}(k-1-m,m;\mathbb{R})\}. \end{aligned} \tag{13.9}$$

13.3 Extrinsic Mean Planar Shapes and Their Estimators

We recall from Chapter 5 that the *extrinsic mean* $\mu_{j,E}(Q)$ of a *nonfocal* probability measure Q on a manifold $\mathcal{M}$ w.r.t. an embedding $j : \mathcal{M} \to \mathbb{R}^N$, when it exists, is the point from which the expected squared (induced Euclidean) distance under Q is minimum.

In this section we will consider extrinsic means associated with embeddings of the two types of manifolds described in Section 13.2.

13.3.1 Mean Planar Direct Similarity Shape

We first describe the extrinsic mean of a probability measure Q on Σ_2^k with respect to the Veronese–Whitney map j given in (13.2). The squared distance in the space $S(k,\mathbb{C})$ of self adjoint matrices is $d_0^2(A,B) = Tr((A-B)(A-B)^*) = Tr((A-B)^2)$.

A probability measure Q on Σ_2^k may be viewed as a distribution of a random shape $[\mathbf{U}]$, where $\mathbf{U}$ is a random preshape (see (3.124)). According to Problem 4.3.6, this probability measure Q is J-nonfocal if the largest eigenvalue of $E(\mathbf{U}\mathbf{U}^*)$ is simple, and in this case $\mu_{J,E}(Q) = [\mu]$, where $\mu \in S(L_k^2)$ is a unit eigenvector corresponding to this largest eigenvalue (see also Bhattacharya and Patrangenaru (2003) [42]).

The extrinsic sample mean direct similarity planar shape $\overline{[\mathbf{X}]}_{J,E}$ of a random sample $[\mathbf{X_r}]$ with preshapes $\mathbf{U_r} = [U_r^1 : \cdots : U_r^k]$, $\mathbf{1}_k^T \mathbf{U_r} = 0$, $\|\mathbf{U_r}\| = 1, r = 1,\dots,n$, from such a nonfocal distribution exists with probability converging to 1 as $n \to \infty$ (see Bhattacharya and Patrangenaru (2003) [42]) and is given by

$$\overline{[\mathbf{X}]}_{J,E} = [\mathbf{U}], \tag{13.10}$$

where $\mathbf{U}$ is a unit eigenvector in $S(L_k^2)$ corresponding to the largest eigenvalue of

$$K := n^{-1}\sum_{r=1}^{n} \mathbf{U_r}\mathbf{U_r^*}. \tag{13.11}$$

We may consider an orthogonal basis over $\mathbb{C}$ and identify $S(L_k^2)$ with the unit sphere in $\mathbb{C}^{k-1}$. W.r.t. this basis the unit vector $\mathbf{U_r}$ corresponds to a random point Z_r on that sphere, and the mean is the shape of the unit eigenvector corresponding to the largest eigenvalue of

$$\widehat{K} = n^{-1}\sum_{i=1}^{n} Z_i Z_i^\star. \tag{13.12}$$

This means that $\overline{[\mathbf{X}]}_{J,E}$ is given by a formula that is similar to the one for the *full Procrustes estimate* of the mean shape in parametric families such as Dryden-Mardia type distributions or complex Bingham type distributions for planar shapes on Σ_2^k (Kent (1992) [181], (1994) [182]). For this reason, the Veronese-Whitney extrinsic sample mean shape may also be called the *Procrustes* mean estimate .

Let $\widehat{\lambda}_1 > \widehat{\lambda}_2 \geq \ldots \geq \widehat{\lambda}_{k-1}$ denote the eigenvalues of the matrix $\widehat{K}$ in 13.12. Write

$$\widehat{R} = \text{diag}\{(\widehat{\lambda}_1 - \widehat{\lambda}_2)^{-1}, \ldots, (\widehat{\lambda}_1 - \widehat{\lambda}_{k-1})^{-1}\}, \tag{13.13}$$

$$\widehat{C} = \frac{1}{n}\sum_{i=1}^{n} |z_i^\star \widehat{m}|^2 \widehat{M}_{k-2} z_i z_i^\star \widehat{M}_{k-2}^\star \tag{13.14}$$

and

$$\widehat{G} = \widehat{R}\widehat{C}\widehat{R}, \tag{13.15}$$

where $\widehat{M}_{k-2}$ is a matrix of dimension $(k-2)\times(k-1)$ such that $\widehat{M}_{k-2}\widehat{m} = 0_{k-2}$, the $(k-2)$-vector of zeros; $\widehat{m}$ is such that $[\widehat{m}]$ is the sample mean shape; and $\widehat{M}_{k-2}\widehat{M}_{k-2}^\star = I_{k-2}$, the $(k-2)\times(k-2)$ identity matrix. An explicit construction for $\widehat{M}_{k-2}$ is given in Amaral et al. (2007) [2]. The pivotal T statistic is defined by

$$T(m) = 2nm^\star \widehat{M}_{k-2}^\star \widehat{G}^{-1} \widehat{M}_{k-2} m \tag{13.16}$$

where $[m]$ is a candidate mean shape, and m a corresponding pre-shape (the particular choice of pre-shape m does not matter). The factor 2 is present due to the standard definition of the complex multivariate normal; see Dryden and Mardia (1998, formula 6.3) [91].

13.3.2 Extrinsic Mean Reflection Shapes

We consider now a random k-ad in general position $\mathbf{X}$, centered as $\mathbf{X_0} = (X^1 - \overline{X}, \ldots, X^k - \overline{X}) \in (\mathbb{R}^m)^k \simeq M(m,k;\mathbb{R})$, and then scaled to U:

$$\mathbf{U} = \mathbf{X_0}/\|\mathbf{X_0}\|. \tag{13.17}$$

Set

$$B = E(\mathbf{U^T U}). \tag{13.18}$$

Obviously $Tr(B) = 1, B1_k = 0, B \geq 0$.

The formula for the extrinsic mean reflection shape of $[\mathbf{X}]_R$, has been found by A. Bhattacharya (2008) [25], and independently by Bandulasiri et al. (2008) [12]. This mean exists if

$$Tr(B - \mathbf{u}^T\mathbf{u})^2 \tag{13.19}$$

has a unique minimizer $\mathbf{u} \in M(m,k;\mathbb{R})$ up to an orthogonal transformation, satisfying the constraints

$$\mathbf{u}1_k = 0, Tr(\mathbf{u}^T\mathbf{u}) = 1. \tag{13.20}$$

THEOREM 13.3.1. *Assume $B = \sum_{i=1}^{k} \lambda_i e_i(B) e_i(B)^T$ is the spectral decomposition of B, then the extrinsic mean reflection shape exists if and only if $\lambda_m > \lambda_{m+1}$. If this is this case, then $\mathbf{u}^T$ can be taken as the matrix*

$$V = (v_1 v_2 \dots v_m), \tag{13.21}$$

whose columns are orthogonal eigenvectors of B corresponding to the largest eigenvalues $\lambda_1 \geq \cdots \geq \lambda_m$ of B, with

$$(a)\ v_j^T 1_k = 0$$
$$and$$
$$(b)\ v_j^T v_j = \lambda_j - \bar{\lambda} + \frac{1}{m}, \forall j = 1, \dots, p, where\ \bar{\lambda} = \frac{1}{m}(\lambda_1 + \cdots + \lambda_m). \tag{13.22}$$

We give here a proof of Theorem 13.3.1 following Bandulasiri et al. (2008) [12]. The extrinsic mean exists if

$$Tr(B - \mathbf{u}^T \mathbf{u})^2 \tag{13.23}$$

has a unique minimizer $\mathbf{u} \in (\mathbb{R}^m)^k$ up to an orthogonal transformation, satisfying the constraints

$$(a) \qquad \mathbf{u} 1_k = 0$$
$$\text{and}$$
$$(b) \qquad Tr(\mathbf{u}^T \mathbf{u}) = 1. \tag{13.24}$$

If v is an eigenvector of B, from $B 1_k = 0$ it follows that $v^T 1_k = 0$. In addition, since $B \geq 0, TrB = 1$, we have $\lambda_j \geq 0, j = 1, \dots, k, \lambda_1 + \cdots + \lambda_k = 1$. Let $e_1, \dots, e_k$ be corresponding unit eigenvectors of B.

Assume $\mathbf{u}$ is minimizer of (13.23) subject to the constraints (13.24). Let $\omega_1 \geq \cdots \geq \omega_k$ be the eigenvalues of $\mathbf{u}^T \mathbf{u}$, and assume $f_1, \dots, f_k$ are corresponding unit eigenvectors. Let E respectively F be the matrices having as columns $e_1, \dots, e_k$ respectively $f_1, \dots, f_k$. If Λ respectively Ω are the matrices having the only possibly nonzero diagonal entries $\lambda_i, i = 1, \dots, k$ respectively $\omega_i, i = 1, \dots, k$, then $B = E \Lambda E^T, \mathbf{u}^T \mathbf{u} = F \Omega F^T$, and $\Psi = Tr(E \Lambda E^T - F \Omega F^T)^2 = Tr(E(\Lambda - E^T F \Omega F^T E) E^T)^2 = TrE(\Lambda - E^T F \Omega F^T E)^2 E^T = Tr((\Lambda - E^T F \Omega F^T E)^2 E^T E) = Tr((\Lambda - E^T F \Omega F^T E)^2)$. If we set $G = E^T F$, we get

$$\Psi = Tr((\Lambda - G \Omega G^T)^2), \tag{13.25}$$

where $G \in O(k)$. For fixed Ω the quantity Ψ is minimized when $G = I_k$, (see Mardia et al. (1979, p.423)). Since $\omega_{m+1} = \cdots = \omega_k = 0$, it follows that $\Psi \geq \sum_{i=1}^{m} (\lambda_i - \omega_i)^2$. On the other hand, from (13.24)(b), it follows that $\omega_1 + \cdots + \omega_m = 1$, so we have to minimize the sum $\sum_{i=1}^{m} (\lambda_i - \omega_i)^2$ subject to the constraint

$\omega_1 + \cdots + \omega_m = 1$. If we use Lagrange multipliers we obtain the minimizers $\omega_j = \lambda_j + \frac{1}{m}(\lambda_{m+1} + \cdots + \lambda_k), \forall j = 1, \ldots, p$. Note that Ψ attains its minimum when $G = I_m$, which is equivalent to $F = E$. It follows that, without loss of generality, we may assume that the eigenvectors of $\mathbf{u}^T\mathbf{u}$ are proportional to the corresponding eigenvectors of B. Therefore, the minimizer $\mathbf{u}$ of Ψ is given by the transposed matrix of the matrix

$$V = (v_1 v_2 \ldots v_m) \tag{13.26}$$

whose columns are orthogonal eigenvectors of B corresponding to the largest eigenvalues $\lambda_1 \geq \cdots \geq \lambda_m$ of B, with

$$v_j^t v_j = \omega_j = \lambda_j + \frac{1}{m}(\lambda_{m+1} + \cdots + \lambda_k) = \lambda_j - \bar{\lambda} + \frac{1}{m}, \forall j = 1, \ldots, p. \tag{13.27}$$

□

COROLLARY 13.3.1. *If $\tilde{\sigma}(\mathrm{U})$ is a Schoenberg non-focal random reflection-shape, and $\tilde{\mu} = E(\mathbf{U}^{\mathbf{T}}\mathbf{U})$, there is a small open neighborhood $\mathcal{N} \subset L_k$ of $\tilde{\mu}$, such that for any $\mu \in \mathcal{N}$, the projection $P = P_J$ is given by*

$$P(\mu) = \sum_{j=1}^{m}(\lambda_j - \bar{\lambda} + \frac{1}{m})W_j W_j^T, \tag{13.28}$$

where $\lambda_1 \geq \cdots \geq \lambda_m > \lambda_{m+1} \geq \cdots \geq \lambda_k$, are the eigenvalues of μ in their non-decreasing order, $\lambda_1 + \cdots + \lambda_k = 1$, and

$$\mu = \sum_{j=1}^{k} \lambda_j W_j W_j^T, \tag{13.29}$$

is the spectral decomposition of μ.

13.4 Asymptotic Distribution of Mean Shapes

13.4.1 Asymptotic Distributions of Veronese–Whitney Sample Mean Direct Similarity Planar Shapes

Theorem 13.4.1 below is from Amaral et al. (2010) [1]. It generalizes the χ^2 limit result of Prentice (1984) and Watson (1984) from real to complex unit vectors. We give sufficient conditions for $T(m)$ in (13.16) to have a χ^2_{2k-4} null asymptotic distribution and cover technical issues not mentioned in these earlier papers, such as sufficient conditions for the relevant covariance matrix to be positive definite. Theorem 13.4.1 is also a version of Theorem 4.2 in Bhattacharya and Patrangenaru (2005), but note that the former differs from the latter as their identification of a shape with a point in $\mathbb{C}P^{k-2}$ is not the same as ours.

As in the introduction, $\Sigma = E(ZZ^\star)$, where Z is a random pre-shape.

THEOREM 13.4.1. *Suppose the underlying population F of pre-shapes Z is such that (i) the largest eigenvalue of $\Sigma = E(ZZ^\star)$ is distinct (so that the corresponding eigenvector is well-defined up to scalar multiplication); (ii) the distribution F has a non-zero component which is absolutely continuous with respect to the uniform distribution on the complex unit sphere; (iii) the density of the absolutely continuous component is strictly positive on an open subset of the complex unit sphere. Then if m is a pre-shape of the population mean shape $[m]$, $T(m)$ has an asymptotic χ^2_{2k-4} distribution.*

We first state two results used in the proof of Theorem 13.4.1. First, by the multivariate central limit theorem for $\widehat{S}$,

$$n^{1/2}(\widehat{S}-\Sigma) \xrightarrow{d} W, \tag{13.30}$$

where W is a $(k-1)\times(k-1)$ Hermitian matrix whose entries are jointly normally distributed with zero mean. Second, by applying the complex versions of the matrix perturbation results in Watson (1983, p.216), using (13.30) and condition (i) in the theorem, we obtain

$$n^{1/2}(\widehat{P}_1 - P_1) \xrightarrow{d} \sum_{k>1} \frac{P_k W P_1 + P_1 W P_k}{\lambda_1 - \lambda_k}, \tag{13.31}$$

where $P_j = m_j m_j^\star$, $\widehat{P}_j = \widehat{m}_j \widehat{m}_j^\star$ and the λ_i are the ordered eigenvalues of Σ, as defined in Section 2 of Amaral et al. (2010). Note that Theorem 13.4.1 assumes that λ_1 is strictly larger than the other eigenvalues. We shall write $m_1 = m$ so that $P_1 = mm^\star$. We define M_{k-2} to be the population analogue of $\widehat{M}_{k-2}$ so that $M_{k-2}m = 0_{k-2}$ and $M_{k-2}M_{k-2}^* = I_{k-2}$, and also, under the non-degeneracy assumption of the theorem, $\widehat{M}_{k-2} \xrightarrow{p} M_{k-2}$.

For simplicity, we shall assume that $\lambda_2 > \ldots > \lambda_{k-1}$, i.e. all the inequalities are strict. However, with a longer proof it can be shown that Theorem A remains true even when some of the $\lambda_2, \ldots, \lambda_{k-1}$ are equal.

Proof of Theorem 13.4.1. This is broken into 4 steps.

Step 1 - Show $||n^{1/2}\widehat{M}_{k-2}m + n^{1/2}M_{k-2}(\widehat{m}-m)|| \xrightarrow{p} 0$ as $n \to \infty$.

We have the identity

$$\begin{aligned} -n^{1/2}\widehat{M}_{k-2}m &= n^{1/2}\widehat{M}_{k-2}(\widehat{m}-m) \\ &= n^{1/2}M_{k-2}(\widehat{m}-m) + n^{1/2}(\widehat{M}_{k-2}-M_{k-2})(\widehat{m}-m). \end{aligned}$$

Thus

$$\begin{aligned} ||n^{1/2}\widehat{M}_{k-2}m &+ n^{1/2}M_{k-2}(\widehat{m}-m)|| \\ &= ||n^{1/2}(\widehat{M}_{k-2}-M_{k-2})(\widehat{m}-m)|| \\ &\leq \text{const.}||\hat{M}_{k-2}-M_{k-2}||.||n^{1/2}(\widehat{m}-m)|| \xrightarrow{p} 0, \end{aligned}$$

using the results that (a) $\widehat{S} \xrightarrow{p} \Sigma$ implies that $\widehat{M}_{k-2} \xrightarrow{p} M_{k-2}$, and (b) as a consequence of (13.31) we have $||n^{1/2}(\widehat{m}-m)|| = O_p(1)$.

Step 2 - Show that

$$n^{1/2}M_{k-2}(\widehat{m}-m) \xrightarrow{d} RM_{k-2}Wm \sim \mathbb{C}N_{k-2}(0_{k-2}, G),$$

where $G = R\mathrm{cov}(M_{k-2}Wm)R$;

$$R = \mathrm{diag}\{(\lambda_1-\lambda_2)^{-1}, \ldots, (\lambda_1-\lambda_{k-1})^{-1}\} \tag{13.32}$$

is the population analogue of $\widehat{R}$ in (13.13); and G, the population analogue of $\widehat{G}$ defined in (13.15), is given by $G = RCR$, where

$$C = E[|z^\star m|^2 M_{k-2}zz^\star M_{k-2}^\star] \tag{13.33}$$

is the population analogue of $\widehat{C}$ in (13.14).

Pre-multiplying the left hand side of (13.31) by M_{k-2} and postmultiplying by m we obtain

$$n^{1/2}M_{k-2}(\widehat{P}_1 - P_1)m \;=\; n^{1/2}M_{k-2}\widehat{m}\widehat{m}^\star m,$$

since $\widehat{P}_1 = \widehat{m}\widehat{m}^\star$ and $M_{k-2}P_1m = 0_{k-2}$ by definition of the quantities involved. Moreover,

$$n^{1/2}M_{k-2}\widehat{m}\widehat{m}^\star m \;=\; n^{1/2}M_{k-2}\widehat{m} + n^{1/2}M_{k-2}(\widehat{m}^\star m - 1)\widehat{m}.$$

Therefore

$$||n^{1/2}M_{k-2}\widehat{m}\widehat{m}^\star m - n^{1/2}M_{k-2}(\widehat{m}-m)|| = ||n^{1/2}M_{k-2}(\widehat{m}^\star m - 1)\widehat{m}|| \xrightarrow{p} 0 \tag{13.34}$$

since $\widehat{m}^\star m \xrightarrow{p} 1$ and $n^{1/2}M_{k-2}\widehat{m} = O_p(1)$.

Now consider the right hand side of (13.31). Pre-multiplying by M_{k-2} and post-multiplying by m we obtain

$$\begin{aligned} M_{k-2}\left[\sum_{j=2}^{k-1}\frac{P_jWP_1+P_1WP_j}{\lambda_1-\lambda_j}\right]m &= M_{k-2}\left[\sum_{j=2}^{k-1}\frac{1}{\lambda_1-\lambda_j}P_j\right]Wm \\ &= RM_{k-2}Wm \sim \mathbb{C}N_{k-2}(0_{k-2}, G), \end{aligned}$$

where G is defined above. We have used the fact that M_{k-2} has the form $\sum_{j=2}^{k-1} r_{j-1}m_j^\star$, where $\{r_j\}_{j=1}^{k-2}$ is a set of orthonormal $(k-2)$-vectors; and consequently

$$\begin{aligned} M_{k-2}\left[\sum_{j=1}^{k-1}\frac{1}{\lambda_1-\lambda_j}P_j\right] &= \sum_{j=2}^{k-1}\frac{1}{\lambda_1-\lambda_j}M_{k-2}P_j \\ &= \sum_{j=2}^{k-1}\frac{1}{\lambda_1-\lambda_j}r_{j-1}m_j^\star m_j m_j^\star \\ &= RM_{k-2}. \end{aligned}$$

Therefore the result follows from the fact that the left hand side of (13.31), pre-multiplied by M_{k-2} and post-multiplied by m, becomes $n^{1/2}M_{k-2}(\widehat{m}-m)$, and the corresponding right hand side is $RM_{k-2}Wm$ which has distribution $CN_{k-2}(0_{k-2},G)$.

Step 3 - Show that $\widehat{G} \xrightarrow{P} G = RCR$, where $\widehat{G}$, R and C are defined in, respectively, (13.15), (13.32) and (13.33).

The (j,h) element of matrix $\widehat{G}$ may be written in the form

$$n^{-1}\widehat{A}_{jh}^{\top}\sum_{i=1}^{n}X_i = \widehat{A}_{jh}^{\top}\bar{X},$$

where X_i is a vector whose components are of the form

$$z_{i\alpha}\bar{z}_{i\beta}z_{i\gamma}\bar{z}_{i\delta} \quad \text{where} \quad \alpha,\beta,\gamma,\delta \in \{1,\ldots,k-1\}$$

and $\widehat{A}_{jh}$ is a vector whose components are polynomial functions of the components of $\widehat{R}$, $\widehat{M}_{k-2}$ and $\widehat{m}$. By the law of large numbers, $\frac{1}{n}\sum X_i$ converges in probability to $E(X_1)=\gamma$, say, where the components of γ are of the form $E(z_{1\alpha}\bar{z}_{1\beta}z_{1\gamma}\bar{z}_{1\delta})$. Moreover, since $\widehat{R}$,$\widehat{M}_{k-2}$ and $\widehat{m}$ converge in probability to their population analogues R, M_{k-2} and m, it follows that $\widehat{A}_{jh} \xrightarrow{P} A_{jh}$, where A_{jh} is obtained from $\widehat{A}_{jh}$ by replacing $\widehat{R}$, $\widehat{M}_{k-2}$ and $\widehat{m}$ by their population values.

Step 4 - G has full rank.

By the Lebesgue decomposition theorem we may write

$$F = \varepsilon F_{ac} + (1-\varepsilon)F_r,$$

where F is the distribution of Z, F_{ac} is the component which is absolutely continuous, in the present context with respect to the uniform distribution on the complex unit sphere, and F_r is the remaining component. By assumption (ii), $\varepsilon \in (0,1]$. Write μ_{ac} and Ω_{ac} for, respectively, the mean and covariance matrix of F_{ac}, with similar definitions for μ_r and Ω_r. Now, writing $\Omega = \mathrm{cov}(Z)$,

$$\begin{aligned}
\Omega &= \varepsilon(\Omega_{ac}+\mu_{ac}\mu_{ac}^{\star})+(1-\varepsilon)(\Omega_r+\mu_r\mu_r^{\star}) \\
&\qquad -(\varepsilon\mu_{ac}+(1-\varepsilon)\mu_r)(\varepsilon\mu_{ac}+(1-\varepsilon)\mu_r)^{\star} \\
&= \varepsilon\Omega_{ac}+(1-\varepsilon)\Omega_r+\varepsilon(1-\varepsilon)(\mu_{ac}-\mu_r)(\mu_{ac}-\mu_r)^{\star}.
\end{aligned}$$

Assumption (iii) ensures that Ω_{ac} is positive definite; see Bhattacharya and Ghosh (1978). Therefore, since the second and third terms are nonnegative definite, it follows that Ω is also positive definite.

Define $a(Z) = m^{\star}ZZ^{\star}m$ and let $Y = M_{k-2}Z$, where $Z \sim F$. Then $a(Z) \in [0,1]$ since m and Z are both complex unit vectors. Note that

$$E[YY^{\star}] = M_{k-2}\Omega M_{k-2}^{\star} = \mathrm{diag}\{\lambda_2,\ldots,\lambda_{k-1}\}$$

which is positive definite since M_{k-2} has full rank and Ω is positive definite. Therefore the result will follow if we can show that, for some $\varepsilon > 0$, $G \geq \varepsilon E[YY^\star]$, where "$\geq$" should be understood in terms of the partial ordering of non-negative definite matrices.

We have, for any $\delta > 0$,

$$\begin{aligned} G &= E[a(Z)YY^\star] \\ &= E\left[a(Z)I(a(Z) > \delta)ZZ^\star\right] + E\left[a(Z)I(a(Z) \leq \delta)YY^\star\right]. \\ &\geq E\left[a(Z)I(a(Z) > \delta)YY^\star\right] \\ &\geq \delta E\left[I(a(Z) > \delta)YY^\star\right]. \end{aligned}$$

But it follows from assumption (*ii*) of the theorem that

$$E[YY^\star] = \lim_{\delta \to 0} E[I(a(Z) > \delta)YY^\star],$$

and so there exists a $\tilde{\delta} > 0$ such that

$$E\left[I(a(Z) > \tilde{\delta})YY^\star\right] \geq \frac{1}{2}E\left[YY^\star\right],$$

from which it follows that

$$G \geq \frac{\tilde{\delta}}{2}E\left[YY^\star\right],$$

and therefore G is positive definite.

Finally, by Step 1 and Step 2,

$$n^{1/2}\widehat{M}_{k-2}m \xrightarrow{d} \mathbb{C}N_{k-2}(0, G),$$

Steps 3 and 4 state that $\widehat{G} \xrightarrow{p} G$ and G has full rank. Thus the inverse of $\widehat{G}$ exists in probability as $n \to \infty$ and the statistic

$$T(m) = 2nm^\star \widehat{M}_{k-2}^\star \widehat{G}^{-1} \widehat{M}_{k-2}m,$$

is well-defined in the limit. Thus

$$T(m) \xrightarrow{d} \chi^2_{2k-4}$$

as required.

Note that we can obtain a slightly more general result by imposing conditions ii) and iii) on the underlying distribution in the shape space, as opposed to the underlying distribution in the pre-shape space. Since our proof does not depend on the particular choice of pre-shape, we obtain the same result under these more general conditions.

The above result enables us to give bootstrap confidence regions for a single population mean shape. When two or more mean shapes are available and we wish to construct multi-sample tests or associated confidence regions based on the bootstrap, then there are various procedures available, e.g. Amaral et al. (2007), Bhattacharya (2008a) [24], Bhattacharya and Bhattacharya (2008), Bhattacharya (2008b), and Bhattacharya and Bhattacharya (2012) [27]. Also note that similar results are available for alternative notions of mean shape, for example extrinsic means from equivariant embeddings (Bandulasiri et al. (2009) [10]; Bhattacharya, 2008b [25]).

13.4.2 Asymptotic Distributions of Schoenberg Sample Mean Reflection Shapes

[Asymptotic Distributions of Schoenberg Sample Mean Reflection Shapes]

Recall that the reflection shape manifold $R\Sigma^k_{m,0}$, can be regarded as a submanifold of $SR\Sigma^k_{m,0}$, as size-and-reflection-shapes of k-ads in general position of size one. Also, recall that its embedding J in (13.3) into the Euclidean vector subspace L_k of $k \times k$ symmetric matrices, is given by

$$J([\mathbf{u}]_R) = \mathbf{u}^T\mathbf{u}.$$

From Theorem 13.2.1, it follows that, if $k \geq p$, the range of J in (13.3) is the set $M_{m,k}$ of trace one $k \times k$ positive semidefinite symmetric matrices A of rank m, with $A\mathbf{1}_k = 0$.

REMARK 13.4.1. *Consider the action of the orthogonal group $O(k)$ on $S(k,\mathbb{R})$:*

$$\beta(S,A) = ASA^T. \tag{13.35}$$

Let O_k be the subgroup of $O(k)$ that leaves invariant the subspace $\mathbb{R}1_k$ of $\mathbb{R}^k$ under the natural action on $\mathbb{R}^k$. It follows that O_k leaves invariant the subspace L_k under the action (13.35)*. The Schoenberg embedding is equivariant with respect to the group actions of O_k, α on $R\Sigma^k_{m,0}$ in equation* (22.27) *given by*

$$\alpha([\mathbf{u}]_R, A) = [\mathbf{u}\mathbf{A}]_R, \tag{13.36}$$

and β on in (13.35)*.*

In this section, we will derive the asymptotics for the extrinsic sample mean reflection shape for samples from a Schoenberg-nonfocal distribution Q on $R\Sigma^k_{m,0}$. For this purpose, we will use the general results for extrinsic means on a manifold in Bhattacharya and Patrangenaru (2005) [43] (Theorem 3.1. and its corollaries). Consider first the map $P = P_J : \mathcal{N}(\tilde{\mu}) \to L_k$, as in Corollary 13.3.1, $P(\mu) = \sum_{j=1}^m (\lambda_j(\mu) - \bar{\lambda}(\mu) + \frac{1}{m})W_j(\mu)W_j(\mu)^T$ as in Theorem 13.3.1. Hence for $\mu \in N(\tilde{\mu}), \lambda_m(\mu) > \lambda_{m+1}(\mu)$. Let $\gamma(t) = \tilde{\mu} + tv$ be a curve in L_k with $\gamma(0) = \tilde{\mu}$, and $\dot{\gamma}(0) = v \in L_k$. Let $\tilde{\mu} = W\Lambda W^T, W = (W_1 \ldots W_k), \Lambda = diag(\lambda_1, \ldots, \lambda_k)$ be a s.v.d. of $\tilde{\mu}$ as in Corollary 13.3.1. Then

$$\gamma(t) = W(\Lambda + tW^T vW)W^T = W\tilde{\gamma}(t)W^T, \tag{13.37}$$

where $\tilde{\gamma}(t) = \Lambda + tW^T vW$. Then $\tilde{\gamma}(t)$ is a curve in L_k starting at Λ, with the derivative $\dot{\tilde{\gamma}}(0) = W^T vW$. From (13.37) and the definition of P, we get that

$$P[\gamma(t)] = WP[\tilde{\gamma}(t)]W^T. \tag{13.38}$$

Differentiate (13.38) at $t = 0$, noting that $(P(\gamma(t)])'(0) = d_{\tilde{\mu}}P(v)$ and $(P(\tilde{\gamma}(t)])'(0) = d_\Lambda P(\tilde{v})$, to get that

$$d_{\tilde{\mu}}P(v) = Wd_\Lambda P(\tilde{v})W^T. \tag{13.39}$$

The equation (13.39) shows that it suffices to compute the differential of P at a matrix $\Lambda \in L_k$ that is diagonal w.r.t. a convenient orthobasis in L_k. If $E_i^j = e_i e_j^T$, where $(e_1, \dots, e_k)$ is the standard basis of $\mathbb{R}^k$, we set $F_j = (1+(j-1)^2)^{-\frac{1}{2}}(E_1^1 + E_2^2 + \dots E_{j-1}^{j-1} - (j-1)E_j^j), j = 2, \dots, k$. An orthonormal basis in L_k is given by

$$\tilde{E} = (F_2, F_3, \dots, F_k, 2^{-\frac{1}{2}}(E_i^j + E_j^i), 1 \leq i < j \leq k). \tag{13.40}$$

Note that the matrix $\Lambda = diag(\lambda_1, \dots, \lambda_k) \in L_k$ has the s.v.d.

$$\Lambda = \sum_{j=1}^{k} \lambda_j e_j e_j^T, \tag{13.41}$$

therefore, by Corollary 13.3.1 its projection on $M_{m,k}$ is also diagonal. From Theorem 13.2.1, it follows that $T_\Lambda M_{m,k}$ is the vector subspace of L_k spanned by the orthobasis

$$e(\Lambda) = (F_2, F_3, \dots, F_m, 2^{-\frac{1}{2}}(E_i^j + E_j^i), 1 \leq i < j \leq p \text{ or } 1 \leq i \leq p < j \leq k-1). \tag{13.42}$$

Since, for any $A \in O(k-1)$, the map $v \to AvA^T$ is an isometry of $Sym(k-1, \mathbb{R})$, an orthobasis in the space $T_{\eta^T\eta}\tilde{M}_{m,k}$ is given then by

$$e(\eta^T\eta) = (F_2^2(\eta), \dots, F_m^m(\eta), 2^{-\frac{1}{2}}(E_i^j(\eta) + E_j^i(\eta)), \tag{13.43}$$

for $1 \leq i < j \leq p$ or $1 \leq i \leq p < j \leq k-1$, where for any symmetric matrix F, $F(\eta) = AFA^T$.

In order to find $(P(\tilde{\gamma}(t)])'(0)$, we first note that since the set of matrices with simple eigenvalues is generic (open and dense), without loss of generality, we may assume that $\lambda_1 > \lambda_2 > \dots > \lambda_k$. Then we can choose a s.v.d. for $\tilde{\gamma}(t) : \tilde{\gamma}(t) = \sum_{j=1}^k \lambda_j(t) w_j(t) w_j(t)^T$ such that $\{\lambda_j(t), w_j(t)\}_{j=1}^k$ are some smooth functions of t satisfying $w_j(0) = e_j$ and $\lambda_j(0) = \lambda_j$, where $\{e_j\}_{j=1}^k$ is the canonical basis for $\mathbb{R}^k$. Since $w_j(t)^T w_j(t) = 1$, we get by differentiating,

$$w_j(t)^T \dot{w}_j(0) = 0, j = 1, \dots, k. \tag{13.44}$$

Also, since $\tilde{\gamma}(t)w_j(t) = \lambda_j(t)w_j(t)$, we get that

$$\tilde{v}e_j + \Lambda\dot{w}_j(0) = \lambda_j(0)\dot{w}_j(0) + \dot{\lambda}_j(0)e_j, j = 1,\dots,k. \tag{13.45}$$

A standard orthonormal basis in the space of matrices $V \in S(m,\mathbb{R}), TrV = 0$, is given by the

$$\begin{aligned}\tilde{E}(m) = (2^{-\frac{1}{2}}(E_1^1 - E_2^2), 6^{-\frac{1}{2}}(E_1^1 + E_2^2 - 2E_3^3),\dots,\\ (1+(m-1)^2)^{-\frac{1}{2}}(E_1^1 + E_2^2 + \dots E_{m-1}^{m-1} - (m-1)E_m^m),\\ 2^{-\frac{1}{2}}(E_i^j + E_j^i), 1 \le i < j \le p),\end{aligned} \tag{13.46}$$

where E_i^j has all entries zero, except for the entry in the ith row and j^{th} column, which equals 1. From (13.9) it follows that $T_\Lambda \tilde{M}_{m,k}$ is spanned by the orthobasis

$$e(\Lambda) = (\tilde{E}(m), 2^{-\frac{1}{2}}(E_i^j + E_j^i), 1 \le i < j \le m \text{ or } 1 \le i \le m < j \le k-1). \tag{13.47}$$

For $j = 2,\dots,p$, let us set $F_j^j = 1 + (j-1)^2)^{-\frac{1}{2}}(E_1^1 + E_2^2 + \dots E_{j-1}^{j-1} - (j-1)E_j^j)$. Since, for any $A \in O(k-1)$, the map $v \to AvA^T$ is an isometry of $Sym(k-1,\mathbb{R})$, an orthobasis in the space $T_{\eta^T\eta}\tilde{M}_{m,k}$ is then given by

$$e(\eta^T\eta) = (F_2^2(\eta),\dots,F_m^m(\eta), 2^{-\frac{1}{2}}(E_i^j(\eta) + E_j^i(\eta)), \tag{13.48}$$

for $1 \le i < j \le m$ or $1 \le i \le m < j \le k-1$, where for any symmetric matrix F, $F(\eta) = AFA^T$.

REMARK 13.4.2. *Since the asymptotic results are often presented in vector notation, it will be useful to order the orthobasis* (13.46) *in the following non-canonical way*

$$\begin{aligned}E = (F_2^2,\dots,F_m^m, 2^{-\frac{1}{2}}(E_i^j + E_j^i), 1 \le i < j \le m \text{ or } 1 \le i \le m < j \le k-1,\\ E_{m+1}^{m+1},\dots,E_{k-1}^{k-1}, 2^{-\frac{1}{2}}(E_i^j + E_j^i), m+1 \le i < j \le k-1).\end{aligned} \tag{13.49}$$

Also note that if $(e_1,\dots,e_{k-1})$ *is the standard basis of* $\mathbb{R}^{k-1}$ *identified with the set of* $m \times 1$ *real matrices, then* $E_i^j = e_j e_j^T$.

From Section 5, it follows that the extrinsic mean $\mu_{J,E}(Q) =$ of a Schoenberg nonfocal probability measure Q on $R\Sigma_{m,0}^k$, given by $\mu_{J,E}(Q) = J^{-1}(P_J(\mu))$, where μ is the mean of $\psi(Q)$ in $S(k,\mathbb{R})$ and P_ψ is the projection on $M_{m,k}$, is the same as the extrinsic mean $\mu_{\psi,E}(Q)$. This will be simply labeled $\mu_E(Q)$ or μ_E.

Following Bhattacharya and Patrangenaru (2005) [43], the extrinsic covariance operator $\Sigma_E = \Sigma_{J,E}$ is the restriction of the self-adjoint linear operator $d_\mu P_\psi \Sigma d_\mu P_\psi^T$ to $T_{P_\psi(\mu)}\tilde{M}_{m,k}$. The extrinsic covariance matrix is the matrix associated to Σ_E with respect to a basis $e_1(P_\psi(\mu)),\dots,e_d(P_\psi(\mu))$ of $T_{P_\psi(\mu)}\tilde{M}_{m,k}, d = \frac{m}{2}(2k-m-1)-1$. We now compute the differential of the projection P_J, following A. Bhattacharya (2008b) [25].

PROPOSITION 13.4.1. *The differential of P_J at μ is given by*

$$d_\mu P(A) = \sum_{i=1}^{m} \sum_{j=m+1}^{k} (\lambda_i - \lambda_j) a_{ij} U F_{ij} U^T, \tag{13.50}$$

where $A = \sum_{1 \le i \le j \le k} a_{ij} U F_{ij} U^T$ and $\{U F_{ij} U^T, 1 \le i \le j \le k\}$ is the orthoframe for $S(k, \mathbb{R})$.

LEMMA 13.4.1. *Assume the mean μ of $\psi(Q)$ is a diagonal matrix Λ. The differential of the projection P_ψ at Λ with respect to the ordered orthobasis (13.49) is given by*

$$d_\Lambda P_\psi(F_i^i) = F_i^i, 2 \le i \le p$$

$$d_\Lambda P_\psi(E_i^i) = 0 \quad i > p,$$

$$d_\Lambda P_\psi(E_j^l + E_l^j) = \begin{cases} E_j^l + E_l^j & j < l \le p \\ \frac{\lambda_j}{\lambda_j - \lambda_l}(E_j^l + E_l^j) & j \le p < l \\ 0 & p < j < l. \end{cases} \tag{13.51}$$

Given the equivariance of the embedding ψ, from Lemma 13.4.1 we obtain the following

PROPOSITION 13.4.2. *If the spectral decomposition of the mean μ of $\psi(Q)$ is $\mu = \sum_{i=1}^{k-1} \lambda_i \tilde{e}_i \tilde{e}_i^t$, with $\lambda_1 \ge \cdots \ge \lambda_m > \lambda_{m+1} \ge \cdots \ge \lambda_{k-1}$, then*
(i) the tangent space $T_{\psi(\mu_E)} M_{m,k} = T_1 \oplus T_2$, where T_1 has the orthobasis

$$2^{-\frac{1}{2}}(\tilde{e}_1\tilde{e}_1^T - \tilde{e}_2\tilde{e}_2^T, \dots, (1 + (m-1)^2)^{-\frac{1}{2}}(\tilde{e}_1\tilde{e}_1^T + \dots + \tilde{e}_{m-1}\tilde{e}_{m-1}^T - (m-1)\tilde{e}_m\tilde{e}_m^T),$$
$$2^{-\frac{1}{2}}(\tilde{e}_i\tilde{e}_j^T + \tilde{e}_j\tilde{e}_i^T), 1 \le i < j \le p), \tag{13.52}$$

and T_2 has the orthobasis

$$(2^{-\frac{1}{2}}(\tilde{e}_j\tilde{e}_l^T + \tilde{e}_l\tilde{e}_j^T), 1 \le j \le p < l \le k-1). \tag{13.53}$$

(ii) Let N be the orthocomplement of $T_{\psi(\mu_E)} M_{m,k}$ Then

$$d_\mu P_\psi | T_1 = Id_{T_1},$$
$$d_\mu P_\psi(\tilde{e}_j\tilde{e}_l^T + \tilde{e}_l\tilde{e}_j^T) = \frac{\lambda_j}{\lambda_j - \lambda_l}(\tilde{e}_j\tilde{e}_l^T + \tilde{e}_l\tilde{e}_j^T), \forall (j,l), 1 \le j \le p < l \le k-1,$$
$$d_\mu P_\psi | N = 0. \tag{13.54}$$

An orthobasis of N in proposition 13.4.2 is

$$2^{-\frac{1}{2}}(\tilde{e}_j\tilde{e}_l^T + \tilde{e}_l\tilde{e}_j^T), p < j < l \le k-1). \tag{13.55}$$

The two orthobases (13.52), (13.53) and (13.55) yield an orthobasis $\mathbf{\tilde{e}}$ of

$Sym(k-1,\mathbb{R})$. From proposition 13.4.2 it follows that the matrix D associated with the differential $d_\mu P_\psi$ relative to the orthobasis $\tilde{\mathbf{e}}$ is diagonal:

$$D = \begin{pmatrix} I_{\frac{m(m-1)}{2}} & \mathbf{0} & \mathbf{0} \\ \mathbf{0} & \Delta_{m(k-m-1)} & \mathbf{0} \\ \mathbf{0} & \mathbf{0} & \mathbf{0} \end{pmatrix}, \tag{13.56}$$

where

$$\Delta_{m(k-m-1)} = \begin{pmatrix} \frac{\lambda_1}{\lambda_1 - \lambda_{m+1}} & \dots & 0 \\ \dots & \dots & \dots \\ 0 & \dots & \frac{\lambda_m}{\lambda_m - \lambda_{k-1}} \end{pmatrix}. \tag{13.57}$$

The space of symmetric matrices $Sym(k-1,\mathbb{R})$, regarded as its own tangent space at μ, splits into three orthogonal subspaces

$$Sym(k-1,\mathbb{R}) = T_1 \oplus T_2 \oplus N, \tag{13.58}$$

leading to a decomposition of the covariance matrix Σ of $\psi(Q)$, with respect to the orthobasis of $Sym(k-1,\mathbb{R})$ obtained by augmenting the orthobasis (13.47) by an orthobasis of N, as follows:

$$\Sigma = \begin{pmatrix} \Sigma_{11} & \Sigma_{12} & \Sigma_{13} \\ \Sigma_{12}^T & \Sigma_{22} & \Sigma_{23} \\ \Sigma_{13}^T & \Sigma_{23}^T & \Sigma_{33} \end{pmatrix}. \tag{13.59}$$

If we change the coordinates in $\mathbb{R}^{k-1}$ by selecting an orthobasis $\tilde{\mathbf{e}}$, the eigenvectors $\tilde{e}_1,\dots,\tilde{e}_{k-1}$ of μ, in such a coordinate system, the mean is a diagonal matrix Λ and the matrix $\Sigma_\mu = D\Sigma D^T$, defined in Bhattacharya and Patrangenaru (2005) [43], is

$$\Sigma_\mu = \begin{pmatrix} \Sigma_{11} & \Sigma_{12}\Delta & 0 \\ \Delta\Sigma_{12}^T & \Delta\Sigma_{22}\Delta & 0 \\ 0 & 0 & 0 \end{pmatrix}, \tag{13.60}$$

and the extrinsic covariance matrix Σ_E defined in Bhattacharya and Patrangenaru (2005), with respect to the basis $d_{\mu_\psi}^{-1}(e(\Lambda))$, where $e(\Lambda)$ is as defined in (13.47), is

$$\Sigma_E = \begin{pmatrix} \Sigma_{11} & \Sigma_{12}\Delta \\ \Delta\Sigma_{12}^T & \Delta\Sigma_{22}\Delta \end{pmatrix}, \tag{13.61}$$

We assume now that $Y_1,\dots,Y_n$ are independent identically distributed random reflection objects from a ψ-nonfocal probability distribution Q on $\Sigma_{m,0}^k$ with $\lambda_m > \lambda_{m+1}$. Let $\widetilde{s\sigma}(\eta)$ be the mean of $\psi(Q)$ and Σ be the covariance matrix of $\psi(Q)$ with respect to the orthobasis $\tilde{\mathbf{e}}$ defined above. Let $\vec{W}$ be the vectorized form of a matrix $W \in Sym(k-1,\mathbb{R})$ with respect to the basis $\tilde{V}$. Assume $tan\vec{W}$ denote the component of $\vec{W}$ tangent to $\vec{\tilde{N}}_{m,k}$ at $\vec{\psi}(\mu_\psi)$.

THEOREM 13.4.2. *(a) The random vector* $n^{\frac{1}{2}}\tan(\vec{\psi}(\overline{Y}_E)-\vec{\psi}(\mu_E))$ *converges weakly to a random vector having a* $N(0,\Sigma_E)$ *distribution, where* Σ_E *is given in* (13.61).
(b) If Σ_E *is nonsingular, then* $n\tan(\vec{\psi}(\overline{Y}_E)-\vec{\psi}(\mu_E))^T\Sigma_E^{-1}tan(\vec{\psi}(\overline{Y}_E)-\vec{\psi}(\mu_E))^T$ *converges weakly to a* $\chi^2_{km-\frac{m(m+1)}{2}}$ *distribution.*

From Theorem 13.4.2 we obtain the following result:

COROLLARY 13.4.1. *Let* G *be a normally distributed matrix in* $Sym(k-1,\mathbb{R})$ *with a weak limit of* $n^{\frac{1}{2}}(\overline{Y}-\mu)$*. Assume the spectral decomposition of* μ *is* $\mu=V\Lambda V^T$*. Set* $G^V=V^TGV=(g^V_{jl})$ *and* $\tilde{G}^V=(\tilde{g}^V_{jl})$ *be determined by*

$$\tilde{g}^V_{jl}=\begin{cases} g^V_{jl} & 1\leq j\leq l\leq m\\ \frac{\lambda_j}{\lambda_j-\lambda_l}g^V_{jl} & 1\leq j\leq m<l\leq m-1\\ 0 & m<j\leq l\leq k-1\end{cases}. \tag{13.62}$$

Then $n^{\frac{1}{2}}(\psi(\overline{Y}_E)-\psi(\mu_E))$ *converges in distribution to the normally distributed random matrix* VG^VV^T.

From Theorem 13.4.2, it follows that the extrinsic mean reflection-shape can be easily estimated using non-pivotal bootstrap. Assume $\{\mathbf{x}_1,\dots,\mathbf{x}_n\}$ is a random sample of configurations $\mathbf{x}_j=(x^1_j,\dots,x^k_j), j=1,\dots,n$. We consider N bootstrap resamples from this sample, where N is a reasonably large number, say $N\geq 500$. For each such resample $\mathbf{x}^*_1,\dots,\mathbf{x}^*_n$, we compute the extrinsic sample mean $\overline{[\mathbf{x}]_{RSE}}^*$. We then use a local parametrization of $SR\Sigma^k_{m,0}$ and find $(1-\alpha)100\%$ Bonferroni simultaneous confidence intervals for the corresponding $km-\frac{m(m+1)}{2}$ local coordinates.

Two sample tests for extrinsic means on can be be derived from the general theory for two sample tests for extrinsic means on manifolds recently developed by A. Bhattacharya (2008) [24].

13.5 A Data Driven Example

In Sharvit et al. (1998) [306], the authors provided matching sketches against a small database consisting of a variety of fish, planes, etc., that were later classified by Klassen et al. (2004) [194] using a geodesic distance on a infinite dimensional space of shapes of contours. We selected an example of a sample of sting rays for display in Figure 13.1. Given that the computations are ultimately discrete, we compute the sample mean for a discrete version of the silhouette of 100 pseudolandmarks, and the nonpivotal bootsrap distribution of mean shapes based on 500 resamples.

Note that while the individual features within the sample are smoothed out in the process, the bootstrap distribution is fairly well centered around the sample

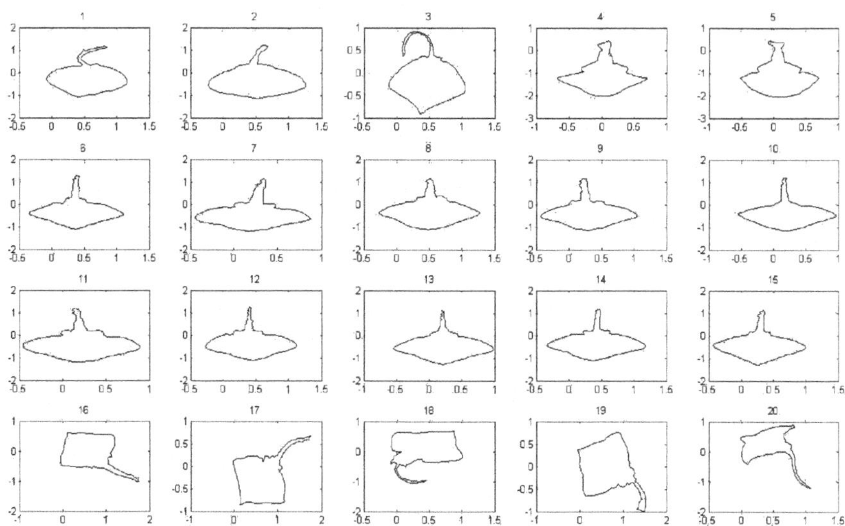

Figure 13.1 *Sample of 20 curves of sting ray fish. (Source: Amaral et al.(2010), Figure 1. Reproduced by permission of Elsevier).*

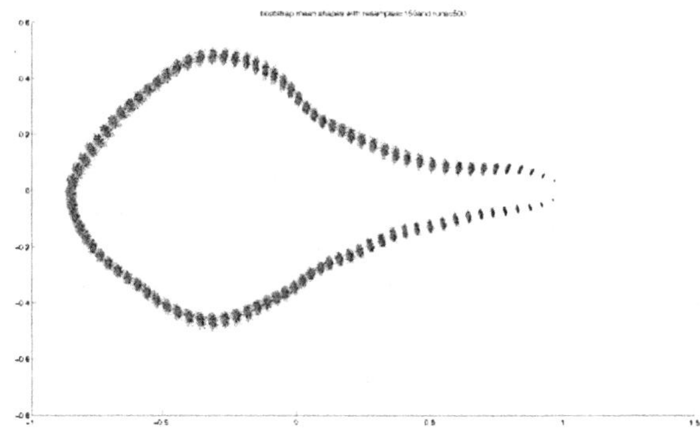

Figure 13.2 *Bootstrap distribution of sample mean shapes of sting ray fish. (Source: Amaral et al.(2010), Figure 2. Reproduced by permission of Elsevier).*

mean shape. One of the main difficulties is that the matching of the pseudolandmarks is automatic, based on the labeling on the preshape of the silhouette.

Chapter 14

Statistics on Grassmannians

According to Section 3.5.7, affine shape analysis of configurations of k-ads in general position utilizes techniques for data analysis on Grassmann manifolds. Therefore, in order to perform an extrinsic analysis, one has to find appropriate chord distances on Grassmannians that are associated with convenient equivariant embeddings. In this chapter, we will explore the theory necessary to do so.

14.1 Equivariant Embeddings of Grassmann Manifolds

Assume $k < m$ and let $G_k(\mathbb{R}^m)$ denote the real Grassmann of k dimensional subspaces in $\mathbb{R}^m$ (see Section 3.5.7). Let $j : G_k(\mathbb{R}^m) \to Sym(m, \mathbb{R})$ be the function which, with respect to a fixed orthonormal basis $\mathbf{e}$ of $\mathbb{R}^m$, maps an element V in $G_k(\mathbb{R}^m)$ to the matrix associated with the orthogonal projection onto V with respect to e. It is known (see Dimitric (1996) [86]) that this map is an embedding, which we will call the *Dimitric embedding* . Dimitric (1996) [86] proved that this embedding is equivariant and embeds the Grassmannian *minimally* into a hypersphere. The Dimitric embedding is an extension of the Veronese-Whitney embedding of projective spaces (see Section 10.2 or Bhattacharya and Patrangenaru (2003) [42]). This was commonly used for axial data (see Mardia and Jupp (2000 ([230]) and for multivariate axial data (Mardia and Patrangenaru (2005) [233]).

Later on, the Dimitric embedding was used in computer vision literature under the name of *projection embedding* (see Helmke et al. (2007) [149], Harandi et al. (2014) [140]). The explicit formula for the Dimitric embedding in terms of a $m \times r$ matrix x with orthogonal norm one columns spanning an r-plane

$V \in G_k(\mathbb{R}^m$, i.e., $x^T x = I_k$, is

$$j(V) = xx^T. \tag{14.1}$$

Obviously, the Dimitric embedding generalizes the VW-embedding of $\mathbb{R}P^{m-1}$, that can be obtained by taking $k = 1$.

Another important embedding can be obtain from classical algebraic geometry, is the *Plücker embedding*, defined below. The Plücker embedding (see Harandi et al. (2014) [140]) is the map j_P defined by

$$\begin{aligned} j_P : G_k(\mathbb{R}^m) &\to P(\wedge^k \mathbb{R}^m) \\ \text{span}(v_1, \ldots, v_k) &\to \mathbb{R}(v_1 \wedge \cdots \wedge v_k) \end{aligned} \tag{14.2}$$

14.2 Derivation of a Formula for the Dimitric Mean of a Random Object on a Grassmann Manifold

Now, let us suppose that a probability measure Q is given on $G_k(\mathbb{R}^m)$. Using the embedding j, we may define the image probability measure $j(Q)$ on $Sym(m,\mathbb{R})$. Let $\mu_{j(Q)}$ be the mean of $j(Q)$ on $Sym(m,\mathbb{R})$. It was first stated in Patrangenaru and Mardia(2003) [274] that the *Dimitric mean*, which is the extrinsic mean with respect to the Dimitric embedding (14.1), is the k-dimensional subspace spanned by the eigenvectors corresponding to the k-largest eigenvalues of $\mu_{j(Q)}$. Note that a projection map of to $j(G_k(\mathbb{R}^m))$ is generically defined on $Sym(m,\mathbb{R})$. This way, the Dimitric mean may be defined as the projection of $\mu_{j(Q)}$ onto $j(G_k(\mathbb{R}^m))$. Our first objective of this section is to prove this claim. The proof is due to Sughatadasa (2006) [322].

For the sake of simplicity, let us denote $\mu_{j(Q)}$ by E. We first note that each orthogonal symmetry matrix is positive semidefinite, hence E is symmetric positive semidefinite. Let A denotes an element of $Sym(m,\mathbb{R})$ in the vicinity of E. Then A is symmetric and positive definite as well, hence its eigenvalues are real and nonnegative, and there exists an orthonormal basis of $\mathbb{R}^m$ consisting of eigenvectors of A. Let us denote the eigenvalues of A as $\lambda_1(A) \geq \lambda_2(A) \cdots \geq \lambda_m(A)$ and the corresponding orthogonal basis of eigenvectors as $\{e_1(A), e_1(A), \cdots, e_m(A)\}$. For the sake of simplicity, we will drop the argument A. The probability measure Q is j-nonfocal iff $\lambda_k > \lambda_{k+1}$ for E and for all elements of $Sym(m,\mathbb{R})$ in a suitably small neighborhood of E (a necessary and sufficient condition to have a unique point on $G_k(\mathbb{R}^m)$ at minimal distance from A).

Let U be such a neighborhood of E. Recall that the standard inner product in $Sym(m,\mathbb{R})$ is given by $< C, D >= \text{Tr (CD)}$, hence the Euclidean norm is $\|C\| = \sqrt{\{\text{Tr (C}^2)\}}$. We claim that the map,

$$\phi : U \to j(G_k(\mathbb{R}^m)), \tag{14.3}$$

where $\phi(A)$ is equal to the orthogonal projection matrix onto the subspace spanned by $e_1(A), \cdots, e_k(A)$, has the property that, for any $A \in U$,

$$\|A - \phi(A)\| \leq \|A - C\| \text{ for all } C \in j(G_k(\mathbb{R}^m))$$

Equality holds if and only if $C = \phi(A)$. Thus, this map is indeed the projection on $j(G_k(\mathbb{R}^m))$ as suggested in [274]. Now, we proceed to prove this claim.

For a fixed $A \in U$, let $\lambda_i,\ i = 1, \cdots m$ and $e_i,\ i = 1, \cdots m$ be the eigenvalues and orthogonal eigenvectors as described above. Let B an arbitrary element of $j(G_k(\mathbb{R}^m))$. So B is an orthogonal projection matrix onto some k dimensional subspace, hence in particular we note that B is positive semi-definite. Now,

$$\begin{aligned}
\|A - B\|^2 &= <A,A> -2<A,B> + <B,B> &(14.4)\\
&= \mathrm{Tr}(A^2) + \mathrm{Tr}(B^2) - 2 - \mathrm{Tr}(BA) &(14.5)\\
&= \sum_{i=1}^{m} \lambda_i^2 + k - 2\sum_{i=1}^{m} \lambda_i < B(e_i), e_i >, &(14.6)
\end{aligned}$$

where the last inner product $< B(e_i), e_i >$ is the standard inner product on $\mathbb{R}^m$. The only term in the right hand side that depends upon B is the third term $\mathrm{Tr(AB)} = \sum_{i=1}^{m} \lambda_i < B(e_i), e_i >$, so we need to show that $\phi(A)$ provides that unique B which maximizes

$$\mathrm{Tr(AB)} = \sum_{i=1}^{m} \lambda_i < B(e_i), e_i > .$$

Let us first notice that when $B = \phi(A)$, $B(e_i) = e_i$ if $i \leq k$ and 0 if $i > k$, hence $\mathrm{Tr(AB)} = \sum_{i=1}^{k} \lambda_i$. Now let us consider an arbitrary $B \in j(G_k(\mathbb{R}^m))$ and write $\alpha_i =< B(e_i), e_i >$. Since B is positive semidefinite $\alpha_i \geq 0$ and by the Cauchy-Schwartz inequality, $\alpha_i \leq 1$. Hence, $0 \leq \alpha_i \leq 1$ for all i. Also

$$\begin{aligned}
\sum_{i=1}^{n} \alpha_i &= \sum_{i=1}^{n} < B(e_i), e_i > &(14.7)\\
&= \mathrm{Tr}\{B(\sum_{i=1}^{n} e_i \otimes e_i')\} &(14.8)\\
&= \mathrm{Tr}\{B(id)\} &(14.9)\\
&= \mathrm{Tr}(B) = k. &(14.10)
\end{aligned}$$

Let $\theta_i = 1 - \alpha_i$. Then from what we have concluded about α_i so far, it follows that $0 \leq \theta_i \leq 1$ and

$$\sum_{i=k+1}^{n} \alpha_i - \sum_{i=1}^{k} \theta_i = \sum_{i=1}^{n} \alpha_1 - k = 0. \qquad (14.11)$$

From

$$\text{Tr(BA)} = \sum_{i=1}^{k} \lambda_i + \sum_{i=k+1}^{n} \lambda_i \alpha_i - \sum_{i=1}^{k} \lambda_i \theta_i$$

and the inequalities $\sum_{i=k+1}^{n} \lambda_i \alpha_i \leq \lambda_{k+1} \sum_{i=k+1}^{n} \alpha_i$ and $\sum_{i=1}^{k} \lambda_i \theta_i \geq \lambda_k \sum_{i=1}^{k} \theta_i$, it follows that Tr(BA) $\leq \sum_{i=1}^{k} \lambda_i$ with equality holding if and only if (since we have assumed $\lambda_k > \lambda_{k+1}$) $\sum_{i=k+1}^{n} \alpha_i = 0 = \sum_{i=1}^{k} \theta_i$. In other words, $\alpha_i = 1$ if $i < k+1$ and 0 if $i > k$. This is precisely the definition of $\phi(A)$. This proves the claim.

14.3 Extrinsic Sample Covariance Matrix on a Grassmann Manifold

Recall that, in general, if $j : \mathcal{M} \to \mathbb{R}^N$ is an embedding of an abstract manifold into $\mathbb{R}^N$ and Q is a j-nonfocal distribution on $\mathcal{M}$, one may introduce a j-extrinsic covariance matrix of Q in terms of a j-adapted frame on $\mathcal{M}$ that is locally defined around the extrinsic mean μ_j (see Chapter 4). Moreover if $\{X_r\}_{r=1,\ldots,n}$ are i.i.d.r.o.'s from such a distribution Q, then the sample extrinsic covariance estimator is defined as the j-extrinsic covariance matrix of the empirical distribution

$$\hat{Q}_n = n^{-1} \sum_{i=1}^{n} \delta_{X_i}.$$

Using this terminology, an explicit formula of the sample Veronese-extrinsic covariance matrix of a Veronese-nonfocal distribution on $\mathbb{R}P^m$ was first obtained by Prentice (1984) [282].

Our goal in this section consists of calculating the extrinsic sample covariance of a j-nonfocal distribution on a Grassmann manifold, thus extending the Prentice's result.

First, let us observe that the tangent space to $j(G_k(\mathbb{R}^m))$ at some $B \in Sym(m,\mathbb{R})$ may be characterized as follows. Let $t \to \sigma(t)$ may be a smooth curve in $j(G_k(\mathbb{R}^m))$ with $\sigma(0) = B$. Let us write $\sigma(t) = B + tC + o(t)$. Then, from the requirements that $\sigma(t)$ is symmetric and a projection matrix of rank k, it follows that

$$BC + CB = C, \tag{14.12}$$
$$C = C^T. \tag{14.13}$$

In other words, C is an eigenvector of the operator

$$L : Sym(m,\mathbb{R}) \to Sym(m,\mathbb{R}),$$
$$L(F) = BF + FB.$$

Now, let $V_B = Range(B)$ and W_B be the orthogonal complement of V_B in $\mathbb{R}^n$.

It is seen easily that, for $a \in V_B$ and $b \in W_B$, $F = ab^T + ba^T$ is symmetric and satisfies $L(F) = F$. As such, it is a tangent vector of $j(G_k(\mathbb{R}^m)))$. Moreover, the vector subspace of $Sym(m, \mathbb{R})$,

$$S_B = \mathrm{span}\{\mathrm{ab}^\mathrm{T} + \mathrm{ba}^\mathrm{T} | \mathrm{a} \in \mathrm{V_B}, \mathrm{b} \in \mathrm{W_B}\}, \tag{14.14}$$

has dimension equal to $k(m-k)$, which is the dimension of $j(G_k(\mathbb{R}^m)))$. Therefore, we conclude that the tangent space $T_B(j(G_k(\mathbb{R}^m))) = S_B$.

It is convenient to use tensor product representation of matrices in the calculations. In what follows, if $\{e_1, \ldots, e_m\}$ is a basis of a vector space, then its dual basis will be denoted by $\{e'_1, \cdots, e'_m\}$. Thus, a linear map from the vector space into itself can be generally represented as $\sum_{i,j} a_{i,j} e_i \otimes e'_j$. This notation will primarily be useful to represent linear maps between vector spaces of matrices since basis elements themselves are denoted using double indices. For example, if $\{F_{\alpha,\beta} | (\alpha, \beta) \in \mathcal{A}\}$ is a basis of a linear space of matrices S, then we may represent a linear map from S into itself as $\sum_{\alpha,\beta,\gamma,\delta} a_{(\alpha,\beta),(\gamma,\delta)} F_{\alpha,\beta} \otimes F'_{\gamma,\delta}$.

Secondarily, we notice that covariance of probability measures transform as a covariant two tensor, and hence may be represented as $\sum_{i,j} g_{i,j} e_i \otimes e_j$ with respect to a given basis $\{e_i\}$. Thus, if the vector space under consideration is a space of matrices, once again the tensor product notation with double suffixes under basis elements will help keep track of how covariance will change under linear maps.

Let us assume that n i.i.d.r.o.'s are taken from a $G_k(\mathbb{R}^m)$ valued distributions. Without loss of generality, we assume that these are given in the form of matrices $Y_r; r = 1, \ldots, n$ where each $Y_r \in Sym(m, \mathbb{R})$ is an $m \times m$ matrix. Then $Y_r = X_r X_r^T$, where X_r is a $m \times k$ matrix whose columns are orthogonal and of norm one. Let us denote the sample mean and covariance of $Y_r, r = 1, \ldots, n$, by $\bar{Y}$ and S_Y respectively. Of course, $\bar{Y} = \frac{1}{n}\sum_{r=1}^n Y_r \in Sym(m, \mathbb{R})$. However, it is convenient for us to treat $\bar{Y}$ as a covariant two tensor on $Sym(m, \mathbb{R})$. Let us express $\bar{Y}$ in terms of components with respect to a particular basis now. Since $\bar{Y}$ is a symmetric matrix, we may write down an orthonormal basis $\{e_1, \ldots, e_m\}$ of $\mathbb{R}^m$ consisting of eigenvalues of $\bar{Y}$. Let us write down the basis of $Sym(m, \mathbb{R})$,

$$F_\alpha^\beta = \begin{cases} e_\alpha \otimes e'_\alpha & \text{if } \alpha = \beta \\ \frac{1}{\sqrt{2}}(e_\alpha \otimes e'_\beta + e_\beta \otimes e'_\alpha) & \text{if } \alpha \neq \beta, \end{cases} \tag{14.15}$$

where $1 \leq \alpha \leq \beta \leq m$. It is clear that this basis is orthonormal with respect to the standard Euclidean inner product of $Sym(m, \mathbb{R})$.

Let us write down the sample covariance tensor S_Y with respect to this basis as

$$S_Y = \sum_{1 \leq \alpha \leq \beta \leq m,\ 1 \leq \gamma \leq \delta \leq m} g_{(\alpha,\beta),(\gamma,\delta)} F_{\alpha,\beta} \otimes F_{\gamma,\delta}, \tag{14.16}$$

where $g_{(\alpha,\beta),(\gamma,\delta)}$ are real numbers for which we will derive formulas now.

Since S_Y is a covariant two tensor, we may write,

$$g_{(\alpha,\beta),(\gamma,\delta)} = \frac{1}{n}\sum_{r=1}^{n} < F_{\alpha}^{\beta}, Y_r - \bar{Y} >< F_{\gamma}^{\delta}, Y_r - \bar{Y} > . \tag{14.17}$$

At this point, we may derive expressions for all $g_{(\alpha,\beta),(\gamma,\delta)}$. However, in the derivation of formulas for the $j-$extrinsic sample covariance we will only need a subset in which $\alpha < \beta$ and $\gamma < \delta$. When this restriction is imposed,

$$\begin{aligned} < F_{\alpha}^{\beta}, \bar{Y} > &= \mathrm{Tr}(F_{\alpha}^{\beta}\bar{Y}) = \mathrm{Tr}\sum_{j=1}^{n} \frac{1}{\sqrt{2}}(e_{\alpha} \otimes e'_{\beta} + e_{\beta} \otimes e'_{\alpha})(\lambda_j e_j \otimes e'_j) \\ &= \mathrm{Tr}\frac{1}{\sqrt{2}}(e_{\alpha} \otimes e'_{\beta} + e_{\beta} \otimes e'_{\alpha}) = 0, \end{aligned} \tag{14.18}$$

where $\{\lambda_j\}$ denotes the eigenvalues of $\bar{Y}$. Hence for $\alpha < \beta$,

$$\begin{aligned} g_{(\alpha,\beta),(\gamma,\delta)} &= \frac{1}{n}\sum_{r=1}^{n} < F_{\alpha}^{\beta}, Y_r >< F_{\gamma}^{\delta}, Y_r > = \frac{1}{\sqrt{2}n}\sum_{r=1}^{n} \mathrm{Tr}(F_{\alpha}^{\beta}Y_r)\frac{1}{\sqrt{2}}\mathrm{Tr}(F_{\gamma}^{\delta}Y_r) \\ &= \frac{1}{2n}\sum_{r=1}^{n}[(Y_r)_{\alpha,\beta} + (Y_r)_{\beta,\alpha}\}\{(Y_r)_{\gamma,\delta} + (Y_r)_{\delta,\gamma}] \\ &= \frac{2}{n}\sum_{r=1}^{n}(Y_r)_{\alpha,\beta}(Y_r)_{\gamma,\delta}, \end{aligned} \tag{14.19}$$

where the notation $(Y_r)_{i,j}$ is used to denote the $(i,j)^{\mathrm{th}}$ element of the matrix Y_r.

Next, we focus on the projection map from a neighborhood of $\bar{Y}$ to $j(G_k(\mathbb{R}^m))$ and how its derivative transforms the sample covariance of Y, S_Y to give the extrinsic sample covariance tensor $\hat{\Sigma}_j$. Here, we make the genericity assumption that the eigenvalues $\{\lambda_j\}$ of $\bar{Y}$ satisfy $\lambda_k > \lambda_{k+1}$, where the eigenvalues are arranged in decreasing order. Let $\overline{Y}_j$ denote the j-extrinsic sample mean of $Y_1, \ldots, Y_n$. Then, from what we have discussed at the beginning, it follows that

$$\overline{Y}_j = \sum_{i=1}^{k} e_i \otimes e_i. \tag{14.20}$$

Recall also that we have an explicit description of the tangent space of $j(G_k(\mathbb{R}^m))$ at $\overline{Y}_j$ as,

$$\begin{aligned} T_{\overline{Y}_j}(j(G_k(\mathbb{R}^m)) &= \mathrm{span}\{e_i \otimes e'_j + e_j \otimes e'_i | 1 \le i \le k, k < j \le m\} \\ &= \mathrm{span}\{F_{\alpha}^{\beta} | \alpha \le k < \beta\}. \end{aligned} \tag{14.21}$$

Let U be a small neighborhood of $\bar{Y}$ in $Sym(m, \mathbb{R})$ on which the projection

$\phi : U \to j(G_k(\mathbb{R}^m))$ is defined. Let $d\phi_{\bar{Y}} : Sym(m,\mathbb{R}) \to T_{\overline{Y}_j}(j(G_k(\mathbb{R}^m)))$ denote the derivative of ϕ at $\bar{Y}$. Let us first show that its tensor coefficients satisfy the following. Here we follow the convention that $\alpha \leq \beta$.

$$d\phi_{\bar{Y}}(F_\alpha^\beta) = \begin{cases} 0 & \text{if } \alpha = \beta \\ 0 & \text{if } \beta \leq \mathrm{k} \\ 0 & \text{if } \alpha > \mathrm{k} \\ \frac{1}{\lambda_\alpha - \lambda_\beta} F_\alpha^\beta & \text{if } \alpha \leq \mathrm{k} < \beta. \end{cases} \tag{14.22}$$

Let us now proceed to prove (14.22). The main idea is to use appropriate curves through $\bar{Y}$ to represent various tangent vectors at $\bar{Y}$, map them by ϕ, and compute the derivatives. It turns out that it is possible to produce curves in the first three cases in such a way that ϕ maps them to the constant point $\overline{Y}_j$. The fourth case is a bit tedious. Here, we have constructed a special curve via a trial and error method that simplify the computation significantly. Below, case numbers 1 through 4 refer to the lines 1 through 4 of (14.22). In all cases, the smooth curves to be constructed will be denoted by $\nu : (-\varepsilon, \varepsilon) \to Sym(m,\mathbb{R})$ with $\nu(0) = \bar{Y}$.

Case 1: $\alpha = \beta$.

Consider the curve $\nu(t) = \bar{Y} + te_\alpha \otimes e'_\alpha$. Notice that $\nu(t)$ has exactly the same eigenvectors as $\bar{Y}$. As far as eigenvalues are concerned, only λ_α has changed to $\lambda_\alpha + t$ and the remaining eigenvalues are unchanged. However, for small values of t, it will still be true that the invariant subspace of the first k eigenvalues will be span$\{\mathrm{e}_1, \cdots, \mathrm{e}_\mathrm{k}\}$. This follows from our assumption that $\lambda_k > \lambda_{k+1}$. Therefore we have $\phi(\nu(t)) = \overline{Y}_j$ for small t, hence $d\phi_{\bar{Y}}(F_\alpha^\alpha) = \frac{d}{dt}\phi(\nu(t))|_{t=0} = 0$.

Case 2: $\alpha < \beta \leq k$

Consider the curve $\nu(t) = \bar{Y} + tF_\alpha^\beta$. Since $\beta < k$, $e_{k+1}, \cdots, e_m$ will remain eigenvectors of $\nu(t)$ and they will be orthogonal to the invariant subspace of the k largest eigenvalues since eigenvalues will be perturbed only slightly, if any. Therefore, $\phi(\nu(t)) = \overline{Y}_j$ for small t and it follows that $d\phi_{\bar{Y}}(F_\alpha^\beta) = 0$.

Case 3: $k < \alpha < \beta$

This case is similar to the previous case. Along the curve $\nu(t) = \bar{Y} + tF_\alpha^\beta$, only the smallest $m - k$ eigenvalues and their eigenspaces get perturbed, which are irrelevant as far as $\phi(\nu(t))$ is concerned.

Case 4: $\alpha \leq k < \beta$

This is an interesting case since it will perturb the invariant subspace of the k largest eigenvalues. Computation of this perturbation is tedious unless a judicious choice of ν is made that will ensure that the perturbations of the eigenspaces can be easily represented, perhaps at the expense of expressions for eigenvalues. Through trial and error, we have come up with the curve

$v(t) = \bar{Y} + \sqrt{2}\theta(t)F_\alpha^\beta$, where

$$\theta(t) = \frac{(\lambda_\alpha - \lambda_\beta)^2 t}{(\lambda_\alpha - \lambda_\beta)^2 - t^2}. \quad (14.23)$$

It is clear that for $s \neq \alpha$ and $s \neq \beta$, λ_s will be an eigenvalue of $v(t)$ with corresponding eigenvector e_s. Thus, there are only two eigenvectors and eigenvalues that gets possibly perturbed; e_α, e_β and $\lambda_\alpha, \lambda_\beta$. Indeed, we claim that

$$\tilde{e}_\alpha = \frac{1}{\sqrt{(\lambda_\alpha - \lambda_\beta)^2 + t^2}}(\lambda_\alpha - \lambda_\beta)e_\alpha + te_\beta \text{ and } \tilde{\lambda} = \lambda,$$

where the tildes denote perturbed values. This is shown by the following explicit computation.

$$\begin{aligned} v(t)[(\lambda_\alpha - \lambda_\beta)e_\alpha + te_\beta] = & \{\lambda_\alpha + \frac{t^2(\lambda_\alpha - \lambda_\beta)}{(\lambda_\alpha - \lambda_\beta)^2 - t^2}\}(\lambda_\alpha - \lambda_\beta)e_\alpha + \\ & \{\lambda_\beta + \frac{(\lambda_\alpha - \lambda_\beta)^3}{(\lambda_\alpha - \lambda_\beta)^2 - t^2}\}te_\beta \quad (14.24) \\ = & \frac{\lambda_\alpha(\lambda_\alpha - \lambda_\beta)^2 - t^2\lambda_\beta}{(\lambda_\alpha - \lambda_\beta)^2 - t^2}[(\lambda_\alpha - \lambda_\beta)e_\alpha + te_\beta]. \quad (14.25) \end{aligned}$$

Here, $\tilde{e}_\alpha = \frac{1}{\sqrt{(\lambda_\alpha - \lambda_\beta)^2 + t^2}}[(\lambda_\alpha - \lambda_\beta)e_\alpha + te_\beta]$ is the perturbation of e_α along the curve v. Therefore,

$$\phi(v(t)) = e_1 \otimes e_1' + \cdots e_{\alpha-1} \otimes e_{\alpha-1}' + \tilde{e}_\alpha \otimes \tilde{e}_\alpha' + e_{\alpha+1} \otimes e_{\alpha+1}' + \cdots + e_k \otimes e_k'. \quad (14.26)$$

Therefore,

$$\begin{aligned} \frac{d}{dt}\phi(v(t))|_{t=0} &= \frac{d}{dt}(\tilde{e}_\alpha \otimes \tilde{e}_\alpha')|_{t=0} \\ &= \frac{1}{\lambda_\alpha - \lambda_\beta}(e_\alpha \otimes e_\beta' + e_\beta \otimes e_\alpha)\frac{\sqrt{2}}{\lambda_\alpha - \lambda_\beta}F_\alpha^\beta. \quad (14.27) \end{aligned}$$

Furthermore, we observe that

$$\frac{d}{dt}v(t)|_{t=0} = \sqrt{2}F_\alpha^\beta. \quad (14.28)$$

The fourth line of (14.22) now follows.

Now, we are in a position to state the Prentice formula for this generalized situation. Recall that F_α^β, $\alpha \leq k < \beta$ forms a basis of $T_{j(\overline{Y}_j)}j(G_k(\mathbb{R}^m))$. The

Prentice formula will state the coefficients of the j extrinsic sample covariance $\hat{\Sigma}_j$ with respect to this basis. Recall that $\hat{\Sigma}_j$ is the projection of S_Y via $d_{\bar{Y}}\phi$. As such, we finally are in the position to give an the explicit Prentice-type of formula for the j extrinsic sample covariance w.r.t. our embedding j:

$$\begin{aligned}
\hat{\Sigma}_j = (d_{\bar{Y}}\phi)S_Y &= \sum_{1\leq\alpha\leq\beta\leq m,\ 1\leq\gamma\leq\delta\leq m} g_{(\alpha,\beta),(\gamma,\delta)}(d_{\bar{Y}}\phi)(F_\alpha^\beta \otimes F_\gamma^\delta) \\
&= \sum_{1\leq\alpha\leq\beta\leq m,\ 1\leq\gamma\leq\delta\leq m} g_{(\alpha,\beta),(\gamma,\delta)}[(d_{\bar{Y}}\phi)F_\alpha^\beta]\otimes[(d_{\bar{Y}}\phi)F_\gamma^\delta] \\
&= \sum_{1\leq\alpha\leq k<\beta\leq m,\ 1\leq\gamma\leq k<\delta\leq m} \frac{1}{\lambda_\alpha-\lambda_\beta}\frac{1}{\lambda_\gamma-\lambda_\delta} g_{(\alpha,\beta),(\gamma,\delta)} F_\alpha^\beta \otimes F_\gamma^\delta \\
&= \sum_{1\leq\alpha\leq k<\beta\leq m,\ 1\leq\gamma\leq k<\delta\leq m} \frac{1}{\lambda_\alpha-\lambda_\beta}\frac{1}{\lambda_\gamma-\lambda_\delta}\frac{2}{n}\sum_{r=1}^{n}(Y_r)_{\alpha,\beta}(Y_r)_{\gamma,\delta} F_\alpha^\beta \otimes F_\gamma^\delta
\end{aligned} \tag{14.29}$$

This yields the following result:

THEOREM 14.3.1. *The extrinsic sample covariance matrix of a nonfocal probability distribution on $G_k(\mathbb{R}^m)$ is given by*

$$\hat{\Sigma}_j = \sum_{1\leq\alpha\leq k<\beta\leq m,\ 1\leq\gamma\leq k<\delta\leq m} \frac{1}{\lambda_\alpha-\lambda_\beta}\frac{1}{\lambda_\gamma-\lambda_\delta}\frac{2}{n}\sum_{r=1}^{n}(X_rX_r^T)_{\alpha,\beta}(X_rX_r^T)_{\gamma,\delta} F_\alpha^\beta \otimes F_\gamma^\delta. \tag{14.30}$$

REMARK 14.3.1. *From Bhattacharya and Patrangenaru (2005) [43], it follows that*

$$n\|\hat{\Sigma}_j^{-\frac{1}{2}} tan_{P_j(\overline{j(X)})}(P_j(\overline{j(X)}) - P_j(\mu))\|^2 \tag{14.31}$$

converges weakly to $\chi^2_{k(m-k)}$. The result will be used in estimation and testing for the extrinsic mean of a population on a Grassmann manifold. In particular, given i.i.r.o.'s from a distribution of affine shapes of k-ads in general position in $\mathbb{R}^m$, the squared norm of the studentized extrinsic sample mean affine shape has asymptotically a $\chi^2_{k(m-k)}$ distribution.

Part III

Applications in Object Data Analysis on Manifolds

Chapter 15

DTI Data Analysis

15.1 Introduction

In Chapter 8, we presented general theory for nonparametric statistical analysis of data from homogeneous Riemannian manifolds . Amongst other applications of this methodology are diffusion tensor imaging (DTI) and cosmic microwave background radiation (CBR) (Schwartzman et al. (2008) [303]). Both of these applications led to the analysis of random objects on the set of positive definite symmetric matrices $Sym^+(m)$, for some dimension $m \geq 2$. In this chapter, we will concentrate on application to DTI data analysis.

Until recently, the main statistical techniques used for DTI data analysis were parametric in nature (Schwartzman (2006) [301], Schwartzman et al. (2008) [303] and Schwartzman et al. (2008a) [304]). Recently, though, Huckemann (2011) [164], and Haff et al. (2011) [134] used new geometric methods for analyzing DTI data, which are related to the approach described here, which follows primarily Osborne et al. (2013) [258] as well as Ellingson et al. (2013) [102].

The application presented here extends from an experiment considered in Schwartzman et al. (2008) [303] in which diffusion direction maps were acquired for two groups of subjects in order to find locations where such maps differ. The objective of that analysis was to find regions of the brain in which the corresponding diffusion directions differ between the groups. That was attained by first computing a test statistic for the difference in direction at every brain location using the Watson model for directional data. While a perfect correspondence between brain locations is impossible, we will assume nevertheless that such a correspondence within and between groups was achieved for at least some region of the brain. As such, the goal for the analysis shown in this chapter is limited to an illustration of the nonparametric inference methods developed in Chapter 8 and is by no means meant to depict discovery for child behavior based on brain imaging.

This chapter is organized as follows. In Section 15.2, we detail the two sample test for generalized Frobenius means on the space of symmetric matrices in terms of Cholesky decompositions, as shown in Theorem 15.2.1 and in Theorem 15.2.2. In Section 15.3, this methodology is applied to a small DTI dataset previously analyzed by Schwartzman et al. (2008) [303], which was kindly provided by the first author of that paper. This analysis consists of a voxelwise comparison of spatially registered DT images belonging to two groups of children, one with normal reading abilities and one with a diagnosis of dyslexia. These methods are illustrated for a single voxel that was found in Schwartzman et al. (2008) [303] to exhibit a strong difference between the two groups using parametric methodology.

15.2 Tests for Equality of Generalized Frobenius Means via Cholesky Decompositions

In this section, we apply the results developed in Section 8.4 to the case of distributions on the Hadamard manifold $(\mathcal{M}, g) = (Sym^+(p), g_F)$ with the simple transitive group action of $\mathcal{G} = T^+(p, \mathbb{R})$ given in Proposition 8.2.2. The Riemannian structure considered here and in Chapter 8 was first used in DTI literature by Fletcher (2004) [119], Moakher (2005) [248], Arsigny et al. (2006,2006a) [5, 6], Deriche et al. (2006) [83], and Pennec et al. (2006) [281]. From equation (8.16), the generalized Frobenius sample mean of a sample $x_1, \dots, x_n$ of matrices on $Sym^+(p)$ is given by

$$\mu_F = \underset{y \in Sym^+(p)}{\arg\min} \frac{1}{n} \sum_{i=1}^{n} [Tr((\log(c(x_i)^{-1} y (c(x_i)^{-1})^T))^2)], \tag{15.1}$$

where $\forall i = 1, \dots, n, x_i = c(x_i)c(x_i)^T$ is the Cholesky decomposition of x_i.

REMARK 15.2.1. *The Cholesky decomposition was used for DTI also by Wang et al. (2004) [331], although they did not use the generalized Frobenius distance in their paper, as Osborne et al. (2013) [258] did.*

From Section 8.4, we see that given two independent populations on $Sym^+(p)$ with generalized Frobenius means $\mu_{a,F}, a = 1, 2$, the testing problem

$$H_0 : \mu_{1,F} = \delta \mu_{2,F} \delta^T$$

versus

$$H_1 : \mu_{1,F} \neq \delta \mu_{2,F} \delta^T \tag{15.2}$$

is equivalent to testing on $\mathcal{G} = T^+(p,\mathbb{R})$

$$H_0 : c(\mu_{1,F})c(\mu_{2,F})^{-1} = \delta$$

versus

$$H_1 : c(\mu_{1,F})c(\mu_{2,F})^{-1} \neq \delta. \tag{15.3}$$

For testing, suppose for $a = 1,2$ we are given the i.i.d.r.o.'s $X_{a,1},\dots,X_{a,n_a} \in Sym^+(p)$ with the total sample size $n = n_1 + n_2$. For $a = 1,2$, the corresponding sample generalized Frobenius means are $\bar{X}_{a,F}$. Note that, in our case, the matrix valued function H in equation (6.23) is given by

$$H(x_1,x_2) = c(x_1)c(x_2)^{-1}. \tag{15.4}$$

A consistent estimator of $H(\mu_{1,F},\mu_{2,F})$ is $T = H(\bar{X}_{1,F},\bar{X}_{2,F})$, and Theorem 6.5.1 becomes

THEOREM 15.2.1. *Assume $X_{a,j_a}, j_a = 1,\dots,n_a$ are i.i.d.r.o.'s from the independent probability measures $Q_a, a = 1,2$ generalized Frobenius moments of order $s, s \leq 4$ on the $Sym^+(p)$. Let $n = n_1 + n_2$ and assume $\lim_{n\to\infty} \frac{n_1}{n} \to q \in (0,1)$. Let $\log : T^+(p,\mathbb{R}) \to T(p,\mathbb{R})$ be the inverse of $\exp$ given in a neighborhood of I_p by*

$$\log(I_p + v) = v - \frac{1}{2}v^2 + \dots + \frac{(-1)^{r+1}}{r}v^r + \dots, \forall v \in T(p,\mathbb{R}), Tr(vv^T) < 1. \tag{15.5}$$

Then under H_0,
i. The sequence of random vectors

$$\sqrt{n}(\log(\delta^{-1}(H(\bar{X}_{1,F},\bar{X}_{2,F})))) \tag{15.6}$$

converges weakly to $\mathcal{N}_{\frac{p(p+1)}{2}}(0_{\frac{p(p+1)}{2}}, \Sigma_F(q))$, for some covariance matrix $\Sigma_F(q)$ that depends linearly on the generalized Frobenius covariance matrices $\Sigma_{a,F}$ of $Q_a, a = 1,2$.
ii. If (i.) holds and Σ_F is positive definite, then the sequence

$$n((\log(\delta^{-1}(H(\bar{X}_{1,F},\bar{X}_{2,F}))))^T \Sigma_F(q)^{-1}(\log(\delta^{-1}(H(\bar{X}_{1,F},\bar{X}_{2,F}))))) \tag{15.7}$$

converges weakly to $\chi^2_{\frac{p(p+1)}{2}}$ distribution.

Theorem 6.6.1, becomes in our case

THEOREM 15.2.2. *Under the hypotheses of Theorem 15.2.1i., assume in addition, that for $a = 1,2$ the distribution of $\mathrm{Exp}^{-1}_{\mu_{F,a}} X_{a,1}$ has an absolutely continuous component, and finite moments of sufficiently high order. Then the joint distribution of*

$$V_\delta = \sqrt{n}(\log(\delta^{-1}H(\bar{X}_{1,F},\bar{X}_{2,F})))$$

can be approximated by the bootstrap joint distribution of

$$V_\delta^* = \sqrt{n}(\log(\delta^{-1}H(\bar{X}_{1,F}^*, \bar{X}_{2,F}^*))) \tag{15.8}$$

with a coverage error $O_p(n^{-\frac{1}{2}})$, *where, for* $a = 1,2$, $\bar{X}_{a,F}^*$ *are the sample generalized Frobenius means of the bootstrap resamples* $X_{a,j_a}^*, j_a = 1,\ldots,n_a$, *from* $X_{a,j_a}, j_a = 1,\ldots,n_a$.

Each bootstrap resample in Theorem 15.2.2 of the total sample is obtained by sampling with replacement from the samples of sizes n_1, n_2 from Q_1 and Q_2, respectively. The matrix valued bootstrap statistics (15.8) are recomputed N times. To insure accuracy, one usually takes N to be at least 5,000. The bootstrap values of V and $\hat{T}$ are given by:

$$\hat{T}^* = \hat{T}_1^* \hat{T}_2^{-1*} \text{ and } V_\delta^* = \log(\delta^{-1}\hat{T}^*), \text{ where} \tag{15.9}$$
$$\hat{T}_a^* = c(\bar{X}_{a,F}^*), a = 1,2.$$

We construct $(1-\alpha)$ bootstrap confidence regions C_α^* for $\tau = c(\mu_{1,F})c(\mu_{2,F})^{-1}$ and $\mathcal{R}(\delta)_\alpha^*$ for $\log(\delta^{-1}\tau)$ based on the bootstrap distributions of $\hat{T}^*$ and V_δ^*.

COROLLARY 15.2.1. *i. We fail to reject* H_0 *in* (15.3) *at level* α *if* $\delta \in C_\alpha^*$ *with an error* $O_P(n^{-\frac{1}{2}})$.
ii. We fail to reject H_0 *in* (15.3) *at level* α *if* $\mathbf{0}_p \in R(\delta)_\alpha^*$ *with an error* $O_P(n^{-\frac{1}{2}})$.

REMARK 15.2.2. *It is known that there are many choices for bootstrap confidence regions (see e.g. Fisher et al.(1996) [116]). Computationally, it is often convenient to use simultaneous confidence intervals, even if the coverage is a bit enlarged. In this case, we reject* H_0 *at level* α *if* $\delta \notin C_\alpha^*$ *or, equivalently,* $\mathbf{0}_p \notin \mathcal{R}(\delta)_\alpha^*$. *That is the same as saying that we will reject* H_0 *if the Bonferroni* $100(1-\alpha)\%$ *simultaneous bootstrap confidence intervals for the nonzero entries* τ_{ij} *of* τ *do not contain 1 in at least one of the diagonal entries (when* $i = j$*) or 0 in at least one of the off-diagonal entries (when* $j < i$*). Similarly, we will reject* H_0 *if the Bonferroni* $100(1-\alpha)\%$ *simultaneous bootstrap confidence intervals for the* $\log(\delta^{-1}\tau)_{ij}$ *does not contain 0 in at least one of the diagonal or off-diagonal entries, when* $j \leq i$*. Such Bonferroni simultaneous bootstrap confidence intervals are formed by cutting off the lower and the upper* $100\left(\frac{\alpha}{2m}\right)\%$ *of the bootstrap distributions of* $\hat{T}_{ij}^*$ *and* V_{ij}^**, respectively. Here* $m = \frac{p(p+1)}{2}$.

15.3 Application to Diffusion Tensor Imaging Data

In this section, we apply the methodology presented in Section 15.2 to a concrete DTI example, using a dataset kindly provided by A. Schwartzman. The data was collected from two populations of children: a group of children with normal reading abilities and a group of children with a diagnosis of dyslexia. For each population, spatially registered DT images were obtained for a sample of 6 children, resulting in a total of 12 observations.

Commonly in DTI group studies, a typical statistical problem is to find regions of the brain whose anatomical characteristics differ between two groups of subjects. Typically, the analysis consists of registering the DT images to a common template so that each voxel corresponds to the same anatomical structure in all the images, and then applying two-sample tests at each voxel. Here, we present the analysis of a single voxel at the intersection of the corpus callosum and corona radiata in the frontal left hemisphere that was found in Schwartzman et al. (2008) [303] to exhibit the strongest difference between the two groups based on a parametric data analysis. Table 15.1 shows the data at this voxel for all 12 subjects. The d_{ij} in the table 15.1 are the entries of the DT on and above the diagonal (the below-diagonal entries would be same since the DTs are symmetric).

Table 15.1 *DTI data in a group of control (columns 1–6) and dyslexia (columns 7–12)*

	1	2	3	4	5	6
d_{11}	0.8847	0.6516	0.4768	0.6396	0.5684	0.6519
d_{22}	0.9510	0.9037	1.1563	0.9032	1.0677	0.9804
d_{33}	0.8491	0.7838	0.6799	0.8265	0.7918	0.7922
d_{12}	0.0448	−0.0392	0.0217	0.0229	−0.0427	0.0269
d_{13}	−0.1168	−0.0631	−0.0091	−0.1961	−0.0879	−0.1043
d_{23}	0.0162	−0.0454	−0.1890	−0.1337	−0.1139	−0.0607

	7	8	9	10	11	12
d_{11}	0.5661	0.6383	0.6418	0.6823	0.6159	0.5643
d_{22}	0.7316	0.8381	0.8776	0.8376	0.7296	0.8940
d_{33}	0.8232	1.0378	1.0137	0.9541	0.9683	0.9605
d_{12}	0.0358	−0.0044	−0.0643	0.0309	−0.0929	−0.0635
d_{13}	−0.2289	−0.2229	−0.1675	−0.2217	−0.1713	−0.1307
d_{23}	−0.1106	−0.0449	−0.0192	−0.0925	−0.0965	−0.1791

For this analysis, the primary goal is to demonstrate that that the nonparametric two-sample testing procedure presented in Section 15.2 is able to detect a significant difference between the generalized Frobenius means of the clinically normal and dyslexia groups without increasing the dimensionality in the process or making any distributional assumptions.

Given two independent populations with i.i.d. samples of random SPD matrices $X_{1,1}, X_{1,2}, \ldots, X_{1,n_1} \in Sym^+(3)$ from the clinically normal population and $X_{2,1}, X_{2,2}, \ldots, X_{2,n_2} \in Sym^+(3)$ from the dyslexia population with sample sizes of $n_1 = 6$ and $n_2 = 6$ and the total sample size $n = n_1 + n_2 = 12$, where, for $a = 1, 2, X_{a,1} \sim \mu_{F,a}$, the sample generalized Frobenius mean for the clinically normal population and dyslexia population is given by

$$\bar{x}_{1,F} = \begin{pmatrix} 0.6318 & 0.0046 & -0.0924 \\ 0.0046 & 0.9863 & -0.0873 \\ -0.0924 & -0.0873 & 0.7803 \end{pmatrix} \text{ and}$$

$$\bar{x}_{2,F} = \begin{pmatrix} 0.6146 & -0.0261 & -0.1910 \\ -0.0261 & 0.8118 & -0.0901 \\ -0.1910 & -0.0901 & 0.9537 \end{pmatrix}.$$

Diffusion tensors are commonly visualized as ellipsoids constructed from their spectral decompositions. Ellipsoids representing the two sample generalized Frobenius means are provided in Figure 15.1. The ellipsoids are shown from three views for both groups to better display the differences between the means.

The values of the test statistics $\hat{T}$ and V are given by

$$\hat{T} = \begin{pmatrix} 0.9862 & 0.0000 & 0.0000 \\ -0.0485 & 0.9067 & 0.0000 \\ -0.1487 & -0.0152 & 1.0781 \end{pmatrix} \text{ and}$$

$$V = \begin{pmatrix} -0.0139 & 0.0000 & 0.0000 \\ -0.0513 & -0.0980 & 0.0000 \\ -0.1446 & -0.0153 & 0.0752 \end{pmatrix}.$$

In addition, let $\hat{t}_{ij}$ and v_{ij} correspond to the entries of the test statistics $\hat{T}$ and V on and below the diagonal (since the test statistics $\hat{T}$ and V are lower triangular matrices).

In order to test hypothesis (15.2) or hypothesis (15.3) for $\delta = I_3$, we repeatedly resample observations from the original data and compute the generalized Frobenius sample mean for each respective group. The generalized Frobenius sample means are computed as described in Section 8.3. Tables 15.2 and 15.3 display a five number summary for the bootstrap distribution of the Generalized Frobenius sample means for the clinically normal and dyslexia groups. To generate the summary statistics presented in tables 15.2 and 15.3, 10,000 bootstrap resamples were used. Figure 15.2 displays a visualization of the bootstrap distributions of the Generalized Frobenius sample means.

In addition, for each bootstrap resample, we calculate the Cholesky decomposition of the bootstrap generalized Frobenius sample mean for each respective group and then proceed to calculate the bootstrap distribution of our test

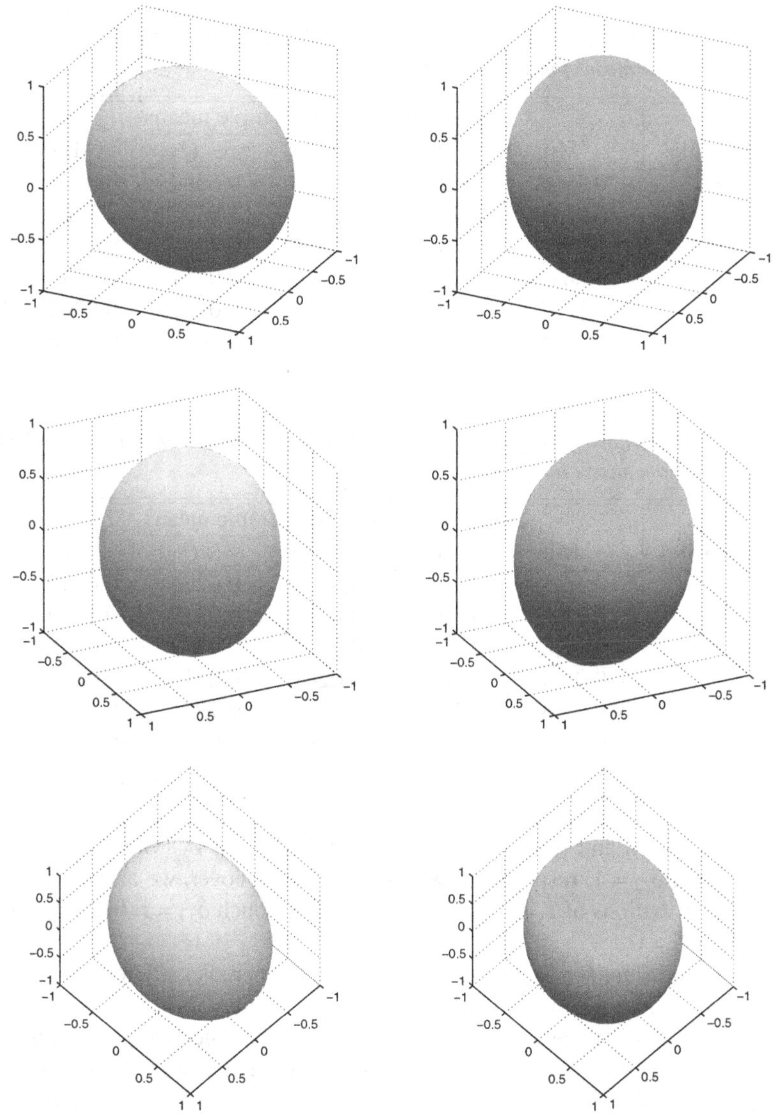

Figure 15.1 *Ellipsoids representing the sample generalized Frobenius means of the control group (left) and the dyslexia group (right).*

statistics $\hat{T}$ and V as described in (15.9). Figures 15.3 and 15.4 display a visualization of the bootstrap distributions of our test statistics $\hat{T}$ and V.

After visually examining Figures 15.3 and 15.4, we informally conclude that there is significant difference between the generalized Frobenius means of the

Table 15.2 *Five Number Summary of the bootstrap distribution of the generalized Frobenius sample means for the clinically normal group*

	Generalized Frobenius sample means				
Quartiles	Min	Q_1	Median	Q_3	Max
d_{11}	0.4902	0.6004	0.6318	0.6662	0.8408
d_{22}	0.9018	0.9608	0.9863	1.0131	1.1402
d_{33}	0.6951	0.7649	0.7819	0.7991	0.8429
d_{12}	−0.0422	−0.0064	0.0046	0.0151	0.0413
d_{13}	−0.1852	−0.1093	−0.0924	−0.0757	−0.0202
d_{23}	−0.1788	−0.1063	−0.0873	−0.0692	0.0058

Table 15.3 *Five Number Summary of the bootstrap distribution of the generalized Frobenius sample means for the dyslexia group*

	Generalized Frobenius sample means				
Quartiles	Min	Q_1	Median	Q_3	Max
d_{11}	0.5626	0.6036	0.6149	0.6276	0.6745
d_{22}	0.7269	0.7935	0.8129	0.8317	0.8940
d_{33}	0.8232	0.9348	0.9561	0.9763	1.0291
d_{12}	−0.0880	−0.0410	−0.0261	−0.0113	0.0358
d_{13}	−0.2289	−0.2012	−0.1910	−0.1809	−0.1307
d_{23}	−0.1791	−0.1025	−0.0896	−0.0767	−0.0192

clinically normal and dyslexia group, since the $\hat{T}^*_{22}$ and V^*_{22} values does not overlap with $\delta_{22} = 1$, respectively with $\mathbf{0}_{3,22} = 0$. Moreover, we also observed that the distributions of $\hat{T}^*_{33}$, V^*_{33} and $\hat{T}^*_{31}$, V^*_{31} barely touch $\delta_{33} = 1$, $\mathbf{0}_{3,33} = 0$ and $\delta_{31} = 0$, $\mathbf{0}_{3,31} = 0$.

These results are formally confirmed at level α; there is sufficient evidence that the clinically normal and dyslexia children display, on average, significantly different DTI responses. The results were obtained by performing a $100(1-\alpha)\%$-simultaneous bootstrap confidence intervals, as described in Remark (15.2.2), for $\hat{T}_{ij}$ and V_{ij}. Tables 15.4 and 15.5 display the results of the Bonferroni $100(1-\alpha)\%$-simultaneous bootstrap confidence intervals for $\hat{T}_{ij}$ and V_{ij} at the following significance levels: $\alpha = 0.06, 0.03$, and 0.006, where for each marginal we consider a $100(1-\frac{\alpha}{12})\%$ confidence interval. The significant differences are marked with an asterisk in those tables.

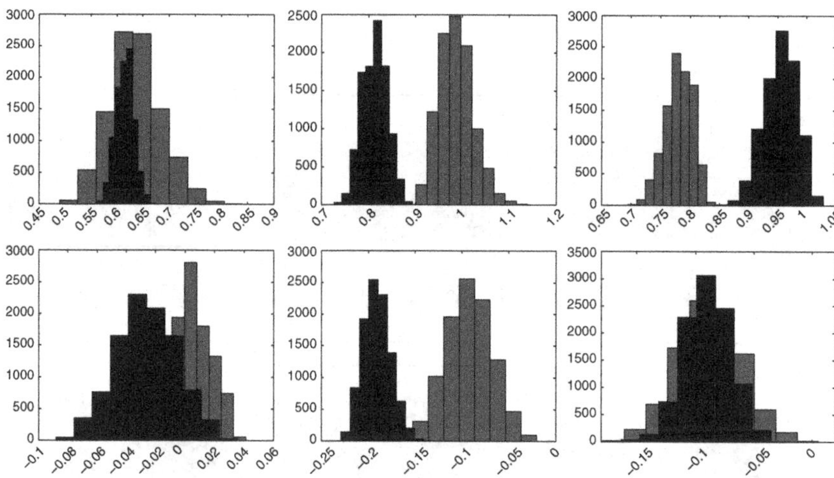

Figure 15.2 *Marginals of the bootstrap distribution for the generalized Frobenius sample means for* d_{11}, d_{22}, d_{33}, d_{12}, d_{13}, *and* d_{23}; *clinically normal (light) vs dyslexia (dark).*

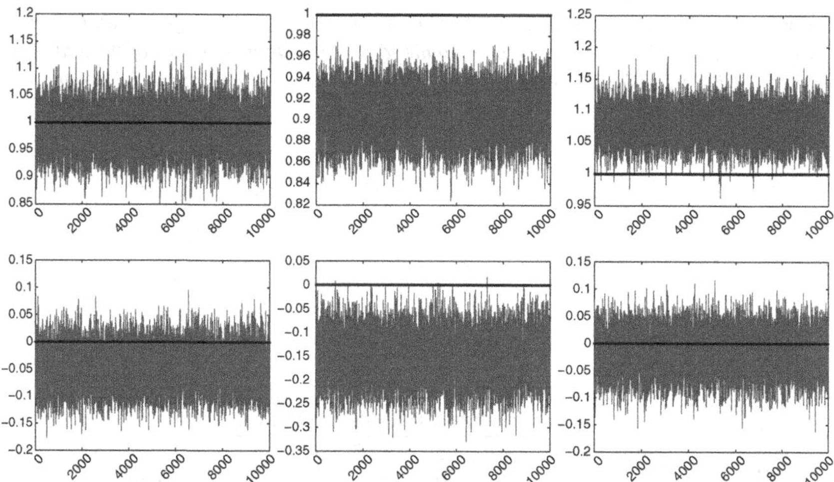

Figure 15.3 *Bootstrap distribution of our test statistics* $\hat{T}$: *The images (1 - 3) in the first row corresponds to the diagonal entries of the matrices* $\hat{T}^*$: t_{11}, t_{22}, t_{33} *and images (4 - 6) in the second row corresponds to the lower triangular off-diagonal entries of the matrices* $\hat{T}^*$: t_{21}, t_{31}, t_{32}. *(Source: Osborne et al.(2013), Figure 2. Reproduced by permission of Elsevier).*

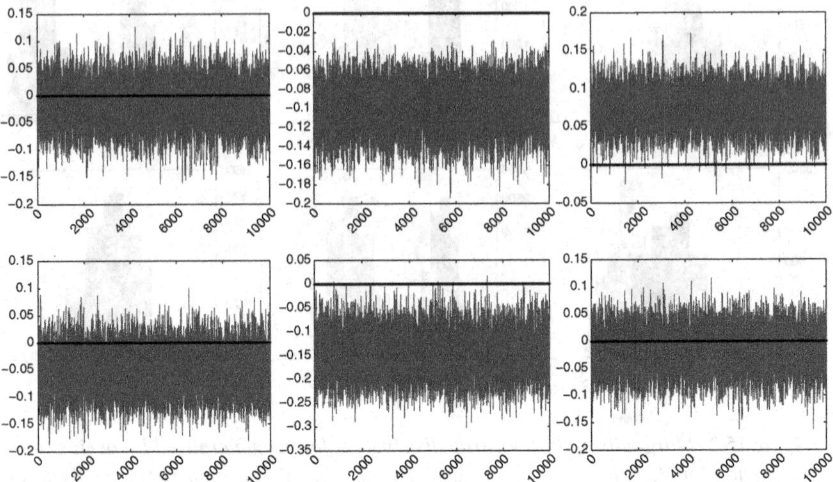

Figure 15.4 *Bootstrap distribution of our test statistics* V*: The images (1 - 3) in the first row corresponds to the diagonal entries of the matrices* V^**:* v_{11}*,* v_{22}*,* v_{33} *and images (4 - 6) in the second row corresponds to the lower off-diagonal entries of the matrices* V^**:* v_{21}*,* v_{31}*,* v_{32}*. (Source: Osborne et al.(2013), Figure 3. Reproduced by permission of Elsevier).*

Table 15.4 *Bonferroni* $100(1-\alpha)\%$ *multiple comparison simultaneous bootstrap confidence intervals for* $\hat{T}_{ij}$

	Bootstrap Confidence Intervals		
	94%	97%	99.4%
t_{11}	$(0.8865, 1.0916)$	$(0.8787, 1.1019)$	$(0.8589, 1.1179)$
t_{22}	$(0.8488, 0.9600)^*$	$(0.8449, 0.9633)^*$	$(0.8352, 0.9701)^*$
t_{33}	$(1.0085, 1.1465)^*$	$(1.0015, 1.1542)^*$	$(0.9844, 1.1712)$
t_{21}	$(-0.1413, 0.0513)$	$(-0.1482, 0.0590)$	$(-0.1619, 0.0748)$
t_{31}	$(-0.2749, -0.0269)^*$	$(-0.2905, -0.0145)^*$	$(-0.3085, 0.0018)$
t_{32}	$(-0.1136, 0.0768)$	$(-0.1214, 0.0859)$	$(-0.1398, 0.0975)$

Table 15.5 *Bonferroni* $100(1-\alpha)\%$ *multiple comparison simultaneous bootstrap confidence intervals for* V_{ij}

	Bootstrap Confidence Intervals		
	94%	97%	99.4%
v_{11}	$(-0.1205, 0.0876)$	$(-0.1293, 0.0970)$	$(-0.1521, 0.1115)$
v_{22}	$(-0.1640\ -0.0409)$*	$(-0.1686, -0.0374)$*	$(-0.1801, -0.0304)$*
v_{33}	$(0.0084, 0.1367)$*	$(0.0015, 0.1434)$*	$(-0.0157, 0.1580)$
v_{21}	$(-0.1501, 0.0539)$	$(-0.1580, 0.0627)$	$(-0.1754, 0.0796)$
v_{31}	$(-0.2584, -0.0272)$*	$(-0.2668, -0.0141)$*	$(-0.2859, 0.0000)$
v_{32}	$(-0.1153, 0.0775)$	$(-0.1241, 0.0843)$	$(-0.1421, 0.0963)$

Chapter 16

Application of Directional Data Analysis: a Bootstrap Approach to Pluto's Origin

16.1 Introduction

The solar nebula theory hypothesizes that planets are formed from an accretion disk of material that, over time, condenses into dust, small planetesimals, and that the planets should have, on average, coplanar, nearly circular orbits. If the orbit of Pluto has a different origin from the other planets in the Solar System, then there will be tremendous repercussions on modeling the spacecrafts for a mission to Pluto. We test here the nebula theory for Pluto. We apply the general theory for asymptotic distributions of extrinsic means on a manifold in Chapter 5 to spherical data analysis. Then we use the derived nonparametric bootstrap based on the large sample distribution of the sample mean direction. Our nonparametric analysis provides very strong evidence that the solar nebula theory does not hold for Pluto.

16.2 The Pluto Controversy

Last century, the solar system was considered to have been divided into *planets*- the big bodies orbiting the Sun, their *satellites*- smaller objects orbiting the planets, *asteroids* - small dense objects orbiting the Sun, and *comets*- small icy objects with highly eccentric orbits. Still the Solar System is more complex than such a simple classification would imply, since there are several moons larger than Pluto, and two larger than Mercury. There are also several small moons considered, that are probably trapped asteroids, etc.

Pluto was considered as the ninth planet since its discovery in 1930. Pluto has always been a kind of a misfit among the other planets in the solar system. It does not fit in the group of Jovian planets, which are large, gaseous, low density worlds. It also does not fit into the Terrestrial group because it lies far away from the sun, as a Jovian planet does.

There were numerous press reports generated in March of 1996 on the eve of the release of new Hubble Space Telescope images of Pluto. Most of these reports focused on the question of whether or not Pluto is a planet.

At the time, there were astronomers who thought Pluto would be better classified as a large asteroid or comet rather than as a planet. Some considered it to be the largest of the Kuiper Belt objects (also known as Trans-Neptunian objects). There is considerable merit to the latter position; Pluto's orbit is highly eccentric, Pluto rotates in the opposite direction from most of the other planets and its orbital inclination is also much higher than the other planets.

However, although Pluto has historically been classified as a planet, it became more and more unlikely to remain classified as such (IAU press release, 2008 [326]). Exploring the contrasts with other planets was not easy, since at the time, Pluto was the only planet that had not been visited by a spacecraft until 2015. The Pluto system (consisting of Pluto and Charon and other moons) was visited by the New Horizons probe for the first time in 2015, when this space probe seemingly took detailed measurements and images of Pluto's neighborhood.

Still, considering that Pluto's orbit is so inclined, reaching Pluto may still appear as a fiction turned to reality scenario. And since Pluto's orbit is much more eccentric, is it a planet in the Solar system, or is it rather an entrapment by the Sun of an extrasolar object? Since the questioning about Pluto's nature started, many other Pluto-like transneptunian objects called *plutoids*, which have a large inclination to Earth's ecliptic orbital plane, have been discovered (see Li et al. (2014) [221]).

In Section 16.3, we give a sketch of the Solar Nebula Theory. In Section 16.4, we applied the results from Chapter 5 to obtain asymptotic and nonparametric bootstrap distributions of extrinsic mean directions, regarded as points on a round sphere of radius one. In Section 16.5, we describe the planet data set used in this chapter to test the null hypothesis that Pluto has the same origin as the other eight planets in the Solar System, as far as its orbit is concerned. Since the sample is small, we also implement in Section 16.5 our new nonparametric approach of Section 16.4. Finally, the chapter ends with a discussion.

16.3 The Solar Nebula Theory

In an attempt to avoid looking into differences between planets, and rather consider what they share in common, we accept the nebula origin theory of the solar system.

According to this theory, the present motions of the planets and of the Sun were inherited from the original motion of the solar nebula. As originally proposed by Kant, Laplace, and others, the solar system is presumed to be formed from a nebula that evolved into a disk. The initial nebula presumably looked like the molecular cloud cores, where we find stars forming today (see Williams and Cieza (2011) [337]).

The solar nebula theory proposes that the Solar System began as a contracting cloud of gas and dust and flattened into a rotating disk. The center of this cloud became the Sun and the planets eventually formed into the disk of the nebula. The nebula contracted, but the tendency to conserve angular momentum caused the gas to spin faster and to flatten, forming – in the case of the Solar System – an accretion disk. In the late stages of accretion, material in the disk condensed into dust. The dust agglomerated into many small planetesimals, most of which grew by collisions into planets. According to this theory, the planets should have, on average, coplanar and nearly circular orbits. Even if one them, Pluto's orbit is more eccentric and more tippy.

16.4 Large Sample and Nonparametric Bootstrap Distributions for the Mean Direction

To each object orbiting around the Sun, we associate a unit vector, that is perpendicular on the orbital plane. In this manner, given the masses gravitating in the Solar System, we obtain a distribution on the round sphere S^2, regarded as a surface in $\mathbb{R}^3$. In general, as a submanifold of $\mathbb{R}^{m+1}$, the sphere S^m can be regarded as a set of zeros of the function $F_1((x^1,\ldots,x^{m+1})) = (x^1)^2 + \cdots + (x^{m+1})^2 - 1$. As shown in Section 5.3, a non-parametric bootstrap method was developed based on large sample theory for extrinsic means of probability distributions on submanifolds $\mathcal{M}$ of $\mathbb{R}^N$ given by implicit equations $F_1(x) = \cdots = F_c(x) = 0$, where $F_1,\ldots,F_c$ are functionally independent and c is the codimension of $\mathcal{M}$ in $\mathbb{R}^N$. It was shown that if H is the projection on the affine subspace $\mu_E(Q) + T_{\mu_E}\mathcal{M}$, the asymptotic distribution of $H(\overline{X})$ given in Theorem 5.3.1 is given by

$$n^{\frac{1}{2}}(H(\overline{X}) - P_{\mathcal{M}}(\mu)) \to_d \mathcal{N}_{N-c}(0, G\Sigma G^t), \tag{16.1}$$

where G is the matrix with entries G_i^j given by (5.17), $P_{\mathcal{M}}$ is the projection on $\mathcal{M}$, μ is the mean vector of the nonfocal distribution Q in the ambient numerical space $\mathbb{R}^N$ and Σ is its covariance matrix.

Note that, in the low sample size case, rather than using the asymptotics in (16.1), Corollary 5.3.1 shows that one may construct an asymptotic $(1-\alpha)$-confidence region for $\mu_E(Q) = P_{\mathcal{M}}\mu$ using the bootstrapped statistic $n^{\frac{1}{2}}(H(\overline{X}^*) - \overline{H}(X))$, where $\overline{H}(X)$ is the projection of $\overline{X}$ on the affine subspace $\overline{X}_E + T_{\overline{X}_E}\mathcal{M}$, and $\overline{X}^*$ is the mean of a random sample with repetition of size n from the empirical $\hat{Q}_n$ considered as a probability measure on $\mathbb{R}^N$.

We now apply the results above to spheres. Let $\mathcal{M} = S^m$, be the unit sphere centered at the origin of $\mathbb{R}^{m+1}$. The probability measure Q is non-focal on S^m, of mean

$$\mu_Q = \int_{\mathbf{R}^{m+1}} xQ(dx) \neq 0. \tag{16.2}$$

If we use the terminology of directional statistics (see Mardia and Jupp (2000) [230]), in this case, the extrinsic mean is the *population mean direction*

$$\mu_E = \frac{1}{\|\mu\|}\mu, \tag{16.3}$$

where μ_E is given by (16.2). Assume $X_1,\dots,X_n$ are independent identically distributed directions (objects on S^m). From (16.3), the extrinsic sample mean direction is

$$\bar{X}_E = \frac{1}{\|\bar{X}\|}\bar{X}, \tag{16.4}$$

and with this preparation it is elementary to show that

$$H(\overline{X}) = \overline{X} - \mu_E - \{(\overline{X} - \mu_E)\cdot\mu_E\}\mu_E. \tag{16.5}$$

Hence we obtain the following result, including a correction of a similar result in Patrangenaru and Mardia (2004) [273]:

THEOREM 16.4.1. *Given i.i.d.r. directions $X_1,\dots,X_n$ from a probability distribution Q on S^m, the quantity $\sqrt{n}(H(\overline{X}) - \mu_E)$ converges in distribution to n m-dimensional multivariate normal distribution supported by the tangent space $T_{\mu_E}S^m$, identified with the linear subspace $\{v \in \mathbb{R}^{m+1} : v^T\mu_E = 0\}$. As a measure on $\mathbb{R}^{m+1}$ this distribution has mean 0_{m+1} and covariance matrix*

$$\Gamma_E = G\Sigma G^T, \tag{16.6}$$

where G is associated with the linear transformation T_G given by

$$T_G(u) = Gu = u - \mu_E^T u\mu_E. \tag{16.7}$$

Here $\Sigma = Cov(X_1)$, with X_1 regarded as a random vector in $\mathbb{R}^{m+1}$.

An asymptotic $(1-\alpha)100\%$-confidence region for μ_E may be constructed using the estimate $\hat{\Sigma}_n$ of Σ and obtained by replacing μ_E by the extrinsic sample mean direction $\bar{X}_E$ in (16.4), thus studentizing $\sqrt{n}(H(\overline{X}) - \mu_E)$ in Theorem 16.4.1. Moreover, if the sample size n is small, one may approximate this asymptotic distribution using the nonparametric bootstrap approximation (see Hall (1997) [137]), which can be formulated as follows.

COROLLARY 16.4.1. *Consider a random sample with repetition $X_1^*, \ldots, X_n^*$ from the random unit normals $X_1, \ldots, X_n$, and let us project the mean $\bar{X}^*$ of $X_1^*, \ldots, X_n^*$ onto the tangent space to the sphere at $\bar{X}_E$. If the number of such resamples is large (thousands), then the distribution of projections of the points $\bar{X}^*$ on this tangent plane, is approximately the same as the distribution of $H(\bar{X})$ in $T_{\mu_E}S^m$. The probability error is of order $O_P(n^{-\frac{1}{2}})$.*

16.5 Implementation of the Nonparametric Approach

If the probability measure of the distribution of the unit normals to the particles' orbits in the solar nebula is Q and

$$\mu_Q = \int_{\mathbb{S}^2} nQ(dn) \tag{16.8}$$

is the mean of this distribution, seen as a distribution in $\mathbb{R}^3$, then we estimate that the mean unit normal is the $\mu = \frac{1}{\|\mu_Q\|}\mu_Q$ by considering the sample mean normal for the nine "planets". Note that if $\mu_E = (v^1 v^2 v^3)^T$, in our case, $m = 2$, the matrix G in (16.7) is given by

$$G = \begin{pmatrix} 1-(v^1)^2 & -v^1v^2 & -v^1v^3 \\ v^1v^2 & 1-(v^2)^2 & -v^2v^3 \\ -v^1v^3 & -v^1v^3 & 1-(v^3)^2 \end{pmatrix}$$

We calculate the directional sample mean normals for the sample of unit normals. Next, we calculate the sample average and the projections of the unit normals to orbital planes on the tangent space to the sample mean. These are displayed in Figure 16.1. We redo these computations for a reduced sample of unit normals by excluding Pluto from the sample. These are displayed in Figure 16.2.

We then randomly resample with replication 5000 times from these projections and then evaluated the bootstrap distribution of the projections of the resamples. We then determined the similar bootstrap distribution in the case of the reduced sample, for the data with Pluto excluded.

The results from Patrangenaru (1998) [266] are displayed in Figure 16.3. Note the elliptic appearance of the bootstrap distributions of means when Pluto is removed is in agreement with Corollary 5.3.1, unlike the situation when Pluto is in the sample, showing that Pluto is too strong of an "outlier" if included in the group of major planets of the solar system. As of 2015, Pluto is classified as the largest *dwarf planet* in the Solar system.

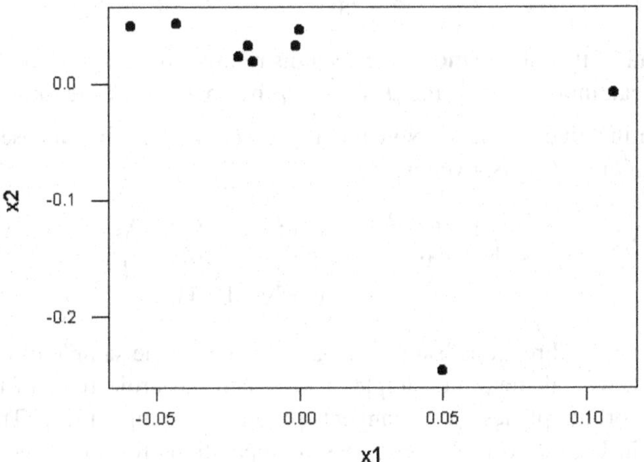

Figure 16.1: *Planet data with Pluto included in the sample.*

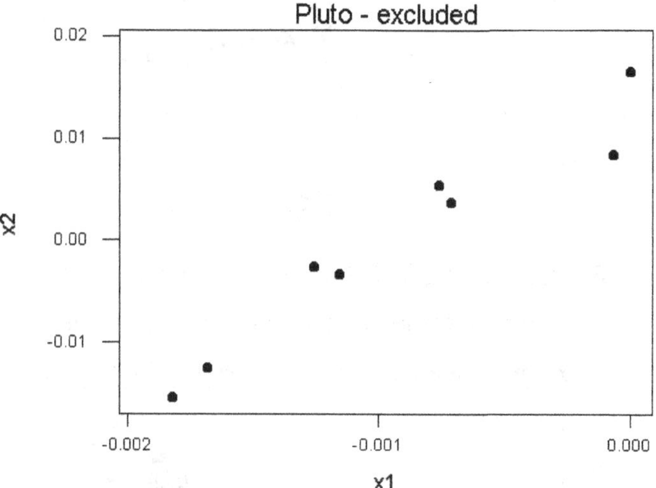

Figure 16.2: *Planet data without Pluto.*

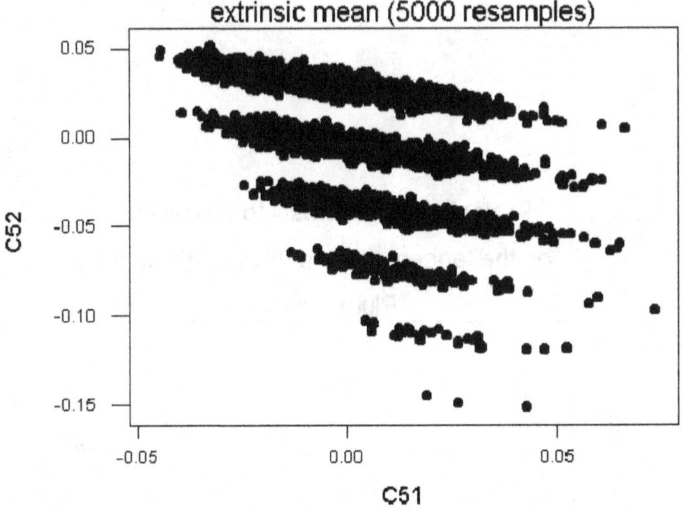

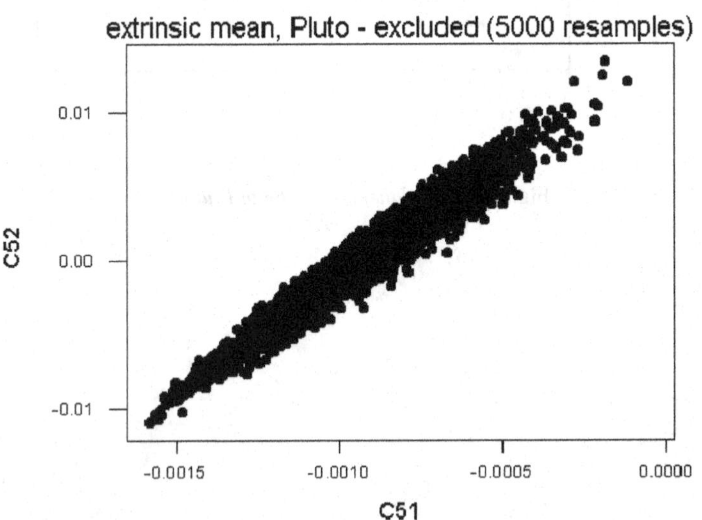

Figure 16.3: *Bootstrap distributions with Pluto in (top) and out (bottom).*

Chapter 17

Applications of Direct Similarity Shape Analysis in Medical Imaging

17.1 Introduction

As an application of the theoretical results in Chapter 13, here we give a large sample confidence region for the extrinsic mean shape of a configuration of eight landmarks obtained from X-rays of midfaces of clinically healthy eight year old children in terms of simultaneous confidence intervals for the affine coordinates of the extrinsic sample mean shape. We also illustrate a test for the difference between 3D mean shapes in a problem involving the detection of glaucomatous mean shape change.

17.2 University School X-ray Data Analysis

As an application of the methodology from Chapter 13, we give a nonpivotal bootstrap confidence region for the mean shape of a group of eight landmarks on the skulls of eight year-old, North American children taken from a growth study of normal children (see Tables 1.4, 1.5, 1.6, and 1.7). The sample used is from the University School data (Bookstein (1991) [51], pp. 400-405). Recall that the data set represents coordinates of anatomical landmarks whose names and positions on the skull are given in Bookstein, op. cit. The corresponding scatterplot is displayed in Figure 17.1 The presentation of raw data is similar to other known shape data displays (see Dryden and Mardia (1998) [91] p.46). The shape of the 8 landmarks on the upper mid face is valued in a planar shape space $\mathbb{C}P^6$, which has real dimension equal to 12. A spherical representation of a shape, in this case, consists of seven marked points. We display such a

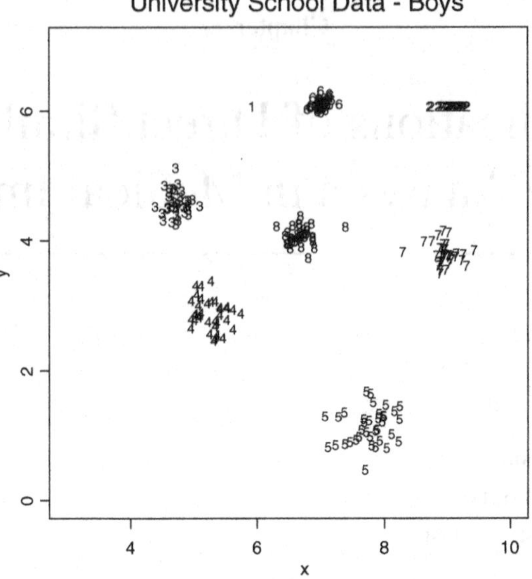

Figure 17.1: *Scatterplot for University School data.*

representation of this data in Figure 17.2. A representative for the extrinsic sample mean (in spherical representation) is

$$\begin{aligned}&(-0.67151+0.66823i, 0.76939+1.05712i, -1.03159-0.15998i,\\&-0.57776-0.87257i, 0.77871-1.36178i, -0.17489+0.82106i,\\&1.00000+0.00000i)\end{aligned}$$

The nonpivotal boostrap distribution was derived using a simple program in S-Plus4.5 that was run for 500 resamples. A spherical representation of the bootstrap distribution of the extrinsic sample means is displayed in Figure 17.3. Here, we added a representative for the last landmark (which is the opposite of the sum of the other landmarks since observations are centered at 0).

Note that the bootstrap distribution of the extrinsic sample mean is very concentrated at each landmark location. This is in agreement with the theory, which predicts, in this case, a spread of about 6 times smaller than the spread of the population. It is also an indication of the usefulness of the spherical coordinates. We determined a confidence region for the extrinsic mean using the six 95% simultaneous bootstrap complex intervals for the complex affine coordinates, and found the following complex intervals (see also Bhattacharya

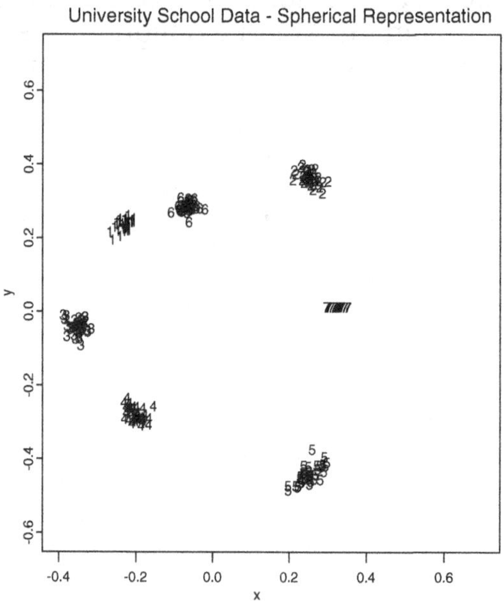

Figure 17.2: *Spherical representation of University School data.*

and Patrangenaru (2005) [43]):

$$
\begin{aligned}
w_1 &: (-0.677268+0.666060i, -0.671425+0.672409i)\\
w_2 &: (0.767249+1.051660i, 0.775592+1.058960i)\\
w_3 &: (-1.036100-0.161467i, -1.029420-0.154403i)\\
w_4 &: (-0.578941-0.875168i, -0.574923-0.871553i)\\
w_5 &: (0.777688-1.366880i, 0.782354-1.358390i)\\
w_6 &: (-0.177261+0.820107i, -0.173465+0.824027i).
\end{aligned}
$$

17.3 LEGS Data Analysis

The data set for this second application consists of a library of Scanning Confocal Laser Tomography (SCLT) images of the complicated ONH topography (see Burgoyne et al. (2000) [56]). Those images are also called *range images*. A range image is, loosely speaking, like a digital camera image except that

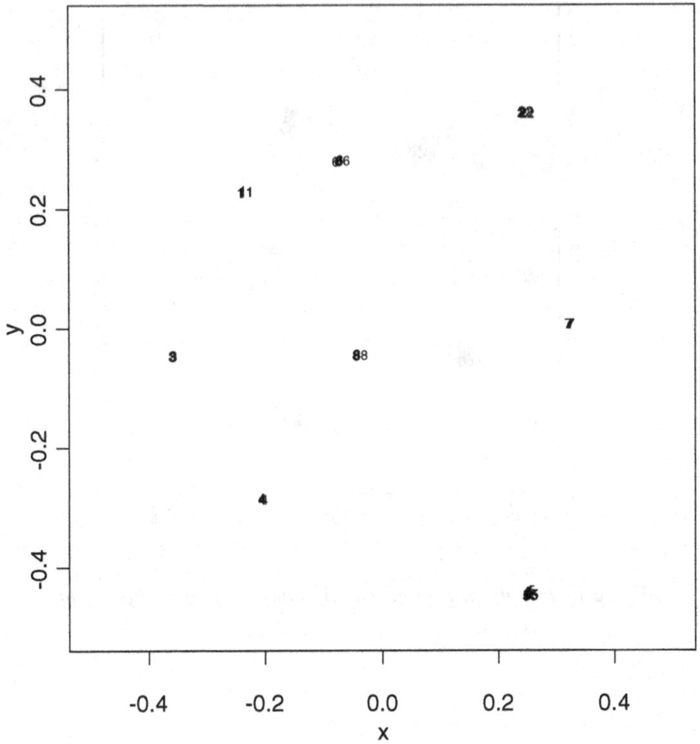

Figure 17.3 *Bootstrap distribution - extrinsic sample mean shape for University School data.*

each pixel stores a depth rather than a color level (see Figure 1.6 in Chapter 1). It can also be seen as a set of points in $3D$. The range data acquired by $3D$ digitizers, such as optical scanners, commonly consist of depths sampled on a regular grid. In the mathematical sense, a range image is a $2D$ array of real numbers which represent those depths. All of the files (observations) are produced by a combination of modules in C++ and SAS that take the raw image output and process it. The 256×256 arrays of height values are the products of this software. Another byproduct is a file which we will refer to as the "abxy" file. This file contains the following information: subjects' names (denoted by:

1c, 1d, 1e, 1f, 1g, 1i, 1j, 1k, 1l, 1n, 1o, 1p), observation points that distinguish the normal and treated eyes, and the 10 or 15 degree fields of view for the imaging. The observation point "03" denotes a 10 degree view of the experimental glaucoma eye, "04" denotes a 15 degree view of the experimental glaucoma eye, "11" and "12" denote respectively the 10 degree and the 15 degree view of the normal eye.

The two-dimensional coordinates of the center (a,b) of the ellipses that bound the ONH region, as well as the sizes of the small and the large axes of the ellipses (x,y), are stored in the so called "abxy"file. To find out more about the LSU study and the image acquisition, see Burgoyne et al. (2000) [56].

The program (created in C++ by Gordana Derado) determines the three-dimensional coordinates of the landmarks for each observation considered in our analysis. It also determines the fifth Bookstein coordinate for each observation. Each image consists of a 256×256 array of elevation values which represent the "depth" of the ONH. By "depth," we mean the distance from an imaginary plane, located approximately at the base of the ONH cup, to the "back of the ONH cup".

To reduce the dimensionality of the shape space to 5, out of five landmarks T, S, N, I, V recorded, only four landmarks $(X_1 = T, X_2 = S, X_3 = N, X_4 = V)$ were considered.

The original data were collected in experimental observations on Rhesus monkeys, and after treatment a healthy eye slowly returns to its original shape. For the purpose of increased intraocular pressure (IOP) increment detection, in this section only the first set of after treatment observations of the treated eye are considered.

This example is relevant in glaucoma detection. Although it is known that IOP may cause a shape change in the eye cup, which is identified with glaucoma, it does not always lead to this shape change. The data analysis presented shows that the device used for measuring the topography of the back of the eye, as reported in Burgoyne et al. (2000) [56], is effective in detecting shape change.

We give a nonpivotal bootstrap confidence region for the mean shape change of the eye cup due to IOP. Glaucoma is an eye disorder caused by IOP that is very high. Due to the increased IOP, as the soft spot where the optic nerve enters the eye is pushed backwards, eventually the optic nerve fibers that spread out over the retina to connect to photoreceptors and other retinal neurons can be compressed and damaged. An important diagnostic tool is the ability to detect increased depth (cupping) of the ONH structures. Depth data is tabulated in tables 1.24, 1.25, and 1.26 in Chapter 1. The processed images of the ONH cup surface from before and after the IOP was increased are shown in Figure 17.4.

The laser image files are, however, huge dimensional vectors, and their sizes

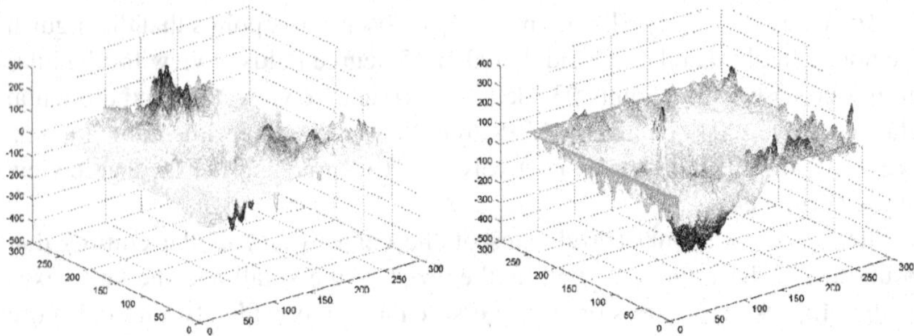

Figure 17.4 *Change in the ONH topography from normal (left) to glaucomatous (right).*

usually differ. Even if we were to restrict the study to a fixed size, there is no direct relationship between the eye cup pictured and the coordinates at a given pixel. A useful data reduction process consists of registering a number of anatomical landmarks that were identified in each of these images. Assume the position vectors of these landmarks are $X_1,\dots,X_k, k \geq 4$. Two configurations of landmarks have the same shape if they can be superimposed after a translation, a rotation, and a scaling. With the notation in Subsection 3.5.3, the shape of the configuration $x = (x_1,\dots,x_k)$ is labeled $[\mathbf{x}]$, a point on the Kendall space Σ_m^k, introduced in that subsection of Chapter 3.

We come back to the shape of an ONH configuration in animal models, described in Section 1.2. This ONH region resembles a "cup" of an ellipsoid and its border has a shape of an ellipse. The first three landmarks, S (superior), T (temporal), and N (nasal), are chosen on this ellipse when referring to the left eye. The last landmark V (vertex) is the point with the largest "depth" inside the ellipse area that determines the border of the ONH. Therefore, in this example, the data analysis is on the Kendall shape space of tetrads Σ_3^4, which is topologically a 5 dimensional sphere (see Kendall et al. (1999) [179], p.38).

On the other hand, it is known that if a probability distribution on Σ_m^k has small support outside a set of singular points, the use of any distance that is compatible with the orbifold topology considered is appropriate in data analysis (Dryden and Mardia (1998) [91], p.65). Our choice of the Riemannian metric chosen in equation (17.4) is motivated by considerations of applicability of results in Chapter 5 and computational feasibility. Dryden and Mardia (1998) [91], pp.78-80) have introduced the following five coordinates defined on the generic subset of Σ_3^4 of shapes of a nondegenerate tetrads that they called

Bookstein coordinates:

$$
\begin{aligned}
v^1 &= (w_{12}w_{13} + w_{22}w_{23} + w_{32}w_{33})/a \\
v^2 &= \left((w_{12}w_{23} - w_{22}w_{13})^2 + (w_{12}w_{33} - w_{32}w_{13})^2 + (w_{22}w_{33} - w_{23}w_{32})^2\right)^{1/2}/a \\
v^3 &= (w_{12}w_{14} + w_{22}w_{24} + w_{32}w_{34})/a \\
v^4 &= (ab^{1/2})^{-1}\left(w_{12}^2(w_{23}w_{24} + w_{33}w_{34}) + w_{22}^2(w_{13}w_{14} + w_{33}w_{34}) + \right. \\
&\quad + w_{32}^2(w_{13}w_{14} + w_{23}w_{24}) - w_{12}w_{13}(w_{22}w_{24} + w_{32}w_{34}) + \\
&\quad \left. - w_{22}w_{32}(w_{23}w_{34} + w_{33}w_{24}) - w_{12}w_{14}(w_{22}w_{23} + w_{32}w_{33})\right) \\
v^5 &= (w_{12}w_{23}w_{34} - w_{12}w_{33}w_{24} - w_{13}w_{22}w_{34} + \\
&\quad + w_{13}w_{32}w_{24} + w_{22}w_{33}w_{14} - w_{32}w_{23}w_{14})/(2ab)^{1/2}
\end{aligned} \tag{17.1}
$$

where

$$
\begin{aligned}
a &= 2(w_{12}^2 + w_{22}^2 + w_{32}^2) \\
b &= w_{12}^2w_{23}^2 + w_{12}^2w_{33}^2 - 2w_{12}w_{13}w_{22}w_{23} + w_{13}^2w_{22}^2 + w_{13}^2w_{32}^2 - \\
&\quad - 2w_{12}w_{13}w_{32}w_{33} + w_{33}^2w_{22}^2 + w_{23}^2w_{32}^2 - 2w_{22}w_{32}w_{23}w_{33},
\end{aligned} \tag{17.2}
$$

where w_{ri} are given by

$$
w_{ri} = x_i^r - (x_1^r + x_2^r)/2 \, \forall i = 2,3,4, \forall r = 1,2,3. \tag{17.3}
$$

These coordinates carry useful geometric information on the shape of the 4-ad; v^1 and v^3 give us information of the appearance with respect to the bisector plane of $[X_1X_2], v^2$ and v^4 give some information about the "flatness" of this 4-ad, and v^5 measures the height of the 4-ad (X_1, X_2, X_3, X_4) relative to the distance $||X_1 - X_2||$. Assume U is the set of Kendall shapes $[\mathbf{X}]$ such that (X_1, X_2, X_3, X_4) is an affine frame in $\mathbb{R}^3$ and $\phi : U \to \mathbb{R}^{3k-7}$ is the map that associates to $[\mathbf{X}]$ its Bookstein coordinates. U is an open dense set in Σ_3^k with the induced topology. In the particular case $k = 4, \Sigma_3^4$ is topologically a 5 dimensional sphere and, by from a classical result of Smale (1961) [309], Σ_3^4 has a differentiable structure diffeomorphic with the sphere S^5. Moreover, if L is a compact subset of U, there is a finite open covering $U_1 = U, \ldots, U_t$ of Σ_3^4 and a partition of unity $\varphi_1, \ldots, \varphi_t$, such that $\varphi_1([\mathbf{X}]) = 1 \forall [\mathbf{X}] \in L$.

We will use the following Riemannian metric on Σ_3^4 : let $(y_1, \ldots, y_5)$ be the Bookstein coordinates of a shape in U_1 and let $g_1 = dy_1^2 + \cdots + dy_5^2$ be a flat Riemannian metric on U_1. For each $j = 2, \ldots, t$, we consider any fixed Riemannian metric g_j on U_j. Let g be the Riemannian metric given by

$$
g = \sum_{j=1}^{t} \varphi_j g_j. \tag{17.4}
$$

Table 17.1: *Abbreviated filenames and w_{ij} -coordinates.*

Filename	w_{12}	w_{13}	w_{14}	w_{22}	w_{23}	w_{24}	w_{32}	w_{33}	w_{34}
1alpn	-610	1830	610	800	800	2400	-69	-124	136
1alp0	-610	1830	610	800	800	2400	-19.5	-267	52
1alcn	-520	1560	520	640	640	1920	-78	-217	108.5
1alc0	-520	1560	520	640	640	1920	-69	-162	150
1aldn	-590	1770	590	900	900	2700	-8	-147	54
1ald0	-600	1800	600	850	850	2550	-24	-100	171
1alen	-650	1950	650	830	830	2490	-15	-138	238
1ale0	-660	1980	660	810	810	2430	-86	-21	241
1alfn	-690	2070	690	880	880	2640	-31	4.5	94
1alf0	-690	2070	690	880	880	2640	-23	-119	108
1algn	-620	1860	620	750	750	2250	-20	-153	199
1alg0	-620	1860	620	750	750	2250	11	-48	166
1alin	-650	1950	650	920	920	2760	-155	-131	174
1ali0	-650	1950	650	880	880	2640	-91	-48	224
1aljn	-580	1740	580	810	810	2430	-187	-175	50.5
1alj0	-600	1800	600	800	800	2400	-200	-90	-38
1alkn	-630	1890	630	810	810	2430	-135	-91	146
1alk0	-630	1890	630	810	810	2430	-33	-77	208
1alln	-570	1710	570	790	790	2370	-84	-174	246
1all0	-570	1710	570	790	790	2370	-122.5	-161	253
1alnn	-660	1980	660	840	840	2520	-120.5	-93	148
1aln0	-660	1980	660	840	840	2520	-78	-115.5	158.5
1alon	-600	1800	600	740	740	2220	-37	-160	129
1alo0	-600	1800	600	740	740	2220	5	-260	95.5

The space (Σ_3^4, ρ_g) is complete and flat in a neighborhood of L. In this example, the two distributions of shapes of tetrads before and after the increase in IOP are close. Hence, L, which contains supports of both distributions, consists of shapes of nondegenerate tetrads only.

Computations for the glaucoma data in Table 1.17 yield the Bookstein coordinates given in the Table 17.1.

The p-value of the test for equality of the intrinsic means was found to be 0.058 based on the bootstrap distribution of the chi-square-like statistic discussed in Remark 5.4.6. The number of bootstrap resamples for this study was 3,000. The chi-square-like density histogram is displayed in Figure 17.5. A matrix plot for the components of the nonpivotal bootstrap distribution of the sample mean differences γ_n^* in Remark 5.4.6 is displayed in Figure 17.6.

The nonpivotal bootstrap 95% confidence intervals for the mean differences

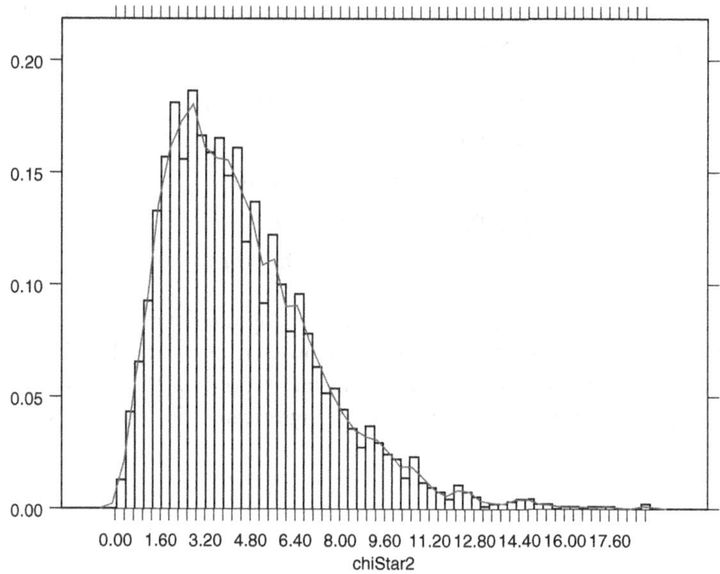

Figure 17.5 *χ^2 like bootstrap distribution for equality of intrinsic mean shapes from glaucoma data.*

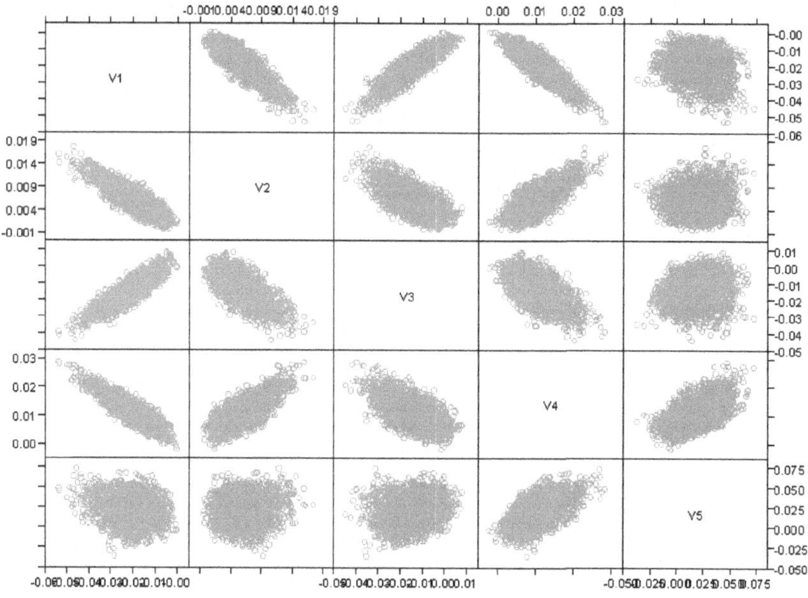

Figure 17.6 *Glaucoma data, matrix plot for the bootstrap mean differences associated with Bookstein coordinates due to increased IOP.*

$\gamma_j, j = 1, \cdots, 5$ components of γ in Remark 5.4.6 associated with the Bookstein coordinates $v_j, j = 1, \cdots, 5$ are:

$$\gamma_1 : (-0.0377073, -0.0058545)$$
$$\gamma_2 : (0.0014153, 0.0119214)$$
$$\gamma_3 : (-0.0303489, 0.0004710)$$
$$\gamma_4 : (0.0031686, 0.0205206)$$
$$\gamma_5 : (-0.0101761, 0.0496181).$$

Note that the individual tests for difference are significant at the 5% level for the first, second and fourth coordinates. However, using the Bonferroni inequality, combining tests for five different shape coordinates each at 5% level leads to a much higher estimated level of significance for the overall shape change.

Chapter 18

Direct Similarity Shape Analysis of Planar Contours

In this chapter, we will discuss problems associated with a nonparametric analysis of direct similarity shapes of planar contours. In addition to the methodology for performing inference, which is a particular case of the material covered in Chapter 11, this chapter will describe some of the computational challenges associated with the study of such data. After detailing a method for addressing these challenges, we will conclude by applying these methods to a problem involving the corpus callosum data presented in Chapter 1. The primary resources for this chapter are Ellingson (2011) [101], Ellingson et al. (2013) [106], Ellingson et al. (2013a) [104], and Qiu et al. (2014) [285].

18.1 Introduction

The study of shape analysis of shapes lying in the plane has progressed considerably differently than functional data analysis. In the latter case, researchers began by studying dense functional data and have since adapted methodology for sparse functional data. On the other hand, planar shape analysis in its modern form originated with D. G. Kendall (1984) [177], which described shapes of finite-dimensional planar configurations and the space of such shapes. Starting midway through the 1990s, many researchers have become interested in studying shapes of planar curves as infinite-dimensional objects. Due to the nature of these configurations, a statistical analysis of such data must draw on methodologies developed for both shape analysis and functional data analysis.

While Kendall (1984) [177] showed that the space Σ_2^k of direct similarity shapes of nontrivial planar configurations of k points is a complex projective space $\mathbb{C}P^{k-2}$, definitions of location and variability parameters for probability distributions were not considered until years later. Kent (1992) [181] defined the full Procrustes estimate of a mean shape and Ziezold (1994) [346] introduced a nonparametric definition of a mean shape based upon the notion of Fréchet population means (see Fréchet (1948) [121] and Ziezold (1977) [345]). This approach was subsequently followed by Ziezold (1998) [347], Le and Kume (2000) [212], Kume and Le (2000 [206], 2003 [207]), Le (2001) [211], Bhattacharya and Patrangnearu (2003) [42], and Huckemann and Ziezold (2006) [166].

While a majority of these methods has defined means in terms of Riemannian distances, as initially suggested by Patrangenaru (1998) [266], a Veronese-Whitney (VW) extrinsic mean similarity shape was also introduced by Patrangenaru (op.cit.) in terms of the VW embedding of $\mathbb{C}P^{k-2}$ into the space $S(k-1,\mathbb{C})$ of selfadjoint $(k-1)\times(k-1)$ matrices. The asymptotic distribution of this extrinsic sample mean shape and the resulting bootstrap distribution are given in Bhattacharya and Patrangenaru (2005) [11], Bandulasiri et al. (2009) [10] and Amaral et al. (2010) [1]. These results are discussed in Chapters 10 and 13 with some applications shown in Chapter 17.

The trend toward considering shapes of infinite-dimensional planar configurations began with Grenander (1993) [132], which was motivated by the pioneering work of Zahn and Roskies (1972) and other applications in object recognition from digital images. A manifold model for direct similarity shapes of planar closed curves, first suggested by Azencott (1994) [7], was then pursued in Azencott et al. (1996) [8] and detailed by Younes (1998 [342], 1999 [343]). This idea further gained ground at the turn of the millennium, with more researchers studying shapes of planar closed curves (for example, see Sebastian et al. (2003) [305]).

Continuing this trend, Klassen et al. (2004) [194], Michor and Mumford (2004) [240], and Younes et al. (2008) [344] followed the methods of Small

(1996) [311] and Kendall by defining a Riemannian structure on a shape manifold. Klassen et al. (op. cit) and the related papers Mio and Srivastava (2004) [245], Mio et al. (2007) [246], Srivastava et al. (2005) [317], Mio et al. (2005) [247], Kaziska and Srivastava (2007) [176], Joshi it et al. (2007) [171], and Kurtek et al. (2012) [209] compute an intrinsic sample mean similarity shape of a closed curve. In many of these cases, the researchers define shape modulo reparameterizations in addition to the traditional similarity transformations, using a Riemannian metric of their preference.

However, these above approaches are focused on mathematical and computational aspects of shape analysis. While a number of them include descriptive statistics, such as sample means, they do not address fundamental definitions of populations of shapes, population means, and population covariance operators. As such, the idea of statistical inference is absent from these papers.

Ellingson (2011) [101] described methodology for a nonparametric statistical analysis of shapes of planar contours and considerations for approximating the contours for computational purposes. Ellingson et al. (2013) [104] refined these ideas and extended the methodology for general Hilbert manifolds, as discussed in Chapter 11. This chapter presents methodology and results that originate from these two sources and is organized as follows.

Section 18.2 presents the definition of the direct similarity shape of planar contours and their space. Following this, Section 18.3 defines the extrinsic mean shape and its sample analogue and presents examples from the Kimia contour database. Section 18.4 describes the asymptotic distribution of the extrinsic sample mean shape and its derivation. Sections 18.5 and 18.6, apply the neighborhood hypothesis test, detailed in Chapter 11, to this problem. Section 18.7 presents confidence regions for the extrinsic mean shape using nonparametric bootstrap with a non-pivotal statistic. Section 18.8 then addresses computational considerations necessary for working with digital imaging data in practice, including the approximation of the contours and the correspondence problem. Lastly, Section 18.9 applies these methodologies to the corpus callosum problem introduced in Chapter 1.

18.2 Similarity Shape Space of Planar Contours

Features extracted from digital images are represented by planar subsets of unlabeled points. If these subsets are uncountable, the labels can be assigned in infinitely many ways. Here, though, we will consider only contours, which are unlabeled boundaries of 2D topological disks in the plane. To keep the data analysis stable, and to assign a *unique* labeling, we make the *generic* assumption that there is a unique point p_0 on such a contour at the maximum distance to its center of mass so that the label of any other point p on the contour is the counterclockwise travel time at a constant speed from p_0 to p. As such, the total time needed to travel from p_0 to itself around the contour once is the length of the contour.

Therefore, we consider direct similarity shapes of nontrivial contours in the plane as described here. A contour $\tilde{\gamma}$ is then regarded as the range of a piecewise differentiable function, which is parameterized by arclength, i.e. $\gamma: [0,L] \to \mathbb{C}$, such that $\gamma(0) = \gamma(L)$, and is one-to-one on $[0,L)$. Recall that the length of a piecewise differentiable curve $\gamma: [a,b] \to \mathbb{R}^2$ is defined as follows:

$$l(\tilde{\gamma}) = \int_a^b \|\frac{d\gamma}{dt}(t)\| dt. \tag{18.1}$$

Its center of mass (the mean of a uniform distribution on $\tilde{\gamma}$) is given by

$$z_{\tilde{\gamma}} = \frac{1}{L}\int_{\gamma} z ds. \tag{18.2}$$

The contour $\tilde{\gamma}$ is said to be regular if γ is a simple closed curve and there is a unique point $z_0 = argmax_{z \in \tilde{\gamma}} \|z - z_{\tilde{\gamma}}\|$.

A direct similarity is a complex polynomial function in one variable of degree one. Two contours $\tilde{\gamma}_1$ and $\tilde{\gamma}_2$ *have the same direct similarity shape* if there is a direct similarity $S : \mathbb{C} \to \mathbb{C}$ such that $S(\tilde{\gamma}_1) = \tilde{\gamma}_2$. The centered contour $\tilde{\gamma}_0 = \tilde{\gamma} - z_{\tilde{\gamma}} = \{z - z_{\tilde{\gamma}}, z \in \tilde{\gamma}\}$ has the same direct similarity shape as $\tilde{\gamma}$.

DEFINITION 18.2.1. *Two regular contours $\tilde{\gamma}_1, \tilde{\gamma}_2$ have the same similarity shape if $\tilde{\gamma}_{2,0} = \lambda \tilde{\gamma}_{1,0}$, where λ is a nonzero complex number.*

In order to construct the space of direct similarity shapes, we note the following.

REMARK 18.2.1. *A function $\gamma: S^1 \to \mathbb{C}$ is centered if $\int_{S^1} \gamma(z) ds = 0$. We consider regular contours since the complex vector space spanned by centered functions γ yielding regular contours $\tilde{\gamma}$ is a pre-Hilbert space. Henceforth, we will be working with the closure of this space. This Hilbert space* **H** *can and will be identified with the space of all measurable square integrable centered functions from S^1 to $\mathbb{C}$.*

Let Σ_2^{reg} be the set of all direct similarity shapes of regular contours, which is the same as the space of all shapes of regular contours centered at zero.

REMARK 18.2.2. *From Definition 18.2.1 and Remark 18.2.1, we associate a unique piecewise differentiable curve γ to a contour $\tilde{\gamma}$ by taking $\gamma(0) = z_0$, the point at the maximum distance to the center of C, and by parameterizing γ using arc length in the counter clockwise direction. Therefore Σ_2^{reg} is a dense and open subset of $P(\mathbf{H})$, the projective space corresponding to the Hilbert space* **H**. *Henceforth, to simplify the notation, we will omit the symbol ˜ in $\tilde{\gamma}$ and identify a regular contour with the associated closed curve, without confusion.*

18.3 The Extrinsic Mean Direct Similarity Shape

The space $P(\mathbf{H})$ is a Hilbert manifold. From Example 11.3.1, this space can be embedded in the Hilbert space $\mathcal{L}_{HS}$ via the Veronese-Whitney (VW) embedding j given by

$$j([\gamma]) = \frac{1}{\|\gamma\|^2}\gamma \otimes \gamma. \tag{18.3}$$

From Proposition 11.3.1, the VW extrinsic mean shape μ_E of a random object $X = [\Gamma]$ in $P(\mathbf{H})$ is $[e_1]$, where e_1 is the eigenvector corresponding to the largest eigenvalue of $\mu = E(\frac{1}{\|\Gamma\|^2}\Gamma \otimes \Gamma)$. From this, the VW extrinsic sample mean shape can be computed similarly.

PROPOSITION 18.3.1. *Given any VW-nonfocal probability measure Q on $P(\mathbf{H})$, then if $\gamma_1, \dots, \gamma_n$ is a random sample from Γ, then the VW sample mean $\hat{\mu}_{E,n}$ is the projective point of the eigenvector corresponding to the largest eigenvalue of $\frac{1}{n}\sum_{i=1}^{n} \frac{1}{\|\gamma_i\|^2}\gamma_i \otimes \gamma_i$.*

To illustrate the behavior of the VW extrinsic sample mean shape, we will now present some examples using selected planar contours from Ben Kimia's contour database. A few samples of contours and their extrinsic sample mean shapes are given below.

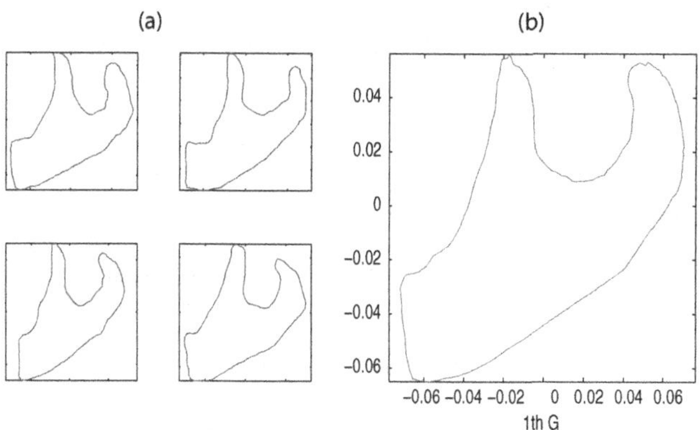

Figure 18.1: *(a) Sample of 4 contours of 't' gestures and (b) the extrinsic mean shape.*

Our computations display many of the characteristics of the VW extrinsic mean shape. Similarly to the standard arithmetic mean, we see that the extrinsic mean provides a summary of the shapes by reducing the variability. This is best shown with the contours of the dogs in Figure 18.2. For this data, there is a large amount of variability in the contours, especially in the legs and tail, resulting in the sample extrinsic mean shape capturing less detail in these regions. This result is also very noticeable with the worm fish (Figure 18.4), the red snapper (Figure 18.5), and the pears (Figure 18.6).

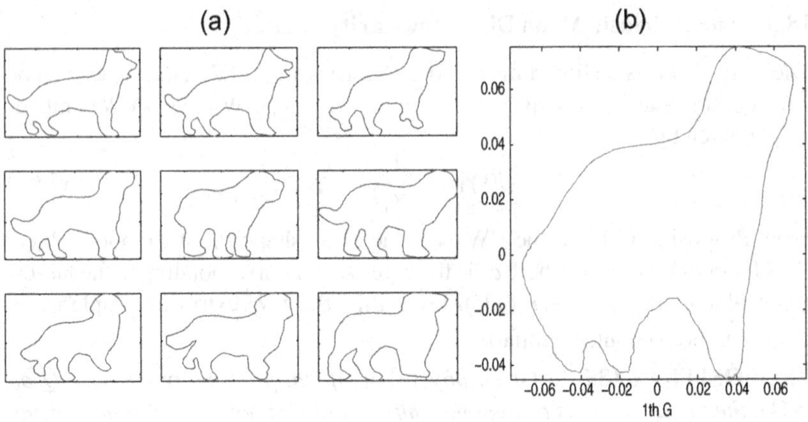

Figure 18.2: *(a) Sample of 9 contours of dogs and (b) the extrinsic mean shape.*

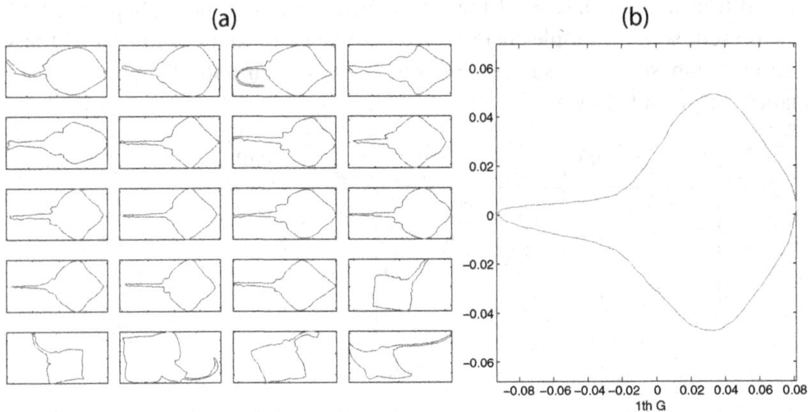

Figure 18.3: *(a) Sample of 20 contours of sting rays and (b) the extrinsic mean shape.*

Also of note is the VW extrinsic mean shape (Figure 18.3 (a)) of the sample of contours of sting rays (Figure 18.3 (b)). In this case, 15 of the 20 sting rays have straight tails. The other 5 have tails that are curved to varying degrees and in different directions. Despite these differences in the tails of the observed contours, the extrinsic mean shape has a straight tail, but is wider in order to account for the curved tails.

It is apparent from these examples that we can describe each sample of planar contours very well by the VW mean shape. Moreover, the extrinsic mean shape preserves geometric characteristics of a random family of shapes.

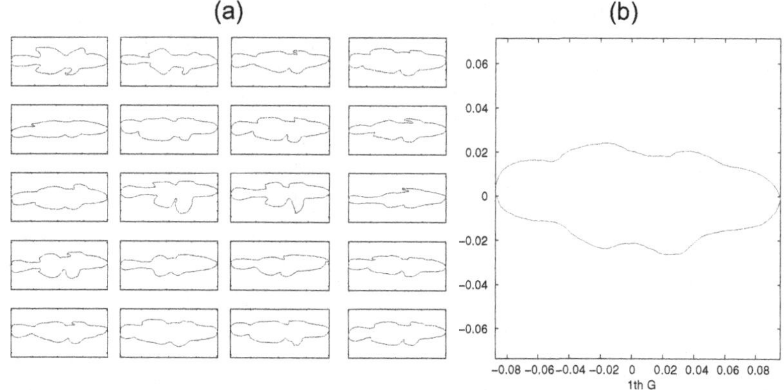

Figure 18.4: *(a) Sample of 20 contours of worm fish and (b) the extrinsic mean shape.*

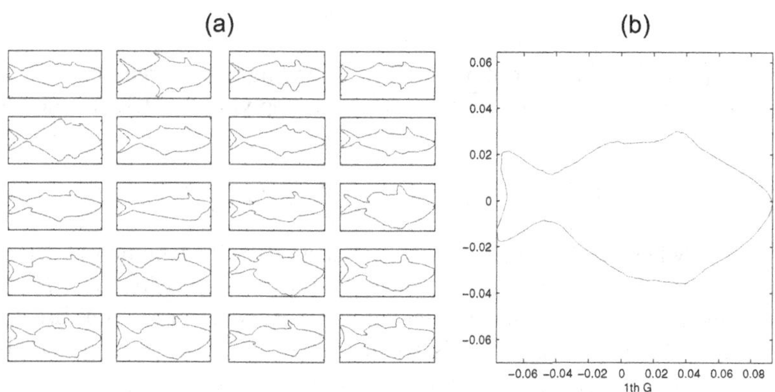

Figure 18.5 *(a) Sample of 20 contours of red snapper fish and (b) the extrinsic mean shape.*

18.4 The Asymptotic Distribution of the Extrinsic Sample Mean Direct Similarity Shape

We can now derive the asymptotic distribution of $\hat{\mu}_{E,n}$ based upon the general formulation specified in Proposition 5.5.1. The asymptotic distribution of $\overline{j(X)}_n$ is as follows:

$$\sqrt{n}(\overline{j(X)}_n - \mu) \to_d \mathcal{G} \text{ as } n \to \infty, \tag{18.4}$$

where $\mathcal{G}$ has a Gaussian distribution $N_{\mathcal{L}_{HS}}(0, \Sigma)$ on $\mathcal{L}_{HS}$ with zero mean and covariance operator Σ. From Proposition **??**, it follows that the projection $P_j : \mathcal{L}_{HS} \to j(P(\mathbf{H})) \subset \mathcal{L}_{HS}$ is given by

$$P_j(A) = \nu_A \otimes \nu_A, \tag{18.5}$$

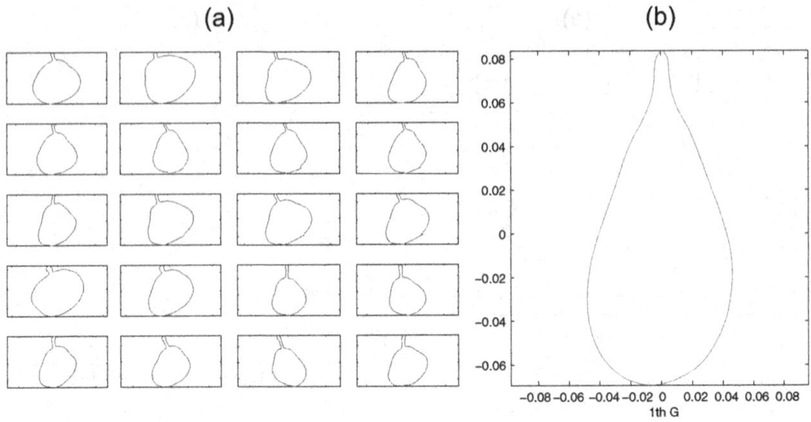

Figure 18.6: *(a) Sample of 20 contours of pears and (b) the extrinsic mean shape.*

where v_A is the eigenvector of norm 1 corresponding to the largest eigenvalue of A, $P_j(\mu) = j(\mu_E)$, and $P_j(\overline{j(X)}_n) = j(\hat{\mu}_{E,n})$. Applying the delta method to (18.4) yields

$$\sqrt{n}(j(\hat{\mu}_{E,n}) - j(\mu_E)) \to_d N_{\mathcal{L}_{HS}}(0, \mathrm{d}_\mu P_j \Sigma (\mathrm{d}_\mu P_j)^T), \tag{18.6}$$

as $n \to \infty$, where $\mathrm{d}_\mu P_j$ denotes the differential, as given in Definition 11.2.1, of the projection P_j, evaluated at μ. It remains to find the expression for $\mathrm{d}_\mu P_j$. To determine the formula for the differential, we must consider the equivariance of the embedding J. Because of this, we may assume without loss of generality that $\mu = diag\{\delta_a^2\}_{a=1,2,3,...}$. As defined previously, the largest eigenvalue of μ is a simple root of the characteristic polynomial with e_1 as the corresponding complex eigenvector of norm 1, where $\mu_E = [e_1]$.

An orthobasis for $T_{[e_1]}P(\mathbf{H})$ is formed by e_a, ie_a, for $a = 2,3,\ldots$, where e_a is the eigenvector over $\mathbb{R}$ that corresponds to the a-th eigenvalue. For any γ which is orthogonal to e_1 w.r.t. the real scalar product, we define the path $\psi_\gamma(t) = [\cos(t)e_1 + \sin(t)\gamma]$. Then $T_{j([e_1])}j(P(\mathbf{H}))$ is generated by the vectors tangent to such paths at $t = 0$. Such vectors have the form $\gamma \otimes e_1 + e_1 \otimes \gamma$. In particular, since the eigenvectors of μ are orthogonal w.r.t. the complex scalar product, we may take $\gamma = e_a, a = 2,3,\ldots$, or $\gamma = ie_a, a = 2,3,\ldots$ to get an orthobasis for $T_{j([e_1])}j(P(\mathbf{H}))$. Normalizing these vectors to have unit lengths, we obtain the following orthonormal frame for $a = 2,3,\ldots$:

$$d_\mu j(e_a) = 2^{-1/2}(e_a \otimes e_1 + e_1 \otimes e_a), \tag{18.7}$$

$$d_\mu j(ie_a) = i2^{-1/2}(e_a \otimes e_1 + e_1 \otimes e_a), \tag{18.8}$$

As stated previously, since the map j is equivariant, we may assume that $\overline{j(X)}_n$ is a diagonal operator D, with the eigenvalues $\delta_1^2 > \delta_2^2 \geq \ldots$ In this case,

$$d_{\mu_E} j(e_a) = 2^{-1/2} E_a^1 = F_a^1, \tag{18.9}$$

$$d_{\mu_E} j(ie_a) = i2^{-1/2} E_a^1 = iF_a^1, \tag{18.10}$$

where E_a^b has all entries zero except those in the positions (a,b) and (b,a) that are all equal to 1. From these formulations and computations of the differential of P_j in the finite dimensional case in Bhattacharya and Patrangenaru (2005), it follows that $\mathrm{d}_D P_j(E_a^b) = 0$, for all values $a \leq b$, except for $a = 1 < b$. In this case,

$$\mathrm{d}_D P_j(F_1^b) = \frac{1}{\delta_1^2 - \delta_b^2} F_1^b, \mathrm{d}_D P_j(iF_1^b) = \frac{1}{\delta_1^2 - \delta_b^2} iF_1^b. \tag{18.11}$$

Equation (18.11) implies that the differential of the projection P_j at μ is the operator Q_1 given by

$$Q_1 = \sum_{k=2}^{\infty} \frac{1}{\delta_1^2 - \delta_k^2} E_k, \tag{18.12}$$

where $\delta_1^2, \delta_2^2, \ldots$ are the eigenvalues of $E(\frac{1}{\|\Gamma\|^2} \Gamma \otimes \Gamma)$ and $E_1, E_2, \ldots$ are the corresponding eigenprojections. Also, in this situation, $\mathcal{G}$ is a normally distributed random element in $\mathcal{L}_{HS}$. This results in the tangential component of the difference between the j - images of the VW sample mean and of the VW mean having an asymptotic normal distribution, albeit with a degenerate covariance operator. From these computations, the asymptotic distribution of this difference can be expressed more explicitly in the following manner.

$$\sqrt{n}(tan(j(\hat{\mu}_{E,n}) - j(\mu_E))) \xrightarrow{d} Q_1 \mathcal{G}, \tag{18.13}$$

where $tan(v)$ is the tangential component of $v \in j(P(\mathbf{H}))$ with respect to the basis $e_a(P_j(\mu)) \in T_{P_j(\mu)} j(P(\mathbf{H}))$, for $a = 2, 3, \ldots$ and is expressed as

$$tan(v) = (e_2(P_j(\mu))^T v, e_3(P_j(\mu))^T v, , \ldots)^T. \tag{18.14}$$

However, this result cannot be used directly because Q_1, which is calculated using the eigenvalues of $E(\frac{1}{\|\Gamma\|^2} \Gamma \otimes \Gamma)$, and μ_E are unknown. This problem is solved by estimating μ_E by $\hat{\mu}_{E,n}$ and Q_1 in the following manner.

$$\hat{Q}_1 = \sum_{k=2}^{\infty} \frac{1}{\hat{\delta}_1^2 - \hat{\delta}_k^2} \hat{E}_k, \tag{18.15}$$

where $\hat{\delta}_1, \hat{\delta}_2, \ldots$ are the eigenvalues of

$$\hat{\mu} = \frac{1}{n} \sum_{i=1}^{n} \frac{1}{\|\gamma_i\|^2} \gamma_i \otimes \gamma_i \tag{18.16}$$

and $\hat{E}_1, \hat{E}_2, \ldots$ are the corresponding eigenprojections. Using this estimation, the asymptotic distribution is as follows:

$$\sqrt{n}(\widehat{tan}(j(\hat{\mu}_{E,n}) - j(\mu_E))) \overset{d}{\approx} \hat{Q}_1 \mathcal{G}, \tag{18.17}$$

where "$\overset{d}{\approx}$" denotes approximate equality in distribution and $\widehat{tan}$ is the tangential component relative to the tangent space of $j(P(\mathbf{H}))$ at $j(\hat{\mu}_{E,n})$ as in (18.14), where μ is replaced with $\hat{\mu}$ in equation (18.16). Applying this result to (18.6), we arrive at the following result.

THEOREM 18.4.1. *If $\Gamma_1, \ldots, \Gamma_n$ are i.i.d.r.o.'s from a VW-nonfocal distribution Q on $P(\mathbf{H})$ with VW extrinsic sample mean $\hat{\mu}_{E,n}$, then*

$$\sqrt{n}(j(\hat{\mu}_{E,n}) - j(\mu_E)) \overset{d}{\approx} d_{\hat{\mu}_n} P_j \mathcal{G} \text{ as } n \to \infty, \tag{18.18}$$

where $\hat{\mu}_n = \overline{j(X)}_n$ is a consistent estimator of μ.

It must be noted that because of the infinite dimensionality of $\mathcal{G}$, in practice, a sample estimate for the covariance that is of full rank cannot be found. Because of this issue, this result cannot be properly studentized. Rather than using a regularizarion technique for the covariance that leads to complicated computations, we will drastically reduce the dimensionality via the use of the neighborhood hypothesis methodology presented in Chapter 11.

18.5 The One-Sample Neighborhood Hypothesis Test for Mean Shape

Suppose that $j : P(\mathbf{H}) \to \mathcal{L}_{HS}$ is the VW embedding in (11.3) and $\delta > 0$ is a given positive number. Using the notation in Chapter 12, we now can apply Theorem 11.4.1 to random shapes of regular contours. Assume $x_r = [\gamma_r], \|\gamma_r\| = 1, r = 1, \ldots, n$ is a random sample from a VW-nonfocal probability measure Q. Then equation (18.18) shows that, asymptotically, the tangential component of the VW-sample mean around the VW-population mean has a complex multivariate normal distribution. Note that such a distribution has a Hermitian covariance matrix (see Goodman, 1963 [128]). Therefore, in this setting, the extrinsic covariance operator and its sample counterpart are infinite-dimensional Hermitian matrices. In particular, if we extend the CLT for VW-extrinsic sample mean Kendall shapes in Bhattacharya and Patrangenaru (2005) to the infinite dimensional case, the j-extrinsic sample covariance operator $S_{E,n}$, when regarded as an infinite Hermitian complex matrix has the following entries

$$S_{E,n,ab} = n^{-1}(\hat{\delta}_1^2 - \hat{\delta}_a^2)^{-1}(\hat{\delta}_1^2 - \hat{\delta}_b^2)^{-1} \tag{18.19}$$

$$\sum_{r=1}^{n} < e_a, \gamma_r >< e_b, \gamma_r >^* | < e_1, \gamma_r > |^2, a, b = 2, 3, \ldots$$

with respect to the complex orthobasis $e_2, e_3, e_4, \ldots$ of unit eigenvectors in the tangent space $T_{\hat{\mu}_{E,n}} P(\mathbf{H})$. Recall that this orthobasis corresponds via the differential $d_{\hat{\mu}_{E,n}}$ with an orthobasis (over $\mathbb{C}$) in the tangent space $T_{j(\hat{\mu}_{E,n})} j(P(\mathbf{H}))$. Therefore, one can compute the components $\hat{v}^a$ of $\hat{v}$ from equation (11.18) with respect to $e_2, e_3, e_4, \ldots$, and derive for s_n^2 in (11.16) the following expression

$$s_n^2 = 4 \sum_{a,b=2}^{\infty} S_{E,n,ab} \hat{v}^a \overline{\hat{v}^b}, \tag{18.20}$$

where the $S_{E,n,ab}$ given in equation (18.19) are regarded as entries of a Hermitian matrix. The test statistic T_n in equation (11.15) is defined on an infinite dimensional Hilbert manifold.

18.6 Application of the One-Sample Test for Mean Shape

The most likely applications involve having a known extrinsic mean shape determined from historical data. In such cases, the hypothesis test can be used to determine whether there is a significant deviation from the historical mean shape. One application in agriculture could be determining whether the use of a new fertilizer treatment results in the extrinsic mean shape of a crop significantly changing from the historical mean. Similarly, this test could be performed for quality control purposes to determine if there is a significant defect in the outline of a produced good.

In practice, δ will be determined by the application and the decision for a test would be reached in the standard fashion. However, for the examples presented here, there is no natural choice for δ, so one can instead consider setting $Z = \xi_{1-\alpha}$ and solving for δ to show what decision would be reached for any value of δ. To do so, it is important to understand the role of δ. The size of the neighborhood around m_0 is completely determined by δ. As such, it follows that smaller values of δ result in smaller neighborhoods. In terms of H_0, this places a greater restriction on M_δ and B_δ, requiring μ_E to have a smaller distance to m_0.

For the examples presented here, the contours are approximated using $k = 300$ sampling times, so the shape space is embedded into $S(300, \mathbb{C})$ to conduct analysis. In this environment, consider having two k-gons that are identical except for at one time. If this exceptional point for the second k-gon differs from the corresponding point in the first k-gon by a difference of 0.01 units, then the distance between the shapes inherited from $S(300, \mathbb{C})$ is approximately 0.0141. For the hypothesis test, if $\delta = 0.0141$, then the neighborhood around m_0 would consist of distances between shapes similar in scope to the situation described above.

First, consider an example for which the one-sample test for extrinsic mean shape is performed for sting ray contours. In this case, the sample extrinsic mean shape for a sample of contours of $n = 10$ sting rays is the shape shown on

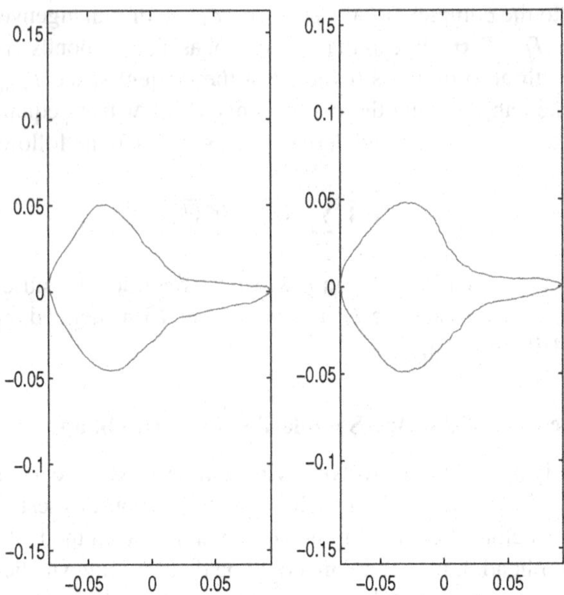

Figure 18.7 *The extrinsic sample mean shape of a sample of 10 sting rays and, respectively, the hypothesized extrinsic mean shape.*

the left hand side in Figure 18.7. After performing the calculations, it was determined that for an asymptotic level 0.05 test, the largest value of δ for which we would reject the null hypothesis is 0.0290. For perspective, this neighborhood has a radius roughly 2 times larger than the example with the nearly identical k-ads described above. This means that we would only reject the null hypothesis if we required the sample extrinsic mean to be nearly identical to the hypothesized mean. It should also be noted here that the sample size is small here, but that the conclusion agrees with intuition based upon a visual inspection of the contours.

Now consider two examples involving contours of pears. In this first case, the sample consists of $n = 87$ pears. The sample extrinsic mean shape and hypothesized extrinsic mean shape are shown in Fig 18.8. It was determined that for an asymptotic level 0.05 test, the maximum value of δ for which we would reject the null hypothesis is 1.2941. This value of δ is almost 92 times greater than the distance between the nearly identical k-ads. This suggests that even if we greatly relax the constraints for similarity, the null hypothesis would still be rejected. This again agrees with intuition.

In this last example, consider another sample of contours of pears. In this scenario, we consider a sample of $n = 83$ pears. The sample extrinsic mean shape and hypothesized extrinsic mean shape are shown in Fig 18.9. After per-

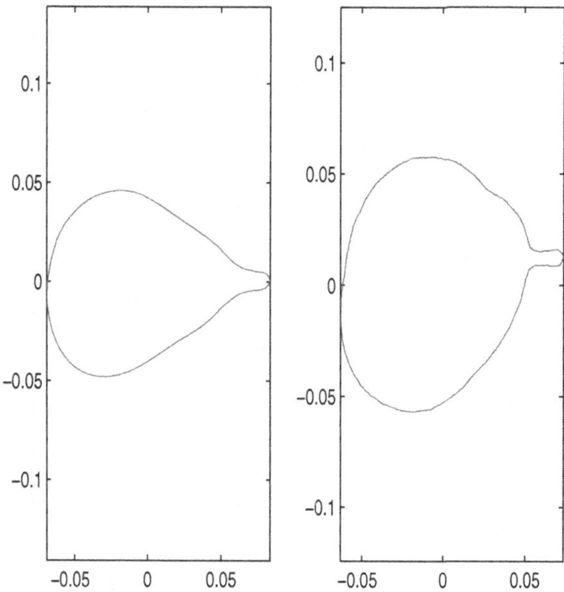

Figure 18.8 *The sample extrinsic mean shape of a sample of 87 pears and, respectively, the hypothesized extrinsic mean shape.*

forming the calculations, we determined that for an asymptotic level 0.05 test, the largest value of δ for which we would reject the null hypothesis is 0.1969, meaning that our procedure does not reject the null hypothesis, unless δ is smaller then 0.1969. For perspective, this neighborhood has a radius nearly 14 times larger than the example with the nearly identical k-ads described above. Unlike in the previous two examples it is unclear whether the null hypothesis should be rejected in this case without having a specific application in mind and, as such, this could be considered a borderline case.

18.7 Bootstrap Confidence Regions for the Sample Mean

Another method for performing inference, which we consider now, is through the use of nonparametric nonpivotal bootstrap. By repeatedly resampling from the available data and computing the distance between each resampled mean and the sample mean, we can obtain a confidence region for the extrinsic mean shape (for the finite-dimensional case, see Bandulasiri et al. (2008) [10] and Amaral et al (2010) [1]).

Below are examples of 95% bootstrap confidence regions for the same sets of contours as provided previously. These regions are based upon 400 resamples from the data. These computations reveal that these regions behave as

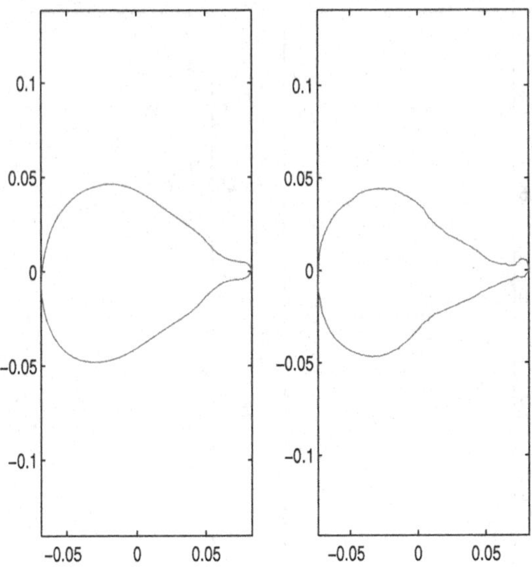

Figure 18.9 *The sample extrinsic mean shape of a sample of 83 pears and, respectively, the hypothesized extrinsic mean shape.*

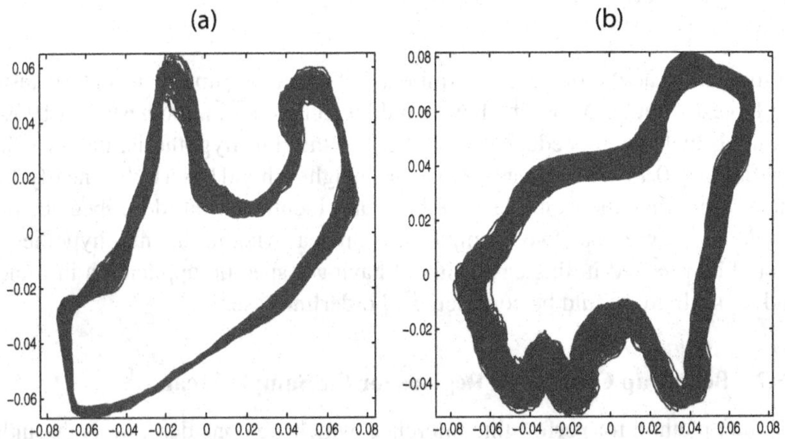

Figure 18.10 *Bootstrap 95% confidence region for the extrinsic mean shape of (a) the "t" hand gesture and (b) the dogs.*

would be expected. For instance, the confidence region for the 't' hand gesture in Figure 18.10(a) is wider in the portions of the shape where there is more variability in the sample (Figure 18.1), such as the knuckle and finger areas, but is narrower in the portions where there is less variability in the sample, such as the wrist. This is also evident in the confidence region for the sting rays 18.11(a) since the bands are thicker in the regions corresponding to the tail of the fish, where, as shown in Figure 18.3, the variability is the greatest.

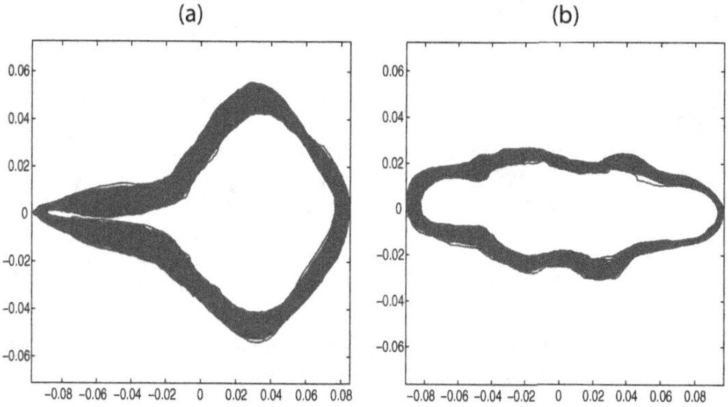

Figure 18.11 *Bootstrap 95% confidence region for the extrinsic mean shape of (a) the sting rays and (b) the wormfish.*

It is also noticeable that the samples with less variability have narrower confidence regions, as well. Comparing the samples of dogs, wormfish, and pears (Figures 18.2, 18.4, & 18.6, respectively), it is easy to see that the contours of the dogs have more variability than the contours of the wormfish, which have more variability than the contours of the pears. This is reflected in the widths of the confidence regions for these three groups of contours, as seen in Figures 18.10(b), 18.11(b) and 18.12(b).

18.8 Approximation of Planar Contours

Ideally, the shapes of planar contours could be studied directly. However, when performing computations, it is necessary to approximate the contour by evaluating the function at only a finite number of times. If k such sampling times are selected, then the linear interpolation of the yielded sampling points is a k-gon z, for which each sampling time is a vertex. As with the contour, z is a one-to-one piecewise differentiable function that can be parametrized by arclength. Let L_k denote the length of the k-gon. For $j = 1, \ldots, k$, let $z(t_j)$ denote the jth ordered vertex, where $t_j \in [0, L_k)$ and $z(t_1) = z(0) = z(L_k)$. It follows that, for $s \in (0,1)$, the k-gon can be expressed as follows:

$$z(sL_k) = \begin{cases} (t_2 - sL_k)z(0) + sL_k z(t_2) & 0 < sL_k \leq t_2 \\ (t_j - sL_k)z(t_{j-1}) + (sL_k - t_{j-1})z(t_j) & t_{j-1} < sL_k \leq t_j \\ (L_k - sL_k)z(t_k) + (sL_k - t_k)z(0) & t_k < sL_k < L_k \end{cases} \quad (18.21)$$

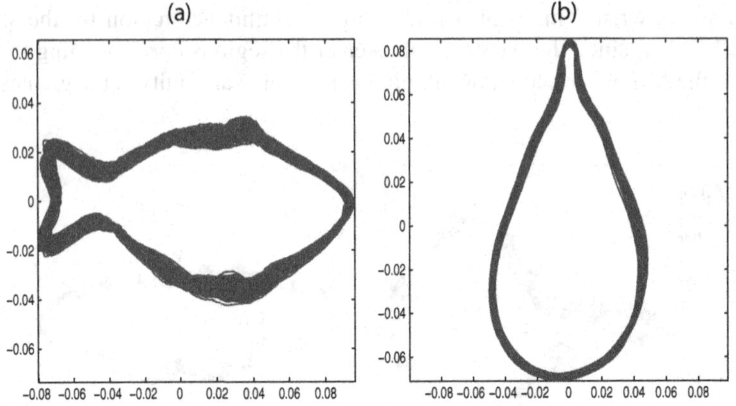

Figure 18.12 *Bootstrap 95% confidence region for the extrinsic mean shape of (a) the red snapper fish and (b) the pears.*

for $j = 3,\ldots,k$. As such, the space of direct similarity shapes of non-self-intersecting regular polygons is dense in the space of direct similarity shapes of regular contours. Therefore, the theory and methodology discussed in sections 3 through 5 hold for the shapes of these functions, as well. For the purposes of inference using the neighborhood hypothesis, then, it suffices to use the test statistic as derived previously. However, when considering these approximations, it is important to choose the sampling times appropriately so that the contour is well approximated by the polygon. Additionally, one must take correspondence across contours into consideration when working with a sample. We will first present an algorithm for choosing sampling times in such a way that the k-gon well represents the contour and converges to it accordingly. Following that, we will address considerations for working with samples.

18.8.1 Random Selection of Sampling Times

To obtain approximations, we propose to randomly select a large number k of sampling times t_j from the uniform distribution over $[0,L)$. By doing so, we insure, on one hand, that we ultimately use a sufficiently large number of vertices so that the k-gon well represents the contour. On the other hand, we ensure the desired density of sampling points.

In order to maintain the within sample matching for a sample of regular contours, we first find the point z_0 at the largest distance from the center of the contour and choose that as $z(0)$. We then randomly select $k-1$ sampling times from the uniform distribution to form the k-gon, making sure to maintain the proper ordering, preventing the k-gon from self-intersecting. This is accomplished by sorting the selected sampling times in increasing order.

It is important to choose an appropriate number of sampling times for the given data. The selected sampling points will be distributed fairly uniformly around the contour for large values of k, ensuring that the curve is accurately represented by the k-gon. However, choosing too many sampling times will needlessly increase the computational cost of performing calculations. This will be most noticeable when utilizing bootstrap techniques to compute confidence regions for the extrinsic mean shape.

Choosing too few sampling times, though, while keeping computational cost down, can be extremely detrimental as the sampling points may not be sufficiently uniform to provide adequate coverage of the contour. This can significantly distort the k-gon, as shown in Figure 18.13. In this particular instance, the 200-gon of the dog includes no information about one of the ears and very little detail about one of the feet. It should be noted that the extraction and

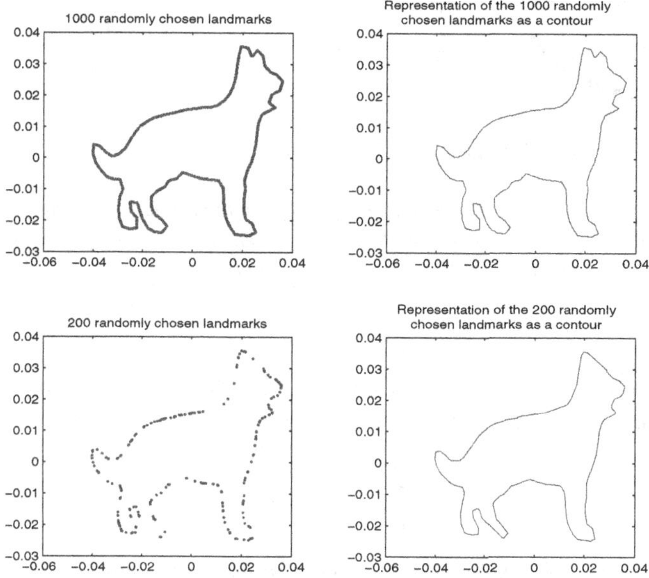

Figure 18.13 *A dog represented by 1000 and, respectively, 200 randomly chosen sampling points.*

approximation of a contour and the previously described eigenanalysis can be automated, or semi-automated, allowing for efficient execution of the methodology. An example of this sampling point selection process from start to finish is shown in Figure 18.14. Please note that the sampling point selection considered here is analogous to the landmark selection method considered by Ellingson (2011) [101] for approximating a contour as a k-ad.

The length of the contour L can be used to assist in determining an appro-

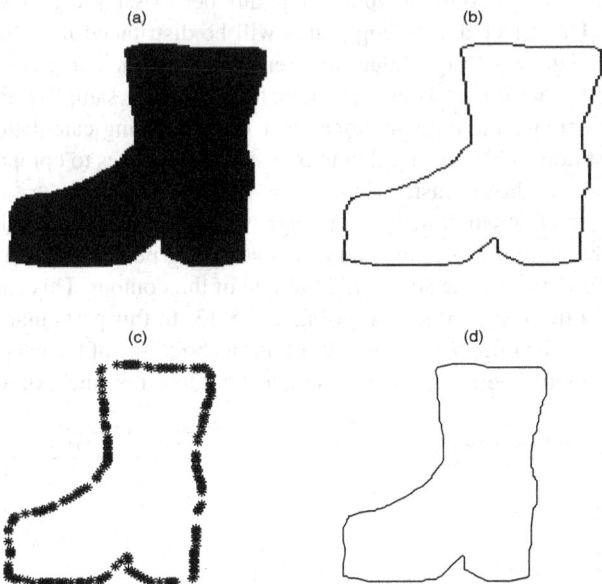

Figure 18.14 *An image of the contour of a boot from the Kimia image database (a) as a silhouette, (b) as a contour, (c) as 200 randomly selected sampling points, and (d) as a contour constructed from the chosen sampling points.*

priate number of sampling points to be chosen. After selecting an initial set of k sampling times as describe above, the length L_k of the k-gon is

$$L_k = \sum_{j=2}^{k+1} \|z(t_j) - z(t_{j-1})\|, \tag{18.22}$$

where $z(t_{k+1}) = z(0)$. An appropriate lower bound for the number of sampling points can be determined by randomly selecting times for various values of k. Compute L_k for each of these k-gons using (18.22) and compute the relative error compared to L. This should be repeated many times to obtain a mean relative error and standard deviation of the relative error for each value of k used. To determine an appropriate number of sampling points to use, compare the mean relative error to a desired threshold. Additionally, the distributions of the relative errors could also be examined. It should be noted, however, that since digital imaging data is discrete by nature, the contour will be represented by K pixels. As such, it is often necessary to replace L by L_K, the length of the closest approximation to the contour, which can be calculated similarly to L_k.

When using this algorithm to select sampling points, it follows that the k-gon will converge to the contour. However, when selecting an additional sampling time $t_{j'}$, care must be taken to properly alter z in such a way that ensures that there is no self-intersection of the k-gon. To do so, simply reorder the sampling times in increasing order and apply the resulting permutation to the sampling points. It follows that, with probability 1, the length of the k-gon between successive sampling points will converge to 0 as the number of sampling times tends to infinity. This can be stated more formally as follows.

LEMMA 18.8.1. *If sampling times $s_1, s_2, \ldots, s_k$ are selected from a uniform distribution over* $[0,1)$, *then*

$$L_{max} = \max_{2,k+1} \|z(s_j L_k) - z(s_{j-1} L_k)\| \xrightarrow{P} 0.$$

Proof.

$$\begin{aligned} P(L_{max} > \varepsilon) &= P(\text{All } k \text{ sampling points are within the remaining } L_k - \varepsilon) \\ &= P\left(\text{All } k \text{ sampling times are within the remaining } 1 - \frac{\varepsilon}{L_k}\right) \end{aligned}$$

We can assume without loss of generality that the section of the k-gon for which the distance between successive sampling times is greater than ε^* is over the interval $(0, \varepsilon)$. In addition, since the sampling times are independently chosen,

$$P(L_{max} > \varepsilon) = \left(F(1) - F\left(\frac{\varepsilon}{L_k}\right)\right)^k = \left(1 - \frac{\varepsilon}{L_k}\right)^k$$

where F is the cdf for the uniform distribution over the interval $[0,1)$. Taking the limit of this expression as $k \to \infty$ results in $L_{max} \xrightarrow{P} 0$ since

$$\lim_{k\to\infty} P(L_{max} > \varepsilon) = \lim_{k\to\infty} \left(1 - \frac{\varepsilon}{L_k}\right)^k = 0$$

□

The center of mass of the k-gon is calculated in the following manner.

$$\bar{z}_k = \frac{1}{2L_k} \sum_{j=2}^{k+1} \|z(t_j) - z(t_{j-1})\| \left(z(t_j) + z(t_{j-1})\right) \tag{18.23}$$

This differs from the Kendall definition of the center of mass for a k-ad, which in this situation, would depend only on the vertices. It follows immediately from the definition that the center of mass of the k-gon converges to the center of mass of the contour.

While the k-gon z converges to the contour γ, it is also of great interest to consider the convergence of $[z]$ to $[\gamma]$. Since z and γ are objects in the same space, the disparity in their shapes can be examined by considering the squared distance $\|j([z]) - j([\gamma])\|^2$ in $\mathcal{L}_{HS}$. However, for the purposes of computational comparisons, it is necessary to evaluate the functions at $m > k$ times using (18.21) and approximate the distance in $S(m, \mathbb{C})$, the space of self-adjoint $m \times m$ matrices. To illustrate, consider the contour considered in Figure 18.15. The

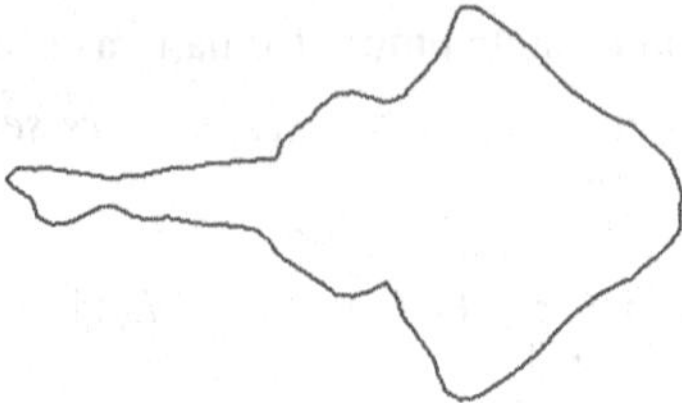

Figure 18.15: *The digital image of the contour of a stingray.*

digital representation of this contour consists of $K = 764$ pixels. sampling times were selected using the above algorithm to form k-gons for $k = 3, \ldots, 763$. Each k-gon was then evaluated at 764 times corresponding to the each of the pixels on the digital image of the contour. As such, squared distances between the k-gons and the contour were computed in $S(764, \mathbb{C})$. After this was repeated 50 times, the means and standard deviations of the squared distances were calculated for each value of k and are shown in Figure 18.16. The mean squared distance to the contour converges quickly, showing that the distance between $[z_k]$ and $[\gamma]$ only diminishes slightly for $k > 100$. Moreover, the variability introduced by selecting the sampling times randomly also rapidly approaches 0. As such, it is clear that $[\gamma]$ is well approximated using $k << K$. However, while the overall shape is well approximated, it is unclear from this alone how well the details of γ are approximated. As such, using the distance between shapes may not be the best indicator for determining a lower bound for k. For this purpose, it may be more helpful to consider $(L_K - L_k)/L_K$, the relative error in the approximation of the length, as described previously. For the contour in Figure 18.15, the relative error in length is shown in Figure 18.17. Here, while the variability approaches 0 quickly, the average relative error approaches 0 at a lower rate. As such, if it is desirable to keep the relative error below 0.05, for this example, no fewer than 300 sampling times should be selected.

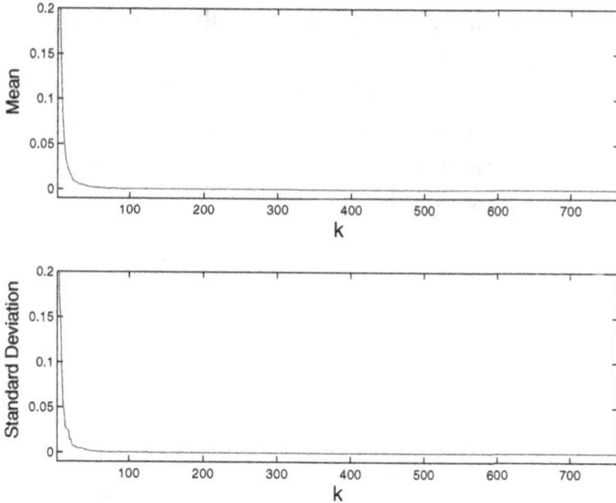

Figure 18.16 *The mean and standard deviation of the squared distance of shapes of k-gons to the shape of the contour, as evaluated in* $S(764, \mathbb{C})$.

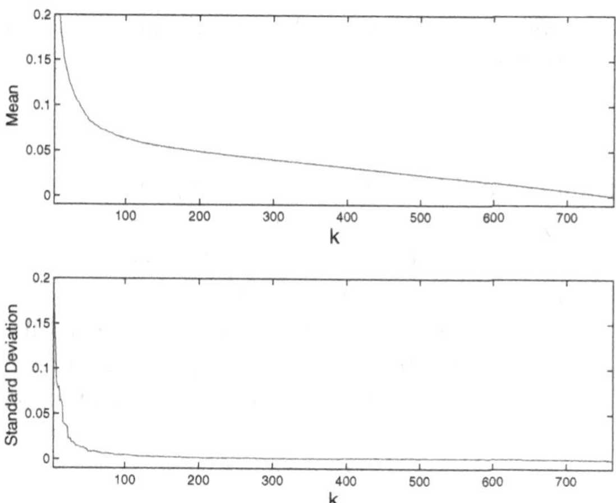

Figure 18.17 *The mean and standard deviation of the relative error of the length of the k-gons.*

18.8.2 Considerations for Samples of Contours

In addition to ensuring that each contour in a sample is well approximated, since each contour must be evaluated at m times for computations, it is necessary that each be evaluated at the same times to maintain correspondence across the n observations. In the ideal scenario, if each contour is well approximated by a k-gon, then select k sampling times $s_1, \ldots, s_k \in [0, 1)$ using the algorithm as described above. For $j = 1,\dots,n$, the sampling points for z_j can then be obtained by evaluating γ_j at times $s_1 \cdot L^j,\dots,s_k \cdot L^j$, where L^j denotes the length of of γ_j. Figure 18.18 shows two examples of utilizing this procedure for samples of contours of hand gestures. Using the same sampling times for each observation, 6 sampling points are highlighted in red to illustrate the correspondence across the sample.

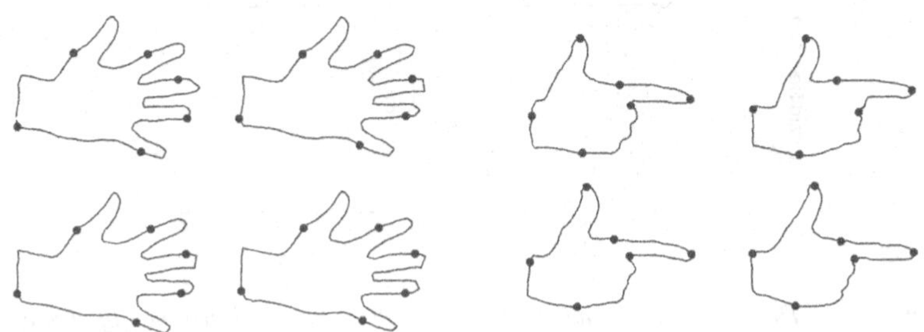

Figure 18.18 *Correspondence of 6 sampling points for contours of hand gestures of (a) the number '5' and (b) the letter 'L'. (Source: Ellingson et al.(2013), Figure 5. Reproduced by permission of Elsevier).*

Alternatively, if contour j requires k_j sampling points for adequate approximation, where $k_i \neq k_j$ for at least one pair i, j, then select sampling times for each contour. Let $\mathcal{T}_j$ denote the set of sampling times that generate the k_j-gon z_j. In order to maintain correspondence, evaluate z_j at the m times contained in $\cup_{i=1}^{n} \mathcal{T}_i$ for $j = 1,\dots,n$. This approach may also be utilized if each contour is approximated using k sampling points, but at different times. Finally, even if the conditions of either of the previous scenarios are met, it may be desired to consistently work within the same shape space when working with multiple samples, so it may be preferred to instead first consider approximating the contours and then approximating each at m subsequently chosen times, thus separating the issues of approximation and correspondence. However, for each of these scenarios, the selection of sampling points, evaluation of the k-gon at m times, and subsequent analysis can be either semi-automated or fully automated, allowing for efficient execution of the methodology.

18.8.3 Approximation of the Sample Mean Shape

Whenever one is dealing with an object that is conceptually of infinite, or very high, dimension, a suitable dimension reduction must inevitably take place to enable computers to handle this object. Because this process is usually a projection from an infinite dimensional sample space of which the original object is an element, onto a finite dimensional subspace, we will for convenience refer to it as a "projection". In the current situation, the infinite dimensional object is the average of projection operators $\hat{\mu}$ in equation (18.16), which is a positive element in the Hilbert space of Hilbert-Schmidt operators. Above, this object has been approximated by rather high-dimensional projection and then successively by projections of lower dimension, in order to arrive at an approximation of sufficiently low dimension that is still a good representative of the original object. What constitutes "good" here has not been established rigorously, but instead primarily on eye ball fitting, which may, in many cases, work rather well.

Ellingson and Prematilake (2013) [105] follows up the above approach with the aim to form a more concrete methodlogy for determining an appropriate approximation for a single contour. However, it is still driven by making a choice about what constitutes a "good" approximation. A more sophisticated, theoretically-based approach seems possible, however, and might be based on a method employed by Gaines (2012) [122] to determine a suitable number of points at which the values of a Brownian motion process should be simulated. This was accomplished by using results from perturbation theory to ensure that the largest eigenvalues of the covariance operator of the projection sufficiently approximated those of the original process. Since $\hat{\mu}_E$ is the largest eigenvalue of $\hat{\Gamma}$, a similar approach should, in principle, be appropriate in the this context.

18.9 Application to Einstein's Corpus Callosum

We will now return to a problem that was introduced in Chapter 1 involving Albert Einstein's corpus callosum. Recall that Einstein's brain was removed and extensively studied after his death. Motivated by Falk et al. (2013) [110] and Man et al. (2014) [228], which showed that the midsatigittal cross section of Einstein's corpus callosum was thicker than average, Qiu et al. (2014) [285] utilized the methodology described above to investigate this problem.

The contours of corpora callosa for Einstein and a sample of similarly aged people from Fletcher (2013) [118] are shown in Figure 1.2 and, respectively, Figure 1.2. Qiu et al. (2014) [285] obtained matched sampling points for the observed contours by using the methodology described in Section 18.8.

Qiu et al. (2014) [285] subsequently performed an analysis using the remaining methodology discussed in this chapter. An icon for the extrinsic sample mean shape is displayed in Figure 18.9, where it is superimposed on the contour for Einstein's corpus callosum. This figure suggests that there may be some differences in shape between these contours.

The two inference procedures discussed in this chapter suggest that the cross-section of Einstein's corpus callosum does indeed differ in shape from the average. Performing the neighborhood hypothesis test revealed that the largest value of δ for which the null hypothesis would be rejected is 0.1367,

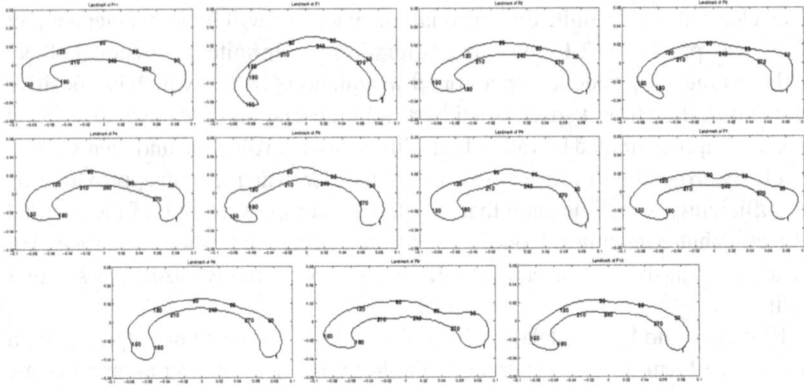

Figure 18.19 *Matched sampling points on the CC contours. Einstein's is the upper left CC.*

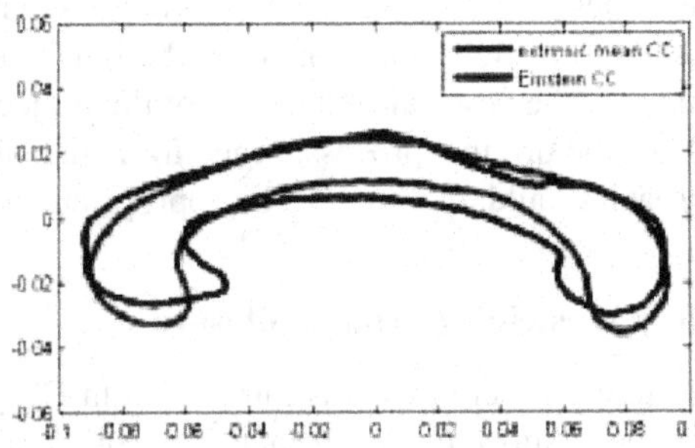

Figure 18.20 *Superimposed icons for the similarity shapes of contours of corpus callosum midsections: sample mean vs Albert Einstein's.*

which is rather large compared to the amount of variability present in the data. This finding is further bolstered by the 95% bootstrap confidence region for the extrinsic mean shape, which is depicted in Figure 18.9. Comparing Figures 18.9 and 18.9 reveals that the Einstein's contour is not contained within the confidence region, thus suggesting that the shape of Einstein's corpus callosum cross section is significantly different than average.

Figure 18.21 *95% bootstrap confidence region for the extrinsic mean CC contour by 1000 resamples.*

Chapter 19

Estimation of Mean Skull Size and Shape from CT Scans, with Applications in Planning Reconstructive Plastic Surgery in Young Adults

In this chapter, we consider an application of size-and-shape analysis arising from images obtained from CT (computed tomography) scans. We return to the data set introduced in Chapter 1 involving CT scans of human skulls. This chapter will discuss image processing techniques necessary for obtaining data in the appropriate form for landmark-based shape analysis and the subsequent nonparametric statistical analysis, as described in Chapter 12. The primary references for this chapter are Osborne (2012) [257], Osborne et al. (2012) [259], and Osborne et al. (2015) [260].

19.1 Introduction

As mentioned in Chapter 1, CT scans are a valuable resource for imaging various parts of the human anatomy. One particular application is in the planning of surgery, where it is appropriate to take the size and shape of the body into account. As such, it is sensible to appeal to statistical methods for analyzing size-and-shape for such purposes. However, it is highly non-trivial to produce useful landmarks from CT data. Here, we will present the methodology detailed in Osborne (2012) [257] to obtain such data for the purposes of studying the size-and-shape of the bones found in the skull.

This procedure can be broken down into a number of steps. First, the bone must be isolated from the CT scans. Then, the bone structures from various scans from a single individual are used to produce a 3D reconstruction of the skull. Subsequently, relevant anatomical landmarks can be obtained from these reconstructions for the purposes of size-and-shape analysis, where the observations are regarded as points on the size-and-reflection-shape space $SR\Sigma^k_{3,0}$, that was introduced in Chapter 12.

The remainder of this chapter will be organized as follows. First, we will present an overview of CT scans. We will then describe the techniques used to isolate the bone structures. After that, we will discuss the reconstruction procedure. We will then conclude with an aplication of the nonparametric techniques described in Chapter 12 to landmarks obtained from this procedure.

19.2 CT Scans

A CT or CAT (computed assisted tomography) scan uses X-rays to make detailed pictures of structures inside of the body. As described in Chapter 1, CT scans can be used to study all parts of the human body. CT scans can also take pictures of the body organs, such as the bladder, liver, lungs, pancreas, intestines, kidneys, and heart. Additionally, they can be used to study the spinal cord, blood vessels, and bones. Because of their great utility, specialized software was designed for retrieval of volumetric data from the CT X-rays.

Figure 19.1 shows a flowchart describing the process involved with obtaining a series of CT scans from acquisition through representation as an image. The value associated with a given voxel in a CT scan, which is typically converted to a grayscale intensity for displaying, represents the material, such as bone, fat, or air, found in the corresponding location within a person's body. Figure 19.2 details the relationship between the materials and their depiction in a CT scan. Good contrast sensitivity is desirable in medical imaging. CT images provide a high level of this, especially for distinguishing bone from soft tissue.

19.3 Bone Surface Segmentation

The data set used in Osborne (2012) [257] contains CT scans of skulls for 20 individuals. Each observation consists of roughly 100 CT slices of one person's head above the mandible, as displayed in Figure 1.20. Since it is necessary to virtually reconstruct the skulls for the size-and-shape analysis, after the observations were obtained, the surfaces of the bone structures had to be extracted.

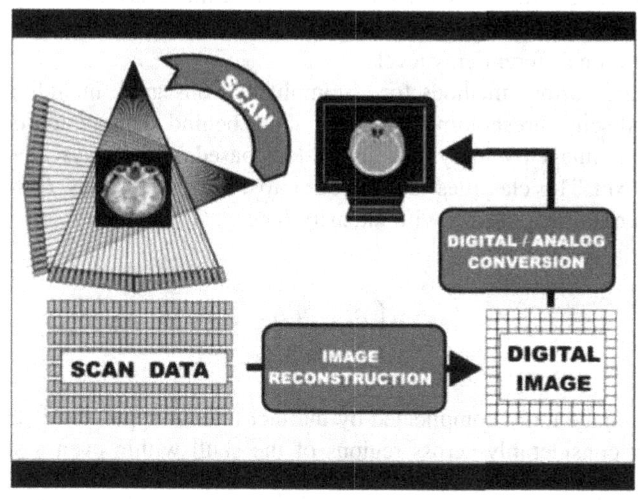

Figure 19.1: *The process of CT image acquisition.*

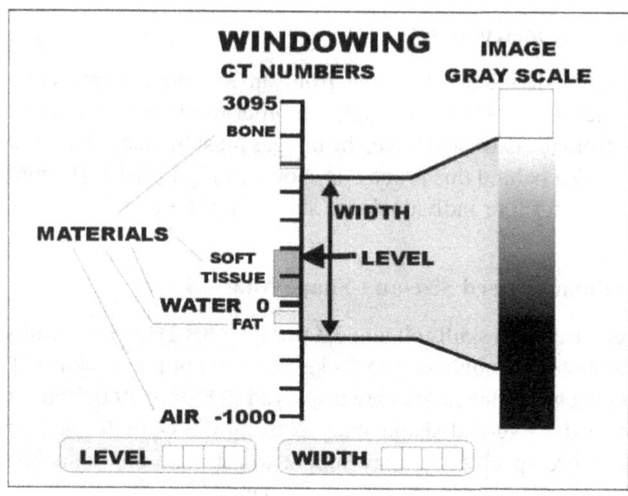

Figure 19.2 *Various materials are represented as a grayscale intensity for a given voxel.*

In other words, the surface is reconstructed from the curves that separate the softer flesh tissue from the hard bone tissue. Since these tissues are represented by different voxel intensities, finding these curves amounts to detecting the edges between different gray levels.

There are various methods for accomplishing this goal, including thresholding and semi-thresholding. The basic ideas behind such an approach is to simplify the image by applying a threshold T based on the grayscale intensity of each voxel. This classifies each voxel x into one of two classes: B (bone) and $\bar{B}$ (non-bone). For a voxel x with intensity $I(x)$,

$$x \in \begin{cases} B, \text{ if } I(x) \geq T \\ \bar{B}, \text{otherwise} \end{cases}$$

This is made more complicated by the fact that an appropriate value of T may vary considerably across regions of the skull within even a single individual. An alternative approach for doing this is based on a segmentation technique introduced by Caselles et al. (1997) [60] called the Geodesic Active Contour model, which is an extension of the snake model of Kass et al.(1988) [175]. The method used by Osborne (2012) [257] was the *total variation segmentation algorithm* of Unger et. al (2008) [325]. Figure 19.3 displays some examples of bone structures extracted from CT slices form a single individual.

19.4 Skull Reconstruction

Because each individual CT image from an individual represents a cross-section of their head at a given height, in order to virtually reconstruct a person's skull from the extracted bone, the images must be stacked atop each other. The general idea behind this process is shown in Figure 19.4. Examples of 3D reconstructions for four individuals are shown in Figure 19.5.

19.5 Landmark-Based Size-and-Shape Analysis

With the reconstructed skulls obtained, Osborne (2012) [257] carefully chose 9 matched landmarks around the eye sockets on each of the 20 skulls. The set of skulls and the chosen landmarks are displayed in Figure 19.6. The researchers then performed a size-and-shape analysis using the k-ads formed from these landmarks. More specifically, they sought to estimate the Schoenberg mean size-and-reflection-shape, as described in Chapter 12.

To do this, they used 500 bootstrap resamples to obtain 90% simultaneous confidence regions for the coordinates of the landmarks with respect to a convenient coordinate system. The lower and upper bounds for the coordinates of these rectangular regions are shown in Table 19.1. To help better visualize

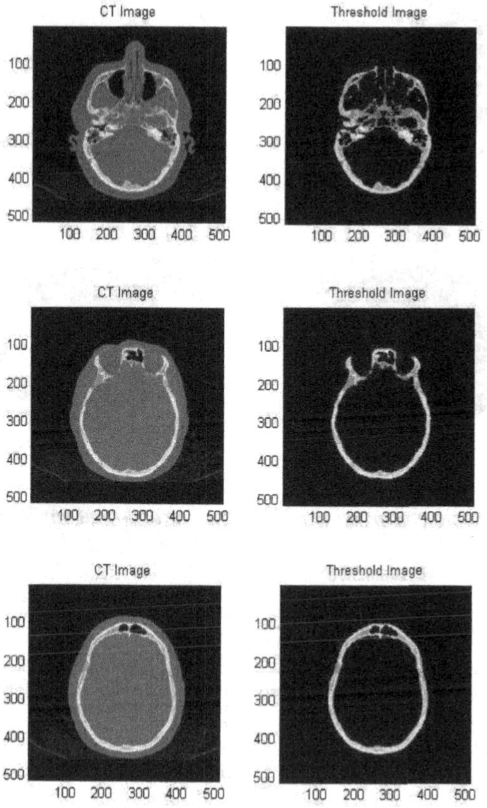

Figure 19.3 *Examples of bone structures extracted from CT slices of a single individual.*

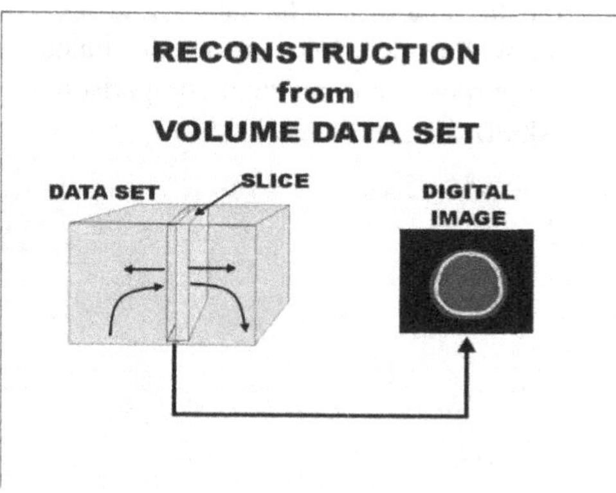

Figure 19.4: *Reconstruction of volume data from CT images.*

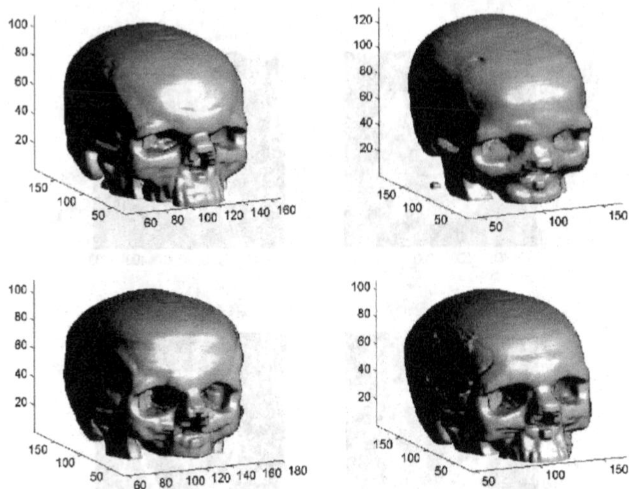

Figure 19.5: *3D Reconstruction of the bone surface-total variation approach.*

Table 19.1 *Simultaneous Bootstrap Confidence Intervals for the Mean Size and Shape of the Distribution of Configurations*

90% Low Confidence Limit for the Bootstrap Distribution of the Sample Mean Configuration									
Landmark No.	1	2	3	4	5	6	7	8	9
x	−45.76	−28.65	−9.75	−32.06	−0.90	7.84	27.15	40.79	26.28
y	10.10	−2.37	−5.91	0.27	−19.20	−3.84	0.86	10.04	−0.06
z	−0.19	9.73	4.24	−11.67	−8.06	3.00	−12.70	−1.70	9.36

90% Upper Confidence Limit for the Bootstrap Distribution of the Sample Mean Configuration									
Landmark No.	1	2	3	4	5	6	7	8	9
x	−42.03	−24.44	−7.27	−28.29	0.41	10.19	30.65	44.52	30.30
y	11.85	−0.93	−3.39	1.82	−15.92	−1.69	2.52	12.83	4.17
z	1.43	11.12	5.93	−10.02	−5.73	4.83	−11.25	−0.32	13.08

these regions, bootstrap distributions for the landmarks of the Schoenberg sample mean are displayed with respect to the same coordinate system are shown in Figure 19.7. For the purpose of convenient comparison, they are displayed alongside a template skull.

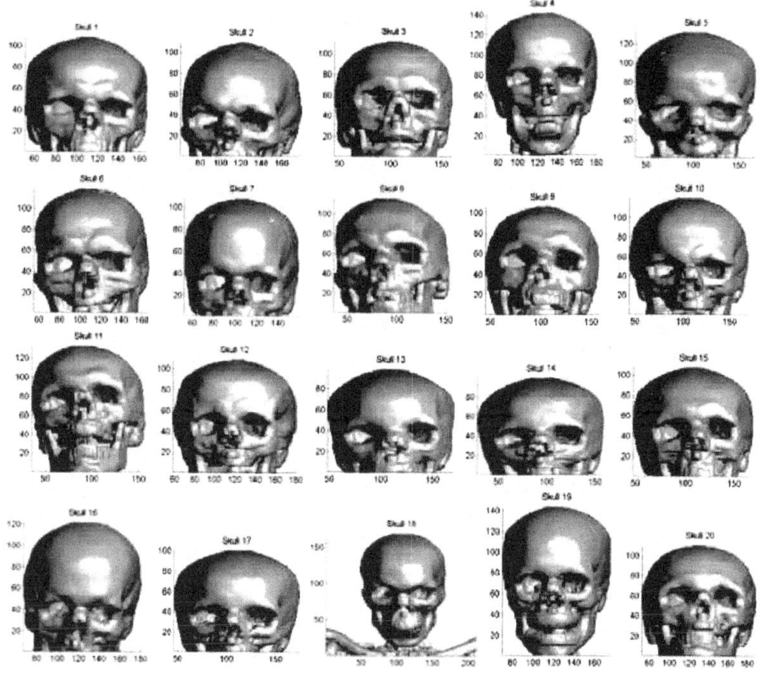

Figure 19.6 *The 20 virtual reconstructions of the skulls with the chosen landmarks labeled.*

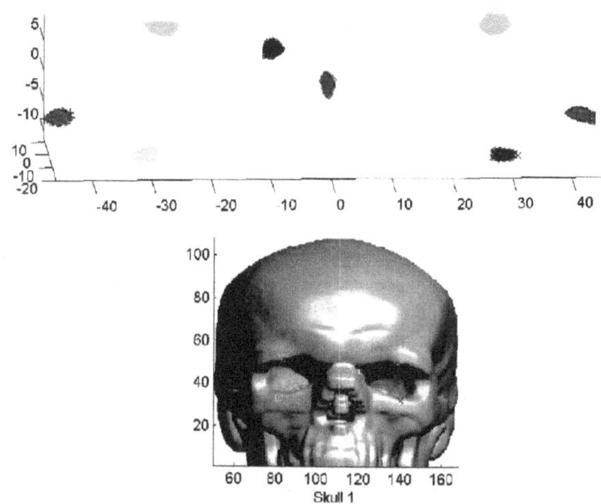

Figure 19.7 *Bootstrap distributions for the landmarks of the Schoenberg sample mean based on 500 resamples.*

Chapter 20

Affine Shape and Linear Shape Applications

20.1 Introduction

A study of affine shape and linear shape is needed in certain problems that arise in in bioinformatics and in pattern recognition . In particular, affine shape analysis is useful in the reconstruction of a larger area from a number aerial images, and linear shape is helpful in the the analysis of 2D electrophoresis images.

A typical example, given in Figure 20.1, is when two aerial images taken by a squadron from different distances of a ground scene need to be merged to create a larger contiguous image that contains the information in both these images.

Another problem consists in identifying or matching configurations of proteins on marked 2D gel electrophoresis. See Figure 20.2. Such problems involve image warping that may be based on *linear shapes* of configurations of selected landmarks. When it comes to affine shape, researchers from computer vision use a different definition than the one in Section 3.5. In Section 20.2, we introduce the computer vision definition due to A. Heyden (1995) [157] and

Figure 20.1: *Aerial photographs of two sections of a ground scene.*

compare it to the one in Subsection 3.5.7, showing that the two definitions are, in fact, equivalent.

As a consequence, either definition allows considering the *affine shape space of configurations of k points in general position* to be regarded as a real Grassmann manifold (see Subsection 3.5.7). Therefore, statistical analysis of affine shape naturally leads to data analysis on a real Grassmann manifold. The preferred data analysis, given its simplicity and computational speed, is an extrinsic analysis based on the Dimitric embedding in (14.1) defined in Section 20.3. In Section 14.2, the Dimitric mean of a distribution Q on $G_k(\mathbb{R}^m)$ was determined as the linear envelope of the eigenvectors corresponding to the k largest eigenvalues of the regular mean regarded as point on $Sym(m,\mathbb{R})$. The Dimitric sample mean is the key statistic considered in this chapter for 2D image reconstruction.

2D-gel electrophoresis (2DGE) is a widely used technique for separating protein molecules for identification purposes. Different proteins have different molecular weights and electrical charges, and hence behave differentially in an applied electric field. In 2DGE protein molecules are allowed to move in a gel medium under an X-Y electric field, and a snapshot is taken once molecules have had sufficient time to move away from the initial point. Due to the variability of gels density and strength of electric fields, a protein molecule is generally considered to be moved to a certain position in each of the perpendicular directions. Their coordinates in two experiments, say (X,Y) and (X',Y') differ via a linear equation

$$X' = aX, Y' = bY, a, b > 0, \tag{20.1}$$

so two configurations of proteins from the same medium, but pictured in different settings, may differ by a linear transformation 20.1. Therefore, up to measurement errors, the linear shape of the configuration of the protein configuration should be independent of experimental settings. The problem of match-

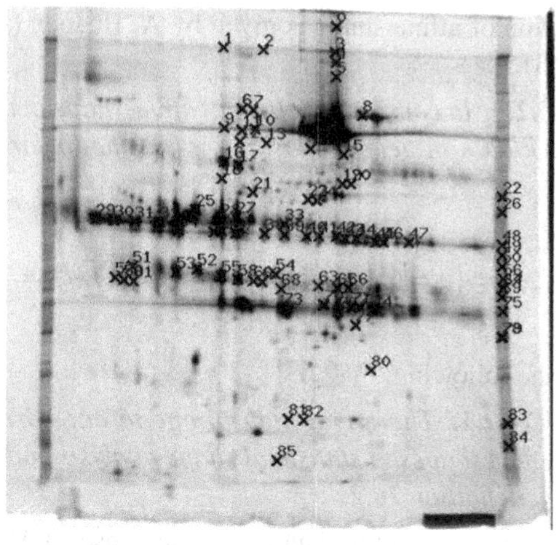

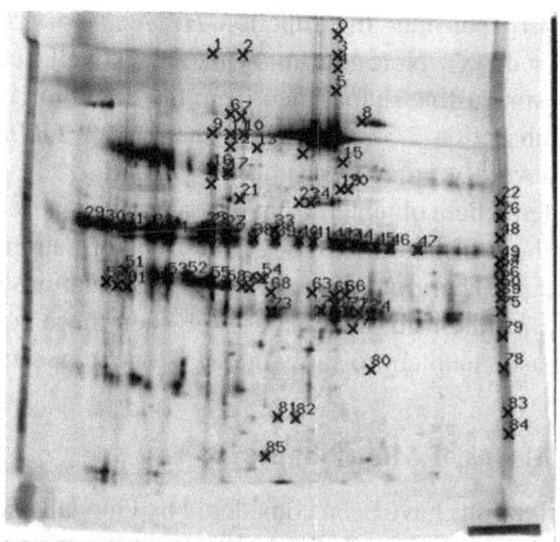

Figure 20.2: *Marked electrophoresis gels.*

ing electrophoresis configurations in 2DGE images, modulo linear transformations, is discussed with an example in Section 20.3.

20.2 The Affine Shape Space in Computer Vision

Recall from subsection 3.5.7 that, for $k \geq m+1$, the reflection-affine shape space $A\Sigma_m^k$ of k-ads in general position in $\mathbb{R}^m$, is in a one to one correspondence with the Grassmann manifold $G_m(\mathbb{R}^{k-1})$. In that section, a reflection-affine shape $a\tilde{\sigma}(\mathbf{x})$ was defined as the orbit of a k-ad $\mathbf{x} = (x_1, \dots, x_k)$ under the diagonal action 3.136 of the affine group $Aff(m)$ on k-ads in $\mathbb{R}^m$.

Another definition of affine shape is given by A. Heyden (1995) [157] and G. Sparr (1998) [315]:

DEFINITION 20.2.1. *In computer vision, an affine shape in the sense of Heyden $\alpha\sigma[\mathbf{x}]$ of the k-ad $\mathbf{x}$ in general position $\mathbb{R}^m$ is the linear subspace of $\mathbb{R}^k$ given by*

$$\alpha\sigma[x_1,\dots,x_k] = \{\xi = (\xi^1,\dots,\xi^k), \sum_{j=1}^{j=k} \xi^j x_j = 0, \sum_{j=1}^{j=k} \xi^j = 0\} \tag{20.2}$$

We can prove the following:

PROPOSITION 20.2.1. *There is a natural one to one correspondence between reflection-affine shapes in statistical shape analysis and affine shapes in computer vision in definition 20.2.1.*

Let $\alpha\Sigma^k_{m,m}$ be the set of affine shapes in the sense of Heyden, as given in definition 20.2.1. Consider the function $H : A\Sigma^k_{m,m} \to \alpha\Sigma^k_{m,m}$, given by $H(\alpha\sigma[x_1,\dots,x_k]) = a\tilde{\sigma}(\mathbf{x})$. Note that this function is well defined. Indeed, if $\mathbf{x}$ and $\mathbf{y}$ have the same affine shape, then from the definition of the reflection-affine shape in Subsection 3.5.7, there is a matrix $A \in GL(m,\mathbb{R})$, such that, $\forall i = 1,\dots,k$, we have $\xi^i A = \xi'^i$, where $(\xi^1 \dots \xi^k)$ and $(\xi'^1 \dots \xi'^k)$ are, respectively, the the centered data obtained from the sampling matrices $\mathbf{x}$ and $\mathbf{x}'$. This is the case since the translation part, b, of the transformation $x' = Ax + b$ disappears via centering. Therefore, if we multiply the equations in (20.2), which are satisfied by ξ, on the right hand side by A, then those equations are also satisfied by ξ'. It is elementary to show that H is one to one and onto.

20.3 Extrinsic Means of Affine Shapes

Affine shape distributions have been considered by Goodall and Mardia (1993) [126], Leung, Burl, and Perona (1998) [220], Berthilsson and Heyden (1999) [23], and others. In view of results in Chapter 14, Dimitric sample means of distributions of reflection affine shapes can be determined using the general approach in Chapter 4. Recall that via Dimitric embedding j, $G_{m-j}(\mathbb{R}^{k-1})$ is mapped to $Sym(k-1)$ by identifying each $m-j$-dimensional vector subspace L with the matrix p_L of the orthogonal projection into L.

Assume a probability distribution Q of affine shapes of configurations in general position is *nonfocal* w.r.t. this embedding j. In this case, from Section 14.2, the *Dimitric mean* of Q is the vector subspace spanned by unit eigenvectors corresponding to the first $m-j$ eigenvalues of $\mu_{j(Q)}$, mean of $j(Q)$. Assume $(\pi_1,\dots,\pi_n)$ is a sample of size n of $m-j$-vector subspaces $\pi_1,\dots,\pi_n$ in $\mathbb{R}^{k-1}$, and the subspace π_r is spanned by the orthonormal unit vectors $\{x_{r,a}\}_{a=1,\dots,m-j}$ and set $x_r = (x_{r,a})_{a=1,\dots,m-j}$. The Dimitric sample mean of this sample, when it exists, is the $m-j$-vector subspace $\overline{\pi}$ generated by the unit

Figure 20.3 *Reconstruction by Gordana Derado of a larger view of the scene in images in Figure 20.1, based on an extrinsic mean affine shape.*

eigenvectors corresponding to the first largest $m - j$ eigenvalues of $\sum_r x_x x_r^t$. This can be compared with the Procrustes sample mean in the sense of Chikuse (1999 [71], 2003 [72])). There are exceptional distributions, such as the uniform distributions (Chikuse and Watson (1995) [73]), for which the extrinsic mean does not exist, and therefore the Procrustes sample mean is inconsistent.

Nevertheless, the condition of existence of the mean of a distribution on a Grassmann manifold is generic and, for a given random sample, the extrinsic mean exists with almost no exception. The extrinsic sample mean is useful in averaging images of remote planar scenes by adapting the standard method of image averaging of Dryden and Mardia (1998) [91] as shown in Mardia et al. (2001) [234]. This method can be used in reconstruction of larger planar scenes, as in Faugeras and Luong (2001) [112], as shown in Figure 3.

Patrangenaru et al. (2012) [269] applied this methodology for the reconstruction of an almost planar ground scene ($2\frac{1}{2}$D scene) by fusing three aerial images of the city Amiens, France, that are given in Figure 20.4.

The reconstruction, including texture, is displayed in Figure 20.5. It clearly shows some partial blurring around the shared portions of the original images,

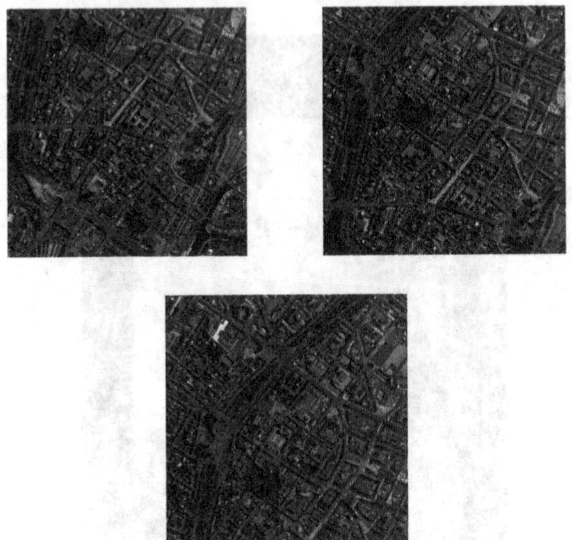

Figure 20.4: *Aerial partially overlapping images of a city.*

essentially due to the camera movement. Note that, given the pixel size (about 10 inches in ground units), the blurring effect could not be removed by any smoothing.

20.4 Analysis of Gel Electrophoresis (2DGE)

Our next focus is on the field of proteomics. Proteomics, or proteome analysis, has become an increasingly important part of the life sciences, especially after the completion of sequencing the human genome. Proteome analysis is the science of separation, identification, and quantization of proteins from biological samples with the purpose of revealing the function of living cells. Applications range from prognosis of virtually all types of cancer over drug development to monitoring environmental pollution.

Currently, one leading technique for protein separation is two-dimensional gel electrophoresis (2DGE), resulting in grey level images showing the separated proteins as dark spots on a bright background (see Figure 20.6 from Sughatadasa (2006) [322]). Such an image can represent thousands of proteins.

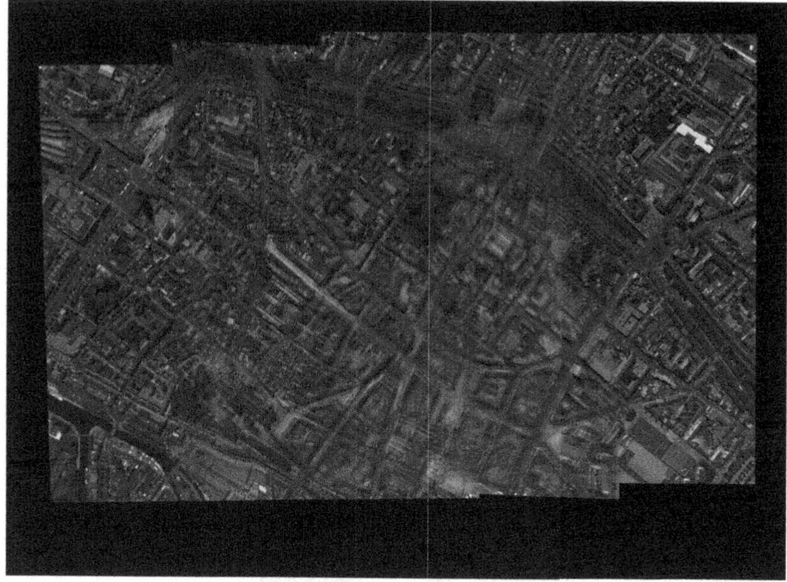

Figure 20.5 *Reconstruction of a* $2\frac{1}{2}D$ *scene from images in Figure 20.4 based on 2D extrinsic mean affine shapes.*

20.4.1 2D Gel Electrophoresis (2DGE)

Gel electrophoresis is an inexpensive technique for separating proteins in a biological sample on a gel. The resulting protein patterns are by captured as digital images of the gel. This image is then analyzed in order to determine the relative amount of each of the proteins in the sample in question or to compare the sample with other sample or a database. The task of analyzing the images can be tedious and is subjective if performed manually.

The most important issues and challenges related to digital image analysis of the gel images are segmentation of the images and the matching the corresponding protein spots. The segmentation of the images is beyond the scope of this dissertation. We will analyze the current matching methods used and will propose a new method.

2DGE enables separation of mixtures of proteins due to differences in their isoelectric points (pI) in the first dimension and, subsequently, by their molecular weight (MWt) in the second dimension (see Berth et al. (2007) [22]).

The change in protein expression, for example in the development of cancer, are subtle: a change in the expression level of a protein of a factor 10 is rare and a factor 5 is uncommon. Furthermore, few proteins change; usually fewer than 200 proteins out of 15,000 would be expected to change by more than a factor of 2.5. Multiple samples need to analyzed because of the natural varia-

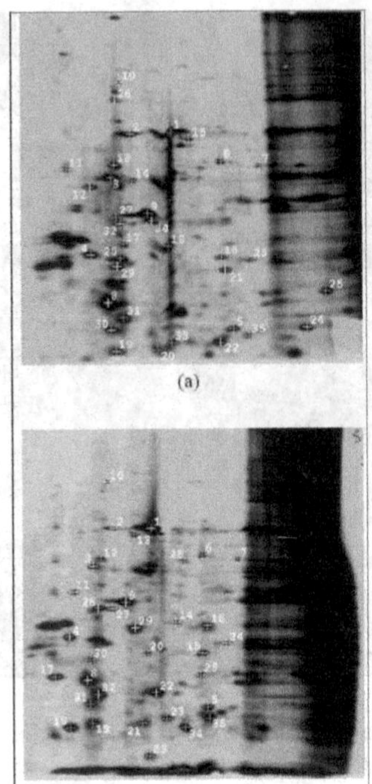

Figure 20.6: *2D gels.*

tion between individuals. Therefore, it is necessary to be able to rely on perfect matching of patterns of the new images.

20.4.2 The Laboratory Process

Even though this book is in the area of statistics, how gel electrophoresis is done in a laboratory is of interest in order to understand the mathematics behind the process. Because of this, we will describe the basic steps for the laboratory process, as it is practiced in CPA, of going from a biological sample of living cells, such as from a biopsy or blood sample, to separated proteins on a gel. Because some steps and details have necessarily been omitted, please refer to Berth et al. (2007) [22] for a detailed description of the procedure.

The procedure described here uses radioactive labeling, IPG for the first dimension, SDS polyacrylamide gels for the second dimension, and phosphor imaging to capture digital images of the protein patterns.

Labeling: A radioactive amino acid is "fed" to the living cells and all the proteins synthesized *de novo* may then contain the radioactive amino acid [^{35}S]-methionine in place of the non-radioactive one. The radioisotope used for the labeling is typically [^{35}S], but other radioisotopes, e.g. [^{32}P] or [^{14}C] can also be used. The radioactive labeling enables detection of the proteins later on. he usual labeling time interval used is 20 hours, but this can be changed for specific purposes or situations.

Solubilization: The cells' structures are broken down and the proteins are dissolved in a detergent lysis buffer. The lysis buffer contains urea, thiourea, detergent (NP40 or CHAPS), ampholytes, and dithiothreitol. All of these have the purpose of dissolving the proteins, unfolding them and preventing proteolysis. The actual procedure used depends on the sample itself and can take from less than 1 minute to 2 days.

1st dimension - isoelectric focusing: On an immobilized pH gradient (IPG) gel, in a glass tube or on a plastic strip, the proteins are separated according to their isoelectric point (pI). An electric field is applied across the gel and the charged proteins start to migrate into the gel. The proteins are differently charged and the electric field will pull them to the point where the pH cast into the IPG gel is the same as the pI of the protein, i.e., the pH value at which the number of positive and negative charges on the protein are the same. At this point, no net electrical force is pulling the protein (See Figure 20.6). Eventually, all proteins will have migrated to their pI state of equilibrium. This process usually lasts between 8–48 hours, depending on the pH range of the IPG gel. For example, it might require 17.5 hours for IPG pH in the range from 4 to 7.

Incubation: In the incubation step, the 1st dimension gel is washed in a detergent ensuring (virtually) the same charge on all proteins per unit length. Proteins are linear chains of amino acids that fold up and can be cross-linked by disulphide bridges. The solutions that are used at CPA contain urea, thiourea and detergents which cause the proteins to unfold into long random-coil chains. This portion of the procedure takes roughly 30 minutes.

2nd dimension - Molecular weight (MWt) separation: The incubated 1st dimensional gel strip is positioned on the upper edge of a polyacrylamide gel slab (See Figure 20.6). The second dimension acts like a molecular sieve so that the small molecules can pass more quickly than the large. Again, an electrical field is applied, but this time in the perpendicular direction, and proteins migrate into the gel. As all proteins have the same charge per unit length now, the same electrical force is pulling them. However, small proteins meet less obstruction in the gel and will migrate with higher velocity through the gel. The larger proteins meet more resistance and migrate slower. Proteins with the same pI will migrate in the same column but will now be separated by molecular weight (MWt). As opposed to the 1st dimension process, the 2nd dimension has no equilibrium state because the proteins keep moving as long as the electric field is applied. The small proteins reach bottom of the gel first and the process has to be halted before they migrate out of the bottom of the gel. This portion of the procedure requires approximately 16 hours.

Drying, etc: The gel is dried on paper support requiring some manual handling, which takes roughly 20 minutes.
Image generation: The dry gel is put in contact with a phosphor plate which is sensitive to emissions. The radiation from the labeled proteins excites the electrons of rare earth atoms in the plate at positions where there is protein present in the gel. The larger amount of protein present at a specific location in the gel, the more electrons in the plate will be excited at that location. The amount of radioactive protein in the samples can be quite small at the picogram level, so the level of radiation is also small and the time required to expose the phosphor plate is long. After exposure, the phosphor plates are read using phosphor imaging technology where a laser beam excites the already excited electrons to an even higher energy state. The electrons return to their normal state while emitting electro-magnetic radiation (light). A CCD chip captures the light and a digital image is generated. The exposure time is usually 5 days, but the capturing of the image capture requires roughly only 1 minute.

20.4.3 Matching Landmarks on 2D-Gel Electrophoresis Image Data Using the Dimitric Distance on the Grassmannian

Here, we discuss the problem of matching proteins from two 2DGE images under the assumption that the image points are matched onto each other via a linear or an affine transformation. A data set consisting two images from the same protein sample consisting of $k = 35$ data points is used to illustrate the methodology discussed. In this data set, the first 10 points have already been already matched.

Let us suppose that two images of 2DGE are to be matched, and the images have been annotated with coordinates of k data pairs $\{x_i = (x_{i,1}, x_{i,2})\}_{i=1}^k$ and $\{y_i = (y_{i,1}, y_{i,2})\}_{i=1}^k$, respectively. Let us assume that the first m data points have been matched. We are seeking a permutation $\sigma \in S_{k-m}$ which yields a match for the remaining $k - m$ points.

Let us consider the group action, $\Sigma \times G \times \mathbb{R}^{2k} \to \mathbb{R}^{2k}$,

$$(\pi \times g).[x_1^T, \cdots, x_k^T]^T = [(x_\pi(1)g)^T, \cdots, (x_\pi(k)g)^T]^T.$$

Recall that, according to Theorem 3.5.1, the linear space $L\Sigma_{2,2}^k$ is the Grassmann manifold $G_2(\mathbb{R}^k)$. In the case when $G = GL(2, \mathbb{R})$, we may identify the G orbit of a point in $(\mathbb{R}^k)^2$ with the two dimensional subspace in $\mathbb{R}^k$ spanned by the vectors $[x_{1,1}, \cdots, x_{k,1}]^T$ and $[x_{1,2}, \cdots, x_{k,2}]^T$.

Now, if we want to considered only *unlabeled landmarks*, we may act by the permutation group Σ_k on this Grassmannian. Thus the matching problem in 2DGE amounts to finding a permutation in the Grassmann so that the distance between the two images, in the sense of the distance in the Grassmannian, is minimized. Let us recall that the distance $\rho(V, W)$ between points V and W of a Grassmann is naturally defined as the distance between the two linear operators corresponding to orthogonal projections in $\Re^k$ onto V and W respectively. Thus we have the following problem.

Given two subspaces V and W in $G_k(\mathbb{R}^m)$ and an integer $m < k$, we want to find the permutation π on the last $k-m$ elements of an ordered set of n elements, such that $\rho(\pi(V), W)$ is minimized.

Now, we describe a numerical scheme which is guaranteed to converge to the permutation that solves our problem. This algorithm appears to scale reasonably well with the size k of the problem. Though a careful analysis of the algorithm has not been done, those experiments that have been done

Let us begin with an orthogonal basis $\{v_1, v_2\}$ of V and $\{w_1, w_2\}$ of W respectively. When π is a permutation, let us denote $\pi(v_i)$ by v_i^π. Let P_V and P_W denotes orthogonal projection operators from $\Re^k$ onto V and W, respectively. Then we have,

$$\begin{aligned}\rho(V,W)^2 &= \|P_V - P_W\|^2 \\ &= \|P_V\|^2 + \|P_W\|^2 - 2 < P_V, P_W > \\ &= 4+4-2\text{trace}(\text{P}_\text{V}\text{P}_\text{W}).\end{aligned}$$

Thus, we are searching for a permutation $\pi*$ such that $Tr(P_{\pi*(V)}P_W) \geq Tr(P_\pi(V)P_W)$ for all allowable permutations π. Since $P_V = [v_1, v_2][v_1, v_2]^T$, we may state

$$TrP_{\pi(V)}P_W = \sum_{i,j=1,2} [(v_i^\pi)^T w_j]^2. \tag{20.3}$$

A straightforward search over all permutations isn't possible since it will involve examining $(k-m)!$ rearrangements. The algorithm proposed by Sughatadasa (2006) [322] is as follows.

1. For i=m+1:k-1
   (a) Set π to be the permutation which transposes $m+1$ and i.
   (b) Use formula (20.3) to find i that maximize $Tr(P_{\pi(V)}P_W) - Tr(P_VP_W)$.
   (c) Interchange rows $m+1$ and i of the vectors v_1 and v_2.
2. Repeat step 1 until no i can be found.
3. Repeat steps 1 and 2 with $m+1$ replaced by $m+2, m+3, \cdots, k-1$.

Observe that each step terminates with a larger value of $Tr(P_{\pi(V)}P_W)$ than before, hence the algorithm proceeds in the right direction. However, due the quadratic nature of the formula (20.3), it may run into a situation in which no transposition will improve the error, yet there are still acceptable permutations that will improve the error. The way out of this is to use a three element permutation to get started again. This follows from the following

LEMMA 20.4.1. *Suppose that there exists a permutation $\pi \in \text{id} \times S_{k-m}$ such that $Tr(P_{\pi(V)}P_W) - Tr(P_VP_W) > 0$. Then, for generic subspaces V and W, π may be found so that it only permutes three elements.*

The key idea is to notice that any permutation may be written as a linear combination of permutations involving only three elements at a time in such a way that the coefficients in the linear combination are nonnegative. This follows by dimension counting. Therefore, if a permutation exists that decrease the matching error, then at least one of the three element permutation also must decrease the error.

20.4.4 Results

The following data set presented by Horgan (1992) [160] was tested using the numerical scheme. For an additional results obtained from a larger data set, we send the reader to Sughatadasa (2006) [322] and the references therein.

Method	Error
Ignore expert match, a match for top 20 matches	0.0903
Holding the top 10 matched by experts and matching additional 10	0.1316
Holding the top 10 matched by experts and matching additional 25	0.1878
Ignore expert match, a match for all 35	0.1401
Ignore expert match, a match for top 10	0.0387
Top 10 matched by experts	0.0927

Chapter 21

Projective Shape Analysis of Planar Contours

21.1 Introduction

In this chapter, the problem of identifying the projective shape of a planar curve is considered as a practical application of the neighborhood hypothesis test presented in Chapter 11, a data-driven example where we determine δ, the smallest radius of the neighborhood hypothesis for which the neighborhood hypothesis is not rejected. The theory is applied to the recognition of the projective shape of a planar curve extracted from digital images of a flat scene.

21.2 Hilbert Space Representations of Projective Shapes of Planar Curves

Planar projective shape analysis of a planar scene is concerned with understanding the scene modulo projective transformations. As described in Ma et al. (2005) [225], well-focused digital camera images may be assumed to have come from an ideal pinhole camera. Recall from Chapter 3 that the real projective plane $\mathbb{R}P^2$ is a geometric model for the set of axes in $\mathbb{R}^3$. This space also models the pinhole camera view. Assume the pinhole is the origin 0 in the ambient Euclidean space $\mathbb{R}^3$. The photographer can take a picture of any point x except of 0. Thus, two points in space yield the same projective point on the film if they are collinear with the origin. The projective point associated with $u \in \mathbb{R}^3\backslash 0$ is $p = [u] = (\mathbb{R}\backslash 0)$. If $u = (x,y,1)^T$ and $u' = (x',y',1)^T$ differ by a projective transformation of the form shown in equation (21.1), then $[u'] = [Au] = \alpha([u])$.

Two images of the same planar scene differ by a composition of two central projections in $\mathbb{R}^3$ from an observed plane to a receiving plane (such as film, retina, etc.) Such a map is therefore a projective transformation, which depends on a nonsingular matrix matrix $A = (a_{ij})$, and is given by

$$x' = \frac{a_{11}x + a_{12}y + a_{10}}{a_{01}x + a_{02}y + a_{00}}$$
$$y' = \frac{a_{21}x + a_{22}y + a_{20}}{a_{01}x + a_{02}y + a_{00}} \tag{21.1}$$

In m dimensions, a projective transformation has the equations $y = f(x)$, with

$$y^j = \frac{\sum_{i=0}^m a_i^j x^i}{\sum_i a_0^j x^i}, \forall j = 1, \dots, m \tag{21.2}$$

where $\det((a_i^j)_{i,j=0,\dots,m}) \neq 0$.

From Subsection 3.5.8, recall that two configurations of points in $\mathbb{R}^m$ have the same the projective shape if they differ by a projective transformation of $\mathbb{R}^m$. Unlike similarities or affine transformations, projective transformations of $\mathbb{R}^m$ do not have a group structure under composition. This is because the domain of definition of a projective transformation depends on the transformation and the maximal domain of a composition has to be restricted accordingly. To avoid such unwanted situations, rather than considering projective shapes of configurations in $\mathbb{R}^m$, one may consider configurations in $\mathbb{R}P^m$ with the projective general linear group action that is described in Subsection 3.5.8. The resulting group action of the projective general linear group $PGL(m,\mathbb{R})$ as a group of transformations of $\mathbb{R}P^m$ is described in Subsection 3.5.8.

Scientists are looking for new computational algorithms, including statistical methods, to deal with digital imaging libraries. Images of approximately planar scenes are very common and they need to be analyzed in their full complexity. Until Munk et al. (2008) [252], only finite configurations were analyzed even though the actual scenes are more complex, often depticting curves and regions bounded by these curves. A toy example of such images, from the so-called "BigFoot" data set is displayed in Chapter 1 in Figure 1.11.

REMARK 21.2.1. *The study of* direct similarity shapes of planar contours *under the group of direct similarities of the Euclidean plane was preceded by that of data analysis of projective shapes of contours, given that data analysis on a a Hilbert space (see Section 2.15) is more elementary than data analysis on a Hilbert manifold (see Chapter 11).*

Our approach to projective shapes of planar closed curves is based on the idea of registration with respect to a projective frame. To keep things simple, assume that, in addition to a closed planar curve, four labeled control points, that yield a projective frame are also known. Two closed curves have the same projective shape if they differ by a planar projective transformation that brings the projective frame in the first configuration into coincidence with the projective frame in the second configuration.

REMARK 21.2.2. *In the context of scene recognition, the frame assumption is natural, given that a scene pictured may contain more information than a curved contour, including featured landmarks that can be spotted in different images of such a scene.*

Assume $x_1,\dots,x_{m+2}$ are points in general position and $x = (x^1,\dots,x^m)$ is an arbitrary point in $\mathbb{R}^m$. Note that, in our notation, the superscripts are reserved for the components of a point, whereas the subscripts are for the labels of points. In order to determine the projective coordinates of $p = [x : 1]$ w.r.t. the projective frame associated with $(x_1,...,x_{m+2})$, we set $\tilde{x} = (x^1,...,x^m,1)^T$ and consider the $(m+1)\times(m+1)$ matrix $U_m = [\tilde{x}_1,...,\tilde{x}_{m+1}]$, whose jth column is $\tilde{x}_j = (x_j,1)^T$, for $j = 1,\dots,m+1$ (see Remark 3.5.2). Also, recall from Subsection 3.5.8 that, if we define an intermediate system of homogeneous coordinates

$$v(x) = U_m^{-1}\tilde{x} \tag{21.3}$$

and write $v(x) = (v^1(x),\dots,v^{m+1}(x))^T$, then we can set

$$z^j(x) = \frac{v^j(x)}{v^j(x_{m+2})}/||\frac{v^j(x)}{v^j(x_{m+2})}||,\ j = 1,...,m+1 \tag{21.4}$$

so that the last point x_{m+2} is now used. Finally, the projective coordinate(s) of x are given by the point $[z^1(x):....:z^{m+1}(x)]$, where $(z^1(x))^2+....+(z^{m+1}(x))^2 = 1$. If $z^{m+1}(x) \neq 0$, the affine representative of this point with respect to the last coordinate is $(\xi^1(x),...,\xi^m(x))$, where

$$\xi^j(x) = \frac{z^j(x)}{z^{m+1}(x)},\ j = 1,...,m. \tag{21.5}$$

Assume $x(t), t \in I$, is a curve in $\mathbb{R}^m$, such that $\forall t \in I, z^{m+1}(x(t)) \neq 0$. Such curves will be said to be in a *convenient position* relative to the projective frame π associated with $(x_1,...,x_{m+2})$.

THEOREM 21.2.1. *There is a one to one correspondence between the set of projective shapes of curves $x(t), t \in I$, in a convenient position relative to π and curves in $\mathbb{R}^m$. In this correspondence, closed curves in a convenient position relative to π correspond to closed curves in $\mathbb{R}^m$.*

We will use the representation Theorem 21.2.1 for projective shapes of closed curves in the projective space that avoid a hyperplane. These correspond to closed curves in the Euclidean space. In particular, in two dimensions, we consider closed curves in the planar projective plane, avoiding a projective line. If we assume that $(x(t), y(t)), t \in [0,1]$ is a closed planar curve, then $[x(t) : y(t) : 1], t \in [0,1]$ is such a projective curve. Using a projective frame π, we associate with this curve the affine affine representative $(\xi(t), \eta(t)), t \in [0,1]$ of its curve of projective coordinates $[x(t) : y(t) : 1]^{\pi}$, which yield another planar curve. If two curves are obtained from a planar curve viewed from different perspective points, then the associated affine curves are the same. This affine representative of the projective curve of a (closed) curve is used in the rmainder of this chapter. Here, we are concerned with recognition of closed curves. That is, we have a closed curve

$$\gamma(t) = (\xi(t), \eta(t)), t \in [0,1], (\xi(0), \eta(0)) = (\xi(1), \eta(1)) \tag{21.6}$$

that is observed with random errors

$$\Gamma(t) = (\xi(t), \eta(t)) + (\varepsilon^X(t), \varepsilon^Y(t)), t \in [0,1], \tag{21.7}$$

where $\varepsilon^X(t), \varepsilon^Y(t)$ are stochastically independent error processes with the restriction that $(\varepsilon^X(0), \varepsilon^Y(0)) = (\varepsilon^X(1), \varepsilon^Y(1))$. As such, the observed curve can, for instance, be considered as a random element in the Hilbert space $\mathbb{H} = L^2(S^1, \mathbb{R}^2)$.

21.3 The One-Sample Problem for Mean Projective Shapes of Planar Curves

The idea of Hilbert space representation of the projective shape with respect to a projective frame is summarized in Theorem 21.2.1. To identify the mean projective shape of a curve, one may now use the neighborhood hypothesis test described in Chapter 11.

This is because, in practice, two curves will not have exactly the same shape even if they should agree according to some theory. In this case, therefore, using the neighborhood hypothesis that states approximate equality of the shapes of the curves seems appropriate.

In our case, the neighborhood hypothesis to be tested is $H_\delta : \mu \in \gamma_0 + B_\delta$, for some $\delta > 0$. In this case, the linear subspace M is the trivial subspace, which is the infinite dimensional analog of the classical null hypothesis $H_0 : \mu = \gamma_0$. The constant $\delta > 0$ in (2.147) is to be determined from the data. We consider an application of this technique to the "Bigfoot" data set and a new image not necessarily belonging to this data set.

21.3.1 Image Processing and Shape Registration Using the Projective Frame

Before the hypothesis test could be performed, however, the data needed to be put into the proper form. Image processing was performed using the Image Processing Toolbox in MATLAB 7.1. and Microsoft©Paint 5.1. After processing the images using the projective coordinates w.r.t. the selected projective frame, the image was registered and the contour of the foot was extracted using the Sobel edge detector. Finally, locations of pixls were put in clockwise order so as to define a piecewise-linear projective curve. The M-file used for this was written by Alister Fong and is available for download from the MATLAB Central File Exchange website.

As an example, Figure 21.1 shows one observation in its original and processed forms. Note that the ordering of pixel locations, in effect, rotates the curve by 90° since the first point on the projective curve corresponds to the first pixel location (when moving from top to bottom and from left to right) lying on the curve in the original image.

The projective curves for the entire sample of "BigFoot" images, as well as their sample mean, are shown in Figure 21.2.

21.3.2 Hypothesis Testing

One of the classical problems in pattern recognition is the identification of a scene for which previous information is known. One typical problem in this area arises when researchers have a number of images from a known planar scene and acquire a new image that may also depict the same scene. In this scenario, the researchers wish to identify whether the new image does, in fact, depict that scene.

As an example, the observations in 1.11 are known to belong to the "BigFoot" data set. An additional image is shown in Figure 21.3. We can use the neighborhood hypothesis test to conclude whether it depicts the same scene as the "BigFoot" data set.

Using the same image processing techniques as described in the previous section, a curve γ_0 was extracted from the new image. This curve is displayed with the sample mean of the "BigFoot" data in Figure 21.4. As a method to explore the feasibility that γ_0 is from the same data set as the other observations, we can consider the null hypothesis $H_0 : \mu \in \gamma_0 + B_\delta$, for some $\delta > 0$, which is equivalent to (2.147).

Similarly to the cases considered in Chapter 18, we can use the notion of a p-value curve to identify the maximum value of δ for which we would reject the null hypothesis. Here, it was determined that if $\delta < 10.7151$, we would reject the approximate equality of the mean projective shape of the "BigFoot" population of curves and the projective shape of the new curve. As this value of δ is quite large, the evidence suggests that the projective mean shape of the first "BigFoot" planar scene is significantly different from that of the curve in Figure 21.3. As a result, it appears that γ_0 depicts a different scene than the original data set.

It should be noted that, since the sample size was small, nonparametric bootstrap could have been used in place of the asymptotic results. Nevertheless, the

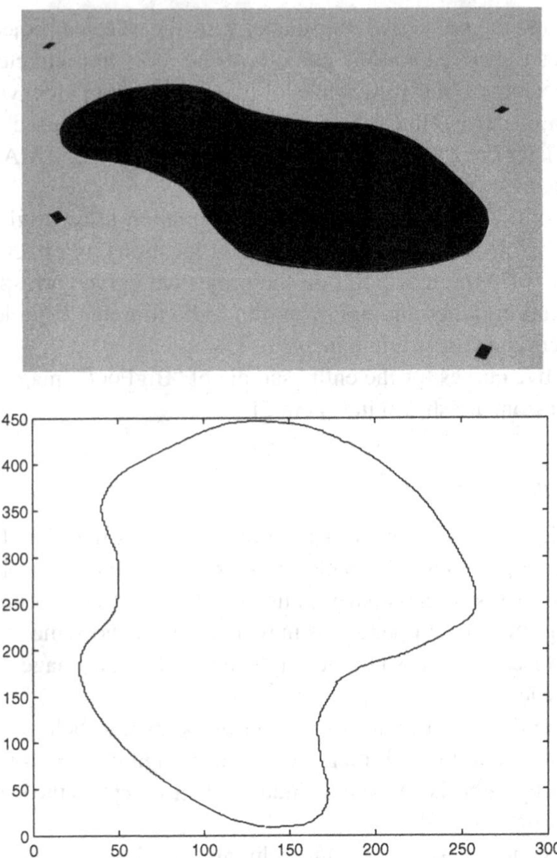

Figure 21.1 *The first "Bigfoot" image in its original form (top) and final processed form (bottom). (Source: Munk et al.(2008), Figure 4 and Figure 5. Reproduced by permission of Elsevier).*

errors are quite insignificant since they depend only on the pose of the scene, which is essentially flat. As such, even for this fairly small sample, the result is reliable.

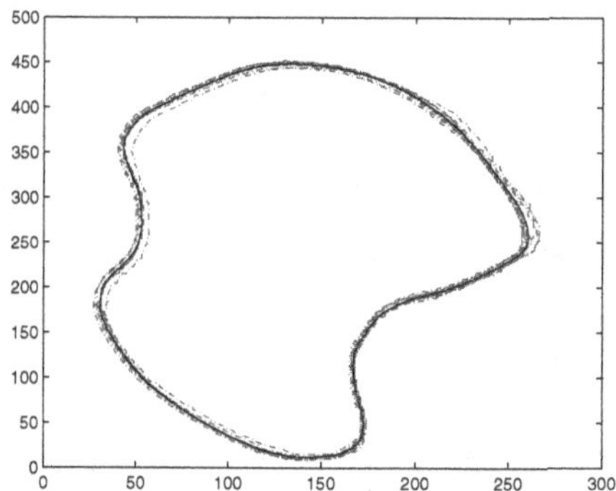

Figure 21.2 *Empirical sample mean of the observations from the first group. (Source: Munk et al.(2008), Figure 7. Reproduced by permission of Elsevier).*

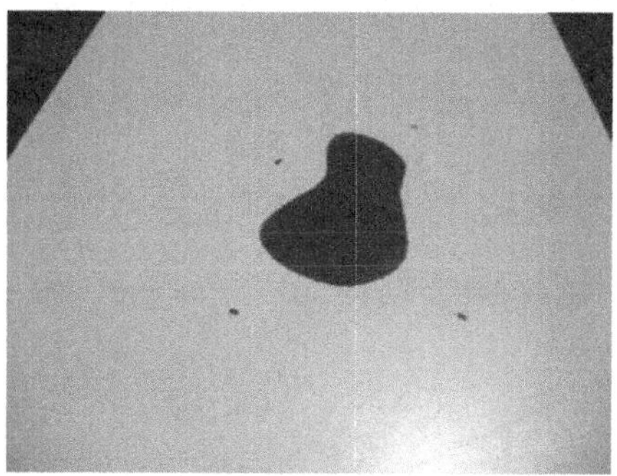

Figure 21.3 *The new image, which includes a natural projective frame and curve, that depicts an unknown scene. (Source: Munk et al.(2008), Figure 8. Reproduced by permission of Elsevier).*

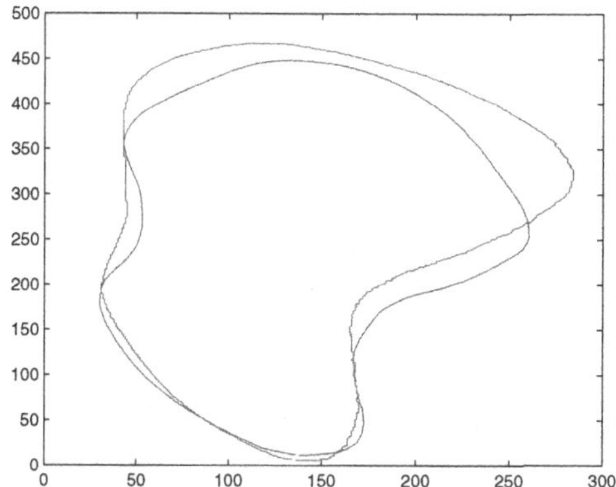

Figure 21.4 *Plot of the mean "Bigfoot" curve and the new image curve. (Source: Munk et al.(2008), Figure 9. Reproduced by permission of Elsevier).*

Chapter 22

Projective Shape Analysis of 3D Scenes from Pinhole Camera Images

22.1 Introduction

Until recently, statistical analysis of similarity shape from images was restricted to a small amount of data since the appearance of similarity shape is relative to the camera position with respect to the scene pictured. In this chapter, we study the shape of a 3D configuration from pictures of this configuration in 2d images without requiring any restriction for the camera positioning respect to the scene pictured. This methodology uses standard reconstruction methods from computer vision. In absence of occlusions, a set of point correspondences in two views can be used to retrieve the 3D configuration of points. A key result due to Faugeras (1992) [111] and Hartley, Gupta, and Chang (1992) [143] states that two such reconstructions differ by a projective transformation in 3D. Sughatadasa (2006) [322] noticed that the object that is recovered without ambiguity is actually the projective shape of the configuration. This cast a new light on the role of projective shape in the identification of a spatial configuration.

Projective shape is a natural approach to shape analysis from digital images since the vast majority of libraries of images are acquired via a central projection from the scene pictured to the black box recording plane. Hartley and Zisserman (2004 [141], p.1) noted that "this often renders classical shape analysis of a spatial scene impossible, since similarity is not preserved when a camera is moving." Advances in statistical analysis of projective shape have been slowed down due to overemphasized importance of similarity shape in image analysis while ignoring the basic principle of image acquisition. Progress was also hampered by the lack of a geometric model for the space of projective shapes, and ultimately by insufficient dialogue between researchers in geometry, computer vision and statistical shape analysis.

For the reasons presented above, projective shapes have been studied only recently, starting with Sughatadasa (2006) [322]. Examples of concrete 3D projective shape analysis are considered in Patrangenaru et al. (2010) [272], Crane and Patrangenaru [78], Qiu et al. (2014) [277], and recent conference presentations that are not yet in refereed publications. However, the remainder of the projective shape literature, including Maybank (1994) [238], Goodall and Mardia (1999) [127], Patrangenaru (2001) [268], Paige et al. (2005) [261], Mardia and Patrangenaru (2005) [233], Kent and Mardia (2006 [184], 2013 [185]) and Munk et al. (2008)[252] has considered linear or planar projective shape data analysis.

Our main goal here is to derive a natural concept of 3D shape that can be extracted from data recorded from regular camera images. This statistical methodology for estimation of a mean 3D projective shape is based on Efron's bootstrap (see Efron [96]). In this chapter, a 3D projective shape is regarded as a random object on a projective shape space. Since sample size for images is typically small, in order to estimate the mean projective shape we use nonparametric bootstrap for the studentized sample mean projective shape on a manifold, as shown in Chapter 5. In the case of projective shapes, the bootstrap distribution was essentially derived in Mardia and Patrangenaru (2005) [233] (also see Chapter 10).

The chapter is organized as follows. Section 22.2 includes a test for coplanarity of a 3D scene. This test is based on the asymptotic distribution of the total sample variance developed in Chapter 5, which is applied to 2D projective shapes. Section 22.3 is devoted to the 3D reconstruction algorithm. Here, we introduce the essential matrix and, respectively, the fundamental matrix, which are associated with a pair of camera views of a 3D scene that is needed in the reconstruction of that scene from both 2D calibrated and noncalibrated camera images. For reconstruction of a configuration of points in space from its views in a pair of images, we refer to the computational algorithms in Ma et al. (2006) [225]. In Section 22.4, we reformulate the Faugeras–Hartley–Chang–

Gupta result on the ambiguity of 3D reconstruction of a scene up to a projective transformation in Theorem 22.3.2. This result opens the statistical shape analysis door to computer vision and pattern recognition of 3D scenes, including face analysis. In Section 22.5 we consider a toy model example for 3D projective shape reconstruction and an empirical principle based example of face identification.

22.2 Test for Coplanarity

Projective shape analysis is concerned with the properties of a configuration of collinear or coplanar points, as they are seen in a central projection by an external observer in what can be considered a simplified analysis of vision in absence of occlusions.

Recall that the projective coordinates p_j^π and the *projective invariants* $(\iota_{sj}), s = 1,...,m, j = m+3,...,k$ of a projective shape due to Goodall and Mardia (1999) [127] determine each other. The relations between the two representations of a projective shape are:

$$p_j^\pi = [\iota_{1j} : ... : \iota_{mj} : 1], \tag{22.1}$$

showing that these invariants are locally the affine coordinates of $(p_j^\pi, j = m+3,...,k)$ whenever the last homogeneous coordinate of each of the $p_j^\pi{}'s$ is not zero.

If no assumption is made on the probability distribution Q on a projective shape space, it is appropriate to use a location parameter, such as a Fréchet mean, associated with the distance induced on $P\Sigma_m^k$.

In 3D projective shape data, sometimes the landmarks pictured are sitting on an almost flat surface. In such a situation, when one registers the landmark coordinates in a configuration extracted from a digital camera, there is little variability in the 2D projective shape of that configuration, which makes it appropriate to regard such observations as coming from a probability distributions of small flat support on $(\mathbb{R}P^2)^{k-4}$.

In this section, we confine ourselves only to probability measures Q with a small flat support on our space. The advantage is that in this case we can use the technique in Section 2. Since any flat metric is locally isometric to an Euclidean metric, we consider a metric on $(\mathbb{R}P^2)^{k-4}$ which is Euclidean in the invariant coordinates on the support of Q, and in particular if x_1, x_2 are two observations from Q with

$$x_r = ([\iota_{r,1j} : ... : \iota_{r,mj} : 1])_{j=m+3,...,k}, \tag{22.2}$$

their distance is

$$d^2(x_1,x_2) = \sum_{j=m+3}^{k} \sum_{s=1}^{m} (\iota_{1,sj} - \iota_{2,sj})^2 \tag{22.3}$$

In this case, if we use the invariant representation for random configurations of k landmarks, the intrinsic sample means and total sample variance are ordinary sample means and total sample variances, for which classical large sample and nonparametric bootstrap theory can be used. This approach requires less computational complexity than for extrinsic means.

Assume $X_1,\ldots,X_n$ are i.i.d.r.v.'s with a common distribution Q of small flat support on $P\Sigma_m^k$. For each $r, r=1,\ldots,n$, assume X_r has the invariant representation

$$X_r = (p_{r,j}^{\pi}), j = m+3,\ldots,k, \tag{22.4}$$

where

$$p_{r,j}^{\pi} = [\iota_{r,1j} : \ldots : \iota_{r,mj} : 1] \tag{22.5}$$

Then the intrinsic sample mean $\overline{X}_{n,I}$ is given in invariant representation by

$$\overline{X}_{n,I} = (p_j^{\pi}), j = m+3,\ldots,k, \tag{22.6}$$

where

$$p_j^{\pi} = [\overline{\iota_{.,1j}} : \ldots : \overline{\iota_{.,mj}} : 1]. \tag{22.7}$$

[267] shows that if an external observer travels in the 3D space and registers images of a configuration of points with respect to a projective frame of marked coplanar points, then this configuration is planar if the projective coordinates are independent of the position of the observer. We assume $m = 2$ and $k = 5$. The corresponding hypothesis testing problem is

$$H_0 : t\Sigma_I = 0 \ vs. \ H_1 : t\Sigma_I \neq 0, \tag{22.8}$$

where $t\Sigma_I$ is the *intrinsic total variance* of a probability measure Q with small flat support on $P\Sigma_m^{m+3}$.

Note that $P\Sigma_m^{m+3}$ is diffeomorphic to $\mathbb{R}P^m$ and, in invariant representation, the random variable associated with Q is

$$X = [\iota_{,1} : \ldots : \iota_{,m} : 1]. \tag{22.9}$$

In this case, the intrinsic mean and intrinsic total variance of X are, respectively, the mean and total variance of the random vector $\iota = (\iota_{,1},\ldots,\iota_{,m})^T$. If $X_1,\ldots,X_n$ are i.i.d.r.v.'s with common distribution Q and we similarly associate X_r with the random vector $\iota_r = (\iota_{r,1},\ldots,\iota_{r,m})^T$, we have

$$\overline{X}_{n,I} = \overline{\iota}_n = (\overline{\iota}_{,1}\overline{\iota}_{,m}) \tag{22.10}$$

and

$$t\hat{\Sigma}_{I,n} = n^{-1}\sum_{j=1}^{m}\sum_{r=1}^{n}(\iota_{r,j} - \overline{\iota_{,j}})^2 \tag{22.11}$$

The value of the $\hat{\Sigma}(V)_4$ given in Corollary 5.2.1 is, in this case, given by

$$\hat{\Sigma}(V)_4 = n^{-1}\sum_{i,j=1}^{m}\sum_{r=1}^{n}(\iota_{r,i}-\overline{\iota_{,i}})^2(\iota_{r,j}-\overline{\iota_{,j}})^2 \tag{22.12}$$

As a consequence, we obtain the following result.

THEOREM 22.2.1. *Assume $x_1,\dots,x_n$ is a large random sample from a probability distribution with small flat support on $P\Sigma_m^{m+3}$. Then we reject the null hypothesis in equation (22.8) at level α if*

$$t\hat{\Sigma}_{I,n} > (S(\hat{\Sigma}(V)_4)^{1/2})z_{\frac{\alpha}{2}}. \tag{22.13}$$

Indeed, it follows that a $100(1-\alpha)\%$ symmetric large sample confidence interval for $t\Sigma_I$ is given by

$$(t\hat{\Sigma}_{i,n}-(S(\hat{\Sigma}(V)_4)^{1/2})z_{\frac{\alpha}{2}}, t\hat{\Sigma}_{i,n}+(S(\hat{\Sigma}(V)_4)^{1/2})z_{\frac{\alpha}{2}}) \tag{22.14}$$

and the condition in the theorem says that 0 is not in this confidence region.

An example of application of Theorem 22.2.1 was given in Patrangenaru and Pruett (1999) [276], where a random sample of 41 scanned images of a large boulder where analyzed to determine whether the scene is actually flat. Five fixed landmarks were identified in each of these images and their coordinates were registered in Adobe Photoshop. One of the images is displayed in figure 1.10 and the recorded landmark coordinates are given in Table 1.20. Computations yielded the following values: $t\hat{\Sigma}_{I,n} = 0.58$ and $S(\hat{\Sigma}(V)_4) = 2.63$. According to Theorem 22.2.1, we fail to reject the flatness of the scene at level $\alpha = 0.05$.

22.3 Basic Projective Geometry for Ideal Pinhole Camera Image Acquisition

Image acquisition from the 3D world to the 2D camera film is based on a central projection principle. Therefore, projective geometry governs the physics of ideal pinhole cameras. A point in the outer space and its central projection via the camera pinhole determine a unique line in space. As such, an image point captured on the camera film can be regarded as a line going through the pinhole, leading to the definition of the real projective plane $\mathbb{R}P^2$ as the space of all lines going through the origin of $\mathbb{R}^3$. Projective geometry also provides a logical justification for the mental reconstruction of a spatial scene from binocular retinal images, playing a central role in vision. In subsection 3.5.8, we reviewed some of the basics of projective geometry that are useful in understanding of image formation and $3D$ scene retrieval from a pair of ideal pinhole camera images.

Consider a matrix $B \in M(m+1, m'+1;\mathbb{R})$ and the linear subspace $K = \{x \in$

$\mathbb{R}^{m'+1}, Bx = 0\}$ of $\mathbb{R}^{m'}$. The *projective map* $\beta : \mathbb{R}P^{m'} \backslash P(K) \to \mathbb{R}P^m$ *associated with* B is defined by $\beta([x]) = [Bx]$. In particular, a *projective transformation* β of $\mathbb{R}P^m$, considered in Subsection 3.5.8, is the projective map associated with a nonsingular matrix $B \in GL(m+1,\mathbb{R})$ and its action on $\mathbb{R}P^m$:

$$\beta([x^1 : \cdots : x^{m+1}]) = [B(x^1,\ldots,x^{m+1})^T]. \tag{22.15}$$

In this chapter, we will use the projective frame approach to projective shape analysis, as given in Subsection 3.5.8. Given a k-ad $u_1,\ldots,u_k$ in general position in $\mathbb{R}^m$, for $k > m+2$, such that $\pi = ([\tilde{u}_1],\ldots,[\tilde{u}_{m+2}])$ is a projective frame, from Remark 3.5.3, the projective shape of the k-ad $([\tilde{u}_1],\ldots,[\tilde{u}_k])$ is uniquely determined by the projective coordinates of the last $k-m-2$ labeled points from k-ad, with respect to π, where the projective coordinate of a projective point w.r.t. of projective frame is given by (3.145). Ideal pinhole camera image acquisition can be thought of in terms of a central projection β from $\mathbb{R}P^3 \backslash \mathbb{R}P^2$ to $\mathbb{R}P^2$, whose representation in conveniently selected affine coordinates $(x,y,z) \in \mathbb{R}^3$, $(u,v) \in \mathbb{R}^2$ is given by

$$\begin{aligned} u &= -f\frac{x}{z} \\ v &= -f\frac{y}{z}, \end{aligned} \tag{22.16}$$

where f is the *focal length*, which is the distance from the *image sensor* or film to the pinhole or *principal plane of the lens* $\mathbb{R}P^2$, complement of the domain of β. In homogeneous coordinates $[x:y:z:w], [u:v:t]$ the *perspective projective map* β can be represented by the matrix $B \in M(3,4;\mathbb{R})$ given by:

$$B = \begin{pmatrix} -f & 0 & 0 & 0 \\ 0 & -f & 0 & 0 \\ 0 & 0 & 1 & 0 \end{pmatrix}. \tag{22.17}$$

In addition to the projective map (22.17), image formation in digital cameras assumes a composition with matrices accounting for camera internal calibration parameters, such as the *pixel aspect ratio, skew parameter, origin of image coordinates in the principal plane (principal point)* and for a change of coordinates between two camera positions involving a roto-translation $(R,t) \in SO(3) \times \mathbb{R}^3$. The projective map of pinhole camera image acquisition $\tilde{\pi}$, in homogeneous coordinates, is associated with the matrix:

$$\tilde{B} = C_{\text{int}}BE = \begin{pmatrix} k_u & k_c & u_0 \\ 0 & k_v & v_0 \\ 0 & 0 & 1 \end{pmatrix} \begin{pmatrix} -f & 0 & 0 & 0 \\ 0 & -f & 0 & 0 \\ 0 & 0 & 1 & 0 \end{pmatrix} \begin{pmatrix} R & t \\ 0_3^T & 1 \end{pmatrix} = NE, \tag{22.18}$$

where k_u and k_v are scale factors of the image plane in units of the focal length f, and $\theta = cot^{-1}k_c$ is the skew, (u_0,v_0) is the *principal point*. The matrix N

contains the internal parameters and the perspective map (22.17), while E contains the external parameters. The matrix $\tilde{B}$ can be decomposed into a 3×3 matrix P and a 3×1 vector p

$$\tilde{B} = (P\ p) \tag{22.19}$$

so that

$$P = AR \text{ and } p = At. \tag{22.20}$$

22.3.1 *Essential and Fundamental Matrices*

We now consider a pair of cameras viewing a point $[u] \in \mathbb{R}P^3$. This point projects onto the two image planes to $m_1 = [u_1] \in \mathbb{R}P^2$ and, respectively, to $m_2 = [u_2] \in \mathbb{R}P^2$. Since we are working in homogeneous coordinates, $[u]$ is represented by a 4×1 column vector and m_1, m_2 are each represented by 3×1 column vectors. If we assume the camera's internal parameters are known (the camera is *calibrated),* then m_1, m_2 are each given with respect to its camera's coordinate frame, so $C_{\text{int}} = I_3$.

DEFINITION 22.3.1. *The epipolar constraint refers to the fact that the vector from the first camera's optical center to the first imaged point, the vector from the second optical center to the second imaged point, and the vector from one optical center to the other are all coplanar.*

If we use only one coordinate system, say the coordinate system of the second camera, the vector from the first camera's optical center to the first imaged point is $t + Ru_1$, the vector from the second optical center to the second imaged point is u_2, and the vector from one optical center to the other is t. Here the change of coordinates between the Euclidean frames of the two cameras is given by a roto-translation $(R,t) \in SO(3) \times \mathbb{R}^3$. The epipolar constraint can be expressed via a zero exterior product $u_2 \wedge (t + Ru_1) \wedge t = 0$, which is equivalent to

$$u_2^T(t \times (Ru_1)) = 0. \tag{22.21}$$

By defining $t_\times$ as the matrix associated with the linear operator $y \rightarrow t \times y$, we can rewrite the equation (22.21) as follows

$$u_2^T(t_\times Ru_1)) = u_2^T E u_1 = 0, \tag{22.22}$$

where $E = t_\times R$ is the so called *essential matrix.* If the cameras are uncalibrated, then the matrices A_1, A_2 from (22.20), which contain the internal parameters of the two cameras, are needed to transform the camera bound Euclidean coordinates into pixel coordinates:

$$\begin{aligned} v_1 &= A_1 u_1 \\ v_2 &= A_2 u_2. \end{aligned}$$

This yields the following equations:

$$(A_2^{-1}v_2)^T(T \times RA_1^{-1}v_1) = v_2^T A_2^{-1}(T \times RA_1^{-1}v_1) = 0, \tag{22.23}$$

and we obtain

$$v_2^T F v_1 = 0, \tag{22.24}$$

where $F = (A_2^{-1})^T E A_1^{-1}$ is the *fundamental matrix*. The fundamental matrix depends only on the relative position of the two cameras and their internal parameters. It has rank two and depends on seven real constants.

22.3.2 *Reconstruction of a 3D Scene from Two of its 2D Images.*

If we conveniently select the coordinates for the first camera, incorporating the internal parameters, we may assume that the matrix associated with $\tilde{\beta}_1$ in (22.18) is $B_1 = (I|0)$ and the fundamental matrix factors as $F = [t]_\times R$, with $B_2 = (R|t)$ corresponding to $\tilde{\beta}_2$, being a realization of the fundamental (or essential) matrix F (here, R is nonsingular and it does not necessarily represent the matrix of a rotation). Let $[u_1], [u_2] \in \mathbb{R}P^2$ be a pair of matched points in the two images. We seek a point $[u] \in \mathbb{R}P^3$ such that $[u_i] = \tilde{\beta}_i[u], i = 1, 2$. From the relation $u_2^T F u_1 = u_2^T t_\times R u_1 = u_2^T(t \times R u_1) = 0$, it follows that u_2, Ru_1, t are linearly dependent. We may assume that $Ru_1 = bu_2 - at$ and, since the position vector u_1 is defined up to a scalar multiple, we may assume that $Ru_1 = u_2 - at$ and define the corresponding landmark position $[u] \in \mathbb{R}P^3$ by $u = (u_1^T, a)^T$. Now, $B_1 u = (I|0)u = u_1$ and $B_2 u = (R|t)u = Ru_1 + at = u_2$, so if β_1, β_2 are the projections associated with B_1, B_2, it follows that $\beta_a([u]) = [u]_a$, for $a = 1, 2$ and $[u]$ is a desired solution to the reconstruction problem. As shown, $[u]$ is determined by the two camera projection matrices B_1 and B_2. If we choose a different pair of camera matrices $B_1 H$ and $B_2 H$ realizing the same fundamental matrix F, then in order to preserve the same pair of matched image points, the point $[u]$ must be replaced by $[H^{-1}u]$.

PROBLEM 22.3.1. *The problem of the reconstruction of a configuration of points in 3D from two ideal noncalibrated camera images with unknown camera parameters is equivalent to the following: given two camera images $\mathbb{R}P_1^2, \mathbb{R}P_2^2$ of unknown relative position and internal camera parameters and two matching sets of labeled points $\{p_{a,1}, \dots, p_{a,k}\} \subset \mathbb{R}P_a^2, a = 1, 2$, find all the sets of points in space $p_1, \dots, p_k$ in such that there exist two positions of the planes $\mathbb{R}P_1^2, \mathbb{R}P_2^2$ and internal parameters of the two cameras $c_a, a = 1, 2$ with the property that the c_a-image of p_j is $p_{a,j}, \forall a = 1, 2, j = 1, \dots, k$.*

The above discussion proves the following theorem (Faugeras (1992) [111], Hartley et al. (1992) [143]):

THEOREM 22.3.2. *The reconstruction problem for two non calibrated camera images has a solution in terms of the realization of the fundamental matrix*

$F = t_{\times}R$. Any two solutions can be obtained from each other by a projective transformation in $\mathbb{R}P^3$.

REMARK 22.3.1. *Note that, although the configurations in correspondence are finite, their size is arbitrarily large, and the assumption of finite matching labeled pairs can be replaced by an assumption of parameterized sets in correspondence. Therefore, in absence of occlusions, a 3D configuration can be reconstructed from 2D images and this reconstruction is unique up a projective transformation.*

22.3.3 Estimation of the Fundamental Matrix

Since equation (22.24) is homogeneous as a linear equation in F and F has rank two, this matrix depends on seven independent parameters. Therefore, in principle, F can be recovered from corresponding configurations of seven points. Due to the fact that the nature of digital imaging data is inherently discrete and other errors in landmark registration, F can be estimated using configurations of eight or more points $p_{a,i}, a = 1,2, i = 1,\dots k, k \geq 8$, whose stacked homogeneous coordinates are the $k \times 3$ matrices $y_a, a = 1,2$. The linear system for $\mathbf{F}$ is

$$y_2^T F y_1 = 0. \tag{22.25}$$

This can also be written as

$$f^T Y = 0, \tag{22.26}$$

where f is a vectorized form of F. If k is large, the linear homogeneous system is over-determined and the optimal estimated solution $\hat{f}$ can be obtained using simple least squares algorithm by minimizing $\|Y^T f\|^2$ subject to $\|f\| = 1$ (see Hartley and Zisserman (2004) [141], p. 593).

22.4 3D Reconstruction and Projective Shape

DEFINITION 22.4.1. *Two configurations of points in $\mathbb{R}^m$ have the same projective shape if they differ by a projective transformation of $\mathbb{R}^m$.*

Unlike similarities or affine transformations, projective transformations of $\mathbb{R}^m$ do not have a group structure under composition (the domain of definition of the composition of two such maps is smaller than the maximal domain of a projective transformation in $\mathbb{R}^m$.) To avoid this complication, rather than considering the projective shapes of configurations in $\mathbb{R}^m$, we consider projective shapes of configurations in $\mathbb{R}P^m$. A *projective shape* of a k-ad (a configuration of k landmarks or labeled points) is the orbit of that k-ad under projective transformations with respect to the diagonal action

$$\alpha_k(p_1,\dots,p_k) = (\alpha(p_1),\dots,\alpha(p_k)). \tag{22.27}$$

Since the action (3.142) of $\beta \in PGL(m)$ on $[x] \in \mathbb{R}P^m$, when expressed in inhomogeneous coordinates (3.141) reduces to (3.143), if two configurations Γ_1, Γ_2 of points in $\mathbb{R}^m$ have the same projective shape, then $h(\Gamma_1), h(\Gamma_2)$ have the same projective shape in $\mathbb{R}P^m$ (h is the affine embedding given by (3.140)).

Patrangenaru (1999 [267], 2001 [268]) considered the set $G(k,m)$ of k-ads $(p_1,...,p_k), k > m+2$ for which $\pi = (p_1,...,p_{m+2})$ is a projective frame. $PGL(m)$ acts simply transitively on $G(k,m)$ and the projective shape space $P\Sigma_m^k$, is the quotient $G(k,m)/PGL(m)$. Using the projective coordinates $(p_{m+3}^\pi, \ldots, p_k^\pi)$ given by (3.145), one can show that $P\Sigma_m^k$ is a manifold diffeomorphic with $(\mathbb{R}P^m)^{k-m-2}$. The projective frame representation is an alternative to the projective invariants-based representation, which was used previously for projective shape analysis by Goodall and Mardia (1999) [127].

The projective frame representation has two useful features. First, the projective shape space has a manifold structure, thus allowing to use the an asymptotic theory for means on manifolds in Bhattacharya and Patrangenaru (2003 [42], 2005 [43]). Secondly, it can be extended to infinite dimensional projective shape spaces, such as projective shapes of curves, as discussed in Chapter 21. This approach has the advantage of being inductive in the sense that each new landmark of a configuration adds an extra marginal axial coordinate, thus allowing to detect its overall contribution to the variability of the configuration and correlation to the other landmarks. The effect of change of projective coordinates due to projective frame selection can be understood via a group of projective transformations, but is beyond the scope of this chapter.

We return to the reconstruction of a spatial configuration. Having in view the definition 22.4.1 of a projective shape of a configuration, Theorem 22.3.2 can be stated as follows:

THEOREM 22.4.1. *A spatial $\mathcal{R}$ reconstruction of a 3D configuration $\mathcal{C}$ can be obtained in absence of occlusions from two of its ideal camera views. Any such 3D reconstruction $\mathcal{R}$ of $\mathcal{C}$, has the same projective shape as $\mathcal{C}$.*

REMARK 22.4.1. *Since the output in a reconstruction algorithm is a projective shape and multiplying by an imposed internal camera parameters matrix keeps the projective shape of the reconstruction unchanged, one may use the essential matrix estimate using the eight point algorithm (see Ma et al. (2006) [225], p. 121) for a conveniently selected internal parameters matrix. Refined eight point algorithms for the estimate $\hat{F}$ of the fundamental matrix, which can be found in Ma et al. (2006) [225] on pages 188 and 395, could also be used, given the projective equivalence of any two 3D reconstructions.*

22.5 Applications

22.5.1 *Estimation of the Mean 3D Projective Shape of a Polyhedral Surface from its Images*

Theorem 22.3.2 lays down the geometric principle of binocular vision that triggers 3D perception in combination with certain neurological mechanisms. The projective ambiguity of the reconstructed scene stated in that theorem was first recognized by Faugeras (1992) [111] and by Hartley et al. (1992) [143]. In this chapter, it is assumed that cameras are *non-calibrated,* meaning that images of the same scene are available and nothing is known about the cameras internal parameters that recorded these images or the camera's relative position. Since the reconstruction algorithms in computer vision were initially designed for calibrated cameras (see Longuet–Higgins (1981) [224]), the projective ambiguity of the reconstructed scene for non-calibrated cameras was perceived as a lack of information and research was directed mostly towards camera calibration.

REMARK 22.5.1. *Theorem 22.4.1 sheds a new light in pattern recognition of a 3D scene imaged by pairs of non-calibrated cameras since the projective shape of a reconstructed configuration does not depend on the reconstruction scheme. As such, the projective shape already provides useful information about the imaged 3D scene, an important point in scene identification that was previously ignored in the literature.*

Due to landmark registration, camera distortion or rounding errors in the reconstruction algorithms, 3D projective shapes from pairs of images can be regarded as random objects on the projective shape space $P\Sigma_3^k$. In our first example, we consider the 3D polyhedral object manufactured from three cubes based on a blueprint displayed on the left hand side of Figure 1.13 in Chapter 1. The object is manufactured from three cubes that sit on the top of each other, which should match the blueprint, whose sides from top to bottom are four, six and ten units.

To see if the projective shape of the object matches the original blueprint, one may take a number of random pictures of the object to estimate its 3D projective shape. We assume the faces of the object are flat and, consequently, its visible surface is determined by the visible corners. On the right hand side of Figure 1.13 is displayed a digital image of such an object with visible corners, taken as landmarks, numbered as specified in Section 1.3. The 2D coordinates of the landmarks selected are and listed in Table 1.22.

Camera images were organized as pairs, each of which were used to reconstruct a 3D object in $\mathbb{R}P^3$, such as the ones displayed in Figure 22.1. Using a reconstruction algorithm (see Ma et al.(2006) [225], p. 121 or p.188) for each pair of corresponding planar configurations, we obtain a $3D$ reconstructed configuration. Homogeneous coordinates of the seven reconstructed objects using the standard reconstruction algorithm are displayed in the Table 22.1 and the sample of 3*D* configurations of points that are joined to reassemble the original object is displayed in Figure 22.1.

Table 22.1: *Reconstructed Object Coordinates of the Eight Configurations*

Landmark No.		1	2	3	4	5	6	7	8	9	10
Reconstructed 3D Config 1	x	−0.09	−0.18	−0.08	0.01	−0.19	−0.09	0.00	−0.24	−0.08	0.05
	y	0.37	0.31	0.25	0.31	0.18	0.12	0.18	0.15	0.06	0.15
	z	2.51	2.44	2.27	2.36	2.53	2.35	2.44	2.49	2.23	2.35
	w	1.00	1.00	1.00	1.00	1.00	1.00	1.00	1.00	1.00	1.00

Landmark No.		11	12	13	14	15	16	17	18	19
Reconstructed 3D Config 1	x	−0.25	−0.09	0.04	−0.35	−0.09	0.13	−0.36	−0.10	0.14
	y	−0.04	−0.12	−0.03	−0.11	−0.22	−0.08	−0.45	−0.53	−0.38
	z	2.62	2.36	2.49	2.55	2.13	2.33	2.78	2.35	2.56
	w	1.00	1.00	1.00	1.00	1.00	1.00	1.00	1.00	1.00

Landmark No.		1	2	3	4	5	6	7	8	9	10
Reconstructed 3D Config 2	x	0.43	0.11	0.27	0.59	0.11	0.26	0.59	−0.05	0.17	0.66
	y	1.39	1.30	1.13	1.22	0.99	0.81	0.91	0.93	0.69	0.81
	z	6.89	6.78	6.52	6.65	6.90	6.64	6.79	6.76	6.47	6.68
	w	1.00	1.00	1.00	1.00	1.00	1.00	1.00	1.00	1.00	1.00

Landmark No.		11	12	13	14	15	16	17	18	19
Reconstructed 3D Config 2	x	−0.07	0.16	0.65	−0.38	−0.00	0.81	−0.39	−0.01	0.80
	y	0.47	0.23	0.35	0.39	−0.02	0.19	−0.36	−0.77	−0.58
	z	6.99	6.69	6.86	6.89	6.32	6.66	7.15	6.61	6.97
	w	1.00	1.00	1.00	1.00	1.00	1.00	1.00	1.00	1.00

Landmark No.		1	2	3	4	5	6	7	8	9	10
Reconstructed 3D Config 3	x	0.22	0.06	0.11	0.27	0.05	0.09	0.26	−0.03	0.03	0.27
	y	0.62	0.61	0.54	0.55	0.45	0.38	0.39	0.43	0.33	0.35
	z	3.28	3.30	3.16	3.16	3.30	3.17	3.15	3.30	3.10	3.08
	w	1.00	1.00	1.00	1.00	1.00	1.00	1.00	1.00	1.00	1.00

Landmark No.		11	12	13	14	15	16	17	18	19
Reconstructed 3D Config 3	x	−0.04	0.02	0.25	−0.21	−0.10	0.29	−0.22	−0.11	0.26
	y	0.20	0.10	0.13	0.18	0.01	0.06	−0.18	−0.35	−0.28
	z	3.30	3.09	3.07	3.29	2.96	2.93	3.31	2.96	2.90
	w	1.00	1.00	1.00	1.00	1.00	1.00	1.00	1.00	1.00

Landmark No.		1	2	3	4	5	6	7	8	9	10
Reconstructed 3D Config 4	x	0.00	−0.31	−0.11	0.20	−0.32	−0.11	0.20	−0.48	−0.16	0.29
	y	1.47	1.36	1.15	1.27	1.03	0.83	0.94	0.96	0.69	0.84
	z	8.10	7.90	7.38	7.61	8.17	7.61	7.80	8.09	7.29	7.56
	w	1.00	1.00	1.00	1.00	1.00	1.00	1.00	1.00	1.00	1.00

Landmark No.		11	12	13	14	15	16	17	18	19
Reconstructed 3D Config 4	x	−0.50	−0.18	0.30	−0.83	−0.27	0.47	−0.87	−0.28	0.51
	y	0.46	0.21	0.36	0.34	−0.06	0.19	−0.57	−0.89	−0.63
	z	8.55	7.70	7.98	8.37	7.10	7.43	9.23	7.69	8.11
	w	1.00	1.00	1.00	1.00	1.00	1.00	1.00	1.00	1.00

Landmark No.		1	2	3	4	5	6	7	8	9	10
Reconstructed 3D Config 5	x	−0.13	−0.43	−0.19	0.10	−0.46	−0.19	0.11	−0.62	−0.23	0.20
	y	1.72	1.69	1.43	1.50	1.36	1.11	1.17	1.30	0.99	1.07
	z	8.10	8.19	7.35	7.51	8.69	7.74	7.76	8.57	7.49	7.44
	w	1.00	1.00	1.00	1.00	1.00	1.00	1.00	1.00	1.00	1.00

Landmark No.		11	12	13	14	15	16	17	18	19
Reconstructed 3D Config 5	x	−0.68	−0.25	0.23	−1.06	−0.31	0.43	−1.19	−0.34	0.51
	y	0.77	0.47	0.59	0.67	0.24	0.42	−0.46	−0.81	−0.52
	z	9.48	8.14	8.22	9.45	7.62	7.54	11.02	9.01	8.88
	w	1.00	1.00	1.00	1.00	1.00	1.00	1.00	1.00	1.00

Landmark No.		1	2	3	4	5	6	7	8	9	10
Reconstructed 3D Config 6	x	0.00	-0.16	-0.07	0.10	-0.19	-0.09	0.07	-0.28	-0.13	0.12
	y	0.52	0.47	0.33	0.38	0.32	0.18	0.24	0.29	0.08	0.17
	z	2.99	2.98	2.83	2.85	3.12	2.97	2.99	3.12	2.89	2.92
	w	1.00	1.00	1.00	1.00	1.00	1.00	1.00	1.00	1.00	1.00

Landmark No.		11	12	13	14	15	16	17	18	19
Reconstructed 3D Config 6	x	-0.32	-0.17	0.08	-0.50	-0.25	0.17	-0.56	-0.31	0.12
	y	0.07	-0.13	-0.04	0.00	-0.33	-0.17	-0.36	-0.70	-0.51
	z	3.31	3.09	3.10	3.31	2.94	2.98	3.64	3.28	3.31
	w	1.00	1.00	1.00	1.00	1.00	1.00	1.00	1.00	1.00

Landmark No.		1	2	3	4	5	6	7	8	9	10
Reconstructed 3D Config 7	x	-0.01	-0.12	0.10	0.20	-0.17	0.05	0.15	-0.23	0.09	0.25
	y	0.80	0.70	0.58	0.68	0.47	0.36	0.45	0.41	0.25	0.40
	z	4.44	4.31	4.06	4.18	4.41	4.14	4.24	4.35	3.95	4.16
	w	1.00	1.00	1.00	1.00	1.00	1.00	1.00	1.00	1.00	1.00

Landmark No.		11	12	13	14	15	16	17	18	19
Reconstructed 3D Config 7	x	-0.31	0.02	0.17	-0.44	0.11	0.35	-0.56	-0.01	0.25
	y	0.06	-0.07	0.08	-0.06	-0.26	-0.01	-0.63	-0.78	-0.51
	z	4.48	4.10	4.25	4.39	3.74	4.01	4.65	3.96	4.21
	w	1.00	1.00	1.00	1.00	1.00	1.00	1.00	1.00	1.00

Landmark No.		1	2	3	4	5	6	7	8	9	10
Reconstructed 3D Config 8	x	-0.04	-0.06	-0.03	-0.01	-0.06	-0.04	-0.02	-0.08	-0.04	-0.00
	y	0.11	0.11	0.11	0.12	0.08	0.08	0.09	0.07	0.08	0.09
	z	0.56	0.56	0.60	0.59	0.59	0.63	0.62	0.59	0.65	0.64
	w	1.00	1.00	1.00	1.00	1.00	1.00	1.00	1.00	1.00	1.00

Landmark No.		11	12	13	14	15	16	17	18	19
Reconstructed 3D Config 8	x	-0.08	-0.04	-0.01	-0.11	-0.04	0.02	-0.12	-0.05	0.02
	y	0.03	0.03	0.04	0.02	0.01	0.04	-0.06	-0.09	-0.05
	z	0.64	0.69	0.68	0.64	0.74	0.72	0.72	0.84	0.81
	w	1.00	1.00	1.00	1.00	1.00	1.00	1.00	1.00	1.00

We selected landmarks 8, 12, 17, 18, and 19 in this order to form a projective frame. The projective coordinates of the remaining landmarks in their original labeling order, with respect to this frame, yield a sample of points in $(RP^3)^{14}$ $(19-3-2=14)$, with each point representing a projective shape in axial representation. The extrinsic mean of the eight projective shapes was computed and is displayed in Figure 22.2(a). The spherical coordinates of the landmarks are given in Table 22.2. Given the large number of covariates in the tangent space (forty two), we display only the "heat map" of the sample extrinsic covariance matrix (10.39), shown here in Figure 22.3.

Table 22.2: *Extrinsic Sample Mean*

Proj. Sp. Copy		1	2	3	4	5	6	7
j_{19} Ext. Samp. Mean	x	0.21	0.29	-0.29	-0.25	0.54	-0.10	-0.07
	y	0.75	0.67	0.46	0.63	0.66	0.57	0.51
	z	-0.41	0.36	0.46	-0.50	0.08	0.10	-0.77
	w	0.47	0.58	0.70	0.54	0.51	0.81	0.38

Proj. Sp. Copy		9	10	11	13	14	15	16
j_{19} Ext. Samp. Mean	x	-0.45	-0.28	0.85	-0.00	0.65	-0.43	-0.32
	y	0.10	0.40	0.41	0.24	0.10	-0.41	0.08
	z	0.64	-0.79	0.02	-0.96	0.71	0.77	-0.93
	w	0.62	0.37	0.32	0.15	0.24	0.22	0.15

We formulate the original question as a hypothesis testing problem

$$H_0 : \mu_{j_{19}} = \mu_0 \text{ vs. } H_1 : \mu_{j_{19}} \neq \mu_0, \tag{22.28}$$

where μ_0 is the projective shape of the blueprint, given in 22.5.1, and $\mu_{j_{19}}$ is the extrinsic mean projective shape of the random 3D configuration of 19

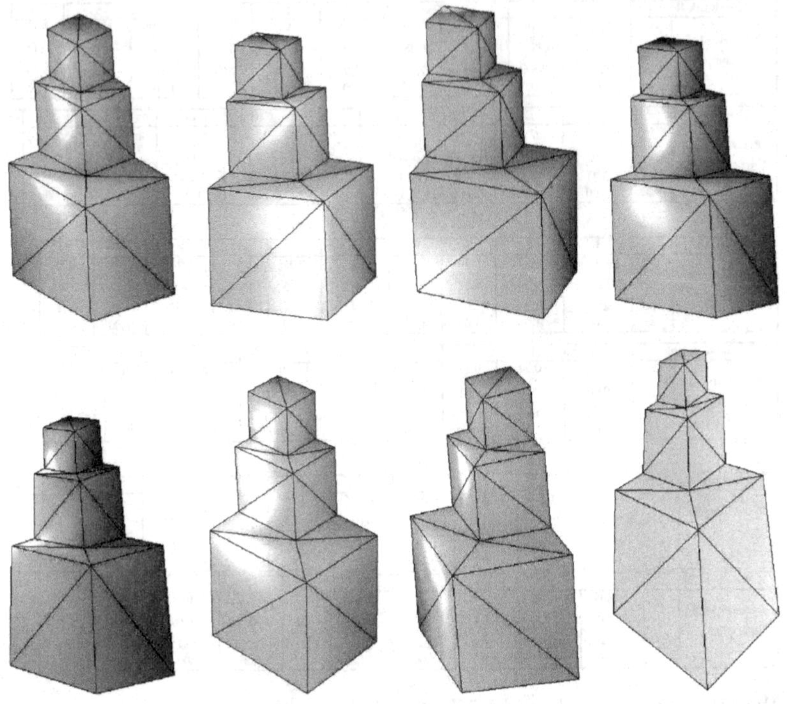

Figure 22.1 *The eight reconstructed 3D projective shapes. (Source: Patrangenaru et al.(2010), Figure 4. Reproduced by permission of Elsevier).*

Table 22.3: *Coordinates of the blueprint*

Landmark No.		1	2	3	4	5	6	7	8	9	10
Blue print	x	10.00	10.00	0.00	10.00	0.00	0.00	4.00	4.00	0.00	4.00
	y	0.00	10.00	10.00	10.00	6.00	0.00	0.00	4.00	4.00	0.00
	z	0.00	0.00	0.00	10.00	10.00	20.00	20.00	20.00	20.00	16.00
	w	1.00	1.00	1.00	1.00	1.00	1.00	1.00	1.00	1.00	1.00

Landmark No.		11	12	13	14	15	16	17	18	19
Blue print	x	4.00	0.00	6.00	6.00	0.00	6.00	6.00	10.00	0.00
	y	4.00	4.00	0.00	6.00	6.00	0.00	6.00	0.00	10.00
	z	16.00	16.00	16.00	16.00	16.00	10.00	10.00	10.00	10.00
	w	1.00	1.00	1.00	1.00	1.00	1.00	1.00	1.00	1.00

vertices on the polyhedral surface of the object as recovered from its pictures, corresponding tho the embedding j_{19}.

It suffices to check that, for a significance level α, μ_0 is in a $(1-\alpha)100\%$ confidence region of μ_{19}. Since the sample size (eight) is so small, the extrinsic sample covariance matrix is a degenerate matrix, so one can not use pivotal bootstrap for the mean projective shape of the entire configuration of nineteen landmarks, as given in Corollary 10.4.4. Nevertheless, we could find

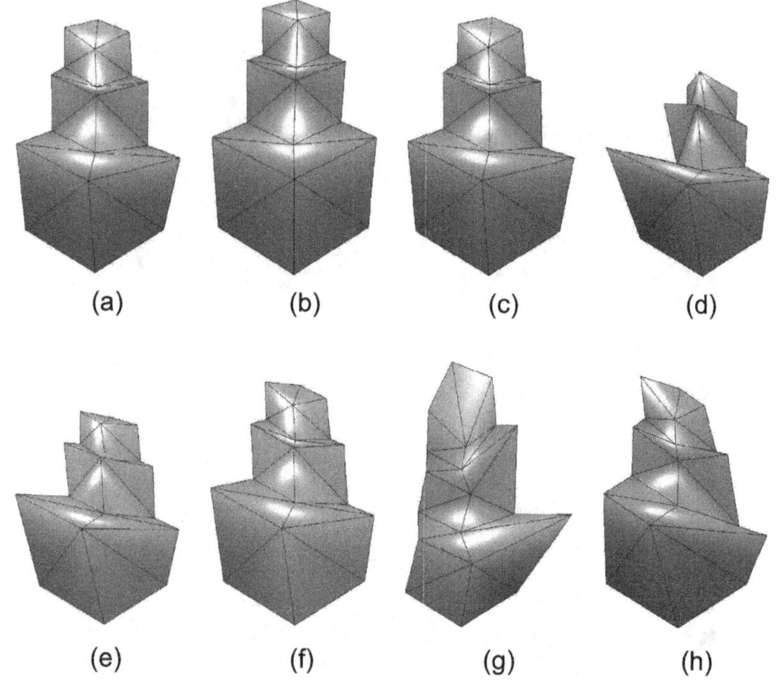

Figure 22.2 *(a) The estimated extrinsic mean shown as a 3D projective shape. (b)-(h) Randomly selected estimated extrinsic means based on bootstrap samples. In each case, a projective transformation is applied to each shape so that landmarks 13, 15, 17, 18, 19 match that of reconstructed 3D images as shown in Figure 22.1. (Source: Patrangenaru et al.(2010), Figure 5. Reproduced by permission of Elsevier).*

$(1-\alpha)100\%$ confidence regions using pivotal bootstrap based on Corollary 10.4.5.

There are $q = 14$ marginal axial distributions corresponding to the fourteen existing landmarks besides those that are part of the projective frame. To achieve a reliable conclusion, we used 20,000 resamples from the original sample. For example, μ_0 is in the 95% confidence region for $\mu_{j_{19}}$ if, for each $s = 1,\ldots,14$, the value of $T_s = T(\overline{Y}^s_{j,7};\mu_{0,s})$ in (10.41) corresponding to the $s-th$ marginal is between the 72nd ranked and the 19928th ranked observation of the corresponding bootstrap distribution (values of degenerate $G^*_{s,7}$ have been omitted). The results are as follows:

- First marginal (Landmark 1): T_1 = 3.0279647168E+00 is between 6301 (T*=3.0243210949E+00) and 6302 (T* = 3.0294218108E+00).
- Second marginal (Landmark 2): T_2 = 2.6459766362E+00 is between 3942 (T*=2.6434475892E+00) and 3943 (T* = 2.6920988816E+00).

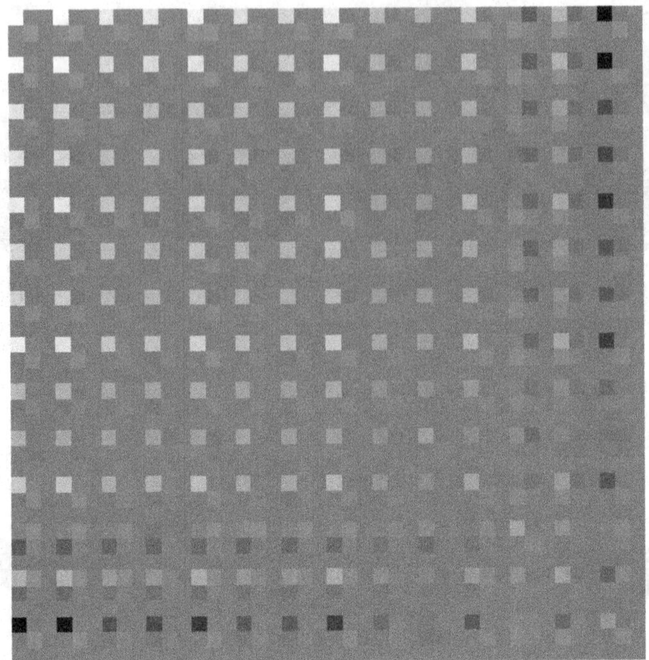

Figure 22.3 *Extrinsic sample covariance matrix shown as an image. (Source: Patrangenaru et al.(2010), Figure 6. Reproduced by permission of Elsevier).*

- Third marginal (Landmark 3): T_3 = 1.5175491E-01 is between 397 (T*=1.4510789E-01) and 398 (T* = 1.5271147E-01).
- Fourth marginal (Landmark 4): T_4 = 3.7407490E+00 is between 7379 (T*=3.7288447E+00) and 7380 (T* = 3.7464786E+00).
- Fifth marginal (Landmark 5): T_5 = 2.6168385E+00 is between 5355 (T*=2.6166643704E+00) and 5356 (T* = 2.6216985741E+00).
- Sixth marginal (Landmark 6): T_6 = 1.7898784E+00 is between 3294 (T*=1.7859106E+00) and 3295 (T* = 1.7914946E+00).
- Seventh marginal (Landmark 7): T_7 = 3.9364703E+00 is between 7194 (T*=3.9191776E+00) and 7195 (T* = 3.9388019E+00).
- Eighth marginal (Landmark 9): T_8 = 1.5700171E+00 is between 4432 (T*=1.5687626E+00) and 4433 (T* = 1.5748148E+00).
- Ninth marginal (Landmark 10): T_9 = 5.0491394E+00 is between 8507 (T*=5.0407173E+00) and 8508 (T* = 5.0521943E+00).
- Tenth marginal (Landmark 11): T_{10} = 1.3735517E+01 is between 15155 (T*=1.3706638E+01) and 15156 (T* = 1.3750192E+01).

Table 22.4: *Projective coordinates of the blueprint*

Landmark No.		1	2	3	4	5	6	7
	x	0.22	0.26	−0.31	−0.26	0.52	−0.09	−0.05
Blue	y	0.77	0.66	0.46	0.66	0.67	0.58	0.58
print	z	−0.33	0.39	0.46	−0.39	0.09	0.13	−0.68
	w	0.50	0.59	0.69	0.59	0.52	0.80	0.45

Landmark No.		9	10	11	13	14	15	16
	x	−0.42	−0.31	0.85	0.00	0.68	−0.47	−0.39
Blue	y	0.11	0.46	0.42	0.31	0.11	−0.47	0.10
print	z	0.64	−0.69	0.00	−0.92	0.68	0.70	−0.89
	w	0.64	0.46	0.32	0.23	0.25	0.26	0.22

- Eleventh marginal (Landmark 13): T_{11} = 2.0352336E+00 is between 4198 (T*=2.0327469E+00) and 4199 (T* = 2.0412641E+00).
- Twelfth marginal (Landmark 14): T_{12} = 3.9488573860E+00 is between 6837 (T*=3.9442688E+00) and 6838 (T* = 3.9573099E+00).
- Thirteenth marginal (Landmark 15): T_{13} = 3.6973595941E+00 is between 7857 (T*=3.6946285E+00) and 7858 (T* = 3.6986865E+00).
- Fourteenth marginal (Landmark 16): T_{14} = 2.8730067E+00 is between 5065 (T*=2.8723605E+00) and 5066 (T* = 2.8770536E+00).

These results show that we fail to reject H_0 for any reasonable level α, thus proving that the projective shape of the object is following the projective shape of the blueprint closely.

Simultaneous confidence intervals for affine coordinates of the extrinsic mean projective shape, using nonpivotal bootstrap, as in Remark 10.4.4, yield similar results, but are less reliable given the poorer coverage error of nonparametric bootstrap. Such results can be provided by the authors by request and can be found in Liu et al. (2007) for a similar data set with different relative sizes of cubes.

22.5.2 *Face Identification Example*

Our second example is in the area of face recognition. This example is based on a data set used in the live BBC program "Tomorrow's World". The example was introduced in Mardia and Patrangenaru (2005) [233], where six landmarks (at the ends of eyes plus ends of lips) have been recorded from fourteen digital images of the same person (an actor posing in different disguises) in fourteen pictures. Face appearance in these pictures may be neither frontal nor lateral. The data was obtained from cropped images are displayed in Figure 1.12 in

Chapter 1. Table 1.21 in that chapter contains the coordinates of the eight landmarks that were used in this analysis.

In Mardia and Patrangenaru (2005) [233], seven of the frontal pictures and, respectively, seven one-quarter pictures were used, for which the coordinates of six anatomically landmarks that are approximately coplanar were recorded (four corners of the eyes canthus and two end points of the lips mouth edge points). Using the four eye-corner landmarks as the projective frame, the landmark coordinates were converted into bivariate axial observations. An empirical test was performed in that paper, which showed evidence that the frontal and one quarter views of the group of landmarks could be of the same person.

In this section, we use two additional landmarks (the "bridge of the nose" and "tip of the nose"). The eight landmarks considered are significantly not coplanar, as shown by Crane (2010) [77]. Therefore, a 3D projective shape analysis is more appropriate for this configuration. If one compares the nonparametric bootstrap distributions of the extrinsic sample mean 2D projective shape of a configuration of 5 points, in one-quarter views versus frontal views of the actor, we notice that even for close-to-coplanar configurations, these regions have only a small overlap. In Figure 22.4, one may notice this effect when the fifth landmark called the "bridge of the nose" is added to a configuration of four coplanar landmarks, showing the limitations of the 2D projective shape analysis of spatial scenes.

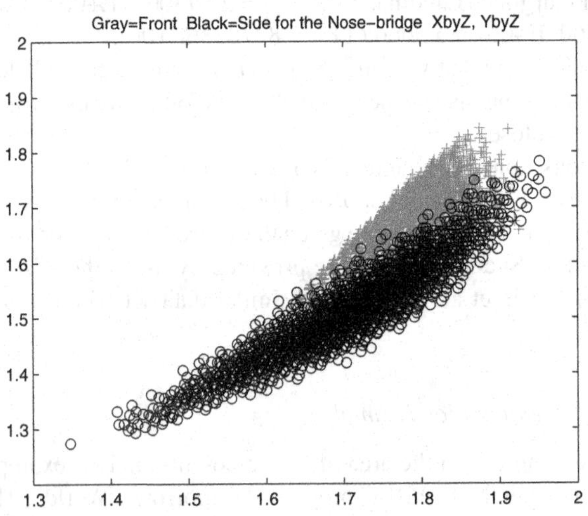

Figure 22.4 *Affine views of the bootstrap distribution of the extrinsic mean axial coordinate corresponding to the "bridge of nose" landmark. Frontal Views = +. One Quarter Views = ○. (Source: Patrangenaru et al.(2010), Figure 8. Reproduced by permission of Elsevier).*

In Figure 1.12, the matched configurations from 1 to 8 are for frontal views and from 9 to 14 for one-quarter views. The 3D reconstruction was obtained for each pair of the images, with 1 and 2 being the first pair, 3 and 4 being the second pair, and so on. In total, there are four reconstructed 3D facial landmark configurations using the front views and three reconstructed 3D faces using the side views. The 3D projective projective frames are given by landmarks 1 to 5.

REMARK 22.5.2. *Note that an asymptotic chi-square two-sample test statistic for the hypothesis testing problem* $\mu_{j_{19},1} = \mu_{j_{19},2}$ *that is similar to the statistic* $T_{n_1 n_2}$ *in formula (3.16) in Bhattacharya and Bhattacharya (2008) [32] for the equality of two extrinsic means on* $\mathbb{C}P^{k-2}$ *may be derived also on* $(\mathbb{R}P^m)^{k-m-2}$*. However, such a test statistic assumes the sample extrinsic covariance matrices are nonsingular, while, in our example, the sample sizes* $n_1 = 4, n_2 = 3$ *are too small to insure this property for these* 9×9 *matrices.*

Therefore, we give only a heuristic computational justification for the equality of the extrinsic mean projective shapes of the reconstructed configurations from frontal images and from one-quarter images, which is similar to the one used in Mardia and Patrangenaru (2005) [233]. We use nonpivotal bootstrap , as mentioned in Remark 10.4.4. The affine coordinates of the extrinsic sample mean 3D projective shapes of the configurations of eight landmarks retrieved from the side images falls inside seven out of nine 95% simultaneous bootstrap confidence intervals for the affine coordinates of the extrinsic mean 3D projective shapes of the corresponding configurations retrieved from the frontal images. The joint 95% confidence regions for the two means overlap.

Simultaneous 95% bootstrap confidence intervals for the affine coordinates of the extrinsic mean 3D projective shapes of spatial configurations of landmarks obtained from pairs of frontal images (red) and, respectively, from pairs of one quarter images (blue) are displayed in Figure 22.5. On the horizontal axis, we have the nine affine coordinates (three for each projective space marginal of the multi-axial representation). For each affine coordinate, the simultaneous confidence intervals are displayed on the vertical of that coordinate. Note that according to Hall (1997, p. 283), if n is the sample size, the number of *atoms for nonparametric bootstrap* is $\binom{2n-1}{n}$. In our example, $n = 4$ for reconstructions from frontal images and we have $\binom{7}{4} = 35$ bootstrap atoms. Similarly, $n = 3$ for reconstructions from frontal images and, for this group, we have only $\binom{5}{3} = 10$ bootstrap atoms. The affine coordinates of bootstrap means in the two groups are displayed in figure 22.5 as small circles and, respectively, as small crosses.

The intersection interval along each of the affine coordinates is given below:

Intersection interval for the 1st coordinate is 0.099812 0.642128 with length 0.542315.

Intersection interval for the 2nd coordinate is -0.160658 0.450349 with length 0.611007.

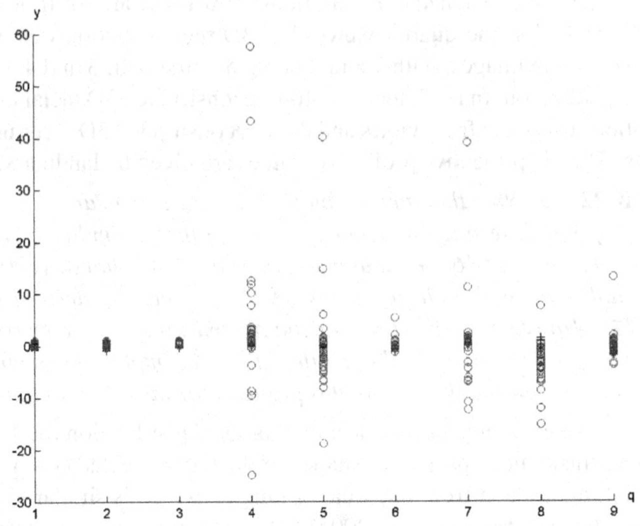

Figure 22.5 *95% simultaneous bootstrap confidence intervals for affine coordinates of mean 3D projective shapes of 8 facial landmarks in front views and one quarter views. (Source: Patrangenaru et al.(2010), Figure 9. Reproduced by permission of Elsevier).*

Intersection interval for the 3rd coordinate is 0.185366 0.292291 with length 0.106926.

Intersection interval for the 4th coordinate is -0.002991 0.657261 with length 0.660252.

Intersection interval for the 5th coordinate is -0.376397 0.983902 with length 1.360299.

Intersection interval for the 6th coordinate is -1.092520 -0.001708 with length 1.090812.

Intersection interval for the 7th coordinate is -0.002792 1.131982 with length 1.134774.

Intersection interval for the 8th coordinate is -0.769301 1.474715 with length 2.244016.

Intersection interval for the 9th coordinate is -1.992983 -0.008885 with length 1.984098.

REMARK 22.5.3. *No matter what model of 3D projective shape space is used, the dimension of that manifold is* $3k-15$. *Therefore, for each additional landmark, ideally one should increase the sample size by at least six additional images. For this reason, Corollary 10.4.5, based on the multivariate axial model, offers a reasonable way of bypassing this high dimensionality - small sample problem, as seen in our application. This is a promising feature of* *our methodology for solving pattern recognition questions from digital images of a scene.*

Chapter 23

Two-Sample Tests for Mean Projective Shapes and Scene Recognition

23.1 Projective Shape Analysis Examples in 1D and 2D

The applications in this chapter are related to a target recognition problem.

EXAMPLE 23.1.1. *The aim of this 1D example, due to Mardia and Patrangenaru (2005) [233], is to distinguish between buildings that have some common architectural features. A number of different views of buildings on the University of Leeds campus in the United Kingdom, including the Education Building, were taken for this goal. Note that, for these illustrative examples, the sample size n is small. Nevertheless, these methods are applicable to larger sample sizes.*

We start with a basic one-sample test before comparing two architectural styles. One set consists of photographs of the Education Building at Leeds University taken at five random angles and the aim is to verify the equi-spacing of the windows. Unlike other examples, which involve larger data sets, for this toy example, image data and landmark coordinates are given here, as opposed to being in the data chapter.

Four consecutive windows and four intersection points of the central part of these four windows (see Figure 23.1) were marked. The problem is also of relevance in machine vision when different types of buildings are to be discriminated (see, for example, Hartley and Zisserman (2004) [141], pp.14-37). Four collinear points p_1, p_2, p_3, p_4 were digitized and the coordinates (x_i, y_i) of p_i were recorded for $i = 1, 2, 3, 4$.

Education building: view 1

Figure 23.1 *One view of a building on the Leeds University campus with four collinear points on consecutive windows marked by white rings at the center of the windows.*

We fix the projective frame $\pi = ([x_1 : 1], [x_2 : 1], [x_3 : 1])$ and determine the cross-ratio c and computed projective coordinate of $[x_4 : 1]^\pi$. Using an angular representation, $[x_4 : 1]^\pi = [\cos\phi : \sin\phi]$ and, doubling the angle, we get a circular representation of $[x_4 : 1]^\pi$ in terms of the angle $\psi = 2\phi$.

Table 23.1 *Horizontal coordinates of collinear points on the Education Building at five different views*

View	x_1	x_2	x_3	x_4
1	22.90	35.7	48.3	61.10
2	23.10	29.1	35.5	45.50
3	41.40	44.3	47.3	50.70
4	39.00	47.0	53.9	60.00
5	42.25	46.9	50.5	53.85

If the landmarks were equidistant, their cross-ratio would be $c = 4/3$ and the corresponding direction is $\psi_0 = 1.287$ rad. Therefore, testing the hypothesis for projective equidistance is equivalent to the problem for the intrinsic mean ψ_I. Here, we consider the testing problem

$$H_0 : \psi_I = \psi_0, \text{ vs. } H_1 : \psi_I \neq \psi_0. \tag{23.1}$$

Given that the sample size $n = 5$ is small, according to Section 2.13, we base our P-value on a nonparametric pivotal bootstrap derived from the studentized large sample distribution of the mean $\bar{\psi}_I$. That is, based on $T(Q,\psi) = n^{\frac{1}{2}} S^{-1}(\bar{\psi}_I - \psi_I)$, where S is the sample standard deviation and $\psi_k, k = 1,\ldots,n$, is a random sample from Q. If $\psi_k^*, k = 1,\ldots,n$ is a random resample with repetition from $\psi_k, k = 1,\ldots,n$, then the bootstrap distribution of

$$T^*(Q^*,\psi^*|\psi) = n^{\frac{1}{2}}(S^*)^{-1}(\bar{\psi}_I^* - \bar{\psi}_I)$$

approximates $T(Q,\mu_I)$ with an error of order $O_p(n^{-1})$ (see Section 2.13), where $\frac{S^*}{\sqrt{n}}$ is the sample standard deviation of $\psi_k^*, k = 1,\ldots,n$. The p-value of the observed statistic $t = n^{\frac{1}{2}}(\bar{\psi}_I - \psi_0)/s$, is $2P^*(T^*(Q^*,\Psi^*|\Psi) < t)$ if $t < 0$. Here, P^* denotes the empirical distribution. That is, for a large number of resamples, half the P-value is the percentage of observed values of T^* less than t. The observed value of t is -1.69. Using 5000 pseudorandom resamples from the directions $\psi_k, k = 1,\ldots,n$, we obtained the pivotal bootstrap distribution of $T^*(Q^*,\psi^*|\psi)$ and ranked the values of $T^*(Q^*,\psi^*|\psi)$. Ultimately, a p-value of 0.105 was obtained. We find that the 95% bootstrap confidence interval based on the observed distribution of T^* is $(-2.133, 6.969)$. In conclusion, we fail to reject the equidistance hypothesis for the landmarks selected on the Education building at the 5% level of significance.

EXAMPLE 23.1.2. *We now compare the styles of two buildings with a similar architecture on the University of Leeds campus; one of which is the Education building described in Example 23.1.1 and another is the Careers building. One of the views of the Careers building is shown in Figure 23.2.*

Two groups of identically positioned noncollinear landmarks A_1, A_2, A_3, A_4, and A_5 were marked on five frontal photographs of the Education Building and four of the Careers Building, so that $n_1 = 5$ for the Education building and $n_2 = 4$ for the Careers building. We selected the projective frame $\pi = ([A_1 : 1],[A_2 : 1],[A_3 : 1],[A_4 : 1])$ and determined the coordinates of the views in the sample using a spherical representation. These spherical coordinates are displayed in Table 23.2.

Here, the extrinsic sample mean projective shapes of views from the Education Building and Careers Building are given in the spherical representation by $Y_{1,E} = [0.8037 : 0.5632 : 0.1922]$ and $Y_{2,E} = [0.7907 : 0.5834 : 0.1855]$, respectively. Now, consider the problem of estimating the distribution of the axis

Careers building: view 1

Figure 23.2 *One view of the Careers Buildings (University of Leeds) with five landmarks.*

Table 23.2 *Spherical coordinates for the Education Building (1 to 5) and the Careers Building (6 to 10)*

View Id	z^1	z^2	z^3
1	0.8142	0.5547	0.1718
2	0.8038	0.5610	0.1977
3	0.8067	0.5591	0.1917
4	0.8150	0.5513	0.1787
5	0.7773	0.5890	0.2211
6	0.7859	0.5768	0.2228
7	0.8170	0.5712	0.0791
8	0.7639	0.6041	0.2268
9	0.7893	0.5766	0.2110

$H(\rho_2)$, as defined in Section 10.5. Since the smaller sample size is 4 and the eigen-analysis has to be repeated for each resample, we limited ourselves to 250 pseudorandom resamples and determined the corresponding nonpivotal bootstrap distribution of $G(\rho_2^*)$. Let $\rho_2 = \rho$ and $\rho_2^* = \rho^*$. The corresponding distribution of $3G(\rho^*)$ is displayed in Figure 23.3, which indicates that the sample mean of $G(\rho)$ is close to $(0,0,0)$.

The rotation that brings $\bar{X}_{1,E}$ and $\bar{X}_{2,E}$ into coincidence is identified with a

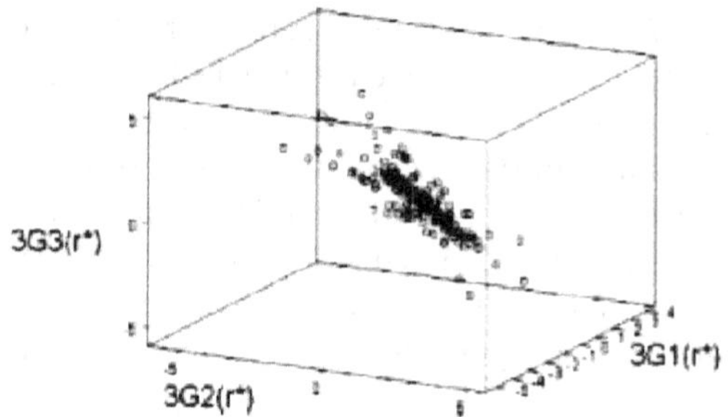

Figure 23.3: *Nonparametric bootstrap distribution of* $3G(\rho)$

4D axis and turns out to be, in spherical coordinates, given by

$$H(\rho) = [.9997 : -0.0077 : 0.0029 : 0.0231].$$

We determined the coordinates of the distribution of $3(G(\rho^*) - G(\rho))$ and, for this distribution, we successively sorted and trimmed the distribution of $3G(\rho^*) = \{3G_1(\rho^*), 3G_2(\rho^*), 3G_3(\rho^*)\}$ and obtain the following 93% simultaneous bootstrap confidence intervals: $[-4.36, 3.02)$ for $3G_1(\rho^*)$, $[-3.59, 2.67)$ for $3G_2(\rho^*)$, $[-2.70, 3.40)$ for $3G_3(\rho^*)$. This analysis shows that $(0,0,0)$ is in the 93% percentile confidence region, which is the identity in the corresponding 93% bootstrap confidence region for r_2 on $SO(3)$. Therefore, we fail to reject that $\mu_{1,E} = \mu_{2,E}$ at the significance level $\alpha = 0.07$.

Mardia and Patrangenaru (2001) [232] showed that the affine shape approach performs more poorly for this data.

23.2 Test for VW Means of 3D Projective Shapes

In this section, we apply the results from Section 10.5, along with previous results in projective shape analysis, to two-sample tests on the projective shape space $P\Sigma_3^k$ in the context of image analysis of 3D scenes.

23.2.1 Two-Sample Tests for VW Mean 3D Projective Shapes from Stereo Images - Matched Pairs

We first consider an application for matched pairs of 3D projective shapes from digital images. The theory for such a two-sample test (a test for mean projective shape change) was developed in Crane and Patrangenaru [78], where it was applied to stereo medical imaging. Here, we consider a toy example consisting of two random samples of polyhedral objects. The first data set (see Figure 1.14 in Chapter 1) was considered in Chapter 22. It consists of 16 digital images of a polyhedral surface taken by a uncalibrated digital camera. For a second data set of 16 digital images, a related polyhedral scene were taken with a different camera that had different internal camera parameters. The second scene was obtained by a slight modification of the first polyhedral object, as displayed in Figure 1.15 of Chapter 1.

Using the algorithm from Hartley and Zisserman (2004) [141] or Ma et al. (2005) [225] (see Chapter 22 for details), the 3D configuration of visible corners was reconstructed from the paired landmark data in Table 1.23. The wire frame representations of these 3D reconstructions are displayed in Figure 23.4. The 3D reconstructions correspond to the similar landmark selection as in Figure 1.13 in Chapter 1.

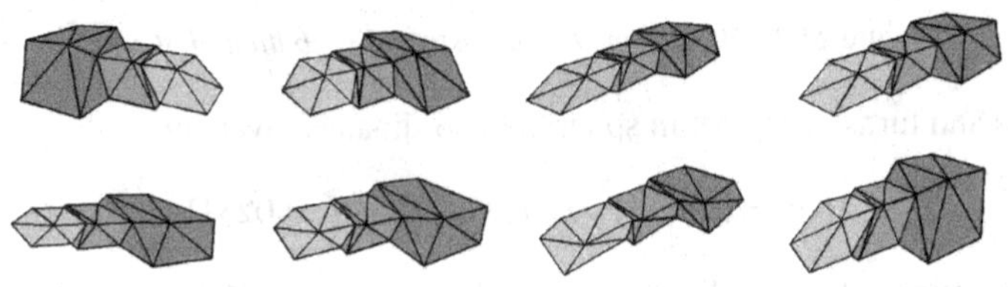

Figure 23.4 *3D Reconstructions of configurations of corners in Figure 1.15. (Source: Patrangenaru et al.(2014), Figure 4. Reproduced by permission of Taylor & Francis).*

From the reconstructed configurations of 19 landmarks (visible corners) carrying the same labels for corresponding corners, we selected a projective frame made of the first five labeled landmarks. Therefore, in this example the projective shape data is carried over to $P\Sigma_3^{19} = (\mathbb{R}P^3)^{14}$, which is a sample space that has a Lie group structure according to Subsection 10.5.1. Given that the two objects are obtained from two lower blocks by adding a third cube at the top, we regard the resulting data as matched pair observations on $(\mathbb{R}P^3)^{14}$. We considered the projective shape change hypothesis in Subsection 10.5.2, and, following the computational steps in that subsection of Chapter 10, from the resulting bootstrap distribution of Veronese-Whitney sample means, we computed the 14 marginal $T_s^*, s = 1, \dots, 14$ statistics on the Lie group $\mathbb{R}P^3$.

For $s = 1, \dots, 14$, the values of the statistics T_s in (10.46), which under the null hypothesis are all larger than the corresponding T_s^* for the 95% simultaneous confidence sets (shown in Figure 23.5 as cutoffs) are:
$T_1 = 1735771.3$, $T_2 = 2234801.4$, $T_3 = 24260037.4$, $T_4 = 949014.2$, $T_5 = 942757.9$, $T_6 = 148967185.2$, $T_7 = 15847127.4$, $T_8 = 3342761.1$, $T_9 = 8042772.6$, $T_{10} = 15528559.7$, $T_{11} = 3800842.3$, $T_{12} = 35097853.3$, $T_{13} = 24107515.0$, $T_{14} = 7085996.9$.

On the other hand, the corresponding values of the bootstrap cutoffs $T_s^*, s = 1, \dots, 14$ are:
$T_1^* = 23.9831$, $T_2^* = 38.9948$, $T_3^* = 441.3134$, $T_4^* = 44.4325$, $T_5^* = 25.1901$, $T_6^* = 305.9000$, $T_7^* = 74.7575$, $T_8^* = 24.2130$, $T_9^* = 35.1296$, $T_{10}^* = 204.4511$, $T_{11}^* = 42.3008$, $T_{12}^* = 40.7353$, $T_{13}^* = 113.6289$, $T_{14}^* = 26.3761$.

The nonexistence of projective shape change from the first object to the second object is therefore rejected at level $\alpha = 0.05$. We infer that, based on the data, the two polyhedral objects are not the same, which is a confirmation of the usefulness of the Lie group methodology.

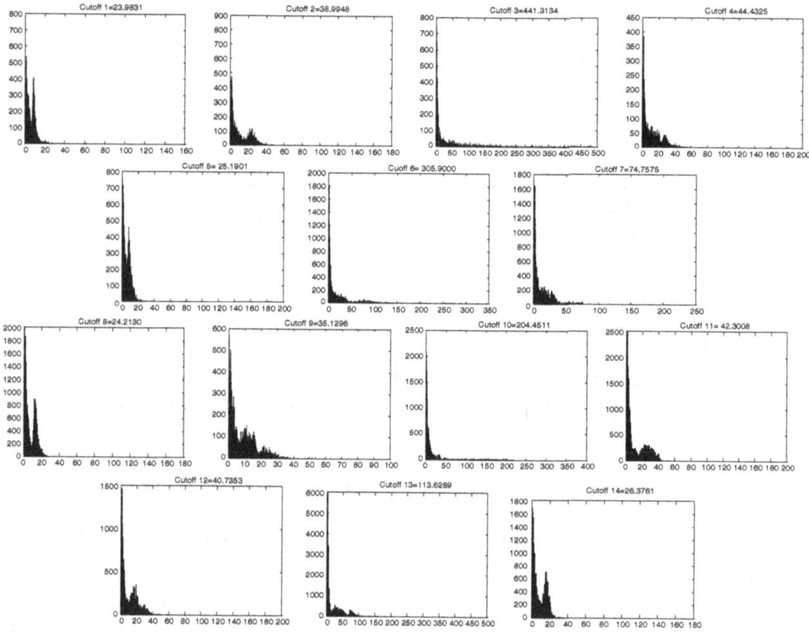

Figure 23.5 *Cutoffs of the T^* marginal bootstrap statistics. (Source: Patrangenaru et al.(2014), Figure 5. Reproduced by permission of Taylor & Francis).*

23.2.2 Example 2 - Two-Sample Test for Means of Independent Pairs

In this example, the data consists of twenty four photos taken of the busts of the Greek philosopher Epicurus, which are displayed in figure 1.16. The number of images from the one-head statue differs from that of the double-head statue. Therefore, a matched pairs test can not be used in this example. The landmark coordinates and the reconstructed 3D configurations obtained from 2D matched configurations in pairs of images are posted at www.stat.fsu.edu/∼vic/NSM. Landmarks 1, 4, 5, 6, 8 were utilized to construct the projective frame. For the confidence region, we computed 2,000,000 bootstrap VW sample means based on landmarks 2, 3, 7, 9.

For the 4 landmarks, the point $(0,0,0)$ is in the 12 confidence intervals of the affine coordinates, so we fail to reject the null hypothesis that there is no change on average from the one-headed statue to the double-headed one. Thus,

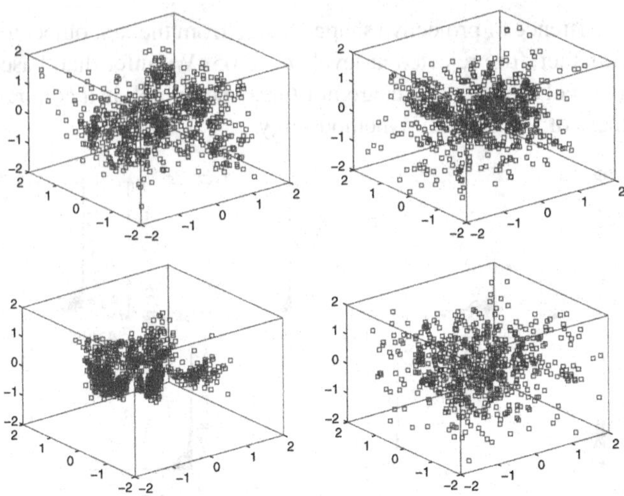

Figure 23.6 *Simultaneous confidence regions for the statue data. (Source: Patrangenaru et al.(2014), Figure 9. Reproduced by permission of Taylor & Francis).*

Table 23.3 *Simultaneous confidence affine intervals for mean projective shape change - statue data*

Bootstrap simultaneous confidence intervals for statue landmarks 2, 3, 7, 9				
Axis	2	3	7	9
x	$(-11.5, 11.6)$	$(-28.6, 30.8)$	$(-2.0, 1.9)$	$(-46.7, 42.6)$
y	$(-11.3, 12.0)$	$(-32.1, 32.6)$	$(-1.8, 1.9)$	$(-47.2, 46.4)$
z	$(-12.3, 12.5)$	$(-24.1, 26.3)$	$(-1.2, 1.4)$	$(-40.1, 37.1)$

the fact that the two statues are portraying the same person is not statistically significant.

Chapter 24

Glaucoma Index from HRT Imaging and Mean Glaucomatous Projective Shape Change Detection from Stereo Pair Images

24.1 Introduction

In Chapter 17 we showed that similarity shape analysis can be used for detection of glaucomatous change from Heidelberg Retina Tomograph (HRT) outputs. It would be nevertheless useful to have a simple geometric marker for detection of the glaucoma onset. Such a measurement, called *glaucoma index* was developed in Derado et al (2004) [82] for HRT data in animal models. This index, which is presented in Section 24.3 was considered by glaucoma experts (see Sanfilippo et al. (2009) [312], Hawker et al. (2007) [144]). One of the problems is that HRT like devices are quite expensive, therefore only a few medical cabinets have them available. For this reason, a different methodology for glaucoma detection, based on inexpensive imaging technology had to be considered as well. That methodology is developed in this chapter (also see Crane and Patrangenaru (2011)[78]). The afferent statistical analysis is on the projective shape space of k-ads in 3D $(\mathbb{R}P^3)^{k-5}$, which has a Lie group structure (see Section 10.5).

Statistical analysis on Lie groups was first considered by Beran (1968) [19] in the context of testing for uniformity of a distribution on a compact homogeneous space. Nonparametric density estimation on Lie groups via deconvolution like techniques and Fourier analysis was developed by Kim (1998 [187], 2000 [188]), Healy et al. (1998) [145], Lesosky et al. (2008) [218] and Koo and Kim (2008)[202]; these have applications in medical imaging, robotics, and polymer science (see Yarman and Yazici(2005) [340], Yarman and Yazici (2003) [339], Koo and Kim (2008a)[203]).

Statistical inference on certain Lie groups in a parametric setting context was considered also in concrete estimation problems in paleomagnetics (see Watson(1983) [333]), plate tectonics (see Chang (1986) [64], Chang (1988) [63], Chang (1989) [65], Chang and Ko (1995) [67], Kirkwood et al. (1999) [193], Rivest and Chang (2006) [290]), biomechanics (see Rivest (2005) [288], Rivest and Chang (2006) [290], Rivest et al. (2008) [289]). The idea of statistical Lie group model is due to Chang (2004) [66].

Correlation of two images of the same object (see Chang and Ko (1995) [67]), comparison of mean axes (see Beran and Fisher (1998) [18]), or of extrinsic mean planar projective shapes (see Mardia and Patrangenaru (2005) [233]), as well as estimation of 3D motion in computer vision (e.g. Tuzel et al. (2005) [324]), leads also to natural inference problems on Lie groups.

In general, if the sampling manifold is a homogeneous space, often times the dimension of the Lie group is higher than the dimension of the space on which the Lie group is acting. Since the matched pairs data analysis is on this Lie group, the increase in dimensionality requires even larger samples. In the other case this kind of data analysis is relevant in problems in medical imaging, face analysis and the like where the sample sizes are often moderate or small. Ideally the dimensions of the Lie group and of the sampling manifold where the group is acting should be the same, which is the case if and only if the sampling manifold with a homogeneous space structure has a Lie group structure in itself. Fortunately, this is the case if a projective shape analysis is pursued for 3D medical imaging data, and luckily again, following the methodology developed in Chapter 22, such 3D projective shape data can be recovered from stereo pairs of images of the eye fundus (see Crane(2010) [77], Crane and Patrangenaru (2011) [78]). Stereo data of the eye is a most common imaging data for eye disease detection and control. The stereo pairs in our analysis are from the Louisiana Experimental Glaucoma Study (LEGS) in Rhesus monkeys (see Burgoyne et al. (2000) [56]). Although the data were collected in the year 2000, they were not analyzed immediately; the analysis was postponed partially due to the hurricane Katrina, when sadly, the animals participating in the study were lost.

Section 24.4 is dedicated to an application of 3D projective shape analysis to mean glaucomatous projective shape change detection, from stereo data of the ONH. Since glaucoma is a disease affecting the 3D appearance of the ONH region due to high IOP, it leads to a change in the 3D projective shape of this region. The results obtained are statistically significant.

24.2 Glaucoma and LEGS Stereo Eye Fundus Data

Glaucoma is a group of eye disorders caused mainly by intraocular IOP that is too high for the maintenance of the normal eye, and in later stages, causes blindness. There are two fluid chambers in the eye, the anterior chamber, most of which we see behind the clear part of the eye (the cornea) in front of the iris (colored part of the externally visible eye), and the posterior chamber. It is in the anterior chamber that aqueous humor is produced, and it is this fluid production, and its slow release from the eye via the trabecular meshwork that maintains the IOP at normal levels. In glaucoma the trabecular meshwork releases less fluid, leading to an increased IOP in the eye. This pressure **pushes against the walls of the eyeball, which is solid, except for its ONH region, where the optic nerve fibers enter the eyeball**. The retinal neurons are thus compressed, damaged and pushed outside the eyeball, a phenomenon leading to a tunneling effect of vision in the diseased eye, leading to gradual loss of the peripheral view. The Louisiana Experimental Glaucoma Study, the ONHs of both eyes of mature Rhesus monkeys were imaged on separate sessions on one hand with an TopSS Scaning Conformal Laser Tomograph (also known as HRT) and a Topcon TRC-WT stereo eye fundus camera. Moderate levels of elevated IOP were then induced by laser treatment of the trabecular meshwork of one eye (the treated eye) of each monkey and both eyes were then imaged a number of times per session, on separate sessions to establish variability in normals. Then experimental glaucoma was induced in one eye of each animal, and the fellow eye was maintained as an untreated control. In this chapter, only a set of after treatment independent observations are considered, for which data was provided. The data set consists of a 15 sets of pairs of stereo images of the complicated ONH topography, that are displayed in Figure 1.7 in Chapter 1.

Recall that during the LEGS experiment, of the two eyes of one animal one was given experimental glaucoma, and the other was left untreated (normal) and imaged over time as a control. The coordinates of nine landmarks on the approximate elliptic contour that determines the ridge of the ONH are recorded, as well as those of certain blood vessels junctions and estimated location of the deepest point are tabulated in Tables 1.24, 1.25, 1.26 in Chapter 1. The nine landmarks considered are S(superior), I(inferior), N(nasal), T(templar), V(vertex-the deepest point of the ONH cup), SM(mid-superior), IM(mid-inferior), NM(mid-nasal) and TM(mid-templar), and their positions in the ONH cup are schematically displayed in the Figure 1.19 in Chapter 1. The landmarks N, T, S, I and V have been used before in statistical analysis for glaucomatous change detection from 3D confocal tomography images, often known as Heidelberg Retina Tomograph (HRT) outputs (see Patrangenaru et al. (2000), Derado et al. (2004), Bhattacharya and Patrangenaru (2005), A. Bhattacharya (2008), Bandulasiri et al. (2009a), Bandulasiri et al. (2009b), A. Bhattacharya (2009)).

24.3 A Shape-based Glaucoma Index for Tomographic Images

From the clinical experience, it is known that the ONH area contains all the relevant information related to glaucoma onset. Figure 1.7 in Chapter 1 shows the relevant area with four landmarks, namely, S for the superior aspect of the retina toward the top of the head, N for the nasal or nose side of the retina, T for temporal, the side of the retinal closest to the temple or temporal bone of the skull and V for the ONH deepest point. That is, the first three are anatomical landmarks and the fourth is a mathematical landmark.

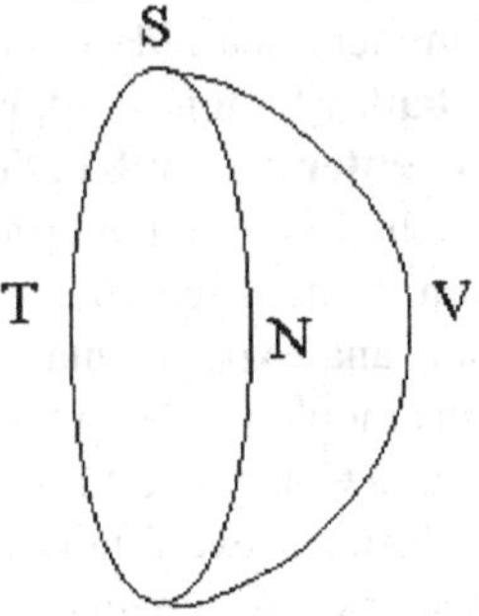

Figure 24.1 *Schematic picture of the four landmarks N, S, T and V in the ONH eye area. (Source: Bandulasiri et al.(2009), Figure 3. Reproduced by permission of Elsevier).*

Figure 24.1 gives a schematic picture depicting the geometry of the four landmarks. Some experimental data on Rhesus monkeys have been available and a shape analysis has been performed by Bhattacharya and Patrangenaru (2005) [43] (also see Chapter 17) using Bookstein coordinates (see Dryden and Mardia (1998) [91]) but they found that the shape change is barely significant. Since the eyeball, and in particular the papilla is rigid, the size of the configuration of the first three landmarks S, N, T remains unchanged during the glaucomatous change. Further, from medical experience, it is known that the physiology of glaucoma triggers a change in shape due to the increased IOP, the soft tissue of eyecup is pushed back, where the ONH enters the eyeball, thus the its depth grows. Therefore, we define the *glaucoma index, G, a shape measure, as the depth of the eyecup normalized by the distance between the two specific landmarks on the papilla.* That is, the ratio of the height from V of the tetrahedron S, T, N and V to its edge ST. Such a quantity has been used in practice (qualitatively) by clinicians.
In order to determine the 3D coordinates (x^1, x^2, x^3) of the four landmarks, we first found the coordinates of the three landmarks N, S, T from the TopSS output, using the knowledge that N, S, T lie on an ellipse. The (x^1, x^2) coordinates of V are straightforward to obtain, and we get its x^3-coordinate as the point of minimum inside the ONH cup. Thus, we get a set of four landmarks. We now

define the glaucoma index G, which was mentioned in the first section. We relabel the four landmarks as $X_1 = T, X_2 = S, X_3 = N, X_4 = V$. Assume the landmark coordinates are given by the 3×1 column vectors $X_j = (x_j^1\, x_j^2\, x_j^3)^T, j = 1, \ldots, 4$.

DEFINITION 24.3.1. *The glaucoma index G is the height from X_4 of the tetrahedron defined by (X_1, X_2, X_3, X_4) relative to the distance $\|X_1 - X_2\|$ (see Figure 24.2).*

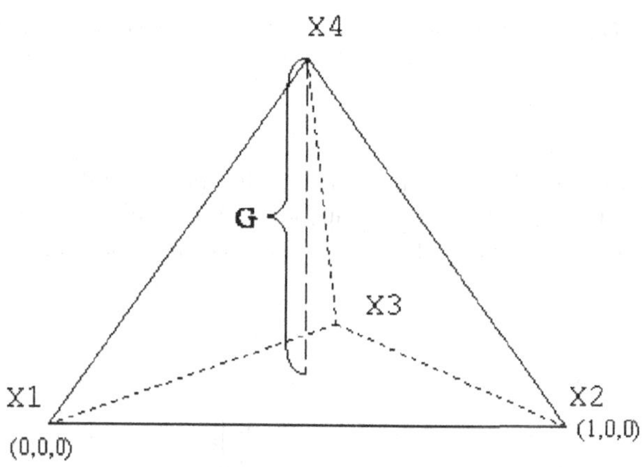

Figure 24.2 *Geometric interpretation of the glaucoma index G where X_1 and X_2 are transformed to $(0,0,0)$ and $(1,0,0)$ and the third coordinate of X_3 is 0. (Source: Derado et al.(2001), Figure 4. Reproduced by permission of Taylor & Francis).*

Explicitly, G is given by the formulas:

$$G = \frac{Y}{\|(X_2 - X_1) \times (X_3 - X_1)\| \|X_2 - X_1\|}, \tag{24.1}$$

where:

$$Y = det(\begin{pmatrix} 1 & 1 & 1 & 1 \\ X_1 & X_2 & X_3 & X_4 \end{pmatrix})$$

Obviously G is a shape coordinate on the space of tetrads (X_1, X_2, X_3, X_4) in general position. Figure 17.4 in Chapter 17 shows the ONH topography for a monkey, before and after treatment. Figure 24.3 gives a schematic diagram of the same information in terms of the landmarks. Thus there is empirical evidence that G does measure the shape change. We proceed with the statistical analysis. Table 1 in Derado et al. (2004) [82] contains before and after the treatment values of the glaucoma index G in the 12 of the monkeys for which the HTR data was available (G_2 denotes before the treatment, and G_1 the after the

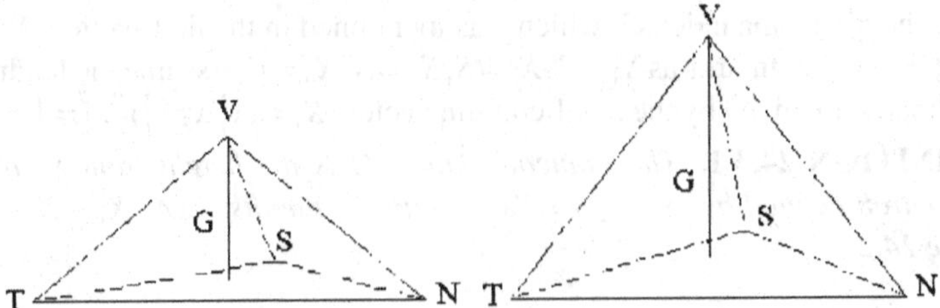

Figure 24.3 *Change in glaucoma index G due to increased internal ocular pressure level. (Source: Derado et al.(2001), Figure 6. Reproduced by permission of Taylor & Francis).*

treatment values) for LEGS. The nonparametric bootstrap distribution T^* of the studentized mean change in the glaucoma index, based on 5,000 resamples with repetition, is displayed in Figure 24.4 below (see Derado et al. (2004) [82] for details).

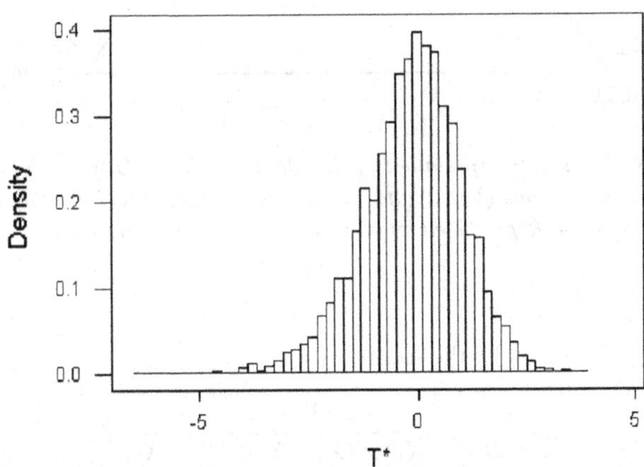

Figure 24.4 *Bootstrapped distribution of the studentized mean change in the glaucoma index G. (Source: Derado et al.(2001), Figure 7. Reproduced by permission of Taylor & Francis).*

The 95% - pivotal bootstrap confidence interval (see Section 2.13) for the mean change in the glaucoma index for the treated eye, with three exact decimals, was found to be (0.018,0.063) which does not contain zero! This data analysis suggests that changes in the G coordinate are a statistically significant measure for the glaucomatous eye, which is in agreement with some clinicians way of detecting the onset of the illness.

24.4 Reconstruction of 3D Landmark Configurations from Stereo Images and Detection of Extrinsic Mean Glaucomatous Projective Shape Change Detection from Stereo Pairs Data

It is no accident that most of the living creatures having the ability to see, from insects and snails to mammals and humans, have binocular vision. Stereopsis helps humans mentally recreating a 3D scene from the two retina 2D views of that scene. Nevertheless, this mental reconstruction is far from being similar to the observed scene, since two observed parallel lines, seem to meet at a point in space in their mental reconstructions. In fact the mentally reconstructed scene, most likely differs from the observed scene by a projective transformation of the surrounding Euclidean space. While the "ancient" mechanism of 3D binocular vision is not yet fully understood, in a nutshell, the results in Chapter 22 mean that **all we see are 3D projective shapes**.

For the mathematical and numerical aspects of the reconstruction of the projective shape of a 3D configuration from the pixel coordinates of two of its digital images, we refer the reader to Chapter 22, and the references therein.

REMARK 24.4.1. *We should point out that extracting the 3D projective shapes from stereo pairs of images is optimal in absence of access to the camera, or to additional useful information, that helps estimate the internal camera parameters. If information, such as a calibration stereo rig is available, the internal camera parameters can be obtained, thus making possible a 3D reflection shape analysis from digital camera images (see Lester et al. (2015) [219]).*

Using a reconstruction algorithm from Ma et al. (2006) [225], p. 121, for each eye stereo pair of corresponding planar configurations in Tables 1.24, 1.25 in Chapter 1, Crane (2010) [77] obtained in MATLAB R2007 a $3D$ reconstructed configurations of landmarks (S, I, N, T, V, SM, IM, NM, TM) in both control and treated eyes. These were the input data for testing if there is any mean glaucomatous projective shape change, based on the bootstrap methodology developed in Section 2.13. Note that the reconstructed configurations are only projectively equivalent to the observed ones. The reconstructions can be found at http:www.stat.fsu.edu/~vic/JMVA-09-179. MATLAB codes for deriving the bootstrap distributions for testing the extrinsic mean projective shape change can be found at the same web address.

In our analysis we selected the projective frame $\pi = (N,T,S,I,V)$ and used the projective coordinates $[h_{1,ji}],[h_{2,ji}],[h_{3,ji}],[h_{4,ji}], j = 1,2, i = 1,\dots,15$ of the other anatomic landmarks with respect to this frame. Then we obtained the sample $(y_{1,i},y_{2,i},y_{3,i},y_{4,i}) \in (\mathbb{R}P^3)^4, i = 1,\dots,15$ from the change in projective shape, using quaternion multiplication:

$$y_{a,i} = [\bar{h}_{a,1i} \cdot h_{a,2i}], a = 1,2,3,4, i = 1,\dots,15. \tag{24.2}$$

The extrinsic sample covariance matrix in Section 10.5 is singular; therefore to estimate the extrinsic sample mean change, we computed the bootstrap $T_s^*, s = 1,2,3,4$, for 20,000 resamples and obtained the following cutoffs upper simultaneous bootstrap .95-quantiles:

- for T_1^* : 20.0663,
- for T_2^* : 16.5607,
- for T_3^* : 25.7405,
- for T_4^* : 21.7158.

The histograms for the bootstrap distributions of $T_s^*, s = 1,2,3,4$ corresponding to the marginal axes are displayed below: The values of the statistics

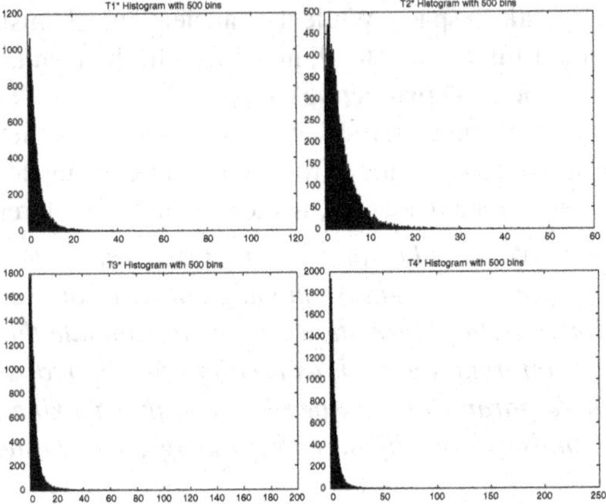

Figure 24.5 *Histograms for the bootstrap distributions of $T_s^*, s = 1,2,3,4$ for 20,000 resamples. (Source: Crane and Patrangenaru(2011), Figure 4. Reproduced by permission of Elsevier).*

$T_s, s = 1,2,3,4$ under the null hypothesis that there is no $3D$ projective shape change are:
$T_1 = 1474.7, T_2 = 2619.9, T_3 = 860.2, T_4 = 1145.7$, and since the T_1, T_2, T_3 and T_4 are much larger than the corresponding cutoffs given above, there is a significant mean projective shape change due to the increased IOP in the treated eye.

REMARK 24.4.2. *If instead of considering the change on a Lie group given by the $C = X^{-1}Y$, we look at $C_1 = YX^{-1}$, we obtain the same conclusion that T_a is larger than the cutoff for T_a^*, for $a = 1,2,3,4$, as expected.*

We close the section by noting that while previous results in literature, which are referred to in this chapter, show that there is mean similarity shape change, or similarity size-and-shape change, as well. However, those studies are based on HRT outputs and most ophthalmologists can not afford an HRT while any ophthalmologist has access to stereo cameras designed for eye fundus imagery.

Chapter 25

Application of Nonparametric Density Estimation on Manifolds

25.1 Introduction

In this section we present results in Lee et al. (2004) [214]. The landmark data reduction approach in high level image analysis has led to significant progress to scene recognition via statistical shape analysis (see Dryden and Mardia (1998) [91]). While a number of families of similarity shape densities have proven useful in landmark based data analysis, parametric models have seldom been considered in the context of projective shape or affine shape (see Mardia and Patrangenaru (2005) [233]). Sample spaces of interest in Statistics (see Section 3.5) that have the geometric structure of symmetric spaces (see Section 3.2) are spheres (as spaces of directions), real projective spaces as spaces of axes, complex projective spaces as planar direct similarity shape spaces (see Kendall (1984) [177]), real Grassmann manifolds as spaces of affine shapes (see Sparr (1992) [314], and products of real projective spaces as spaces of projective shapes of configurations of points in general position (see Patrangenaru (2001) [268], Mardia and Patrangenaru (2005) [233]). Therefore, density estimation of distributions, regarded as points on such symmetric spaces and arising from directional data or from digitizing landmarks in images, was necessary.

Henry and Rodriguez (2009) [156] used the Pelletier estimator for in directional paleomagnetism data and Di Marzio et al. (2011)[236] used kennel density estimators for data on the torus.

In Section 25.4, following Pelettier (2005) [278], we consider kernel density estimation on general Riemannian manifolds. Note that those results however hold only in homogeneous spaces (for a definition of a homogeneous space see Chapter 3). This is good enough for image analysis, since any sample spaces considered there are symmetric spaces, and a symmetric space is homogeneous. *Pelletier density estimators* generalize the density estimators on certain homogeneous spaces introduced by Ruymgaart (1989), by H. Hendriks et al. (1993) and by Lee and Ruymgaart (1998). In this chapter, we propose a class of adjusted Pelletier density estimators on homogeneous spaces that converge uniformly and almost surely at the same rate as naive kernel density estimators on Euclidean spaces. A concrete detailed example of projective shape density estimation of 6-ads arising from digitized images of the "actor" data set in Table 1.21.

25.2 Pelletier Density Estimators on Homogeneous Spaces

Given that no chapter was devoted to density estimation in Part II, in this section, we first consider some necessary definitions and results on this subject.

A kernel density estimator of a probability distribution on an arbitrary complete Riemannian manifold $(\mathcal{M}, g)$ of dimension m was introduced in Pelletier (2005) [278]. Assume ρ_g is the geodesic distance on $\mathcal{M}$ associated with the Riemannian structure g (see (3.58)) and let $\mathcal{B}$ denote the Borel $\sigma-$field of M. Let $(\Omega, \mathcal{F}, \mathbf{P})$ denote an underlying probability space and let X be a random object on $\mathcal{M}$, and let Q be the probability measure on M associated with X. If $dV = \sqrt{|g|}\lambda_m(dx)$, is the *volume measure* associated with the invariant Riemannian metric g on $\mathcal{M}$ (see (25.6)) and Q has a probability density f w.r.t. the volume measure dV, Pelletier defines the density estimator $f_{n,K}$ of f as follows: let $K : \mathbb{R}_+ \to \mathbb{R}$ be a nonnegative map with support in $[0,1]$ such that $K(\| \times \|$ is a density with finite moments of order $p \leq 2$. Then $f_{n,K}$ is defined by

$$f_{n,K}(p) = \frac{1}{n}\sum_{i=1}^{n} |g(p)|^{-\frac{1}{2}} \frac{1}{r^m} K(\frac{\rho_g(p, x_i)}{r}), \tag{25.1}$$

where $|g(p)|$ is the determinant of the metric, and r is the radius considered above.

Equation (25.1) needs to be used with caution since the coefficients of the Riemannian metric depend on the choice of a coordinate system around p. If in addition we assume that the $(\mathcal{M}, g)$ is homogeneous, then for any pair of indices i, j there is an isometry $h : \mathcal{M} \to \mathcal{M}$, with $h(x_i) = x_j$, which insures that $f_{n,K}$ is a density.

REMARK 25.2.1. *If M is an arbitrary compact Riemannian manifold, one may modify equation* (25.1) *as follows and get a probability density function: for each $i = 1, \ldots, n$ one should consider $g_i(p)$ instead of $g(p)$, where g_i is the determinant of g w.r.t. the the log-chart centered at the observation x_i. The log-chart is the inverse of the exponential map at x_i considered in* (3.70).

To assess the quality of this estimator we shall consider the MSE

$$MSE(x) = \mathbf{E}\{f_{n,K}(x) - f(x)\}^2 = Var(f_{n,K})(x) + \{f_{n,r}(x) - f(x)\}^2, \qquad (25.2)$$

and the mean integrated squared error on the Riemannian manifold $\mathcal{M}$ (MISE)

$$\text{MISE} = \int_{\mathcal{M}} MSE(x)dV(x). \qquad (25.3)$$

that is a generalization to the nonlinear case of the MISE in (2.111), Section 2.14. The following result due to Pelletier (2005) [278] extends a result of Ruymgaart (2004) [298]:

THEOREM 25.2.1. *If f is of class C^2 on M and $f_{n,K}$ is the density estimator in* (25.1), *then $MISE \leq C_f(\frac{1}{nr^m} + r^4)$. Consequently for $r = O(n^{-\frac{1}{m+4}})$ we have $MISE = O(n^{-\frac{4}{m+4}})$*

For a proof, see Pelletier (2005) [278].

25.3 Density Estimation on Symmetric Spaces

If $(\mathcal{M}, g)$ is a simply connected compact Riemannian symmetric space, with the isometry group $\mathcal{G}$ equipped with a $\mathcal{G}$ invariant measure, it is known that $\mathcal{M}$ factors as a product irreducible symmetric spaces (see Kobayashi and Nomizu (1963) [197]). Therefore, any symmetric space $\mathcal{M}$ is locally isometric to a direct product $\mathcal{M}_1 \times \cdots \times \mathcal{M}_q$ of irreducible symmetric spaces. Assume $p = (p_1, \dots, p_q)$ is a fixed point on such a product. A vector u tangent to $\mathcal{M}$ at p can be represented as $u = (u_1, ..., u_q), u_a \in T_{p_a}M_a \forall a = 1, ..., q$, and with these identifications, the exponential map at p is given by $Exp_p(u) = (Exp_{p_1}(u_1), ..., Exp_{p_q}(u_q))$. Since exponential maps are easy to compute in irreducible symmetric spaces, for such products, rather than using caps $C_r(p)$, it is more convenient to use products of caps $C_{r_a}(p_a)$ in the irreducible factors $M_a, a = 1, ..., q$. Also, instead of using the density estimator in (25.1), we use an estimator $\widehat{f}_{n,K}(p)$ that is compatible with the decomposition in irreducible factors. If the random sample is $X = (x_1, ..., x_n)$, where $x_i = (x_{i1}, ..., x_{iq}), \forall i = 1, ..., n$, and $p = (p_1, ..., p_q)$ is an arbitrary point as above then

$$\widehat{f}_{n,K}(p) = \frac{1}{n}\sum_{i=1}^{n} \Pi_{a=1}^{q} |g_{ia}(p_a)|^{-\frac{1}{2}} \frac{1}{r_a^{m_a}} K(\frac{\rho_{g_a}(p_a, x_{ia})}{r_a}). \qquad (25.4)$$

In (25.4), $|g_{ia}(p_a)|$ is the determinant of the metric tensor of M_a at $Log_{x_{ia}} p_a$, and m_a is the dimension of M_a. For convenience, one may take equal radii $r_a, a = 1, ..., q$. The density estimator $\widehat{f}_{n,K}(p)$ has the same asymptotic order of error as $f_{n,K}(p)$ mentioned in Theorem 25.2.1.

25.4 An Example of Projective Shape Density Estimation

Mardia and Patrangenaru (2005) [233] have shown that the projective shape space $P\Sigma_m^k$, of projective k–ads with first $m+2$ points of the k-ad in general position in P^m is a manifold diffeomorphic with a direct product of $q = k-m-2$ copies of P^m (see Subsection 3.5.8). Using the projective frame approach they considered the so called spherical representation of a shape of such a k-ad. When a distribution on the $P\Sigma_m^k$ is concentrated, one may simply regard it as a distribution on a product of q unit spheres S^m, given that the projective space $\mathbb{R}P^m$ is obtained from S^m by identifying antipodal points (see Chapter 3). The planar case (m=2), is relevant in high level image analysis. In this particular case $q = k-4$. Caution should be taken in the selection of landmarks: while the actual physical landmarks of the scene pictured are in $3D$, one should select a group of landmarks that are approximately coplanar.

Recall from Chapter 1, that the "actor" library is a data set of images of an individual who appears in different disguises in front of a camera. A small number of such images have been digitized and six approximately coplanar landmarks (ends of eyes and lips–see coordinates of landmarks (3) to (8) in Table 1.21) have been used in a planar projective shape density estimation example, using Ruymgaart's spherical caps (see Ruymgaart (2004) [298]) where K is a step function. Figure 25.1 displays graphs of smoothed histogram estimators of type $f_{n,r}$ for the distribution of spherical coordinates, of the two spherical marginals associated with the k-ads, for $k = 6$ is given by

$$f_{n,r}(\phi,\theta) = \frac{1}{nc_n(r)}\sum_{i=1}^{n} 1_{B_r(\phi,\theta)}(x_i), x \in \mathbb{R}^2 \tag{25.5}$$

where $B_r(\phi,\theta)$ is a small disc of area $c_n(r)$ with center $(\phi,\theta) \in \mathbb{R}^2$ and radius (bandwidth) r. Here we consider a small geodesic ball of radius $0 < r < \frac{diam\mathbb{R}P^2}{2}$ and refer to is as the *cap* $C_r(x)$, with center $x \in \mathbb{R}P^2$ and radius r. The volume of such a cap has the form

$$|C_r(x)| = c_n(r) = O(r^2). \tag{25.6}$$

In fact the formula of the area of the cap is known from elementary geometry to be $c_n(r) = 2\pi rh$, where h is the height of the corresponding cap on S^2. In our actor data example given that $q = 2$, there are two axial marginals. The marginal densities are pictured in Figure 25.1

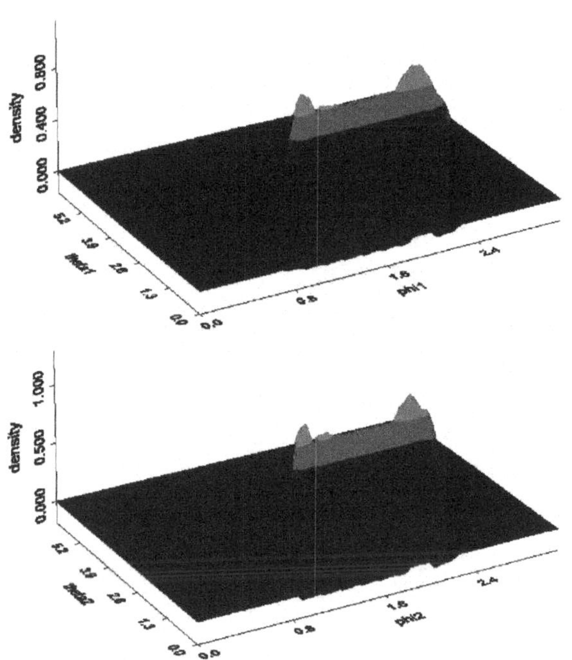

Figure 25.1 *Actor data: marginal densities of kernel estimates in spherical coordinates.*

Part IV

Additional Topics

Chapter 26

Persistent Homology

26.1 Introduction

This section is essentially due to Bubenik et al. (2010) [53], who gracefully introduced us to the fascinating new area of *topological data analysis* (TDA). The prerequisites for this TDA section are Chapter 3, with an emphasis on Subsection 3.4.3.

Big and complex data raise new challenges for quantitative scientists of diverse backgrounds who are being asked to apply the techniques of their specialty to data which is greater in both size and complexity than that which has been studied previously. Massive, data sets on manifolds, for which traditional linear methods are inadequate, pose challenges in representation, visualization, interpretation and analysis. A common finding is that these massive multivariate data sets require the development of new statistical methodology and that these advances are dependent on increasing technical sophistication. Two such data-analytic techniques that have recently come to the fore are computational algebraic topology and statistics on manifolds.

Commonly, one starts with data obtained from some induced geometric structure, such as a curved submanifold of a numerical space, or, even a *singular algebraic variety*, set of zeroes of a polynomial function in more variables. The observed data is obtained as a random sample from this space, and the objective is to statistically recover features of the underlying space.

In *computational algebraic topology*, one attempts to recover qualitative global features of the underlying data, such as connectedness, or the number of holes, or the existence of obstructions to certain constructions, based upon the random sample. In other words, one hopes to recover the underlying topology. An advantage of topology is that it is stable under deformations and thus can potentially lead to robust statistical procedures. A combinatorial construction such as the alpha complex or the Čech complex, see for example Zomorodian (2005) [348], converts the data into an object for which it is possible to compute the topology. However, it is quickly apparent that such a construction and its calculated topology depend on the scale at which one considers the data. A multi–scale solution to this problem is the technique of *persistent homology*. It quantifies the persistence of topological features as the scale changes. Persistent homology is useful for visualization, feature detection and object recognition. Applications of persistent topology include protein structure analysis (see Sacan et al. (2007) [299]) and sensor networks (see de Silva and Ghrist (2007) [81]). In an application to brain image data (see Chung et al. (2009) [74]), a demonstration of persistent topology in discriminating between two populations was exhibited.

In geometric statistics, one uses the underlying Riemannian structure to recover quantitative information concerning the underlying probability distribution and functionals thereof. The idea is to extend statistical estimation techniques to functions over Riemannian manifolds, utilizing the Riemannian structure. One then considers the magnitude of the statistical accuracy of these estimators. Considerable progress has been achieved in terms of optimal estimation which include works by Hendriks (1990) [150], Efromovich (2000) [94], Kim and Koo (2005) [190], Pelletier (2005, 2006) [278], [279], Koo and Kim (2008) [202], and Kim, Koo and Luo (2009) [191]. Other related works include Rooij and Ruymgaart (1991) [328], Ruymgaart (1993) [296], Mair and Ruymgaart (1996) [227], Angers and Kim (2005) [4], and Bissantz, Hohage, Munk and Ruymgaart (2007) [50]. There is also a growing interest in function estimation over manifolds in the learning theory literature. Some papers that have direct relevance are Cucker and Smale (2002a) [79], Smale and Zhou (2004) [310], and Belkin and Niyogi (2004) [15]; see also the references cited therein.

Although computational algebraic topology and *geometric statistics* appear dissimilar and seem to have different objectives, it has recently been noticed that they share a commonality through statistical sampling. In particular, a pathway between them can be established by using elements of Morse theory. This is achieved through the fact that persistent homology can be applied to Morse functions and comparisons between two Morse functions can be assessed by a metric called the *bottleneck distance*. Furthermore, the bottleneck distance is bounded by the sup–norm distance between the two Morse functions on some underlying manifold. This framework thus provides just enough structure for a statistical interpretation. Indeed, consider a nonparametric regression problem on some manifold. Given data in this framework one can construct a nonparametric regression function estimator such that the persistent homology associated with this estimated regression function is an estimator for the persistent homology of the true regression function, as assessed by the bottleneck distance. Since this will be bounded by the sup-norm loss, by providing a sharp sup–norm minimax estimator of the regression function, we can effectively bound the expected bottleneck distance between the estimated persistent homology and the true persistent homology. Consequently, by showing consistency in the sup-norm risk, we can effectively show consistency in the bottleneck risk for persistent homology which is what we will demonstrate. Let us again emphasize that the pathway that allows us to connect computational algebraic topology with geometric statistics is Morse theory. This is very intriguing in that a pathway between the traditional subjects of geometry and topology is also Morse theory.

We now summarize this chapter. In Section 26.1.1 we will lay down the topological preliminaries needed to state our main results. In Section 26.2, we go over the preliminaries needed for nonparametric regression on a Riemannian manifold. Section 26.3 states the main results where sharp sup-norm minimax bounds consisting of constant and rate, and sharp sup-norm estimators are presented. The connection to bounding the persistent homology estimators thus ensues. Following this in Section 26.4, a brief discussion of the implementation is given. Proofs to the main results are collected in Section 26.5.

26.1.1 Persistent Topology

Edelsbrunner et al. (2001) [92] and Zomorodian and Carlsson (2005) [348] provide a computational procedure for determining how the homology persists as the level r changes. In the example in Section 3.4 there are two persistent homology classes (defined below). One class is born when $r = 1.1$, the first sublevel set that has two holes, and dies at $r = 1.4$ the first sublevel set for which the second hole disappears. The other class is born at $r = 0$ and persists until $r = 2$. Thus the persistent homology can be completely described by the two ordered pairs $\{(1.1, 1.4), (0, 2)\}$. This is called the reduced persistence diagram (defined below) of f, denoted $\bar{\mathcal{D}}(f)$. For a persistent homology class described by (a, b), call $b - a$ its lifespan. From the point of view of an experimentalist, a long-lived persistent homology is evidence of a significant feature in the data, while a short-lived one is likely to be an artifact.

A precise definition follows:

DEFINITION 26.1.1. *Let k be a nonnegative integer. Given $f : \mathcal{M} \to \mathbb{R}$ and $a \leq b \in \mathbb{R}$ the inclusion of sublevel sets $i_a^b : \mathcal{M}_{f \leq a} \hookrightarrow \mathcal{M}_{f \leq b}$ induces a map on homology*

$$H_k(i_a^b) : H_k(\mathcal{M}_{f \leq a}) \to H_k(\mathcal{M}_{f \leq b}).$$

The image of $H_k(i_a^b)$ is the persistent homology group from a to b. Let β_a^b be its dimension. This counts the independent homology classes which are born by time a and die after time b.

Call a real number a a homological critical value of f if for all sufficiently small $\varepsilon > 0$ the map $H_k(i_{a-\varepsilon}^{a+\varepsilon})$ is not an isomorphism. Call f tame if it has finitely many homological critical values, and for each $a \in \mathbb{R}$, $H_k(\mathcal{M}_{f \leq a})$ is finite dimensional. In particular, any Morse function on a compact manifold is tame.

Assume that f is tame. Choose ε smaller than the distance between any two homological critical values. For each pair of homological critical values $a < b$, we define their multiplicity μ_a^b which we interpret as the number of independent homology classes that are born at a and die at b. We count the homology classes born by time $a + \varepsilon$ that die after time $b - \varepsilon$. Among these subtract those born by $a - \varepsilon$ and subtract those that die after $b + \varepsilon$. This double counts those born by $a - \varepsilon$ that die after $b + \varepsilon$, so we add them back. That is,

$$\mu_a^b = \beta_{a+\varepsilon}^{b-\varepsilon} - \beta_{a-\varepsilon}^{b-\varepsilon} - \beta_{a+\varepsilon}^{b+\varepsilon} + \beta_{a-\varepsilon}^{b+\varepsilon}.$$

The persistent homology of f may be encoded as follows. The reduced persistence diagram of f, $\bar{\mathcal{D}}(f)$, is the multiset of pairs (a, b) together with their multiplicities μ_a^b. We call this a diagram since it is convenient to plot these points on the plane. We will see that it is useful to add homology classes which are born and die at the same time. Let the persistence diagram of f, $\mathcal{D}(f)$, be given by the union of $\bar{\mathcal{D}}(f)$ and $\{(a, a)\}_{a \in \mathbb{R}}$ where each (a, a) has infinite multiplicity.

26.1.2 Bottleneck Distance

Cohen–Steiner, Edelsbrunner and Harer (2005) [75] introduced the following metric on the space of persistence diagrams. This metric is called the bottleneck distance and it bounds the Hausdorff distance. It is given by

$$d_B(\mathcal{D}(f), \mathcal{D}(g)) = \inf_{\gamma} \sup_{p \in \mathcal{D}(f)} \|p - \gamma(p)\|_\infty, \tag{26.1}$$

where the infimum is taken over all bijections $\gamma : \mathcal{D}(f) \to \mathcal{D}(g)$ and $\|\cdot\|_\infty$ denotes supremum–norm over sets.

For example, let f be the function considered at the start of this section. Let g be a unimodal, radially-symmetric function on the same domain with maximum 2.2 at the origin and minimum 0. We showed that

$\bar{\mathcal{D}}(f) = \{(1.1, 1.4), (0, 2)\}$. Similarly, $\bar{\mathcal{D}}(g) = (0, 2.2)$. The bottleneck distance is achieved by the bijection γ which maps $(0,2)$ to $(0,2.2)$ and $(1.1,1.4)$ to $(1.25,1.25)$ and is the identity on all 'diagonal' points (a,a). Since the diagonal points have infinite multiplicity this is a bijection. Thus, $d_B(\mathcal{D}(f), \mathcal{D}(g)) = 0.2$.

In Cohen–Steiner, Edelsbrunner and Harer (2005) [75], the following result is proven:

$$d_B(\mathcal{D}(f), \mathcal{D}(g)) \leq \|f - g\|_\infty \tag{26.2}$$

where $f, g : \mathcal{M} \to \mathbb{R}$ are tame functions and $\|\cdot\|_\infty$ denotes sup–norm over functions.

26.1.3 Connection to Statistics

It is apparent that most articles on persistent topology do not as of yet incorporate statistical foundations although they do observe them heuristically. One approach that combines topology and statistics is that of Niyogi, Smale and Weinberger (2008) [255]. They calculate how much data is needed to guarantee recovery of the underlying topology of the manifold. A drawback of their technique is that it supposes that the size of the smallest features of the data is known a priori. To date the most comprehensive parametric statistical approach is contained in Bubenik and Kim (2007) [54]. In this chapter, the unknown probability distribution is assumed to belong to a parametric family of distributions. The data is then used to estimate the level so as to recover the persistent topology of the underlying distribution.

As far as we are aware no statistical foundation for the nonparametric case has been formulated although Cohen–Steiner et al.(2005) [75] provide the topological machinery for making a concrete statistical connection. In particular, persistent homology of a function is encoded in its reduced persistence diagram. A metric on the space of persistence diagrams between two functions is available which bounds the Hausdorff distance and this in turn is bounded by the sup–norm distance between the two functions. Thus by viewing one function as the parameter, while the other is viewed as its estimator, the asymptotic sup–norm risk bounds the expected Hausdorff distance thus making a formal nonparametric statistical connection. This in turn lays down a framework for topologically classifying clusters in high dimensions.

26.2 Nonparametric Regression on Manifolds

In general, given two embedded manifolds $j_a : \mathcal{M}_a \to \mathbb{R}^{d_a}, a = 1, 2$, and a random pair (X_1, X_2) on $\mathcal{M} \times \mathcal{N}$, we assume that $\forall x_1 \in \mathcal{M}_1$, the r.o. $X_2|X_1 = x_1$ is j_2-nonfocal. The regression function of predictor X_1 and response X_2 is the function $f_{12} : \mathcal{M}_1 \to \mathcal{M}_2$ given by

$$f_{12}(x_1) = E_{j_2}(X_2|X_1 = x_1), \tag{26.1}$$

where the right hand side of (26.1) is the conditional extrinsic mean of X_2, for $X_1 = x_1$. In particular if $\mathcal{M}_2 = \mathbb{R}^{d_2}$, the right hand side is the conditional mean vector of X_2, for $X_1 = x_1$. In this later case, we may assume in a different formulation, that

$$X_2 = f_{12}(x_1) + \varepsilon, \tag{26.2}$$

where ε has mean vector 0_{d_2}

Consider in particular the following nonparametric regression problem

$$y = f(x) + \varepsilon, \quad x \in \mathcal{M}, \tag{26.3}$$

where $\mathcal{M}$ is a d–dimensional compact Riemannian manifold, $f : \mathcal{M} \to \mathbb{R}$ is the regression function and ε is a normal random variable with mean zero and variance $\sigma^2 > 0$.

For a given sample $(y_1, x_1), \dots, (y_n, x_n)$, let $\tilde{f}$ be an estimator of f based on the regression model (26.3). We will assess the estimator's performance by the sup–norm loss:

$$\| \tilde{f} - f \|_\infty = \sup_{x \in \mathcal{M}} |\tilde{f}(x) - f(x)|. \tag{26.4}$$

Furthermore, we will take as the parameter space, $\Lambda(\beta, L)$, the class of Hölder functions

$$\Lambda(\beta, L) = \{ f : \mathcal{M} \to \mathbb{R} \mid |f(x) - f(z)| \leq L\rho(x,z)^\beta, x, z \in \mathcal{M} \}, \tag{26.5}$$

where $0 < \beta \leq 1$ and ρ is the Riemannian metric on $\mathcal{M}$, i.e., $\rho(x,z)$ is the geodesic length (determined by the metric tensor) between $x, z \in \mathcal{M}$.

For $w(u)$, a continuous non-decreasing function which increases no faster than a power of its argument as $u \to \infty$ with $w(0) = 0$, we define the *sup-norm minimax risk* by

$$r_n(w, \beta, L) = \inf_{\tilde{f}} \sup_{f \in \Lambda(\beta, L)} \mathbb{E} w(\psi_n^{-1} \| \tilde{f} - f \|_\infty), \tag{26.6}$$

where the $\psi_n \to 0$ is the sup–norm minimax rate, as $n \to \infty$, and $\mathbb{E}$ denotes expectation with respect to (26.3) where ε is normally distributed.

26.2.1 Asymptotic Equidistance on Manifolds

Consider a set of points $z_i \in \mathcal{M}$, $i = 1, \cdots, m$. We will say that the set of points is asymptotically equidistant if

$$\inf_{i \neq j} \rho(z_i, z_j) \sim \frac{(\operatorname{vol} \mathcal{M})^{1/d}}{m} \tag{26.7}$$

as $m \to \infty$ for all $i, j = 1, \dots, m$, where for two real sequences $\{a_m\}$ and $\{b_m\}$, $a_m \sim b_m$ will mean $|a_m / b_m| \to 1$ as $m \to \infty$, this implies that

$$\frac{\max_j \min_{i \neq j} \rho(z_i, z_j)}{\min_j \min_{i \neq j} \rho(z_i, z_j)} \sim 1 , \tag{26.8}$$

as $m \to \infty$. It will be assumed throughout that the manifold admits a collection of asymptotically equidistant points. This is certainly true for the sphere (in any dimension), and will be true for all compact Riemannian manifolds since the injectivity radius is strictly positive. We note that Pelletier [279] makes use of this condition as well.

We will need the following constants

$$C_0 = L^{d/(2\beta+d)} \left(\frac{\sigma^2 \mathrm{vol}\,\mathcal{M}\ (\beta+d)d^2}{\mathrm{vol}\,\mathbb{S}^{d-1}\beta^2} \right)^{\beta/(2\beta+d)}, \tag{26.9}$$

$$\psi_n = \left(\frac{\log n}{n} \right)^{\beta/(2\beta+d)}, \tag{26.10}$$

and "vol" denotes the volume of the object in question, where $\mathbb{S}^{d-1}$ is the $(d-1)$–dimensional unit sphere with vol $\mathbb{S}^{d-1} = 2\pi^{d/2}/\Gamma(d/2)$ and Γ is the gamma function.

Define the geodesic ball of radius $r > 0$ centered at $z \in \mathcal{M}$ by

$$B_z(r) = \{x \in \mathcal{M} \,|\, \rho(x,z) \le r\}. \tag{26.11}$$

We have the following result whose proof will be detailed in Section 26.5.1

LEMMA 26.2.1. *Let $z_i \in \mathcal{M}, i = 1, \cdots, m$, be asymptotically equidistant. Let $\lambda = \lambda(m)$ be the largest number such that $\bigcup_{i=1}^m \overline{B}_{z_i}(\lambda^{-1}) = \mathcal{M}$, where $\overline{B}_{z_i}(\lambda^{-1})$ is the closure of the geodesic ball of radius λ^{-1} around z_i. Then there is a $C_1 > 0$ such that* $\limsup_{m\to\infty} m\lambda(m)^{-d} \le C_1$.

26.2.2 An Estimator

Fix a $\delta > 0$ and let

$$m = \left[C_1 \left(\frac{L(2\beta+d)}{\delta C_0 d \psi_n} \right)^{d/\beta} \right],$$

where C_1 is a sufficiently large constant from Lemma 26.2.1, hence $m \le n$ and $m \to \infty$ when $n \to \infty$ and for $s \in \mathbb{R}$, $[s]$ denotes the greatest integer part.

For the design points $\{x_i : i = 1, \ldots, n\}$ on $\mathcal{M}$, assume that $\left\{x_{i_j} \in \mathcal{M}, j = 1, \ldots, m\right\}$ is an asymptotically equidistant subset on $\mathcal{M}$. Let $A_j, j = 1, \ldots, m$, be a partition of $\mathcal{M}$ such that A_j is the set of those $x \in \mathcal{M}$ for which x_{i_j} is the closest point in the subset $\{x_{i_1}, \ldots, x_{i_m}\}$. Thus, for $j = 1, \ldots, m$,

$$A_j = \left\{ x \in \mathcal{M} \mid \rho(x_{i_j}, x) = \min_{k=1,\ldots,m} \{\rho(x_{i_k}, x)\} \right\}. \tag{26.12}$$

Let A_j, $j = 1, \ldots, m$ be as in (26.12) and define $1_{A_j}(x)$ to be the indicator function on the set A_j and consider the estimator

$$\hat{f}(x) = \sum_{j=1}^{m} \hat{a}_j 1_{A_j}(x), \tag{26.13}$$

where for $L > 0$, $0 < \beta \leq 1$,

$$\hat{a}_j = \frac{\sum_{i=1}^{n} K_{\kappa, x_{i_j}}(x_i) y_i}{\sum_{i=1}^{n} K_{\kappa, x_{i_j}}(x_i)},$$

$$K_{\kappa, x_{i_j}}(\omega) = \left(1 - (\kappa \rho(x_{i_j}, \omega))^{\beta}\right)_+,$$

$$\kappa = \left(\frac{C_0 \psi_n}{L}\right)^{-1/\beta},$$

and $s_+ = \max(s, 0)$, $s \in \mathbb{R}$. We remark that when m is sufficiently large hence κ is also large, the support set of $K_{\kappa, x_{i_j}}(\omega)$ is the closed geodesic ball $\overline{B}_{x_{i_j}}(\kappa^{-1})$ around x_{i_j} for $j = 1, \ldots, m$.

26.3 Main Results

We now state the main results of this chapter. The first result provides an upper bound for the estimator (26.13), where the function $w(u)$ satisfies $w(0) = 0$, $w(u) = w(-u)$, $w(u)$ does not decrease, and $w(u)$ increases not faster than a power as $u \to \infty$.

THEOREM 26.3.1. *For the regression model* (26.3) *and the estimator* (26.13), *we have*

$$\sup_{f \in \Lambda(\beta, L)} \mathbb{E} w\left(\psi_n^{-1} \|\hat{f} - f\|_{\infty}\right) \leq w(C_0),$$

as $n \to 0$, *where* $\psi_n = (n^{-1} \log n)^{\beta/(2\beta + d)}$.

We have the asymptotic minimax result for the sup–norm risk .

THEOREM 26.3.2. *For the regression model* (26.3)

$$\lim_{n \to \infty} r_n(w, \beta, L) = w(C_0).$$

In particular, we have the immediate result.

COROLLARY 26.3.1. *For the regression model* (26.3) *and the estimator* (26.13),

$$\sup_{f \in \Lambda(\beta, L)} \mathbb{E} \|\hat{f} - f\|_{\infty} \sim C_0 \left(\frac{\log n}{n}\right)^{\beta/(2\beta + d)}$$

as $n \to \infty$.

We note that the above generalizes earlier one-dimensional results of Korostelev (1993) [204], and Korostelev and Nussbaum (1996) [205], where the domain is the unit interval. Klemelä (2000) [195] generalizes this result to higher dimensional unit spheres.

Now that a sharp sup–norm minimax estimator has been found we would like to see how we can use this for topological data analysis. The key is the sup–norm bound on the bottleneck distance for persistence diagrams. In particular, for the regression function f in (26.3) and $\hat{f}$ the estimator (26.13), we have the persistence diagram $\mathcal{D}(f)$ as well as an estimator of the persistence diagram $\mathcal{D}(\hat{f})$. Using the results of Section 26.1.2, and in particular (26.2), we have

$$d_B\left(\mathcal{D}(\hat{f}),\mathcal{D}(f)\right) \leq \left\|\hat{f}-f\right\|_{\infty}. \tag{26.1}$$

Let $\Lambda_t(\beta,L)$ denote the subset of tame functions in $\Lambda(\beta,L)$. By Corollary 26.3.1, the following result is immediate.

COROLLARY 26.3.2. *For the nonparametric regression model* (26.3), *let* $\hat{f}$ *be defined by* (26.13). *Then for* $0<\beta\leq 1$ *and* $L>0$,

$$\sup_{f\in\Lambda_t(\beta,L)} \mathbb{E}d_B\left(\mathcal{D}(\hat{f}),\mathcal{D}(f)\right) \leq L^{d/(2\beta+d)}\left(\frac{\sigma^2\mathrm{vol}\,\mathcal{M}\;(\beta+d)d^2}{\mathrm{vol}\,\mathbb{S}^{d-1}\beta^2}\;\frac{\log n}{n}\right)^{\beta/(2\beta+d)}$$

as $n\to 0$.

26.4 Discussion

To calculate the persistence diagrams of the sublevel sets of $\widehat{f}$, we suggest that because of the way $\widehat{f}$ is constructed, we can calculate its persistence diagrams using a triangulation, $\mathcal{T}$ of the manifold in question.

We can then filter $\mathcal{T}$ using $\widehat{f}$ as follows. Let $r_1\leq r_2\leq\ldots\leq r_m$ be the ordered list of values of $\widehat{f}$ on the vertices of the triangulation. For $1\leq i\leq m$, let $\mathcal{T}_i$ be the subcomplex of $\mathcal{T}$ containing all vertices v with $\widehat{f}(v)\leq r_i$ and all edges whose boundaries are in $\mathcal{T}_i$ and all faces whose boundaries are in $\mathcal{T}_i$. We obtain the following filtration of $\mathcal{T}$,

$$\phi=\mathcal{T}_0\subseteq\mathcal{T}_1\subseteq\mathcal{T}_2\subseteq\cdots\subseteq\mathcal{T}_m=\mathcal{T}.$$

Because the critical points of $\widehat{f}$ only occur at the vertices of $\mathcal{T}$, Morse theory guarantees that the persistent homology of the sublevel sets of $\widehat{f}$ equals the persistent homology of the above filtration of $\mathcal{T}$.

Using the software Plex, one calculates the persistent homology, in degrees 0, 1, 2, ..., d of the triangulation $\mathcal{T}$ filtered according to the estimator. Since the data will be d–dimensional, we do not expect any interesting homology in higher degrees, and in fact, most of the interesting features would occur in the lower degrees.

A demonstration of this is provided in Chung et al. (2009) [74] for brain image data, where the topology of cortical thickness in an autism study takes place. The persistent homology, in degrees 0, 1 and 2 is calculated for 27 subjects. Since the data is two-dimensional, we do not expect any interesting homology in higher degrees. For an initial comparison of the autistic subjects and control subjects, we take the union of the persistence diagrams, see Figure 4 in Chung et al. (2009) [74] page 392. We note the difference in the topological structures as seen through the persistent homologies between the autistic and control group, particularly, as we move away from the diagonal line. A test using concentration pairings reveal group differences.

26.5 Proofs

The proofs will use the ideas from Klemelä (2000) [195] and Korostelev (1993) [204].

26.5.1 *Upper Bound*

We first prove the earlier lemma.

Proof [Proof of Lemma 26.2.1]. Let $(\mathcal{U},(x^i))$ be any normal coordinate chart centered at x_i, then the components of the metric at x_i are $g_{ij} = \delta_{ij}$, so $\sqrt{|g_{ij}(x_i)|} = 1$, see [215]. Consequently,

$$\begin{aligned}
\mathrm{vol}(\overline{B}_{x_i}(\lambda^{-1})) &= \int_{B(\lambda^{-1})} \sqrt{|g_{ij}(\exp_{x_i}(x))|}dx \\
&= \sqrt{|g_{ij}(\exp_{x_i}(t))|}\int_{B(\lambda^{-1})} dx \\
&\sim \mathrm{vol}(\mathbb{B}(\lambda^{-1})) \\
&= \mathrm{vol}(\mathbb{B}(1))\lambda^{-d} \\
&= \mathrm{vol}(\mathbb{S}^{d-1})\lambda^{-d}/d\,.
\end{aligned}$$

The first line uses the integration transformation, where $\exp_{x_i}: B(\lambda^{-1}) \to \overline{B}_{x_i}(\lambda^{-1})$ is the exponential map from the tangent space $\mathrm{T}\mathcal{M}_{x_i} \to \mathcal{M}$. The second line uses the integral mean value theorem and r is the radius from the origin to point x in the Euclidean ball $\mathbb{B}(\lambda^{-1})$. The third line is asymptotic as $\lambda \to \infty$ and uses the fact that $|g_{ij}(\exp_{x_i}(t))| \to 1$ when $\lambda \to \infty$. In the fourth line $\mathrm{vol}(\mathbb{B}(1))$ is the volume of d-dimensional Euclidean unit ball. The last line uses the fact $\mathrm{vol}(\mathbb{B}(1)) = \mathrm{vol}(\mathbb{B}^{d-1})/d$.

Let $\lambda' = \lambda'(m) > 0$ be the smallest number such that $\overline{B}_{x_i}((\lambda')^{-1})$ are disjoint. Then

$$\lambda^{-1} = c(m) \times (\lambda')^{-1}$$

where $c(m) > 1$ and $c(m) \to 1$ as $m \to \infty$. Consequently

$$\text{vol}(\mathcal{M}) \geq \sum_{i=1}^{m} \text{vol}(\overline{B}_{x_i}((\lambda')^{-1})) \sim m\text{vol}(\mathbb{S}^{d-1})(\lambda')^{-d}/d.$$

Thus

$$\limsup_{m\to\infty} m\lambda(m)^{-d} = \limsup_{m\to\infty} c(m)^d m(\lambda')^{-d} \leq \frac{d\text{vol}(\mathcal{M})}{\text{vol}(\mathbb{S}^{d-1})}.$$

We now calculate the asymptotic variance of $\hat{a}_j$ for $j = 1, \ldots, m$. Let

$$M = \left\lceil \frac{n\text{vol}(\overline{B}_{x_{i_j}}(\kappa^{-1}))}{\text{vol}(\mathcal{M})} \right\rceil$$

$$\begin{aligned}
\text{var}(\hat{a}_j) &= \frac{\sigma^2 \sum_{i=1}^n K^2_{\kappa,x_{i_j}}(x_i)}{(\sum_{i=1}^n K_{\kappa,x_{i_j}}(x_i))^2} \\
&\sim \frac{\sigma^2 \text{vol}(\overline{B}_{x_{i_j}}(\kappa^{-1})) \int_{\overline{B}_{x_{i_j}}(\kappa^{-1})} (1-(\kappa\rho(x_{i_j},\omega))^\beta)^2 d\omega}{M(\int_{\overline{B}_{x_{i_j}}(\kappa^{-1})}(1-(\kappa\rho(x_{i_j},\omega))^\beta)d\omega)^2} \\
&= \frac{\sigma^2 \text{vol}(\overline{B}_{x_{i_j}}(\kappa^{-1})) \int_{B(\kappa^{-1})}(1-(\kappa r)^\beta)^2 \sqrt{|g_{ii_j}(\exp_{x_{i_j}}(x))|}dx}{M\,(\int_{B(\kappa^{-1})}(1-(\kappa r)^\beta)\sqrt{|g_{ii_j}(\exp_{x_{i_j}}(x)))|}dx)^2} \\
&= \frac{\sigma^2 \text{vol}(\overline{B}_{x_{j_i}}(\kappa^{-1}))\sqrt{|g_{ii_j}(\exp_{x_{i_j}}(t))|} \int_0^{\kappa^{-1}} \int_0^{\pi}\cdots\int_0^{\pi}\int_0^{2\pi}(1-(\kappa r)^\beta)^2 r^{d-1} dr d\sigma_{d-1}}{M\,|g_{ii_j}(\exp_{x_{i_j}(t')})|(\int_0^{\kappa^{-1}}\int_0^{\pi}\cdots\int_0^{\kappa^{-1}}\int_0^{\pi}\cdots\int_0^{\pi}\int_0^{2\pi}(1-(\kappa r)^\beta)^2 r^{d-1} dr d\sigma_{d-1})^2} \\
&\sim \frac{\sigma^2 \text{vol}(\overline{B}_{x_{i_j}}(\kappa^{-1}))d\text{vol}(\mathbb{B}^d)\int_0^{\kappa^{-1}}(1-(\kappa r)^\beta)^2 r^{d-1}dr}{M\,d^2\text{vol}(\mathbb{B}^d)^2(\int_0^{\kappa^{-1}}(1-(\kappa r)^\beta)r^{d-1}dr)^2} \\
&= \sigma^2\kappa^d \frac{\text{vol}(\mathcal{M})2d(\beta+d)}{n\text{vol}(\mathbb{S}^{d-1})(2\beta+d)}
\end{aligned}$$

as $n \to \infty$, where $d\sigma_{d-1}$ is the spherical measure on $\mathbb{S}^{d-1}$.

LEMMA 26.5.1.

$$\lim_{n\to\infty} \mathbb{P}\left(\psi_n^{-1} \parallel \hat{f}_n - \mathbb{E}\hat{f}_n \parallel_\infty > (1+\delta)C_0\frac{2\beta}{2\beta+d}\right) = 0$$

Proof. Denote $Z_n(x) = \hat{f}_n(x) - \mathbb{E}\hat{f}_n(x)$. Define

$$D_n^2 = \text{var}(\psi_n^{-1}Z_n(x_j)) = \psi_n^{-2}\text{var}(\hat{a}_j) \sim \frac{2\beta^2 C_0^2}{d(2\beta+d)\log n}.$$

Denote $y = (1+\delta)C_0 2\beta/(2\beta+d)$. Then

$$\frac{y^2}{D_n^2} = \frac{2d(1+\delta)^2\log n}{2\beta+d}.$$

For sufficiently large n, $Z_n(x_j) \sim N(0, \psi_n^2 D_n^2)$, hence as $n \to \infty$,

$$
\begin{aligned}
\mathbb{P}\left(\| \psi_n^{-1} Z_n \|_\infty > y \right) &\leq \mathbb{P}\left(\max_{i=1,\cdots,m} \psi_n^{-1} |Z_n(x_j)| > y \right) \\
&\leq m\mathbb{P}\left(D_n^{-1} \psi_n^{-1} |Z_n(x_j)| > \frac{y}{D_n} \right) \\
&\leq m\exp\left\{ -\frac{1}{2} \frac{y^2}{D_n^2} \right\} \\
&= m\exp\left\{ -\frac{d(1+\delta)^2 \log n}{2\beta + d} \right\} \\
&= n^{-d((1+\delta)^2-1)/(2\beta+d)} (\log n)^{-d/(2\beta+d)} D_n \left(\frac{L(2\beta+d)}{\delta C_0 d} \right)^{d/\beta}.
\end{aligned}
$$

LEMMA 26.5.2.

$$
\limsup_{n\to\infty} \sup_{f \in \Lambda(\beta, L)} \psi_n^{-1} \| f - \mathbb{E}\hat{f}_n \|_\infty \leq (1+\delta) C_0 \frac{d}{2\beta + d}
$$

Proof. We note that

$$
\begin{aligned}
\| f - \mathbb{E}\hat{f} \|_\infty &= \max_{j=1,\ldots,m} \sup_{x \in A_j} |f(x) - \mathbb{E}\hat{f}(x)| \\
&\leq \max_{j=1,\ldots,m} \sup_{x \in A_j} \left(|f(x) - f(x_j)| + |\mathbb{E}\hat{f}(x_j) - f(x_j)| \right) \\
&\leq \max_{j=1,\ldots,m} \left(|\mathbb{E}\hat{f}(x_j) - f(x_j)| + L \sup_{x \in A_j} \rho(x, x_j)^\beta \right).
\end{aligned}
$$

When m is sufficiently large, $A_j \subset \overline{B}_{x_j}(\lambda^{-1})$, hence by Lemma 26.2.1

$$
\limsup_{n\to\infty} \sup_{x \in A_j} \rho(x, x_j) \leq \limsup_{n\to\infty} \lambda^{-1} \leq \limsup_{n\to\infty} \left(\frac{C_1}{m} \right)^{1/d}.
$$

Thus

$$
\limsup_{n\to\infty} \sup_{x \in A_j} \psi_n^{-1} \rho(x, x_j)^\beta \leq \limsup_{n\to\infty} \psi_n^{-1} \left(\frac{C_1}{m} \right)^{\beta/d} \leq \frac{\delta C_0 d}{L(2\beta + d)}.
$$

For $j = 1, \cdots, m$,

$$
\begin{aligned}
|\mathbb{E}\hat{f}(x_{i_j}) - f(x_{i_j})| &= |E\hat{a}_j - f(x_{i_j})| \\
&= \left| \frac{\sum_{j=1}^{m} K_{\kappa,x_i}(x_{i_j}) f(x_{i_j})}{\sum_{j=1}^{m} K_{\kappa,x_i}(x_{i_j})} - f(x_i) \right| \\
&\leq \frac{\sum_{j=1}^{m} K_{\kappa,x_i}(x_{i_j}) |f(x_{i_j}) - f(x_i)|}{\sum_{j=1}^{m} K_{\kappa,x_i}(x_{i_j})} \\
&\leq \frac{L \int_{\overline{B}_{x_i}(\kappa^{-1})} (1 - (\kappa\rho(x_i,\omega))^{\beta}) \rho(x_i,\omega))^{\beta} d\omega}{\int_{\overline{B}_{x_i}(\kappa^{-1})} (1 - (\kappa\rho(x_i,\omega))^{\beta}) d\omega} \\
&\sim \frac{L}{\kappa^{\beta}} \frac{d}{2\beta + d} \\
&= C_0 \psi_n \frac{d}{2\beta + d}
\end{aligned}
$$

as $n \to \infty$.

Proof [Proof of the upper bound].

$$
\begin{aligned}
&\lim_{n\to\infty} \mathbb{P}\left(\psi_n^{-1} \parallel \hat{f} - f \parallel_{\infty} > (1+\delta)C_0\right) \\
&\leq \lim_{n\to\infty} \mathbb{P}\left(\psi_n^{-1} \parallel \hat{f} - \mathbb{E}\hat{f} \parallel_{\infty} + \psi_n^{-1} \parallel \mathbb{E}\hat{f} - f \parallel_{\infty} > (1+\delta)C_0\right) \\
&\leq \lim_{n\to\infty} \mathbb{P}\left(\psi_n^{-1} \parallel \hat{f} - \mathbb{E}\hat{f} \parallel_{\infty} + (1+\delta)C_0 \frac{d}{2\beta+d} > (1+\delta)C_0\right) \\
&= \lim_{n\to\infty} \mathbb{P}\left(\psi_n^{-1} \parallel \hat{f} - \mathbb{E}\hat{f} \parallel_{\infty} > (1+\delta)C_0 \frac{2\beta}{2\beta+d}\right) = 0
\end{aligned}
$$

the second inequality uses Lemma 26.5.2 and the last line uses Lemma 26.5.1.

Let g_n be the density function of $\psi_n^{-1} \parallel \hat{f} - f \parallel_{\infty}$, then

$$
\begin{aligned}
&\limsup_{n\to\infty} \mathbb{E} w^2(\psi_n^{-1} \parallel \hat{f}_n - f \parallel_{\infty}) \\
&= \limsup_{n\to\infty} \left(\int_0^{(1+\delta)C_0} w^2(x) g_n(x) dx + \int_{(1+\delta)C_0}^{\infty} w^2(x) g_n(x) dx \right) \\
&\leq w^2((1+\delta)C_0) + \limsup_{n\to\infty} \int_{(1+\delta)C_0}^{\infty} x^{\alpha} g_n(x) dx \\
&= w^2((1+\delta)C_0) \\
&\leq B < \infty,
\end{aligned}
$$

where the constant B does not depend on f, the third lines uses the assumption on the power growth and non-decreasing property of the loss function $w(u)$.

Using the Cauchy–Schwartz inequality, we have

$$\begin{aligned}
&\limsup_{n\to\infty} \mathbb{E}w(\psi_n^{-1} \parallel \hat{f}_n - f \parallel_\infty) \\
&\le w((1+\delta)C_0) \limsup_{n\to\infty} \mathbb{P}\left(\psi_n^{-1} \parallel \hat{f} - f \parallel_\infty \le (1+\delta)C_0\right) \\
&+ \limsup_{n\to\infty} \left\{\mathbb{E}w^2(\psi_n^{-1} \parallel \hat{f}_n - f \parallel_\infty)\mathbb{P}(\psi_n^{-1} \parallel \hat{f} - f \parallel_\infty > (1+\delta)C_0)\right\}^{1/2} \\
&= w((1+\delta)C_0).
\end{aligned}$$

26.5.2 *The Lower Bound*

We now prove the lower bound result on $\mathcal{M}$.

LEMMA 26.5.3. *For sufficiently large κ, let $N = N(\kappa)$ be such that $N \to \infty$ when $\kappa \to \infty$ and $x_i \in \mathcal{M}, i = 1, \cdots, N$, be such that x_i are asymptotically equidistant,and such that $\overline{B}_{x_i}(\kappa^{-1})$ are disjoint. There is a constant $0 < D < \infty$ such that*

$$\liminf_{\kappa\to\infty} N(\kappa)\kappa^{-d} \ge D. \tag{26.1}$$

Proof. Let $\kappa' > 0$ be the largest number such that $\bigcup_{i=1}^{N(\kappa)} \overline{B}_{x_i}((\kappa')^{-1}) = \mathcal{M}$. Then

$$(\kappa')^{-1} = c(\kappa) \times \kappa^{-1}$$

where $c(\kappa) > 1$ and $c(\kappa) \to$ const. ≥ 1 as $\kappa \to \infty$.

$$\text{vol}(\mathcal{M}) \le \sum_{i=1}^{N} \text{vol}(\overline{B}_{x_i}((\kappa')^{-1})) \sim N \text{vol}(\mathbf{S}^{d-1})(\kappa')^{-d}/d$$

Thus

$$\liminf_{\kappa\to\infty} N(\kappa)\kappa^{-d} = \liminf_{\kappa\to\infty} c(\kappa)^{-d} N(\kappa')^{-d} \ge \text{const.} \times \frac{d\text{vol}(\mathcal{M})}{\text{vol}(\mathbf{S}^{d-1})}.$$

Let $J_{\kappa,x} : \mathcal{M} \to \mathbb{R}$, and

$$J_{\kappa,x} = L\kappa^{-\beta} K_{\kappa,x}(x) = L\kappa^{-\beta}(1 - (\kappa d(x,x))^{\beta})_+,$$

where $\kappa > 0, x \in \mathcal{M}$. Let $N = N(\kappa)$ be the greatest integer such that there exists observations $x_i \in \mathcal{M}, i = 1, \cdots, N$ (with possible relabeling) in the observation set $\{x_i, i = 1, \cdots, n\}$ such that the functions J_{κ,x_i} have disjoint supports. From (26.1)

$$\liminf_{\kappa\to\infty} N(\kappa)\kappa^{-d} \ge \text{const.}$$

Let

$$\mathcal{C}(\kappa, \{x_i\}) = \left\{ \sum_{i=1}^{N} \theta_i J_{\kappa,x_i} : |\theta_i| \le 1, i = 1, \cdots, N \right\},$$

where $\mathcal{C}(\kappa, \{x_i\}) \subset \Lambda(\beta, L)$ when $0 < \beta \leq 1$. The complete class of estimators for estimating $f \in \mathcal{C}(\kappa, \{x_i\})$ consists of all of the form

$$\hat{f}_n = \sum_{i=1}^{N} \hat{\theta}_i J_{\kappa, x_i} \tag{26.2}$$

where $\hat{\theta}_i = \delta_i(z_1, \cdots, z_N), i = 1, \cdots, N$, and

$$z_i = \frac{\sum_{j=1}^{n} J_{\kappa, x_i}(x_j) y_j}{\sum_{j=1}^{n} J_{\kappa, x_i}^2(x_j)}.$$

When $\hat{f}_n$ is of the form (26.2) and $f \in \mathcal{C}(\kappa, \{x_i\})$ then

$$\begin{aligned} \| \hat{f}_n - f \|_\infty &\geq \max_{i=1,\cdots,N} |\hat{f}_n(x_i) - f(x_i)| \\ &= |J_{\kappa, x_1}(x_1)| \, \| \hat{\theta} - \theta \|_\infty \\ &= L\kappa^{-\beta} \, \| \hat{\theta} - \theta \|_\infty \end{aligned}$$

Hence

$$\begin{aligned} r_n &\geq \inf_{\hat{f}_n} \sup_{f \in \mathcal{C}(\kappa, \{x_i\})} \mathbb{E} w(\psi_n^{-1} \| \hat{f}_n - f \|_\infty) \\ &\geq \inf_{\hat{\theta}} \sup_{|\theta_i| \leq 1} \mathbb{E} w(\psi_\varepsilon^{-1} L\kappa^{-\beta} \| \hat{\theta} - \theta \|_\infty), \end{aligned}$$

where the expectation is with respect to a multivariate normal distribution with mean vector θ and the variance-covariance matrix $\sigma_N^2 \mathbf{I}_N$, where $\mathbf{I}_N$ is the $N \times N$ identity matrix and $\sigma_N^2 = \text{var}(z_1) = \sigma^2 / \sum_{j=1}^{N} J_{\kappa, x_i}^2(x_j)$.

Fix a small number δ such that $0 < \delta < 2$ and

$$C_0' = L^{d/(2\beta+d)} \left(\frac{(2-\delta)\text{vol}(\mathcal{M})(\beta+d)d^2}{2\text{vol}(\mathbf{S}^{d-1})\beta^2} \right)^{\beta/(2\beta+d)}$$

and

$$\kappa = \left(\frac{C_0' \psi_\varepsilon}{L} \right)^{-1/\beta}.$$

Since

$$\begin{aligned}
\sigma_N^{-1} &= \sigma^{-1}\sqrt{\sum_{j=1}^{N} J^2_{\kappa,x_i}(x_j)} \\
&\sim \sqrt{\frac{(2-\delta)d}{2\beta+d}\log n} \\
&\leq \sqrt{(2-\delta)(\log(\log n/n)^{-d/(2\beta+d)})} \\
&= \sqrt{2-\delta}\sqrt{\log(\text{cons}\times\kappa^d)} \\
&= \sqrt{2-\delta}\sqrt{\log N}
\end{aligned}$$

by (26.1), it follows that if

$$\sigma_N^{-1} \leq \sqrt{2-\delta}\sqrt{\log N}$$

for some $0 < \delta < 2$, then

$$\inf_{\hat{\theta}} \sup_{|\theta_i|\leq 1} \mathbb{E}w(\| \hat{\theta} - \theta \|_\infty) \to w(1),$$

as $N \to \infty$, but

$$\psi_n^{-1} L\kappa^{-\beta} = C_0'.$$

By the continuity of the function w, we have

$$\inf_{\hat{\theta}} \sup_{|\theta_i|\leq 1} \mathbb{E}w(\psi_n^{-1} L\kappa^{-\beta} \| \hat{\theta} - \theta \|_\infty) \to w(C_0'),$$

when $N \to \infty$. Since δ was chosen arbitrarily, the result follows.

Chapter 27

Further Directions in Nonparametric Statistics on Manifolds

27.1 Introduction

Throughout this text, a number of locally homogeneous spaces (see Patrangenaru (1994) [264]) have been discussed that arise as sample spaces in data analysis. These manifolds are smooth, including $\mathbb{R}^p$ for multivariate analysis and the spheres S^{p-1} for directional data analysis. Certain Lie groups, such as the special orthogonal groups for the analysis of the movement of tectonic plates (see Chang (1988) [63]) and the group of positive definite symmetric matrices for DTI analysis (see Osborne et al. (2013) [258]), also fall into this category. Additionally, real and complex Grassmann manifolds arise for the analysis of affine shape spaces (see Patrangenaru and Mardia (2002) [274]), Kendall's similarity shape spaces (see Kendall (1984) [177]), and in signal tracking problems. Products of real projective spaces are found in projective shape analysis (see Mardia and Patrangenaru (2005) [233]).

Due to the increase of the use of nonparametric statistics on manifolds, including those discussed above, the growth of both methodology and applications of such methods prevents us from exhaustively presenting the entire field in this manuscript. However, we wish to mention some recent developments in a number of these areas in the second section of this chapter. In the third section, which follows from Bhattacharya et al. (2012) [37]), we will touch upon some computational issues that arise when analyzing data.

27.2 Additional Topics

As has been discussed throughout this text, nonparametric methods for statistical analysis on data with manifold sample spaces have tended to develop in at least one of two manners. First, many of these methods serve to generalize standard methods for Euclidean data analysis for sample spaces that have been previously studied. Secondly, the methods are often developed to extend nonparametric statistics on manifolds to new types of data. In this section, we will first discuss two developments that fall into the first category, those being principal component analysis and spatial statistics. We will then consider shape analysis of 3D surfaces, which fits into the second category. Finally, we will discuss a further generalization of statistical methods to spaces with a manifold stratification.

27.2.1 Principal Component Analysis

For multivariate data, principal component analysis (PCA) greatly aids in understanding the primary directions of variation. These can, in turn, be used to perform dimension reduction by considering only those principal components that contribute greatly to the variability. For data on manifolds, understanding the principal directions of variation is of great use to help visualize the variability, especially due to the abstract nature of the sample spaces.

Dryden and Mardia (1998) [91] and Fletcher et al. (2004) [120] suggest to perform Euclidean PCA for data on manifolds by, respectively, projecting data onto the tangent space of the extrinsic mean or using the inverse Riemannian exponential map. However, in recent years, the focus has shifted to performing analogues of PCA directly on the manifold. Huckemann and Hotz (2009) [165] and Huckemann et al. (2010) present notions of geodesic principal component analysis for, resepctively, planar shape spaces under a Riemannian metric and general Riemannian manifolds. Jung et al. (2012) [172] introduced the idea of principal nested spheres as a decomposition of S^p to capture non-geodesic variation on a unit sphere using nested sub-manifolds of decreasing dimension. Principal nested spheres are also shown to be applicable to Kendall's shape space under some modifications.

27.2.2 Spatial Statistics

Spatial statistics is an area of study largely concerned with spatial characterization and spatial dependency of data. Spatial methods have useful applications in geography and meteorology, among many others. These two are of particular note here, though, as data points tend to lie on the surface of the earth, and as such, can be modelled as though they are on a sphere. For applications in these areas in which the data is located within a region well approximated using a plane, traditional methods are suitable. However, when this is not the case, the standard methods may not be suitable due to the curvature of the earth's surface. As such, techniques for statistics on manifolds may be of great use. Recently, Hitczenko and Stein (2012) [158] considered such a problem and utilize sphericial harmonics to model Gaussian processes on a sphere.

27.2.3 Shape Analysis of Surfaces

Planar shape analysis developed from initially being concerned with shapes of finite dimensional configurations to those of contours. Similarly, theory and methodology have been developed for shapes of configurations in higher dimensions. While most of this work has been done for either finite dimensional configurations or curves, recent years have given rise to the study of shapes of surfaces in three dimensions. Of particular note, Kurtek et al. (2012) [208] uses a parametrization-invariant Riemannian metric to analyze shapes of parametrized surfaces.

27.2.4 Statistics on Stratified Spaces

In recent years, many researchers have also become interested in studying data that arise on spaces that are not manifolds, but rather are stratified spaces, where each stratum is a manifold. The Data Analysis on Sample Spaces with a Manifold Stratification working group from the 2010-2011 SAMSI Program on Analysis of Object Data delved into problems arising in this area. Bhattacharya et al. (2011) [34] presents an overview of a number of problems arising in this area. One specific area of interest that this group studied is that for a number of these spaces, Fréchet sample means exhibit a property known as *stickiness*. In short, stickiness refers to the occurrence of a Fréchet sample mean remaining at a given location when the underlying probability distribution is perturbed. Hotz et al. (2013) [162] presents asymptotic results on a particular space known as an open book. Ellingson et al. (2014) [103] discusses stickiness on sample spaces that are low-dimensional stratified spaces.

A specific type of stratified space that has garnered great interest is the space of phylogenetic trees, which was introduced in Billera et al. (2001) [47] and is discussed in the papers referenced above. Barden et al. (2013) [13] presents asymptotic results for Fréchet sample means on the space of phylogenetic trees.

Among other topics, Feragen et al. (2013) [114] presents methodology for performing principal component analysis and a permutation test for problems involving tree-valued data. Skwerer et al. (2014) [308] applies ideas developed for phylogenetic trees to the structure of brain arteries.

27.3 Computational Issues

As seen throughout this text, Fréchet means have been shown to be very important for nonparametric analysis of data arising from sample spaces that are manifolds. The asymptotic distributions of Fréchet sample means for a random sample from a probability measure on a smooth manifold were derived by Bhattacharya and Patrangenaru (2005) [43], as described in Chapter 5, providing methodology for inference. However, computations of Cartan means (or intrinsic sample means) for a given geodesic distance are, in general, based on iterative algorithms, making evaluation of Cartan means time consuming.

While manifolds like the sphere, the torus, and the surface of a pretzel are easy to comprehend as submanifolds of Euclidean spaces, other manifolds naturally arising in statistics have abstract descriptions, such as the spaces of axes and shape spaces. For the latter case, these manifolds have to be *embedded* into a numerical space $\mathbb{R}^N$ for a better understanding. Whitney showed that any smooth m-dimensional manifold can be embedded in $\mathbb{R}^{2m+1}$. A manifold that is embedded in $\mathbb{R}^N$ naturally inherits a Riemannian structure. Therefore, an embedded manifold automatically inherits two distances: *a chord distance* and a *Riemannian distance*. The pioneers of differential geometry (Gauss, Riemann, Cartan) considered geometric properties of a surface, meaning a 2D manifold that is embedded in the Euclidean space, to be *extrinsic* if they are derived from the chord distance or *intrinsic* if they are derived from the restriction of the infinitesimal chord distance, also known as infinitesimal geodesic or arc distance.

Following from these classical ideas, a Fréchet statistic (mean, mean set, total variance, etc.) that is associated with a random sample on a manifold $\mathcal{M}$ is said to be an *extrinsic statistic* (mean, mean set, total variance, etc.) if the distance on the manifold is the chord distance associated with an embedding of the manifold in $\mathbb{R}^N$. Respectively, a Fréchet statistic associated with a manifold $\mathcal{M}$ is said to be an *intrinsic statistic* (mean, mean set, total variance etc) if the distance on the manifold is the geodesic distance associated with the induced Riemannian structure on that manifold.

The notions of intrinsic and extrinsic means on abstract manifolds were first introduced by Patrangenaru (1998) [266]. Extrinsic means extend the notion of the means of distributions on a sphere or a real projective space, using the method of the center of mass and, respectively, the method of moments of inertia in directional data analysis (see Watson (1983) [333]). Means of random vectors on submanifolds, as defined by Hendricks and Landsman (1998) [154], are extrinsic means.

Note that Riemannian structures on an abstract manifold can be also obtained via a partition of the unity, leading to the question: Are there Riemannian manifolds that cannot be isometrically embedded in some Euclidean space? The answer to this question is *negative* and is due to J. F. Nash (1956) [253], who considered this result more important than his other achievements for which he was awarded the Nobel Prize in Economics. Moreover, any homogeneous Riemannian manifold, and in particular any symmetric space, can be equivariantly embedded in an Euclidean space (Moore, 1976 [249]). This shows that there are both infinitely many extrinsic means and infinitely many intrinsic means that can be associated with a probability measure on a manifold. This is true even on a homogeneous space, given the metric classification of Riemannian homogeneous spaces (Patrangenaru 1994, 1996 [264, 265]).

Despite all of the above, many years after the well established isometric embedding theorem of Nash, confusion seems to persist in some circles about the roles of extrinsic and intrinsic means in statistics. It is sometimes argued that intrinsic analysis, based on *"the"* Riemannian distance, is to be preferred to extrinsic analysis because the latter is based on an arbitrary embedding among infinitely many possible embeddings (See, for example, Srivastava and Klassen (2002) [319]). The implication in this argument that there is a unique Riemannian metric tensor on a manifold is, of course, false; indeed, there are in general infinitely many metric tensors on a manifold. The argument of a unique Riemannian metric on a manifold was nevertheless taken at face value by many computational scientists, who prefer performing an intrinsic analysis for the above reasons. The idea that intrinsic means (or their local version, *Karcher means* (Srivastava and Klassen, 2004 [320]) are "more important" has made swift gains among large segments of electrical engineers, computer scientists, and statisticians.

As the above arguments show, other things being equal, intrinsic and extrinsic means have comparable theoretical statures for statistical inference. *But other things are not equal*! Unfortunately, there are no verifiable general criteria for the existence of a unique intrinsic mean, with the exception of those requiring a small support of the underlying distribution (see Karcher (1977) [174] and Kendall et al. (1999) [179]). Even in the case of the circle S^1, there is no simple broad criterion for uniqueness. Furthermore, when distributions have small support, intrinsic and extrinsic sample means are generally indistinguishable (see Bhattacharya and Patrangenaru (2005) [43] and Bhattacharya and Bhattacharya (2008, 2009) [26, 33]).

As such, intrinsic means inherit the poor properties of arbitrary Fréchet means and the computation of Cartan means can be very slow, which can limit their utility in data analysis on manifolds. This is especially true when resampling methods, such as the nonparametric bootstrap, are employed. Cartan means have the additional disadvantage that their sample counterparts depend on the choice of the initial point used in the iterative computational algorithm. Due to this, it is possible for the algorithm to converge to a critical point of the Fréchet function that is not the intrinsic mean. Indeed, such an algorithm may, in some cases, converge even to a point of maximization of the Fréchet function.

By contrast, the necessary and sufficient condition for the existence of the extrinsic mean of random object X on a manifold $\mathcal{M}$ embedded via $j : \mathcal{M} \to \mathbb{R}^N$ is that the mean vector of $j(X)$ is a *nonfocal* point of $j(\mathcal{M})$ (see Chapter 4 or Bhattacharya and Patrangenaru (2003) [42]). Additionally, extrinsic means can be computed easily.

One should note, though, that in Physics there are sometimes considerations of Mechanics which dictate the use of a particular metric tensor and the corresponding intrinsic mean. However, this is infrequently the case in statistics, where a main problem is discrimination among different distributions.

Throughout the remainder of this section, we present examples of various types of data analysis on manifolds. For each type of data analysis, a methodology is described for the calculation of an extrinsic mean and for an intrinsic mean. Examples of computations are then performed using both methodologies for a given sample and are presented with the required computational time. For timing purposes, all computations were performed using MATLAB on a machine running Windows XP on an Intel Core 2 Duo processor running at 2.33 GHz.

It should be noted, though, that there certainly exist alternative algorithms for computing intrinsic means, such as the one given in Groisser (2004) [133]. As other algorithms may result in somewhat reduced computational costs, the computational times provided for these examples are intended solely for illustrative purposes.

27.3.1 Directional Data

We first consider directional data analysis. Given a random vector X on the unit sphere $S^m \subset \mathbb{R}^{m+1}$, $X^T X = 1$, and the extrinsic mean is $\mu_E = \frac{1}{\|E(X)\|} E(X)$. For this example, $m = 2$. For observations $x_1, \ldots, x_n \in S^2$, the extrinsic sample mean $\hat{\mu}_E$ is calculated using the formula:

$$\hat{\mu}_E = \frac{\bar{x}}{\|\bar{x}\|}, \tag{27.1}$$

where $\bar{x}$ is the usual Euclidean mean and $\|\cdot\|$ is the standard norm. This mean is based upon the chord distance between points. For observations $x_1, \ldots, x_n \in S^2$, the Cartan mean $\hat{\mu}_I$, which is the minimizer of the Fréchet function:

$$F(p) = \sum_{i=1}^{n} \rho_1^2(x_i, p), \tag{27.2}$$

where $\rho_1(x,y)$ is the arc distance between points x and y on the round sphere of radius 1 (and sectional curvature 1 as well), has no closed form. Instead, an iterative algorithm must be used. Such an algorithm is given as follows:

1. Make an initial estimate $\tilde{\mu}_I$ of $\hat{\mu}_I$; i.e. Use $\hat{\mu}_E$
2. For each x_i, compute

$$v_i = u_i * \arccos \frac{x_i \tilde{\mu}_I^T}{\sqrt{u_i u_i^T}},$$

   where $u_i = x_i - (x_i \tilde{\mu}_I^T)\tilde{\mu}_I^T$
3. Compute $\bar{v}$ and update $\tilde{\mu}_I$ in the following manner:

$$\tilde{\mu}_{I,new} = \cos(\varepsilon\sqrt{\bar{v}\bar{v}^T})\tilde{\mu}_I + \sin(\varepsilon\sqrt{\bar{v}\bar{v}^T})\frac{\bar{v}}{\sqrt{\bar{v}\bar{v}^T}},$$

   where $\varepsilon > 0$.
4. Repeat until $\sqrt{\bar{v}\bar{v}^T} < \varepsilon$.

To illustrate these computations and the time required to perform such tasks, we consider a set of wind direction data from Fisher, Lewis, and Embleton (1987, p.308) [117]. The data consist of 29 observations and is provided as pairs of colatitude and longitude. Figure 1.1 displays this data on S^2 in four views.

The extrinsic and intrinsic means were calculated using the methods described above and are shown in Figure 27.1, though in order to calculate the extrinsic mean, it was first necessary to convert to Cartesian coordinates. Additionally, the Fréchet function of which the intrinsic mean is the minimizer is displayed for a grid of values of colatitude and longitude in Figure 27.1. The amount of time required to compute $\hat{\mu}_E$ was : 9.531×10^{-5} seconds. The amount of time required to compute $\hat{\mu}_I$ was 10.88 seconds.

To further illustrate the disparity in computational cost, we consider the calculation of bootstrap means. Using the same methodology as previously, 200 resamples were taken and both types of means were obtained for each. To obtain the bootstrap means, as displayed in Figure 27.3.1, the methodology was applied again on the sample of means. For the extrinsic mean, the computational time required was 0.016762 seconds. For the intrinsic mean, the computational time required was 1572 seconds.

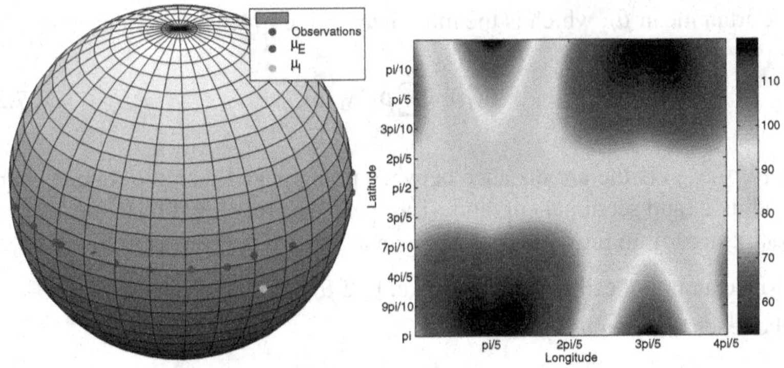

Figure 27.1 *(a) The extrinsic and intrinsic means for the data (b) The Fréchet function using arc distance as a function of colatitude and longitude. (Source: Bhattacharya et al.(2012), Figure 5, p.227. Reproduced by permission of John Wiley & Sons LTD).*

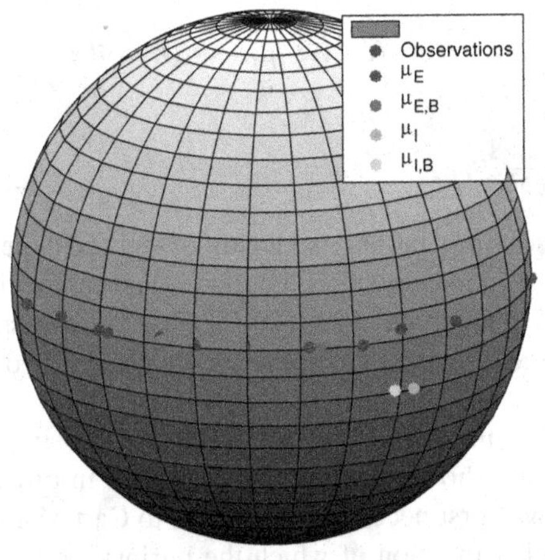

Figure 27.2 *The extrinsic and intrinsic bootstrap means means for the data compared to the extrinsic and intrinsic means. (Source: Bhattacharya et al.(2012), Figure 6, p.228. Reproduced by permission of John Wiley & Sons LTD).*

27.3.2 Congruences of k-ads in 3D

We now consider data for which we are interested in analyzing both the shape and size of objects. For such data, each observation is represented as a k-ad, an ordered set of k points. A k-ad in $\mathbb{R}^p$ is said to be in general position if the k-ad spans $\mathbb{R}^p$. For our purposes, we consider the case that p=3.

Let $\{\mathbf{x}_1,\ldots,\mathbf{x}_n\}$, where $\mathbf{x}_j = (x_j^1,\ldots,x_j^k), j = 1,\ldots,n$, be a sample of k-ads in general position in $\mathbb{R}^3$. The *extrinsic sample mean reflection size-and-shape* is $\overline{[\mathbf{x}]}_E = [\hat{\xi}]_{RS}$, where $\hat{\xi}$ is given by the eigenvectors corresponding to the 3 largest eigenvalues of

$$\hat{C} = \frac{1}{n}\sum_{j=1}^{n} \xi_j^T \, \xi_j$$

assuming that $\hat{\lambda}_3$ is strictly greater than $\hat{\lambda}_4$, where $\hat{\lambda}_1 \geq \cdots \geq \hat{\lambda}_k$ are the eigenvalues of $\hat{C}$, and $\xi_j = \mathbf{x} - \bar{\mathbf{x}}$ (Bandulasiri et al (2009) [10]). If $\hat{\lambda}_3 = \hat{\lambda}_4$, which occurs with probability 0, then there is not a unique extrinsic mean.

The Cartan mean size-and-shape $\hat{\mu}_I$ is the minimizer of the Fréchet function:

$$F(\mathbf{p}) = \sum_{i=1}^{n} d^2(\mathbf{x}_i,\mathbf{p}) = \sum_{i=1}^{n} \inf_{\Gamma_i \in SO(3)} \|\mathbf{p} - \mathbf{x_i}\Gamma_i\|^2, \tag{27.3}$$

where $\|\cdot\|$ is the standard norm in $\mathbb{R}^3$ and Γ_i is a special orthogonal matrix. As with the spherical data, there is no closed form solution for the intrinsic mean size-and-shape. Instead, the following iterative algorithm is used (Dryden and Mardia (1998) [91]):

1. Make an initial estimate $\tilde{\mu}_I$ of $\hat{\mu}_I$; i.e. Use $\hat{\mu}_E$
2. For each $\mathbf{x}_i$, find the optimal "rotation" matrix Γ_i using Procrustes alignment and compute
$$V_i = \xi_i\Gamma_i - \tilde{\mu}_I, \tag{27.4}$$
3. Compute $\bar{V}$ and update $\tilde{\mu}_I$ in the following manner:
$$\tilde{\mu}_{I,new} = \tilde{\mu}_I + \varepsilon\bar{V}, \tag{27.5}$$
where $\varepsilon > 0$.
4. Repeat until $\|\bar{V}\| < \varepsilon$.

To demonstrate, we performed the above computations for a data set consisting of 4 protein binding sites. The binding sites, obtained from the RCSB Protein Data Bank and shown in Figure 12.1, are found in the proteins 1phf, 1phg, 2cpp, and 1m85 and bind to the ligand heme. As shown in Figure 12.2, the extrinsic mean size-and-shape, obtained after atoms matching, is visually indistinguishable from the intrinsic mean-size-and-shape.

To detail the computational speeds of the two types of analysis, bootstrap means are computed, similarly to the wind direction data. To examine the effect of sample size on the computational cost, these calculations were performed for samples of size 4, 5, 6, 8, 12, and 16. For the samples of size greater than 4, the observed data was simulated based upon the original sample. The times, in

Table 27.1 *The times required to compute the mean size-and-shape for various sample sizes*

Sample Size	Extrinsic	Intrinsic	Ratio
4	15.9	29.3	1.84
5	15.3	32.0	2.09
6	15.2	36.2	2.38
8	14.8	60.2	4.07
12	15.3	92.1	6.01
16	16.5	123.6	7.49

seconds, required for these computations are shown in Table 27.1. Increasing the sample size has no significant effect on the computational cost for calculating the extrinsic mean-size-and-shape. However, increasing the sample size has a large effect on the computational cost for calculating the intrinsic mean-size-and-shape.

27.3.3 Direct Similarity Shapes of Planar Contours

In this section, we consider data analysis of shapes of planar contours (See Chapter 18). A digital representation of contour is represented as a k-gon, where each vertex on the object is a point in $\mathbb{C}$. Let $\zeta_1,\ldots,\zeta_n$ be a sample of *centered* k-gons. The Veronese–Whitney extrinsic sample mean shape is the unit eigenvector corresponding to the largest eigenvalue of $\sum_{i=1}^{n}\frac{1}{\|\zeta_i\|^2}\zeta_i\zeta_i^*$ assuming that the largest eigenvalue is simple.

The recently developed elastic framework for shape analysis of planar curves utilizes an intrinsic analysis (See Joshi et al. (2007) [171]). As with the previously discussed types of data, there is no closed form for the intrinsic mean shape, so an iterative algorithm similar in concept to that used for the wind direction data must be used to perform computations.

Computations for both approaches were performed on a sample of 4 observations of contours of the "l" hand gesture, shown in Figure 1.3. To illustrate the difference in computational cost, 95% bootstrap confidence intervals were computed for both the extrinsic mean shape and the intrinsic mean shape using 400 resamples and 300 randomly chosen stopping times (vertices). These confidence regions are shown in Figure 27.3. For the extrinsic mean shape, these calculations required 48 seconds to complete. However, for the intrinsic mean shape, these calculations required 47.9 hours.

As a second example, these methods were also performed on a sample of 4 observations of contours of dogs, which is shown in 27.4. Again, 95% bootstrap confidence regions were computed for both approaches, using 300 resamples, where each contour is provided as 100 evenly spaced landmarks. These

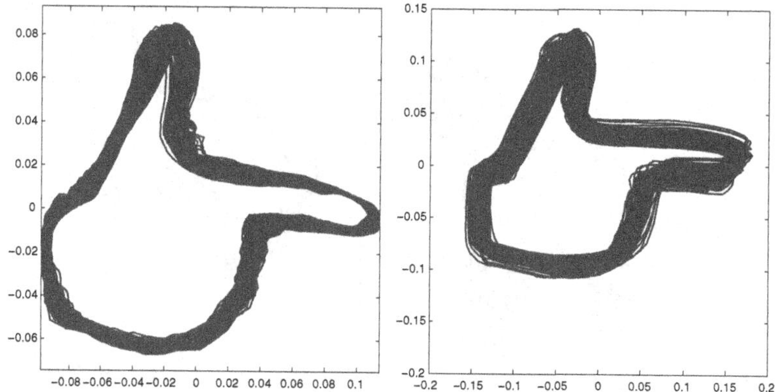

Figure 27.3 *Bootstrap 95% confidence regions using 400 resamples for (a) the extrinsic mean shape of the "l" hand gesture and (b) the intrinsic mean shape of the "l" hand gesture. (Source: Bhattacharya et al.(2012), Figure 10, p.230. Reproduced by permission of John Wiley & Sons LTD).*

confidence regions are shown in Figure 27.3. For the extrinsic mean shape, these calculations required 5.6 seconds to complete. However, for the intrinsic mean shape, these calculations required 8.9 hours.

Figure 27.4 *9 observations of contours of a side view of a dog. (Source: Bhattacharya et al.(2012), Figure 11, p.231. Reproduced by permission of John Wiley & Sons LTD).*

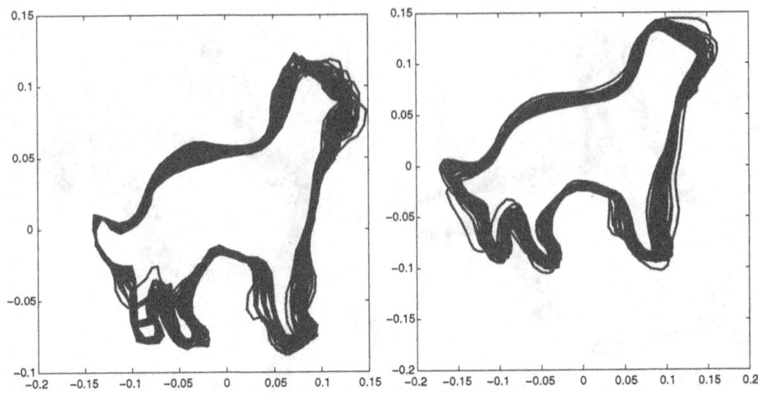

Figure 27.5 *Bootstrap 95% confidence regions using 400 resamples for (a) the extrinsic mean shape of the 4 dogs and (b) the intrinsic mean shape of the 4 dogs. (Source: Bhattacharya et al.(2012), Figure 12, p.231. Reproduced by permission of John Wiley & Sons LTD).*

27.4 Summary

Much of modern data analysis consists of problems involving data on manifolds. Among the fields in which such data occurs are medical imaging, directional data analysis, pattern recognition, and visual quality control. To conduct a statistical analysis for this data, either extrinsic or intrinsic analysis may be used.

The computational cost of performing extrinsic analysis on manifold data is, as shown here, often substantially less than the computational cost of performing intrinsic analysis for the same data. This is especially noticeable when working with large data sets and/or performing analysis requiring large numbers of repetitions, as with nonparametric bootstrap techniques. As shown with the protein data, in many cases, the extrinsic and intrinsic means are indistinguishable from each other despite the difference in computational time, providing strong support for the use of extrinsic analysis in such situations.

In other scenarios, one must look to the requirements for the application at hand. For instance, extrinsic means exist outside a negligible singular set, whereas intrinsic means are not, in general, guaranteed to exist. Furthermore, for a given embedding J, the extrinsic mean is the projection on $J(\mathcal{M})$, but there is no natural description of the intrinsic mean. Due to these considerations, extrinsic analysis is often preferable for statistical analysis.

Bibliography

[1] G. J. A. Amaral, I. L. Dryden, V. Patrangenaru, and A. T. A. Wood. Bootstrap confidence regions for the planar mean shape. *Journal of Statistical Planning and Inference*, 140:3026–3034, 2010.

[2] G. J. A. Amaral, I. L. Dryden, and A. T. A. Wood. Pivotal bootstrap methods for k-sample problems in directional statistics and shape analysis. *J. Amer. Statist. Assoc.*, 102:695–707, 2007.

[3] S.-I. Amari. *Differential-Geometrical Methods in Statistics*. Springer, New York. Lecture Notes in Statist. 28, 1985.

[4] J. F. Angers and P. T. Kim. Multivariate Bayesian function estimation. *Ann. Statist.*, 33:2967–2999, 2005.

[5] V. Arsigny, P. Fillard, X. Pennec, and N. Ayache. Geometric means in a novel vector space structure on symmetric positive-definite matrices. *SIAM Matrix Anal. Appl.*, 29:328–347, 2006.

[6] V. Arsigny, P. Fillard, X. Pennec, and N. Ayache. Log-Euclidean metrics for fast and simple calculus on diffusion tensors. *Magnetic Resonance in Medicine*, 56:411–421, 2006.

[7] R. Azencott. Deterministic and random deformations applications to shape recognition. *Conference at HSSS workshop in Cortona, Italy*, 1994.

[8] R. Azencott, F. Coldefy, and L. Younes. A distance for elastic matching in object recognition. *Proceedings of 12th ICPR*, pages 687–691, 1996.

[9] G. J. Babu and K. Singh. On one term Edgeworth correction by Efron's bootstrap. *Sankhya Ser. A.*, 46:219–232, 1984.

[10] A. Bandulasiri, R. N. Bhattacharya, and V. Patrangenaru. Nonparametric inference for extrinsic means on size-and-(reflection)-shape manifolds with applications in medical imaging. *Journal of Multivariate Analysis*, 100:1867–1882, 2009.

[11] A. Bandulasiri and V. Patrangenaru. Algorithms for nonparametric inference on shape manifolds. *Proc. of JSM 2005, Minneapolis, MN*, pages 1617–1622, 2005.

[12] A. Bandulasiri, V. Patrangenaru, J. Su, and J. Zhang. Applications

of nonparametric statistics on reflection shape manifolds and reflection size-and-shape manifolds. *Proc. of JSM 2008, Denver, CO.*, pages 2769–2776, 2009.

[13] D. Barden, H. Le, and M. Owen. Central limit theorems for Fréchet means in the space of phylogenetic trees. *Electron. J. Probab.*, 18:no. 25, 1–25, 2013.

[14] P. J. Basser and C. Pierpaoli. Microstructural and physiological features of tissues elucidated by quantitative-diffusion-tensor MRI. *J. Magn. Reson. B*, 111:209–219, 1996.

[15] M. Belkin and P. Niyogi. Semi-supervised learning on Riemannian manifolds. *Mach. Learn.*, 56:209–239, 2004.

[16] R. Beran. Prepivoting to reduce level error of confidence sets. *Biometrika*, 74:457–468, 1987.

[17] R. Beran. Prepivoting test statistics: a bootstrap view of asymptotic refinements. *J. Amer. Statist. Assoc.*, 83:687–697, 1988.

[18] R. Beran and N.I. Fisher. Nonparametric comparison of mean axes. *Ann. Statist.*, 26:472–493, 1998.

[19] R. J. Beran. Testing for uniformity on a compact homogeneous space. *J. Appl. Probability*, 5:177–195, 1968.

[20] J. O. Berger and M. Delampady. Testing precise hypotheses. *Statist. Sci.*, 2:317–352, 1987.

[21] D. Bernoulli. Recherches physiques et astronomiques, sur le probléme propospour la second fois par l'Académie Royale des Sciences de Paris. *Recueuil des piéces qui ont remport le prix de l'Académie Royale des Sciences Tome III*, pages 95–134, 1734.

[22] M. Berth, F. M. Moser, M. Kolbe, and J. Bernhardt. The state of the art in the analysis of two-dimensional gel electrophoresis images. *Appl. Microbiol. Biotechnol.*, 76:1223–1243, 2007.

[23] R. Berthilsson and A. Heyden. Recognition of planar objects using the density of affine shape. *Computer Vision and Image Understanding*, 76:135–145, 1999.

[24] A. Bhattacharya. Nonparametric statistics on manifolds with applications to shape spaces. *Thesis, Uni. of Arizona*, 2008.

[25] A. Bhattacharya. Statistical analysis on manifolds: A nonparametric approach for inference on shape spaces. *Sankhya*, 70-A:223–266, 2008.

[26] A. Bhattacharya and R. N. Bhattacharya. Statistics on Riemannian manifolds: Asymptotic distribution and curvature. *Proceedings of the American Mathematical Society*, pages 2957–2967, 2008.

[27] A. Bhattacharya and R. N. Bhattacharya. *Nonparametric inference*

on manifolds. With applications to shape spaces., volume 2. Institute of Mathematical Statistics (IMS) Monographs. Cambridge University Press, Cambridge, 2012.

[28] A. Bhattacharya and D. B. Dunson. Nonparametric Bayesian density estimation on manifolds with applications to planar shapes. *Biometrika*, 97:851–865, 2010.

[29] R. N. Bhattacharya. Refinements of the multidimensional central limit theorem and applications. *Ann. Probability*, 5:1–27, 1977.

[30] R. N. Bhattacharya. *New perspectives in Theoretical and Applied Statistics (M.L. Puri and J.P. Vilaplana and W. Wertz Eds.)*, chapter Some aspects of Edgeworth expansions in statistics and probability. (M.L. Puri, J.P. Vilaplana and W. Wertz, Eds.) (Wiley, NY) 157-171., pages 157–171. Wiley, NY, 1987.

[31] R. N. Bhattacharya. *A Course in Large Sample Theory*. Class Notes, Indiana University, 1997.

[32] R. N. Bhattacharya and A. Bhattacharya. Nonparametric statistics on manifolds with applications to shape spaces. *Pushing the Limits of Contemporary Statistics: Contributions in honor of J.K. Ghosh(S.Ghoshal and B. Clarke eds.). IMS Collections*, 3:282–301, 2008.

[33] R. N. Bhattacharya and A. Bhattacharya. *Perspectives in Mathematical Sciences I: Probability and Mathematics*, chapter Statistics on Manifolds with Applications to Shape Spaces, pages 41–70. Indian Statistical Institute, Bangalore. Eds: N. S. Narasimha Sastry, T.S.S.R.K. Rao, Mohan Delampady and B. Rajeev, 2009.

[34] R. N. Bhattacharya, M. Buibas, I. L. Dryden, L. A. Ellingson, D. Groisser, H. Hendriks, S. Huckemann, H. Le, X. Liu, J. S. Marron, D. E. Osborne, V. Patrangenaru, A. Schwartzman, H. W. Thompson, and A. T. A. Wood. Extrinsic data analysis on sample spaces with a manifold stratification. *Advances in Mathematics, Invited Contributions at the Seventh Congress of Romanian Mathematicians, Brasov, 2011*, pages 241–252, 2011.

[35] R. N. Bhattacharya and N. H. Chan. Comparisons of chisquare, Edgeworth expansions and bootstrap approximations to the distribution of the frequency chi-square. *Sankhya Ser. A*, 58:57–68, 1996.

[36] R. N. Bhattacharya and M. Denker. *Asymptotic Statistics*, volume 14. Birkhäuser, DMV Seminar, 1990.

[37] R. N. Bhattacharya, L. Ellingson, X. Liu, V. Patrangenaru, and M. Crane. Extrinsic analysis on manifolds is computationally faster than intrinsic analysis with applications to quality control by machine vision. *Appl. Stochastic Models Bus. Ind.*, 28:222–235, 2012.

[38] R. N. Bhattacharya and J. K. Ghosh. On the validity of the formal Edgeworth expansion. *Ann. Statist.*, 6:434–451, 1978.

[39] R. N. Bhattacharya and L. Lin. A central limit theorem for Fréchet means. *arXiv.org*, arXiv:1306.5806:1–8, 2011.

[40] R. N. Bhattacharya, L. Lin, and V. Patrangenaru. *A Course in Mathematical Statistics and Large Sample Theory-in preparation.* Springer Series of Statistics, New York, 2014.

[41] R. N. Bhattacharya and V. Patrangenaru. Non parametric estimation of location and dispersion on Riemannian manifolds. *J. Statist. Plan. Infer. Volume in honor of the 80th birthday of professor C.R.Rao*, 108:23–35, 2002.

[42] R. N. Bhattacharya and V. Patrangenaru. Large sample theory of intrinsic and extrinsic sample means on manifolds - part i. *Annals of Statistics*, 31(1):1–29, 2003.

[43] R. N. Bhattacharya and V. Patrangenaru. Large sample theory of intrinsic and extrinsic sample means on manifolds-part ii. *Annals of Statistics*, 33(3):1211–1245, 2005.

[44] R. N. Bhattacharya and V. Patrangenaru. Statistics on manifolds and landmarks based image analysis: A nonparametric theory with applications. *J. of Statist. Planning and Inference*, 145:1–22, 2014.

[45] R. N. Bhattacharya and M. Qumsiyeh. Second order and l^p-comparisons between the bootstrap and empirical Edgeworth expansion methodologies. *Ann. Statist.*, 17:160–169, 1989.

[46] R. N. Bhattacharya and R. Ranga Rao. *Normal Approximation and Asymptotic Expansions.* Wiley, 1976.

[47] L. Billera, S. Holmes, and K. Vogtmann. Geometry of the space of phylogenetic trees. *Adv. in Appl. Math.*, 27:733–767, 2001.

[48] P. Billingsley. *Probability and measure. Third edition.* Wiley Series in Probability and Mathematical Statistics., 1995.

[49] L. Birgé. Approximation dans les espaces metriques et theorie de l'estimation (in french). *Z. Wahrsch. verw Gebiete*, 65:181–237, 1983.

[50] N. Bissantz, T. Hohage, A. Munk, and F. Ruymgaart. Convergence rates of general regularization methods for statistical inverse problems and applications. *SIAM J. Numerical Analysis*, 45:2610–2636, 2007.

[51] F. L. Bookstein. *Morphometric Tools for Landmark Data, Geometry and Biology.* Cambridge University Press, Cambridge, 1991.

[52] T. Bröcker and T. tom Diek. *Representations of Compact Lie Groups.* Springer-Verlag. New York, 1985.

[53] P. Bubenik, G. Carlsson, P. T. Kim, and Z.-M. Luo. *Algebraic methods*

in statistics and probability II, Contemp. Math., 516, chapter Statistical topology via Morse theory persistence and nonparametric estimation., pages 75–92. Amer. Math. Soc., Providence, RI, 2010, 2010.

[54] P. Bubenik and P. T. Kim. A statistical approach to persistent homology. *Homology Homotopy and Applications*, 9:337–362, 2007.

[55] M. Buibas, M. Crane, L. Ellingson, and V. Patrangenaru. A projective frame based shape analysis of a rigid scene from noncalibrated digital camera imaging outputs. *JSM Proceedings, 2011, Miami, FL. Institute of Mathematical Statistics*, pages 4730–4744, 2012.

[56] C. F. Burgoyne, H. W. Thompson, D. E. Mercante, and R. Amin. Basic issues in the sensitive and specific detection of Optic Nerve Head surface change within longitudinal LDT TOPSS images. *The Shape of Glaucoma, Quantitative Neural Imaging Techniques*, pages 1–37, 2000.

[57] É. Cartan. Sur certaines formes riemanniennes remarquables des géométries *à* groupe fondamental simple. (french). *Ann. Sci. École Norm. Sup.*, 3(44):345–467, 1927.

[58] É. Cartan. *Léçons sur la Géométrie des Espaces de Riemann (in French)*. Gauthier-Villars, Paris, 1928.

[59] É. Cartan. *Léçons sur la Géométrie des Espaces de Riemann (in French)*. Gauthier-Villars, Paris, 1946.

[60] V. Caselles, R. Kimmel, and G. Sapiro. Geodesic active contours. *Intl. Journ. of Computer Vision*, 22(1):61–79, 1997.

[61] L. Cavalier and A. Tsybakov. Random change on a lie group and mean glaucomatous projective shape change detection from stereo pair images. *Probab. Theory Relat. Fields*, 123:323–354, 2002.

[62] T. K. Chandra and J. K. Ghosh. Valid asymptotic expansion for likelihood ratio statistic and other perturbed chisquare variables. *Sankhya*, 41:22–47, 1979.

[63] T. Chang. Estimating the relative rotation of two tectonic plates from boundary crossings. *J. Amer. Statist. Assoc.*, 83:1178–1184, 1979.

[64] T. Chang. Spherical regression. *Ann. Statist.*, 14:907–924, 1986.

[65] T. Chang. Spherical regression with errors in variables. *Ann. Statist.*, 17:293–306, 1989.

[66] T. Chang. Spatial statistics. *Statistical Science*, 19:624–635, 2004.

[67] T. Chang and D. Ko. M-estimates of rigid body motion on the sphere and in Euclidean space. *Ann. Statist.*, 23:1823–1847, 1995.

[68] I. Chavel. *Riemannian Geometry. A Modern Introduction*. Cambridge University Press, Cambridge, 1993.

[69] I. Chavel. *Riemannian Geometry. A Modern Introduction, Second Edi-*

tion. Cambridge University Press, Cambridge, 2006.

[70] Y. Chikuse. The matrix angular central Gaussian distribution. *J. Multivariate Anal.*, 33:265–274, 1990.

[71] Y. Chikuse. Procrustes analysis on some special manifolds. statistical inference and data analysis. *Comm. Statist. Theory Methods*, 28:885–903, 1999.

[72] Y. Chikuse. *Statistics on Special Manifolds*. Lecture Notes in Statistics. New York: Springer-Verlag, 2003.

[73] Y. Chikuse and G. S. Watson. Large sample asymptotic theory of tests for uniformity on the grassmann manifold. *J. Multivariate Anal.*, 54:18–31, 1995.

[74] M. K. Chung, P. Bubenik, and P. T. Kim. Persistence diagrams of cortical surface data. *LNCS: Proceedings of IPMI 2009*, 5636:386–397, 2009.

[75] D. Cohen-Steiner, H. Edelsbrunner, and J. Harer. Stability of persistence diagrams. *SCG '05: Proceedings of the twenty-first annual symposium on Computational geometry*, pages 263–271, 2005.

[76] H. Cramér. *Mathematical Methods of Statistics*. Princeton University Press. Princeton, New Jersey, 1946.

[77] M. Crane. Nonparametric estimation of three dimensional projective shapes with applications in medical imaging and in pattern recognition. *Ph.D. Dissertation, The Florida State University*, 2010.

[78] M. Crane and V. Patrangenaru. Random change on a lie group and mean glaucomatous projective shape change detection from stereo pair images. *Journal of Multivariate Analysis*, 102:225–237, 2011.

[79] F. Cucker and S. Smale. Best choices for regularization parameters in learning theory: on the bias-variance problem. *Found. Comput. Math.*, 2:413–428, 2002.

[80] J. Dauxois, A. Prousse, and Y. Romain. Asymptotic theory for the principal component analysis of a vector random function: some applications to statistical inference. *J. Multivariate Anal.*, 12:136–154, 1982.

[81] V. de Silva and R. Ghrist. Homological sensor networks. *Notic. Amer. Math. Soc.*, 54:10–17, 2007.

[82] G. Derado, K. V. Mardia, V. Patrangenaru, and H. W. Thompson. A shape based glaucoma index for tomographic images. *Journal of Applied Statistics*, 31:1241–1248, 2004.

[83] R. Deriche, D. Tschumperlé, C. Lenglet, and M. Rousson. *Handbook of mathematical models in computer vision*, chapter Variational approaches to the estimation, regularization and segmentation of diffu-

sion tensor images, pages 517–530. Springer, 2006.

[84] H. Dette and A. Munk. Validation of linear regression models. *Ann. Statist.*, 26:778–800, 1998.

[85] H. Dette and A. Munk. Some methodological aspects of validation of models in nonparametric regression. *Statistica Neerlandica*, 57:207–244, 2003.

[86] I. Dimitric. A note on equivariant embeddings of Grassmannians. *Publ. Inst. Math. (Beograd) (N.S.)*, 59:131–137, 1996.

[87] M. P. do Carmo. *Riemannian Geometry*. Birkhäuser, Boston, 1992.

[88] T. D. Downs. Orientation statistics. *Biometrika*, 59:665–676, 1972.

[89] I. L. Dryden, A. Koloydenko, and D. Zhou. Non-Euclidean statistics for covariance matrices, with applications to diffusion tensor imaging. *Ann. Appl. Stat.*, 3:1102–1123, 2009.

[90] I. L. Dryden, A. Kume, H. Le, and A. T. A. Wood. A multi-dimensional scaling approach to shape analysis. *Biometrika*, 95:779–798, 2008.

[91] I. L. Dryden and K. V. Mardia. *Statistical Shape Analysis*. Wiley, Chichester., 1998.

[92] H. Edelsbrunner, D. Letscher, and A. Zomorodian. Topological persistence and simplification. *Discrete Comput. Geom.*, 28:511–533, 2001.

[93] F.Y. Edgeworth. The law of error. *Proceedings, Cambridge Philos. Soc.*, 20:36–65, 1905.

[94] S. Efromovich. On sharp adaptive estimation of multivariate curves. *Math. Methods Statist.*, 9:117–139, 2000.

[95] B. Efron. Defining the curvature of a statistical problem (with applications to second order efficiency) (with discussion). *Annals of Statistics*, 3:1189–1242, 1975.

[96] B. Efron. Bootstrap methods: Another look at the jackknife. *Annals of Statistics*, 7:1–26, 1979.

[97] B. Efron. *The Jackknife, the Bootstrap and Other Resampling Plans*. CBMS-NSF Regional Conference Series in Applied Mathematics, 38, SIAM, 1982.

[98] B. Efron. Second thoughts on the bootstrap. Silver anniversary of the bootstrap. *Statist. Sci.*, 18:135–140, 2003.

[99] B. Efron and R. J. Tibshirani. *An introduction to the bootstrap*. Monographs on Statistics and Applied Probability, 57. Chapman and Hall, New York, 1993.

[100] S. Eilenberg and N. Steenrod. *Foundations of algebraic topology*. Princeton, 1952.

[101] L. Ellingson. Statistical shape analysis on manifolds with applications to planar contours and structural proteomics. *PhD Dissertation, Florida State University*, 2011.

[102] L. Ellingson, D. Groisser, D. Osborne, V. Patrangenaru, and A. Schwartzman. Nonparametric bootstrap of sample means of positive-definite matrices with an application to diffusion tensor imaging data analysis. *to appear*, TBA.

[103] L. Ellingson, V. Patrangenaru, H. Hendriks, and P. S. Valentin. *Topics in Nonparametric Statistics. Editors: M.G. Akritas, S.N. Lahiri and D. N. Politis*, chapter CLT on Low Dimensional Stratified Spaces, pages 227–240. Springer, 2014.

[104] L. Ellingson, V. Patrangenaru, and F. Ruymgaart. Nonparametric estimation of means on Hilbert manifolds and extrinsic analysis of mean shapes of contours. *Journal of Multivariate Analysis*, 122:317–333, 2013a.

[105] L. Ellingson and C. Prematilake. Problems in approximating shapes of planar contours. *Proceedings of the 2013 Joint Statistical Meetings*, 2013.

[106] L. Ellingson, F. H. Ruymgaart, and V. Patrangenaru. *Statistical Models and Methods for non-Euclidean Data with Current Scientific Applications, The 32nd Leeds Annual Statistical Research Workshop 2nd-4th July 2013*, chapter Data analysis on Hilbert manifolds and shapes of planar contours, pages 23–27. Leeds University Press, 2013.

[107] L. Ellingson and J. Zhang. Protein surface matching by incorporating local and global geometric information. *PLoS ONE*, 7(7), 2012.

[108] M. Émery and G. Mokobodzki. *Séminaire de Probabilités*, volume Lecture Notes in Math., 1485, chapter Sur le barycentre d'une probabilité dans une variété., pages 220–233. Springer, Berlin, 1991.

[109] V. A Epanechnikov. Nonparametric estimation of a multidimensional probability density. *Theory of Probability and Applications*, 14:156–162, 1969.

[110] D. Falk, F. E. Lepore, and A. Noe. The cerebral cortex of Albert Einstein: a description and preliminary analysis of unpublished photographs. *Brain*, pages 1304–1327, 2012.

[111] O. Faugeras. What can be seen in three dimensions with an uncalibrated stereo rig? *Proc. European Conference on Computer Vision, LNCS 588*, pages 563–578, 1992.

[112] O. Faugeras and Q. T. Luong. *The geometry of multiple images. With contributions from Theo Papadopoulo.* MIT Press, Cambridge, MA, 2001.

[113] H. D. Fegan. *Introduction to Compact Lie Groups*. Singapore: World Scientific, 1991.

[114] A. Feragen, M. Owen, J. Petersen, M. M. W. Wille, L. H. Thomsen, A. Dirksen, and M. de Bruijne. Tree-space statistics and approximations for large-scale analysis of anatomical trees. *Information Processing in Medical Imaging, Lecture Notes in Computer Science*, 7917:74–85, 2013.

[115] T. Ferguson. *Large Sample Theory*. Chapman & Hall/CRC, 1996.

[116] N. I. Fisher, P. Hall, B. Y. Jing, and A. T. A. Wood. Properties of principal component methods for functional and longitudinal data analysis. *J. Amer. Statist. Assoc.*, 91:1062–1070, 1996.

[117] N. I. Fisher, T. Lewis, and B. J. J. Embleton. *Statistical Analysis of Spherical Data*. Cambridge University Press, Cambridge, 1987.

[118] P. T. Fletcher. Geodesic regression and the theory of least squares on Riemannian manifolds. *Int. J. Comput. Vis.*, 105:171–185, 2004.

[119] P. T. Fletcher. Statistical variability in nonlinear spaces: Application to shape analysis and DT-MRI. *Ph.D. Thesis, University of North Carolina*, 2004.

[120] P. T. Fletcher, C. Lu, S. M. Pizer, and S. C. Joshi. Principal geodesic analysis for the study of nonlinear statistics of shape. *IEEE Trans. Med. Imaging*, 23(8):995–1005, 2004.

[121] M. Fréchet. Les élements aléatoires de nature quelconque dans un espace distancié (in French). *Ann. Inst. H. Poincaré*, 10:215–310, 1948.

[122] Gaines. *Random perturbation of a self-adjoint operator with a multiple eigenvalue*. PhD Dissertation, Texas Tech University, 2012.

[123] K. F. Gauss. *Disquisitiones generales circa superficies curvas (in Latin)*. Dieterich, Göttingen, 1828.

[124] E. M. Giné. The addition formula for the eigenfunctions of the Laplacian. *Adv. in Math.*, 18:102–107, 1975.

[125] C. Goodall. Procrustes methods in the statistical analysis of shape. *J. Roy. Statist. Soc. Ser. B*, 53:285–339, 1991.

[126] C. Goodall and K. V. Mardia. Multivariate aspects of shape theory. *Ann. Statist.*, 21:848–866, 1993.

[127] C. R. Goodall and K. V. Mardia. Projective shape analysis. *J. Graphical and Computational Statist.*, 8:143–168, 1999.

[128] N. R. Goodman. Statistical analysis based on a certain multivariate complex Gaussian distribution. (an introduction). *Ann. Math. Statist.*, 34:152–177, 1963.

[129] C. Goutis and C. Robert. Model choice in generalised linear models:

A Bayesian approach via Kullback-Leibler projections. *Biometrika*, 85:29–37, 1998.

[130] J. C. Gower. Some distance properties of latent root and vector methods in multivariate analysis. *Biometrika*, 53:315 – 328, 1966.

[131] M. J. Greenberg and J. H. Harper. *Algebraic Topology. A first course*. Mathematics Lecture Note Series, 58. Benjamin/Cummings Publishing Co., 1981.

[132] U. Grenander. *General Pattern Theory*. Oxford Univ. Press, 1993.

[133] D. Groisser. Newtons method, zeroes of vector fields, and the Riemannian center of mass. *Advances in Applied Mathematics*, 33:95–135, 2004.

[134] L. R. Haff, P. T. Kim, J. Y. Koo, and D. St. P. Richards. Minimax estimation for mixtures of Wishart distributions. *Ann. Statist.*, 39:3417–3440, 2011.

[135] P. Hall. On the bootstrap and likelihood-based confidence regions. *Biometrika*, 74:481–493, 1987.

[136] P. Hall. Theoretical comparison of bootstrap confidence intervals. *Ann. Statist.*, 16:927–985, 1988.

[137] P. Hall. *The Bootstrap and Edgeworth Expansion*. Springer Series in Statistics, Springer-Verlag, NewYork, 1997.

[138] P. Hall and J. D. Hart. Bootstrap test for difference between means in nonparametric regression. *J. Amer. Stat. Asso.*, 85:1039–1049, 1990.

[139] P. Hall, H. G. M*ü*ller, and J. L. Wang. Properties of principal component methods for functional and longitudinal data analysis. *Ann. Statist.*, 34:1493–1517, 2006.

[140] M. T. Harandi, M. Salzmann, S. Jayasumana, R. I. Hartley, and L. Hongdong. Expanding the family of Grassmannian kernels: An embedding perspective. *arXiv:1407.1123*, 2014.

[141] R. Hartley and A. Zisserman. *Multiple view Geometry in computer vision; 2nd edition*. Cambridge University Press, 2004.

[142] R. I. Hartley. Projective reconstruction and invariants from multiple images. *IEEE Transactions on Pattern Analysis and Machine Intelligence*, 16:1036–1041, 1994.

[143] R. I. Hartley, R. Gupta, and T. Chang. Stereo from uncalibrated cameras. *Proceedings IEEE Conference on Computer Vision and Pattern Recognition*, pages 761 – 764, 1992.

[144] M. J. Hawker, S.A. Vernon, C. L. Tattersall, and H. S. Dua. Linear regression modeling of rim area to discriminate between normal and glaucomatous optic nerve heads - the Bridlington Eye Assessment project.

Journal OF GLAUCOMA, 16:345–351, 2007.

[145] D. M. Healy and P. T. Kim. An empirical Bayes approach to directional data and efficient computation on the sphere. *Ann Statist*, 24:232–254, 1996.

[146] S. Helgason. *Differential Geometry and Symmetric Spaces*. Academic Press, Inc., 1962.

[147] S. Helgason. *Differential Geometry, Lie groups, and Symmetric Spaces*. New York: Academic Press, 1978.

[148] S. Helgason. *Groups and Geometric Analysis*. Orlando: Academic Press, 1984.

[149] U. Helmke, K. Hüper, and J. Trumpf. Newtonss method on Grassmann manifolds. *arXiv:0709.2205*, 2007.

[150] H. Hendriks. Nonparametric estimation of a probability density on a Riemannian manifold using fourier expansions. *Ann. Statist.*, 18:832–849, 1990.

[151] H. Hendriks. Sur le cut-locus d'une sous-variété de l'espace euclidien. (french). *C. R. Acad. Sci. Paris Sér. I Math.*, 311:637–639, 1990.

[152] H. Hendriks and Z. Landsman. Asymptotic behavior of sample mean location for manifolds. *Statist. Probab. Lett.*, 26:169–178, 1996.

[153] H. Hendriks and Z. Landsman. Asymptotic tests for mean location on manifolds (french). *C. R. Acad. Sci. Paris Sér. I Math*, 322:773–778, 1996.

[154] H. Hendriks and Z. Landsman. Mean location and sample mean location on manifolds: asymptotics, tests, confidence regions. *J. Multivariate Anal.*, 67:227–243, 1998.

[155] H. Hendriks and Z. Landsman. Asymptotic data analysis on manifolds. *Ann. Statist.*, 35:109–131, 2007.

[156] G. Henry and D. Rodriguez. Kernel density estimation on Riemannian manifolds: asymptotic results. *J. Math. Imaging Vision*, 34:235–239, 2009.

[157] A. Heyden. Geometry and algebra of multiple projective transformations. *PhD Dissertation, University of Lund, Sweden*, 1995.

[158] M Hitczenko and M. L. Stein. Some theory for anisotropic processes on the sphere. *Statistical Methodology*, 9:211–227, 2012.

[159] J. L. Hodges and E. L. Lehmann. Testing the approximate validity of statistical hypotheses. *J. R. Statist. Soc. B*, 16:261–268, 1954.

[160] G. W. Horgan, A. Creasey, and B. Fenton. Superimposing two dimensional gels to study genetic variation in malaria parasites. *Electrophoresis*, 13:871–875, 1992.

[161] H. Hotelling. The generalization of student's ratio. *Ann. Math. Statist.*, 2:360–364, 1931.

[162] T. Hotz, S. Huckemann, H. Le, J. S. Marron, J. Mattingly, E. Miller, J. Nolen, M. Owen, V. Patrangenaru, and S. Skwerer. Sticky central limit theorems on open books. *Annals of Applied Probability*, 23:2238–2258, 2013.

[163] P. J. Huber. *Robust statistics*. Wiley Series in Probability and Mathematical Statistics, 1981.

[164] S. Huckemann. Inference on 3d procrustes means: tree bole growth, rank deficient diffusion tensors and perturbation models. *Scand. J. Stat.*, 38:424–446, 2011.

[165] S. Huckemann and T. Hotz. Principal component geodesics for planar shape spaces. *J. Multivariate Anal.*, 100(4):699–714, 2009.

[166] S. Huckemann and H. Ziezold. Principal component analysis for Riemannian manifolds, with an application to triangular shape spaces. *Adv. in Appl. Probab.*, 38(2):299–319, 2006.

[167] Brouwer L. E. J. Beweis der invarianz des n-dimensionalen gebiets. *Mathematische Annalen*, 71:305–315, 1912.

[168] Brouwer L. E. J. Beweis der invarianz des n-dimensionalen gebiets. *Mathematische Annalen*, 72:55–56, 1913.

[169] A. T. James. Normal multivariate analysis and the orthogonal group. *Ann. Math. Statist.*, 25:40–75, 1954.

[170] R. A. Johnson and D. W. Wichern. *Applied Multivariate Statistical Analysis, 6th Edition*. Pearson Prentice Hall, Upper Saddle River, New Jersey, 2007.

[171] S. Joshi, A. Srivastava, E. Klassen, and I. Jermyn. Removing shape-preserving transformations in square-root elastic (sre) framework for shape analysis of curves. *Workshop on Energy Minimization Methods in CVPR*, 2007.

[172] S. Jung, I. L. Dryden, and J. S. Marron. Analysis of principal nested spheres. *Biometrika*, 99(3):551 – 568, 2012.

[173] P. E. Jupp and K. V. Mardia. A unified view of the theory of directional statistics, 1975–1988. *International Statistical Review*, 57:261–294, 1989.

[174] H. Karcher. Riemannian center of mass and mollifier smoothing. *Comm. Pure Appl. Math.*, 30:509–554, 1977.

[175] M. Kass, A. Witkin, and D. Terzopoulos. Snakes: Active contour models. *Intl. Journ. of Computer Vision*, 1(4):321–331, 1988.

[176] D. Kaziska and A. Srivastava. Gait-based human recognition by classifi-

cation of cyclostationary processes on nonlinear shape manifolds. *JASA*, 102(480):1114–1124, 2007.

[177] D. G. Kendall. Shape manifolds, procrustean metrics, and complex projective spaces. *Bull. London Math. Soc.*, 16:81–121, 1984.

[178] D. G. Kendall. How to look at objects in a five-dimensional shape space: looking at geodesics. *Adv. in Appl. Probab.*, 27:35–43, 1995.

[179] D. G. Kendall, D. Barden, T. K. Carne, and H. Le. *Shape and Shape theory*. Wiley, New York, 1999.

[180] W. S. Kendall. Probability, convexity, and harmonic maps with small image. i. uniqueness and the fine existence. *Proc. London Math. Soc.*, 61:371–406, 1990.

[181] J. T. Kent. *The Art of Statistical Science, A Tribute to G. S. Watson*, volume 32, chapter New directions in shape analysis, pages 115–127. John Wiley and Sons, 1992.

[182] J. T. Kent. The complex Bingham distribution. *J. Roy. Statist. Soc. Ser. B*, 56:285–299, 1994.

[183] J. T. Kent and K. V. Mardia. Consistency of procrustes estimators. *J. Roy. Statist. Soc. Ser. B*, 59:281 – 290, 1997.

[184] J. T. Kent and K. V. Mardia. A new representation for projective shape. *Interdisciplinary Statistics and Bioinformatics, Leeds, Leeds University Press.*, S. Barber, P. D. Baxter, K. V. Mardia, & R. E. Walls (Eds.).:75–78, 2006.

[185] J. T. Kent and K. V. Mardia. A geometric approach to projective shape and the cross ratio. *Biometrika*, 99:833–849, 2012.

[186] C. G. Khatri and K. V. Mardia. The von Mises-Fisher matrix distribution in orientation statistics. *J. Royal Statist. Soc. B*, 39(1):95–106, 1977.

[187] P. T. Kim. Deconvolution density estimation on SO(N). *Ann Statist*, 26:1083–1102, 1998.

[188] P. T. Kim. *High dimensional probability, II (Seattle, WA, 1999), Progr. Probab., 47*, chapter On the characteristic function of the matrix von Mises-Fisher distribution with application to SO(N)-deconvolution, pages 477–492. Birkh*ä*user: Boston, 2000.

[189] P. T. Kim and J. Y. Koo. Optimal spherical deconvolution. *J. Multivariate Anal.*, 80:21–428, 2002.

[190] P. T. Kim and J. Y. Koo. Statistical inverse problems on manifolds. *Fourier Anal. Appl.*, 11:639–653, 2005.

[191] P. T. Kim, J. Y. Koo, and Z. Luo. Weyl eigenvalue asymptotics and sharp adaptation on vector bundles. *J. Multivariate Anal.*, 100:1962–1978, 2009.

[192] Y. T. Kim and H. S. Park. Geometric structures arising from kernel density estimation on Riemannian manifolds. *J. Multivariate Anal.*, 114:112–126, 2013.

[193] B. Kirkwood, J.-Y. Royer, T. Chang, and R. Gordon. Statistical tools for estimating and combining finite rotations and their uncertainties. *Geophysical J. Internat.*, 137:408–428, 1999.

[194] E. Klassen, A. Srivastava, W. Mio, and S. Joshi. Analysis of planar shapes using geodesic paths on shape spaces. *IEEE Transactions on Pattern Analysis and Machine Intelligence*, 26(3):372–383, 2004.

[195] J. Klemelä. Estimation of densities and derivatives of densities with directional data. *J Multivariate Anal.*, 73:18–40, 2000.

[196] S. Kobayashi. Isometric imbeddings of compact symmetric spaces. *Tôhoku Math. J.*, 20:21–25, 1968.

[197] S. Kobayashi and K. Nomizu. *Foundations of Differential Geometry. Vol. I.* John Wiley & Sons, 1963.

[198] S. Kobayashi and K. Nomizu. *Foundations of Differential Geometry. Vol. II.* John Wiley & Sons, 1969.

[199] S. Kobayashi and K. Nomizu. *Foundations of Differential Geometry. Vol.* **I, II**. *Reprint of the 1969 original.* Wiley Classics Library, 1996.

[200] J. Y. Koo. Optimal rates of convergence for nonparametric statistical inverse problems. *Ann Statist*, 21:590–599, 1993.

[201] J. Y. Koo and H. Y. Chung. Log-density estimation in linear inverse problems. *Ann Statist*, 26:335–362, 1998.

[202] J. Y. Koo and P. T. Kim. Asymptotic minimax bounds for stochastic deconvolution over groups. *IEEE Trans. Inform. Theory*, 54:289–298, 2008.

[203] J. Y. Koo and P. T. Kim. Sharp adaptation for spherical inverse problems with application to medical imaging. *J. Multivariate Anal.*, 99:165–190, 2008a.

[204] A. P. Korostelev. An asymptotically minimax regression estimator in the uniform norm up to exact constant. *Theory Probab. Appl.*, 38:737–743, 1993.

[205] A. P. Korostelev and M. Nussbaum. The asymptotic minimax constant for sup-norm loss in nonparametric density estimation. *Bernoulli*, 5:1099–1118, 1996.

[206] A. Kume and H. Le. Estimating Fréchet means in Bookstein's shape space. *Adv. in Appl. Probab.*, 32:663–674, 2000.

[207] A. Kume and H. Le. On Fréchet means in simplex shape spaces. *Adv. in Appl. Probab.*, 35:885–897, 2003.

[208] S. Kurtek, E. Klassen, J. Gore, Z. Ding, and A. Srivastava. Elastic geodesic paths in shape space of parametrized surfaces. *IEEE Transactions on Pattern Analysis and Machine Intelligence*, 34(9):1717–1730, 2012.

[209] S. Kurtek, A. Srivastava, E. Klassen, and Z. Ding. Statistical modeling of curves using shapes and related features. *Journal of the American Statistical Association*, 107(499):1152–1165, 2012.

[210] H. Le. On the consistency of procrustean mean shapes. *Adv. in Appl. Probab.*, 30:53–63, 1998.

[211] H. Le. Locating Fréchet means with application to shape spaces. *Adv. in Appl. Probab.*, 33:324–338, 2001.

[212] H. Le and A. Kume. The Fréchet mean shape and the shape of the means. *Adv. in Appl. Probab.*, 32:101–113, 2000.

[213] D. LeBihan, J.-F. Mangin, C. Poupon, C. A. Clark, S. Pappata, N. Molko, and H. Chabriat. Diffusion tensor imaging: Concepts and applications. *J. Magn. Reson. Imaging*, 13:534–546, 2001.

[214] J. L. Lee, R. Paige, V. Patrangenaru, and F. Ruymgaart. Nonparametric density estimation on homogeneous spaces in high level image analysis. *Bioinformatics, Images, and Wavelets. Department of Statistics, University of Leeds*, pages 37–40, 2004.

[215] J. M. Lee. *Riemannian Manifolds: An introduction to curvature*. Graduate Texts in Mathematics, 176, Springer-Verlag, New York, 1997.

[216] E. L. Lehmann. *Elements of Large Sample Theory*. Springer Texts in Statistics, Springer-Verlag, New York, 1999.

[217] S. Lele. Euclidean distance matrix analysis (edma): estimation of mean form and mean form difference. *Math. Geol.*, 25:573–602, 1993.

[218] M. Lesosky, Peter T. Kim, and D.W. Kribs. Regularized deconvolution on the 2D-Euclidean motion group. *Inverse Problems, no.5, 055017*, 24:15 pp, 2008.

[219] D. Lester, V. Patrangenaru, and R. Guo. Mean reflection shapes of landmark configurations in large and small acrosterigma magnum clamshells. *Department of Statistics-Florida State University. Tech report M1009. http://www.stat.fsu.edu/techreports.php*, 2013.

[220] T. Leung, M.C. Burl, and P. Perona. Probabilistic affine invariants for recognition. *Proceedings Conf. Computer Vision and Pattern Recognition, June 1998, Santa Barbara, California*, pages 678–684, 1998.

[221] J. Li, L.-Y. Zhou, and Y.-S. Sun. A study of the high-inclination population in the Kuiper belt – 1. the plutinos. *Monthly Notices of the Royal Astronomical Society*, 437:215–226, 2014.

[222] J. Liu and B. Lindsay. Building and using semiparametric tolerance regions for parametric multinomial models. *Technical report, Penn State University*, 2005.

[223] M Loève. *Probability Theory*. Springer-Verlag, Berlin, 4th edition, 1977.

[224] C. Longuet-Higgins. A computer algorithm for reconstructing a scene from two projections. *Nature*, 293:133–135, 1981.

[225] Y. Ma, A. Soatto, J. Košecká, and S. Sastry. *An Invitation to 3-D Vision: From Images to Geometric Models*. Springer, 2005.

[226] S. Mac Lane. *Categories for the working mathematician. Second Edition*. Graduate Texts in Mathematics, 5. Springer-Verlag, New York, 1998.

[227] B. A. Mair and F. H. Ruymgaart. Statistical inverse estimation in Hilbert spaces. *SIAM J Appl Math*, 56:1424–1444, 1996.

[228] W. Man, D. Falk, T. Sun, W. Chen, J. Li, D. Yin, L. Zang, and M. Fan. The corpus callosum of Albert Einsteins brain: another clue to his high intelligence? *Brain*, pages 1–8, 2013.

[229] K. V. Mardia. Statistics of directional data (with discussion). *JR Statist Soc B*, 37:349–393, 1975.

[230] K. V. Mardia and P. E. Jupp. *Directional Statistics*. Chichester: Wiley, 2000.

[231] K. V. Mardia, J. T. Kent, and J. M. Bibby. *Multivariate Analysis*. Academic Press, 1979.

[232] K. V. Mardia and V. Patrangenaru. *Proceedings of the 20th LASR Workshop, Leeds University Press*, chapter On Affine and Projective Shape Data Analysis, pages 39–45. Leeds University Press, 2001a.

[233] K. V. Mardia and V. Patrangenaru. Directions and projective shapes. *Ann. Statist.*, 33:1666–1699, 2005.

[234] K. V. Mardia, V. Patrangenaru, G. J. Davis, and G. Derado. *Proceedings of the 20th LASR Workshop, Leeds University Press*, chapter Averaging side view images of almost flat spatial scenes, pages 47–54. Leeds University Press, 2001.

[235] K. V. Mardia, V. Patrangenaru, and S. Sughatadasa. *The Proceedings of the 24th LASR Workshop*, chapter Protein gels matching, pages 163–165. Leeds University Press, 2005.

[236] M. Di Marzio, A. Panzera, and C.C. Taylor. Kernel density estimation on the torus. *J. Statist. Plann. Inference*, 141:2156–2173, 2011.

[237] C. Mattos, D. Rasmussen, X. Ding, G. A. Petsko, and D. Ringe. Analogous inhibitors of elastase do not always bind analogously. *Nat. Struct.*

Biol., 1:55–58, 1994.

[238] S. J. Maybank. *Applications of Invariance in Computer Vision. Lecture Notes in Comput. Sci. Eds: J.L. Mundy and A. Zisserman and D. Forsyth*, volume 825, chapter Classification based on the cross ratio, pages 433–472. Springer, Berlin, 1994.

[239] H. G. M*ü*ller, U. Stadtm*ü*ller, and F. Yao. Functional variance processes. *J. Amer. Statist. Assoc.*, 101:1007–1018, 2006.

[240] P. W. Michor and D. Mumford. Riemannian geometries on spaces of plane curves. *Journal of the European Mathematical Society*, 8:1–48, 2004.

[241] J. Milnor. On axiomatic homology theory. *Pacific J. of Math.*, 12:337–341, 1962.

[242] J. Milnor. *Morse Theory*. Princeton University Press, 1963.

[243] J. W. Milnor and J. D. Stasheff. *Characteristic classes*. Annals of Mathematics Studies, No. 76. Princeton University Press, Princeton, 1974.

[244] S. Minakshisundaram and A. Pleijel. Some properties of the eigenfunctions on the Laplace operator on Riemannian manifolds. *Can. J. Math*, 1:242–256, 1949.

[245] W. Mio and A. Srivastava. Elastic-string models for representation and analysis of planar shapes. *Proceedings of the IEEE Computer Society International Conference on CVPR*, pages 10–15, 2004.

[246] W. Mio, A. Srivastava, and S. Joshi. On the shape of plane elastic curves. *International Journal of Computer Vision*, 73:307–324, 2005.

[247] W. Mio, A. Srivastava, and E. Klassen. Interpolation with elasticae in Euclidean spaces. *Quarterly of Applied Math*, 62:359–378, 2004.

[248] M. Moakher. A differential geometric approach to the geometric mean of symmetric positive-definite matrices. *SIAM J. Matrix Anal. Appl.*, 26:735–747, 2005.

[249] D. Moore. Equivariant embeddings of Riemannian homogeneous spaces. *Indiana Univ. Math. J.*, 25:271–279, 1976.

[250] C. Müller. *Analysis of Spherical Symmetries in Euclidean Spaces*. New York: Springer, 1998.

[251] A. Munk and H. Dette. Nonparametric comparison of several regression functions: exact and asymptotic theory. *Ann. Statist.*, 26:2339–2368, 1998.

[252] A. Munk, R. Paige, J. Pang, V. Patrangenaru, and F. Ruymgaart. The one and multisample problem for functional data with applications to projective shape analysis. *J. of Multivariate Anal.*, 99:815–833, 2008.

[253] J. Nash. The imbedding problem for Riemannian manifolds. *Ann. of*

Math., 63:20–63, 1956.

[254] L. Nicolaescu. *An invitation to Morse theory. Second edition.* Universitext. Springer, New York, 2011.

[255] P. Niyogi, S. Smale, and S. Weinberger. Finding the homology of submanifolds with high confidence from random samples. *Discrete and Computational Geometry*, 39:419–441, 2008.

[256] J. M. Oller and J. M. Corcuera. Intrinsic analysis of statistical estimation. *Ann. Statist.*, 23:1562–1581, 1995.

[257] D. Osborne. Nonparametric data analysis on manifolds with applications in medical imaging. *PhD Dissertation, Florida State University*, 2012.

[258] D. Osborne, V. Patrangenaru, L. Ellingson, D. Groisser, and A. Schwartzman. Nonparametric two-sample tests on homogeneous Riemannian manifolds, Cholesky decompositions and diffusion tensor image analysis. *Journal of Multivariate Analysis*, 119:163–175, 2013.

[259] D. Osborne, V. Patrangenaru, X. Liu, and H. W. Thompson. 3D size-and-reflection shape analysis for planning reconstructive surgery of the skull. *JSM Proceedings, 2011, Miami, FL. Institute of Mathematical Statistics*, pages 4838–4850, 2012.

[260] D. Osborne, V. Patrangenaru, M. Qiu, and H. W. Thompson. *Festschrift in honor of Kantilal V. Mardia. Eds: I.L. Dryden and J.T. Kent*, chapter Nonparametric Data Analysis Methods in Medical Imaging, pages 183–203. Wiley, 2015.

[261] R. Paige, V. Patrangenaru, F. Ruymgaart, and W. Wang. Analysis of projective shapes of curves using projective frames. *Quantitative Biology, Shape Analysis, and Wavelets*, pages 71–74, 2005.

[262] E. Parzen. On estimation of a probability density function and mode. *Ann. Math. Stat.*, 33(3):1065–1–76, 1962.

[263] V. Patrangenaru. S-manifolds as hypersurfaces in Euclidean spaces. *Revue Roumaine de Mathematiques Pures et Appliqués.*, 29:341 – 348, 1984.

[264] V. Patrangenaru. Locally homogeneous Riemannian manifolds and Cartan triples. *Geom. Dedicata*, 50:143–164, 1994.

[265] V. Patrangenaru. Classifying 3- and 4-dimensional homogeneous Riemannian manifolds by Cartan triples. *Pacific J. Math.*, 173:511–532, 1996.

[266] V. Patrangenaru. Asymptotic Statistics on Manifolds. *PhD Dissertation, Indiana University*, 1998.

[267] V. Patrangenaru. Moving projective frames and spatial scene identifica-

tion. In K. V. Mardia, R. G. Aykroyd, and I. L. Dryden, editors, *Proceedings of the 18th LASR Workshop.*, pages 53–56, Leeds, 1999. Leeds University Press.

[268] V. Patrangenaru. New large sample and bootstrap methods on shape spaces in high level analysis of natural images. *Communications in Statistics Theory and Methods*, 30:1675–1695, 2001.

[269] V. Patrangenaru, M. A. Crane, X. Liu, X. Descombes, G. Derado, W. Liu, V. Balan, V. P. Patrangenaru, and H. W. Thompson. Methodology for 3D scene reconstruction from digital camera images. *BSG Proceedings*, 19:110–124, 2012.

[270] V. Patrangenaru, H. Hendricks, M. Qiu, and K. D. Yao. *Book of Abstracts, June, 12th - 16th, 2014, Cadiz, Spain*, chapter Cartan means and Cartan anti-means on Stratified Spaces, page 196. 2nd Conference of ISNPS. Edited by Universidad Carlos III and The International Society for NonParametric Statistics, 2014a.

[271] V. Patrangenaru, K.D.Yao, and R. Guo. Means and Antimeans. *to appear*, page TBA, 2015a.

[272] V. Patrangenaru, X. Liu, and S. Sugathadasa. Nonparametric 3D projective shape estimation from pairs of 2D images - i, in memory of W. P. Dayawansa. *Journal of Multivariate Analysis*, 101:11–31, 2010.

[273] V. Patrangenaru and K. V. Mardia. A bootstrap approach to Pluto's origin. *Journal of Applied Statistics*, 29:935–943, 2002.

[274] V. Patrangenaru and K. V. Mardia. *Proceedings of the Leeds Annual Statistics Research Workshop*, chapter Affine Shape Analysis and Image Analysis, pages 57–62. Leeds University Press, 2003.

[275] V. Patrangenaru and V. P. Patrangenaru. Mean shapes, image fusion and scene reconstruction. In G. Tsagas, editor, *Proceedings of The Conference of Applied Differential Geometry - Aristotle University of Thessaloniki, Greece, 2002*, pages 230–242. http://www.mathem.pub.ro/proc/bsgp-11/bsgp-11.htm, 2004.

[276] V. Patrangenaru and R. Pruett. *Nonparametric Image Analysis in Planar Scene identification of Sope Creek stone data*, page 13. Abstracts of IPMAS, Georgia Insitute of Technology, Sept. 10-12 Atlanta, GA, 1999.

[277] V. Patrangenaru, M. Qiu, and M. Buibas. Two sample tests for mean 3D projective shapes from digital camera images. *Methodology and Computing in Applied Probability*, 16:485–506, 2014.

[278] B. Pelletier. Kernel density estimation on Riemannian manifolds. *Stat. Prob. Letter.*, 73:297–304, 2005.

[279] B. Pelletier. Non-parametric regression estimation on a closed Riemannian manifold. *J. Nonparametric Statist.*, 18:57–67, 2006.

[280] X. Pennec and N. Ayache. Uniform distribution, distance and expectation problems for geometric features processing. *Journal of Mathematical Imaging and Vision*, 9:49–67, 1998.

[281] X. Pennec, P. Fillard, and N. Ayache. A Riemannian framework for tensor computing. *Int. J. Comput. Vision*, 66:41–66, 2006.

[282] M. J. Prentice. A distribution-free method of interval estimation for unsigned directional data. *Biometrika*, 71:147–154, 1984.

[283] M. J. Prentice. Orientation statistics without parametric assumptions. *J. Royal Statist. Soc. B*, 48:214–222, 1986.

[284] M. J. Prentice. Spherical regression on matched pairs of orientation statistics. *J. Royal Statist. Soc. B*, 51:241–248, 1989.

[285] M. Qiu, V. Patrangenaru, and L. Ellingson. *Proceedings of COMPSTAT2014, The 21st International Conference on Computational Statistics, August 19-22, 2014, Geneva, Switzerland. Eds:Manfred Gilli and Gil Gonzalez-Rodriguez and Alicia Nieto-Reyes*, chapter How far is the Corpus Callosum of an Average Individual from Albert Einsteins?, pages 403–410. The International Statistical Institute/International Association for Statistical Computing, 2014.

[286] J. Ramsay and B. W. Silverman. *Functional Data Analysis*. Springer, New York, 1997.

[287] C. R. Rao. Information and the accuracy attainable in the estimation of statistical parameters. *Bull. Calcutta Math. Soc.*, 37:81–91, 1945.

[288] L.-P. Rivest. A correction for axis misalignment in the joint angle curves representing knee movement in gait analysis. *Journal of Biomechanics*, 38:1604–1611, 2005.

[289] L.-P. Rivest, S. Baillargeon, and M. Pierrynowski. A directional model for the estimation of the rotation axes of the ankle joint. *J. Amer. Statist. Assoc.*, 103:1060–1069, 2008.

[290] L.-P. Rivest and T. Chang. Regression and correlation for 3×3 rotation matrices. *Canadian J. of Statist.*, 34:184–202, 2006.

[291] J. Rohlf. Morphometrics at S.U.N.Y. Stony Brook. *http://life.bio.sunysb.edu/morph/index.html*, 2010.

[292] M. Rosenblatt. Remarks on some nonparametric estimates of a density function. *Ann. Math. Stat.*, 27(3):832, 1956.

[293] S. N. Roy and R.C. Bose. Simultaneous confidence interval estimation. *Ann. Math. Statist.*, 24:513 – 536, 1953.

[294] F. H. Ruymgaart. Strong uniform convergence of density estimators on spheres. *J. Statist. Plann. Inference*, 23:45–52, 1989.

[295] F. H. Ruymgaart and S. Yang. Some applications of watson's pertur-

bation approach to random matrices. *J. Multivariate Anal.*, 60:48 – 60, 1997.

[296] F.H. Ruymgaart. A unified approach to inversion problems in statistics. *Math Methods Statist*, 2:130–146, 1993.

[297] F.H. Ruymgaart. *Mathematical Statistics* . Manuscript. Class Notes. Texas Tech University., 2001.

[298] F.H. Ruymgaart. Statistics on products of spheres. *unpublished manuscript*, 2004.

[299] A. Sacan, O. Ozturk, H. Ferhatosmanoglu, and Y. Wang. Lfm-pro: A tool for detecting significant local structural sites in proteins. *Bioinformatics*, 6:709–716, 2007.

[300] I. J. Schoenberg. Remarks to Maurice Fréchet's article " sur la definition axiomatique d'une classe d'espace distancies vectoriellement appicable sur l'espace de Hilbert". *Ann. Math. Statist.*, 36:724–732, 1935.

[301] A. Schwartzman. Random ellipsoids and false discovery rates: statistics for diffusion tensor imaging data. *PhD Thesis, Stanford University*, 2006.

[302] A. Schwartzman. Lognormal distributions and geometric averages of positive definite matrices. *arXiv.org*, arXiv:1407.6383, 2014.

[303] A. Schwartzman, R. F. Dougherty, and J. E. Taylor. False discovery rate analysis of brain diffusion direction maps. *Ann. Statist.*, 36:153–175, 2008.

[304] A. Schwartzman, W. F. Mascarenhas, and J. E. Taylor. Inference for eigenvalues and eigenvectors of Gaussian symmetric matrices. *Ann. Statist.*, 36:2886–2919, 2008.

[305] T. B. Sebastian, P. N. Klein, and B. B. Kimia. On aligning curves. *IEEE Trans. Pattern Analysis and Machine Intelligence*, 25(1):116–125, 2003.

[306] D. Sharvit, J. Chan, H. Tek, and B. B. Kimia. Symmetry-based indexing of image databases. *Journal of Visual Communication and Image Representation*, pages 366–380, 1998.

[307] K. Singh. On the asymptotic accuracy of Efron's bootstrap. *Ann. Statist.*, 9:1187 – 1195, 1981.

[308] S. Skwerer, E. Bullitt, S. Huckemann, E. Miller, I. Oguz, M. Owen, and V. Patrangenaru. Tree-oriented analysis of brain artery structure. *J. Math. Imaging Vision*, 50:126–143, 2014.

[309] S. Smale. The generalized poincaré conjecture in dimensions greater than four. *Ann. Math. Statist.*, 74:391–406, 1961.

[310] S. Smale and D. Zhou. Shannon sampling and function reconstruction from point values. *Bull. Amer. Math. Soc. (NS)*, 41:279–305, 2004.

[311] C. G. Small. *The Statistical theory of shape.* Springer-Verlag, New York, 1996.

[312] P. G. Sanfilippo snd A. Cardini, A.W. Hewitt, J. G. Crowston, and D.A. Mackey. Optic disc morphology - rethinking shape. *PROGRESS IN RETINAL AND EYE RESEARCH*, 28(4):227–248, 2009.

[313] E. H. Spanier. *Algebraic topology. Corrected reprint.* Springer-Verlag, New York-Berlin, 1981.

[314] G. Sparr:. Depth-computations from polihedral images. *Image and Vision Computing*, 10:683–688, 1992.

[315] G. Sparr. Euclidean and affine structure/motion for uncalibrated cameras from affine shape and subsidiary information. In *Proc. SMILE Workshop on Structure from Multiple Images, Freiburg*, 1998.

[316] M. Spivak. *A Comprehensive Introduction to Differential Geometry, Vol. I-II. Second edition.* Publish or Perish, Inc., Wilmington, Del., 1979.

[317] A. Srivastava, S. Joshi, W. Mio, and X. Liu. Statistical shape analysis: Clustering, learning and testing. *IEEE Transactions on Pattern Analysis and Machine Analysis*, 27:590–602, 2005.

[318] A. Srivastava and E. Klassen. Monte-Carlo extrinsic estimators of manifold-values parameters. *IEEE Trans. Signal Processing*, 20:100–111, 2001.

[319] A. Srivastava and P. E. Klassen. Monte Carlo extrinsic estimators for manifold-values parameters. *IEEE Transactions on Signal Processing*, 50:299–308, 2002.

[320] A. Srivastava and P. E. Klassen. Bayesian and geometric subspace tracking. *Adv. Appl. Prob. (SGSA)*, 36:43 – 56, 2004.

[321] C.J. Stone. Optimal rates of convergence for nonparametric estimators. *Ann. Statist.*, 8:1348–1360, 1980.

[322] S. M. Sughatadasa. Affine and projective shape analysis with applications. *Ph.D. Dissertation, Texas Tech University*, 2006.

[323] A. Tsybakov. *Introduction to Nonparametric Estimation.* Springer, New York, 2009.

[324] O. Tuzel, R. Subbarao, and P. Meer. Simultaneous multiple 3d motion estimation via mode finding on lie groups. *10th IEEE International Conference on Computer Vision, Beijing, China, October 2005*, I:18–25, 2005.

[325] M. Unger, T. Pock, W. Trobin, D. Cremers, and H. Bischof. Tvseg - interactive total variation based image segmentation. *Proceedings of the British Machine Vision Conference (BMVC)*, 2008.

[326] International Astronomical Union. Plutoid cho-

sen as name for solar system objects like pluto. *http://www.iau.org/static/archives/releases/pdf/iau0804.pdf*, 2008.

[327] A. Van der Vaart. *Asymptotic Statistics.* Cambridge University Press, Cambridge., 1998.

[328] A.C.M. van Rooij and F. H. Ruymgaart. *Nonparametric Functional Estimation and Related Topics (G. Roussas, ed.)*, chapter Regularized deconvolution on the circle and the sphere, pages 679–690. Kluwer, Amsterdam, 1991.

[329] A.C.M. van Rooij and F. H. Ruymgaart. Abstract inverse estimation with application to deconvolution on locally compact abelian groups. *Ann. Inst Statist Math.*, 53:781–798, 2001.

[330] G. Wahba. Practical approximate solutions to linear operator equations when the data are noisy. *SIAM: J. Numer. Anal.*, 14:651–667, 1977.

[331] Z. Wang, B. Vemuri, Y. Chen, and T. Mareci. A constrained variational principle for direct estimation and smoothing of the diffusion tensor field from complex dwi. *IEEE Trans. Med. Imaging*, 23:930–939, 2004.

[332] F. W. Warner. The conjugate locus of a Riemannian manifold. *IEEE Trans. Med. Imaging*, 87:575–604, 1965.

[333] G. S. Watson. *Statistics on Spheres.* Lecture Notes in the Mathematical Sciences, John Wiley & Sons, 1983.

[334] G. S. Watson. Limit theorems in high dimensional spheres and stiefel manifolds. *Studies in Econometrics, Time Series, and Multivariate Statistics.*, 1983b.

[335] H. Weyl. Das asymptotische verteilungsgesetz der eigenwerte linearer partieller differentialgleichungen (mit einer anwendung auf die theorie der hohlraumstrahlung). (in German). *Math. Ann.*, 71:441–479, 1912.

[336] H. Whitney. The self-intersections of a smooth n-manifold in 2n-space. *Ann. of Math.*, 45:220–246, 1944.

[337] J. P. Williams and L. A. Cieza. Protoplanetary disks and their evolution. *Annual Review of Astronomy and Astrophysics*, 49:67–117, 2011.

[338] T. Willmore. *Riemannian Geometry.* Oxford University Press, Oxford, 1993.

[339] C. E. Yarman and B. Yazici. Radon transform inversion via wiener filtering over the Euclidean motion group. *Proc. IEEE Int. Conf. Acoust., Speech and Signal Process., Mar. 2005*, 3:811–814, 2003.

[340] C. E. Yarman and B. Yazici. Radon transform inversion based on harmonic analysis of the Euclidean motion group. *Proc. IEEE Int. Conf. Acoust., Speech and Signal Process., Mar. 2005*, 2:481–484, 2005.

[341] Y. G. Yatracos. A lower bound on the error in nonparametric regression

type problems. *Ann. Statist.*, 16:1180–1187, 1988.

[342] L. Younes. Computable elastic distance between shapes. *SIAM Journal of Applied Mathematics*, 58:565–586, 1998.

[343] L. Younes. Optimal matching between shapes via elastic deformations. *Journal of Image and Vision Computing*, 17:381–389, 1999.

[344] L. Younes, P. W. Michor, J. Shah, and D. Mumford. A metric on shape space with explicit geodesics. *Rend. Lincei Mat. Appl.*, 19:25–57, 2008.

[345] H. Ziezold. On expected figures and a strong law of large numbers for random elements in quasi-metric spaces. *Transactions of the Seventh Prague Conference on Information Theory, Statistical Decision Functions, Random Processes and of the Eighth European Meeting of Statisticians*, A:591–602, 1977.

[346] H. Ziezold. Mean figures and mean shapes applied to biological figure and shape distributions in the plane. *Biometrical J.*, 36(4):491–510, 1994.

[347] H. Ziezold. *Numbers, information and complexity*. Kluwer Acad. Publ., Boston, MA, 1998.

[348] A. Zomorodian and G. Carlsson. Computing persistent homology. *Discrete Comput. Geom.*, 33:249–274, 2005.

Index

Printed in the United States
by Baker & Taylor Publisher Services